Wnt Signaling

A subject collection from *Cold Spring Harbor Perspectives in Biology*

OTHER SUBJECT COLLECTIONS FROM *COLD SPRING HARBOR PERSPECTIVES IN BIOLOGY*

Protein Synthesis and Translational Control
The Synapse
Extracellular Matrix Biology
Protein Homeostasis
Calcium Signaling
The Golgi
Germ Cells
The Mammary Gland as an Experimental Model
The Biology of Lipids: Trafficking, Regulation, and Function
Auxin Signaling: From Synthesis to Systems Biology
The Nucleus
Neuronal Guidance: The Biology of Brain Wiring
Cell Biology of Bacteria
Cell–Cell Junctions
Generation and Interpretation of Morphogen Gradients
Immunoreceptor Signaling
NF-κB: A Network Hub Controlling Immunity, Inflammation, and Cancer
Symmetry Breaking in Biology
The Origins of Life
The p53 Family

SUBJECT COLLECTIONS FROM *COLD SPRING HARBOR PERSPECTIVES IN MEDICINE*

Parkinson's Disease
Type 1 Diabetes
Angiogenesis: Biology and Pathology
HIV: From Biology to Prevention and Treatment
The Biology of Alzheimer Disease

Wnt Signaling

A subject collection from *Cold Spring Harbor Perspectives in Biology*

EDITED BY

Roel Nusse
Stanford University

Xi He
Harvard Medical School

Renée van Amerongen
Netherlands Cancer Institute

COLD SPRING HARBOR LABORATORY PRESS
Cold Spring Harbor, New York • www.cshlpress.org

Wnt Signaling

A Subject Collection from *Cold Spring Harbor Perspectives in Biology*
Articles online at www.cshperspectives.org

Printed in the United States of America

Executive Editor	Richard Sever
Managing Editor	Maria Smit
Project Manager	Barbara Acosta
Permissions Administrator	Carol Brown
Production Editor	Diane Schubach
Production Manager/Cover Designer	Denise Weiss
Publisher	John Inglis

Front cover artwork: Green (GFP-labeled) neural cells activated by Wnt signaling, coming from a radial glial stem cell in the developing nervous system. Figure kindly provided by Angela Bowman, Renée van Amerongen, and Roel Nusse (Stanford University Medical School).

Library of Congress Cataloging-in-Publication Data

Wnt signaling : a subject collection from Cold Spring harbor perspectives in biology / edited by Roel Nusse, Xi He, Renee van Amerongen.
p. cm.
Includes bibliographical references and index.
ISBN 978-1-936113-23-1 (harccover : alk. paper)
1. Wnt proteins. 2. Wnt genes. 3. Developmental cytology. 4. Cellular signal transduction. I. Nusse, Roel, 1950- II. He, Xi, 1962- III. Amerongen, Renee van, 1976-

QP552.W58W68 2012
572′.64--dc23

2012016545

10 9 8 7 6 5 4 3 2 1

All World Wide Web addresses are accurate to the best of our knowledge at the time of printing.

For a complete catalog of all Cold Spring Harbor Laboratory Press publications, visit our website at www.cshlpress.org.

Contents

SECTION III: STEM CELL BIOLOGY AND DIFFERENTIATION

SECTION IV: DISEASE AND INTERVENTION

Preface

Signaling by the Wnt family of secreted proteins has come to be recognized as one of the most fundamental and conserved regulatory systems in animals. Wnt signaling controls many critical aspects of development and adult tissue homeostasis and its deregulation is intimately linked to a broad spectrum of human diseases, ranging from congenital defects to osteoporosis and cancer. As such, Wnt-related research has become a major area for targeted drug discovery and therapeutic development.

Since the identification of the first Wnt gene 30 years ago, research in the Wnt field has taken flight—as reflected by more than 13,000 publications in Medline to date—and has progressed into numerous exciting areas. Wnt-related investigations continue to reveal fascinating principles of embryonic patterning, cell growth and differentiation, the wiring of the nervous system, the pathogenic mechanisms underlying cancer as well as degenerative disease, stem cell and regenerative biology, and potential therapeutic applications. This book on Wnt signaling is therefore both timely and long overdue.

When the plan for this book was first conceived, we felt that it should cover the scope of the Wnt field as broadly as we could envision. We have therefore organized the various chapters to include the biochemistry and molecular mechanisms that control Wnt signal transduction (Section I), in parallel to the evolutionary origin of Wnt signaling and its functions in the development of model organisms (Section II). A major focus lies on the role of Wnt signaling in stem cell biology and differentiation (Section III), as our knowledge in this area has increased tremendously over the last decade or so. Finally, the implications of Wnt signal transduction for pathogenesis and treatment of human diseases are presented in Section IV. As this book is published in the same year as the 30th anniversary of the discovery of the *int1/wnt1* gene, we have included a separate chapter describing the history of the Wnt field, in the form of a personal perspective by Roel Nusse and Harold Varmus. This chapter was previously published by *EMBO J* (doi:10.1038/emboj.2012.146) and is reprinted with kind permission from the editor.

The pace of progress in the Wnt field has been astonishing, and several chapters had to be revised even at the production stage to include the latest exciting discoveries that just appeared in the literature. However, in spite of its broad range, this book does not pretend to cover all of the fascinating insights we have gained in Wnt biology over the years. It is our hope that this volume serves as a stepping-stone for the reader to guide and encourage further exploration and, perhaps, to open up novel avenues of investigation, particularly applications in the fields of bioengineering, regenerative medicine, and cancer treatment.

A big thank you goes to all of the authors who contributed timely chapters so enthusiastically to this project. We are very grateful to Richard Sever and in particular to Barbara Acosta at CSHL Press, who guided and coordinated the entire process and who showed tremendous patience in managing our deadlines.

Roel Nusse
Xi He
Renée van Amerongen

Three Decades of Wnts: A Personal Perspective on How a Scientific Field Developed[1]

Roel Nusse[2] and Harold Varmus[3]

[2]Department of Developmental Biology, Howard Hughes Medical Institute, Stanford University School of Medicine, Stanford, California 94305

[3]National Cancer Institute, National Institutes of Health, Bethesda, Maryland 20892

Correspondence: rnusse@stanford.edu

Wnt genes and components of Wnt signalling pathways have been implicated in a wide spectrum of important biological phenomena, ranging from early organismal development to cell behaviours to several diseases, especially cancers. Emergence of the field of Wnt signalling can be largely traced back to the discovery of the first mammalian *Wnt* gene in 1982. In this essay, we mark the thirtieth anniversary of that discovery by describing some of the critical scientific developments that led to the flowering of this field of research.

INTRODUCTION: HOW WE LEARN ABOUT WHAT WE KNOW

Knowledge differs from the growth of knowledge. To learn the facts about a subject, such as *Wnt* genes or the Wnt signalling pathway, one can consult an encyclopaedia, a textbook, or a conventional review article. To understand how those facts were unearthed and assembled into coherent concepts, it is necessary to probe the history of a field—to learn about the sequence of events, the logical and illogical connections between those events, and the people who participated in them.

We have approached this essay with more attention to historical development than to a full repertoire of facts. While interesting experimental results about *Wnt* genes and their effects on cells and organisms continue to appear at an accelerating pace, we believe that there is a great deal to learn about the scientific enterprise more broadly by looking back on the unusual way in which the knowledge about those genes has grown over the past three decades. Some understanding of Wnt signalling is now required of those who aspire to succeed in many prominent fields of biology—including organismal development, cancer research, and stem-cell biology. There is also a sizeable subset of biologists who define themselves primarily as students of *Wnt* genes, while also aligning themselves with cancer, developmental, or stem-cell biologists or with communities devoted to the fruit fly, worm, amphibians, mouse, or Homo sapiens. This cannot be said of those working on many other genes or signalling pathways, raising the interesting questions of how and why scientists

[1]First published in *EMBO J* **31**: 2670–2684 (22 May 2012) doi:10.1038/emboj.2012.146, and reprinted with express permission from the authors, Roel Nusse and Harold Varmus.

organize themselves in unusual ways and view their subjects through certain kinds of lenses.

We have composed this essay on the occasion of the thirtieth anniversary of the published report that announced our discovery of what proved to be the first mammalian *Wnt* gene (Nusse and Varmus 1982). Taking that report as an arbitrary starting point in the history of this field, we have tried to highlight the most significant ways in which the field increased in knowledge, enlarged in scope, and grew in disciples. As we emphasize, some of these advances were logical and straightforward, others were technically difficult and protracted, and yet others were serendipitous and surprising. We also note more briefly how the shape of the field was determined by certain beneficial attitudes and behaviours that may be worthy of emulation.

THE "PRE-HISTORY": MOUSE MODELS FOR BREAST CANCER AND CANCER-CAUSING RETROVIRUSES PRECEDED KNOWLEDGE OF WNT GENES

All modern science is built on earlier science. Accordingly, the discovery that launched the intense study of *Wnt* genes 30 years ago depended on at least two earlier and closely related lines of enquiry: mouse models of cancer and oncogenic retroviruses.

It had been known since the 1930s that certain strains of laboratory mice are highly susceptible to breast cancer, and that the disease is usually transmitted from mothers to offspring mice through the milk (Bittner 1936; Korteweg 1936). Later, the tumour-inducing activity was purified from the milk (Lyons and Moore 1962), and the milk-transmitted factor was shown to be a morphologically atypical retrovirus, called the Mouse Mammary Tumour Virus or MMTV.

Although the study of oncogenic retroviruses can be tracked to the first decade of the 20th century, the basis of their cancer-causing properties came into focus only in the century's second half. The first great advances came from tissue culture assays for viral growth and cell transforming capacities and from the biochemical and genetic analysis of the RNA genomes of retroviruses isolated from chickens, mice, rats, and other experimental animals. These methods led to the discovery of distinct viral oncogenes, such as Src, Myc, and Ras, and their cellular precursors, called proto-oncogenes, a term denoting any cellular genes that could be converted to active cancer-causing genes (Bishop and Varmus 1985). Conversion to oncogenicity could occur by the mechanisms that produced highly oncogenic retroviruses or, as was shown in time and described later in this essay, by a variety of other mechanisms, most commonly somatic mutations of several types.

The proto-oncogenes initially discovered by tracing viral oncogenes to their cellular origins were generally not related to each other, but they shared several properties. They had been conserved during evolution and were converted to cancer-causing genes by gain-of-function mutations—as first shown for retroviral oncogenes and later for activated cellular oncogenes found in human cancers. The rapid onset of cancer caused by many retroviruses reflected the ability of active viral oncogenes to transform many infected cells (Bishop and Varmus 1985).

Before RN came to UCSF in 1980 to work with HV, we had both been interested in breast cancer in the mouse and in the general properties of retroviruses isolated from animals. Such viruses often cause haematological cancers and sarcomas, relatively infrequent types of cancers in human beings; only rarely do they induce epithelial carcinomas, the most common human cancers. So, we were curious about the mechanism by which MMTV might cause carcinoma of the breast. Unlike the most intensively studied cancer-causing retroviruses, however, there was no readily identifiable oncogene in the viral genome. But, like other retroviruses, MMTV was known to insert a DNA copy of its RNA genome into the host cell genome during infection. Moreover, MMTV caused cancer slowly, over the course of several months, unlike oncogene-containing retroviruses, suggesting that viral infection alone, while required for tumour induction, was not sufficient to transform a host cell—a property that had also been recognized in the study of certain leukaemia-inducing retroviruses found in birds and mice (Teich et al. 1982).

It was known, however, that tumours induced by such viruses were composed of cell clones defined by shared proviral integration sites (Cohen et al. 1979; Payne et al. 1981) implying that tumours consisted of the descendants of a single infected cell and were thus the outcome of rare events. The pattern of clonality raised the possibility that one of many infected cells had randomly acquired a provirus that could initiate tumourigenesis. For instance, the insertion of viral DNA might cause a mutation of a gene in the vicinity of the integration site, and the change might confer a growth advantage to that cell. While it was possible that such mutations caused loss-of-function mutations, by disrupting a host cell gene, an alternative and more attractive model was that a host cell gene was transcriptionally activated by the incoming provirus. The latter model was inspired by findings in yeast and prokaryotes, where transposable elements, which are often functionally and structurally similar to retroviral proviruses (Brown and Varmus 1989), were known to activate residing host cell genes (Errede et al. 1980).

The model of gene activation by proviral insertion gained momentum when Bill Hayward's group showed that B-cell lymphomas caused by the slowly oncogenic Avian Leukosis Virus (ALV) commonly contained proviruses inserted near the *c-myc* gene and that c-myc was overexpressed in those tumours (Hayward et al. 1981). Initially, activation of c-myc appeared to be the consequence of a transcriptional promoter present in an ALV provirus inserted upstream of the protein- coding exons of c-myc in the same transcriptional orientation (Hayward et al. 1981), but research in HV's group showed that proviruses could also enhance c-myc expression from the endogenous c-myc promoter when inserted downstream of the gene or upstream in either orientation (Payne et al. 1982).

The discoveries concerning c-myc and ALV were of critical importance in cancer research, as it was the first time that a cellular homologue of a viral oncogene was shown to be activated in cancer by mutation. Soon thereafter, many kinds of mutations other than viral insertion mutations—gene amplifications, chromosomal translocations, and point mutations—were shown to activate c-myc and many other progenitors of retroviral oncogenes, in human as well as in animal tumours, without the intervention of retroviruses (Weinberg 1983). The results with ALV and c-myc also set the stage for looking for a similar mechanism in tumours caused by other retroviruses, such as MMTV, with the prospect of finding proto-oncogenes that are not related to known viral oncogenes.

MOLECULAR CLONING OF INT1, THE FIRST NOVEL PROTO-ONCOGENE IDENTIFIED BY PROVIRAL TAGGING

To seek host cell genes that are activated by insertions of MMTV proviruses, we decided to do a systematic screen rather than to look for activation of only those proto-oncogenes known at the time, such as c-myc. We assumed, perhaps naively, that only one or perhaps a few genes in the mouse genome would confer tumourigenic growth when activated by MMTV in mammary cancer. This assumption implied that by comparing MMTV proviruses in multiple different tumours we would find common integration sites, or at least common regions, near those unusual genes. This segment of the genome would represent a so-called "common integration site." But the commonality would not reflect a predisposition to integrate MMTV at preferred locations in the mouse genome—all data, then and now, suggest that retroviral integration occurs quasi-randomly in host genomes—but instead would result from selection of cells that acquired a growth advantage when a provirus activated a nearby proto-oncogene.

The goal then became to isolate an integrated MMTV provirus plus its adjacent host cell DNA from a tumour by molecular cloning. We reasoned that we should start from a tumour with just one MMTV insertion. Most tumours in the strain we were using (C3H) carried multiple newly acquired proviruses, in addition to a few copies of inherited "endogenous" MMTV DNA that were invariant among the arising tumours because they were products of ancient infections of the mouse germ line (Bentvelzen et al. 1970; Cohen and Varmus 1979). Since it was difficult to know which of the multiple new proviruses would be functionally important, we looked for tumours carrying only one new provirus; this single insert should, according to the model, be near the relevant gene. Restriction fragments of host DNA derived from the vicinity of the provirus from that tumour would then be used as probes for two purposes: to screen other tumours for integrations in the same domain and to measure transcription of local genes, both in tumours and in normal mammary tissue. Of course, today, with complete sequences of normal mouse genomes and powerful methods for sequencing tumour DNA in hand, precise mapping of insertion sites has become a relatively simple exercise. But at the time, these tools were not available, so a more cumbersome method was required.

We initiated the screen in October 1980, gathering over 30 MMTV-infected C3H mice with mammary tumours, and we found just one tumour (#18) with a single new provirus. Happily, that single tumour proved to suffice for our purposes. From the genome of this tumour, we cloned a "junctional" fragment, containing a part of an MMTV provirus joined to adjacent chromosomal DNA, then derived smaller fragments containing only host cell DNA. With those fragments as probes, we "walked" along the chromosome in either direction from the original proviral insertion site to obtain more probes for a broader region of the mouse genome (Fig. 1). These were then used to examine DNA from other tumours for disruptions caused by MMTV proviruses. Indeed,

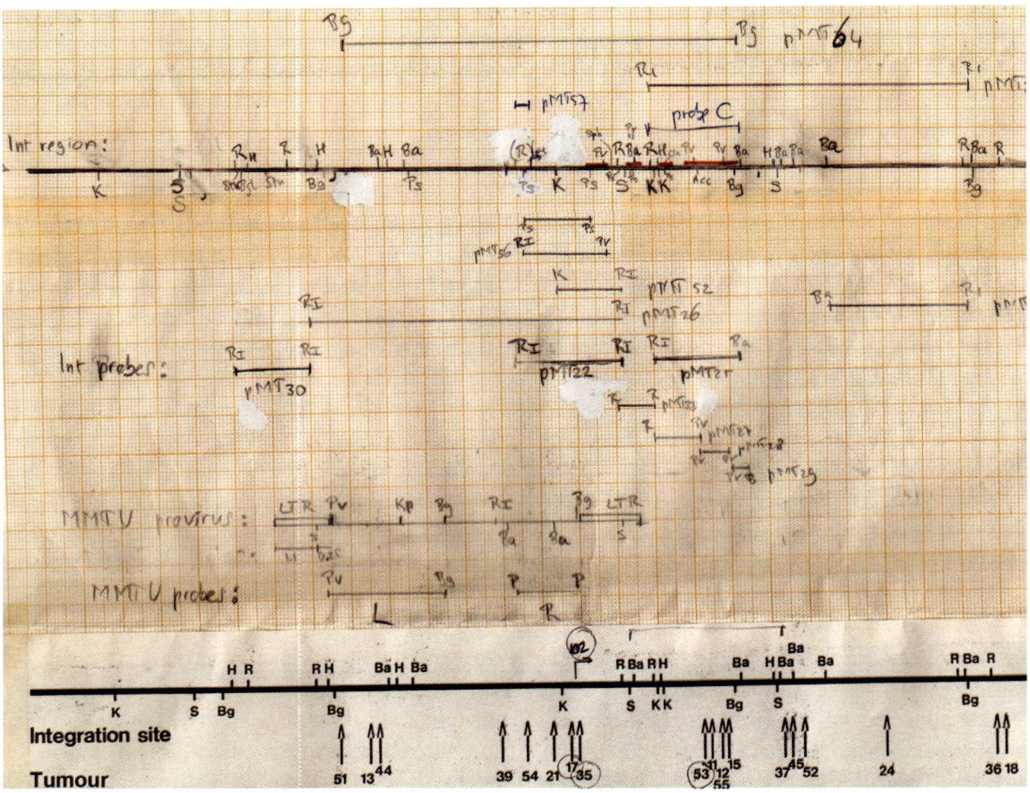

Figure 1. A working map of the mouse int1 locus as drawn by RN and used from 1982 to 1984, with the position of various cloned genomic restriction fragments. Red lines indicate the presence of int1 exons, mapped in 1984. Note the location of Probe C, a genomic fragment hybridizing with int1 mRNA in mouse mammary tumours. At the *bottom*, the position of MMTV proviruses mapped in different tumours, with the position of the provirus in tumour #18, the starting point of the cloning of the locus, indicated at the *right* hand end.

many tumours in our collection had evidence for MMTV DNA insertions in the region occupied by the single provirus in tumour #18; these were manifested by novel restriction fragments that hybridized with both MMTV and host cell probes. We assembled a map of the relevant region of the mouse genome, with the various proviral insertions in different tumours scattered throughout a region spanning ~30 KB (Fig. 1). It turned out that the original insertion site we cloned from tumour #18 was at one end of the cluster.

But where was the cellular gene that, according to our hypothesis, should have been activated as a result of these integrations? One of the probes (probe C, Fig. 1) proved to be particularly helpful. It picked up integration events in many individual tumours and mapped approximately to the middle of the cluster of integrations. More importantly, the same probe also detected a tumour-specific mRNA—that is, a species not found in normal mammary gland tissue—on northern blots. This was the evidence we were looking for—a host gene, a putative proto-oncogene, had been transcriptionally activated by MMTV proviruses in multiple, independent tumours (Nusse and Varmus 1982). We called the gene *int1* (to denote the first common integration site) and promptly submitted a manuscript that was published in the journal *Cell* in November 1982 (Nusse and Varmus 1982). The method we used, called proviral tagging, is now widely used to discover proto-oncogenes in cancers induced in mice by retroviruses or transposable elements and in cancers whose growth is accelerated in transgenic animals by these elements (Kool and Berns 2009; Copeland and Jenkins 2010).

The paper describing int1 was well received, but its impact at that time was overshadowed by an avalanche of incredibly exciting developments in cancer research. In the same year, 1982, human cancers were found to have mutations in cellular RAS genes, and other human tumours were shown to contain chromosome translocations directly affecting c-myc (Bishop 1983). Soon thereafter, equally explosive findings were announced, identifying oncogenic proteins as known factors governing growth control: the erbB oncogene was discovered to be derived from the EGF receptor gene, and both encoded protein-tyrosine kinases; and the precursor to the v-sis oncogene was shown to be the gene encoding PDGF, a secreted growth factor (Bishop and Varmus 1985).

THE DIFFICULTIES OF DEFINING THE MECHANISM OF ACTION OF THE INT1 GENE AND ITS ENCODED PROTEIN

In the midst of all of the excitement about retroviral oncogenes and their progenitors, int1 attracted relatively little attention beyond the confines of our laboratories, despite the many open questions that seemed important and interesting to us. Initially, in the absence of a nucleotide sequence, we had no clue how the gene would function. Work done in both our laboratories (HV at UCSF; and RN after his return to the Netherlands Cancer Institute in Amsterdam in 1982) elucidated the structure and sequence of the *int1* gene (van Ooyen and Nusse 1984) and its cDNA (Fung et al. 1985), revealing no homology with any other gene or protein known at the time. We did notice, however, that the predicted protein sequence started with a signal sequence, indicating that the int1 protein would be secreted. This opened up the exciting possibility that this protein might be an extracellular growth factor. But direct proof of this prediction turned out to be very hard to obtain. For many years, no one was able to produce or isolate significant quantities of the int1 protein, a problem that was not solved until 2003 (see below). To make matters worse, generating useful antibodies to int1 was an equally frustrating enterprise; in fact, detecting int1 protein in cells or tissues remains an elusive goal even today. These problems, in particular the lack of active protein for experimental use, precluded conventional signalling assays in cell culture. As a result, indirect assays, such as those dependent on gene transfer, had to be used to study signalling events in cells expressing int1. In particular, the identification of specific int1 cell surface receptors by binding assays was not possible until much later.

On the other hand, we did establish that expression of the *int1* gene could affect cell behaviour in a fashion that resembled conventional transformation and provided a biological assay: various mammary epithelial cell lines could be morphologically altered by overexpression of int1, albeit in a subtle way and rarely leading to formation of cells capable of growing into a tumour (Brown et al. 1986; Rijsewijk et al. 1987b). More dramatically, Ann Tsukamoto in HV's laboratory was able to recapitulate the oncogenic effect of int1 in mice without resorting to virus infection: mice expressing an int1 transgene under the influence of an MMTV transcriptional regulator developed cancer in the mammary gland within about 6 months of age (Tsukamoto et al. 1988). This established that int1 is a *bona fide* proto-oncogene. These mice—and several others expressing int1 under the control of inducible promoters—have since become widely used mouse models for studying breast carcinogenesis and for finding genes that can cooperate with the int1 transgene during oncogenesis (MacArthur et al. 1995).

DEVELOPMENTAL GENETICS HELPED TO REVEAL THE FUNCTION OF INT1 WHEN THE GENE WAS DISCOVERED TO BE THE HOMOLOGUE OF THE *DROSOPHILA* SEGMENT POLARITY GENE, WINGLESS

How could the function of int1 protein and the components of its signalling pathway be deciphered without a direct, convenient biochemical or cell-based assay? Fortunately, int1 did have one advantage: a high degree of conservation across species. The human int1 protein sequence turned out to be almost completely (99%) identical to that of the mouse homologue (Van Ooyen et al. 1985). Moreover, int1-related sequences appeared to be present in the DNA of *Drosophila melanogaster* as judged from molecular hybridization (Nusse et al. 1984).

At the time, the new molecular methods that were responsible for unveiling the genes central to cancer research had also re-energized efforts to understand embryogenesis, using the rich treasury of *Drosophila* developmental mutants. During the 1970s, genetic screens in *Drosophila* had unveiled a set of genes that were essential for the development of the embryo. Nüsslein-Volhard and Wieschaus (1980) showed specific patterning defects, ranging from abnormal segment numbers to polarity changes, in mutants for many of these genes. They coined the term "segment polarity genes" for one class of mutants that shared a similar patterning phenotype during embryogenesis. One of the genes in this group was called *Wingless;* others, *Armadillo* and *Arrow.* The *Wingless* gene had actually been identified earlier as a weak mutant allele leading to loss of wing tissue, hence the name Wingless (Sharma and Chopra 1976). Subsequent to the genetic screens, many of the segmentation genes were molecularly cloned, generating a treasure trove of reagents to study developmental mechanisms. For example, the expression patterns of these genes often produced stripes corresponding to body segments. By examining the expression pattern of one gene in the background of mutations in other genes, hierarchies of genetic interactions were uncovered, providing unparalleled insights into how embryos develop (Ingham 1988).

RN and his colleagues cloned the *Drosophila* int1 homologue and used polytene chromosome mapping (the genomics technology at the time) to locate the gene. It turned out to map close to Wingless, one of the segment polarity genes; a striped expression pattern observed with a *Drosophila* int1 probe also suggested a role in segmentation (Rijsewijk et al. 1987a). Around the same time, Baker (1987) had cloned the *Wingless* gene by a P-element transposon tagging, a method akin to the proviral tagging methods we had used for int1. The gene he cloned had restriction maps matching our *Drosophila* int1 clone. The genes were identical; the int1 homologue in *Drosophila* was Wingless, one of the first examples of a gene involved in development and also activated in cancer (Rijsewijk et al. 1987a). This was an exciting discovery in its own right. In addition, the membership of Wingless in the segment polarity group promised to open doors to discovering the mechanisms of action of int1/Wingless, since it seemed likely that other genes in the group would interact with int1/Wingless genetically and biochemically. As we now understand, the core of its signalling pathway is indeed based on genetic relationships between segment polarity genes, with a key role for *Armadillo* (see below) (Fig. 2).

It had long been argued by some that cancer can be considered akin to a developmental abnormality, a disease caused by cells that have escaped from the normal developmental constraints on proliferation and differentiation (Boveri 2008). Implicated in cancer as well as in embryogenesis, int1 became a poster child for these connections. Interest in the roles of int1 in development was further strengthened by reports by McMahon and Moon (1989) showing that int1 was implicated in embryonic axis formation in another animal, *Xenopus laevis.* Xenopus had been used classically to study morphological changes in development; more recently, injection methods had been used to perturb development by introduction of wild-type and mutant genes and proteins into early stage Xenopus embryos. McMahon and Moon (1989) found that ectopic expression of int1 duplicated the dorsal axis, suggesting a role for int1 as an organizer and an embryonic "inducer," a signal between germ layers that would lead to pattern formation. Because of the speed and clarity of the embryological assays, Xenopus embryology became a mainstay of research on int1 and other developmental signals.

In other organisms, int1 expression patterns in embryos and other tissues suggested a diversity of functions in the development of tissues. For instance, int1 was initially found to be expressed mainly near the midbrain and cerebellum at the mid-stage of mouse development and in a specific stage of male germ cell maturation (Shackleford and Varmus 1987; Wilkinson et al. 1987).

INT1 WAS ONE OF THE FIRST GENES TO BE KNOCKED OUT BY HOMOLOGOUS GENE TARGETING IN MICE, PRODUCING A DEVELOPMENTAL PHENOTYPE

During the late 1980s, Mario Capecchi and Oliver Smithies developed the groundbreaking technique for generating gene-specific mouse mutants by homologous recombination. After learning about the highly restricted pattern of expression of int1, Capecchi's group chose the gene as one of its initial targets for application of this amazing technology. After breeding to homozygosity, disruption of int1 produced a dramatically diminished cerebellum, accompanied by severe ataxia (Thomas and Capecchi 1990). In McMahon's laboratory, int1 null mutants caused an embryonic lethal phenotype in the

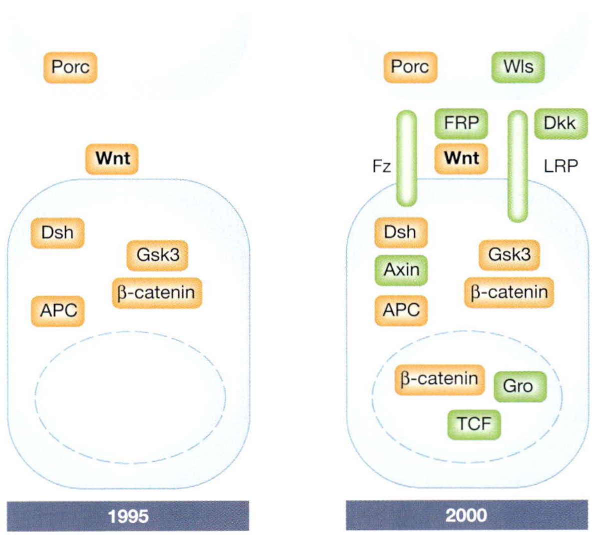

Figure 2. Wnt signalling components as known in 1995 and 2000.

mid-brain also affecting the development of the cerebellum (McMahon and Bradley 1990). Soon thereafter, it was also shown that a classical mouse mutation, Swaying (Lane 1967), was an allele of the *int1* gene; when int1 was cloned from Swaying mice, it proved to have a frameshift mutation (Thomas et al. 1991).

These several discoveries about int1 mutant organisms in flies and mice were very striking. In a review article published in 1992 (Nusse and Varmus 1992), we wrote: "With the benefit of hindsight, we now recognize that phenomena studied for several decades are the consequences of *Wnt* gene mutations. Viral insertion mutations regularly promote mammary tumours in laboratory mice (Bittner 1936; Korteweg 1936), a spontaneous frameshift mutation of mice (swaying) impairs cerebellar structure and function (Lane 1967; Thomas et al. 1991) and wingless mutations in *Drosophila* can transform a wing to a notum or disrupt segment polarity (Sharma and Chopra 1976; Nüsslein-Volhard and Wieschaus 1980)."

RENAMING INT1 AS WNT1 AND RECOGNITION OF A WNT GENE FAMILY

Around 1990, it became clear that the int nomenclature had become inadequate and confusing. For example, additional screens for MMTV proviral insertion sites in tumours had yielded other activated genes, called *int2, int3,* and *int4* (Dickson et al. 1984; Gallahan and Callahan 1987; Roelink et al. 1990). But by sequence comparisons, these genes were not usually related to int1. One MMTV target gene, initially called *int4,* did prove to be related to int1 (Roelink et al. 1990). But the frequently activated *int2* gene, first identified by Clive Dickson and Gordon Peters (Dickson et al. 1984), turned out to be a member of the FGF family (Dickson and Peters 1987). Interestingly, FGF genes were also implicated in normal development at about this time, sometimes in coordination with int1-related genes, as in mesoderm formation (Kimelman et al. 1992). Moreover, int1 and int2 are sometimes co-activated in MMTV-induced breast tumours (Peters et al. 1986). int3 was shown to be related to Notch, another important developmental regulator (Gallahan and Callahan 1997).

At the same time that the *int* gene nomenclature was becoming unworkable, various experiments, including PCR-based homology screens, had revealed a large family of genes related to int1 (Gavin et al. 1990). It would have been confusing to christen all these genes with the term "int," whether or not they had been activated by proviral integration. To avoid further confusion, all those working on int1 and its relatives, including Wingless, consented to a new hybrid name "Wnt" (for Wingless-related integration site) to denote genes belonging to the int1/Wingless family, with int1, now called Wnt1, as the founding member (Nusse et al. 1991). (In accord with other recognized relationships, *int2* is now called FGF3, the *int3* gene is Notch4, and *int4* is Wnt3A.)

THE WNT FAMILY AS A VANTAGE POINT TO STUDY GENE EVOLUTION AND DEVELOPMENT IN METAZOANS

With the complete sequences of the genomes of many multicellular animals in hand, we now realize that vertebrates contain a family of 19 Wnt-related genes; pairs of these genes can often be placed in subfamilies that are highly similar to each other, perhaps reflecting gene duplications relatively recently in evolution (Gavin et al. 1990). Each of these genes seems likely to have a specific role in development or other processes; they are generally expressed in different cells and at different times in maturation (Gavin et al. 1990). Many of the *Wnt* genes have now been deliberately mutated in the mouse, almost always leading to striking phenotypes, including limb polarity defects and sexual dimorphic abnormalities (Parr and McMahon 1995, 1998).

Even representatives of the earliest branches of the animal kingdom, such as Hydra and Nematostella, have the same number of Wnt subfamilies as vertebrates (Kusserow et al. 2005). Readily recognized orthologues of specific *Wnt* genes—for example, Wnt1—have been found throughout the entire animal kingdom, often expressed in tantalizing patterns (Kusserow et al.

2005). As a result, evolutionary biologists speculate that the early amplification and diversification of the Wnt family were at the roots of the increased complexity of animal body plans (Sidow 1992; Holstein 2012). It appears from such findings that *Wnt* genes were probably present in genomes prior to the split of the animal kingdom into protostomes and deuterostomes, are therefore at least 600 M years old, and may have a universal role in setting up the primary axis of animals (Petersen and Reddien 2009; Niehrs 2010; Holstein 2012). However, it is also clear that single-cell organisms do not contain *Wnt* genes, nor do plants.

In the midst of all of these genes and families, it remains striking that Wnt1 is one of the key Wnt family members and may have been the primordial one. Wnt1 is the true orthologue of *Wingless,* a gene in *Drosophila* with numerous functions in later development as well as early embryogenesis. While there are six other Wnts in *Drosophila,* the others each play a minor role compared with Wingless. Moreover, there are very useful temperature-sensitive alleles of Wingless to study its numerous functions. As a result of these attributes, Wingless has been a rich source for understanding developmental processes. In other organisms that have multiple Wnts, Planaria in particular, the true Wnt1 orthologue also has a special place because of the requirement for it in regeneration (Gurley et al. 2010). While we do not understand why Wnt1 is the most frequently activated gene in MMTV-induced breast cancer, it should be noted that we now know that Wnt1 is closely linked to another *Wnt* gene that is often insertionally activated, Wnt10B (Lee et al. 1995). Thus, some MMTV inserts may have activated both genes, providing a greater growth advantage.

UNEXPECTED FINDINGS REVEAL THE IMPORTANCE OF THE WNT PATHWAY IN HUMAN CANCERS

From the time that *Wnt1* was discovered as an initiating gene in mouse mammary carcinogenesis, it remained of great interest and importance to establish whether *Wnt* genes were involved in any human cancer. Early tests for aberrant expression of *Wnt* genes in human breast cancer gave ambiguous results, at best, and no *Wnt* genes appeared to be mutated in any kinds of human tumours by DNA rearrangements or (as more recently documented by next generation sequencing of whole exomes or whole genomes) by point mutations (http://www.sanger.ac.uk/genetics/CGP/cosmic).

But work from a different angle changed our perceptions about the role of *Wnt* genes in human cancer, demonstrating that downstream components of the Wnt signalling pathway, rather than the *Wnt* genes themselves at the upstream end of the pathway, were commonly altered in several types of human cancer. The first news came from the study of colon cancer, and it was dramatic.

Around 1990, significant advances had been made in positional cloning of inherited human disease genes, including genes predisposing to several types of cancer. Among the hereditary forms of human cancer, adenomatous polyposis coli (APC), a trait associated with multiple polyps in the colon, often leads to colon cancer at a relatively early age. The corresponding mutations—often nonsense or frameshift mutations that produce truncated proteins—were found in an enormous gene called *APC,* which was cloned from human chromosome 5 in 1991 (Groden et al. 1991; Kinzler et al. 1991). Identifying the human *APC* gene led to the cloning of a mouse homologue, subsequently shown by Bill Dove's group to be mutated in a mouse strain called Min (Multiple intestinal neoplasia) (Su et al. 1993). Just as in human families, the cancer trait in the Min mouse is produced by an APC truncating mutation, inherited in an autosomal dominant manner. But despite the new genetic insights into intestinal cancer, the function of the large APC protein posed a biochemical mystery.

Soon thereafter, the groups of Paul Polakis, Bert Vogelstein, and Ken Kinzler established that APC interacted in cells with a protein called β-catenin (Rubinfeld et al. 1993; Su et al. 1993). At the time, β-catenin had just been characterized by Masatoshi Takeichi and Rolf Kemler as a protein binding to the cytoplasmic domain of the adhesion molecule

E-cadherin (Ozawa et al. 1989; Takeichi 1990). Intriguingly, Pierre McCrea and Barry Gumbiner had found that β-*catenin* gene was a vertebrate homologue of the segment polarity gene, *Armadillo* (McCrea et al. 1991), while Peifer and Wieschaus (1990) had established similarity between Plakoglobin (a β-catenin-related adhesion complex member) and Armadillo. Together, these findings suggested that APC and β-catenin/Armadillo were involved in regulating adhesion between vertebrate cells. Given the role of Armadillo in segment polarity, a function shared with Wingless, a model emerged in which Wnt/Wingless signalling controlled cell adhesion in development (Peifer and Wieschaus 1990; Peifer et al. 1993), and the same adhesion-based mechanism could control growth of cells in tissues and cause cancer when misregulated.

While this was tantalizing, there were also reports that β-catenin/Armadillo was present in the nucleus, as well as at the cell membrane (Funayama et al. 1995). Other publications mentioned that injection of antibodies to β-catenin/Armadillo could induce dorsal axis duplication in Xenopus (McCrea et al. 1993), possibly by stabilizing the β-catenin/Armadillo protein; yet others claimed that depletion of maternal β-catenin/Armadillo could eliminate the dorsal axis (Heasman et al. 1994). These observations would ultimately all make sense: β-catenin is a key participant of Wnt signalling, but the molecular mechanisms remained unexplained until a few years later.

In deciphering the cascade of events between the Wnt signal and the role of β-catenin/Armadillo, the genetic interactions between a protein kinase called Glycogen Synthase Kinase 3 (GSK3) and other Wnt components proved to be of critical importance. Norbert Perrimon and colleagues showed that GSK3 (the *Drosophila* homologue was called zeste-white 3, also known as shaggy) was a negative regulator of the pathway; at the genetic level, Wnt/Wingless acted as a GSK3 inhibitor (Siegfried et al. 1992). Until then, GSK3 was known for its role in glucose metabolism (Dent et al. 1989), so its newly discovered role in developmental signalling was certainly surprising.

A WNT SIGNALLING CASCADE FROM THE CELL SURFACE TO THE NUCLEUS, AN UNUSUAL PATHWAY EXPERIMENTALLY ASSEMBLED FROM SEVERAL DIFFERENT MODELS

By 1995, the combined results from fly and mouse genetics, Xenopus embryology, and fly and mammalian cell culture experiments had generated an outline of a Wnt signalling pathway (Fig. 2). It became clear that Wnt signalling was unusual compared with the other pathways known at the time: those consisted mostly of successions of protein phosphorylations, with protein associations based on recognition of phosphorylated domains. In Wnt signalling, the most upstream known component was a cytoplasmic protein of uncertain biochemical function, Dishevelled, which then was proposed to inhibit the abundant GSK3 protein kinase (Peifer et al. 1991; Siegfried et al. 1992, 1994; Noordermeer et al. 1994). GSK3 was known to be a negative regulator of β-catenin/Armadillo and was found in a complex with β-catenin/Armadillo, together with the APC protein (Rubinfeld et al. 1996). The role of the GSK3 kinase activity as a suppressor of Wnt action was confirmed by injecting dominant-negative (kinase-dead) mutants of GSK3 into early Xenopus embryos; this maneuver produced a phenocopy of the effect of Wnt1, duplication of the dorsal axis, that had been reported previously by Moon and MacMahon (Dominguez et al. 1995; He et al. 1995; Pierce and Kimelman 1996).

A critical next step was to determine how phosphorylation by GSK3 governed β-catenin. This seemed likely to occur by control of the level of β-catenin protein. In cells activated by Wnt, levels of β-catenin are commonly increased (Riggleman et al. 1990; Peifer et al. 1994; Van Leeuwen et al. 1994) by stabilizing the β-catenin protein, not by an increase in its synthesis. Several highly conserved Ser/Thr phosphorylation sites near the amino terminus of β-catenin were proposed as possible targets for phosphorylation by GSK3 (Peifer et al. 1994). As shown by Rolf Kemler's and Randall Moon's groups, phosphorylated β-catenin is targeted for degradation by the ubiquitination/proteasome pathway

(Yost et al. 1996; Aberle et al. 1997; Orford et al. 1997), with a critical role for F-box proteins (Jiang and Struhl 1998). Eliminating one or more of the N-terminal phosphorylation sites stabilizes β-catenin, producing abundant protein highly active in Xenopus axis formation assays (Yost et al. 1996). As a result, there are striking parallels between the Wnt, Hedgehog, and NF-κB signalling pathways; in all three cases, regulated signalling depends on degradation of a key pathway component by the ubiquitination/proteasome pathway after phosphorylation (Jiang and Struhl 1998; Maniatis 1999).

The study of the molecular pathology of colon cancers then offered a remarkable example of the predictive power of knowledge about signalling. As mentioned earlier, inherited mutations in the *APC* gene were known to cause the familial disease APC, and somatic mutations in APC were found in most (ca. 85%) but not in all sporadic colorectal cancers. Why not all? Could mutations affecting other components of the Wnt signalling pathway substitute for APC mutations? Or did other mutant signalling pathways drive those tumours? Paul Polakis, Bert Vogelstein, Hans Clevers, and their colleagues looked specifically for altered β-*catenin* genes in APC wild-type tumours, on the supposition that β-catenin protein could be stabilized by mutations affecting the N-terminus as well as by loss of APC. Indeed they found that about 5–10% of sporadic colon cancers had mutations, often short deletions, that removed or changed the phosphorylation sites that target β-catenin for degradation (Korinek et al. 1997; Morin et al. 1997; Rubinfeld et al. 1997). Subsequently, mutations have also been found in another component of the degradation complex, Axin, in colorectal and other types of cancers (Satoh et al. 2000).

This other component of the β-catenin/Armadillo/GSK3/APC complex has an interesting history. Cloned by Frank Costantini as a mouse developmental mutant called fused (Zeng et al. 1997), the *Axin* gene encodes a protein that shares homology with Dishevelled, suggesting possible participation in the Wnt pathway. This turned out to be the case. Axin is now known to participate in the β-catenin destruction complex, together with APC and GSK3 (Behrens et al. 1998; Ikeda et al. 1998). Using a novel cell-free system for studying β-catenin degradation, Marc Kirschner and colleagues showed that Axin is the rate-determining component of the complex, even though it was the most recently identified (Lee et al. 2003). Axin has a similar role in intact mammalian cells (Li et al. 2012).

After these several elements were implicated in Wnt signalling, two apparent and important gaps in the pathway remained, one at each end of the pathway (Fig. 2). On the upstream end, there were no proteins known to recognize extracellular Wnt proteins and transmit a signal to the cell's interior (Wnt receptors). On the downstream end, the anticipated effects on gene expression through transcriptional control could not be explained because β-catenin does not have the expected physical attributes of a transcription factor, and no established transcription factor was known to partner with it. Then, in a single year, 1996, these gaps were closed, generating excitement in the growing Wnt field (Fig. 2).

WNT RECEPTOR PROTEINS WERE KNOWN IN OTHER CONTEXTS BEFORE THEIR ROLES IN WNT SIGNALLING WERE UNCOVERED

After many trial-and-error searches, one class of the elusive Wnt receptors was identified: the Frizzled transmembrane proteins. Originally found in mutant screens by Calvin Bridges (Bridges and Brehme 1944), *Frizzled* had been identified in *Drosophila* as a gene required for planar polarity (Gubb and Garcia-Bellido 1982; Vinson and Adler 1987), the orientation of cells in tissues. Paul Adler and colleagues showed Frizzled to encode a seven-pass transmembrane protein (Vinson et al. 1989). Genetically, at least, Frizzled interacted with Dishevelled, which was shown by Perrimon and Mahowald (1987) to be involved in Wingless signalling as well. While this suggested that Frizzled could mediate Wingless signalling, the absence of an embryonic segment polarity phenotype in Frizzled mutants indicated otherwise.

Here serendipity stepped in. Jeremy Nathans and his colleagues found a Frizzled homologue among components of a human retinal cDNA library that had been made to pursue their interests in the molecular biology of vision. When the cDNA was used to seek homologues in a library of *Drosophila* DNA, a second *Drosophila* Frizzled gene (*Dfz2*) was cloned, and the *Dfz2* gene displayed a striped pattern of gene expression in the embryo, implying that it might be directly involved in segment polarity. A collaboration between the Nathans and RN laboratories revealed that the Wingless protein, which RN's laboratory had solubilized at the time, could bind to Dfz2 and, more weakly, to Frizzled itself (Bhanot et al. 1996). Moreover, in cultured *Drosophila* cells that did not express *Frizzled* genes, transfection of an expression vector containing *Frizzled* genes conferred active signalling, as demonstrated by an increase in Armadillo (β-catenin) levels (Bhanot et al. 1996). Genetic and other interactions between Frizzleds and Wnts were also reported by the groups of Randall Moon and Robert Horvitz (Sawa et al. 1996; Yang-Snyder et al. 1996).

Just like Wnts, Frizzleds form a large gene family in all branches of metazoan animals. Genetically, *Frizzled* genes are often redundant and display phenotypes only when mutated in combination with other family members (Ye et al. 2011). In *Drosophila,* this was shown for Frizzled and Dfz2 using dsRNA interference technology (Kennerdell and Carthew 1998). (Interestingly, this occurred in the same year that this revolutionizing method to inhibit gene expression was first reported by Andy Fire and Craig Mello; Fire et al. 1998). By using loss-of-function mutations in Frizzled and Dfz2, Eric Wieschaus, Gary Struhl, Ken Cadigan, and Krishna Bhat uncovered a segment polarity phenotype indistinguishable from phenotypes characteristic of the other genes in that class—but only as double mutants, explaining why these receptor genes were not in the original Nüsslein-Volhard/Wieschaus collection (Bhat 1998; Bhanot et al. 1999; Chen and Struhl 1999; Muller et al. 1999). Despite the evidence for redundancy, Frizzled proteins have different affinities for different Wnts (Rulifson et al. 2000), indicating a high degree of specificity in their interactions. However, persistent experimental problems with the biochemistry of Wnt proteins have hampered systematic surveys of the interactions.

To complement the Wnt receptor story, the *Drosophila* gene *Arrow,* one of the last segment polarity genes to be identified, was cloned by Stephen DiNardo and colleagues a few years later. Arrow proved to be a member of the Low density lipoprotein receptor-Related Protein (LRP) family of receptors (Wehrli et al. 2000). Based on additional genetic data from Bill Skarnes, who made LRP mouse mutants, and biochemical experiments from Xi He's laboratory, a model emerged in which Arrow/LRP is a co-receptor for Wnts, physically adjacent to Frizzleds in the cell membrane (Pinson et al. 2000; Tamai et al. 2000). When signalling to downstream components, however, Arrow/LRP may be the key player. Its cytoplasmic tail is phosphorylated as a consequence of Wnt binding and interacts directly with GSK3 and Axin (Mao et al. 2001b; Tamai et al. 2004; Davidson et al. 2005; Zeng et al. 2005) and Frizzled's intracellular role in signalling may be limited to binding Dishevelled (Macdonald et al. 2009). Arrow/LRP is also the target of several Wnt antagonists including the protein Dickkopf, isolated by Christof Niehrs (Glinka et al. 1998; Bafico et al. 2001; Mao et al. 2001a; Semenov et al. 2001). The Dickkopf-Wnt antagonism is conserved across many animal phyla (Guder et al. 2006), illustrating the ancient nature of Wnt signalling in animal development and evolution.

Eddy De Robertis, Jeremy Nathans, Jeff Rubin, and their co-workers have uncovered several other Wnt antagonists, in addition to Dickkopf. These are secreted molecules usually consisting of Wnt receptor domains that bind to Wnt itself. Some of these molecules have names such as FRP or FRZB, reflecting their similarity to the Frizzled receptor (Finch et al. 1997; Leyns et al. 1997; Rattner et al. 1997). Others, such as the WIF protein (Hsieh et al. 1999), are unrelated to Frizzled. These proteins are likely involved in fine-tuning the concentration of active Wnt outside cells.

The tandem arrays of Frizzled and LRP are not the only Wnt receptors, as there are various members of the trans-membrane tyrosine kinase family that serve to receive Wnts; these include the ROR (Oishi et al. 2003; Mikels and Nusse 2006) and Derailed/RYK (Yoshikawa et al. 2003) proteins. Interestingly, these two classes of molecules have different Wnt binding modules: the RORs contain a CRD domain similar to the Frizzled CRD, while Derailed/RYK is related to the WIF protein mentioned above (Patthy, 2000). Wnt interactions with these receptors often lead to effects in cells that are unrelated to β-catenin, possibly mediating "non-canonical Wnt signalling" (van Amerongen et al. 2008).

ANOTHER UNEXPECTED BUT PREVIOUSLY WELL-KNOWN PROTEIN, TCF/LEF1, EXPLAINS THE ROLE OF β-CATENIN IN THE WNT SIGNALLING PATHWAY

The TCF/Lef1 protein proved to be the long-sought Wnt transcription factor in the nucleus Discovery of the critical interaction between this protein and β-catenin highlights one of the themes of this essay: historically, the map of Wnt signalling was assembled by merging evidence from several different cell types and organisms. TCF/Lef1, an HMG box-containing transcription factor, was first implicated in immune T-cell gene expression (Travis et al. 1991; van de Wetering et al. 1991; Waterman et al. 1991) without any evident link to Wnt signalling. Working separately on *C. elegans,* Jim Priess and colleagues identified an HMG-box family member, POP1, involved in mesoderm specification in the worm embryo (Lin et al. 1995), initially also without connections to the Wnt pathway.

Soon thereafter, a surprising discovery was reported: TCF/Lef1 could interact with β-catenin, considered at that time to be an adhesion molecule. Hans Clevers extended his earlier work on TCF/Lef1 to make this finding (Molenaar et al. 1996), while Walter Birchmeier (Behrens et al. 1996) and Rolf Kemler (Huber et al. 1996) started from β-catenin to establish binding to TCF/Lef1. Using *Drosophila,* Mariann Bienz and Rudi Grosschedl (Riese et al. 1997) found that Wingless signalling was mediated by TCF/Lef1 while Konrad Basler used mutagenesis screens to find a gene called *Pangolin* (Brunner et al. 1997), the single TCF/Lef1 homologue in *Drosophila.* Around the same time, continued investigations into *C. elegans* embryogenesis by Jim Priess, Craig Mello, and Bruce Bowerman unveiled that the set of MOM genes implicated in lineage choices were members of the Wnt pathway, including Wnt itself (MOM2), Porcupine (MOM1), and Frizzled (MOM5). All of these MOMs converged on POP1 as a transcription factor and WRM1 as a β-catenin-related gene (Rocheleau et al. 1997, 1999; Thorpe et al. 1997).

In many contexts, TCF/Lef1 can switch between two states. When bound to Groucho, it acts as a repressor of target genes; but when Groucho is displaced by β-catenin, the same target genes are transcriptionally activated (Cavallo et al. 1998; Daniels and Weis, 2005). Crystallographic studies by Bill Weis and Wenqing Xu revealed the molecular details of the binding between TCF, β-catenin, and other proteins. The structure of β-catenin contains a groove made by the "Armadillo" repeats in the protein, explaining how β-catenin can interact with several different partners, including TCF/ Lef1, E-cadherin, and APC (Huber et al. 1997; Graham et al. 2000).

These discoveries about the interaction between β-catenin and TCF/Lef1 were in more than one respect very significant, as they not only closed the gaps in the Wnt pathway but also provided unparalleled tools for experiments. TCF/Lef1 recognizes a well-defined DNA binding site. By multimerizing this sequence, Hans Clevers and colleagues generated very convenient luciferase-based Wnt reporters, called Top-FLASH and now widely used in the Wnt field to measure signalling (Korinek et al. 1997). There are now numerous genes known to have TCF/Lef1 binding sites in their promoters and hence likely to be transcriptional targets for the Wnt signalling pathway in at least some cell types; among these genes are several implicated in cancer, such as c-myc (He et al. 1998).

AT LAST, THE PURIFICATION OF ACTIVE WNT PROTEIN

As we have recounted, by the end of the 20th century we had a blueprint of the Wnt signalling pathway and a readout for the pathway, both of which were missing in the two previous decades in which *Wnt* genes were intensively studied (Fig. 2). Still lacking, however, was the purification of any active Wnt protein, a problem that we and many other researchers had been working on since the initial cloning of Wnt1. Why were Wnt proteins so much more refractory to biochemical purification than many other secretory proteins? Were they modified in a fashion that rendered them insoluble or highly adherent?

It was known from Norbert Perrimon's work and *Drosophila* genetics that export likely required a specialized protein encoded by Porcupine. Evidence that Porcupine encoded a putative acyltransferase (Kadowaki et al. 1996; Hofmann 2000) suggested that detergents might be needed to keep lipid-modified Wnt proteins soluble during extraction from cells. With the help of assays that judged Wnt activity in extracts based on increased β-catenin levels in cells, Karl Willert in RN's laboratory finally managed to break through the purification barriers (Willert et al. 2003). (Coincidentally, Willert had obtained his PhD working with HV at UCSF.) Wnts were found to be indeed covalently attached to lipids, explaining to some extent their resistance to biochemical manipulation (Willert et al. 2003). It is now recognized that secretion of Wnt proteins is a complex process, involving a dedicated enzyme (Porcupine; Kadowaki et al. 1996) and secretory proteins that are specific for Wnt signals. Among these is the multiple-pass transmembrane protein Wntless/Evi, identified by the groups of Konrad Basler and Michael Boutros, once again using the *Drosophila* genetic resources that have over the years been so instrumental in understanding the details of Wnt signalling (Banziger et al. 2006; Bartscherer et al. 2006; Port and Basler 2010).

A GROWING AND VERY ACTIVE FIELD OF WNT SIGNALLING

The availability of Wnt proteins and more quantitative reporters (e.g., TopFlash) as reliable end points for signalling have simplified the study of Wnt signalling in cell culture, attracting many new investigators. The generation of Wnt reporter mice, initially Top-gal animals from Elaine Fuchs (DasGupta and Fuchs 1999) and later animals with transgenic markers driven by Axin 2 promoters (Lustig et al. 2002; van Amerongen et al. 2012), provided yet more experimental opportunities. One can now trace Wnt-responding cells in any tissue of the mouse, examine the origin of these cells, and follow their fate in normal settings or after injury (Barker et al. 2007, 2010; van Amerongen et al. 2012). These new experimental tools have led to a rapidly growing list of Wnt signalling components, built on the core pathway.

Wnt signalling is clearly complicated and unusual when compared with other growth factor cascades. At various nodes in the Wnt pathway, there are links to cyto-architectural proteins, such as those involved in adhesion and cell polarity (Nelson and Nusse 2004). The Wnt pathway is clearly important for cell fate changes and the control of gene expression, but Wnt signalling can also influence how cells are shaped and polarized and how they divide (Veeman et al. 2003). Hitoshi Sawa, Bruce Bowerman, and Craig Mello have provided conclusive evidence for a major role for Wnt signalling in asymmetric cell division in *C. elegans* and the annelid worm *Platynereis dumerilii* (Rocheleau et al. 1999; Schneider and Bowerman 2007; Sugioka et al. 2011). Given the multiple roles of the Wnt pathway in development, these cell biological phenotypes are perhaps not surprising and have opened fertile ground for further research.

SOCIOLOGY OF THE FIELD OF WNT SIGNALLING: ANNUAL MEETINGS, SHARING INFORMATION, AND A DEDICATED WEBSITE

A history of Wnt signalling would be incomplete without a few comments on the sociology of the field, which, we believe, has several unusual aspects. Those features have contributed to one of the overarching characteristics of this field: the propensity of investigators working on *Wnt* genes and Wnt signalling to identify themselves, at least in part and often primarily, as students of Wnts, regardless of whether they are cancer biologists, developmental biologists, or biochemists. First and foremost among the unifying activities are the annual (or sometimes biannual) Wnt meetings. These are organized by working scientists in the field, not by institutions or meetings specialists, in a very informal, low-cost, but effective way. No one, except a keynote speaker, receives an invitation accompanied by a promise of reimbursements; all others are expected to get there, find food and lodging, and arrange to cover expenses. Nevertheless, the meetings are well attended by many principal investigators, not by just trainees. The meetings started in on a small scale in 1990 as regular gatherings of our two nearby laboratories, after RN had moved to Stanford from Amsterdam. We then asked members of other laboratories to attend as well; Figure 3 shows most of the attendees at the 1991 meeting at UCSF. After HV moved to the NIH in 1993, we continued the gatherings; attendance soon grew to ~300 people, as our own trainees started their own laboratories to work on *Wnt* genes, and others joined the field. Many of the discoveries we have presented in this essay were first made public during Wnt meetings, including the identification of Frizzleds as Wnt receptors and TCFs/Lefs as transcription factors in Wnt signalling, both at the 1996 meeting at Stanford (Fig. 4); and the Arrow/LRP findings were first reported at another Wnt meeting, at Stanford in 1999. The meetings have covered an increasingly wide range of subjects and biological systems related to Wnt signalling, and their popularity attests to the loyalty of Wnt researchers to the subject matter in its many manifestations.

Wnt meetings have also helped to establish a culture of sharing information and reagents. An example of the congenial relationship was the unanimous and friction-free acceptance of the new nomenclature (from int to Wnt) when it was felt that this would benefit the field and the general comprehension of its work. In parallel to the informal Wnt meetings in the United States, there have been numerous other Wnt conferences throughout the world, perhaps a bit more official but still in the same spirit of open exchange and camaraderie.

Figure 3. Participants of the 1991 Wnt meeting at UCSF. From *left*: RN, Andrew McMahon, Arend Sidow, Vladimir Pecenka, John Mason, Lee Fradkin, HV, Henk Roelink, Jasprina Noordermeer, Supriya Shivakumar, Frank van Leeuwen, Cindy Harryman, Jean-Paul Vincent, Jackie Papkoff, two unidentified people, Tony Brown, a third unidentified person, Helen Kwan. *Top* row, from *left*: Karl Willert, Neil Parkin, and Jan Kitajewski.

Figure 4. Celebration of discovery of Frizzleds as Wnt receptors at the 1996 Wnt meeting at Stanford. From *left*: Jeremy Nathans, Matthew Scott, RN, and HV.

In 1996, around the time that the Internet became easier to navigate with browsers such as Netscape, RN started and still maintains a website, the Wnt homepage (http://wnt.stanford.edu). The various pages list genes and signalling diagrams, in many cases linked to other genome databases. But the site is also used to announce meetings and to supply information on Wnt technology and reagents, and has become a popular resource for the world-wide Wnt community and for others seeking information about *Wnt* genes, as appreciation of their importance has expanded.

CONTEMPORARY WNT SIGNALLING SYSTEMS, INCLUDING STEM CELLS, AND AN OUTLOOK

It is now clear that Wnt signalling is widely implicated in diverse biological processes. For instance, the large majority of developmental decisions that cells make during embryogenesis and thereafter appears to be coordinated, in large part, by Wnts. Scientists are beginning to understand how organs grow and regenerate after injury, and it is clear that Wnt signalling has major functions in these processes as well. A particularly prominent example of the centrality of Wnt signalling is the recently recognized role of Wnts in maintaining stem cells. The choices that stem cells make to self-renew or to differentiate are very much dependent on extrinsic signalling factors coming from a niche. Wnt signals are widely active as niche factors, as illustrated by the identification of the LGR5 receptor as a Wnt target gene in many different kinds of adult stem cells (Barker et al. 2007) and the requirement for TCF4 to maintain stem cells in the intestine (Korinek et al. 1998).

The discoveries related to stem cells have also illuminated the connections between Wnt and cancer: in a simple but likely correct view, stem cells are normally dependent on external Wnts for self-renewal, but when a negative Wnt regulator such as APC is mutated in stem cells, cells with markers of early lineage development proliferate in an uncontrolled manner, producing cancers of the colon and other organs. In the mammary gland, where we first identified Wnt1 as an oncogene, stem cells are also Wnt dependent (Shackleton et al. 2006; Zeng and Nusse 2010; van Amerongen et al. 2012), and Wnt1-induced tumours bear hallmarks of normal mammary stem cells (Li et al. 2003).

Cite this article as *Cold Spring Harb Perspect Biol* doi: 10.1101/cshperspect.a007856

Wnt signalling mutations have also been implicated in a growing list of degenerative diseases. Important among these are bone density abnormalities with dysfunctional LRP receptors (Gong et al. 2001) and retinal degeneration with Frizzled mutations (Robitaille et al. 2002). Some metabolic disorders, including diabetes mellitus, have been associated with alterations in Wnt pathway genes (Grant et al. 2006).

It has been gratifying to witness growth of the Wnt field, from the finding of a single cancer gene in a mouse model to a rich system branching out to influence so many aspects of metazoan biology and human disease. While outsiders may be intimidated by the current size of the field and the biochemical complexities of Wnt signalling, we suggest that there are still many fundamental aspects of Wnt-related biology to be discovered, understood, and exploited. Increasingly, structures of Wnt signalling components are being elucidated, often in complexes with their partners (Dann et al. 2001; Schwarz-Romond et al. 2007; Ahn et al. 2011; Chen et al. 2011; Cheng et al. 2011); but important aspects of the pathway's molecular machinery and biochemical regulators remain incompletely defined.

Progress in the Wnt field is much more rapid today than it was in the early history of this field, thanks to more sophisticated tools: Wnt-specific assays and materials and more general methods in structural biology, genetics, and cell biology. For instance, recognition of the role of Wnt signalling in stem-cell regulation has already led to the use of Wnt proteins or Wnt agonists to expand stem cells in culture (Sato et al. 2010; Zeng and Nusse, 2010; ten Berge et al. 2011). On the other hand, despite the evidence for widespread involvement of Wnt signaling in human carcinogenesis, the kinds of targeted cancer therapies that are now being developed against components of several other signaling pathways have not yet been produced to interfere with the Wnt pathway. Among the most significant challenges in future research in the Wnt field is the identification of effective and specific Wnt pathway inhibitors for use in cancer and other diseases. We expect that further understanding of the intricacies and varieties of Wnt signaling will help to achieve these important goals.

ACKNOWLEDGMENTS

We thank Thomas Schwarz-Romond and Renee van Amerongen for inviting us to write this essay. We thank Elizabeth Matthews for comments on various drafts and for suggesting the name Wnt during the nomenclature discussions over 20 years ago.

CONFLICT OF INTEREST

The authors declare that they have no conflict of interest.

REFERENCES

*Reference is also in this collection.

Aberle H, Bauer A, Stappert J, Kispert A, Kemler R. 1997. β-Catenin is a target for the ubiquitin-proteasome pathway. *EMBO J* **16:** 3797–3804.

Ahn VE, Chu ML, Choi HJ, Tran D, Abo A, Weis WI. 2011. Structural basis of Wnt signaling inhibition by Dickkopf binding to LRP5/6. *Dev Cell* **21:** 862–873.

Bafico A, Liu G, Yaniv A, Gazit A, Aaronson SA. 2001. Novel mechanism of Wnt signalling inhibition mediated by Dickkopf-1 interaction with LRP6/Arrow. *Nat Cell Biol* **3:** 683–686.

Baker NE. 1987. Molecular cloning of sequences from wingless, a segment polarity gene in Drosophila: The spatial distribution of a transcript in embryos. *EMBO J* **6:** 1765–1773.

Banziger C, Soldini D, Schutt C, Zipperlen P, Hausmann G, Basler K. 2006. Wntless, a conserved membrane protein dedicated to the secretion of Wnt proteins from signaling cells. *Cell* **125:** 509–522.

Barker N, Huch M, Kujala P, van de Wetering M, Snippert HJ, van Es JH, Sato T, Stange DE, Begthel H, van de Born M, et al. 2010. Lgr5+ve stem cells drive self-renewal in the stomach and build long-lived gastric units *in vitro*. *Stem Cell* **6:** 25–36.

Barker N, van Es JH, Kuipers J, Kujala P, van de Born M, Cozijnsen M, Haegebarth A, Korving J, Begthel H, Peters PJ, et al. 2007. Identification of stem cells in small intestine and colon by marker gene Lgr5. *Nature* **449:** 1003–1007.

Bartscherer K, Pelte N, Ingelfinger D, Boutros M. 2006. Secretion of Wnt ligands requires Evi, a conserved transmembrane protein. *Cell* **125:** 523–533.

Behrens J, Jerchow BA, Wurtele M, Grimm J, Asbrand C, Wirtz R, Kuhl M, Wedlich D, Birchmeier W. 1998. Functional interaction of an axin homolog, conductin, with β-catenin, APC, and GSK3β. *Science* **280:** 596–599.

Behrens J, von Kries JP, Kuhl M, Bruhn L, Wedlich D, Grosschedl R, Birchmeier W. 1996. Functional interaction of β-catenin with the transcription factor LEF-1. *Nature* **382:** 638–642.

Bentvelzen P, Daams JH, Hageman P, Calafat J. 1970. Genetic transmission of viruses that incite mammary tumor in mice. *Proc Natl Acad Sci USA* **67:** 377–384.

Bhanot P, Brink M, Samos CH, Hsieh JC, Wang Y, Macke JP, Andrew D, Nathans J, Nusse R. 1996. A new member of the frizzled family from *Drosophila* functions as a Wingless receptor. *Nature* **382:** 225–230.

Bhanot P, Fish M, Jemison JA, Nusse R, Nathans J, Cadigan KM. 1999. Frizzled and Dfrizzled-2 function as redundant receptors for Wingless during Drosophila embryonic development. *Development* **126:** 4175–4186.

Bhat KM. 1998. Frizzled and frizzled 2 play a partially redundant role in wingless signaling and have similar requirements to wingless in neurogenesis. *Cell* **95:** 1027–1036.

Bishop JM. 1983. Cellular oncogenes and retroviruses. *Annu Rev Biochem* **52:** 301–354.

Bishop JM, Varmus H. 1985. Functions and origins of retroviral transforming genes. In *RNA tumor viruses* (ed. Weis R, Teich N, Varmus H, Coffin JM), pp. 249–356. Cold Spring Harbor Laboratory Press, Cold Spring Harbor, NY.

Bittner J. 1936. Some possiūle effects of nursing on the mammary. gland tumor incidence in mice. *Science* **84:** 162.

Boveri T. 2008. *Concerning the origin of malignant tumours.* The Company of Biologists Limited and Cold Spring Harbor Laboratory Press, Cold Spring Harbor, NY.

Bridges C, Brehme K. 1944. *The mutants of Drosophila melanogaster.* Carnegie Institute, Washington, DC.

Brown AM, Wildin RS, Prendergast TJ, Varmus HE. 1986. A retrovirus vector expressing the putative mammary oncogene int-1 causes partial transformation of a mammary epithelial cell line. *Cell* **46:** 1001–1009.

Brown P, Varmus H. 1989. Retroviruses. In *Mobile DNA* (ed. Berg D, Howe M). American Society for Microbiology Press, Washington, DC.

Brunner E, Peter O, Schweizer L, Basler K. 1997. Pangolin encodes a Lef-1 homologue that acts downstream of Armadillo to transduce the Wingless signal in *Drosophila. Nature* **385:** 829–833.

Cavallo RA, Cox RT, Moline MM, Roose J, Polevoy GA, Clevers H, Peifer M, Bejsovec A. 1998. Drosophila Tcf and Groucho interact to repress Wingless signalling activity. *Nature* **395:** 604–608.

Chen CM, Struhl G. 1999. Wingless transduction by the Frizzled and Frizzled2 proteins of Drosophila. *Development* **126:** 5441–5452.

Chen S, Bubeck D, Macdonald BT, Liang WX, Mao JH, Malinauskas T, Llorca O, Aricescu AR, Siebold C, He X, et al. 2011. Structural and functional studies of LRP6 ectodomain reveal a platform for Wnt signaling. *Dev Cell* **21:** 848–861.

Cheng Z, Biechele T, Wei Z, Morrone S, Moon RT, Wang L, Xu W. 2011. Crystal structures of the extracellular domain of LRP6 and its complex with DKK1. *Nat Struct Mol Biol* **18:** 1204–1210.

Cohen JC, Shank PR, Morris VL, Cardiff R, Varmus HE. 1979. Integration of the DNA of mouse mammary tumor virus in virus-infected normal and neoplastic tissue of the mouse. *Cell* **16:** 333–345.

Cohen JC, Varmus HE. 1979. Endogenous mammary tumour virus DNA varies among wild mice and segregates during inbreeding. *Nature* **278:** 418–423.

Copeland NG, Jenkins NA. 2010. Harnessing transposons for cancer. gene discovery. *Nat Rev Cancer* **10:** 696–706.

Daniels DL, Weis WI. 2005. β-Catenin directly displaces Groucho/TLE repressors from Tcf/Lef in Wnt-mediated transcription activation. *Nat Struct Mol Biol* **12:** 364–371.

Dann CE, Hsieh JC, Rattner A, Sharma D, Nathans J, Leahy DJ. 2001. Insights into Wnt binding and signalling from the structures of two Frizzled cysteine-rich domains. *Nature* **412:** 86–90.

DasGupta R, Fuchs E. 1999. Multiple roles for activated LEF/TCF transcription complexes during hair follicle development and differentiation. *Development* **126:** 4557–4568.

Davidson G, Wu W, Shen J, Bilic J, Fenger U, Stannek P, Glinka A, Niehrs C. 2005. Casein kinase 1γ couples Wnt receptor activation to cytoplasmic signal transduction. *Nature* **438:** 867–872.

Dent P, Campbell DG, Hubbard MJ, Cohen P. 1989. Multisite phosphorylation of the glycogen-binding subunit of protein phosphatase-1G by cyclic AMP-dependent protein kinase and glycogen synthase kinase-3. *FEBS Lett* **248:** 67–72.

Dickson C, Peters G. 1987. Potential oncogene product related to growth factors [letter]. *Nature* **326:** 833.

Dickson C, Smith R, Brookes S, Peters G. 1984. Tumorigenesis by mouse mammary tumor virus: Proviral activation of a cellular gene in the common integration region int-2. *Cell* **37:** 529–536.

Dominguez I, Itoh K, Sokol SY. 1995. Role of glycogen synthase kinase 3β as a negative regulator of dorsoventral axis formation in Xenopus embryos. *Proc Natl Acad Sci USA* **92:** 8498–8502.

Errede B, Cardillo TS, Sherman F, Dubois E, Deschamps J, Wiame JM. 1980. Mating signals control expression of mutations resulting from insertion of a transposable repetitive element adjacent to diverse yeast genes. *Cell* **22:** 427–436.

Finch PW, He X, Kelley MJ, Uren A, Schaudies RP, Popescu NC, Rudikoff S, Aaronson SA, Varmus HE, Rubin JS. 1997. Purification and molecular cloning of a secreted, Frizzled-related antagonist of Wnt action. *Proc Natl Acad Sci USA* **94:** 6770–6775.

Fire A, Xu S, Montgomery MK, Kostas SA, Driver SE, Mello CC. 1998. Potent and specific genetic interference by double-stranded RNA in Caenorhabditis elegans. *Nature* **391:** 806–811.

Funayama N, Fagotto F, McCrea P, Gumbiner BM. 1995. Embryonic axis induction by the armadillo repeat domain of β-Catenin: Evidence for intracellular signaling. *J Cell Biol* **128:** 959–968.

Fung YK, Shackleford GM, Brown AM, Sanders GS, Varmus HE. 1985. Nucleotide sequence and expression *in vitro* of cDNA derived from mRNA of int-1, a provirally activated mouse mammary oncogene. *Mol Cell Biol* **5:** 3337–3344.

Gallahan D, Callahan R. 1987. Mammary tumorigenesis in feral mice-identification of a new int locus in mouse mammary tumor virus (Czech II)-induced mammary tumors. *J Virol* **61:** 66–74.

Gallahan D, Callahan R. 1997. The mouse mammary tumor associated gene INT3 is a unique member of the NOTCH gene family (NOTCH4). *Oncogene* **14:** 1883–1890.

Gavin BJ, McMahon JA, McMahon AP. 1990. Expression of multiple novel Wnt-1/int-1-related genes during fetal and adult mouse development. *Genes Dev* **4:** 2319–2332.

Glinka A, Wu W, Delius H, Monaghan AP, Blumenstock C, Niehrs C. 1998. Dickkopf-1 is a member of a new family of secreted proteins and functions in head induction. *Nature* **391:** 357–362.

Gong Y, Slee RB, Fukai N, Rawadi G, Roman-Roman S, Reginato AM, Wang H, Cundy T, Glorieux FH, Lev D, et al. 2001. LDL receptor- related protein 5 (LRP5) affects bone accrual and eye development. *Cell* **107:** 513–523.

Graham TA, Weaver C, Mao F, Kimelman D, Xu W. 2000. Crystal structure of a β-catenin/Tcf complex. *Cell* **103:** 885–896.

Grant SF, Thorleifsson G, Reynisdottir I, Benediktsson R, Manolescu A, Sainz J, Helgason A, Stefansson H, Emilsson V, Helgadottir A, et al. 2006. Variant of transcription factor 7-like 2 (TCF7L2) gene confers risk of type 2 diabetes. *Nat Genet* **38:** 320–323.

Groden J, Thliveris A, Samowitz W, Carlson M, Gelbert L, Albertsen H, Joslyn G, Stevens J, Spirio L, Robertson M, et al. 1991. Identification and characterization of the familial adenomatous polyposis coli gene. *Cell* **66:** 589–600.

Gubb D, Garcia-Bellido A. 1982. A genetic analysis of the determination of cuticular polarity during development in Drosophila melanogaster. *J Embryol Exp Morphol* **68:** 37–57.

Guder C, Pinho S, Nacak TG, Schmidt HA, Hobmayer B, Niehrs C, Holstein TW. 2006. An ancient Wnt-Dickkopf antagonism in Hydra. *Development* **133:** 901–911.

Gurley KA, Elliott SA, Simakov O, Schmidt HA, Holstein TW, Alvarado AS. 2010. Expression of secreted Wnt pathway components reveals unexpected complexity of the planarian amputation response. *Dev Biol* **347:** 24–39.

Hayward WS, Neel BG, Astrin SM. 1981. Activation of a cellular onc gene by promoter insertion in ALV-induced lymphoid leukosis. *Nature* **290:** 475–480.

He X, Saint-Jeannet JP, Woodgett JR, Varmus HE, Dawid IB. 1995. Glycogen synthase kinase-3 and dorsoventral patterning in Xenopus embryos. *Nature* **374:** 617–622.

He TC, Sparks AB, Rago C, Hermeking H, Zawel L, da Costa LT, Morin PJ, Vogelstein B, Kinzler KW. 1998. Identification of c-MYC as a target of the APC pathway. *Science* **281:** 1509–1512.

Heasman J, Crawford A, Goldstone K, Garner-Hamrick P, Gumbiner B, McCrea P, Kintner C, Noro CY, Wylie C. 1994. Overexpression of cadherins and underexpression of β-catenin inhibit dorsal mesoderm induction in early Xenopus embryos. *Cell* **79:** 791–803.

Hofmann K. 2000. A superfamily of membrane-bound *O*-acyltransferases with implications for wnt signaling. *Trends Biochem Sci* **25:** 111–112.

* Holstein TW. 2012. The evolution of the Wnt pathway. *Cold Spring Harb Perspect Biol* doi: 10.1101/cshperspect.a007922.

Hsieh JC, Kodjabachian L, Rebbert ML, Rattner A, Smallwood PM, Samos CH, Nusse R, Dawid IB, Nathans J. 1999. A new secreted protein that binds to Wnt proteins and inhibits their activities. *Nature* **398:** 431–436.

Huber O, Korn R, McLaughlin J, Ohsugi M, Herrmann BG, Kemler R. 1996. Nuclear localization of β-catenin by interaction with transcription factor LEF-1. *Mech Dev* **59:** 3–10.

Huber AH, Nelson WJ, Weis WI. 1997. Three-dimensional structure of the armadillo repeat region of β-catenin. *Cell* **90:** 871–882.

Ikeda S, Kishida S, Yamamoto H, Murai H, Koyama S, Kikuchi A. 1998. Axin, a negative regulator of the Wnt signaling pathway, forms a complex with GSK-3β and β-catenin and promotes GSK-3β-dependent phosphorylation of β-catenin. *EMBO J* **17:** 1371–1384.

Ingham PW. 1988. The molecular genetics of embryonic pattern formation in Drosophila. *Nature* **335:** 25–34.

Jiang J, Struhl G. 1998. Regulation of the Hedgehog and Wingless signalling pathways by the F-box/WD40-repeat protein Slimb. *Nature* **391:** 493–496.

Kadowaki T, Wilder E, Klingensmith J, Zachary K, Perrimon N. 1996. The segment polarity gene porcupine encodes a putative multitransmembrane protein involved in Wingless processing. *Genes Dev* **10:** 3116–3128.

Kennerdell JR, Carthew RW. 1998. Use of dsRNA-mediated genetic interference to demonstrate that frizzled and frizzled 2 act in the wingless pathway. *Cell* **95:** 1017–1026.

Kimelman D, Christian JL, Moon RT. 1992. Synergistic principles of development: Overlapping patterning systems in Xenopus mesoderm induction. *Development* **116:** 1–9.

Kinzler KW, Nilbert MC, Su LK, Vogelstein B, Bryan TM, Levy DB, Smith KJ, Preisinger AC, Hedge P, McKechnie D, et al. 1991. Identification of FAP locus genes from chromosome 5q21. *Science* **253:** 661–665.

Kool J, Berns A. 2009. High-throughput insertional mutagenesis screens in mice to identify oncogenic networks. *Nat Rev Cancer* **9:** 389–399.

Korinek V, Barker N, Morin PJ, van Wichen D, de Weger R, Kinzler KW, Vogelstein B, Clevers H. 1997. Constitutive transcriptional activation by a β-catenin-Tcf complex in $APC^{-/-}$ colon carcinoma. *Science* **275:** 1784–1787.

Korinek V, Barker N, Moerer P, van Donselaar E, Huls G, Peters PJ, Clevers H. 1998. Depletion of epithelial stem-cell compartments in the small intestine of mice lacking Tcf-4. *Nat Genet* **19:** 379–383.

Korteweg R. 1936. On the manner in which the disposition to carcinoma of the mammary gland is inherited in mice. *Genetics* **18:** 350–371.

Kusserow A, Pang K, Sturm C, Hrouda M, Lentfer J, Schmidt HA, Technau U, von Haeseler A, Hobmayer B, Martindale MQ, et al. 2005. Unexpected complexity of the Wnt gene family in a sea anemone. *Nature* **433:** 156–160.

Lane P. 1967. *Mouse Newsletter* **36:** 40.

Lee FS, Lane TF, Kuo A, Shackleford GM, Leder P. 1995. Insertional mutagenesis identifies a member of the Wnt gene family as a candidate oncogene in the mammary epithelium of int-2/Fgf-3 transgenic mice. *Proc Natl Acad Sci USA* **92:** 2268–2272.

Lee E, Salic A, Kruger R, Heinrich R, Kirschner MW. 2003. The roles of APC and Axin derived from experimental and theoretical analysis of the Wnt pathway. *PLoS Biol* **1:** E10.

Leyns L, Bouwmeester T, Kim SH, Piccolo S, De Robertis EM. 1997. Frzb-1 is a secreted antagonist of Wnt signaling expressed in the Spemann organizer. *Cell* **88:** 747–756.

Li Y, Welm B, Podsypanina K, Huang S, Chamorro M, Zhang X, Rowlands T, Egeblad M, Cowin P, Werb Z, et al. 2003. Evidence that transgenes encoding components of the Wnt signaling pathway preferentially induce mammary cancers from progenitor cells. *Proc Natl Acad Sci USA* **100:** 15853–15858.

Li V, Ng SS, Boersema P, Low T, Karthaus W, Gerlach J, Mohammed S, Hweck A, Maurice M, Mahmoudi T, et al. 2012. Wnt signaling inhibits proteasomal β-catenin degradation within a compositionally intact Axin1 complex. *Cell* **149:** 1245–1256.

Lin R, Thompson S, Priess JR. 1995. pop-1 encodes an HMG box protein required for the specification of a mesoderm precursor in early C. elegans embryos. *Cell* **83:** 599–609.

Lustig B, Jerchow B, Sachs M, Weiler S, Pietsch T, Karsten U, van de Wetering M, Clevers H, Schlag PM, Birchmeier W, et al. 2002. Negative feedback loop of Wnt signaling through upregulation of conductin/axin2 in colorectal and liver tumors. *Mol Cell Biol* **22:** 1184–1193.

Lyons MJ, Moore DH. 1962. Purification of the mouse mammary tumour virus. *Nature* **194:** 1141–1142.

MacArthur CA, Shankar DB, Shackleford GM. 1995. Fgf-8, activated by proviral insertion, cooperates with the Wnt-1 transgene in murine mammary tumorigenesis. *J Virol* **69:** 2501–2507.

Macdonald BT, Tamai K, He X. 2009. Wnt/β-catenin signaling: Components, mechanisms, and diseases. *Dev Cell* **17:** 9–26.

Maniatis T. 1999. A ubiquitin ligase complex essential for the NF-κB, Wnt/Wingless, and Hedgehog signaling pathways. *Genes Dev* **13:** 505–510.

Mao B, Wu W, Li Y, Hoppe D, Stannek P, Glinka A, Niehrs C. 2001a. LDL-receptor-related protein 6 is a receptor for Dickkopf proteins. *Nature* **411:** 321–325.

Mao J, Wang J, Liu B, Pan W, Farr GH III, Flynn C, Yuan H, Takada S, Kimelman D, Li L, et al. 2001b. Low-density lipoprotein receptor-related protein-5 binds to Axin and regulates the canonical Wnt signaling pathway. *Mol Cell* **7:** 801–809.

McCrea PD, Turck CW, Gumbiner B. 1991. A homolog of the armadillo protein in *Drosophila* (plakoglobin) associated with E-cadherin. *Science* **254:** 1359–1361.

McCrea PD, Brieher WM, Gumbiner BM. 1993. Induction of a secondary body axis in Xenopus by antibodies to β-catenin. *J Cell Biol* **123:** 477–484.

McMahon AP, Bradley A. 1990. The Wnt-1 (int-1) proto-oncogene is required for development of a large region of the mouse brain. *Cell* **62:** 1073–1085.

McMahon AP, Moon RT. 1989. Ectopic expression of the protooncogene int-1 in Xenopus embryos leads to duplication of the embryonic axis. *Cell* **58:** 1075–1084.

Mikels AJ, Nusse R. 2006. Purified Wnt5a protein activates or inhibits β-catenin-TCF signaling depending on receptor context. *PLoS Biol* **4:** e115.

Molenaar M, van de Wetering M, Oosterwegel M, Peterson-Maduro J, Godsave S, Korinek V, Roose J, Destree O, Clevers H. 1996. XTcf-3 transcription factor mediates β-catenin-induced axis formation in Xenopus embryos. *Cell* **86:** 391–399.

Morin PJ, Sparks AB, Korinek V, Barker N, Clevers H, Vogelstein B, Kinzler KW. 1997. Activation of β-catenin-Tcf signaling in colon cancer by mutations in β-catenin or APC. *Science* **275:** 1787–1790.

Muller HA, Samanta R, Wieschaus E. 1999. Wingless signaling in the *Drosophila* embryo: Zygotic requirements and the role of the frizzled genes. *Development* **126:** 577–586.

Nelson WJ, Nusse R. 2004. Convergence of Wnt, β-catenin, and cadherin pathways. *Science* **303:** 1483–1487.

Niehrs C. 2010. On growth and form: A Cartesian coordinate system of Wnt and BMP signaling specifies bilaterian body axes. *Development (Cambridge, England)* **137:** 845–857.

Noordermeer J, Klingensmith J, Perrimon N, Nusse R. 1994. *dishevelled* and *armadillo* act in the *wingless* signalling pathway in *Drosophila*. *Nature* **367:** 80–83.

Nusse R, Varmus HE. 1992. Wnt genes. *Cell* **69:** 1073–1087.

Nusse R, Varmus HE. 1982. Many tumors induced by the mouse mammary tumor virus contain a provirus integrated in the same region of the host genome. *Cell* **31:** 99–109.

Nusse R, van Ooyen A, Cox D, Fung YK, Varmus HE. 1984. Mode of proviral activation of a putative mammary oncogene (int-1) on mouse chromosome 15. *Nature* **307:** 131–136.

Nusse R, Brown A, Papkoff J, Scambler P, Shackleford G, McMahon A, Moon R, Varmus H. 1991. A new nomenclature for int-1 and related genes: The Wnt gene family. *Cell* **64:** 231.

Nüsslein-Volhard C, Wieschaus E. 1980. Mutations affecting segment number and polarity in *Drosophila*. *Nature* **287:** 795–801.

Oishi I, Suzuki H, Onishi N, Takada R, Kani S, Ohkawara B, Koshida I, Suzuki K, Yamada G, Schwabe GC, et al. 2003. The receptor tyrosine kinase Ror2 is involved in non-canonical Wnt5a/JNK signalling pathway. *Genes Cells* **8:** 645–654.

Orford K, Crockett C, Jensen JP, Weissman AM, Byers SW. 1997. Serine phosphorylation-regulated ubiquitination and degradation of β-catenin. *J Biol Chem* **272:** 24735–24738.

Ozawa M, Baribault H, Kemler R. 1989. The cytoplasmic domain of the cell adhesion molecule uvomorulin associates with three independent proteins structurally related in different species. *EMBO J* **8:** 1711–1717.

Parr BA, McMahon AP. 1995. Dorsalizing signal Wnt-7a required for normal polarity of D-V and A-P axes of mouse limb. *Nature* **374:** 350–353.

Parr BA, McMahon AP. 1998. Sexually dimorphic development of the mammalian reproductive tract requires Wnt-7a. *Nature* **395:** 707–710.

Patthy L. 2000. The WIF module. *Trends Biochem Sci* **25:** 12–13.

Payne GS, Courtneidge SA, Crittenden LB, Fadly AM, Bishop JM, Varmus HE. 1981. Analysis of avian leukosis virus DNA and RNA in bursal tumours: Viral gene expression is not required for maintenance of the tumor state. *Cell* **23:** 311–322.

Payne GS, Bishop JM, Varmus HE. 1982. Multiple arrangements of viral DNA and an activated host oncogene in bursal lymphomas. *Nature* **295:** 209–214.

Peifer M, Orsulic S, Pai LM, Loureiro J. 1993. A model system for cell adhesion and signal transduction in *Drosophila. Development* (Suppl 1)**:** 163–176.

Peifer M, Wieschaus E. 1990. The segment polarity gene armadillo encodes a functionally modular protein that is the *Drosophila* homolog of human plakoglobin. *Cell* **63:** 1167–1176.

Peifer M, Rauskolb C, Williams M, Riggleman B, Wieschaus E. 1991. The segment polarity gene armadillo interacts with the wingless signaling pathway in both embryonic and adult pattern formation. *Development* **111:** 1029–1043.

Peifer M, Pai LM, Casey M. 1994. Phosphorylation of the *Drosophila* adherens junction protein Armadillo: Roles for wingless signal and zeste-white 3 kinase. *Dev Biol* **166:** 543–556.

Perrimon N, Mahowald AP. 1987. Multiple functions of segment polarity genes in Drosophila. *Dev Biol* **119:** 587–600.

Peters G, Lee AE, Dickson C. 1986. Concerted activation of two potential proto-oncogenes in carcinomas induced by mouse mammary tumour virus. *Nature* **320:** 628–631.

Petersen CP, Reddien PW. 2009. Wnt signaling and the polarity of the primary body axis. *Cell* **139:** 1056–1068.

Pierce SB, Kimelman D. 1996. Overexpression of Xgsk-3 disrupts anterior ectodermal patterning in Xenopus. *Dev Biol* **175:** 256–264.

Pinson KI, Brennan J, Monkley S, Avery BJ, Skarnes WC. 2000. An LDL-receptor-related protein mediates Wnt signalling in mice. *Nature* **407:** 535–538.

Port F, Basler K. 2010. *Wnt trafficking: New insights into Wnt maturation, secretion and spreading*. Traffic, Copenhagen, Denmark.

Rattner A, Hsieh JC, Smallwood PM, Gilbert DJ, Copeland NG, Jenkins NA, Nathans J. 1997. A family of secreted proteins contains homology to the cysteine-rich ligand-binding domain of frizzled receptors. *Proc Natl Acad Sci USA* **94:** 2859–2863.

Riese J, Yu X, Munnerlyn A, Eresh S, Hsu SC, Grosschedl R, Bienz M. 1997. LEF-1, a nuclear factor coordinating signaling inputs from wingless and decapentaplegic. *Cell* **88:** 777–787.

Riggleman B, Schedl P, Wieschaus E. 1990. Spatial expression of the Drosophila segment polarity gene armadillo is posttranscriptionally regulated by wingless. *Cell* **63:** 549–560.

Rijsewijk F, Schuermann M, Wagenaar E, Parren P, Weigel D, Nusse R. 1987a. The *Drosophila* homolog of the mouse mammary oncogene int-1 is identical to the segment polarity gene wingless. *Cell* **50:** 649–657.

Rijsewijk F, van Deemter L, Wagenaar E, Sonnenberg A, Nusse R. 1987b. Transfection of the int-1 mammary oncogene in cuboidal RAC mammary cell line results in morphological transformation and tumorigenicity. *EMBO J* **6:** 127–131.

Robitaille J, MacDonald ML, Kaykas A, Sheldahl LC, Zeisler J, Dube MP, Zhang LH, Singaraja RR, Guernsey DL, Zheng B, et al. 2002. Mutant frizzled-4 disrupts retinal angiogenesis in familial exudative vitreoretinopathy. *Nat Genet* **32:** 326–330.

Rocheleau CE, Downs WD, Lin R, Wittmann C, Bei Y, Cha YH, Ali M, Priess JR, Mello CC. 1997. Wnt signaling and an APC-related gene specify endoderm in early C. elegans embryos. *Cell* **90:** 707–716.

Rocheleau CE, Yasuda J, Shin TH, Lin R, Sawa H, Okano H, Priess JR, Davis RJ, Mello CC. 1999. WRM-1 activates the LIT-1 protein kinase to transduce anterior/posterior polarity signals in C. elegans. *Cell* **97:** 717–726.

Roelink H, Wagenaar E, Lopes da Silva S, Nusse R. 1990. Wnt-3, a gene activated by proviral insertion in mouse mammary tumors, is homologous to int-1/Wnt-1 and is normally expressed in mouse embryos and adult brain. *Proc Natl Acad Sci USA* **87:** 4519–4523.

Rubinfeld B, Souza B, Albert I, Müller O, Chamberlain SH, Masiarz FR, Munemitsu S, Polakis P. 1993. Association of the APC gene product with β-catenin. *Science (New York, NY)* **262:** 1731–1734.

Rubinfeld B, Albert I, Porfiri E, Fiol C, Munemitsu S, Polakis P. 1996. Binding of GSK3β to the APC-β-catenin complex and regulation of complex assembly. *Science (New York, NY)* **272:** 1023–1026.

Rubinfeld B, Robbins P, El-Gamil M, Albert I, Porfiri E, Polakis P. 1997. Stabilization of β-catenin by genetic defects in melanoma cell lines. *Science* **275:** 1790–1792.

Rulifson EJ, Wu CH, Nusse R. 2000. Pathway specificity by the bifunctional receptor frizzled is determined by affinity for wingless. *Mol Cell* **6:** 117–126.

Sato T, van Es JH, Snippert HJ, Stange DE, Vries RG, van den Born M, Barker N, Shroyer NF, van de Wetering M, Clevers H. 2010. Paneth cells constitute the niche for Lgr5 stem cells in intestinal crypts. *Nature* **469:** 415–418.

Satoh S, Daigo Y, Furukawa Y, Kato T, Miwa N, Nishiwaki T, Kawasoe T, Ishiguro H, Fujita M, Tokino T, et al. 2000. AXIN1 mutations in hepatocellular carcinomas, and growth suppression in cancer cells by virus-mediated transfer of AXIN1. *Nat Genet* **24:** 245–250.

Sawa H, Lobel L, Horvitz HR. 1996. The Caenorhabditis elegans gene lin-17, which is required for certain asymmetric cell divisions, encodes a putative seven-transmembrane protein similar to the Drosophila frizzled protein. *Genes Dev* **10:** 2189–2197.

Schneider SQ, Bowerman B. 2007. β-Catenin asymmetries after all animal/vegetal-oriented cell divisions in Platynereis dumerilii embryos mediate binary cell-fate specification. *Dev Cell* **13:** 73–86.

Schwarz-Romond T, Fiedler M, Shibata N, Butler PJ, Kikuchi A, Higuchi Y, Bienz M. 2007. The DIX domain of Dishevelled confers Wnt signaling by dynamic polymerization. *Nat Struct Mol Biol* **14:** 484–492.

Semenov MV, Tamai K, Brott BK, Kuhl M, Sokol S, He X. 2001. Head inducer Dickkopf-1 is a ligand for Wnt coreceptor LRP6. *Curr Biol* **11:** 951–961.

Shackleford GM, Varmus HE. 1987. Expression of the proto-oncogene int-1 is restricted to postmeiotic male germ cells and the neural tube of mid-gestational embryos. *Cell* **50:** 89–95.

Shackleton M, Vaillant F, Simpson KJ, Stingl J, Smyth GK, Asselin- Labat ML, Wu L, Lindeman GJ, Visvader JE. 2006. Generation of a functional mammary gland from a single stem cell. *Nature* **439:** 84–88.

Sharma RP, Chopra VL. 1976. Effect of the Wingless (wg1) mutation on wing and haltere development in *Drosophila* melanogaster. *Dev Biol* **48:** 461–465.

Sidow A. 1992. Diversification of the *Wnt* gene family on the ancestral lineage of vertebrates. *Proc Natl Acad Sci USA* **89:** 5098–5102.

Siegfried E, Chou TB, Perrimon N. 1992. Wingless signaling acts through zeste-white 3, the Drosophila homolog of glycogen synthase kinase-3, to regulate engrailed and establish cell fate. *Cell* **71:** 1167–1179.

Siegfried E, Wilder EL, Perrimon N. 1994. Components of wingless signalling in Drosophila. *Nature* **367:** 76–80.

Su LK, Kinzler KW, Vogelstein B, Preisinger AC, Moser AR, Luongo C, Gould KA, Dove WF. 1992. Multiple intestinal neoplasia caused by a mutation in the murine homolog of the APC gene. *Science* **256:** 668–670.

Su LK, Vogelstein B, Kinzler KW. 1993. Association of the APC tumor suppressor protein with catenins. *Science* **262:** 1734–1737.

Sugioka K, Mizumoto K, Sawa H. 2011. Wnt regulates spindle asymmetry to generate asymmetric nuclear β-catenin in C. elegans. *Cell* **146:** 942–954.

Takeichi M. 1990. Cadherins: A molecular family important in selective cell-cell adhesion. *Annu Rev Biochem* **59:** 237–252.

Tamai K, Semenov M, Kato Y, Spokony R, Liu C, Katsuyama Y, Hess F, Saint-Jeannet JP, He X. 2000. LDL-receptor-related proteins in Wnt signal transduction. *Nature* **407:** 530–535.

Tamai K, Zeng X, Liu C, Zhang X, Harada Y, Chang Z, He X. 2004. A mechanism for Wnt coreceptor activation. *Mol Cell* **13:** 149–156.

Teich N, Wyke J, Mak T, Bernstein A, Hardy W. 1982. Pathogenesis of Retrovirus-indiced disease. In *RNA tumor viruses* (ed. Weis R, Teich N, Varmus H, Coffin JM), pp. 785–998. Cold Spring Harbor Laboratory Press, Cold Spring Harbor, NY.

ten Berge D, Kurek D, Blauwkamp T, Koole W, Maas A, Eroglu E, Siu RK, Nusse R. 2011. Embryonic stem cells require Wnt proteins to prevent differentiation to epiblast stem cells. *Nat Cell Biol* **13:** 1070–1075.

Thomas KR, Capecchi MR. 1990. Targeted disruption of the murine int-1 proto-oncogene resulting in severe abnormalities in mid-brain and cerebellar development. *Nature* **346:** 847–850.

Thomas KR, Musci TS, Neumann PE, Capecchi MR. 1991. Swaying is a mutant allele of the proto-oncogene Wnt-1. *Cell* **67:** 969–976.

Thorpe CJ, Schlesinger A, Carter JC, Bowerman B. 1997. Wnt signaling polarizes an early C. elegans blastomere to distinguish endoderm from mesoderm. *Cell* **90:** 695–705.

Travis A, Amsterdam A, Belanger C, Grosschedl R. 1991. LEF-1, a gene encoding a lymphoid-specific protein with an HMG domain, regulates T-cell receptor α enhancer function [corrected]. *Genes Dev* **5:** 880–894.

Tsukamoto AS, Grosschedl R, Guzman RC, Parslow T, Varmus HE. 1988. Expression of the int-1 gene in transgenic mice is associated with mammary gland hyperplasia and adenocarcinomas in male and female mice. *Cell* **55:** 619–625.

van Amerongen R, Mikels A, Nusse R. 2008. Alternative wnt signaling is initiated by distinct receptors. *Sci Signal* **1:** re9.

van Amerongen R, Bowman A, Nusse R. 2012. Developmental stage and time dictate the fate of Wnt/β-catenin responsive stem cells in the mammary gland. *Cell Stem Cell* (in press).

van de Wetering M, Oosterwegel M, Dooijes D, Clevers H. 1991. Identification and cloning of TCF-1, a T lymphocyte-specific transcription factor containing a sequence-specific HMG box. *EMBO J* **10:** 123–132.

Van Leeuwen F, Harryman Samos C, Nusse R. 1994. Biological activity of soluble wingless protein in cultured Drosophila imaginal disc cells. *Nature* **368:** 342–344.

van Ooyen A, Nusse R. 1984. Structure and nucleotide sequence of the putative mammary oncogene int-1; proviral insertions leave I the protein-encoding domain intact. *Cell* **39:** 233–240.

van Ooyen A, Kwee V, Nusse R. 1985. The nucleotide sequence of the human int-1 mammary oncogene; evolutionary conservation of coding and non-coding sequences. *EMBO J* **4:** 2905–2909.

Veeman MT, Axelrod JD, Moon RT. 2003. A second canon. Functions and mechanisms of β-catenin-independent Wnt signaling. *Dev Cell* **5:** 367–377.

Vinson CR, Adler PN. 1987. Directional non-cell autonomy and the transmission of polarity information by the frizzled gene of Drosophila. *Nature* **329:** 549–551.

Vinson CR, Conover S, Adler PN. 1989. A Drosophila tissue polarity locus encodes a protein containing seven potential transmembrane domains. *Nature* **338:** 263–264.

Waterman ML, Fischer WH, Jones KA. 1991. A thymus-specific member of the HMG protein family regulates the human T cell receptor C α enhancer. *Genes Dev* **5:** 656–669.

Wehrli M, Dougan ST, Caldwell K, O'Keefe L, Schwartz S, Vaizel-Ohayon D, Schejter E, Tomlinson A, DiNardo S. 2000. *arrow* encodes an LDL-receptor-related protein essential for Wingless signalling. *Nature* **407:** 527–530.

Weinberg RA. 1983. A molecular basis of cancer. *Sci Am* **249:** 126–142.

Wilkinson DG, Bailes JA, McMahon AP. 1987. Expression of the proto-oncogene int-1 is restricted to specific neural cells in the developing mouse embryo. *Cell* **50:** 79–88.

Wnt Proteins

Karl Willert[1] and Roel Nusse[2]

[1]Department of Cellular and Molecular Medicine, University of California, San Diego, California 92093

[2]Department of Developmental Biology, Howard Hughes Medical Institute, Stanford University School of Medicine, Stanford, California 94305

Correspondence: kwillert@ucsd.edu

Wnt proteins comprise a major family of signaling molecules that orchestrate and influence a myriad of cell biological and developmental processes. Although our understanding of the role of Wnt signaling in regulating development and affecting disease, such as cancer, has been ever increasing, the study of the Wnt proteins themselves has been painstaking and slow moving. Despite advances in the biochemical characterization of Wnt proteins, many mysteries remain unsolved. In contrast to other developmental signaling molecules, such as fibroblast growth factors (FGF), transforming growth factors (TGFβ), and Sonic hedgehog (Shh), Wnt proteins have not conformed to many standard methods of protein production, such as bacterial overexpression, and analysis, such as ligand–receptor binding assays. The reasons for their recalcitrant nature are likely a consequence of the complex set of posttranslational modifications involving several highly specialized and poorly characterized processing enzymes. With the recent description of the first Wnt protein structure, the time is ripe to uncover and possibly resolve many of the remaining issues surrounding Wnt proteins and their interactions. Here we describe the process of maturation of Wnt from its initial translation to its eventual release from a cell and interactions in the extracellular environment.

Since the isolation of the first *Wnt* gene in 1982, then called *int-1* (Nusse and Varmus 1982), the study of Wnt has impacted virtually all aspects of developmental biology, from establishing the polarity of a single cell within a tissue to specifying the entire body axis of an organism. At the cellular level, Wnts have been described to regulate stem cell self-renewal, apoptosis, and cell motility. During development, the requirement for Wnt can be observed at stages as early as the first cleavages of the zygote all the way to adulthood, where Wnts regulate tissue homeostasis, such as of the skin and intestine, and, when dysregulated, lead to multiple disorders, such as cancer and premature aging.

Even in light of the vast literature on Wnt, currently totaling approximately 2000 PubMed citations per year, our understanding of the gene products orchestrating these diverse biological processes—the Wnt proteins—is still quite rudimentary. Early analysis of overexpressed *Wnt* genes, in particular *Wnt1* (renamed from *int-1*) (Nusse et al. 1991), revealed important properties of Wnts, including their secretion from cells, glycosylation, and tight association with the cell surface and extracellular matrix. Purification of the first Wnt protein

Cite this article as *Cold Spring Harb Perspect Biol* doi: 10.1101/cshperspect.a007864

(Willert et al. 2003) led to the identification of covalent protein acylation and provided insight into the poor solubility of these secreted glycosylated lipid-modified signaling molecules. However, many mysteries still surround Wnt protein biochemistry and consequently their mode of action. For example, very little is known regarding Wnt binding affinities and specificities to its many substrates, including other secreted molecules, cell surface receptors, and proteoglycans. The recent description of a high-resolution crystal structure of a Wnt protein in complex with a Frizzled receptor binding domain will surely provide many important insights into how these potent growth factors engage their cognate receptors.

Here we describe the current state of our understanding of Wnt proteins from their initial translation, transit through the secretory pathway, where Wnts are extensively modified and processed, to their secretion from the cell and subsequent interactions with components of the extracellular environment. We also highlight some of the major tools available to manipulate Wnt proteins and interrogate their signaling pathways.

THE BASICS: WNT GENES AND PREDICTED PROTEIN PRODUCTS

All metazoan species express *Wnt* genes, with the genome of *Hydra vulgaris* carrying 13 and mice and humans carrying 19 independent genes (additional information can be found on the Wnt homepage: http://wnt.stanford.edu). Based on their primary amino acid sequence, all *Wnt* genes are predicted to encode secreted proteins. The defining property of Wnt proteins is a nearly invariant positioning of 22 cysteine residues, most of which are postulated to form disulfide bridges that maintain a globular secondary structure. Some studies have suggested that Wnt proteins exist as dimers (Burrus and McMahon 1995; Cha et al. 2008) maintained by intermolecular disulfide linkages; however, this has not been observed to be the case for purified and biologically active Wnt proteins. In addition, the recent high-resolution structure of a Wnt protein suggests that all conserved cysteine residues are occupied in intramolecular rather than intermolecular disulfide bridges (Janda et al. 2012). Wnt proteins carry several stretches of highly charged amino acids and have a predicted isoelectric point of nearly 9, which, in combination with multiple glycosylation events, would lead one to expect that Wnt proteins are readily soluble in an aqueous environment.

The primary amino acid sequence of Wnt shows several hallmarks of secreted proteins, most notably a signal sequence for secretion, a stretch of approximately 20 hydrophobic amino acids. Cleavage of this signal peptide can be predicted using several computer algorithms; however, the true first amino acid of a Wnt protein was identified by amino-terminal sequencing of a purified Wnt protein (Willert et al. 2003). Interestingly, in the case of Wnt5a, the amino-terminal residue was found to be located 62 amino acids from the predicted translational start site (Mikels and Nusse 2006). This observation serves as a cautionary note for those who wish to append amino-terminal tags (His, HA, or GFP) onto Wnts, because such tags may be cleaved from the mature protein upon signal sequence cleavage. In addition, with only a few exceptions, tagged *Wnt* genes produce proteins with significantly lower activity (K Willert and R Nusse, unpubl.), and caution should be exercised with the interpretation of any data using tagged *Wnt* genes.

Aside from targeting Wnt proteins for secretion, the amino terminus may harbor additional critical biological functions. A survey of isoform and alternative splicing databases reveals that multiple *Wnt* genes carry distinct 5′ untranslated regions (UTRs) and are predicted to encode distinct amino termini. In the case of Wnt16, two isoforms with distinct 5′ UTRs are expressed from alternative promoters (Fear et al. 2000). Although little is known regarding the biological significance of these two isoforms, it is intriguing that one of the two isoforms has a broad expression pattern, whereas the other is restricted to the pancreas. Changes in the amino terminus of Wnt proteins may represent a common mechanism by which signaling activity can be affected, as indicated by the studies on *Tiki*, a gene encoding a transmembrane protein that antagonizes Wnt signaling (Zhang et al. 2012). Tiki protein acts cell-autonomously as a protease to cleave eight amino terminal residues from a Wnt protein, thereby reducing receptor-binding and signaling activities. Such observations suggest that the Wnt family of proteins is significantly more complex and diverse than expected for 19 genes.

It should be noted that in contrast to Wnt signaling, which has been conveniently—and perhaps inappropriately—categorized as either canonical or noncanonical, no sequence or structural basis for this distinction has been identified in Wnt proteins. Although many studies make reference to either canonical or noncanonical Wnts, this difference is most likely conferred by cellular context as determined by the expressed repertoire of receptors and signal transducers rather than by an intrinsic property of the Wnt proteins. The hypothesis that Wnt signaling activity is conferred by cellular context rather than by the Wnt protein sequence is supported by the observations that a so-called noncanonical Wnt5a can act "canonically" by activating β-catenin signaling in certain contexts (He et al. 1997; Mikels and Nusse 2006). Furthermore, maternal Wnt11, which has been largely studied for its roles in noncanonical Wnt signaling, specifies the dorsal axis in *Xenopus* by localizing β-catenin to dorsal nuclei, thus promoting a canonical signaling pathway (Tao et al. 2005).

INTRACELLULAR PROCESSING OF WNT

Upon translation and targeting to the endoplasmic reticulum (ER), Wnt proteins associate with multiple processing enzymes that chaperone Wnts on their journey to the extracellular space. The intermediate vesicle compartments through which Wnt passes and the order in which these posttranslational modifications are added have not been precisely elucidated. The two most prominent modifications on Wnt are glycosylation and acylation. The number of glycosylation attachments varies significantly between Wnts, for example, Wnt1 carries four and Wnt3a carries two *N*-linked glycosylations, whereas the most distantly related Wnt homolog, *Drosophila* WntD, appears to be devoid of any glycosylations or any modification because its mass is identical to the predicted mass based on amino acid content (Ching et al. 2008). When overexpressed, Wnts often resolve as multiple bands in immunoblots, which return to a single species upon treatment with glycosidases or site-directed mutagenesis of the individual predicted glycosylation sites. Interestingly, in the rare cases in which Wnt proteins have been detected in soluble form (Wnt1, Wnt3a, Wnt5a, Wingless [Wg], and WntD), the protein migrates as a single species, suggesting that upon complete intracellular processing, the composition of Wnt is quite homogeneous. Site-directed mutagenesis of individual glycosylation sites has minor effects on Wnt1 activity in overexpression assays (Mason et al. 1992), whereas secretion of Wnt3a and Wnt5a glycosylation mutants is significantly impaired (Komekado et al. 2007; Kurayoshi et al. 2007), suggesting that glycosylation likely plays a critical role in control over Wnt folding and subsequent secretion.

In contrast to glycosylation, acylation is absolutely essential for Wnt activity. Although early studies suggested that glycosylation is not critical for Wnt signaling (Mason et al. 1992), subsequent studies have argued that nonglycosylated Wnt is not acylated and consequently not secreted (Komekado et al. 2007). Initial mass spectrometry studies of purified Wnt3a identified two types of covalently attached acyl groups: a palmitic acid linked via a thioester to a conserved cysteine residue (Willert et al. 2003) and a palmitoleic acid linked via an oxyester to a conserved serine residue (Takada et al. 2006). However, the recently published crystal structure of *Xenopus* Wnt8 in complex with a Frizzled CRD indicates that only the conserved serine residue is lipidated, whereas the conserved cysteine is occupied in a disulfide linkage, thus arguing that Wnt proteins are modified by a single lipid moiety (Janda et al. 2012). Mutagenesis of the serine lipid modification sites (S209 in mouse Wnt3a) yielded nonfunctional and poorly secreted protein (Takada et al. 2006). In contrast, Wg carrying a mutation of the predicted acyl-modified serine (S239) is secreted and shows poor signaling activity (Franch-Marro et al. 2008a). Despite this apparent discrepancy between Wnt3a and Wg, the emerging model is that acylation of Wnt is required for proper intracellular processing and signaling activity.

The mechanism by which a lipid is attached to the Wnt polypeptide backbone has not been elucidated at the biochemical level. However, the critical role for the resident ER protein Porcupine (Porcn) in Wnt processing (van den Heuvel et al. 1993; Kadowaki et al. 1996) and its homology to membrane-bound *O*-acyltransferases (MBOAT) (Hofmann 2000) have led to the hypothesis that Porcn catalyzes the transfer of a lipid to serine. The requirement for acylation in Wnt function is highlighted by the fact that mutations in *Porcn* ablate all Wnt signaling and result in early embryonic lethality in mice (Barrott et al. 2011; Biechele et al. 2011) with embryos failing to gastrulate and form mesoderm. Mutations in human *PORCN*, an X-linked gene, lead to a rare genetic disorder, called focal dermal hypoplasia (FDH), characterized by skin abnormalities and a host of developmental malformations and defects (Grzeschik et al. 2007; Wang et al. 2007). Mutations in *PORCN* are lethal in males, consistent with the early embryonic lethality observed in mouse knockouts. In contrast, females survive, albeit with a host of defects, owing to random X-inactivation, which produces mosaic PORCN expression with cells that express PORCN rescuing those that do not.

Although glycosylation and acylation appear to occur on all Wnt proteins (with the exception of the distantly related *Drosophila* WntD) (Ching et al. 2008), several other post-translational modifications, which were not revealed in the high-resolution structure of *Xenopus* Wnt8, may be restricted to specific Wnts. For example, tyrosine sulfation has been shown to regulate Wnt5a and 11 hetero-oligomer formation, and the resulting Wnt multimer shows higher signaling activity than either individual Wnt (Cha et al. 2009). Using antisense oligonucleotides to deplete specific maternal mRNAs, the investigators of this study showed that the enzyme tyrosyl-protein sulfotransferase-1 (TPST-1) is required for *Xenopus* dorsal axis formation and for *O*-sulfation of specific tyrosine residues on Wnt5a and 11.

Additionally, the ER-resident Oto protein may promote the addition of glycophosphatidylinositol (gpi)-like anchors to Wnt1 and 3a, thereby increasing their hydrophobicity and ER retention (Zoltewicz et al. 2009), although the site of this modification is not known. Overexpression of gpi-specific phospholipase D (GPI-PLD) leads to the release of Wnt protein, presumably by cleavage of the gpi anchor. Interestingly, the phospholipases D1 and D2 are direct targets of Wnt/β-catenin signaling and act in a positive-feedback loop to increase signaling in cancer (Kang and Min do 2010; Kang et al. 2010). A possible mechanism for this increase in Wnt signaling may be through the release of ER-retained Wnt proteins. These studies also provide a possible mechanistic basis for poor Wnt secretion.

STRUCTURE OF WNT PROTEINS

While it required 20 years to obtain a pure and biologically active Wnt protein, it took another 10 years to achieve its crystallization and provide a high-resolution structure of a Wnt protein. The feat was accomplished in Chris Garcia's laboratory and has revealed several interesting properties of Wnt proteins (Janda et al. 2012). The 3.25 Å crystal structure was determined for *Xenopus* Wnt8 (XWnt8) in complex with the CRD of mouse Frizzled8 and reveals a highly unusual two-domain structure with amino-terminal and carboxy-terminal domains (NTD and CTD) forming a protein fold previously not identified in any other protein structure (Fig. 1). The NTD is composed of a cluster of α-helices with 10 of the conserved cysteine residues forming five disulfide bridges, whereas the CTD is dominated by two β-sheets and maintained by six disulfide bridges.

In this structure, Wnt extends a thumb from the NTD and an index finger from the CTD to grasp the globular Frizzled CRD. Interestingly, the thumb extends a lipid at serine 187, the highly conserved residue previously identified to carry a covalently attached palmitoleic acid

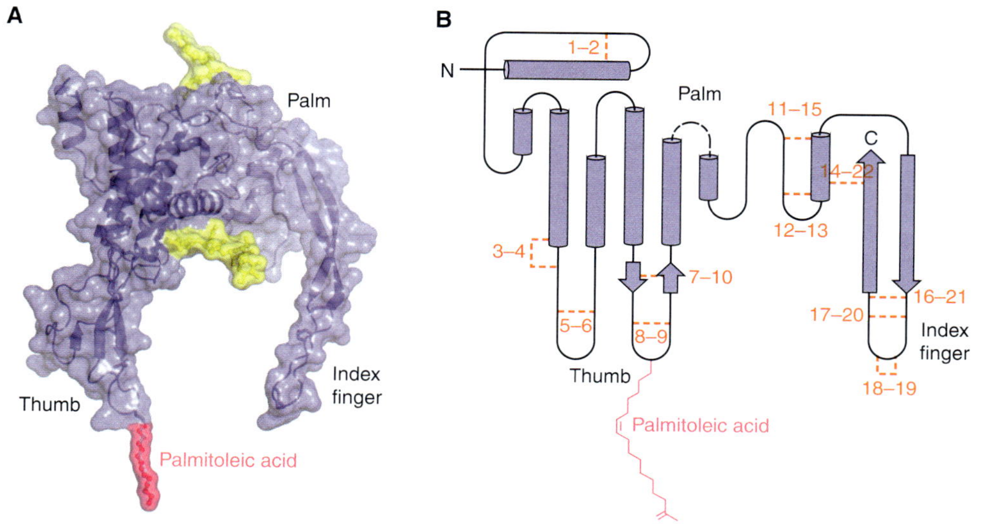

Figure 1. Structure of Wnt. (*A*) Space-filling model of XWnt8. The Frizzled CRD structure has been removed. (Yellow clusters) The *N*-linked glycosylations in XWnt8. (*B*) Secondary structure for Wnt. (Orange) The conserved 22 cysteine residues are numbered to indicate the pairs that form disulfide bridges. *N*-linked glycosylations are not shown because the numbers and positions of *N*-linked glycosylations are highly variable among Wnts. (Dashed line) The approximate position of the linker region where Wg carries an insert of about 80 amino acids. (Figures were generated with the kind assistance of C. Janda and C. Garcia, Stanford University.)

(Takada et al. 2006). This protein–lipid thumb structure is nestled in a deep hydrophobic groove of the CRD, where it makes multiple contacts to completely traverse the CRD surface. As is the case with the thumb, the points of contact of the index finger in the CTD with the CRD are also dominated by hydrophobic and highly conserved residues. Some of the contact points on the Frizzled8 CRD are substituted in other Frizzled CRDs, thus providing a possible mechanism to influence Wnt–Frizzled binding specificities. To further extend the hand analogy for the Wnt structure, the region between the thumb and the index finger is akin to the palm, where the NTD and CTD are intimately associated. Interestingly, the solvent-exposed linker region between NTD and CTD corresponds to the region with the greatest flexibility among Wnt proteins; in particular, Wg carries a large insert of 80 amino acids not present in other Wnts. This "Wg insert" has been used to generate arguably the best antisera to any Wnt protein (van den Heuvel et al. 1989, 1993), further supporting the model that this nonconserved linker region is solvent exposed and likely not involved in binding to the Frizzled CRD.

This Wnt structure will enable a more rational approach to interrogate Wnt–Frizzled interactions and design Wnt agonists and antagonists. A remaining unanswered question concerns the structure of an uncomplexed Wnt protein. The covalently attached lipid necessitates some type of interaction, either with a carrier protein, such as Swim (Mulligan et al. 2012), or with membranous domains, to shield this hydrophobic moiety in a largely aqueous environment. The highly accessible presentation of the lipid moiety on Wnt makes the interaction with carrier molecules or the plasma membrane a likely scenario. An alternative possibility is that the hydrophobic portions of the thumb and index finger fold in to form a fist; however, the current crystal structure does not provide evidence for such intramolecular folding.

SECRETION OF WNT

As ER and Golgi-processed Wnt protein is targeted for secretion (Fig. 2), a sorting receptor encoded by the *Wntless* gene (Wls, also known as Eveness interrupted/Evi, Sprinter, MIG-14, and Gpr177), a multispan transmembrane protein, binds and accompanies Wnt to the cell surface (Banziger et al. 2006; Bartscherer et al. 2006; Goodman et al. 2006). Wls binding of Wnt3a requires acylation on Ser-209, indicating that it acts downstream from Porcn. A comprehensive mutational analysis indicated that acylation of the Ser equivalent of Wg^{S239} is required for the interaction of Wls with all Wnts, except WntD (Herr and Basler 2011). Furthermore, by using a membrane-tethered Wg protein, WgNRT (Zecca et al. 1996), Herr and Basler provided evidence that mere membrane association is not sufficient for Wg association with Wls, thus suggesting that acylation by Porcn enables Wnt's functional interaction with Wls. In addition, in the case of Wg, glycosylation on conserved residues does not affect the dependence of Wnt on Wls.

Vacuolar acidification is required for release of Wnt protein (Coombs et al. 2010), and small drug inhibition of the V-ATPase, a proton pump required for vacuolar acidification, prevents Wls from releasing Wnt so that Wnt–Wls complexes accumulate both in cells and at the plasma membrane (Coombs et al. 2010). However, although essential, a decrease in pH is not sufficient to dissociate the Wnt–Wls complex.

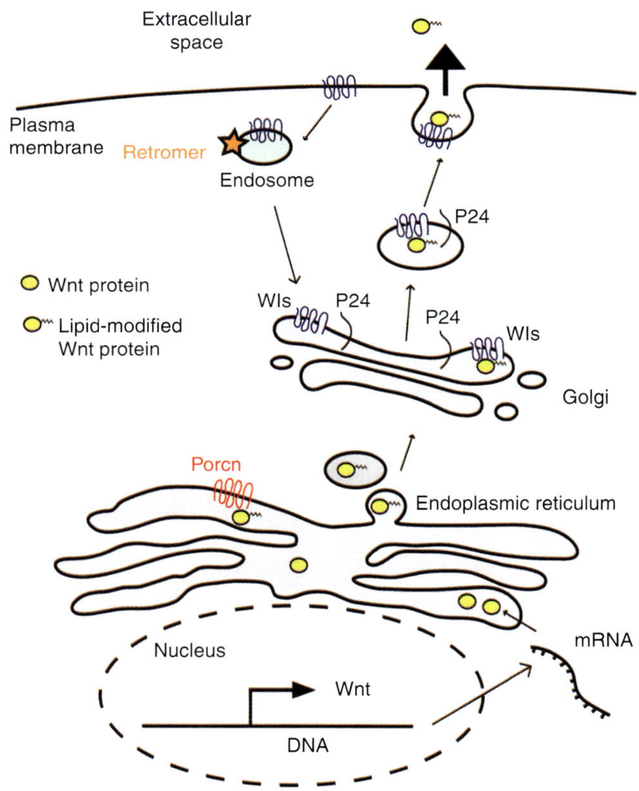

Figure 2. Wnt secretion. Upon translation, Wnt proteins (yellow ovals) undergo a series of modifications as they transit through the secretory pathway and associate with several proteins, including Porcn in the endoplasmic reticulum and Wls in the Golgi apparatus. Efficient secretion of Wnt also requires the recycling of Wls via the retromer complex.

 Cite this article as *Cold Spring Harb Perspect Biol* doi: 10.1101/cshperspect.a007864

RNA interference screens in two separate laboratories identified members of the p24 protein family as cargo receptors for Wnt in anterograde transport and secretion (Buechling et al. 2011; Port et al. 2011). Although the two groups differ slightly on which of the nine fly p24 family members are involved in Wg secretion, they agree in the basic finding that these cargo proteins specifically regulate Wg secretion and that secretion of other signaling molecules (e.g., Decapentaplegic, Hedgehog, and Unpaired) is unaffected. Therefore, Wnt protein does not exit a cell through passive transport or bulk flow but requires specific cargo proteins, such as the p24 family of highly conserved transmembrane receptors, to exit from the ER.

Once Wnt is released from the cell, Wls is recycled via endosomes and the retromer complex to the Golgi, where it acts again to escort a newly processed Wnt protein to the cell surface (Coudreuse et al. 2006; Prasad and Clark 2006; Belenkaya et al. 2008; Franch-Marro et al. 2008b; Port et al. 2008; Yang et al. 2008). As a result, interfering with either Wls expression or its recycling via the retromer inhibits Wnt secretion. In contrast to the classical retromer complex, which involves the sorting nexins SNX1–SNX2 and SNX5–SNX6 (referred to as SNX-BAR sorting nexins) and cargo-selective VPS26, VPS29, and VPS35, Wls recycling requires the distantly related sorting nexin SNX3 (Harterink et al. 2011; Zhang et al. 2011). Therefore, retrograde transport of Wls is distinct from other recycled cargo, raising the possibility of specifically interfering with Wnt secretion through targeted disruption of Wls recycling via SNX3.

If retromer components Vps35 and Vps26 are mutant, Wls is targeted for degradation rather than for endosomal recycling to the Golgi so that it is no longer available to facilitate Wnt secretion. Such defects in the retromer complex can be rescued by overexpression of Wls. In *Caenorhabditis elegans*, it has been shown that the recycling of the Wls homolog MIG-14 from the cell surface to the Golgi requires the MTM-6/MTM-9 myotubularin complex, which dephosphorylates PIP_3, a central regulator of endosomal trafficking (Silhankova et al. 2010). A conserved endocytosis motif required for Wls recycling has been identified in the third intracellular loop of Wls; its mutation results in Wls accumulation on the cell surface and impairs Wg secretion and signaling (Gasnereau et al. 2011).

Although it is clear that the Wls/retromer system is essential for Wnt secretion, studies performed in flies and worm have reached apparently contradicting conclusions with respect to the signaling range of Wnt secreted from retromer-defective cells. Long-range, but not short-range action of Egl-20 (a worm Wnt) was impaired in retromer mutants, suggesting that the retromer complex is important for packaging Wnt for long-range signaling (Coudreuse et al. 2006). In contrast, flies carrying mutations in Vps35 are defective in short-range signaling as evidenced by strong reduction in expression of the Wg target *senseless* (Franch-Marro et al. 2008b).

In motor neurons in *Drosophila*, Wls not only transports Wnt from the Golgi to the plasma membrane, but also functions to shuttle Wnt across the synaptic cleft of the neuromuscular junction in exosome-like vesicles (Korkut et al. 2009). Furthermore, postsynaptically in the muscle, Wls also guides Wg-activated Frizzled-2 trafficking before Frizzled-2 is proteolytically cleaved and its carboxyl terminus is imported into the nucleus (Mathew et al. 2005; Ataman et al. 2006; Korkut et al. 2009).

Structural modeling suggests that the ER luminal portion of Wls contains a lipocalin-family fold (Coombs et al. 2010), which has been shown to interact with lipids. Interestingly, a lipocalin fold has also been predicted by NMR analysis of the Wnt inhibitory factor (WIF) domain (Liepinsh et al. 2006; Malinauskas 2008), which is found in the secreted Wnt antagonist WIF1 and the Wnt receptor Ryk. Furthermore, a fly lipocalin, named Swim (secreted Wnt interacting molecule) has been proposed to act as Wnt chaperone and shield the hydrophobic moieties on Wnt, thereby enabling efficient Wnt diffusion or transport in a largely aqueous environment (Mulligan et al. 2012). Expression of certain lipid-binding proteins, including Lipocalin2 and FABP5, is up-regulated in cells overexpressing Wnt1 (Ziegler et al. 2005,

2007). Additionally, Wls/Gpr177 is also a target of Wnt signaling in mouse cells (Fu et al. 2009), suggesting the presence of feed-forward regulation that ensures sufficient chaperones are available to usher Wnt through the secretory pathway. However, in flies, Wls appears not to be a target of canonical Wnt signaling (Herr and Basler 2011). Alternatively, it is tempting to speculate that Wnt expression triggers a signaling system similar to that of the unfolded protein response, thereby activating expression of genes that encode fatty acid-binding proteins, such as lipocalins, which facilitate the export of hydrophobic Wnt proteins from the ER and to the cell surface. During the transit from the Golgi to the plasma membrane, Wnt protein is passed from Wls to a lipocalin, which then accompanies it in the extracellular space and may regulate its distribution. However, it has been shown that the secretion of the Swim protein is independent of Wg (Mulligan et al. 2012).

Following release from the secretory machinery, it has been suggested that Wnts become tethered to the plasma membrane via their lipid moieties, a feasible model given the lipid's accessibility as revealed by the protein structure (Janda et al. 2012). Although this is a formal possibility, it is clear that a significant proportion of Wnt (e.g., Wnt3a and Wnt5a) can be purified in a biologically active form from conditioned media. Alternatively, Wnt proteins may become incorporated into lipoprotein complexes. In *Drosophila*, Wg (and Hh) was found to colocalize and associate with the lipoprotein Lipophorin, a particle with similar characteristics to ApoB-based lipoproteins (Panakova et al. 2005). Knockdown of Lipophorin expression by RNA interference significantly reduced the signaling range of both Wg and Hh, suggesting that lipoprotein complexes regulate long-range Wnt signaling. In mammalian cells, Wnt3a was shown to cofractionate by KBr isopycnic density centrifugation with the lipoprotein marker hApoB100 and was associated with both high- and low-density lipoproteins (HDL and LDL) (Neumann et al. 2009). Furthermore, addition of HDL, but not LDL, supported Wnt3a solubility in medium containing delipidated fetal calf serum. Taken together with its hydrophobic properties, Wnt solubility in the extracellular environment necessitates molecules or complexes that interact with hydrophobic moieties.

WNTs IN THE EXTRACELLULAR ENVIRONMENT

As they reach the cell surface, Wnt proteins encounter a myriad of interacting molecules. Early biochemical studies revealed a high affinity of Wnts for polyanionic compounds, including heparin and suramin (Brown et al. 1987; Papkoff et al. 1987; Bradley and Brown 1990; Burrus and McMahon 1995). Multiple biochemical and genetic studies since then have underscored the importance of these interactions between Wnt and extracellular carbohydrate chains. Multiple genes involved in heparan sulfate (HS) synthesis, including *sugarless*, *sulfateless*, *tout-velu*, *sister of tout-velu*, and *brother of tout-velu*, regulate Wg signaling (Binari et al. 1997; Hacker et al. 1997; Haerry et al. 1997; Lin and Perrimon 1999; Bornemann et al. 2004; Han et al. 2004; Takei et al. 2004). Although mutations in these genes do not completely block signaling, clear *wg*-related phenotypes are observed, suggesting an important modulatory role for these enzymes in Wnt signaling.

Biochemical studies have revealed that components of the extracellular space, such as Glypicans (Capurro et al. 2005a,b) and Biglycan (Berendsen et al. 2011), modulate Wnts–receptor interactions and can significantly influence signaling output and strength. Mutations of zebrafish Knypek, a heparan sulfate proteoglycan (HSPG), leads to defects in convergent extension movements, a process regulated by Wnt11 (Topczewski et al. 2001) and controlling the location of the centrosome/microtubule-organizing center (MTOC) relative to the cell nucleus and the body axes (Sepich et al. 2011). HSPGs have also been found to maintain solubility and activity of Wnt protein and prevent their aggregation in low serum conditions (Fuerer et al. 2010). Furthermore, the degree of *O*-sulfation of HS has been shown to regulate Wnt signaling activity (Ai et al. 2003). Desulfation of heparin and HS chains of Glypican1 by the cell surface sulfatase QSulf1 reduces Wnt binding, thus producing a low-affinity HS–Wnt complex that can more effectively engage their receptors and initiate Wnt signal transduction.

In addition to their complex interactions with glycans in the extracellular space, Wnts encounter multiple protein-binding partners, including WIF, Sfrp, and cell surface receptors. Subsequent articles in this collection address these interactions in detail.

TOOLS AND METHODS TO STUDY WNT PROTEINS AND SIGNALING

Wnt Proteins

The purification of Wnt eluded many attempts for two decades. Initial purification strategies focused on the founding Wnt protein, Wnt1 (initially referred to as Int-1), and found it to be poorly soluble and largely associated with the extracellular matrix and the cell surface (Brown et al. 1987; Papkoff et al. 1987). The first cell-free and biologically active Wnt proteins were Wg (van Leeuwen et al. 1994) and mouse Wnt3a (Shibamoto et al. 1998). These soluble Wnt proteins have been used extensively in the form of conditioned medium harvested from cells overexpressing Wnt. Although convenient, these Wnt-conditioned media harbor countless contaminating proteins and activities that may complicate the interpretation of any biological assay.

Despite extremely low expression levels ($\sim$100 μg Wnt3a/L), a sensitive and reliable activity assay involving β-catenin stabilization enabled the purification of the first biologically active Wnt protein, Wnt3a (Willert et al. 2003). This purification scheme has been applied to other Wnt proteins with varying degrees of success. Whereas Wnt3a and Wnt5a (Mikels and Nusse 2006) were readily purified using nearly identical protocols, purification of other Wnt proteins has proven more difficult because they tend to form protein aggregates lacking biological activity (K Willert and R Nusse, unpubl.). At present, the reasons for such differences in the biochemical properties of Wnts are unknown, leaving us with the current conclusion that not all Wnt proteins are created equal. Further advances in Wnt protein biochemistry are certain to come, and the recently published high-resolution structure of a Wnt protein (Janda et al. 2012) will likely provide insight into many remaining mysteries in Wnt biochemistry.

Wnt Agonists and Antagonists (Proteins)

Aside from Wnts, several other proteins with Wnt signaling activities have been identified, including Norrin (Xu et al. 2004; Junge et al. 2009) and R-spondin (Kazanskaya et al. 2004; Kim et al. 2006; Nam et al. 2006). Recent studies identified the receptor for R-spondin to be encoded by the Lgr5 homologs (Carmon et al. 2011; de Lau et al. 2011; Glinka et al. 2011). Interestingly, Lgr5, which was previously identified as a Wnt target gene and stem cell marker in intestinal crypts (Barker et al. 2007), associates with the Frizzled/Lrp Wnt–receptor complexes. Therefore, the emerging model involves a complex of cell surface molecules in which Wnt signals acting through Frizzled/Lrp are potentiated by an R-spondin/Lgr receptor complex. Removal of R-spondin1 or mutation of Lgr4/5 in cultured crypt organoids can be rescued by small molecule Wnt pathway agonists (see below), suggesting that the R-spondin/Lgr signaling axis serves to enhance Wnt signaling. Taken together, although R-spondins by themselves have no known intrinsic Wnt signaling activity, they may serve to elevate and thereby uncover endogenous Wnt signaling activities.

Several genetic and biochemical studies have identified multiple secreted proteins that potently antagonize Wnt signaling, including DKK, Sfrp, WIF (for review, see Cruciat and Niehrs 2012), and Klotho (Liu et al. 2007). The protein Wise/Sclerostin can act both as an activator or inhibitor depending on cellular context (Itasaki et al. 2003). These proteins and their mode of action are discussed in detail in a subsequent article. Additionally, several Wnt antagonists have been engineered by fusing Wnt-binding domains, such as the cysteine-rich domain (CRD) of Frizzled, to an affinity tag, such as the constant region of human immunoglobulin heavy chain.

Wnt Agonists and Antagonists (Small Molecules)

The most widely used class of small molecule agonists of Wnt signaling inhibits GSK3. Most prominent among these is lithium, which at millimolar concentrations potently inhibits GSK3 and leads to β-catenin stabilization (Klein and Melton 1996). Several other small molecule GSK3 inhibitors with IC_{50} values in the nanomolar to micromolar ranges have been identified (Bregman et al. 2004; Sato et al. 2004) and are commercially available. These GSK3 inhibitors potently activate the Wnt pathway, but their use can be problematic because GSK3 represents a major signaling hub for multiple other signaling pathways and its inhibition will most certainly affect other pathways in addition to Wnt.

A chemical library screen led to the identification of two potent and highly specific inhibitors of Wnt signaling, called IWP and IWR (Chen et al. 2009). IWR interacts and stabilizes Axin protein, thereby accelerating β-catenin degradation. IWP interacts with PORCN and inhibits its acyl transferase activity. Because PORCN activity is essential for processing of all Wnts, its inhibition will likely block secretion of all Wnts, thus producing the equivalent of an "all Wnt mutant phenotype." Importantly, inhibition of endogenous Wnt signaling with this drug in cell culture can be readily rescued by providing exogenous Wnt protein (e.g., see Ten Berge et al. 2011). In addition, because PORCN function is dedicated to Wnt processing, it is likely that treatment of cells with IWP will not affect any other processes besides Wnt signal transduction.

CONCLUDING REMARKS

With their diverse functions and roles in countless biological processes, isolation and further characterization of Wnt proteins will certainly provide insights into mechanisms and modes of action. Its interactions with receptors, coreceptors, proteoglycans, and secreted proteins underscore the critical role of Wnt proteins as major players in regulating and controlling cell behavior and fate. Many obstacles still need to be overcome to provide us with a better understanding of Wnts' actions. For example, measuring affinities with which each of the 19 Wnts interacts with the various components it encounters in the extracellular environment will lead to a better understanding of how signaling specificities are achieved. In addition, identification and characterization of Wnt-containing complexes (e.g., Wnt–Rspondin–Fzd–Lgr) is critical in controlling Wnt activities, because formation of such complexes likely affects signaling outputs. Finally, with the first crystal structure of a Wnt protein in hand, we will now be able to take a more rational approach in elucidating and dissecting Wnt interactions and binding activities.

ACKNOWLEDGMENTS

We are grateful to C.Y. Janda and K.C. Garcia for their kind assistance in generating the space-filling model of Wnt shown in Figure 1. Research on Wnt proteins in our laboratories is supported by grants from the California Institute of Regenerative Medicine (RB1-010406 to K.W. and RC1-00133-1 to R.N.). K.W. has support from the National Institutes of Health and start-up funding from the UCSD Stem Cell Program. R.N. is a Howard Hughes Investigator.

REFERENCES

**Reference is also in this collection.*

Ai X, Do AT, Lozynska O, Kusche-Gullberg M, Lindahl U, Emerson CP Jr. 2003. QSulf1 remodels the 6-*O* sulfation states of cell surface heparan sulfate proteoglycans to promote Wnt signaling. *J Cell Biol* **162:** 341–351.

Ataman B, Ashley J, Gorczyca D, Gorczyca M, Mathew D, Wichmann C, Sigrist SJ, Budnik V. 2006. Nuclear trafficking of *Drosophila* Frizzled-2 during synapse development requires the PDZ protein dGRIP. *Proc Natl Acad Sci* **103:** 7841–7846.

Banziger C, Soldini D, Schutt C, Zipperlen P, Hausmann G, Basler K. 2006. Wntless, a conserved membrane protein dedicated to the secretion of Wnt proteins from signaling cells. *Cell* **125:** 509–522.

Barker N, van Es JH, Kuipers J, Kujala P, van den Born M, Cozijnsen M, Haegebarth A, Korving J, Begthel H, Peters PJ, et al. 2007. Identification of stem cells in small intestine and colon by marker gene *Lgr5*. *Nature* **449:** 1003–1007.

Barrott JJ, Cash GM, Smith AP, Barrow JR, Murtaugh LC. 2011. Deletion of mouse Porcn blocks Wnt ligand secretion and reveals an ectodermal etiology of human focal dermal hypoplasia/Goltz syndrome. *Proc Natl Acad Sci* **108:** 12752–12757.

Bartscherer K, Pelte N, Ingelfinger D, Boutros M. 2006. Secretion of Wnt ligands requires Evi, a conserved transmembrane protein. *Cell* **125:** 523–533.

Belenkaya TY, Wu Y, Tang X, Zhou B, Cheng L, Sharma YV, Yan D, Selva EM, Lin X. 2008. The retromer complex influences Wnt secretion by recycling Wntless from endosomes to the *trans*-Golgi network. *Dev Cell* **14:** 120–131.

Berendsen AD, Fisher LW, Kilts TM, Owens RT, Robey PG, Gutkind JS, Young MF. 2011. Modulation of canonical Wnt signaling by the extracellular matrix component biglycan. *Proc Natl Acad Sci* **108:** 17022–17027.

Biechele S, Cox BJ, Rossant J. 2011. Porcupine homolog is required for canonical Wnt signaling and gastrulation in mouse embryos. *Dev Biol* **355:** 275–285.

Binari RC, Staveley BE, Johnson WA, Godavarti R, Sasisekharan R, Manoukian AS. 1997. Genetic evidence that heparin-like glycosaminoglycans are involved in wingless signaling. *Development* **124:** 2623–2632.

Bornemann DJ, Duncan JE, Staatz W, Selleck S, Warrior R. 2004. Abrogation of heparan sulfate synthesis in *Drosophila* disrupts the Wingless, Hedgehog and Decapentaplegic signaling pathways. *Development* **131:** 1927–1938.

Bradley RS, Brown AM. 1990. The proto-oncogene *int-1* encodes a secreted protein associated with the extracellular matrix. *EMBO J* **9:** 1569–1575.

Bregman H, Williams DS, Atilla GE, Carroll PJ, Meggers E. 2004. An organometallic inhibitor for glycogen synthase kinase 3. *J Am Chem Soc* **126:** 13594–13595.

Brown AM, Papkoff J, Fung YK, Shackleford GM, Varmus HE. 1987. Identification of protein products encoded by the proto-oncogene *int-1*. *Mol Cell Biol* **7:** 3971–3977.

Buechling T, Chaudhary V, Spirohn K, Weiss M, Boutros M. 2011. p24 proteins are required for secretion of Wnt ligands. *EMBO Rep* **12:** 1265–1272.

Burrus LW, McMahon AP. 1995. Biochemical analysis of murine Wnt proteins reveals both shared and distinct properties. *Exp Cell Res* **220:** 363–373.

Capurro MI, Shi W, Sandal S, Filmus J. 2005a. Processing by convertases is not required for glypican-3-induced stimulation of hepatocellular carcinoma growth. *J Biol Chem* **280:** 41201–41206.

Capurro MI, Xiang YY, Lobe C, Filmus J. 2005b. Glypican-3 promotes the growth of hepatocellular carcinoma by stimulating canonical Wnt signaling. *Cancer Res* **65:** 6245–6254.

Carmon KS, Gong X, Lin Q, Thomas A, Liu Q. 2011. R-spondins function as ligands of the orphan receptors LGR4 and LGR5 to regulate Wnt/β-catenin signaling. *Proc Natl Acad Sci* **108:** 11452–11457.

Cha SW, Tadjuidje E, Tao Q, Wylie C, Heasman J. 2008. Wnt5a and Wnt11 interact in a maternal Dkk1-regulated fashion to activate both canonical and non-canonical signaling in *Xenopus* axis formation. *Development* **135:** 3719–3729.

Cha SW, Tadjuidje E, White J, Wells J, Mayhew C, Wylie C, Heasman J. 2009. Wnt11/5a complex formation caused by tyrosine sulfation increases canonical signaling activity. *Curr Biol* **19:** 1573–1580.

Chen B, Dodge ME, Tang W, Lu J, Ma Z, Fan CW, Wei S, Hao W, Kilgore J, Williams NS, et al. 2009. Small molecule-mediated disruption of Wnt-dependent signaling in tissue regeneration and cancer. *Nat Chem Biol* **5:** 100–107.

Ching W, Hang HC, Nusse R. 2008. Lipid-independent secretion of a *Drosophila* Wnt protein. *J Biol Chem* **283:** 17092–17098.

Coombs GS, Yu J, Canning CA, Veltri CA, Covey TM, Cheong JK, Utomo V, Banerjee N, Zhang ZH, Jadulco RC, et al. 2010. WLS-dependent secretion of WNT3A requires Ser209 acylation and vacuolar acidification. *J Cell Sci* **123:** 3357–3367.

Coudreuse DY, Roel G, Betist MC, Destree O, Korswagen HC. 2006. Wnt gradient formation requires retromer function in Wnt-producing cells. *Science* **312:** 921–924.

* Cruciat C-M, Niehrs C. 2012. Secreted and transmembrane Wnt inhibitors and activators. *Cold Spring Harb Perspect Biol* doi: 10.1101/cshperspect.a015081.

de Lau W, Barker N, Low TY, Koo BK, Li VS, Teunissen H, Kujala P, Haegebarth A, Peters PJ, van de Wetering M, et al. 2011. Lgr5 homologues associate with Wnt receptors and mediate R-spondin signalling. *Nature* **476:** 293–297.

Fear MW, Kelsell DP, Spurr NK, Barnes MR. 2000. Wnt-16a, a novel Wnt-16 isoform, which shows differential expression in adult human tissues. *Biochem Biophys Res Commun* **278:** 814–820.

Franch-Marro X, Wendler F, Griffith J, Maurice MM, Vincent JP. 2008a. In vivo role of lipid adducts on Wingless. *J Cell Sci* **121:** 1587–1592.

Franch-Marro X, Wendler F, Guidato S, Griffith J, Baena-Lopez A, Itasaki N, Maurice MM, Vincent JP. 2008b. Wingless secretion requires endosome-to-Golgi retrieval of Wntless/Evi/Sprinter by the retromer complex. *Nat Cell Biol* **10:** 170–177.

Fu J, Jiang M, Mirando AJ, Yu HM, Hsu W. 2009. Reciprocal regulation of Wnt and Gpr177/mouse Wntless is required for embryonic axis formation. *Proc Natl Acad Sci* **106:** 18598–18603.

Fuerer C, Habib SJ, Nusse R. 2010. A study on the interactions between heparan sulfate proteoglycans and Wnt proteins. *Dev Dyn* **239:** 184–190.

Gasnereau I, Herr P, Chia PZ, Basler K, Gleeson PA. 2011. Identification of an endocytosis motif in an intracellular loop of Wntless, essential for its recycling and the control of Wnt signalling. *J Biol Chem* **286:** 43324–43333.

Glinka A, Dolde C, Kirsch N, Huang YL, Kazanskaya O, Ingelfinger D, Boutros M, Cruciat CM, Niehrs C. 2011. LGR4 and LGR5 are R-spondin receptors mediating Wnt/β-catenin and Wnt/PCP signalling. *EMBO Rep* **12:** 1055–1061.

Goodman RM, Thombre S, Firtina Z, Gray D, Betts D, Roebuck J, Spana EP, Selva EM. 2006. Sprinter: A novel transmembrane protein required for Wg secretion and signaling. *Development* **133:** 4901–4911.

Grzeschik KH, Bornholdt D, Oeffner F, Konig A, del Carmen Boente M, Enders H, Fritz B, Hertl M, Grasshoff U, Hofling K, et al. 2007. Deficiency of PORCN, a regulator of Wnt signaling, is associated with focal dermal hypoplasia. *Nat Genet* **39:** 833–835.

Hacker U, Lin X, Perrimon N. 1997. The *Drosophila sugarless* gene modulates Wingless signaling and encodes an enzyme involved in polysaccharide biosynthesis. *Development* **124:** 3565–3573.

Haerry TE, Heslip TR, Marsh JL, O'Connor MB. 1997. Defects in glucuronate biosynthesis disrupt Wingless signaling in *Drosophila*. *Development* **124:** 3055–3064.

Han C, Belenkaya TY, Khodoun M, Tauchi M, Lin X. 2004. Distinct and collaborative roles of *Drosophila* EXT family proteins in morphogen signalling and gradient formation. *Development* **131:** 1563–1575.

Harterink M, Port F, Lorenowicz MJ, McGough IJ, Silhankova M, Betist MC, van Weering JR, van Heesbeen RG, Middelkoop TC, Basler K, et al. 2011. A SNX3-dependent retromer pathway mediates retrograde transport of the Wnt sorting receptor Wntless and is required for Wnt secretion. *Nat Cell Biol* **13:** 914–923.

He X, Saint-Jeannet JP, Wang Y, Nathans J, Dawid I, Varmus H. 1997. A member of the Frizzled protein family mediating axis induction by Wnt-5A. *Science* **275:** 1652–1654.

Herr P, Basler K. 2011. Porcupine-mediated lipidation is required for Wnt recognition by Wls. *Dev Biol* **361:** 392–402.

Hofmann K. 2000. A superfamily of membrane-bound O-acyltransferases with implications for Wnt signaling. *Trends Biochem Sci* **25:** 111–112.

Itasaki N, Jones CM, Mercurio S, Rowe A, Domingos PM, Smith JC, Krumlauf R. 2003. Wise, a context-dependent activator and inhibitor of Wnt signalling. *Development* **130:** 4295–4305.

Janda CY, Waghray D, Levin AM, Thomas C, Garcia KC. 2012. Structural basis of Wnt recognition by Frizzled. *Science* doi: 10.1126/science.1222879.

Junge HJ, Yang S, Burton JB, Paes K, Shu X, French DM, Costa M, Rice DS, Ye W. 2009. TSPAN12 regulates retinal vascular development by promoting Norrin- but not Wnt-induced FZD4/β-catenin signaling. *Cell* **139:** 299–311.

Kadowaki T, Wilder E, Klingensmith J, Zachary K, Perrimon N. 1996. The segment polarity gene porcupine encodes a putative multitransmembrane protein involved in Wingless processing. *Genes Dev* **10:** 3116–3128.

Kang DW, Min do S. 2010. Positive feedback regulation between phospholipase D and Wnt signaling promotes Wnt-driven anchorage-independent growth of colorectal cancer cells. *PLoS ONE* **5:** e12109.

Kang DW, Lee SH, Yoon JW, Park WS, Choi KY, Min do S. 2010. Phospholipase D1 drives a positive feedback loop to reinforce the Wnt/β-catenin/TCF signaling axis. *Cancer Res* **70:** 4233–4242.

Kazanskaya O, Glinka A, del Barco Barrantes I, Stannek P, Niehrs C, Wu W. 2004. R-Spondin2 is a secreted activator of Wnt/β-catenin signaling and is required for *Xenopus* myogenesis. *Dev Cell* **7:** 525–534.

Kim KA, Zhao J, Andarmani S, Kakitani M, Oshima T, Binnerts ME, Abo A, Tomizuka K, Funk WD. 2006. R-Spondin proteins: A novel link to β-catenin activation. *Cell Cycle* **5:** 23–26.

Klein PS, Melton DA. 1996. A molecular mechanism for the effect of lithium on development. *Proc Natl Acad Sci* **93:** 8455–8459.

Komekado H, Yamamoto H, Chiba T, Kikuchi A. 2007. Glycosylation and palmitoylation of Wnt-3a are coupled to produce an active form of Wnt-3a. *Genes Cells* **12:** 521–534.

Korkut C, Ataman B, Ramachandran P, Ashley J, Barria R, Gherbesi N, Budnik V. 2009. Trans-synaptic transmission of vesicular Wnt signals through Evi/Wntless. *Cell* **139:** 393–404.

Kurayoshi M, Yamamoto H, Izumi S, Kikuchi A. 2007. Post-translational palmitoylation and glycosylation of Wnt-5a are necessary for its signalling. *Biochem J* **402:** 515–523.

Liepinsh E, Banyai L, Patthy L, Otting G. 2006. NMR structure of the WIF domain of the human Wnt-inhibitory factor-1. *J Mol Biol* **357:** 942–950.

Lin X, Perrimon N. 1999. Dally cooperates with *Drosophila* Frizzled 2 to transduce Wingless signalling. *Nature* **400:** 281–284.

Liu H, Fergusson MM, Castilho RM, Liu J, Cao L, Chen J, Malide D, Rovira II, Schimel D, Kuo CJ, et al. 2007. Augmented Wnt signaling in a mammalian model of accelerated aging. *Science* **317:** 803–806.

Malinauskas T. 2008. Docking of fatty acids into the WIF domain of the human Wnt inhibitory factor-1. *Lipids* **43:** 227–230.

Mason JO, Kitajewski J, Varmus HE. 1992. Mutational analysis of mouse Wnt-1 identifies two temperature-sensitive alleles and attributes of Wnt-1 protein essential for transformation of a mammary cell line. *Mol Biol Cell* **3:** 521–533.

Mathew D, Ataman B, Chen J, Zhang Y, Cumberledge S, Budnik V. 2005. Wingless signaling at synapses is through cleavage and nuclear import of receptor DFrizzled2. *Science* **310:** 1344–1347.

Mikels AJ, Nusse R. 2006. Purified Wnt5a protein activates or inhibits β-catenin–TCF signaling depending on receptor context. *PLoS Biol* **4:** e115.

Mulligan K, Fuerer C, Ching W, Fish M, Willert K, Nusse R. 2012. The Wnt/Wingless binding protein Swim promotes long-range signaling by maintaining Wingless solubility. *Proc Natl Acad Sci* **109:** 370–377.

Nam JS, Turcotte TJ, Smith PF, Choi S, Yoon JK. 2006. Mouse cristin/R-spondin family proteins are novel ligands for the Frizzled 8 and LRP6 receptors and activate β-catenin-dependent gene expression. *J Biol Chem* **281:** 13247–13257.

Neumann S, Coudreuse DY, van der Westhuyzen DR, Eckhardt ER, Korswagen HC, Schmitz G, Sprong H. 2009. Mammalian Wnt3a is released on lipoprotein particles. *Traffic* **10:** 334–343.

Nusse R, Varmus HE. 1982. Many tumors induced by the mouse mammary tumor virus contain a provirus integrated in the same region of the host genome. *Cell* **31:** 99–109.

Nusse R, Brown A, Papkoff J, Scambler P, Shackleford G, McMahon A, Moon R, Varmus H. 1991. A new nomenclature for *int-1* and related genes: The Wnt gene family. *Cell* **64:** 231.

Panakova D, Sprong H, Marois E, Thiele C, Eaton S. 2005. Lipoprotein particles are required for Hedgehog and Wingless signalling. *Nature* **435:** 58–65.

Papkoff J, Brown AM, Varmus HE. 1987. The *int-1* proto-oncogene products are glycoproteins that appear to enter the secretory pathway. *Mol Cell Biol* **7:** 3978–3984.

Port F, Kuster M, Herr P, Furger E, Banziger C, Hausmann G, Basler K. 2008. Wingless secretion promotes and requires retromer-dependent cycling of Wntless. *Nat Cell Biol* **10:** 178–185.

Port F, Hausmann G, Basler K. 2011. A genome-wide RNA interference screen uncovers two p24 proteins as regulators of Wingless secretion. *EMBO Rep* **12:** 1144–1152.

Prasad BC, Clark SG. 2006. Wnt signaling establishes anteroposterior neuronal polarity and requires retromer in *C. elegans*. *Development* **133:** 1757–1766.

Sato N, Meijer L, Skaltsounis L, Greengard P, Brivanlou AH. 2004. Maintenance of pluripotency in human and mouse embryonic stem cells through activation of Wnt signaling by a pharmacological GSK-3-specific inhibitor. *Nat Med* **10:** 55–63.

Sepich DS, Usmani M, Pawlicki S, Solnica-Krezel L. 2011. Wnt/PCP signaling controls intracellular position of MTOCs during gastrulation convergence and extension movements. *Development* **138:** 543–552.

Shibamoto S, Higano K, Takada R, Ito F, Takeichi M, Takada S. 1998. Cytoskeletal reorganization by soluble Wnt-3a protein signalling. *Genes Cells* **3:** 659–670.

Silhankova M, Port F, Harterink M, Basler K, Korswagen HC. 2010. Wnt signalling requires MTM-6 and MTM-9 myotubularin lipid-phosphatase function in Wnt-producing cells. *EMBO J* **29:** 4094–4105.

Takada R, Satomi Y, Kurata T, Ueno N, Norioka S, Kondoh H, Takao T, Takada S. 2006. Monounsaturated fatty acid modification of Wnt protein: Its role in Wnt secretion. *Dev Cell* **11:** 791–801.

Takei Y, Ozawa Y, Sato M, Watanabe A, Tabata T. 2004. Three *Drosophila* EXT genes shape morphogen gradients through synthesis of heparan sulfate proteoglycans. *Development* **131:** 73–82.

Tao Q, Yokota C, Puck H, Kofron M, Birsoy B, Yan D, Asashima M, Wylie CC, Lin X, Heasman J. 2005. Maternal wnt11 activates the canonical Wnt signaling pathway required for axis formation in *Xenopus* embryos. *Cell* **120:** 857–871.

Ten Berge D, Kurek D, Blauwkamp T, Koole W, Maas A, Eroglu E, Siu RK, Nusse R. 2011. Embryonic stem cells require Wnt proteins to prevent differentiation to epiblast stem cells. *Nat Cell Biol* **13:** 1070–1075.

Topczewski J, Sepich DS, Myers DC, Walker C, Amores A, Lele Z, Hammerschmidt M, Postlethwait J, Solnica-Krezel L. 2001. The zebrafish glypican knypek controls cell polarity during gastrulation movements of convergent extension. *Dev Cell* **1:** 251–264.

van den Heuvel M, Nusse R, Johnston P, Lawrence PA. 1989. Distribution of the *wingless* gene product in *Drosophila* embryos: A protein involved in cell–cell communication. *Cell* **59:** 739–749.

van den Heuvel M, Harryman-Samos C, Klingensmith J, Perrimon N, Nusse R. 1993. Mutations in the segment polarity genes *wingless* and *porcupine* impair secretion of the Wingless protein. *EMBO J* **12:** 5293–5302.

van Leeuwen F, Samos CH, Nusse R. 1994. Biological activity of soluble wingless protein in cultured *Drosophila* imaginal disc cells. *Nature* **368:** 342–344.

Wang X, Reid Sutton V, Omar Peraza-Llanes J, Yu Z, Rosetta R, Kou YC, Eble TN, Patel A, Thaller C, Fang P, et al. 2007. Mutations in X-linked PORCN, a putative regulator of Wnt signaling, cause focal dermal hypoplasia. *Nat Genet* **39:** 836–838.

Willert K, Brown JD, Danenberg E, Duncan AW, Weissman IL, Reya T, Yates JR 3rd, Nusse R. 2003. Wnt proteins are lipid-modified and can act as stem cell growth factors. *Nature* **423:** 448–452.

Xu Q, Wang Y, Dabdoub A, Smallwood PM, Williams J, Woods C, Kelley MW, Jiang L, Tasman W, Zhang K, et al. 2004. Vascular development in the retina and inner ear: Control by Norrin and Frizzled-4, a high-affinity ligand–receptor pair. *Cell* **116:** 883–895.

Yang PT, Lorenowicz MJ, Silhankova M, Coudreuse DY, Betist MC, Korswagen HC. 2008. Wnt signaling requires retromer-dependent recycling of MIG-14/Wntless in Wnt-producing cells. *Dev Cell* **14:** 140–147.

Zecca M, Basler K, Struhl G. 1996. Direct and long-range action of a Wingless morphogen gradient. *Cell* **87:** 833–844.

Zhang P, Wu Y, Belenkaya TY, Lin X. 2011. SNX3 controls Wingless/Wnt secretion through regulating retromer-dependent recycling of Wntless. *Cell Res* **21:** 1677–1690.

Zhang X, Abreu JG, Yokota C, MacDonald BT, Singh S, Coburn KL, Cheong S-M, Zhang MM, Ye Q-Z, Hang HC, et al. 2012. Tiki1 is required for head formation via Wnt cleavage-oxidation and inactivation. *Cell* **149:** 1565–1577.

Ziegler S, Rohrs S, Tickenbrock L, Moroy T, Klein-Hitpass L, Vetter IR, Muller O. 2005. Novel target genes of the Wnt pathway and statistical insights into Wnt target promoter regulation. *FEBS J* **272:** 1600–1615.

Ziegler S, Rohrs S, Tickenbrock L, Langerak A, Chu ST, Feldmann I, Jakubowski N, Muller O. 2007. Lipocalin 24p3 is regulated by the Wnt pathway independent of regulation by iron. *Cancer Genet Cytogenet* **174:** 16–23.

Zoltewicz JS, Ashique AM, Choe Y, Lee G, Taylor S, Phamluong K, Solloway M, Peterson AS. 2009. Wnt signaling is regulated by endoplasmic reticulum retention. *PLoS ONE* **4:** e6191.

Secreted and Transmembrane Wnt Inhibitors and Activators

Cristina-Maria Cruciat[1] and Christof Niehrs[1,2]

[1]Division of Molecular Embryology, DKFZ-ZMBH Alliance, DKFZ, Im Neuenheimer Feld 280, D-69120 Heidelberg, Germany

[2]Institute of Molecular Biology, D-55128 Mainz, Germany

Correspondence: niehrs@dkfz-heidelberg.de

Signaling by the Wnt family of secreted glycoproteins plays important roles in embryonic development and adult homeostasis. Wnt signaling is modulated by a number of evolutionarily conserved inhibitors and activators. Wnt inhibitors belong to small protein families, including sFRP, Dkk, WIF, Wise/SOST, Cerberus, IGFBP, Shisa, Waif1, APCDD1, and Tiki1. Their common feature is to antagonize Wnt signaling by preventing ligand–receptor interactions or Wnt receptor maturation. Conversely, the Wnt activators, R-spondin and Norrin, promote Wnt signaling by binding to Wnt receptors or releasing a Wnt-inhibitory step. With few exceptions, these antagonists and agonists are not pure Wnt modulators, but also affect additional signaling pathways, such as TGF-β and FGF signaling. Here we discuss their interactions with Wnt ligands and Wnt receptors, their role in developmental processes, as well as their implication in disease.

Wnt signaling is regulated at different levels by a wide range of effectors. These effectors function as agonists or antagonists and act either intracellularly to modulate components of the signal transduction machinery or extracellularly to modulate ligand–receptor interactions. Antagonists and agonists are of great importance, as they control the fine-tuning of Wnt signaling and inhibit or activate Wnt-regulated developmental processes, such as anterior–posterior (AP) axial patterning, somitogenesis, angiogenesis, vasculogenesis, and limb, bone, tooth, and eye formation, and they are implicated in pathological events, including cancer and bone disease. Six families of secreted and four families of transmembrane Wnt antagonists are known to date: the Dickkopf proteins (Dkks), secreted Frizzled-related proteins (sFRPs), Wnt-inhibitory factor 1 (WIF-1), Wise/SOST, Cerberus, insulin-like growth-factor binding protein 4 (IGFBP-4), Shisa, Wnt-activated inhibitory factor 1 (Waif1/5T4), adenomatosis polyposis coli down-regulated 1 (APCDD1), and Tiki1, the latter four being transmembrane. Among them, the Dkk protein family is best characterized. Two families of growth factors are known to activate Wnt signaling besides Wnts, Norrin, and R-spondins (Rspo). These protein families are not related to each other, and some of them act specifically on canonical Wnt signaling, whereas others affect both canonical and noncanonical Wnt pathways. This

Cite this article as *Cold Spring Harb Perspect Biol* doi: 10.1101/cshperspect.a015081

review describes the individual families and their members, with emphasis on their structure, mode of action, and role in development and disease. Related reviews examining Dkks (Niehrs 2006), sFRPs (Jones and Jomary 2002; Kawano and Kypta 2003; Bovolenta et al. 2008), Cerberus (Belo et al. 2009), and Rspo family members (Yoon and Lee 2012) are available.

SECRETED WNT INHIBITORS

Dkk Protein Family

Physical Properties and Structure

Dkks represent a small family of evolutionarily conserved secreted glycoproteins. The founding member of the family, Dkk1, was identified as embryonic head inducer and Wnt antagonist in *Xenopus* (Glinka et al. 1998). Since then, Dkks were identified in other vertebrates, including humans, as well as in invertebrates such as *Dictyostelium*, cnidarians, urochordates, and ascidians, but not in *Drosophila* and *Caenorhabditis elegans*. For a more comprehensive overview on the topic, the reader is referred to Niehrs (2006). In vertebrates, the Dkk family comprises four members, Dkk1–4. They consist of 255–350 amino acids and share two conserved cysteine-rich domains (CRDs) (Fig. 1). Whereas the amino-terminal CRD, DKK_N, is unique to the Dkks, the carboxy-terminal CRD shows homology with the colipase fold, a domain found in a wide range of functionally unrelated proteins, including, for example, colipases, toxins, protease inhibitors, and prokineticins (Niehrs 2006). In addition to the two CRDs, an sgy domain is found only in Dkk3 (see below).

Dkk3 appears to be a divergent member of the Dkk family. In contrast to *Dkk1*, *-2*, and *-4*, which are more related to each other than they are to *Dkk3* (Glinka et al. 1998), *Dkk3* shares sequence homology with *soggy* (*sgy*), a distant *Dkk* family member, also called Dickkopf-like protein 1 (Krupnik et al. 1999). In addition, Cnidaria, *Hydra*, and *Nematostella* each have only two *Dkk* genes, one related to vertebrate *Dkk1*, *-2*, and *-4* (Guder et al. 2006) and one related to vertebrate *Dkk3* (Fedders et al. 2004). Furthermore, human *Dkk1*, *-2*, and *-4* are located on the same chromosome 4/5/8/10 paralogy group, genes of which duplicated early in vertebrate evolution (Pollard and Holland 2000; Luke et al. 2003), but Dkk3 is not part of this group.

Little is known about Soggy outside of a potential role in spermatogenesis and its homology with Dkk3 (Kaneko and DePamphilis 2000; Kohn et al. 2005).

Mechanism of Action

Of the various signaling pathways activated by Wnts, Dkks specifically inhibit the Wnt/β-catenin signaling cascade. Dkk1 and Dkk2 bind to low-density lipoprotein receptor-related protein (LRP) 5/6 with high affinity and an apparent K_d in the range of 0.3 to 0.7 nM (Bafico et al. 2001; Mao et al. 2001; Semenov et al. 2001). As for Dkk1 and Dkk2, a functional interaction with LRP6 has been shown for Dkk4. In contrast, Dkk3 does not bind LRP6 and does not affect Wnt signaling (Mao et al. 2001; Mao and Niehrs 2003), but rather regulates transforming growth factor-β (TGF-β) signaling (Pinho and Niehrs 2007). Within LRP6, the last two YWTD–epidermal growth factor (EGF) repeat domains mediate binding to Dkk1 (Mao et al. 2001). Within Dkk1, the colipase fold, but not the DKK_N domain, is sufficient for LRP6 binding and Wnt inhibition (Brott and Sokol 2002; Li et al. 2002; Mao and Niehrs 2003), and mutation of the conserved Cys-220 of the colipase fold abolishes the interaction (Semenov et al. 2001).

Genetic evidence for a Dkk1–LRP6 interaction is provided by the rescue of severe developmental defects of *LRP6*- and *Dkk1*-null mice observed in *LRP6*/*Dkk1* double-knockout mice (MacDonald et al. 2004).

In addition to LRP5/6, Dkks bind with high affinity to another class of receptors, Kremen1 and 2 (Krm1/2), evolutionarily conserved single-pass transmembrane proteins. They contain a Kringle, WSC, and CUB domain, which are all required for Dkk1 interaction, and an intracellular domain without obvious sequence motifs (Mao et al. 2002). Kremens bind both Dkk1 and Dkk2 but not Dkk3 with an apparent K_d in the nanomolar range (Mao et al. 2002). Similar to

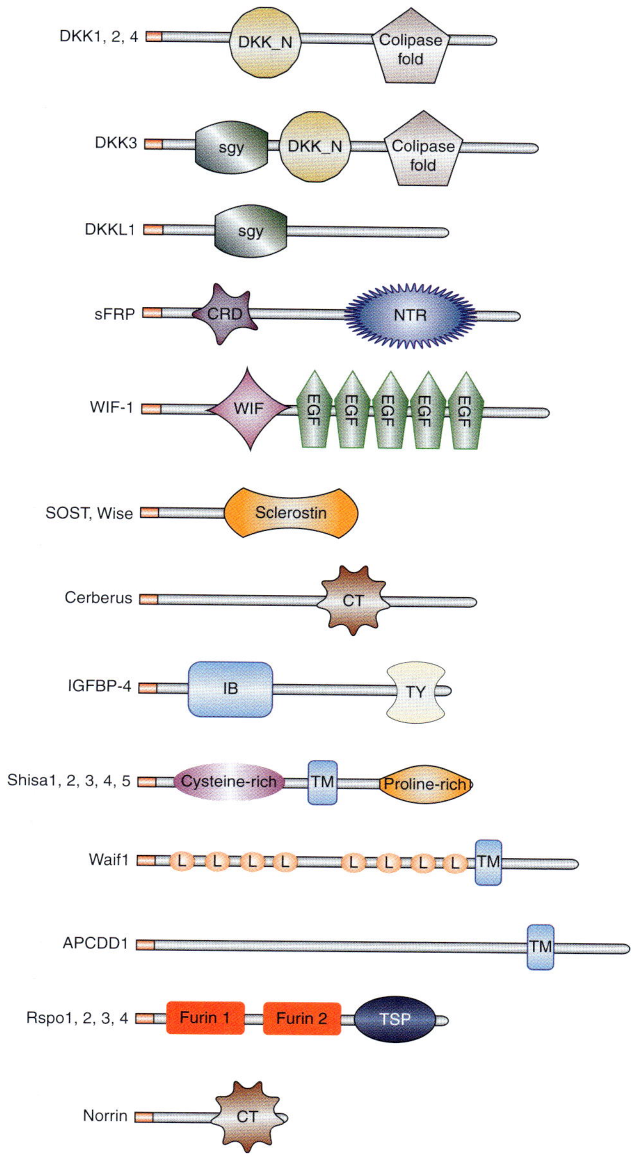

Figure 1. Domain structure of Wnt antagonists and agonists. Signal peptides are shown in red. NTR, Netrin-related motif; WIF, Wnt-inhibitory factor 1 domain; EGF, epidermal growth factor-like domain; CT, cystine knot-like domain; IB, insulin growth factor binding protein domain; TY, thyroglobulin type-1 domain; TM, transmembrane domain; L, leucine-rich repeats; TSP, thrombospondin type-1 domain.

the Dkk1–LRP6 interaction, it is the colipase fold of Dkk1, which is necessary and sufficient for Kremen binding (Mao and Niehrs 2003). Krm1/2 greatly potentiate the ability of Dkk1 to inhibit Wnt signaling, and membrane attachment of the Krm2 extracellular domain is critical for its function. Evidence for a functional interaction of these proteins in vivo is provided by loss-of-function experiments. Morpholino-mediated knockdown of Krm1/2 in *Xenopus* embryos induces head defects, which can be rescued by *Dkk1* mRNA overexpression. In addition, there is strong enhancement of head defects when both *Dkk1* and *Krm1/2* are inhibited (Davidson et al. 2002). Furthermore, in mice, these proteins genetically interact during limb development, as triple-mutant *Krm1*$^{-/-}$/*Krm2*$^{-/-}$/*Dkk1*$^{+/-}$ mice show enhanced formation of ectopic digits (Ellwanger et al. 2008).

The dominant mode by which Dkk1 acts is to prevent the Wnt–LRP6 interaction and disrupt the Wnt-induced Fz8–LRP6 complex formation (Fig. 2A) (Semenov et al. 2001). In addition, when Kremen is present, Dkk1 can form a ternary complex with Krm2 and LRP6 and induce rapid endocytosis and removal of LRP6 from the plasma membrane (Fig. 2B) (Mao et al. 2002). Recent biochemical studies (Semenov et al. 2008; Wang et al. 2008) and the characterization of *Krm1*$^{-/-}$/*Krm2*$^{-/-}$ double-mutant mice (Ellwanger et al. 2008) indicate that Kremens are not universally required for Dkk1 function, but rather this interaction is physiologically relevant only in certain tissues, and that the ability of Dkk1 to prevent Wnt–LRP6 interaction may be sufficient for effective Wnt antagonism in many cells.

Dual Role of Dkk2

Unlike Dkk1, which is a pure inhibitor of Wnt/β-catenin signaling, Dkk2 is able to act either as inhibitor or activator of the pathway, depending on the cellular context. When overexpressed in *Xenopus*, Dkk2 synergizes with coexpressed Fz8 (Wu et al. 2000) or LRP6 (Brott and Sokol 2002) to induce Wnt/β-catenin signaling. In contrast, in HEK293T or NIH3T3 cells, Dkk2 inhibits Wnt/β-catenin signaling when cotransfected with Wnt and Fz, and activates the pathway when cotransfected with LRP5/6 (Wu et al. 2000; Li et al. 2002; Mao and Niehrs 2003). The colipase fold of Dkk2 is sufficient for this synergy (Li et al. 2002). In the presence of Krm2, though, Dkk2 acts as Wnt inhibitor (Mao and Niehrs 2003), indicating that receptor context may determine its action.

Structure–function analysis revealed that the isolated colipase fold of Dkk1 is sufficient to synergize with LRP6 in activating Wnt signaling, whereas its DKK_N domain is important to impart the inhibitory effect. The DKK_N domain of Dkk2 is neutral in this regard (Brott and Sokol 2002). Thus, whereas the colipase fold of Dkk1/2 mediates binding of Dkks to LRPs, the DKK_N domain modulates the outcome of this interaction (Brott and Sokol 2002).

Role of Dkks in Embryonic Development and Disease

Given the paramount role that Wnts play during embryonic development, it is not surprising that one main function of Dkks is to control cell fate in vertebrates as they show highly regionalized expression (Grotewold et al. 1999; Monaghan et al. 1999; Hashimoto et al. 2000; Shinya et al. 2000; Chapman et al. 2004; Diep et al. 2004; Idkowiak et al. 2004; Fjeld et al. 2005; Nie 2005; Nie et al. 2005). *Dkk1*, *-2*, and *-3* mouse mutants are available and a summary of their phenotypes is presented in Table 1.

Dkk1 and AP Axial Patterning. Wnt/β-catenin signaling plays an important role in AP patterning of the primary embryonic axis. Together with bone morphogenetic protein (BMP) and Nodal signaling, Wnt/β-catenin signaling inhibits anterior development in general, and in particular the anterior central nervous system. This inhibition is prevented by the Spemann organizer in *Xenopus*, the zebrafish shield, and the mouse anterior mesendoderm, which secretes a cocktail of growth factor antagonists (De Robertis and Kuroda 2004; Niehrs 2004, 2010).

Dkk1 was initially identified as a gene conferring Spemann's head organizer activity in *Xenopus* embryos. *Dkk1* is specifically expressed in the anterior mesendoderm and induces entire ectopic heads when coexpressed with BMP inhibitors, whereas injection of anti-Dkk1 antibodies yields microcephalic or even headless *Xenopus* embryos (Glinka et al. 1998). In zebrafish, Dkk1 is also expressed in organizer derivatives of gastrulating embryos (Hashimoto et al. 2000; Shinya et al. 2000) and promotes, when overexpressed, anterior neural development (Hashimoto et al. 2000). These data support the model that Wnt/β-catenin signaling inhibits anterior embryonic development, which is prevented by Dkk1.

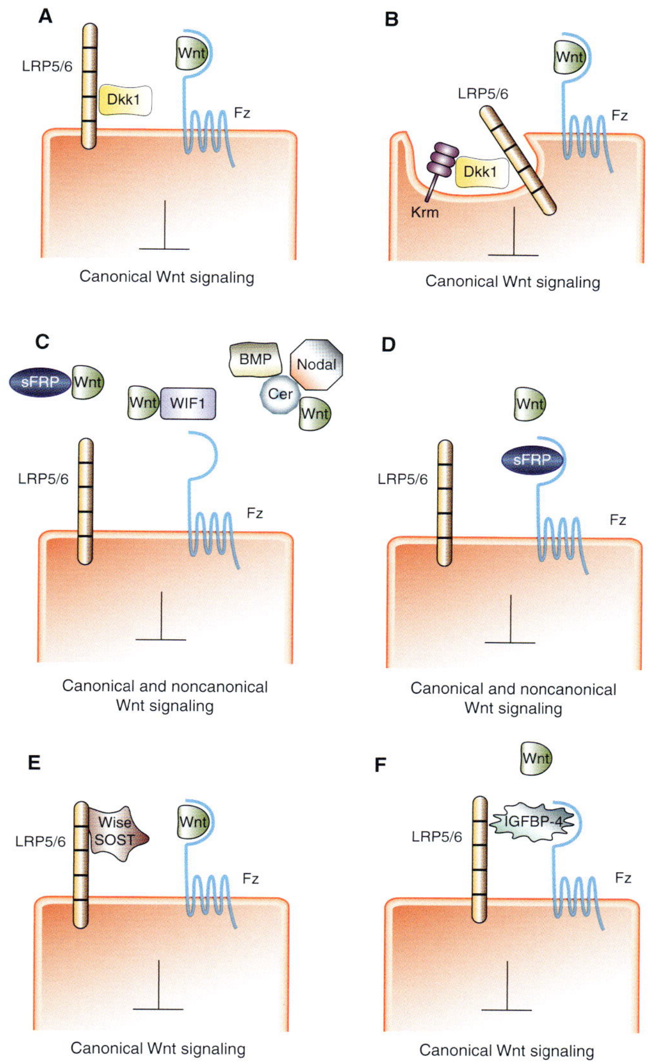

Figure 2. Models of Wnt signaling inhibition. (*A*,*B*) Dkk1 binding to LRP6 disrupts the Wnt-induced Fz–LRP6 complex formation (*A*) and/or induces LRP6 endocytosis in the presence of its coreceptor Kremen (*B*). (*C*) sFRPs, WIF-1, and Cerberus sequester Wnt, thereby inhibiting both canonical and noncanonical Wnt signaling. (*D*) sFRPs may also inhibit canonical and noncanonical Wnt signaling by binding to Fz. (*E*) Wise/SOST binding to LRP6 blocks Wnt-induced Fz–LRP6 complex formation. (*F*) IGFBP-4 binds to LRP6 and Fz, thereby preventing signal transduction by Wnt.

In mouse embryos, *Dkk1* is expressed from gastrula to neurula in the anterior visceral endoderm, anterior mesendoderm, and foregut endoderm, respectively, tissues which are all associated with anterior specification. Mice mutant for *Dkk1* lack head structures anterior to the mid-hindbrain boundary, and already at embryonic day 7.5 (E7.5) the anterior marker *Hesx1* fails to be expressed (Mukhopadhyay et al. 2001). In the transgene-induced mouse mutant *doubleridge*, which harbors a hypomorphic *Dkk1* allele ($Dkk1^{d}$) with 10% normal expression levels, head development is normal. However, $Dkk1^{d/-}$ compound heterozygotes on a

Table 1. Mouse mutant phenotypes

Mouse mutant	Phenotype	References
Dkk1$^{-/-}$	Embryonic lethal Absence of anterior head structures Postaxial polysyndactyly Fused vertebrae	Mukhopadhyay et al. 2001
Dkk1$^{d/d}$	Viable and fertile Postaxial polysyndactyly	MacDonald et al. 2004
Dkk2$^{-/-}$	Viable and fertile Osteopenia Blindness, cornea transformation	Li et al. 2005a; Mukhopadhyay et al. 2006
Dkk3$^{-/-}$	Viable and fertile Hyperactive Reduced lung ventilation and elevated hematocrit Elevated IgM and natural killer cell levels	Barrantes Idel et al. 2006
Sfrp1$^{-/-}$	Viable and fertile Normal and healthy	Satoh et al. 2006
Sfrp2$^{-/-}$	Viable and fertile Hindlimb syndactyly at lower frequency	Satoh et al. 2006
Sfrp5$^{-/-}$	Viable and fertile Normal and healthy	Leaf et al. 2006; Satoh et al. 2006
Sfrp1$^{-/-}$/*Sfrp2*$^{-/-}$	Embryonic lethal Severe shortening of the thoracic region Craniofacial defects Limb outgrowth defects, extra digits Defects in gonad morphology, gonad positioning, and reproductive tract maturation	Satoh et al. 2006; Warr et al. 2009
Wif-1$^{-/-}$	Normal development Susceptible to spontaneous and radiation-induced osteosarcoma	Kansara et al. 2009
Wise$^{-/-}$	Viable and fertile Defects in tooth development	Kassai et al. 2005; Yanagita et al. 2006; Ahn et al. 2010
Sost$^{-/-}$	Viable and fertile Increased bone density, bone volume, bone formation, and bone strength	Li et al. 2008a
Cerl-1$^{-/-}$	Viable and fertile	Simpson et al. 1999; Belo et al. 2000; Shawlot et al. 2000; Stanley et al. 2000
Igfb4$^{-/-}$	Viable and fertile Modest growth deficit	Ning et al. 2006
Shisa$^{-/-}$	Viable and fertile Dwarf (in two-thirds of mutants)	Furushima et al. 2007
Shisa3$^{-/-}$ *Shisa4*$^{-/-}$ *Shisa5*$^{-/-}$	Normal and fertile	Furushima et al. 2007
Waif1/5T4$^{-/+}$	Viable Hydrocephalus	Southgate et al. 2010
Rspo1$^{-/-}$ *XX*	Viable Partial sex-reversed phenotype Masculinized features	Chassot et al. 2008; Tomizuka et al. 2008

Continued

 Cite this article as *Cold Spring Harb Perspect Biol* doi: 10.1101/cshperspect.a015081

Table 1. *Continued*

Mouse mutant	Phenotype	References
Rspo2$^{-/-}$ *Rspo2*$^{Tg/Tg}$	Mortality Defects: craniofacial, skeleton, limb, digits, tail, renal, cardiovascular, respiratory, reproductive	Nam et al. 2007a; Aoki et al. 2008; Bell et al. 2008; Yamada et al. 2009; Jin et al. 2011
Rspo3$^{-/-}$	Embryonic lethal Defective angiogenesis in placenta and yolk sac	Aoki et al. 2007; Kazanskaya et al. 2008
Norrin$^{-/-}$	Abnormal vasculature in cochlea and retina	Richter et al. 1998; Rehm et al. 2002

C57BL/6J genetic background show a variety of head defects, from hydrocephaly with micrognathia to loss of anterior head structures and eyes (MacDonald et al. 2004). Whereas *Dkk1* heterozygous mice are normal and fertile, mice double heterozygous for *Dkk1* and the BMP inhibitor *Noggin* show head defects similar to those of *Dkk1* homozygous mutants (del Barco Barrantes et al. 2003), supporting the model that double inhibition of BMP and Wnt signals is required for head development (Glinka et al. 1997).

Dkk1 plays also a role in pregastrulation at E5.5–E6, when the initial proximal–distal axis of the mouse embryo is converted to AP polarity, by a process involving coordinated cell migrations (Kimura-Yoshida et al. 2005). Similarly, *LRP5*$^{-/-}$/*LRP6*$^{-/-}$ double mutants fail to induce and/or maintain the precursors of the primitive streak in the proximal epiblast of the pregastrula embryo (Kelly et al. 2004).

Role of Dkk1 in Limb, Bone, and Eye Formation. During early limb development, Wnt/β-catenin signaling by Wnt3a is required for formation of the apical ectodermal ridge (AER), a signaling center that controls limb growth (Barrow et al. 2003; Soshnikova et al. 2003). Similarly, overexpression of *Dkk1* in chicks induces limb truncation, and this is accompanied by apoptosis (Mukhopadhyay et al. 2001; Grotewold and Ruther 2002). Conversely, overactivation of Wnt signaling by expression of a gain-of-function mutation of *β-catenin* results in expansion of the AER (Soshnikova et al. 2003). An expansion of the AER is observed in both *Dkk1*-null and hypomorphic *Dkk1*$^{d/d}$ mutants, suggesting a genetic interaction with *Wnt3a* (Mukhopadhyay et al. 2001; Adamska et al. 2004; MacDonald et al. 2004). In addition to AER expansion, *Dkk1*-null and *Dkk1*$^{d/d}$ mutants display postaxial polysyndactyly in the forelimbs (Mukhopadhyay et al. 2001; MacDonald et al. 2004), and normal digit numbers are restored in *Dkk1*$^{d/d}$/*LRP6*$^{+/-}$ mice (MacDonald et al. 2004). Polydactyly in *Dkk1* mutants can be reversed by the simultaneous loss of Wnt7a (Adamska et al. 2004). Thus, Dkk1 can control different steps involving Wnt3 or Wnt7a signaling during mouse limb development and digit patterning.

Wnt/β-catenin signaling plays a central role in mammalian bone density regulation (Krishnan et al. 2006). Loss-of-function mutations of *LRP5* are associated with the recessive familial osteoporosis–pseudoglioma syndrome (Gong et al. 2001), whereas gain-of-function mutations of *LRP5* are associated with diseases of high bone mass (Boyden et al. 2002). Interestingly, for one particular *LRP5* gain-of-function mutation, G171V, decreased inhibition of LRP5 by Dkk1 was suggested to account for increased LRP5 signaling (Boyden et al. 2002; Ai et al. 2005). Similar to humans, heterozygous *Dkk1*$^{+/-}$ mice show an increase in bone mineral density (Morvan et al. 2006; MacDonald et al. 2007). Conversely, transgenic mice overexpressing Dkk1 in bone develop osteopenia (Li et al. 2006), suggesting that Dkk1 antagonizes Wnt–LRP5 signaling to regulate physiological levels of bone mass.

Wnt/β-catenin signaling plays an important role during eye development, and Dkk1, -2, and -3 are highly expressed in this organ (for details, see Niehrs 2006). The severity of head defects in *Dkk1*-null mutants precludes analysis of the role that Dkk1 may play in the developing eye. In contrast, *Dkk2* knockout mice are viable, but blind, because of the development of a keratinized corneal epithelium and skin appendages as a result of constitutive Wnt/β-catenin signaling in the cornea (Mukhopadhyay et al. 2006).

Dkks and Disease. A number of studies have implicated Dkk family members in human disease based on transcriptional profiling and epigenetic studies (for details, see Niehrs 2006). In particular, alterations of Dkk expression have been observed in a number of cancer models, which is not surprising given the importance of Wnt signaling in cancer biology. In colon cancer, for example, the observed epigenetic silencing of *Dkk1* and other *Dkk* family genes is likely to contribute further to the activation of the Wnt/β-catenin pathway (Aguilera et al. 2006; Sato et al. 2007). Dkk1 expression itself can be directly upregulated by Wnt/β-catenin signaling in a T-cell factor (TCF)-dependent manner, but this mechanism appears to be lost through silencing in colon cancer (Gonzalez-Sancho et al. 2005).

Dkk1 has also been implicated in neurodegenerative processes and induction of apoptosis after neuronal injury (Caricasole et al. 2004; Cappuccio et al. 2005).

sFRP Protein Family

Physical Properties and Structure

The sFRPs represent the largest family of secreted Wnt inhibitors and resemble the ligand-binding CRD domain of the Frizzled family of Wnt receptors. The founding member of the family, Frzb (for Frizzled motif associated with bone development), was first purified as a chondrogenic factor from bovine cartilage extracts (Hoang et al. 1996). Shortly after, Frzb was isolated during a differential screen for cDNAs enriched in the Spemann organizer of *Xenopus* embryos (Bouwmeester et al. 1996) and shown to act as a Wnt antagonist (Leyns et al. 1997; Wang et al. 1997). Subsequently, additional members of the family were identified (Finch et al. 1997; Melkonyan et al. 1997; Rattner et al. 1997).

In humans, the sFRP family comprises five members, sFRP1–5, with sFRP3 being the ortholog of Frzb, and orthologous genes have been identified in all vertebrates analyzed so far, as well as in invertebrates, but not in *Drosophila* (Bovolenta et al. 2008). On the basis of sequence homology and phylogenetic analysis, *sFRP1*, *-2*, and *-5* form a subgroup that diverges from the one formed by *sFRP3* and *-4* (Jones and Jomary 2002; Kawano and Kypta 2003; Bovolenta et al. 2008). This clustering also reflects a different genomic organization. Similar to human *Dkk1*, *-2*, and *-4*, *sFRP1*, *-2*, and *-5* are located within the same chromosome 4/5/8/10 paralogy goup, whereas *sFRP3* and *-4* are not part of this group. A third subgroup of sFRPs has been identified in *Xenopus*, chick, and zebrafish, but not in mammals. Members of this subgroup are *Sizzled*, *Sizzled2*, and *Crescent*, and they share sequence similarities with the *sFRP1*, *-2*, and *-5* subgroup (Pfeffer et al. 1997; Salic et al. 1997; Bradley et al. 2000; Pera and De Robertis 2000; Houart et al. 2002).

Human sFRPs consist of 295–346 amino acids and share at the amino terminus a CRD domain (Fig. 1). The CRDs of sFRPs show 30%–50% sequence similarity with those of Fz receptors and contain 10 conserved cysteine residues (Rehn et al. 1998), which form a pattern of disulfide bridges (Chong et al. 2002). The carboxy-terminal part of sFRPs contains a Netrin-related motif (NTR), which is found in a number of unrelated proteins, including the axon guidance protein netrin 1, tissue inhibitors of metalloproteinases (TIMPs), type-1 procollagen C-proteinase enhancer proteins (PCOLCEs), and complement proteins (Fig. 1) (Banyai and Patthy 1999). The NTR domain is characterized by segments of positively charged residues that appear to confer heparin-binding properties (Uren et al. 2000) and by six cysteine residues that form three disulfide bridges (Chong et al. 2002).

Mechanism of Action

Initial biochemical and functional analysis using *Xenopus* embryos and cultured cells showed that Frzb/sFRP3 binds to Wnt1 and XWnt8 and inhibits Wnt/β-catenin signaling (Leyns et al. 1997; Lin et al. 1997; Wang et al. 1997). These results favor a model wherein sFRPs inhibit Wnt signaling by sequestering Wnts away from active receptor complexes (Fig. 2C). The CRD domain of Frzb seemed to be necessary and sufficient for both activities, Wnt binding and inhibition (Lin et al. 1997). In contrast, a mutant of sFRP1 lacking the CRD domain retained the ability to bind to Wingless, the *Drosophila* Wnt homolog (Uren et al. 2000). Another study of sFRP1 structure and function using Wnt reporter assays indicates that both protein domains, CRD and NTR, are important for optimal Wnt inhibition (Bhat et al. 2007). Furthermore, a recent study on sFRP1 combining biochemical and functional assays in cell culture and medakafish embryos shows that its NTR domain mimics the function of the entire molecule in binding Wnt8 and inhibiting Wnt signaling (Lopez-Rios et al. 2008). These conflicting results might imply that sFRPs have multiple Wnt-binding sites and/or sFRP–Wnt pairs associate with different affinities. Indeed, whereas sFRP1–4 have been shown to bind to Wnt3a with affinities in the nanomolar range, sFRP1 and -2, but not sFRP3 and -4, bound to Wnt5a. Moreover, only sFRP1 and -2 could block Wnt3a signaling in L cells (Wawrzak et al. 2007). Several other studies have described possible biochemical and/or functional specificity of the sFRPs–Wnt interactions in different developmental models, including the neural tube (Galli et al. 2006), somites (Lee et al. 2000), vascular endothelium (Dennis et al. 1999), and developing heart (Schneider and Mercola 2001), as well as during AP axial patterning of *Xenopus* embryos (Bradley et al. 2000; Pera and De Robertis 2000).

The crystal structures of the CRDs from mouse Fz8 and mouse sFRP3 have been determined and revealed the potential for the CRDs to dimerize (Dann et al. 2001). This possibility has been supported by the ability of sFRPs and Fz proteins to form homo- and heteromeric complexes through the CRD domain (Bafico et al. 1999; Rodriguez et al. 2005). These findings suggest an alternative model of how sFRPs could inhibit Wnt signaling, by forming nonfunctional complexes with Fz receptors (Fig. 2D).

Unlike Dkks, which specifically inhibit Wnt/β-catenin signaling, sFRPs can also inhibit noncanonical Wnt/PCP (planar cell polarity) signaling (Li et al. 2008b; Satoh et al. 2008; Matsuyama et al. 2009; Sugiyama et al. 2010), which is not surprising, as they bind to both types of Wnts. Inhibition of PCP signaling, which antagonizes Wnt/β-catenin signaling (Yan et al. 2001; Schwarz-Romond et al. 2002; Simons et al. 2005; Li et al. 2011), may also explain why sFRPs can activate Wnt/β-catenin signaling (Swain et al. 2005). Interestingly, in *Xenopus*, Frzb and Crescent can activate canonical Wnt signaling by promoting the diffusion of Wnt8 and Wnt11 (Mii and Taira 2009). In addition to Wnt signaling, sFRPs can regulate other signaling cascades. sFRP1, for example, binds to RANKL, a member of the tumor necrosis factor family, thereby inhibiting osteoclast formation (Hausler et al. 2004). sFRP1 also interacts with and inhibits the metalloprotease ADAM10, thus acting as a negative modulator of Notch signaling to regulate retinal neurogenesis (Esteve et al. 2011). Sizzled binds to and inhibits the activity of BMP1/Tolloid, a metalloprotease that cleaves the BMP antagonist chordin, thereby inhibiting BMP signaling (Lee et al. 2006).

Role of sFRPs in Embryonic Development and Pathological Events

The expression pattern of several sFRPs has been analyzed in different developmental systems, including medaka fish, *Xenopus*, chick, and mouse embryos (for details, refer to Jones and Jomary 2002; Bovolenta et al. 2008). sFRPs show a dynamic, partly overlapping, but also distinct expression pattern. To examine their roles in embryogenesis, single *Sfrp1*, *Sfrp2*, and *Sfrp5* knockout mice (Leaf et al. 2006; Satoh et al. 2006), as well as different combinations of double- and triple-mutant mouse lines carrying homozygous or heterozygous mutations of two or all three genes, have been generated (Table 1) (Satoh et al. 2008). Apart from *Sfrp1*$^{-/-}$/*Sfrp2*$^{-/-}$ double-mutant and *Sfrp1*$^{-/-}$/*Sfrp2*$^{-/-}$/*Sfrp5*$^{-/-}$ triple-mutant mice, the other mutant mice are viable and fertile and have no obvious abnormalities in gross morphology, indicative of the redundant role of *Sfrps* in embryonic development. In contrast, inactivation of *Sfrp1* plus *Sfrp2* is embryonic lethal as a result of severe shortening of the AP axis and incomplete somite segmentation, and this phenotype becomes more severe when *Sfrp5* is also inactivated. Genetic analysis combining *Sfrp* triple-knockout mice and *Loop-tail* mice, which carry a mutation in *Stbm/Vangl2*, revealed the involvement of sFRPs in convergent extension through the regulation of the PCP pathway (Satoh et al. 2008). In addition, *Dkk1*-deficient embryos carrying *Sfrp1* homozygous and *Sfrp2* heterozygous mutations display irregular somites and indistinct intersomitic boundaries, indicating that sFRP-mediated inhibition of Wnt/β-catenin signaling is required for somitogenesis (Satoh et al. 2008).

In addition to the role in regulating AP-axis elongation and somitogenesis during mouse development, *Sfrp1* and *-2* exert redundant roles also in embryonic organogenesis, as both genes are required for normal male sexual development in mice (Warr et al. 2009).

In line with their function as Wnt signaling antagonists, sFRPs appear to act as tumor suppressor genes, as loss or down-regulation of sFRP expression has been observed in a variety of invasive carcinomas. Epigenetic silencing of sFRP1 and -2 by hypermethylation seems to occur in basically all tumor types, and the *sFRP1* promoter methylation status has been proposed to serve as a biomarker for cancer detection and progression (for details, see Bovolenta et al. 2008; Esteve and Bovolenta 2010). sFRPs are also implicated in other pathological events, such as osteolysis and heterotopic ossification (Gordon et al. 2007) and during photoreceptor degeneration (Hackam 2005).

WIF-1

Wnt-inhibitory factor 1 was first identified as an expressed sequence tag from the human retina (Hsieh et al. 1999), and is present in fish, amphibians, and mammals. *Wif-1* is expressed in a variety of tissues, being most abundant in brain, lung, retina, and cartilage (Hsieh et al. 1999; Hunter et al. 2004; Hu et al. 2008; Surmann-Schmitt et al. 2009). WIF-1 is a protein of 379 amino acids with a unique and highly conserved WIF domain, five EGF-like repeats, and a hydrophilic tail (Fig. 1). Interestingly, the WIF domain is also found in the extracellular domain of RYK receptor tyrosine kinase. The phenotype induced by overexpression of WIF-1 in *Xenopus* embryos, namely, induction of a secondary axis and abnormal somitogenesis, together with its ability to block XWnt8 activity, suggested it plays a role in Wnt signaling (Hsieh et al. 1999). Indeed, WIF-1 binds to XWnt8 and Wingless and inhibits the interaction between XWnt8 and *Drosophila* Fz2 (Hsieh et al. 1999). This result suggests that similar to sFRPs, WIF-1 prevents Wnt from binding to its receptors, thus affecting canonical and noncanonical pathways (Fig. 2C). Recently, WIF-1 has been shown to bind to both types of Wnts, canonical and noncanonical, including Wnt3a, Wnt4, Wnt5a, Wnt7a, Wnt9a, and Wnt11, and to regulate Wnt activity during cartilage development (Surmann-Schmitt et al. 2009).

The mechanism of how WIF-1 regulates Wnt signaling is not completely understood, but the observed silencing of WIF-1 in different tumors suggests that like Dkks and sFRPs, this secreted factor plays an important role in cancer and other biological processes related to Wnt signaling (Chien et al. 2009; Kansara et al. 2009; Elston and Clifton-Bligh 2010).

Wise and SOST

Wise, also known as SOSTDC1 (sclerostin domain–containing 1), Ectodin, and USAG-1 (uterine sensitization-associated gene-1), has been isolated by a functional screen for activities that alter the AP character of neuralized *Xenopus* tissues (Itasaki et al. 2003). Sequence and structural analysis revealed that Wise forms a cystine knot (Ellies et al. 2006; Lintern et al. 2009) and belongs together with SOST to the CAN subfamily of cystine knot–containing BMP antagonists (Avsian-Kretchmer and Hsueh 2004). In *Xenopus* Wise appears to be a context-dependent regulator of Wnt signaling; it may inhibit or activate Wnt signaling in different assays (Itasaki et al. 2003). In cell culture Wise blocks Wnt1 and Wnt3a activities in reporter assays (Yanagita et al. 2004; Blish et al. 2008). It interacts with LRP6 through one of the three loops formed by the cystine knot (Lintern et al. 2009) and is able to compete with Wnt8 for binding to LRP6 (Fig. 2E) (Itasaki et al. 2003). Not only secreted Wise is able to inhibit Wnt signaling, but endoplasmic reticulum (ER)-retained Wise is also able to do so by reducing cell surface LRP6 (Guidato and Itasaki 2007). Wise also binds the LRP4 receptor (Ohazama et al. 2008), which is able to modulate Wnt signaling mediated by LRP5 and -6 (Ohazama et al. 2008; Li et al. 2010).

Wise is expressed in various tissues, including the surface ectoderm of the posterior axis, branchial arches, dermal papilla in hair follicles, vibrissae, tooth cusps, rat endometrium, developing testis, and kidney (Lintern et al. 2009). *Wise*-null mice are viable and fertile and apparently healthy, except that they exhibit tooth abnormalities, including supernumerary incisors and molars, fused molars, and cusp defects (Table 1) (Kassai et al. 2005; Yanagita et al. 2006; Ahn et al. 2010). This abnormal tooth development seems to be caused by elevated Wnt/β-catenin signaling. Indeed, in *Wise*$^{-/-}$/*LRP5*$^{-/-}$/*LRP6*$^{+/-}$ triple-mutant mice, most of the different tooth defects of *Wise*-null mutants are rescued, indicating that inhibition of Wnt/β-catenin by Wise controls tooth development (Ahn et al. 2010).

Wise shares 38% identity and a cystine knot domain with SOST (Fig. 1). SOST is highly expressed in osteoblasts and osteocytes, and loss-of-function mutations or down-regulation of SOST are responsible for two rare forms of autosomal recessive severe craniotubular hyperostoses, sclerosteosis (Balemans et al. 2001; Brunkow et al. 2001), and van Buchem disease (Balemans et al. 2002; Staehling-Hampton et al. 2002), which are characterized by overgrowth of bone tissue. Similarly, *SOST*-null mutant mice have a high-bone-mass phenotype characterized by increase in bone mass density, bone volume, bone formation, and bone strength (Table 1) (Li et al. 2008a).

SOST antagonizes Wnt signaling by binding to the first two YWTD–EGF repeat domains of LRP5 and LRP6 (Li et al. 2005b; Semenov et al. 2005). Like Dkk1, SOST has the ability to disrupt Wnt1-induced Fz8–LRP6 complex formation, suggesting a potential mechanism of action (Fig. 2E) (Semenov et al. 2005). Thus, the loss of SOST function likely leads to the hyperactivation of Wnt signaling that underlies bone overgrowth seen in sclerosteosis patients. Interestingly, sclerosteosis shares remarkable similarities with diseases of high bone mass caused by gain-of-function mutations in the *LRP5* gene, such as the G171V point mutation (Boyden et al. 2002; Little et al. 2002), and this mutation as well as an analogous mutation in LRP6 (G158V) blocks binding of SOST to LRP5 and LRP6 (Ellies et al. 2006; Semenov and He 2006). Evidence for a genetic interaction between SOST and LRP6 is provided by rescue experiments. In *SOST*$^{-/-}$/*LRP6*$^{-/-}$ double-mutant mice, the hand and foot defects observed in *LRP6*$^{-/-}$ mutant mice are rescued (Collette et al. 2010).

Wise and SOST belong to a subfamily of cystine knot–containing proteins, members of which inhibit BMP signaling (Avsian-Kretchmer and Hsueh 2004). In cell culture models Wise and SOST inhibit BMP signaling (Laurikkala et al. 2003; Winkler et al. 2003); however, the physiological role of this inhibition is still unclear.

Cerberus

Cerberus was isolated from *Xenopus* as an abundant organizer-specific gene, capable of inducing ectopic heads, when overexpressed in *Xenopus* embryos (Bouwmeester et al. 1996). Cerberus-related proteins have been identified in other vertebrates, such as zebrafish, chick, and mouse (Belo et al. 2009). They share a cystine knot domain at the carboxyl terminus (Fig. 1) and undergo proteolytic cleavage after secretion (Piccolo et al. 1999). *Xenopus* Cerberus binds to Nodal, BMP, and Wnt proteins via independent sites and inhibits all three signaling pathways, which leads to simultaneous head formation and trunk inhibition (Piccolo et al. 1999). Conversely, Morpholino-mediated knockdown of Cerberus in *Xenopus* embryos impairs head induction (Silva et al. 2003; Kuroda et al. 2004).

Unlike its *Xenopus* counterpart, mouse Cerl-1 does not bind Wnt and inhibit Wnt signaling (Belo et al. 2000) and is not essential for mouse head formation (Table 1) (Simpson et al. 1999; Belo et al. 2000; Shawlot et al. 2000; Stanley et al. 2000). Rather, Cerberus-related proteins are key regulators of Nodal signaling and play an important role in the establishment of left–right asymmetry in the vertebrate embryo (Rodriguez Esteban et al. 1999; Yokouchi et al. 1999; Zhu et al. 1999; Hashimoto et al. 2004; Marques et al. 2004; Tavares et al. 2007; Vonica and Brivanlou 2007).

IGFBP-4

IGFBP-4 belongs to the family of insulin-like growth factor binding proteins (IGFBPs), which modulate the actions of insulin-like growth factors (IGFs) (Firth and Baxter 2002). IGFBP-4 was identified in a screen for factors able to induce cardiomyocyte differentiation of P19CL6 cells (Zhu et al. 2008a). It promotes cardiogenesis in an IGF-independent fashion, namely, by antagonizing Wnt/β-catenin signaling (Zhu et al. 2008a). It binds directly to LRP6 and Fz8 via the carboxy-terminal thyroglobulin domain and blocks binding of Wnt3a to the receptors (Fig. 2F). Morpholino-mediated knockdown of IGFBP-4 in *Xenopus* embryos leads to cardiac defects, similar to those induced by overexpression of XWnt8. These defects were rescued by coexpression of a dominant–negative form of LRP6, suggesting that the cardiogenic effect of IGFBP-4 is mediated by inhibition of Wnt/β-catenin signaling.

Of the six IGFBP protein family members, IGFBP-1, -2, and -6, but not IGFBP-3 and -5, are also able to bind directly to LRP6 and Fz8 and to inhibit Wnt/β-catenin signaling, albeit with a lower efficiency compared with IGFBP-4 (Zhu et al. 2008a). Thus, the lack of cardiac phenotypes in *Igfb-4*-null mice or *Igfb-3*, *-4*, and *-5* triple-knockout mice (Ning et al. 2006) may be caused by genetic redundancies between IGFBP-4 and other IGFBPs.

IGFBP-4-mediated inhibition of Wnt/β-catenin signaling might have some implications for cancer biology. Treatment with IGFBP-4 reduces cell proliferation in some cancer cell lines in vitro, and overexpression of IGFBP-4 attenuates the growth of prostate cancer in vivo (Durai et al. 2006).

TRANSMEMBRANE WNT INHIBITORS

Shisa Protein Family

The Shisa protein family comprises five subfamilies in vertebrates (Furushima et al. 2007). Recently, in a search for sequence similarity to Shisa proteins, four additional subfamilies were found in vertebrates (Pei and Grishin 2012). Shisa proteins share a signal peptide, an amino-terminal CRD, a predicted transmembrane segment, and a carboxy-terminal proline-rich region (Fig. 1) (Pei and Grishin 2012).

The founding member of this family, *Xenopus Shisa1*, was identified as a gene specifically expressed in the prospective head ectoderm and the Spemann organizer (Yamamoto et al. 2005). In *Xenopus* embryos, overexpressed Shisa1 expands anterior neural structures, and together with a BMP inhibitor induces a secondary head. Conversely, knockdown of Shisa1 by Morpholino injection in *Xenopus* embryos suppresses anterior neural marker expression and head structures, indicating a requirement for head formation. Shisa1 does not inhibit BMP and Nodal signaling, but functions in head formation as an inhibitor of Wnt and fibroblast growth factor (FGF) signaling. It functions cell-autonomously in the ER, where it traps Fz and the FGF receptor and prevents their maturation (Yamamoto et al. 2005).

In *Xenopus*, three paralogous genes to *XShisa1* were identified: *XShisa2*, *-3*, and *-4* (Nagano et al. 2006; Silva et al. 2006). *XShisa2* is expressed in somitic mesoderm and like Shisa1 inhibits both Wnt and FGF signaling, thereby regulating segmental patterning during somitogenesis (Nagano et al. 2006). *Shisa* homologs were also reported in zebrafish, rat, mouse, chicken, and human (Katoh and Katoh 2005; Furushima et al. 2007; Hedge and Mason 2008; Zhu et al. 2008b).

In mouse, four Shisa family members are present: mShisa and mShisa3, -4, and -5 (Furushima et al. 2007). The expression of *mShisa* is similar to that of *XShisa1* and *-2*. It is expressed in anterior visceral endoderm, anterior mesendoderm, anterior neuroectoderm, and somitic mesoderm (Furushima et al. 2007). Like *XShisa1* and *-2*, overexpressed *mShisa* inhibits Wnt signaling in *Xenopus* and mammalian cells. However, *mShisa*$^{-/-}$ mutant mice had no phenotype in either head development or somitogenesis, and double-knockout mutants of *mShisa* with *mShisa3*, *-4*, or *-5* displayed no additional phenotype (Furushima et al. 2007). These results suggest that Shisa loss-of-function in mouse head and somites development may be compensated by other Wnt antagonists, such as Dkk, sFRP, and Cerberus.

Waif1/5T4

Waif1a was identified in gene expression profiling of zebrafish embryos as a novel direct Wnt/β-catenin target and functionally characterized in zebrafish, *Xenopus*, mouse, and mammalian cells (Kagermeier-Schenk et al. 2011). Waif1a is a single-pass plasma membrane protein with several leucine-rich repeats in the extracellular part and a short carboxyl terminus without conserved motifs (Fig. 1). In zebrafish and *Xenopus*, *Waif1a* is expressed during early embryogenesis in regions of active Wnt/β-catenin signaling, in particular the dorsolateral marginal epiblast during gastrulation, and its expression is directly regulated by Wnt8. Waif1a inhibits Wnt/β-catenin signaling in zebrafish and *Xenopus*. Likewise, mouse and human Waif1/5T4 antagonizes Wnt signaling in cultured cells. Zebrafish Waif1a acts as a direct feedback inhibitor of Wnt8-mediated mesoderm and neuroectoderm patterning during zebrafish gastrulation. At the molecular level Waif1a binds to LRP6 and inhibits both Wnt3a- and Dkk1-induced LRP6 internalization without affecting Wnt3a-induced LRP6 phosphorylation, by a mechanism that needs further elucidation. At the same time Waif1a activates noncanonical Wnt/PCP signaling in zebrafish embryos and *Xenopus* explants by promoting the ability of Dkk1 to activate Wnt/PCP signaling.

The elevated expression of Waif1/5T4 in human carcinomas, which correlates with poor survival prognosis, highlights its relevant role in cancer biology (Starzynska et al. 1994; Naganuma et al. 2002).

APCDD1

Adenomatosis polyposis coli down-regulated 1 (APCDD1) was initially isolated in a search for genes that were down-regulated after induction of APC in SW480 colon cancer cells (Takahashi et al. 2002). APCDD1 is a membrane-bound glycoprotein, conserved throughout vertebrate evolution. In mouse embryos, *APCDD1* is expressed in the nervous and vascular system, the inner ear, and the mesenchyme of several developing organs (Jukkola et al. 2004). Abundant expression of *APCDD1* is observed in mouse and human hair follicles (Jukkola et al. 2004; Shimomura et al. 2010). A point mutation (L9R) in *APCDD1* causes hereditary hypotrichosis simplex, a rare autosomal dominant form of hair loss (Shimomura et al. 2010). This mutation is located in the signal sequence and impairs protein transport to the plasma membrane. The finding that APCDD1 is a direct Wnt/β-catenin target (Takahashi et al. 2002), and its similarity in expression with Wise (O'Shaughnessy et al. 2004), encouraged Shimomura et al. to analyze if APCDD1 may function as a Wnt inhibitor. They found that soluble APCDD1 binds to Wnt3a and to the extracellular domain of LRP5 in vitro, and it inhibits Wnt/β-catenin signaling in cultured cells, possibly by preventing Fz from binding to Wnt. In vivo, APCDD1 inhibits Wnt signaling during the generation of neurons from progenitors in the developing chick nervous system and during axis specification in *Xenopus* embryos. Given its broad expression in various cell types, APCDD1 may regulate several other biological processes controlled by Wnt signaling.

Tiki1

Tiki1 was identified by functional cDNA screening as an organizer-specific gene, required for anterior neural development in Xenopus (Zhang et al. 2012). Tiki1 encodes an evolutionary conserved transmembrane metalloprotease that inhibits Wnt signaling by removing eight amino-terminal residues from Wnt itself, thereby leading to formation of oxidized Wnt oligomers with impaired receptor-binding capability (Zhang et al. 2012).

WNT ACTIVATORS

R-Spondin Protein Family

Physical Properties and Structure

R-Spondins (Rspo1 to -4) are a small family of four secreted growth factors, which in addition to Wnts potently activate β-catenin signaling. The first identified member of this family was human R-spondin 3 (Chen et al. 2002). Subsequently, mouse *R-spondin1* (roof plate-specific spondin) was discovered as a gene specifically expressed in the roof plate of the neural tube and the dorsal part of telencephalon (Kamata et al. 2004), and *Xenopus R-spondin2* was isolated in a functional screen for its property to activate Wnt/β-catenin signaling (Kazanskaya et al. 2004). Rspos are evolutionarily conserved and are present also in invertebrates, such as hemichordates, chordates, and echinoderms, but not in *Drosophila* and *C. elegans* (Kim et al. 2006; Yoon and Lee 2012). In mammals, Rspo1 to -4 show about 60% overall sequence homology and share a signal peptide, two amino-terminal furinlike CRDs, followed by a thrombospondin type-1 domain and a positively charged carboxy-terminal region (Fig. 1) (Kazanskaya et al. 2004).

Although Rspos contain an amino-terminal signal peptide, secreted proteins are barely detectable in the medium of transfected cells (Kazanskaya et al. 2004). Treatment of cells with soluble heparin or sodium chlorate, an inhibitor of sulfation, increases the level of Rspo proteins in the conditioned medium, suggesting that Rspos may bind to heparin sulfate proteoglycans, including syndecans and glypicans (Nam et al. 2006). Indeed, Rspo3 binds syndecan 4 (Sdc4) (Ohkawara et al. 2011) (see below).

Mechanism of Action

The characteristic feature of all four Rspos is their ability to activate canonical Wnt signaling. Rspos synergize with Wnts (Kazanskaya et al. 2004; Kim et al. 2005; Nam et al. 2006; Wei et al. 2007; Kim et al. 2008) and require the presence of Wnts to activate β-catenin signaling (Binnerts et al. 2007). The precise mechanism of Rspo-driven β-catenin activation is poorly understood, yet recent discoveries started to shed light on it.

For a couple of years the Rspo receptor turned out to be a matter of controversy. Initially, it was suggested that Rspos bind to Fz8 (Nam et al. 2006), but this interaction was not confirmed (Wei et al. 2007; Ohkawara et al. 2011). In another study Kremen was reported to bind Rspo1 (Binnerts et al. 2007), but this interaction may not be physiologically relevant, because *Kremen1* and *-2* double-knockout mice are viable, unlike most *Rspo* single-mutant mice, and fibroblasts isolated from these *Kremen*-deficient mice respond normally to Rspo (Ellwanger et al. 2008). Alternatively, LRP6 has been proposed as Rspo receptor (Wei et al. 2007), but this interaction also was not confirmed by others (Binnerts et al. 2007; Glinka et al. 2011). Of note, none of these studies addressed whether Fz8, LRP6, or Kremen was required for Rspo signaling.

Recently, several groups showed that the leucine-rich repeat containing G-protein-coupled receptor 5 (LGR5) and its homolog LGR4 are Rspo receptors. LGR5 marks proliferative stem cells in several Wnt-dependent compartments, such as small intestine, colon, stomach, and hair follicle, whereas LGR4 expression is much broader (Barker and Clevers 2010). LGR4 and -5 are required for Rspo signaling, and they synergize with Rspo and Wnt3a in Wnt/β-catenin signaling. Rspo1 to -4 bind with high affinity to LGR4 and -5 and their apparent K_d was determined to be in the nanomolar range (Carmon et al. 2011; de Lau et al. 2011; Glinka et al. 2011). Both Rspo binding to LGR4/5 and its function are mediated by the furin domains (Kazanskaya et al. 2004; Glinka et al. 2011). Moreover, Rspo3–LGR4 signaling requires clathrin-mediated endocytosis, which is essential for β-catenin activation (Glinka et al. 2011). Requirement of Rspo1/Wnt3a-induced clathrin-mediated endocytosis of LGR5 for Wnt/β-catenin signaling was not observed by others (Carmon et al. 2012), which might be due to artificial overexpression of LGR5 together with LRP6 and Fz5.

What is the mechanism by which Rspo and LGR4/5 potentiate Wnt/β-catenin signaling? Cong and his coworkers provide insight into this mechanism. In a search for novel negative Wnt regulators, they identified transmembrane E3 ubiquitin ligase ZNRF3 and its homolog ring finger 43 (RNF43) (Hao et al. 2012). ZNRF3 inhibits Wnt signaling by promoting the turnover of Fz receptors and LRP6. Rspo binds to the extracellular part of ZNRF3 and promotes the interaction between ZNRF3 and LGR4, which results in ZNRF3 clearance from the membrane. These results suggest a model wherein Rspo activates Wnt signaling by inhibiting ZNRF3 in an LGR4-dependent manner, resulting in the accumulation of Wnt receptors at the cell surface (Fig. 3A). The ZNRF3-mediated suppression of Wnt/β-catenin signaling is crucial for lens development in mice (Hao et al. 2012).

Apart from their function in canonical Wnt signaling, Rspos also amplify Wnt/PCP signaling. In particular, Rspo3 binds Sdc4 through its thrombospondin type-1 domain, and they functionally interact during *Xenopus* gastrulation and head cartilage morphogenesis, two

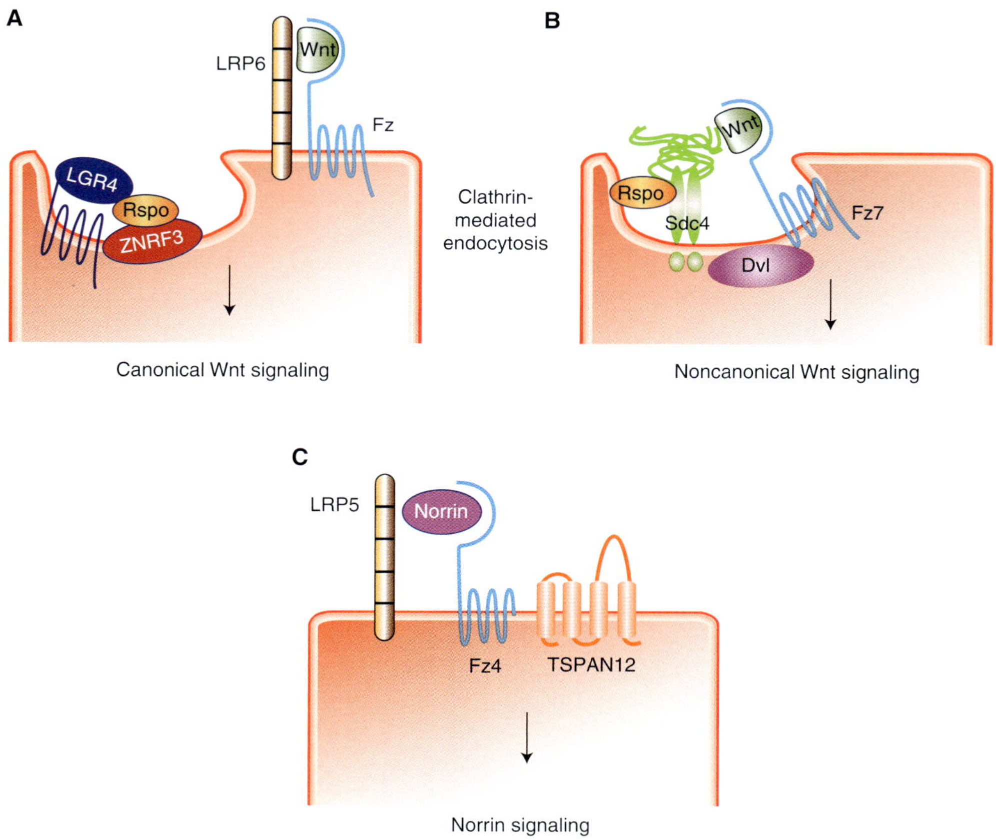

Figure 3. Models of Wnt signaling activation. (*A*) Rspo binds to ZNRF3 and LGR4 and induces clathrin-mediated receptor endocytosis. Internalized ZNRF3 is unable to ubiquitinate and target Wnt receptors for degradation. As a consequence, Fz and LRP6 accumulate on the plasma membrane and transmit canonical Wnt signals. (*B*) Rspo3 binding to Sdc4 promotes clathrin-mediated endocytosis of the Wnt receptor complex and thereby activates noncanonical Wnt signaling. (*C*) TSPAN12 is part of the Norrin/Fz4/LRP5 signaling complex and promotes Norrin/β-catenin signaling.

Wnt/PCP-driven processes (Ohkawara et al. 2011). Rspo3/Sdc4 signaling requires Wnt5a, Fz7, and Dvl and activates JNK (Ohkawara et al. 2011). Moreover, Rspo3 binding to Sdc4 promotes clathrin-mediated endocytosis, which is essential for PCP signal transduction (Fig. 3B). Not surprisingly, LGR4/5 are also essential for Rspo3-induced Wnt/PCP signaling in *Xenopus* (Glinka et al. 2011), and the interaction of ZNRF3 with Rspo and LGR4 suggests that this complex functions in Wnt/PCP signaling as well. Further studies are needed to clarify the interplay between Rspo/Sdc4 and Rspo/LGR/ZNRF3 in PCP signaling, and to determine the factor(s) that confer specificity of Rspo to signal either by β-catenin or the PCP pathway.

Role of Rspos in Embryonic Development and Disease

Loss-of-function analyses in *Xenopus* and mice as well as recent genetic studies in humans underline the role of Rspos in embryonic development and disease. In *Xenopus* and mouse, *Rspo* family members show differential expression in a broad range of embryonic tissues, such as dorsal neural tube, somites, tailbud, AER of the limb, and developing brain (Kazanskaya et al. 2004; Nam et al. 2007b). They are coexpressed with several *Wnt* genes and their expression is regulated by *Wnts* (Kazanskaya et al. 2004).

Role of Rspo1 in Female Sex Determination. Mutations in human *Rspo1* were identified in individuals with female-to-male sex reversal (Parma et al. 2006) or XX true hermaphroditism (Tomaselli et al. 2008), a disorder of gonadal development characterized by the presence of both ovarian and testicular tissue. These mutations result either in a stop codon immediately after the signal peptide or in a truncated protein lacking the first furinlike domain, and thus are linked to the inability of Rspo1 to activate Wnt signaling. The potential role of *Rspo1* as a candidate female-determining gene is confirmed in *Rspo1* knockout XX mice, which display masculinized ovaries (ovotestis) with epididymis and vas deferens-like structures (Chassot et al. 2008; Tomizuka et al. 2008). The observed reduction of *Wnt4* expression and the impaired Wnt/β-catenin activity in these mice, as well as the sex reversal phenotype observed in $Wnt4^{-/-}$ animals (Vainio et al. 1999), suggest a crucial role for Rspo1/Wnt4/β-catenin signaling in ovary development.

Mutations in human *Rspo1* are also responsible for palmoplantar hyperkeratosis and predisposition to squamous cell carcinoma of the skin (Parma et al. 2006), suggesting that *Rspo1* might act as a tumor suppressor gene. Indeed, suppression of *Rspo1* expression by hypermethylation of its promoter is found in leukemia cell lines and leukemia patients (Kuang et al. 2008).

Role of Rspo2 in Limb, Craniofacial, and Lung Morphogenesis and Myogenesis. *Footless* is a mouse mutant carrying a hypomorphic allele of *Rspo2* ($Rspo2^{Tg/Tg}$) (Bell et al. 2008). *Rspo2* knockout mice were generated (Nam et al. 2007a; Aoki et al. 2008; Bell et al. 2008; Yamada et al. 2009; Jin et al. 2011). All these animals immediately die after birth and display hindlimb defects, severe laryngeal and tracheal malformations, lung hypoplasia, and branching defects. In particular, abnormal limb development is manifested through a delayed maturation of the AER. Expression of *Axin2* and the *TopGAL* transgene is significantly reduced in the AER of *Rspo2* mutant mice, indicating that Rspo2 activity in the AER is mediated by Wnt/β-catenin signaling (Nam et al. 2007a; Bell et al. 2008). Furthermore, double-knockout mutants of *Rspo2* and *LRP6* ($Rspo2^{Tg/Tg}/LRP6^{-/-}$) showed much more severe limb defects than the single-knockouts, indicating that both genes functionally interact to regulate limb development (Bell et al. 2008).

In *Xenopus*, Rspo2 through Wnt/β-catenin signaling is required for the expression of myogenic marker genes, *myf5* and *myoD*, and later muscle development (Kazanskaya et al. 2004). Consistently, limb-specific *myf5* expression is down-regulated in *Rspo2* knockout mice (Han et al. 2011), supporting a role of Rspo2 in myogenesis.

Role of Rspo3 in Vasculogenesis and Angiogenesis. In *Xenopus* embryos, Rspo3 regulates the balance between hematopoietic and endothelial differentiation by promoting angioblast specification and inhibiting blood cell specification. It does so by promoting Wnt/β-catenin signaling, which is required for expression of *VEGF* (Kazanskaya et al. 2008). In mice, targeted deletion of *Rspo3* is embryonic lethal because of angiogenesis defects in placenta and yolk sac (Aoki et al. 2007; Kazanskaya et al. 2008). As in *Xenopus*, mouse Rspo3 is required for Wnt/β-catenin-mediated induction of vascular endothelial growth factor (VEGF). Moreover, recombinant Rspo3 promotes proliferation and angiogenesis in endothelial cell lines (Kazanskaya et al. 2008).

Role of Rspo4 in Nail Development. Mutations in *Rspo4* cause anonychia (Bergmann et al. 2006; Blaydon et al. 2006; Bruchle et al. 2008; Ishii et al. 2008), an autosomal recessive disorder characterized by partial or complete loss of finger- and toenails (Baran and Kechijian 2001). *Rspo4* mutations are found either in the furinlike domains, or predict truncated versions lacking any feature of an Rspo protein.

Role of Rspos in Cancer. In addition to the possible role of Rspo1 as tumor suppressor, Rspo2 and -3 promote tumorigenesis in vivo. *Rspo2* was identified as a common integration site for the mouse mammary tumor virus (MMTV) in mouse mammary tumors (Lowther et al. 2005), as well as a candidate colorectal cancer gene (Starr et al. 2009). In a similar screen for MMTV insertion sites, *Rspo3* was identified as a candidate breast cancer gene, which can strongly enhance oncogenicity (Theodorou et al. 2007).

Norrin

The Norrie disease pseudoglioma (*NDP*) or *Norrin* gene encodes a small secreted factor containing a cystine knot motif (Fig. 1) and a tertiary structure similar to that of TGF-β (Meitinger et al. 1993). Mutations in *Norrin* cause Norrie disease (Berger et al. 1992; Chen et al. 1992; Meindl et al. 1992), an X-linked disorder characterized by vascular abnormalities in the eye and blindness, often accompanied by progressive hearing loss and mental retardation (Berger 1998).

The nearly identical vascular defects in the retina and inner ear observed in *Fz4*$^{-/-}$ mice (Xu et al. 2004) and in *Ndp* knockout mice (Richter et al. 1998; Rehm et al. 2002) encouraged Nathans and his coworkers to analyze whether Norrin and Fz4 constitute a ligand–receptor pair. Indeed, Norrin is a high-affinity ligand for Fz4. It binds to the CRD of Fz4 with nanomolar affinity and cooperates with Fz4 to activate lymphoid enhancer factor (LEF)/TCF-mediated transcription in an LRP5/6-dependent manner (Xu et al. 2004). Thus, although structurally unrelated to Wnts, Norrin functions like a Wnt. Norrin function requires three pairs of cysteines that form the conserved trio of disulfide bonds shared among all cystine knot proteins (Smallwood et al. 2007).

Evidence for a functional interaction between *Norrin*, *Fz4*, and *LRP5* in vivo is provided by the finding that in humans, mutations in each gene cause familial exudative vitreoretinopathy (Chen et al. 1993; Robitaille et al. 2002; Toomes et al. 2004a,b), an inherited disease characterized by peripheral retinal avascularity (Warden et al. 2007). Consistent with the findings in human patients, targeted inactivation of *Norrin*, *Fz4*, or *LRP5* in mice causes similar alterations in the retinal vascularization (Xu et al. 2004; Luhmann et al. 2005a; Xia et al. 2008).

Among the 10 mammalian Fzs, Fz4 is the only Norrin receptor (Smallwood et al. 2007). Yet Fz4 can also transduce the signal of Wnts, raising the question of how Fz4/LRP5 can respond to different types of ligands to activate β-catenin-dependent transcription. The identification of TSPAN12, a tetraspanin family member, as a component of the Norrin/Fz4/LRP5 complex provides an answer to this question. TSPAN12 specifically promotes Fz4/LRP5 signaling induced by Norrin, but not by Wnt ligands, to regulate retinal vascular development in mice (Fig. 3C) (Junge et al. 2009).

The expression of mouse Norrin in the developing retina, neural tube, and brain and its expression in the adult central nervous system, uterus, and decidua (Luhmann et al. 2005b; Ye et al. 2011) suggest that Norrin may have developmental functions beyond those described in the eye.

CONCLUDING REMARKS

Considerable progress has been made in our effort to understand the molecular mechanism by which Wnt inhibitors and activators regulate Wnt signaling. These molecules have emerged as key regulators of Wnt signaling–controlled processes during embryogenesis, including head, limb, bone, heart, somites, and vasculature development, and have implications for pathological events, including cancer and bone disease. It is not surprising that agents targeting Dkk1 and sclerostin are currently in clinical trials. BHQ880, an anti-Dkk1 neutralizing antibody, is being investigated as an osteoblastogenesis stimulator for the potential treatment of bone lesions in multiple myeloma patients (Fulciniti et al. 2009), and anti-sclerostin monoclonal antibody AMG 785 is being examined for the potential treatment of diseases associated with bone loss, including osteoporosis (Lewiecki 2011; Padhi et al. 2011). Moreover, because of its potent proliferative effect on the intestinal epithelium in vivo and in vitro, Rspo1 holds promise as therapeutic target for gastrointestinal diseases and regenerative medicine (Kim et al. 2005; Ootani et al. 2009; Sato et al. 2009).

It has become evident that beyond the regulation of Wnt signaling, Wnt inhibitors and activators do modulate other signaling cascades as well, and we will benefit from future research to fully understand their Wnt-dependent and -independent roles.

ACKNOWLEDGMENTS

We thank B. Engelhardt for artwork. This work is supported by the Deutsche Forschungsgemeinschaft.

REFERENCES

Adamska M, MacDonald BT, Sarmast ZH, Oliver ER, Meisler MH. 2004. En1 and Wnt7a interact with Dkk1 during limb development in the mouse. *Dev Biol* **272:** 134–144.

Aguilera O, Fraga MF, Ballestar E, Paz MF, Herranz M, Espada J, Garcia JM, Munoz A, Esteller M, Gonzalez-Sancho JM. 2006. Epigenetic inactivation of the Wnt antagonist *DICKKOPF-1* (*DKK-1*) gene in human colorectal cancer. *Oncogene* **25:** 4116–4121.

Ahn Y, Sanderson BW, Klein OD, Krumlauf R. 2010. Inhibition of Wnt signaling by Wise (Sostdc1) and negative feedback from Shh controls tooth number and patterning. *Development* **137:** 3221–3231.

Ai M, Holmen SL, Van Hul W, Williams BO, Warman ML. 2005. Reduced affinity to and inhibition by DKK1 form a common mechanism by which high bone mass-associated missense mutations in LRP5 affect canonical Wnt signaling. *Mol Cell Biol* **25:** 4946–4955.

Aoki M, Mieda M, Ikeda T, Hamada Y, Nakamura H, Okamoto H. 2007. R-Spondin3 is required for mouse placental development. *Dev Biol* **301:** 218–226.

Aoki M, Kiyonari H, Nakamura H, Okamoto H. 2008. R-Spondin2 expression in the apical ectodermal ridge is essential for outgrowth and patterning in mouse limb development. *Dev Growth Differ* **50:** 85–95.

Avsian-Kretchmer O, Hsueh AJ. 2004. Comparative genomic analysis of the eight-membered ring cystine knot-containing bone morphogenetic protein antagonists. *Mol Endocrinol* **18:** 1–12.

Bafico A, Gazit A, Pramila T, Finch PW, Yaniv A, Aaronson SA. 1999. Interaction of frizzled related protein (FRP) with Wnt ligands and the frizzled receptor suggests alternative mechanisms for FRP inhibition of Wnt signaling. *J Biol Chem* **274:** 16180–16187.

Bafico A, Liu G, Yaniv A, Gazit A, Aaronson SA. 2001. Novel mechanism of Wnt signalling inhibition mediated by Dickkopf-1 interaction with LRP6/Arrow. *Nat Cell Biol* **3:** 683–686.

Balemans W, Ebeling M, Patel N, Van Hul E, Olson P, Dioszegi M, Lacza C, Wuyts W, Van Den Ende J, Willems P, et al. 2001. Increased bone density in sclerosteosis is due to the deficiency of a novel secreted protein (SOST). *Hum Mol Genet* **10:** 537–543.

Balemans W, Patel N, Ebeling M, Van Hul E, Wuyts W, Lacza C, Dioszegi M, Dikkers FG, Hildering P, Willems PJ, et al. 2002. Identification of a 52 kb deletion downstream of the *SOST* gene in patients with van Buchem disease. *J Med Genet* **39:** 91–97.

Banyai L, Patthy L. 1999. The NTR module: Domains of netrins, secreted frizzled related proteins, and type I procollagen C-proteinase enhancer protein are homologous with tissue inhibitors of metalloproteases. *Protein Sci* **8:** 1636–1642.

Baran R, Kechijian P. 2001. Understanding nail disorders. *Eur J Dermatol* **11:** 159–162.

Barker N, Clevers H. 2010. Leucine-rich repeat-containing G-protein-coupled receptors as markers of adult stem cells. *Gastroenterology* **138:** 1681–1696.

Barrantes Idel B, Montero-Pedrazuela A, Guadano-Ferraz A, Obregon MJ, Martinez de Mena R, Gailus-Durner V, Fuchs H, Franz TJ, Kalaydjiev S, Klempt M, et al. 2006. Generation and characterization of *dickkopf3* mutant mice. *Mol Cell Biol* **26:** 2317–2326.

Barrow JR, Thomas KR, Boussadia-Zahui O, Moore R, Kemler R, Capecchi MR, McMahon AP. 2003. Ectodermal *Wnt3*/*β-catenin* signaling is required for the establishment and maintenance of the apical ectodermal ridge. *Genes Dev* **17:** 394–409.

Bell SM, Schreiner CM, Wert SE, Mucenski ML, Scott WJ, Whitsett JA. 2008. R-Spondin 2 is required for normal laryngeal-tracheal, lung and limb morphogenesis. *Development* **135:** 1049–1058.

Belo JA, Bachiller D, Agius E, Kemp C, Borges AC, Marques S, Piccolo S, De Robertis EM. 2000. Cerberus-like is a secreted BMP and nodal antagonist not essential for mouse development. *Genesis* **26:** 265–270.

Belo JA, Silva AC, Borges AC, Filipe M, Bento M, Goncalves L, Vitorino M, Salgueiro AM, Texeira V, Tavares AT, et al. 2009. Generating asymmetries in the early vertebrate embryo: The role of the Cerberus-like family. *Int J Dev Biol* **53:** 1399–1407.

Berger W. 1998. Molecular dissection of Norrie disease. *Acta Anat (Basel)* **162:** 95–100.

Berger W, Meindl A, van de Pol TJ, Cremers FP, Ropers HH, Doerner C, Monaco A, Bergen AA, Lebo R, Warburgh M, et al. 1992. Isolation of a candidate gene for Norrie disease by positional cloning. *Nat Genet* **2:** 84.

Bergmann C, Senderek J, Anhuf D, Thiel CT, Ekici AB, Poblete-Gutierrez P, van Steensel M, Seelow D, Nurnberg G, Schild HH, et al. 2006. Mutations in the gene encoding the Wnt-signaling component R-spondin 4 (RSPO4) cause autosomal recessive anonychia. *Am J Hum Genet* **79:** 1105–1109.

Bhat RA, Stauffer B, Komm BS, Bodine PV. 2007. Structure-function analysis of secreted frizzled-related protein-1 for its Wnt antagonist function. *J Cell Biochem* **102:** 1519–1528.

Binnerts ME, Kim KA, Bright JM, Patel SM, Tran K, Zhou M, Leung JM, Liu Y, Lomas WE III, Dixon M, et al. 2007. R-Spondin1 regulates Wnt signaling by inhibiting internalization of LRP6. *Proc Natl Acad Sci* **104:** 14700–14705.

Blaydon DC, Ishii Y, O'Toole EA, Unsworth HC, Teh MT, Ruschendorf F, Sinclair C, Hopsu-Havu VK, Tidman N, Moss C, et al. 2006. The gene encoding R-spondin 4 (*RSPO4*), a secreted protein implicated in Wnt signaling, is mutated in inherited anonychia. *Nat Genet* **38:** 1245–1247.

Blish KR, Wang W, Willingham MC, Du W, Birse CE, Krishnan SR, Brown JC, Hawkins GA, Garvin AJ, D'Agostino RB Jr, et al. 2008. A human bone morphogenetic protein antagonist is down-regulated in renal cancer. *Mol Biol Cell* **19:** 457–464.

Bouwmeester T, Kim S, Sasai Y, Lu B, De Robertis EM. 1996. Cerberus is a head-inducing secreted factor expressed in the anterior endoderm of Spemann's organizer. *Nature* **382:** 595–601.

Bovolenta P, Esteve P, Ruiz JM, Cisneros E, Lopez-Rios J. 2008. Beyond Wnt inhibition: New functions of secreted Frizzled-related proteins in development and disease. *J Cell Sci* **121:** 737–746.

Boyden LM, Mao J, Belsky J, Mitzner L, Farhi A, Mitnick MA, Wu D, Insogna K, Lifton RP. 2002. High bone density due to a mutation in LDL-receptor-related protein 5. *N Engl J Med* **346:** 1513–1521.

Bradley L, Sun B, Collins-Racie L, LaVallie E, McCoy J, Sive H. 2000. Different activities of the frizzled-related proteins *frzb2* and *sizzled2* during *Xenopus* anteroposterior patterning. *Dev Biol* **227:** 118–132.

Brott BK, Sokol SY. 2002. Regulation of Wnt/LRP signaling by distinct domains of Dickkopf proteins. *Mol Cell Biol* **22:** 6100–6110.

Bruchle NO, Frank J, Frank V, Senderek J, Akar A, Koc E, Rigopoulos D, van Steensel M, Zerres K, Bergmann C. 2008. *RSPO4* is the major gene in autosomal-recessive anonychia and mutations cluster in the furin-like cysteine-rich domains of the Wnt signaling ligand R-spondin 4. *J Invest Dermatol* **128:** 791–796.

Brunkow ME, Gardner JC, Van Ness J, Paeper BW, Kovacevich BR, Proll S, Skonier JE, Zhao L, Sabo PJ, Fu Y, et al. 2001. Bone dysplasia sclerosteosis results from loss of the *SOST* gene product, a novel cystine knot-containing protein. *Am J Hum Genet* **68:** 577–589.

Cappuccio I, Calderone A, Busceti CL, Biagioni F, Pontarelli F, Bruno V, Storto M, Terstappen GT, Gaviraghi G, Fornai F, et al. 2005. Induction of Dickkopf-1, a negative modulator of the Wnt pathway, is required for the development of ischemic neuronal death. *J Neurosci* **25:** 2647–2657.

Caricasole A, Copani A, Caraci F, Aronica E, Rozemuller AJ, Caruso A, Storto M, Gaviraghi G, Terstappen GC, Nicoletti F. 2004. Induction of Dickkopf-1, a negative modulator of the Wnt pathway, is associated with neuronal degeneration in Alzheimer's brain. *J Neurosci* **24:** 6021–6027.

Carmon KS, Gong X, Lin Q, Thomas A, Liu Q. 2011. R-Spondins function as ligands of the orphan receptors LGR4 and LGR5 to regulate Wnt/β-catenin signaling. *Proc Natl Acad Sci* **108:** 11452–11457.

Carmon KS, Lin Q, Gong X, Thomas A, Liu Q. 2012. LGR5 interacts and cointernalizes with Wnt receptors to modulate Wnt/β-catenin signaling. *Mol Cell Biol* **32:** 2054–2064.

Chapman SC, Brown R, Lees L, Schoenwolf GC, Lumsden A. 2004. Expression analysis of chick Wnt and Frizzled genes and selected inhibitors in early chick patterning. *Dev Dyn* **229:** 668–676.

Chassot AA, Ranc F, Gregoire EP, Roepers-Gajadien HL, Taketo MM, Camerino G, de Rooij DG, Schedl A, Chaboissier MC. 2008. Activation of β-catenin signaling by Rspo1 controls differentiation of the mammalian ovary. *Hum Mol Genet* **17:** 1264–1277.

Chen ZY, Hendriks RW, Jobling MA, Powell JF, Breakefield XO, Sims KB, Craig IW. 1992. Isolation and characterization of a candidate gene for Norrie disease. *Nat Genet* **1:** 204–208.

Chen ZY, Battinelli EM, Fielder A, Bundey S, Sims K, Breakefield XO, Craig IW. 1993. A mutation in the Norrie disease gene (NDP) associated with X-linked familial exudative vitreoretinopathy. *Nat Genet* **5:** 180–183.

Chen JZ, Wang S, Tang R, Yang QS, Zhao E, Chao Y, Ying K, Xie Y, Mao YM. 2002. Cloning and identification of a cDNA that encodes a novel human protein with thrombospondin type I repeat domain, hPWTSR. *Mol Biol Rep* **29:** 287–292.

Chien AJ, Conrad WH, Moon RT. 2009. A Wnt survival guide: From flies to human disease. *J Invest Dermatol* **129:** 1614–1627.

Chong JM, Uren A, Rubin JS, Speicher DW. 2002. Disulfide bond assignments of secreted Frizzled-related protein-1 provide insights about Frizzled homology and netrin modules. *J Biol Chem* **277:** 5134–5144.

Collette NM, Genetos DC, Murugesh D, Harland RM, Loots GG. 2010. Genetic evidence that SOST inhibits WNT signaling in the limb. *Dev Biol* **342:** 169–179.

Dann CE, Hsieh JC, Rattner A, Sharma D, Nathans J, Leahy DJ. 2001. Insights into Wnt binding and signalling from the structures of two Frizzled cysteine-rich domains. *Nature* **412:** 86–90.

Davidson G, Mao B, Del Barco Barrantes I, Niehrs C. 2002. Kremen proteins interact with Dickkopf1 to regulate anteroposterior CNS patterning. *Development* **129:** 5587–5596.

de Lau W, Barker N, Low TY, Koo BK, Li VS, Teunissen H, Kujala P, Haegebarth A, Peters PJ, van de Wetering, et al. 2011. Lgr5 homologues associate with Wnt receptors and mediate R-spondin signalling. *Nature* **476:** 293–297.

del Barco Barrantes I, Davidson G, Grone HJ, Westphal H, Niehrs C. 2003. *Dkk1* and *noggin* cooperate in mammalian head induction. *Genes Dev* **17:** 2239–2244.

Dennis S, Aikawa M, Szeto W, d'Amore PA, Papkoff J. 1999. A secreted Frizzled related protein, FrzA, selectively associates with Wnt-1 protein and regulates Wnt-1 signaling. *J Cell Sci* **112:** 3815–3820.

De Robertis EM, Kuroda H. 2004. Dorsal-ventral patterning and neural induction in *Xenopus* embryos. *Annu Rev Cell Dev Biol* **20:** 285–308.

Diep DB, Hoen N, Backman M, Machon O, Krauss S. 2004. Characterisation of the Wnt antagonists and their response to conditionally activated Wnt signalling in the developing mouse forebrain. *Brain Res Dev Brain Res* **153:** 261–270.

Durai R, Davies M, Yang W, Yang SY, Seifalian A, Goldspink G, Winslet M. 2006. Biology of insulin-like growth factor binding protein-4 and its role in cancer (review). *Int J Oncol* **28:** 1317–1325.

Ellies DL, Viviano B, McCarthy J, Rey JP, Itasaki N, Saunders S, Krumlauf R. 2006. Bone density ligand, Sclerostin, directly interacts with LRP5 but not LRP5G171V to modulate Wnt activity. *J Bone Miner Res* **21:** 1738–1749.

Ellwanger K, Saito H, Clement-Lacroix P, Maltry N, Niedermeyer J, Lee WK, Baron R, Rawadi G, Westphal H, Niehrs C. 2008. Targeted disruption of the Wnt regulator Kremen induces limb defects and high bone density. *Mol Cell Biol* **28:** 4875–4882.

Elston MS, Clifton-Bligh RJ. 2010. Identification of Wnt family inhibitors: A pituitary tumor directed whole genome approach. *Mol Cell Endocrinol* **326:** 48–54.

Esteve P, Bovolenta P. 2010. The advantages and disadvantages of Sfrp1 and Sfrp2 expression in pathological events. *Tohoku J Exp Med* **221:** 11–17.

Esteve P, Sandonis A, Cardozo M, Malapeira J, Ibanez C, Crespo I, Marcos S, Gonzalez-Garcia S, Toribio ML, Arribas J, et al. 2011. SFRPs act as negative modulators of ADAM10 to regulate retinal neurogenesis. *Nat Neurosci* **14:** 562–569.

Fedders H, Augustin R, Bosch TC. 2004. A *Dickkopf-3*-related gene is expressed in differentiating nematocytes in the basal metazoan *Hydra*. *Dev Genes Evol* **214:** 72–80.

Finch PW, He X, Kelley MJ, Uren A, Schaudies RP, Popescu NC, Rudikoff S, Aaronson SA, Varmus HE, Rubin JS. 1997. Purification and molecular cloning of a secreted, Frizzled-related antagonist of Wnt action. *Proc Natl Acad Sci* **94:** 6770–6775.

Firth SM, Baxter RC. 2002. Cellular actions of the insulin-like growth factor binding proteins. *Endocr Rev* **23:** 824–854.

Fjeld K, Kettunen P, Furmanek T, Kvinnsland IH, Luukko K. 2005. Dynamic expression of Wnt signaling-related Dickkopf1, -2, and -3 mRNAs in the developing mouse tooth. *Dev Dyn* **233:** 161–166.

Fulciniti M, Tassone P, Hideshima T, Vallet S, Nanjappa P, Ettenberg SA, Shen Z, Patel N, Tai YT, Chauhan D, et al. 2009. Anti-DKK1 mAb (BHQ880) as a potential therapeutic agent for multiple myeloma. *Blood* **114:** 371–379.

Furushima K, Yamamoto A, Nagano T, Shibata M, Miyachi H, Abe T, Ohshima N, Kiyonari H, Aizawa S. 2007. Mouse homologues of Shisa antagonistic to Wnt and Fgf signalings. *Dev Biol* **306:** 480–492.

Galli LM, Barnes T, Cheng T, Acosta L, Anglade A, Willert K, Nusse R, Burrus LW. 2006. Differential inhibition of Wnt-3a by Sfrp-1, Sfrp-2, and Sfrp-3. *Dev Dyn* **235:** 681–690.

Glinka A, Wu W, Onichtchouk D, Blumenstock C, Niehrs C. 1997. Head induction by simultaneous repression of Bmp and Wnt signalling in *Xenopus*. *Nature* **389:** 517–519.

Glinka A, Wu W, Delius H, Monaghan AP, Blumenstock C, Niehrs C. 1998. Dickkopf-1 is a member of a new family of secreted proteins and functions in head induction. *Nature* **391:** 357–362.

Glinka A, Dolde C, Kirsch N, Huang YL, Kazanskaya O, Ingelfinger D, Boutros M, Cruciat CM, Niehrs C. 2011. LGR4 and LGR5 are R-spondin receptors mediating Wnt/β-catenin and Wnt/PCP signalling. *EMBO Rep* **12:** 1055–1061.

Gong Y, Slee RB, Fukai N, Rawadi G, Roman-Roman S, Reginato AM, Wang H, Cundy T, Glorieux FH, Lev D, et al. 2001. LDL receptor-related protein 5 (LRP5) affects bone accrual and eye development. *Cell* **107:** 513–523.

Gonzalez-Sancho JM, Aguilera O, Garcia JM, Pendas-Franco N, Pena C, Cal S, Garcia de Herreros A, Bonilla F, Munoz A. 2005. The Wnt antagonist *DICKKOPF-1* gene is a downstream target of β-catenin/TCF and is downregulated in human colon cancer. *Oncogene* **24:** 1098–1103.

Gordon A, Southam L, Loughlin J, Wilson AG, Stockley I, Hamer AJ, Eastell R, Wilkinson JM. 2007. Variation in the secreted frizzled-related protein-3 gene and risk of osteolysis and heterotopic ossification after total hip arthroplasty. *J Orthop Res* **25:** 1665–1670.

Grotewold L, Ruther U. 2002. The Wnt antagonist Dickkopf-1 is regulated by Bmp signaling and c-Jun and modulates programmed cell death. *EMBO J* **21:** 966–975.

Grotewold L, Theil T, Rüther U. 1999. Expression pattern of *Dkk-1* during mouse limb development. *Mech Dev* **89:** 151–153.

Guder C, Pinho S, Nacak TG, Schmidt HA, Hobmayer B, Niehrs C, Holstein TW. 2006. An ancient Wnt-Dickkopf antagonism in *Hydra*. *Development* **133:** 901–911.

Guidato S, Itasaki N. 2007. Wise retained in the endoplasmic reticulum inhibits Wnt signaling by reducing cell surface LRP6. *Dev Biol* **310:** 250–263.

Hackam AS. 2005. The Wnt signaling pathway in retinal degenerations. *IUBMB Life* **57:** 381–388.

Han XH, Jin YR, Seto M, Yoon JK. 2011. A WNT/β-catenin signaling activator, R-spondin, plays positive regulatory roles during skeletal myogenesis. *J Biol Chem* **286:** 10649–10659.

Hao HX, Xie Y, Zhang Y, Charlat O, Oster E, Avello M, Lei H, Mickanin C, Liu D, Ruffner H, et al. 2012. ZNRF3 promotes Wnt receptor turnover in an R-spondin-sensitive manner. *Nature* **485:** 195–200.

Hashimoto H, Itoh M, Yamanaka Y, Yamashita S, Shimizu T, Solnica-Krezel L, Hibi M, Hirano T. 2000. Zebrafish Dkk1 functions in forebrain specification and axial mesendoderm formation. *Dev Biol* **217:** 138–152.

Hashimoto H, Rebagliati M, Ahmad N, Muraoka O, Kurokawa T, Hibi M, Suzuki T. 2004. The Cerberus/Dan-family protein Charon is a negative regulator of Nodal signaling during left-right patterning in zebrafish. *Development* **131:** 1741–1753.

Hausler KD, Horwood NJ, Chuman Y, Fisher JL, Ellis J, Martin TJ, Rubin JS, Gillespie MT. 2004. Secreted frizzled-related protein-1 inhibits RANKL-dependent osteoclast formation. *J Bone Miner Res* **19:** 1873–1881.

Hedge TA, Mason I. 2008. Expression of Shisa2, a modulator of both Wnt and Fgf signaling, in the chick embryo. *Int J Dev Biol* **52:** 81–85.

Hoang B, Moos M Jr, Vukicevic S, Luyten FP. 1996. Primary structure and tissue distribution of FRZB, a novel protein related to *Drosophila* Frizzled, suggest a role in skeletal morphogenesis. *J Biol Chem* **271:** 26131–26137.

Houart C, Caneparo L, Heisenberg C, Barth K, Take-Uchi M, Wilson S. 2002. Establishment of the telencephalon during gastrulation by local antagonism of Wnt signaling. *Neuron* **35:** 255–265.

Hsieh JC, Kodjabachian L, Rebbert ML, Rattner A, Smallwood PM, Samos CH, Nusse R, Dawid IB, Nathans J. 1999. A new secreted protein that binds to Wnt proteins and inhibits their activities. *Nature* **398:** 431–436.

Hu YA, Gu X, Liu J, Yang Y, Yan Y, Zhao C. 2008. Expression pattern of *Wnt inhibitor factor 1* (*Wif1*) during the development in mouse CNS. *Gene Expr Patterns* **8:** 515–522.

Hunter DD, Zhang M, Ferguson JW, Koch M, Brunken WJ. 2004. The extracellular matrix component WIF-1 is expressed during, and can modulate, retinal development. *Mol Cell Neurosci* **27:** 477–488.

Idkowiak J, Weisheit G, Plitzner J, Viebahn C. 2004. Hypoblast controls mesoderm generation and axial patterning in the gastrulating rabbit embryo. *Dev Genes Evol* **214:** 591–605.

Ishii Y, Wajid M, Bazzi H, Fantauzzo KA, Barber AG, Blaydon DC, Nam JS, Yoon JK, Kelsell DP, Christiano AM. 2008. Mutations in R-spondin 4 (RSPO4) underlie inherited anonychia. *J Invest Dermatol* **128:** 867–870.

Itasaki N, Jones CM, Mercurio S, Rowe A, Domingos PM, Smith JC, Krumlauf R. 2003. Wise, a context-dependent activator and inhibitor of Wnt signalling. *Development* **130:** 4295–4305.

Jin YR, Turcotte TJ, Crocker AL, Han XH, Yoon JK. 2011. The canonical Wnt signaling activator, R-spondin2, regulates craniofacial patterning and morphogenesis within the branchial arch through ectodermal-mesenchymal interaction. *Dev Biol* **352:** 1–13.

Jones SE, Jomary C. 2002. Secreted Frizzled-related proteins: Searching for relationships and patterns. *Bioessays* **24:** 811–820.

Jukkola T, Sinjushina N, Partanen J. 2004. *Drapc1* expression during mouse embryonic development. *Gene Expr Patterns* **4:** 755–762.

Junge HJ, Yang S, Burton JB, Paes K, Shu X, French DM, Costa M, Rice DS, Ye W. 2009. TSPAN12 regulates retinal vascular development by promoting Norrin- but not Wnt-induced FZD4/β-catenin signaling. *Cell* **139:** 299–311.

Kagermeier-Schenk B, Wehner D, Ozhan-Kizil G, Yamamoto H, Li J, Kirchner K, Hoffmann C, Stern P, Kikuchi A, Schambony A, et al. 2011. Waif1/5T4 inhibits Wnt/β-catenin signaling and activates noncanonical Wnt pathways by modifying LRP6 subcellular localization. *Dev Cell* **21:** 1129–1143.

Kamata T, Katsube K, Michikawa M, Yamada M, Takada S, Mizusawa H. 2004. *R-spondin*, a novel gene with thrombospondin type 1 domain, was expressed in the dorsal neural tube and affected in *Wnt*s mutants. *Biochim Biophys Acta* **1676:** 51–62.

Kaneko KJ, DePamphilis ML. 2000. *Soggy*, a spermatocyte-specific gene, lies 3.8 kb upstream of and antipodal to *TEAD-2*, a transcription factor expressed at the beginning of mouse development. *Nucleic Acids Res* **28:** 3982–3990.

Kansara M, Tsang M, Kodjabachian L, Sims NA, Trivett MK, Ehrich M, Dobrovic A, Slavin J, Choong PF, Simmons PJ, et al. 2009. Wnt inhibitory factor 1 is epigenetically silenced in human osteosarcoma, and targeted disruption accelerates osteosarcomagenesis in mice. *J Clin Invest* **119:** 837–851.

Kassai Y, Munne P, Hotta Y, Penttila E, Kavanagh K, Ohbayashi N, Takada S, Thesleff I, Jernvall J, Itoh N. 2005. Regulation of mammalian tooth cusp patterning by ectodin. *Science* **309:** 2067–2070.

Katoh Y, Katoh M. 2005. Comparative genomics on Shisa orthologs. *Int J Mol Med* **16:** 181–185.

Kawano Y, Kypta R. 2003. Secreted antagonists of the Wnt signalling pathway. *J Cell Sci* **116:** 2627–2634.

Kazanskaya O, Glinka A, Del Barco Barrantes I, Stannek P, Niehrs C, Wu W. 2004. R-Spondin2 is a secreted activator of Wnt/β-catenin signaling and is required for *Xenopus* myogenesis. *Dev Cell* **7:** 525–534.

Kazanskaya O, Ohkawara B, Heroult M, Wu W, Maltry N, Augustin HG, Niehrs C. 2008. The Wnt signaling regulator R-spondin 3 promotes angioblast and vascular development. *Development* **135:** 3655–3664.

Kelly OG, Pinson KI, Skarnes WC. 2004. The Wnt co-receptors Lrp5 and Lrp6 are essential for gastrulation in mice. *Development* **131:** 2803–2815.

Kim KA, Kakitani M, Zhao J, Oshima T, Tang T, Binnerts M, Liu Y, Boyle B, Park E, Emtage P, et al. 2005. Mitogenic influence of human R-spondin1 on the intestinal epithelium. *Science* **309:** 1256–1259.

Kim KA, Zhao J, Andarmani S, Kakitani M, Oshima T, Binnerts ME, Abo A, Tomizuka K, Funk WD. 2006. R-Spondin proteins: A novel link to β-catenin activation. *Cell Cycle* **5:** 23–26.

Kim KA, Wagle M, Tran K, Zhan X, Dixon MA, Liu S, Gros D, Korver W, Yonkovich S, Tomasevic N, et al. 2008. R-Spondin family members regulate the Wnt pathway by a common mechanism. *Mol Biol Cell* **19:** 2588–2596.

Kimura-Yoshida C, Nakano H, Okamura D, Nakao K, Yonemura S, Belo JA, Aizawa S, Matsui Y, Matsuo I. 2005. Canonical Wnt signaling and its antagonist regulate anterior-posterior axis polarization by guiding cell migration in mouse visceral endoderm. *Dev Cell* **9:** 639–650.

Kohn MJ, Kaneko KJ, DePamphilis ML. 2005. DkkL1 (Soggy), a Dickkopf family member, localizes to the acrosome during mammalian spermatogenesis. *Mol Reprod Dev* **71:** 516–522.

Krishnan V, Bryant HU, Macdougald OA. 2006. Regulation of bone mass by Wnt signaling. *J Clin Invest* **116:** 1202–1209.

Krupnik VE, Sharp JD, Jiang C, Robison K, Chickering TW, Amaravadi L, Brown DE, Guyot D, Mays G, Leiby K, et al. 1999. Functional and structural diversity of the human Dickkopf gene family. *Gene* **238:** 301–313.

Kuang SQ, Tong WG, Yang H, Lin W, Lee MK, Fang ZH, Wei Y, Jelinek J, Issa JP, Garcia-Manero G. 2008. Genome-wide identification of aberrantly methylated promoter associated CpG islands in acute lymphocytic leukemia. *Leukemia* **22:** 1529–1538.

Kuroda H, Wessely O, De Robertis EM. 2004. Neural induction in *Xenopus*: requirement for ectodermal and endomesodermal signals via Chordin, Noggin, β-Catenin, and Cerberus. *PLoS Biol* **2:** E92.

Laurikkala J, Kassai Y, Pakkasjarvi L, Thesleff I, Itoh N. 2003. Identification of a secreted BMP antagonist, ectodin, integrating BMP, FGF, and SHH signals from the tooth enamel knot. *Dev Biol* **264:** 91–105.

Leaf I, Tennessen J, Mukhopadhyay M, Westphal H, Shawlot W. 2006. *Sfrp5* is not essential for axis formation in the mouse. *Genesis* **44:** 573–578.

Lee CS, Buttitta LA, May NR, Kispert A, Fan CM. 2000. SHH-N upregulates *Sfrp2* to mediate its competitive interaction with WNT1 and WNT4 in the somitic mesoderm. *Development* **127:** 109–118.

Lee HX, Ambrosio AL, Reversade B, De Robertis EM. 2006. Embryonic dorsal-ventral signaling: Secreted frizzled-related proteins as inhibitors of tolloid proteinases. *Cell* **124:** 147–159.

Lewiecki EM. 2011. Sclerostin monoclonal antibody therapy with AMG 785: A potential treatment for osteoporosis. *Expert Opin Biol Ther* **11:** 117–127.

Leyns L, Bouwmeester T, Kim SH, Piccolo S, De Robertis EM. 1997. Frzb-1 is a secreted antagonist of Wnt signaling expressed in the Spemann organizer. *Cell* **88:** 747–756.

Li L, Mao J, Sun L, Liu W, Wu D. 2002. Second cysteine-rich domain of Dickkopf-2 activates canonical Wnt signaling pathway via LRP-6 independently of Dishevelled. *J Biol Chem* **277:** 5977–5981.

Li X, Liu P, Liu W, Maye P, Zhang J, Zhang Y, Hurley M, Guo C, Boskey A, Sun L, et al. 2005a. Dkk2 has a role in terminal osteoblast differentiation and mineralized matrix formation. *Nat Genet* **37:** 945–952.

Li X, Zhang Y, Kang H, Liu W, Liu P, Zhang J, Harris SE, Wu D. 2005b. Sclerostin binds to LRP5/6 and antagonizes canonical Wnt signaling. *J Biol Chem* **280:** 19883–19887.

Li J, Sarosi I, Cattley RC, Pretorius J, Asuncion F, Grisanti M, Morony S, Adamu S, Geng Z, Qiu W, et al. 2006. Dkk1-mediated inhibition of Wnt signaling in bone results in osteopenia. *Bone* **39:** 754–766.

Li X, Ominsky MS, Niu QT, Sun N, Daugherty B, D'Agostin D, Kurahara C, Gao Y, Cao J, Gong J, et al. 2008a. Targeted deletion of the sclerostin gene in mice results in increased bone formation and bone strength. *J Bone Miner Res* **23:** 860–869.

Li Y, Rankin SA, Sinner D, Kenny AP, Krieg PA, Zorn AM. 2008b. Sfrp5 coordinates foregut specification and morphogenesis by antagonizing both canonical and noncanonical Wnt11 signaling. *Genes Dev* **22:** 3050–3063.

Li Y, Pawlik B, Elcioglu N, Aglan M, Kayserili H, Yigit G, Percin F, Goodman F, Nurnberg G, Cenani A, et al. 2010. LRP4 mutations alter Wnt/β-catenin signaling and cause limb and kidney malformations in Cenani-Lenz syndrome. *Am J Hum Genet* **86:** 696–706.

Li S, Esterberg R, Lachance V, Ren D, Radde-Gallwitz K, Chi F, Parent JL, Fritz A, Chen P. 2011. Rack1 is required for Vangl2 membrane localization and planar cell polarity signaling while attenuating canonical Wnt activity. *Proc Natl Acad Sci* **108:** 2264–2269.

Lin K, Wang S, Julius MA, Kitajewski J, Moos MJ, Luyten FP. 1997. The cysteine-rich frizzled domain of Frzb-1 is required and sufficient for modulation of Wnt signaling. *Proc Natl Acad Sci* **94:** 11196–11200.

Lintern KB, Guidato S, Rowe A, Saldanha JW, Itasaki N. 2009. Characterization of Wise protein and its molecular mechanism to interact with both Wnt and BMP signals. *J Biol Chem* **284:** 23159–23168.

Little RD, Carulli JP, Del Mastro RG, Dupuis J, Osborne M, Folz C, Manning SP, Swain PM, Zhao SC, Eustace B, et al. 2002. A mutation in the LDL receptor-related protein 5 gene results in the autosomal dominant high-bone-mass trait. *Am J Hum Genet* **70:** 11–19.

Lopez-Rios J, Esteve P, Ruiz JM, Bovolenta P. 2008. The Netrin-related domain of Sfrp1 interacts with Wnt ligands and antagonizes their activity in the anterior neural plate. *Neural Dev* **3:** 19.

Lowther W, Wiley K, Smith GH, Callahan R. 2005. A new common integration site, Int7, for the mouse mammary tumor virus in mouse mammary tumors identifies a gene whose product has furin-like and thrombospondin-like sequences. *J Virol* **79:** 10093–10096.

Luhmann UF, Lin J, Acar N, Lammel S, Feil S, Grimm C, Seeliger MW, Hammes HP, Berger W. 2005a. Role of the Norrie disease pseudoglioma gene in sprouting angiogenesis during development of the retinal vasculature. *Invest Ophthalmol Vis Sci* **46:** 3372–3382.

Luhmann UF, Meunier D, Shi W, Luttges A, Pfarrer C, Fundele R, Berger W. 2005b. Fetal loss in homozygous mutant Norrie disease mice: A new role of Norrin in reproduction. *Genesis* **42:** 253–262.

Luke GN, Castro LF, McLay K, Bird C, Coulson A, Holland PW. 2003. Dispersal of NK homeobox gene clusters in amphioxus and humans. *Proc Natl Acad Sci* **100:** 5292–5295.

MacDonald BT, Adamska M, Meisler MH. 2004. Hypomorphic expression of *Dkk1* in the *doubleridge* mouse: Dose dependence and compensatory interactions with *Lrp6*. *Development* **131:** 2543–2552.

MacDonald BT, Joiner DM, Oyserman SM, Sharma P, Goldstein SA, He X, Hauschka PV. 2007. Bone mass is inversely proportional to *Dkk1* levels in mice. *Bone* **41:** 331–339.

Mao B, Niehrs C. 2003. Kremen2 modulates Dickkopf2 activity during Wnt/LRP6 signaling. *Gene* **302:** 179–183.

Mao B, Wu W, Li Y, Hoppe D, Stannek P, Glinka A, Niehrs C. 2001. LDL-receptor-related protein 6 is a receptor for Dickkopf proteins. *Nature* **411:** 321–325.

Mao B, Wu W, Davidson G, Marhold J, Li M, Mechler B, Delius H, Hoppe D, Stannek P, et al. 2002. Kremens are novel Dickkopf receptors that regulate Wnt/β-catenin signalling. *Nature* **417:** 664–667.

Marques S, Borges AC, Silva AC, Freitas S, Cordenonsi M, Belo JA. 2004. The activity of the Nodal antagonist Cerl-2 in the mouse node is required for correct L/R body axis. *Genes Dev* **18:** 2342–2347.

Matsuyama M, Aizawa S, Shimono A. 2009. Sfrp controls apicobasal polarity and oriented cell division in developing gut epithelium. *PLoS Genet* **5:** e1000427.

Meindl A, Berger W, Meitinger T, van de Pol D, Achatz H, Dorner C, Haasemann M, Hellebrand H, Gal A, Cremers F, et al. 1992. Norrie disease is caused by mutations in an extracellular protein resembling C-terminal globular domain of mucins. *Nat Genet* **2:** 139–143.

Meitinger T, Meindl A, Bork P, Rost B, Sander C, Haasemann M, Murken J. 1993. Molecular modelling of the Norrie disease protein predicts a cystine knot growth factor tertiary structure. *Nat Genet* **5:** 376–380.

Melkonyan HS, Chang WC, Shapiro JP, Mahadevappa M, Fitzpatrick PA, Kiefer MC, Tomei LD, Umansky SR. 1997. SARPs: A family of secreted apoptosis-related proteins. *Proc Natl Acad Sci* **94:** 13636–13641.

Mii Y, Taira M. 2009. Secreted Frizzled-related proteins enhance the diffusion of Wnt ligands and expand their signalling range. *Development* **136:** 4083–4088.

Monaghan AP, Kioschis P, Wu W, Zuniga A, Bock D, Poustka A, Delius H, Niehrs C. 1999. Dickkopf genes are co-ordinately expressed in mesodermal lineages. *Mech Dev* **87:** 45–56.

Morvan F, Boulukos K, Clement-Lacroix P, Roman Roman S, Suc-Royer I, Vayssiere B, Ammann P, Martin P, Pinho S, Pognonec P, et al. 2006. Deletion of a single allele of the *Dkk1* gene leads to an increase in bone formation and bone mass. *J Bone Miner Res* **21:** 934–945.

Mukhopadhyay M, Shtrom S, Rodriguez-Esteban C, Chen L, Tsukui T, Gomer L, Dorward DW, Glinka A, Grinberg A, Huang SP, et al. 2001. *Dickkopf1* is required for embryonic head induction and limb morphogenesis in the mouse. *Dev Cell* **1:** 423–434.

Mukhopadhyay M, Gorivodsky M, Shtrom S, Grinberg A, Niehrs C, Morasso MI, Westphal H. 2006. Dkk2 plays an essential role in the corneal fate of the ocular surface epithelium. *Development* **133:** 2149–2154.

Nagano T, Takehara S, Takahashi M, Aizawa S, Yamamoto A. 2006. Shisa2 promotes the maturation of somitic precursors and transition to the segmental fate in *Xenopus* embryos. *Development* **133:** 4643–4654.

Naganuma H, Kono K, Mori Y, Takayoshi S, Stern PL, Tasaka K, Matsumoto Y. 2002. Oncofetal antigen 5T4 expression as a prognostic factor in patients with gastric cancer. *Anticancer Res* **22:** 1033–1038.

Nam JS, Turcotte TJ, Smith PF, Choi S, Yoon JK. 2006. Mouse cristin/R-spondin family proteins are novel ligands for the Frizzled 8 and LRP6 receptors and activate β-catenin-dependent gene expression. *J Biol Chem* **281:** 13247–13257.

Nam JS, Park E, Turcotte TJ, Palencia S, Zhan X, Lee J, Yun K, Funk WD, Yoon JK. 2007a. Mouse R-spondin2 is required for apical ectodermal ridge maintenance in the hindlimb. *Dev Biol* **311:** 124–135.

Nam JS, Turcotte TJ, Yoon JK. 2007b. Dynamic expression of R-spondin family genes in mouse development. *Gene Expr Patterns* **7:** 306–312.

Nie X. 2005. Dkk1, -2, and -3 expression in mouse craniofacial development. *J Mol Histol* **36:** 367–372.

Nie X, Luukko K, Fjeld K, Kvinnsland IH, Kettunen P. 2005. Developmental expression of *Dkk1-3* and *Mmp9* and apoptosis in cranial base of mice. *J Mol Histol* **36:** 419–426.

Niehrs C. 2004. Regionally specific induction by the Spemann-Mangold organizer. *Nat Rev Genet* **5:** 425–434.

Niehrs C. 2006. Function and biological roles of the Dickkopf family of Wnt modulators. *Oncogene* **25:** 7469–7481.

Niehrs C. 2010. On growth and form: A Cartesian coordinate system of Wnt and BMP signaling specifies bilaterian body axes. *Development* **137:** 845–857.

Ning Y, Schuller AG, Bradshaw S, Rotwein P, Ludwig T, Frystyk J, Pintar JE. 2006. Diminished growth and enhanced glucose metabolism in triple knockout mice containing mutations of insulin-like growth factor binding protein-3, -4, and -5. *Mol Endocrinol* **20:** 2173–2186.

Ohazama A, Johnson EB, Ota MS, Choi HY, Porntaveetus T, Oommen S, Itoh N, Eto K, Gritli-Linde A, Herz J, et al. 2008. Lrp4 modulates extracellular integration of cell signaling pathways in development. *PLoS ONE* **3:** e4092.

Ohkawara B, Glinka A, Niehrs C. 2011. Rspo3 binds syndecan 4 and induces Wnt/PCP signaling via clathrin-mediated endocytosis to promote morphogenesis. *Dev Cell* **20:** 303–314.

Ootani A, Li X, Sangiorgi E, Ho QT, Ueno H, Toda S, Sugihara H, Fujimoto K, Weissman IL, Capecchi MR, et al. 2009. Sustained in vitro intestinal epithelial culture within a Wnt-dependent stem cell niche. *Nat Med* **15:** 701–706.

O'Shaughnessy RF, Yeo W, Gautier J, Jahoda CA, Christiano AM. 2004. The WNT signalling modulator, Wise, is expressed in an interaction-dependent manner during hair-follicle cycling. *J Invest Dermatol* **123:** 613–621.

Padhi D, Jang G, Stouch B, Fang L, Posvar E. 2011. Single-dose, placebo-controlled, randomized study of AMG 785, a sclerostin monoclonal antibody. *J Bone Miner Res* **26:** 19–26.

Parma P, Radi O, Vidal V, Chaboissier MC, Dellambra E, Valentini S, Guerra L, Schedl A, Camerino G. 2006. R-

Spondin1 is essential in sex determination, skin differentiation and malignancy. *Nat Genet* **38:** 1304–1309.

Pei J, Grishin NV. 2012. Unexpected diversity in Shisa-like proteins suggests the importance of their roles as transmembrane adaptors. *Cell Signal* **24:** 758–769.

Pera EM, De Robertis EM. 2000. A direct screen for secreted proteins in *Xenopus* embryos identifies distinct activities for the Wnt antagonists Crescent and Frzb-1. *Mech Dev* **96:** 183–195.

Pfeffer PL, De Robertis EM, Izpisua-Belmonte JC. 1997. *Crescent*, a novel chick gene encoding a Frizzled-like cysteine-rich domain, is expressed in anterior regions during early embryogenesis. *Int J Dev Biol* **41:** 449–458.

Piccolo S, Agius E, Leyns L, Bhattacharyya S, Grunz H, Bouwmeester T, De Robertis EM. 1999. The head inducer Cerberus is a multifunctional antagonist of Nodal, BMP and Wnt signals. *Nature* **397:** 707–710.

Pinho S, Niehrs C. 2007. Dkk3 is required for TGF-β signaling during *Xenopus* mesoderm induction. *Differentiation* **75:** 957–967.

Pollard SL, Holland PW. 2000. Evidence for 14 homeobox gene clusters in human genome ancestry. *Curr Biol* **10:** 1059–1062.

Rattner A, Hsieh J-C, Smallwood PM, Gilbert D, Copeland NG, Jenkins NA, Nathans J. 1997. A family of secreted proteins contains homology to the cysteine-rich ligand binding domain of frizzled receptors. *Proc Natl Acad Sci* **94:** 2859–2863.

Rehm HL, Zhang DS, Brown MC, Burgess B, Halpin C, Berger W, Morton CC, Corey DP, Chen ZY. 2002. Vascular defects and sensorineural deafness in a mouse model of Norrie disease. *J Neurosci* **22:** 4286–4292.

Rehn M, Pihlajaniemi T, Hofmann K, Bucher P. 1998. The frizzled motif: In how many different protein families does it occur? *Trends Biochem Sci* **23:** 415–417.

Richter M, Gottanka J, May CA, Welge-Lussen U, Berger W, Lutjen-Drecoll E. 1998. Retinal vasculature changes in Norrie disease mice. *Invest Ophthalmol Vis Sci* **39:** 2450–2457.

Robitaille J, MacDonald ML, Kaykas A, Sheldahl LC, Zeisler J, Dube MP, Zhang LH, Singaraja RR, Guernsey DL, Zheng B, et al. 2002. Mutant frizzled-4 disrupts retinal angiogenesis in familial exudative vitreoretinopathy. *Nat Genet* **32:** 326–330.

Rodriguez Esteban C, Capdevila J, Economides AN, Pascual J, Ortiz A, Izpisua Belmonte JC. 1999. The novel Cer-like protein Caronte mediates the establishment of embryonic left–right asymmetry. *Nature* **401:** 243–251.

Rodriguez J, Esteve P, Weinl C, Ruiz JM, Fermin Y, Trousse F, Dwivedy A, Holt C, Bovolenta P. 2005. SFRP1 regulates the growth of retinal ganglion cell axons through the Fz2 receptor. *Nat Neurosci* **8:** 1301–1309.

Salic AN, Kroll KL, Evans LM, Kirschner MW. 1997. Sizzled: A secreted Xwnt8 antagonist expressed in the ventral marginal zone of *Xenopus* embryos. *Development* **124:** 4739–4748.

Sato H, Suzuki H, Toyota M, Nojima M, Maruyama R, Sasaki S, Takagi H, Sogabe Y, Sasaki Y, Idogawa M, et al. 2007. Frequent epigenetic inactivation of *DICKKOPF* family genes in human gastrointestinal tumors. *Carcinogenesis* **28:** 2459–2466.

Sato T, Vries RG, Snippert HJ, van de Wetering M, Barker N, Stange DE, van Es JH, Abo A, Kujala P, Peters PJ, et al. 2009. Single Lgr5 stem cells build crypt-villus structures in vitro without a mesenchymal niche. *Nature* **459:** 262–265.

Satoh W, Gotoh T, Tsunematsu Y, Aizawa S, Shimono A. 2006. *Sfrp1* and *Sfrp2* regulate anteroposterior axis elongation and somite segmentation during mouse embryogenesis. *Development* **133:** 989–999.

Satoh W, Matsuyama M, Takemura H, Aizawa S, Shimono A. 2008. Sfrp1, Sfrp2, and Sfrp5 regulate the Wnt/β-catenin and the planar cell polarity pathways during early trunk formation in mouse. *Genesis* **46:** 92–103.

Schneider VA, Mercola M. 2001. Wnt antagonism initiates cardiogenesis in *Xenopus laevis*. *Genes Dev* **15:** 304–315.

Schwarz-Romond T, Asbrand C, Bakkers J, Kuhl M, Schaeffer HJ, Huelsken J, Behrens J, Hammerschmidt M, Birchmeier W. 2002. The ankyrin repeat protein Diversin recruits Casein kinase Iε to the β-catenin degradation complex and acts in both canonical Wnt and Wnt/JNK signaling. *Genes Dev* **16:** 2073–2084.

Semenov MV, He X. 2006. LRP5 mutations linked to high bone mass diseases cause reduced LRP5 binding and inhibition by SOST. *J Biol Chem* **281:** 38276–38284.

Semenov MV, Tamai K, Brott BK, Kuhl M, Sokol S, He X. 2001. Head inducer Dickkopf-1 is a ligand for Wnt coreceptor LRP6. *Curr Biol* **11:** 951–961.

Semenov M, Tamai K, He X. 2005. SOST is a ligand for LRP5/LRP6 and a Wnt signaling inhibitor. *J Biol Chem* **280:** 26770–26775.

Semenov MV, Zhang X, He X. 2008. DKK1 antagonizes Wnt signaling without promotion of LRP6 internalization and degradation. *J Biol Chem* **283:** 21427–21432.

Shawlot W, Min Deng J, Wakamiya M, Behringer RR. 2000. The *cerberus*-related gene, *Cerr1*, is not essential for mouse head formation. *Genesis* **26:** 253–258.

Shimomura Y, Agalliu D, Vonica A, Luria V, Wajid M, Baumer A, Belli S, Petukhova L, Schinzel A, Brivanlou AH, et al. 2010. APCDD1 is a novel Wnt inhibitor mutated in hereditary hypotrichosis simplex. *Nature* **464:** 1043–1047.

Shinya M, Eschbach C, Clark M, Lehrach H, Furutani-Seiki M. 2000. Zebrafish Dkk1, induced by the pre-MBT Wnt signaling, is secreted from the prechordal plate and patterns the anterior neural plate. *Mech Dev* **98:** 3–17.

Silva AC, Filipe M, Kuerner KM, Steinbeisser H, Belo JA. 2003. Endogenous Cerberus activity is required for anterior head specification in *Xenopus*. *Development* **130:** 4943–4953.

Silva AC, Filipe M, Vitorino M, Steinbeisser H, Belo JA. 2006. Developmental expression of Shisa-2 in *Xenopus laevis*. *Int J Dev Biol* **50:** 575–579.

Simons M, Gloy J, Ganner A, Bullerkotte A, Bashkurov M, Kronig C, Schermer B, Benzing T, Cabello OA, Jenny A, et al. 2005. Inversin, the gene product mutated in nephronophthisis type II, functions as a molecular switch between Wnt signaling pathways. *Nat Genet* **37:** 537–543.

Simpson EH, Johnson DK, Hunsicker P, Suffolk R, Jordan SA, Jackson IJ. 1999. The mouse *Cer1* (*Cerberus related* or *homologue*) gene is not required for anterior pattern formation. *Dev Biol* **213:** 202–206.

Smallwood PM, Williams J, Xu Q, Leahy DJ, Nathans J. 2007. Mutational analysis of Norrin-Frizzled4 recognition. *J Biol Chem* **282:** 4057–4068.

Soshnikova N, Zechner D, Huelsken J, Mishina Y, Behringer RR, Taketo MM, Crenshaw EB III, Birchmeier W. 2003. Genetic interaction between Wnt/β-catenin and BMP receptor signaling during formation of the AER and the dorsal–ventral axis in the limb. *Genes Dev* **17:** 1963–1968.

Southgate TD, McGinn OJ, Castro FV, Rutkowski AJ, Al-Muftah M, Marinov G, Smethurst GJ, Shaw D, Ward CM, Miller CJ, et al. 2010. CXCR4 mediated chemotaxis is regulated by 5T4 oncofetal glycoprotein in mouse embryonic cells. *PLoS ONE* **5:** e9982.

Staehling-Hampton K, Proll S, Paeper BW, Zhao L, Charmley P, Brown A, Gardner JC, Galas D, Schatzman RC, Beighton P, et al. 2002. A 52-kb deletion in the *SOST-MEOX1* intergenic region on 17q12-q21 is associated with van Buchem disease in the Dutch population. *Am J Med Genet* **110:** 144–152.

Stanley EG, Biben C, Allison J, Hartley L, Wicks IP, Campbell IK, McKinley M, Barnett L, Koentgen F, Robb L, et al. 2000. Targeted insertion of a *lacZ* reporter gene into the mouse *Cer1* locus reveals complex and dynamic expression during embryogenesis. *Genesis* **26:** 259–264.

Starr TK, Allaei R, Silverstein KA, Staggs RA, Sarver AL, Bergemann TL, Gupta M, O'Sullivan MG, Matise I, Dupuy AJ, et al. 2009. A transposon-based genetic screen in mice identifies genes altered in colorectal cancer. *Science* **323:** 1747–1750.

Starzynska T, Marsh PJ, Schofield PF, Roberts SA, Myers KA, Stern PL. 1994. Prognostic significance of 5T4 oncofetal antigen expression in colorectal carcinoma. *Br J Cancer* **69:** 899–902.

Sugiyama Y, Stump RJ, Nguyen A, Wen L, Chen Y, Wang Y, Murdoch JN, Lovicu FJ, McAvoy JW. 2010. Secreted frizzled-related protein disrupts PCP in eye lens fiber cells that have polarised primary cilia. *Dev Biol* **338:** 193–201.

Surmann-Schmitt C, Widmann N, Dietz U, Saeger B, Eitzinger N, Nakamura Y, Rattel M, Latham R, Hartmann C, von der Mark H, et al. 2009. Wif-1 is expressed at cartilage-mesenchyme interfaces and impedes Wnt3a-mediated inhibition of chondrogenesis. *J Cell Sci* **122:** 3627–3637.

Swain RK, Katoh M, Medina A, Steinbeisser H. 2005. *Xenopus frizzled-4S*, a splicing variant of *Xfz4* is a context-dependent activator and inhibitor of Wnt/β-catenin signaling. *Cell Commun Signal* **3:** 12.

Takahashi M, Fujita M, Furukawa Y, Hamamoto R, Shimokawa T, Miwa N, Ogawa M, Nakamura Y. 2002. Isolation of a novel human gene, *APCDD1*, as a direct target of the β-catenin/T-cell factor 4 complex with probable involvement in colorectal carcinogenesis. *Cancer Res* **62:** 5651–5656.

Tavares AT, Andrade S, Silva AC, Belo JA. 2007. Cerberus is a feedback inhibitor of Nodal asymmetric signaling in the chick embryo. *Development* **134:** 2051–2060.

Theodorou V, Kimm MA, Boer M, Wessels L, Theelen W, Jonkers J, Hilkens J. 2007. MMTV insertional mutagenesis identifies genes, gene families and pathways involved in mammary cancer. *Nat Genet* **39:** 759–769.

Tomaselli S, Megiorni F, De Bernardo C, Felici A, Marrocco G, Maggiulli G, Grammatico B, Remotti D, Saccucci P, Valentini F, et al. 2008. Syndromic true hermaphroditism due to an R-spondin1 (*RSPO1*) homozygous mutation. *Hum Mutat* **29:** 220–226.

Tomizuka K, Horikoshi K, Kitada R, Sugawara Y, Iba Y, Kojima A, Yoshitome A, Yamawaki K, Amagai M, Inoue A, et al. 2008. R-Spondin1 plays an essential role in ovarian development through positively regulating Wnt-4 signaling. *Hum Mol Genet* **17:** 1278–1291.

Toomes C, Bottomley HM, Jackson RM, Towns KV, Scott S, Mackey DA, Craig JE, Jiang L, Yang Z, Trembath R, et al. 2004a. Mutations in *LRP5* or *FZD4* underlie the common familial exudative vitreoretinopathy locus on chromosome 11q. *Am J Hum Genet* **74:** 721–730.

Toomes C, Bottomley HM, Scott S, Mackey DA, Craig JE, Appukuttan B, Stout JT, Flaxel CJ, Zhang K, Black GC, et al. 2004b. Spectrum and frequency of *FZD4* mutations in familial exudative vitreoretinopathy. *Invest Ophthalmol Vis Sci* **45:** 2083–2090.

Uren A, Reichsman F, Anest V, Taylor WG, Muraiso K, Bottaro DP, Cumberledge S, Rubin JS. 2000. Secreted frizzled-related protein-1 binds directly to Wingless and is a biphasic modulator of Wnt signaling. *J Biol Chem* **275:** 4374–4382.

Vainio S, Heikkila M, Kispert A, Chin N, McMahon AP. 1999. Female development in mammals is regulated by Wnt-4 signalling. *Nature* **397:** 405–409.

Vonica A, Brivanlou AH. 2007. The left–right axis is regulated by the interplay of Coco, Xnr1 and derrière in *Xenopus* embryos. *Dev Biol* **303:** 281–294.

Wang S, Krinks M, Lin K, Luyten FP, Moos M Jr. 1997. Frzb, a secreted protein expressed in the Spemann organizer, binds and inhibits Wnt-8. *Cell* **88:** 757–766.

Wang K, Zhang Y, Li X, Chen L, Wang H, Wu J, Zheng J, Wu D. 2008. Characterization of the Kremen-binding site on Dkk1 and elucidation of the role of Kremen in Dkk-mediated Wnt antagonism. *J Biol Chem* **283:** 23371–23375.

Warden SM, Andreoli CM, Mukai S. 2007. The Wnt signaling pathway in familial exudative vitreoretinopathy and Norrie disease. *Semin Ophthalmol* **22:** 211–217.

Warr N, Siggers P, Bogani D, Brixey R, Pastorelli L, Yates L, Dean CH, Wells S, Satoh W, Shimono A, et al. 2009. *Sfrp1* and *Sfrp2* are required for normal male sexual development in mice. *Dev Biol* **326:** 273–284.

Wawrzak D, Metioui M, Willems E, Hendrickx M, de Genst E, Leyns L. 2007. Wnt3a binds to several sFRPs in the nanomolar range. *Biochem Biophys Res Commun* **357:** 1119–1123.

Wei Q, Yokota C, Semenov MV, Doble B, Woodgett J, He X. 2007. R-Spondin1 is a high affinity ligand for LRP6 and induces LRP6 phosphorylation and β-catenin signaling. *J Biol Chem* **282:** 15903–15911.

Winkler DG, Sutherland MK, Geoghegan JC, Yu C, Hayes T, Skonier JE, Shpektor D, Jonas M, Kovacevich BR, Staehling-Hampton K, et al. 2003. Osteocyte control of bone formation via sclerostin, a novel BMP antagonist. *EMBO J* **22:** 6267–6276.

Wu W, Glinka A, Delius H, Niehrs C. 2000. Mutual antagonism between *dickkopf1* and *dickkopf2* regulates Wnt/β-catenin signalling. *Curr Biol* **10:** 1611–1614.

Xia CH, Liu H, Cheung D, Wang M, Cheng C, Du X, Chang B, Beutler B, Gong X. 2008. A model for familial exudative vitreoretinopathy caused by LPR5 mutations. *Hum Mol Genet* **17:** 1605–1612.

Xu Q, Wang Y, Dabdoub A, Smallwood PM, Williams J, Woods C, Kelley MW, Jiang L, Tasman W, Zhang K, et al. 2004. Vascular development in the retina and inner ear: Control by Norrin and Frizzled-4, a high-affinity ligand-receptor pair. *Cell* **116:** 883–895.

Yamada W, Nagao K, Horikoshi K, Fujikura A, Ikeda E, Inagaki Y, Kakitani M, Tomizuka K, Miyazaki H, Suda T, et al. 2009. Craniofacial malformation in R-spondin2 knockout mice. *Biochem Biophys Res Commun* **381:** 453–458.

Yamamoto A, Nagano T, Takehara S, Hibi M, Aizawa S. 2005. Shisa promotes head formation through the inhibition of receptor protein maturation for the caudalizing factors, Wnt and FGF. *Cell* **120:** 223–235.

Yan D, Wallingford JB, Sun TQ, Nelson AM, Sakanaka C, Reinhard C, Harland RM, Fantl WJ, Williams LT. 2001. Cell autonomous regulation of multiple Dishevelled-dependent pathways by mammalian Nkd. *Proc Natl Acad Sci* **98:** 3802–3807.

Yanagita M, Oka M, Watabe T, Iguchi H, Niida A, Takahashi S, Akiyama T, Miyazono K, Yanagisawa M, Sakurai T. 2004. USAG-1: A bone morphogenetic protein antagonist abundantly expressed in the kidney. *Biochem Biophys Res Commun* **316:** 490–500.

Yanagita M, Okuda T, Endo S, Tanaka M, Takahashi K, Sugiyama F, Kunita S, Takahashi S, Fukatsu A, Yanagisawa M, et al. 2006. Uterine sensitization-associated gene-1 (USAG-1), a novel BMP antagonist expressed in the kidney, accelerates tubular injury. *J Clin Invest* **116:** 70–79.

Ye X, Smallwood P, Nathans J. 2011. Expression of the Norrie disease gene (*Ndp*) in developing and adult mouse eye, ear, and brain. *Gene Expr Patterns* **11:** 151–155.

Yokouchi Y, Vogan KJ, Pearse RV II, Tabin CJ. 1999. Antagonistic signaling by *Caronte*, a novel *Cerberus*-related gene, establishes left–right asymmetric gene expression. *Cell* **98:** 573–583.

Yoon JK, Lee JS. 2012. Cellular signaling and biological functions of R-spondins. *Cell Signal* **24:** 369–377.

Zhang X, Abreu JG, Yokota C, Macdonald BT, Singh S, Coburn KL, Cheong SM, Zhang MM, Ye QZ, Hang HC, et al. 2012. tiki1 is required for head formation via Wnt cleavage-oxidation and inactivation. *Cell* **149:** 1565–1577.

Zhu L, Marvin MJ, Gardiner A, Lassar AB, Mercola M, Stern CD, Levin M. 1999. *Cerberus* regulates left–right asymmetry of the embryonic head and heart. *Curr Biol* **9:** 931–938.

Zhu W, Shiojima I, Ito Y, Li Z, Ikeda H, Yoshida M, Naito AT, Nishi J, Ueno H, Umezawa A, et al. 2008a. IGFBP-4 is an inhibitor of canonical Wnt signalling required for cardiogenesis. *Nature* **454:** 345–349.

Zhu Y, Tsuchida A, Yamamoto A, Furukawa K, Tajima O, Tokuda N, Aizawa S, Urano T, Kadomatsu K. 2008b. Expression and roles of a *Xenopus* head-forming gene homologue in human cancer cell lines. *Nagoya J Med Sci* **70:** 73–82.

 Cite this article as *Cold Spring Harb Perspect Biol* doi: 10.1101/cshperspect.a015081

Frizzled and LRP5/6 Receptors for Wnt/β-Catenin Signaling

Bryan T. MacDonald and Xi He

The F. M. Kirby Neurobiology Center, Boston Children's Hospital, Boston, Massachusetts 02115; Department of Neurology, Harvard Medical School, Boston, Massachusetts 02115

Correspondence: xi.he@childrens.harvard.edu

Frizzled and LRP5/6 are Wnt receptors that upon activation lead to stabilization of cytoplasmic β-catenin. In this study, we review the current knowledge of these two families of receptors, including their structures and interactions with Wnt proteins, and signaling mechanisms from receptor activation to the engagement of intracellular partners Dishevelled and Axin, and finally to the inhibition of β-catenin phosphorylation and ensuing β-catenin stabilization.

The Wnt/β-catenin pathway, or canonical Wnt pathway as it is often referred to, is an ancient and conserved signaling cascade involving β-catenin acting as a transcriptional coactivator (Logan and Nusse 2004). The pathway is best understood when considered in a two-state model of OFF (without Wnt) and ON (with Wnt). In the OFF state, cytoplasmic β-catenin is constitutively targeted for degradation by two multidomain scaffolding proteins, Axin and Adenomatous polyposis coli (APC), which facilitate the amino-terminal phosphorylation of β-catenin via the kinases GSK3 and CKIα. Phosphorylated β-catenin is recognized by the E3 ubiquitin ligase β-Trcp and is thus ubiquitinated and degraded by the proteasome, thereby maintaining low levels of free β-catenin in the cytoplasm and nucleus (MacDonald et al. 2009). In the ON state, a Wnt ligand binds to the seven-pass transmembrane receptor Frizzled (FZD) and the single-pass low-density lipoprotein receptor-related protein 5 or 6 (LRP5/6) (He et al. 2004). The Wnt–FZD–LRP5/6 trimeric complex recruits Dishevelled (DVL) and Axin through the intracellular domains of FZD and LRP5/6, resulting in inhibition of β-catenin phosphorylation and thus ensuing β-catenin stabilization. The rise of cytoplasmic and nuclear levels of β-catenin levels promotes β-catenin partnering with the TCF/LEF transcription factors for activation of Wnt-responsive gene expression.

Wnt/β-catenin signaling controls cell proliferation and differentiation and is a key regulatory mechanism for stem cells. As such, mutations in the Wnt pathway cause many diseases including cancer (Clevers 2006). All of the above "core" Wnt/β-catenin signaling components are present in vertebrates and the well-studied fruit fly *Drosophila* and are also encoded in sequenced genomes of radial symmetric Cnidarians *Hydra magnipapillata* and *Nematostella vectensis* (Guder et al. 2006) and of the primitive metazoan sponge *Amphimedon queenslandica*

Cite this article as *Cold Spring Harb Perspect Biol* doi: 10.1101/cshperspect.a007880

(Adamska et al. 2010). Gene loss is prevalent in the genome of nematodes, such as the popular model organism *Caenorhabditis elegans*, which appears to lack an ortholog of LRP5/6 (Phillips and Kimble 2009). As a result, the mechanisms of Wnt receptor activation are likely divergent in nematodes, whose Wnt pathways are discussed in Jackson and Eisenmann (2012). We focus on the vertebrate (and *Drosophila*) Wnt/FZD/LRP5/6 pathway, which is characterized by its absolute requirement for both FZD and LRP5/6 receptors. We note that FZD is also required for other "noncanonical" or "alternative" Wnt pathways that are independent of β-catenin (van Amerongen et al. 2008), and that in some cases such as in planar cell polarity (PCP) signaling in *Drosophila* (Bayly and Axelrod 2011), FZD may signal in a Wnt-independent manner.

WNT–RECEPTOR INTERACTIONS: OUTSIDE THE CELL

FZD Receptors

Calvin Bridges discovered a recessive fly mutant, which he called *frizzled* (*fz*), with irregularly orientated hairs and ommatidia (compound eye) and an occasionally slight reduction in wing size (Bridges and Brehme 1944). It became clear subsequently that *fz* plays an important role in planar cell polarity (PCP) (Bayly and Axelrod 2011). *fz* was shown to encode a protein with an extracellular cysteine-rich domain and a predicted topology of seven transmembrane helices resembling a G-protein-coupled receptor (GPCR) (Vinson et al. 1989). Epistasis and similarities between the *fz* and *dishevelled* (*dsh*) mutants (Wong and Adler 1993) suggested a link of the two genes, although in PCP signaling initially. With the fledgling *wingless* (*wg*, Wnt)–*dsh* (DVL)–*zeste white3* (*zw3*, GSK3)–*armadillo* (*arm*, β-catenin) signaling pathway important for embryonic patterning and wing development (Klingensmith and Nusse 1994; Krasnow et al. 1995), came along the identification of the second fly Frizzled gene *Dfz2* and the finding that Wg is a ligand for Dfz2 and Fz (Bhanot et al. 1996). *fz*(also called *Dfz1*) and *Dfz2* have redundant functions, although *Dfz2* has a more prominent role, in Wg signaling, whereas *fz*/*Dfz1* has a unique role in the PCP pathway (Kennerdell and Carthew 1998; Chen and Struhl 1999; Boutros et al. 2000; Rulifson et al. 2000).

In human, there are 10 FZD genes, numbered FZD1 through 10. Phylogenetic analysis of the mature FZD proteins generates the following five subgroups: FZD1/2/7, FZD3/6, FZD5/8, FZD9/10, and FZD4 (Fig. 1A). *Dfz2* groups closely with FZD5/8 and *fz*/*Dfz1* with FZD3/6. The Hedgehog (another family of secreted signaling protein) pathway protein Smoothened (SMO) is distantly related to FZD (Schulte and Bryja 2007) in their membrane topologies and amino-terminal cysteine-rich domains (CRDs); however, FZD proteins are distinguished from SMO by a conserved juxtamembrane KTxxxW motif in the carboxy tail region necessary for signaling (Fig. 1B).

THE FZD EXTRACELLULAR DOMAIN AND WNT BINDING: THE WNT8-FZD8CRD CO-CRYSTAL STRUCTURE

FZD contains a conserved 120-amino-acid cysteine-rich domain (CRD) at the amino terminus, which is connected to the first transmembrane helix through a variable 70- to 120-amino-acid linker region (Fig. 1B). Wnt/Wg ligands bind to CRD with high affinity (K_d of 1–10 nM) (Hsieh et al. 1999; Rulifson et al. 2000; Wu and Nusse 2002), although fz/Dfz1 PCP signaling may not involve any Wnt ligands. Deletion of the CRD prevents Wnt/Wg binding, but the CRD can be replaced with other heterologous Wnt-binding domains to generate a functional FZD receptor (Povelones and Nusse 2005; Mulligan et al. 2012). Crystal structures of CRDs from mouse FZD8 and Sfrp3 (secreted FZD-related protein 3) reveal a compact, predominantly α-helical regions held in place by disulfide bonds among 10 invariable cysteines (Fig. 1B,C) (Dann et al. 2001).

How a Wnt ligand engages the FZD receptor is revealed by the co-crystal structure of Xenopus Wnt8 in complex with mouse FZD8CRD (Fig. 1D) (Janda et al. 2012). In the complex, Wnt8 forms an unusual structure resembling a human hand with a central "palm" that extends

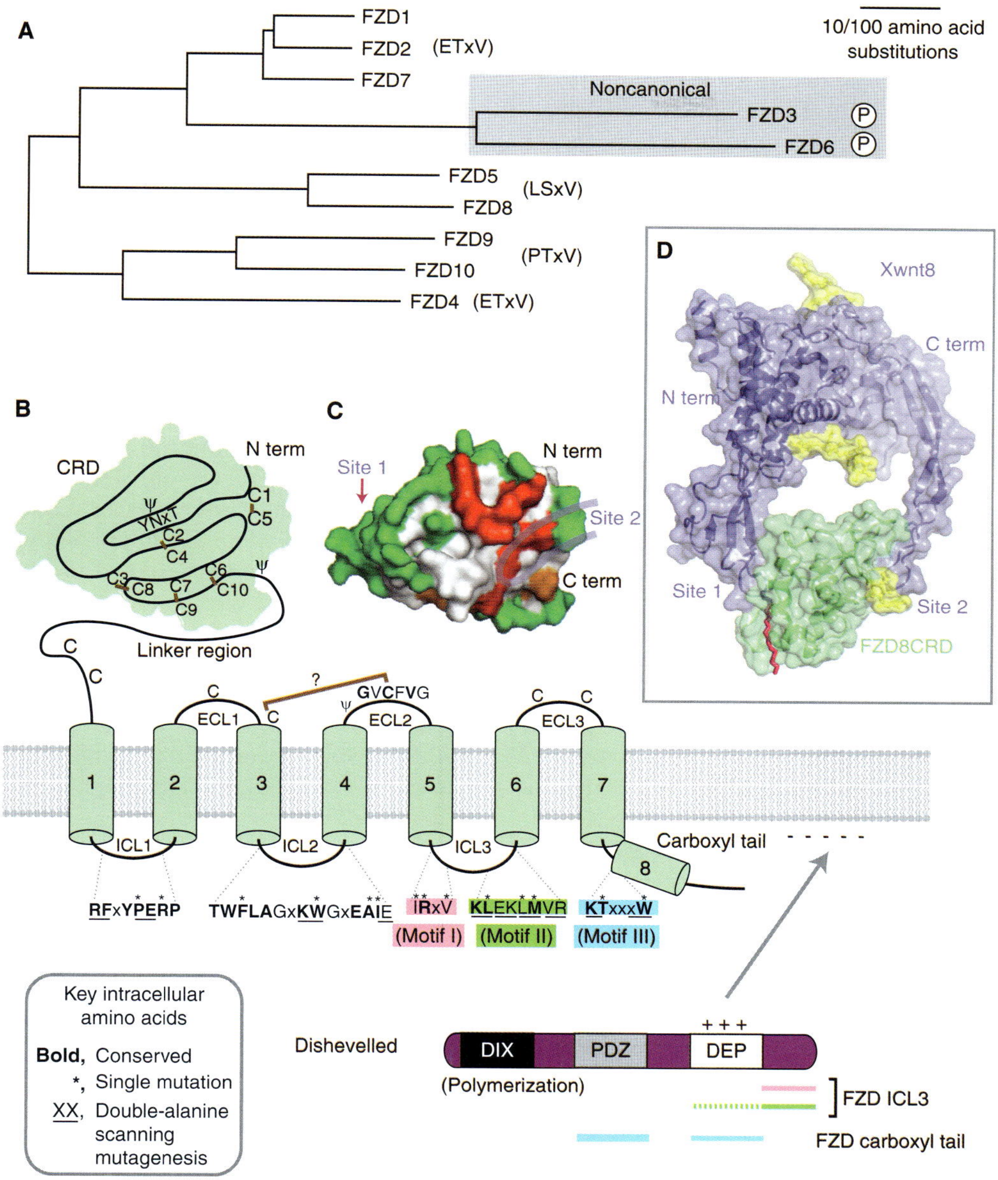

Figure 1. Frizzled (FZD) and Dishevelled (DVL). (*A*) Phylogeny comparing human FZD proteins. (*B*) Topology of a generic FZD on the plasma membrane. The shape of CRD is shaded in green, with the 10 invariable cysteine (C) residues forming five disulfide bonds highlighted. (ψ) Potential *N*-glycosylation sites. Additional conserved cysteine and other residues in the linker and ECL1-3 in the extracellular space are indicated. Conserved residues in ICL1-3 and the carboxy-terminal domain in the intracellular space are also indicated, with invariable residues among all FZD proteins in bold. (* above the letter) Missense mutations found in *Drosophila* Dfz1 (Povelones et al. 2005) and human FZD4 (Robitaille et al. 2002); (underlined) residues tested via double alanine substitution scanning mutagenesis in FZD5 (Cong et al. 2004b). DVL protein is also shown schematically with DIX, PDZ, and DEP domains and their interaction partners highlighted. Two discontinuous regions of FZD ICL3 (motif I [pink]; motif II [green]) bind to the carboxy-terminal region of DVL. The FZD carboxyl tail containing the KTxxxW motif (blue) interacts strongly with the DVL PDZ domain and also the DEP domain, which also shows some interaction with motif II in ICL3. (*See following page for legend.*)

a "thumb" plus an "index finger" to grab/pinch the FZD8CRD globular structure on two opposite sides, without changing FZD8CRD conformation (Fig. 1D). The amino terminal two-thirds of Wnt8 give rise to the palm—which consists mostly of α-helixes and intervening loops—and the thumb that consists of two anti-parallel β-strands and a connecting loop rigidified by a pair of disulfide bonds (Fig. 1D) (Janda et al. 2012). Most strikingly, a fatty acid adduct covalently attached to a serine at the tip of the thumb loop, likely a palmitoleic acid as seen in Wnt3a (Takada et al. 2006), inserts into a hydrophobic groove of FZD8CRD, constituting much of the Wnt8-FZD8CRD binding "site 1" that features extensive hydrophobic interactions between the lipid and apolar residues of FZD8CRD, "lipid-in-groove" fashion (Fig. 1D). The remaining portion of site 1 is contributed by protein–protein contacts between residues of the Wnt8 thumb loop and FZD8CRD (Fig. 1D). The carboxyl one-third of Wnt8 makes up the index finger featuring two anti-parallel β-strands and a long intervening loop, which is also rigidified by several disulfide bonds and engages in hydrophobic contacts within a depression of FZD8CRD, "knob-in-hole" fashion (Fig. 1D). This constitutes Wnt8-FZD8CRD interaction "site 2," which roughly corresponds to a Wnt-binding surface encompassing residues near the second cysteine (C2) of FZD8CRD as suggested by scanning mutagenesis (Fig. 1B,C) (Hsieh et al. 1999; Dann et al. 2001). Both site 1 and site 2 are dominated by hydrophobic contacts (lipid-in-groove and knob-in-hole, respectively), which are mostly mediated by conserved residues of Wnt8 and FZD8CRD, suggesting an explanation for broad or relatively promiscuous specificity of Wnt–FZD relationships, i.e., a single Wnt can often engage multiple FZD proteins and vice versa. However, at both site 1 and site 2, Wnt8 also exhibits protein–protein interactions with FZD8CRD residues that are conserved in some but altered in other FZD proteins, implying certain selectivity on top of broad specificity in Wnt-FZD interactions (Janda et al. 2012).

A YNxT motif at site 2 is conserved in all FZD proteins and is a predicted *N*-glycosylation site, and is indeed glycosylated in the Wnt8CRD crystal (Fig. 1B,C,D). Another predicted *N*-glycosylation site is at the end of the CRD in most FZDs, with the exception of the FZD3/6 group, which has its own unique predicted glycosylation site in extracellular loop 2 (ECL2) (Fig. 1B). *N*-glycans in both Wnt8 and FZD8CRD in the crystal structure are solvent-exposed and do not appear to contribute directly to Wnt–FZD interaction (Fig. 1D) (Janda et al. 2012). FZD *N*-glycosylation appears to be required for receptor maturation and can be regulated by the ER-resident protein Shisa, which binds preferentially to the immature form of FZD for ER trapping and thus antagonizes Wnt signaling (Yamamoto et al. 2005).

Beyond the Wnt–CRD interaction, little is known regarding the function of other FZD extracellular regions. FZD receptors are structurally analogous to GPCRs, and there is evidence that FZDs can interact and signal through G-proteins (see below), although this issue remains debated. Nonetheless, some common GPCR structural elements that are also found in FZD may suggest other potentially important regions. Many small GPCR ligands bind to the extracellular loop regions or between

Figure 1. (*Continued*) (*C*) Crystal structure of FZD8CRD. Residues in red (and to a lesser degree those in orange) are suggested to be a Wnt-binding interface from an alanine scanning mutagenesis (Hsieh et al. 1999; Dann et al. 2001), which partially overlaps with site 2 indentified in the Wnt8–FZD8CRD co-crystal structure. Residues in green when altered did not affect Wnt activity and areas in gray were not tested. Site 1 and site 2, which mediate contacts between Wnt8 and FZD8CRD in the crystal structure, are labeled. A shade of Wnt8 index finger contacting site 2 is sketched. (Panel *C* is derived from Fzd8 1IJY [PDB doi: 10.2210/pdb1ijy/pdb].) (*D*) Wnt8–FZD8CRD co-crystal structure, shown as a ribbon diagram superimposed on surface representation (Janda et al., 2012). Palmitoleic acid adduct from the Wnt8 thumb at site 1 (red); *N*-glycans of Wnt8 and FZD8CRD (yellow). Note that the FZD8CRD in *C* and *D* is viewed in different angles. N term, amino terminal; C term, carboxyl terminal.

the transmembrane helices at the extracellular face (Tebben and Schnur 2011), causing a conformation change for GPCR activation. Ligand binding and relative orientation of the transmembrane helices are influenced by inter- and intraloop disulfide bonds in the extracellular regions/loops of GPCRs. Aside from the 10 invariable cysteines within the CRD, there are additional conserved cysteines in FZD extracellular regions: two in the linker between the CRD and the first transmembrane helix, two in extracellular loop 1 (ECL1), one in ECL2, and two in ECL3 (Fig. 1B). A common extracellular disulfide bond in GPCRs is between the top of transmembrane helix 3 and ECL2 and is thought to be important for orienting and stabilizing the transmembrane helices (Peeters et al. 2010). This disulfide bond is potentially present in all FZDs (Fig. 1B). In addition, the residues flanking the cysteine in ECL2 are conserved (GVCFV) and are predicted to form a β-strand. There are also examples of an intraloop disulfide bond in ECL3 and disulfide bonds connecting with the amino-terminal region (Wheatley et al. 2012). Smo, a distant member of the FZD family, may share these GPCR features in that mutations of cysteines in ECL1 and 2 or a partial deletion of ECL2 renders a constitutively active or less active receptor (Carroll et al. 2012). Furthermore, a fly *fz* allele with a mutation in the ECL2 cysteine causes a partial loss of function (Povelones et al. 2005). Studies are required to elucidate the role of these ECLs in Wnt-induced FZD activation and signaling.

LRP5/6 AND ARROW

Fly mutants for *arrow* were first reported by Nüsslein-Volhard and Wieschaus from the famous genetic screen for embryonic lethal mutants (Nusslein-Volhard et al. 1984). Fly embryo segments normally develop a stripe of anterior denticles that can be distinguished from the naked posterior region of the segment. Mutants for *arrow* contained extra bands of denticles that were more prominent in the midline, causing the denticle stripes to look like arrows. Elimination of maternal and zygotic *arrow* resulted in a phenotype identical to that of *wg* mutants, and molecular cloning revealed that Arrow is homologous to LRP5 and LRP6 (Wehrli et al. 2000), which had been cloned as members of the LDLR family (Brown et al. 1998; Hey et al. 1998). With the evidence that the Lrp6 mouse mutant phenotypically resembles a composite of several Wnt mutants (Pinson et al. 2000), that LRP5 and LRP6 display critical roles in Wnt/β-catenin signaling in *Xenopus* (Tamai et al. 2000), and that Wnt1 can bridge a complex formation between the extracellular domains of FZD and LRP6 (Tamai et al. 2000), LRP5/6 and Arrow were established as coreceptors for the Wnt/β-catenin pathway.

THE LRP5/6 EXTRACELLULAR DOMAIN

LRP5 and LRP6 have more than 1600 amino acids and represent a unique group of the LDLR family. LRP5 and LRP6 proteins are 70% identical, and each is 45% identical to Arrow (He et al. 2004). These single transmembrane receptors have an extracellular domain containing four tandem β-propeller/epidermal growth factor (EGF) repeats followed by three LDLR type A repeats (Fig. 2A). The β-propeller domain was first proposed by Springer, who predicted a layout of a six-bladed propeller, with each blade consisting of four short β-strands and a YWTD motif in strand 2 that acts to stabilize the neighboring β-sheets and is key to the blade structure (Springer 1998). The structure for the single LDLR β-propeller-EGF (PE) unit was solved confirming the predicted six-bladed propeller, which intimately interacts with the EGF repeat (of 50 amino acids) with six cysteines in a disulfide bond pattern of C1–C3, C2–C4, and C5–C6 (Jeon et al. 2001).

Earlier mapping studies divided the LRP5/6 extracellular domain into three segments: β-propeller-EGFs 1 and 2 (P1E1-P2E2), β-propeller-EGFs 3 and 4 (P3E3–P4E4), and the three LDLR type A repeats (Fig. 2A) (He et al. 2004). Deletion of a single β-propeller in LRP5/6 typically interferes with receptor biogenesis, and better stability and recombinant protein production were achieved using tandem pairs of P1E1–P2E2 and P3E3–P4E4, but not P2E2–P3E3 (Liu et al. 2009; Bourhis et al. 2010).

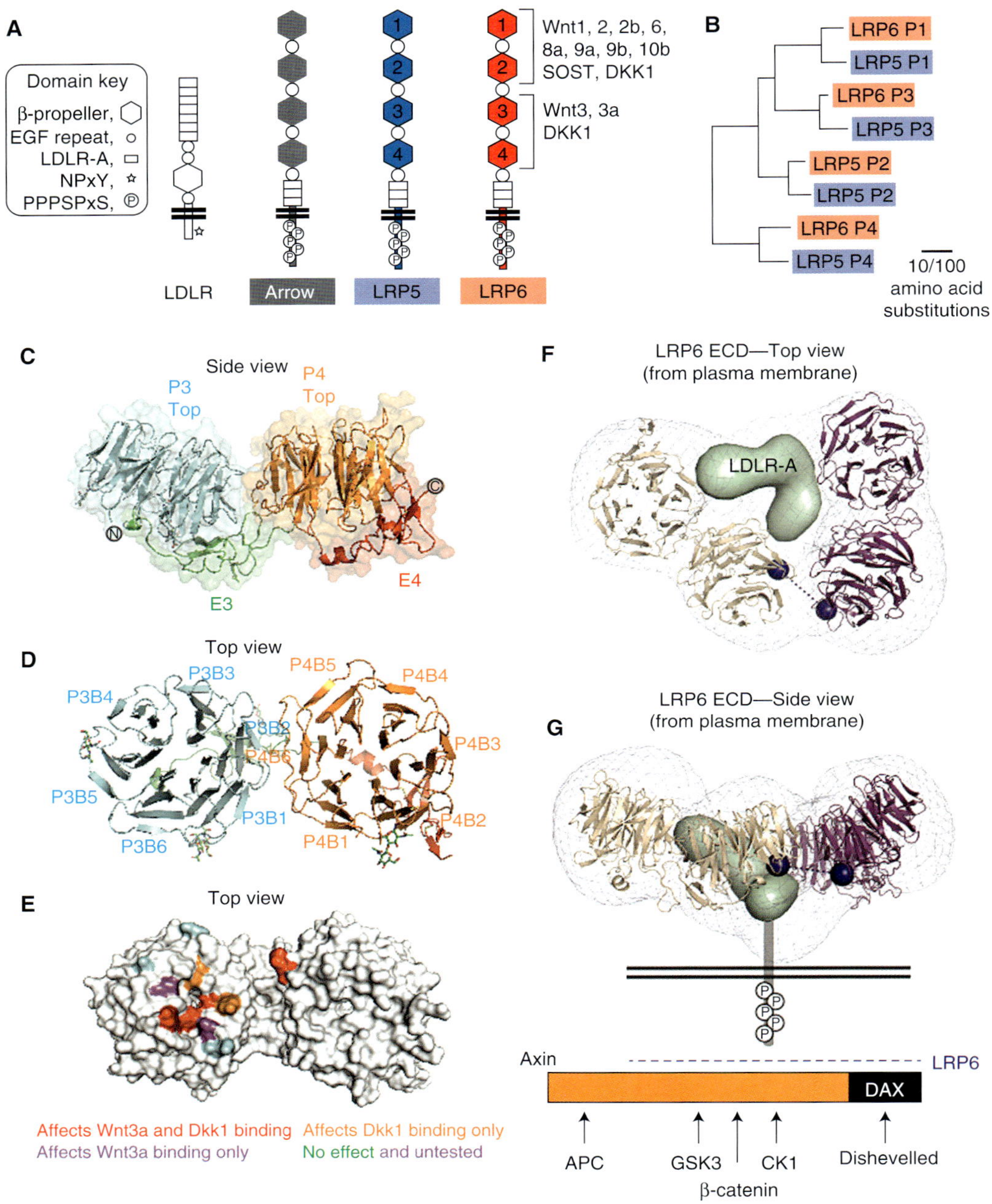

Figure 2. LRP5/6 and Axin. (*A*) Arrow, LRP5, and LRP6 are shown schematically together with LDLR. LRP6 binding to different Wnt proteins and antagonists SOST and DKK1 are shown. (*B*) Phylogeny of the four β-propellers (P1–4) in LRP5 and LRP6. (*C,D*) *Side* and *top* views of the atomic structure of LRP6 P3E3–P4E4 (Chen et al. 2011). (*E*) Wnt3a and DKK1 share an overlapping binding surface on the top of LRP6 P3. (*F,G*) *Top* and *side* views of a model of the entire LRP6 extracellular domain derived from electron microscopic staining (Chen et al. 2011). Orientation and relationship to the plasma membrane are speculative. Axin is shown schematically. (Dashed line) Indicates the unknown nature of the domain involved in LRP6-binding. Axin domains that interact with DVL, APC, β-catenin, GSK3, and CK1α are indicated. (DAX) The DIX domain of Axin. (*C–G*, From Chen et al. 2011; reprinted, with permission, from the author.)

Recent crystal structures of LRP6 P1E1–P2E2 and P3E3–P4E4 highlighted the tandem nature of these repeating units (Fig. 2C,D) (Ahn et al. 2011; Chen et al. 2011; Cheng et al. 2011). Although conforming to the prototypic PE structure of LDLR, extensive interface interactions between P1E1 and P2E2, and between P3E3 and P4E4 were observed, for example, between P3 and P4 and between E3 and P4 (Fig. 2C,D), and these inter-PE interactions were shown to be critical for LRP6 biogenesis and maturation to the plasma membrane. The interface between P2E2 and P3E3 is unresolved but is likely to be different. A low-resolution electron microscopic structure of the entire LRP6 extracellular domain suggested a compact horseshoe platform configuration (Fig. 2F,G), which may be consistent with the notion of a P2E2–P3E3 interface that is distinct from those of P1E1–P2E2 and of P3E3–P4E4 (Chen et al. 2011). A platform configuration of a different kind was also suggested for the LRP6 extracellular domain (lacking the LDLR-A repeats) via a low-resolution small-angle X-ray scattering analysis (Ahn et al. 2011).

The endoplasmic reticulum chaperone protein MESD is required for proper folding of the LRP5/6 extracellular domain (Culi and Mann 2003; Hsieh et al. 2003). MESD also facilitates the folding of other LDLR family members and interacts with multiple β-propellers of LRP5/6 (Culi et al. 2004; Lighthouse et al. 2011). Examination of LRP5/6 via western blotting often reveals two bands, a lower (faster migrating) immature form and a higher fully glycosylated form. Mest (also known as Paternally expressed gene 1, or Peg1), a multispan transmembrane protein that resides in the ER and contains an α/β hydrolase domain, modifies LRP6 glycosylation resulting in less mature LRP6 at the plasma membrane, thereby modulating LRP6 in a manner that appears to be analogous to Shisa inhibition of FZD receptors (Jung et al. 2011).

TWO OR MORE WNT-BINDING SURFACES OF LRP6

Early studies suggested that Wnt1 bridges extracellular domains of a FZD (mFZD8CRD) and LRP6 into a receptor complex via direct binding (Tamai et al. 2000). Consistent with this notion, an LRP6 mutant that lacks the cytoplasmic domain (and is thus membrane bound) or lacks transmembrane plus cytoplasmic domains (and is thus secreted) behaves as a dominant-negative receptor for Wnt/β-catenin signaling (Fig. 3A), presumably via binding to Wnt ligands (Tamai et al. 2000; He et al. 2004). Wnt1 signaling via LRP6 is mediated through the PE domains (P1E1–P4E4) (Mao et al. 2001a). A recent study showed a K_d between the LRP6 extracellular domain and Wnt3a or Wnt9b to be ~10 nM (Bourhis et al. 2010). Unexpectedly, however, Wnt3a and Wnt9b were shown to preferentially interact with P3E3–P4E4 and P1E1–P2E2, respectively, and a Wnt3a–Wnt9b–LRP6 (extracellular domain) complex could be detected in vitro (Bourhis et al. 2010), suggesting a possibility that a single LRP6 may engage two different Wnt proteins simultaneously. Furthermore, functional-blocking monoclonal antibodies (mAbs) against epitopes in P1 and P3, respectively, show a distinct inhibition profile toward different Wnt proteins, presumably via blocking Wnt binding to either P1 or P3 (Ettenberg et al. 2010; Gong et al. 2010). These data infer that many Wnts (Wnt1, Wnt2, Wnt2b, Wnt6, Wnt8a, Wnt9a, Wnt9b, and Wnt10a) interact with P1, whereas Wnt3 and Wnt3a prefer P3 (Fig. 2A). Other Wnts (Wnt7a, Wnt7b, and Wnt10a) could not be assigned into either group because they were not inhibited by the mAbs alone or in combination (Gong et al. 2010), raising the possibility that these Wnts may bind to regions outside P1 and P3. The remaining Wnts have not been tested in these assays. The top surfaces of P1 and P3 (and P2 and P4 as well) do not harbor *N*-glycosylation sites (Ahn et al. 2011; Bourhis et al. 2011; Chen et al. 2011; Cheng et al. 2011) and are likely the Wnt-binding interface. Indeed, missense substitutions of multiple top surface residues of P3 of LRP6, designed based on the crystal structure, diminish or enhance Wnt3a binding and signaling (Fig. 2E) but show minimal effect on signaling by Wnt1 (Chen et al. 2011), which prefers to bind to P1. Therefore, an emerging model is that LRP6, and likely LRP5, engage different Wnts via multiple ligand interfaces

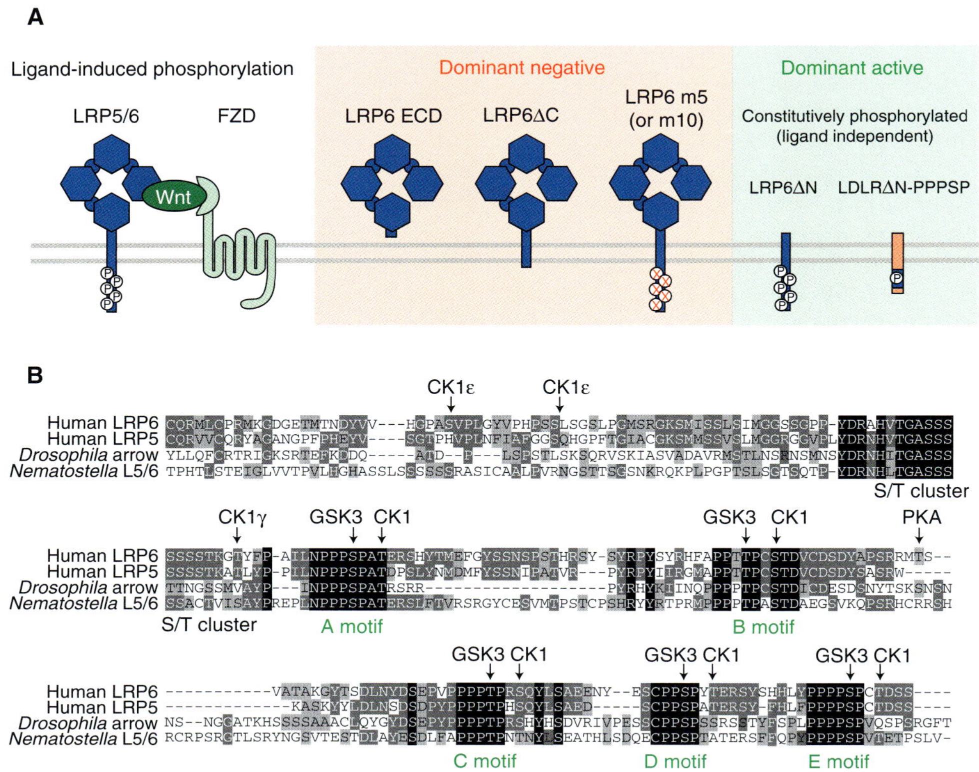

Figure 3. LRP6 phosphorylation and phosphorylation sites. (*A*) Wnt induces LRP6 phosphorylation in the FZD–LRP6 complex. Dominant-negative LRP6 mutants are generated by deleting the cytoplasmic domain or mutating all five PPPSPxS motifs (S to A). Constitutively activated LRP6ΔN and LDLRΔN-PPPSP are constitutively phosphorylated. (*B*) Alignment of the cytoplasmic domain of LRP6, LRP5, Arrow, and the *Nematostella* LRP5/6 homolog. Identified phosphorylation sites by various kinases in LRP6 are indicated.

(Fig. 2A). The LRP6 platform configuration seems to suit this model (Fig. 2F,G). Sequence analysis of LRP5 and LRP6 reveals that P1, P2, and P3 are most related by homology, but P4 is more divergent (Fig. 2B). Further studies of LRP5/6 will be needed to parse out different binding sites for Wnts and other ligands, including whether P2 or P4 also represents a Wnt docking site. The role of three LDLR-A repeats in LRP5/6 remains unknown.

THE WNT–FZD–LRP5/6 COMPLEX

A distinguishing feature of the Wnt/β-catenin pathway is the requirement for both FZD and LRP5/6 receptors (He et al. 2004). Extracellular domains of LRP6 and FZD8 were shown to form a complex in vitro in the presence of Wnt1 (Tamai et al. 2000). A similar trimeric complex was confirmed using recombinant Wnt3a and extracellular domains of FZD8 and LRP6 (Bourhis et al. 2010). These results suggest that Wnts can interact with both receptors simultaneously, perhaps via different parts of the Wnt molecule, although this important question has not been addressed (because of the prior lack of Wnt structural information and the difficulty in generating soluble Wnt protein fragments necessary for mapping studies). The Wnt8–FZD8CRD structure (Janda et al. 2012) will help resolve this issue. Also not addressed is the stoichiometry of the putative Wnt–FZD–LRP6 (or LRP5) complex, which is commonly drawn at a 1:1:1 ratio in models. The observation in vitro that an LRP6 can simultaneously bind two different Wnt proteins suggests the possibility that various combinations of Wnt–FZD–LRP6 complexes may exist in vivo. An LRP6 platform model (Fig. 2F,G) speculates that β-propeller top surfaces/ligand binding regions may position at roughly the same height relative to the plasma membrane, potentially facilitating engagement of the CRD domain of a single FZD or more FZDs (Ahn et al. 2011; Chen et al. 2011). Interestingly, the Wnt8–FZD8CRD structure suggests a possibility of asymmetric Wnt–FZD oligomers that are formed to fully shield the palmitoleic acid adduct from aqueous solvent (Janda et al. 2012). Whether such asymmetric Wnt–FZD oligomer formation occurs in vivo and has a role in signaling deserves further investigation.

Another limiting factor for understanding the Wnt–FZD–LRP5/6 complex is our poor knowledge of Wnt–FZD and Wnt–LRP5/6 binding specificity and affinity, because most Wnts are not available in soluble forms (Mulligan et al. 2012). FZD8 appears to bind Wnt3a with 2× –3× stronger affinity than LRP6 does (Bourhis et al. 2010), seemingly consistent with some anecdotal experience that Wnt–LRP6 binding appears to be weaker than that of Wnt–FZD (He et al. 2004). In contrast, the reverse is true for Wnt9b, which displays weak binding to FZD8 but binds to LRP6 comparably to how Wnt3a does (Bourhis et al. 2010). It remains possible, however, that Wnt9b may prefer a FZD or FZDs other than FZD8. Some of the questions on the Wnt–FZD–LRP6 complex may have to wait until more recombinant Wnt proteins become available and a high-resolution structure of a Wnt protein in complex with an FZD plus LRP6 is achieved.

WNT RECEPTOR SIGNAL TRANSDUCTION: INSIDE THE CELL

Dishevelled and Axin

FZD and LRP5/6 transduce Wnt signal via engaging downstream cytoplasmic components, among which two scaffolding proteins, Dishevelled and Axin, have prominent roles.

Dishevelled (DVL1-3 in human, Dsh in *Drosophila*, and Xdsh in *Xenopus*) is a multifunctional protein that serves as a hub for canonical and noncanonical Wnt signaling. First identified in *Drosophila*, most fly *dsh* mutants are embryonic lethal because of their loss in wg signaling (Perrimon and Mahowald 1987). DVL proteins are about 700 amino acids in length and contain three main domains of about 80–90 amino acids each: DIX (Dishevelled, Axin), PDZ (Postsynaptic density 95, discs large, zona occludens-1), and DEP (Dishevelled, Egl-10, Pleckstrin) (Fig. 1B). The PDZ domain is essential for binding to a juxtamembrane KTxxxW motif (Umbhauer et al. 2000) in the FZD carboxyl cytoplasmic region (Wong et al. 2003). The first dsh mutant identified, dsh^1, displays a PCP phenotype (Fahmy and Fahmy 1959), and is a missense mutation (K417M) in the DEP domain (Axelrod et al. 1998; Boutros et al. 1998). Further studies have shown that the DEP domain has a positively charged surface that may interact with phospholipids in the plasma membrane (Wong et al. 2000; Simons et al. 2009) and additional surfaces for interacting proteins including the endocytic adaptor protein 2 (AP-2) complex (Yu et al. 2010). A recent study further suggests that DEP plus the carboxyl region of DVL interact with FZD ICL3 (via the so-called motif I and motif II) (Fig. 1B) and the KTxxxW region (so-called motif III) (Fig. 1B), thereby facilitating DVL association with a discontinuous cytoplasmic surface of FZD (Tauriello et al. 2012). Thus, PDZ and DEP domains have roles in recruiting DVL to FZD at the plasma membrane (Fig. 1B). The DIX domain shows an interesting property of head-to-tail polymerization (Schwarz-Romond et al. 2007a). This property correlates well with DVL aggregates under the overexpression condition, and it has been argued that the endogenous DVL may form such aggregates, which are highly dynamic (Schwarz-Romond et al. 2005). DIX oligomerization/polymerization is proposed to provide a DVL platform for dynamic assembly of protein–protein interactions of low avidity, such as between DVL and Axin (Schwarz-Romond et al. 2007b), and is a main underpinning for the receptor "signalosome" hypothesis (Bilic et al. 2007) (see below). However, some have suggested that the DVL "aggregates" (or "dots" under microscopes) represent endocytic vesicles (Capelluto et al. 2002; Taelman et al. 2010). The simplest model states that DIX and PDZ domains, but not DEP, are required for Wnt/β-catenin signaling, as seen in some overexpression studies. However, evidence exists that the DEP domain influences β-catenin signaling (Wong et al. 2000; Simons et al. 2009; Tauriello et al. 2012). DVL associates with kinases such as CK1ε and CK2, and becomes hyperphosphorylated upon Wnt signaling (Willert et al. 1997; Peters et al. 1999; Kishida et al. 2001; Cong et al. 2004a; Klein et al. 2006). These kinases, in particular CK1ε, have been documented to have activating roles in Wnt/β-catenin signaling, but the role of DVL phosphorylation remains unclear.

Axin was identified as the gene mutated in *Fused* mice (Zeng et al. 1997), which display a dominant phenotype of vertebral fusions or kinked tails, with homozygous mutants showing more severe phenotypes including axis duplication of the tail (Reed 1937). Axin has about 850 amino acids and has a close homolog in vertebrates (Behrens et al. 1998; Yamamoto et al. 1998). Axin is a key negative component of the β-catenin pathway, owing to its scaffolding function in promoting β-catenin phosphorylation and degradation (Behrens et al. 1998; Ikeda et al. 1998; Kishida et al. 1998; Liu et al. 2002). Axin directly binds to β-catenin, GSK3, CK1α, and APC (Fig. 2), and additional proteins in the assembly of the "β-catenin destruction complex," in which β-catenin phosphorylation (and degradation) is performed (MacDonald et al. 2009; Stamos and Weis 2012). In relevance to Wnt receptor and DVL functions, Axin has a carboxy-terminal DIX domain, which was also called for convenience DAX to distinguish it from the DVL DIX domain (Schwarz-Romond et al. 2007b). DAX shows an oligomerization/polymerization property similar to but less dynamic than DIX (Schwarz-Romond et al. 2007a). Studies in mammalian and *Drosophila* models argue that DAX oligomerization is critical for Axin function in β-catenin phosphorylation/degradation (Kishida et al. 1999b; Fiedler et al. 2011). Axin binds directly with DVL and is recruited into DVL aggregates (under overexpression) via, in part, DAX–DIX interaction (Kishida et al. 1999b; Schwarz-Romond et al. 2007b), which may represent one mechanism by which DVL inhibits Axin (Kishida et al. 1999b; Fiedler et al. 2011). But it is unclear whether/how Wnt signaling regulates DVL–Axin interaction.

FZD AND DVL INTERACTION

The characterized FZD–DVL binding involves PDZ interaction with the juxtamembrane KTxxxW motif in the FZD carboxyl terminus (Umbhauer et al. 2000; Wong et al. 2003), DEP interaction with the same or an overlapping KTxxxW motif, and the interaction of the DVL carboxyl region with FZD ICL3 (motif I and motif II) (Fig. 1B) (Tauriello et al. 2012). Thus, multiple DVL domains/regions appear to engage a discontinuous cytoplasmic surface of FZD, and FZD–DVL association, which derives from a sum of multiple relatively weak interactions, may be further stabilized by the DEP–phospholipids (in the plasma membrane) interaction (Fig. 1B). It is unknown how PDZ and DEP may bind to the same or overlapping KTxxxW region, or whether each binding may be exclusive of the other and related to FZD activation status. A common feature of GPCRs is an intracellular helical region in the carboxyl tail, sometimes referred to as helix 8 taking into account the seven transmembrane helices. Crystal structures commonly show helix 8 to be perpendicular to TM7 and parallel with the membrane and often amphipathic to enable an interaction with the lipid membrane. In the case of FZD, a predicted helix 8 begins at the KTxxxW motif and extends for 14 amino acids in most receptors (Fig. 1B). Given the importance of the KTxxxW region for DVL (PDZ and/or DEP) binding (Punchihewa et al. 2009; Tauriello et al. 2012), one potential mechanism for FZD activation would be a Wnt-induced movement of TM7 to expose the key FZD–DVL interaction site, if we assume that Wnt regulates FZD–DVL interaction. It should be noted that FZD–DVL interaction has thus far been studied

 Cite this article as *Cold Spring Harb Perspect Biol* doi: 10.1101/cshperspect.a007880

either in vitro or under overexpression conditions in vivo, and it remains unknown whether this interaction is under Wnt regulation. However, because Dfz1/Dsh PCP signaling in *Drosophila* appears to be Wnt independent (Bayly and Axelrod 2011), it is conceivable that FZD–DVL interaction can occur without Wnt, at least in some circumstances.

In addition to the helix 8 region, many GPCRs use a region in intracellular loop (ICL) 2 that serves as an interaction site for the Gα protein. However, ICLs are divergent between GPCRs and FZD. Comparison of the 10 FZDs reveals conserved residues in the ICLs (Fig. 1B). The importance of these residues for signaling has been shown by fly Fz/Dfz1 mutants, patient mutations in FZD4, and site-directed mutagenesis studies (Robitaille et al. 2002; Cong et al. 2004b; Povelones et al. 2005; Zeng et al. 2008; Nikopoulos et al. 2010). Some of these findings can potentially be explained by DVL interaction with FZD ICL3 (Tauriello et al. 2012). Whether ICL1 and ICL2 residues are also involved in interaction with DVL, or G-proteins (see below), or unidentified proteins remains to be investigated.

Most FZD receptors, except for the noncanonical FZD3/6 group, contain a carboxy-terminal S/TxV motif (Fig. 1A), which is commonly involved in PDZ binding (Hering and Sheng 2002) and may serve as an accessory site to augment FZD clustering by perhaps PDZ proteins. Phosphorylation of the FZD carboxyl region has been observed and is associated with down-regulation of FZD/PCP signaling, such as Fz/Dfz1 phosphorylation by aPKC around the DVL-interacting SxKTxxSW motif in *Drosophila* (Djiane et al. 2005), and Dvl-dependent FZD3 phosphorylation, which inhibits FZD3 endocytosis and signaling, in axon guidance (Shafer et al. 2011) and possibly in neural crest induction (Yanfeng et al. 2006). However, it is unknown whether phosphorylation regulates FZD function in β-catenin signaling.

LRP6 PHOSPHORYLATION AND AXIN RECRUITMENT

The LRP5/6 cytoplasmic region has about 200 amino acids and contains five signature PPPSPxS motifs named from A to E, which are conserved from invertebrates to human (Fig. 3B). LRP6 mutants that lack the entire cytoplasmic region (LRP6ΔC), or have all five PPPSPxS motifs changed to PPPAPxA (LRP6m10), or have all five PPPSP motifs changed to PPPAP (LRP6m5), behave each as a loss-of-function and, in fact, dominant-negative mutant (Fig. 3A) (Tamai et al. 2000, 2004; Zeng et al. 2005). Conversely, LRP6 (and LRP5) mutants that lack the extracellular domain (LRP6ΔN) behave as constitutively activated Wnt receptors (Mao et al. 2001a,b; Tamai et al. 2004). Most tellingly, a single PPPSPxS motif is sufficient to transfer Wnt signaling activation function to a heterologous receptor (Fig. 3A) (Tamai et al. 2004; MacDonald et al. 2008). These results highlight the critical importance of the PPPSPxS motif.

The PPPSPxS motif is dually phosphorylated in the endogenous LRP6 in response to Wnt but is constitutively phosphorylated in either LRP6ΔN or in any single PPPSPxS motif that is transferred to a heterologous receptor (Fig. 3A) (Tamai et al. 2004; Zeng et al. 2005; MacDonald et al. 2008), correlating fully with the signaling activity. The phosphorylated PPPSPxS motif, but not the unphosphorylated one, is a docking site for Axin (Tamai et al. 2004), explaining, in part, Wnt-induced recruitment of Axin via LRP5/6 to the plasma membrane (Mao et al. 2001b). Axin binding to the phosphorylated PPPSPxS motif is likely direct (Tamai et al. 2004; Zeng et al. 2005; MacDonald et al. 2008, 2011), but it remains unclear which part of Axin is involved in the binding, with the earlier mapping studies implicating a broad segment of Axin (Fig. 2) (Mao et al. 2001b). This critical but unresolved issue has fueled an alternative model in which LRP6–Axin interaction may be indirectly mediated by other factors such as GSK3 (Piao et al. 2008).

LRP6 PHOSPHORYLATION: KINASES AND REGULATION

Of the kinases that have been implicated in LRP6 phosphorylation, the GSK3 and CK1 families are the most prominent in their phosphorylation of PPPSP and xS, respectively (Fig.

3B) (Davidson et al. 2005; Zeng et al. 2005). PPPSP phosphorylation has been mostly studied using Ab1490, which corresponds to motif A (Tamai et al. 2004), and studies of other PPPSP motifs (C and E) suggested similar regulation (MacDonald et al. 2008, 2011). PPPSP is phosphorylated by GSK3, which primes xS phosphorylation by CK1 (Zeng et al. 2005). This dual kinase model contrasts that for β-catenin, whose phosphorylation by CK1α primes phosphorylation by GSK3 (Liu et al. 2002). In many cases such as in β-catenin phosphorylation, GSK3 requires a priming kinase, but in its PPPSP phosphorylation, there does not seem to be a priming requirement (Zeng et al. 2005). GSK3 phosphorylation and activation of LRP6, which also contrasts its phosphorylation and degradation of β-catenin, is supported by two key pieces of evidence: (1) Wnt-induced LRP6 PPPSP phosphorylation does not occur in cells that lack both Gsk3α and Gsk3β; and (2) inhibition of GSK3 by a plasma membrane-tethered inhibitor blocks Wnt signaling, whereas the same inhibitor in the cytoplasm activates Wnt signaling (Zeng et al. 2005, 2008). Such dichotomic roles of GSK3 are not without parallel: in *Drosophila*, protein kinase A (PKA) antagonizes Hedgehog (Hh) signaling by phosphorylating and inhibiting the Ci transcription factor in the absence of Hh but activates Hh signaling by phosphorylating and activating Smo in response to Hh (Jiang and Hui 2008).

FZD is required for Wnt-induced PPPSP phosphorylation by GSK3, and forced FZD–LRP6 association is sufficient to induce PPPSP phosphorylation (Zeng et al. 2008), consistent with the importance of Wnt-induced FZD–LRP6 complex formation in Wnt signaling. Mutations in FZD intracellular loops or the carboxyl tail that prevent FZD–DVL interaction (Wong et al. 2003; Cong et al. 2004b) abolish the ability of FZD to promote LRP6 phosphorylation, implying that DVL is required to mediate FZD regulation of LRP6 PPPSP phosphorylation (Fig. 4), a notion that is supported by Dvl depletion experiments (Zeng et al. 2008). The recent observation in *Drosophila* that Dsh acts upstream of Arrow further corroborates this biochemical relationship (Metcalfe et al. 2010).

Unexpectedly, Axin is also required for Wnt-induced LRP6 PPPSP phosphorylation, and Axin overexpression enhances PPPSP phosphorylation (Tamai et al. 2004; Yamamoto et al. 2006), but only if Axin is not impaired in GSK3 binding (Zeng et al. 2008). These findings led to a model that DVL binding to and recruitment of Axin into the FZD–LRP6 complex initiates GSK3 phosphorylation of LRP6 on PPPSP motifs (Zeng et al. 2008). This model posits that the Axin–GSK3 complex phosphorylates β-catenin in the absence of Wnt but switches to phosphorylate LRP6 under the FZD/DVL control. DIX/DAX-mediated DVL and Axin oligomerization is compatible with this model.

Wnt-induced CK1 phosphorylation of LRP6 presents a more complicated picture. Firstly, there are CK1 phosphorylation sites within and outside the PPPSPxS motif, such as the conserved region that precedes the first PPPSPxS motif and contains residues "YDRxH" and a block of serine and threonine residues, referred to as the S/T cluster (Fig. 3B). This region is likely phosphorylated by CK1, including the nearby T1479 by CK1γ (Davidson et al. 2005). The functional significance of this S/T cluster is unclear, although one study suggests it as a GSK3-binding site upon CK1 phosphorylation (Piao et al. 2008). Secondly, there exist seven mammalian CK1s (α, β, γ1, γ2, γ3, δ, and ε) (Price 2006). CK1α and CK1ε/δ are associated with Axin and DVL, respectively, whereas CK1γ is membrane tethered via isoprenylation (Price 2006; MacDonald et al. 2009), and therefore these CK1s are each in proximity to LRP6 in the receptor complex. xS phosphorylation by CK1 is proceeded/primed by PPPSP phosphorylation (Zeng et al. 2005), and therefore appears to be secondary to regulation of PPPSP phosphorylation by Wnt through FZD, DVL, and Axin, although independent layers of CK1 regulation may exist. Indeed, the transmembrane protein TMEM198 enhances LRP6 signaling and phosphorylation by CK1 through its binding to CK1α, ε, and γ (Liang et al. 2011). Efforts have been taken to distinguish roles of different CK1s in multiple steps in Wnt signaling (Del Valle-Perez et al. 2011), but the available

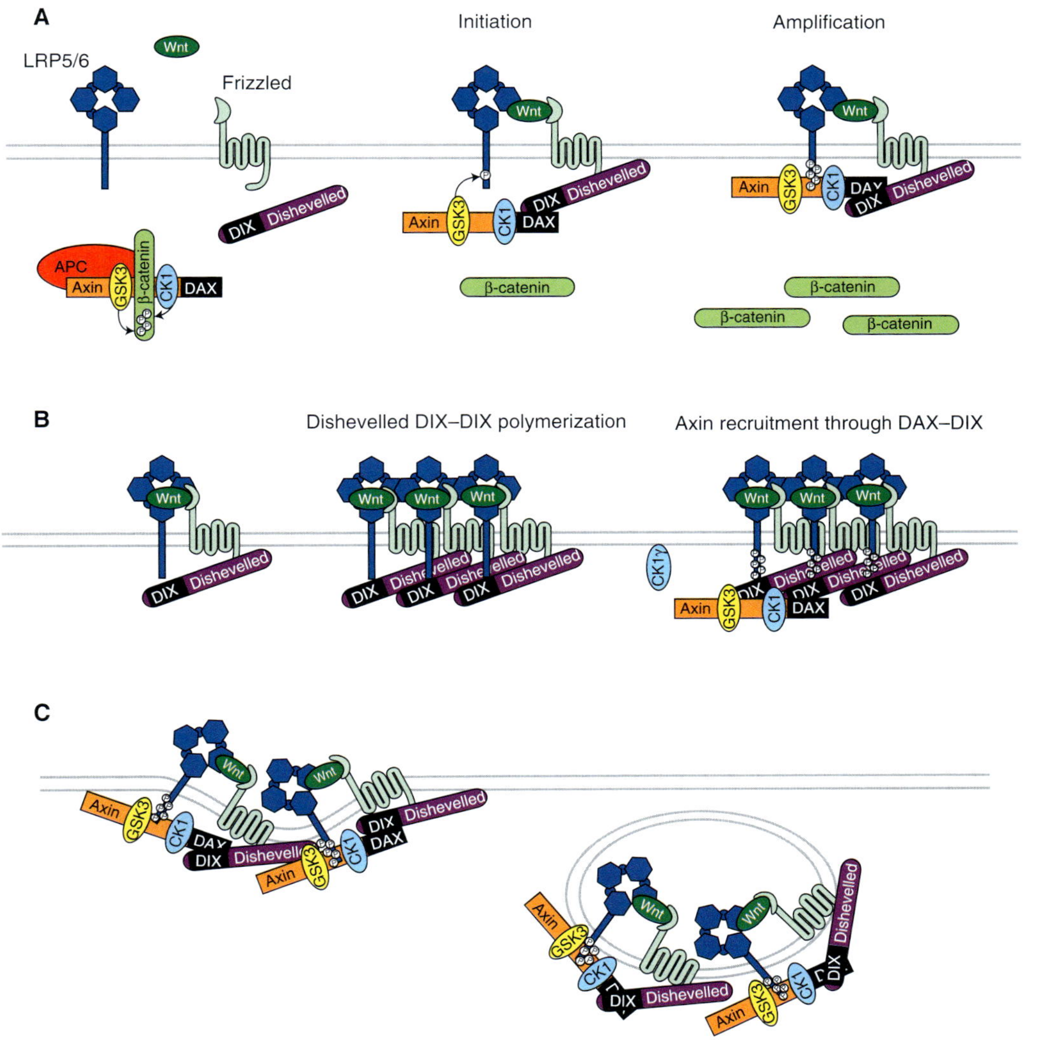

Figure 4. Models for LRP6 phosphorylation and signaling. (*A*) Initiation-amplification model. (*B*) Signalosome model. Only one Axin molecule is drawn for clarity. (*C*) Endosomal signaling model.

evidence suggests overlapping functions of CK1s in LRP6 phosphorylation.

LRP6 PHOSPHORYLATION: ROLES OF PIP2, G-PROTEIN, AND OTHER KINASES

Two lipid kinases, phosphatidylinositol 4-kinase type II (PI4KIIα) and phosphatidylinositol-4-phosphate 5-kinase type I (PIP5KI), which are bound and activated by DVL, mediate Wnt-induced PIP2 (phosphatidylinositol 4,5-bisphosphate) production that promotes LRP6 phosphorylation by GSK3 and CK1 (Pan et al. 2008; Qin et al. 2009). One model suggests that PIP2 enhances LRP6 aggregation/signalosome formation (see below) and thereby phosphorylation (Pan et al. 2008). An alternative mechanism has been suggested by a study of WTX/Amer1, which is an X-linked tumor suppressor in Wilms tumor. WTX binds to Axin and promotes β-catenin degradation in absence of Wnt (Major et al. 2007). However, upon Wnt stimulation, WTX via its PIP2-binding motif recruits Axin into the Wnt receptor complex, where PIP2 is high, thereby facilitating LRP6 phosphorylation by GSK3 and CK1γ (Regimbald-Dumas and He 2011; Tanneberger et al. 2011). WTX, which is vertebrate specific, thus has dichotomic roles in Wnt signaling, analogous to those for Axin and GSK3.

The FZD topology invites GPCR comparisons, and FZD coupling to trimeric G-proteins in Wnt/β-catenin signaling has received certain support from pharmacological and genetic studies (Ahumada et al. 2002; Katanaev et al. 2005; Liu et al. 2005; Schulte and Bryja 2007), although this remains a debated issue. Although Gαo and Gαq are suggested to regulate Axin–GSK3 interaction (Liu et al. 2005), Gβγ is shown to be in complex with LRP6, DVL, and Axin and to recruit GSK3 to phosphorylate LRP6 (Jernigan et al. 2010). Thus, multiple routes through DVL, G-protein, and WTX promote LRP6 phosphorylation, and DVL is essential in this capacity (Bilic et al. 2007; Zeng et al. 2008).

The SP motif is preferred by the so-called proline-directed kinases, which include GSK3, MAPK, and CDK subfamilies (Manning et al. 2002). PFTK, a CDK, and several MAPKs have been suggested to phosphorylate PPPSP in LRP6 in contexts in which cross-regulation with Wnt signaling may occur. A study showed that the PFTK/CyclinY pair phosphorylates LRP6 in G_2/M of the cell cycle, and thus "poises" LRP6 for Wnt signaling to peak at G_2/M (Davidson et al. 2009). This notion, however, contrasts the conventional view that proliferative signaling pathways commonly act during G_1/S transition (to regulate DNA synthesis). Indeed, a recent study suggests that Wnt signaling is highest at G_1/S but lowest at G_2/M (Hadjihannas et al. 2012). MAPK phosphorylation of LRP6 PPPSP was also suggested and may underlie synergy between FGF and Wnt signaling (Cervenka et al. 2011). This is likely context dependent, however, because antagonism between Wnt and MAPK signaling pathways has been observed in development (Szuts et al. 1997; Freeman and Bienz 2001) and in cancer such as melanoma (Biechele et al. 2012). GRK5/6 kinases, which are mostly associated with down-regulation of GPCR signaling, can phosphorylate LRP6 on PPPSP motifs and other sites in vitro (Chen et al. 2009), although the significance in vivo is unclear. Finally, evidence also exists that PKA can phosphorylate LRP6 at a site between PPPSPxS motifs (Fig. 3B) in response to activation of PTHR, a GPCR, and this event correlates with β-catenin signaling (Wan et al. 2011).

LRP6 SIGNAL INITIATION-AMPLIFICATION, SIGNALOSOMES, AND ENDOSOMAL SIGNALING

Several models have been elaborated based on Wnt-induced LRP6 phosphorylation/activation. FZD/DVL recruitment of Axin–GSK3 initiates PPPSPxS phosphorylation, which, in turn, reinforces Axin interaction, thereby forming a local positive feed-forward loop that promotes further PPPSPxS phosphorylation and enhances further Axin recruitment (Fig. 4A) (MacDonald et al. 2008; Zeng et al. 2008). This feed-forward loop between LRP6 and Axin may act in *cis* (among five PPPSPxS motifs within an LRP6 molecule) or in *trans* (among PPPSPxS motifs of different LRP6 molecules if LRP6 forms higher orders of oligomers), and has inspired the Wnt receptor "initiation- amplification" model (MacDonald et al. 2008; Zeng et al. 2008), which is consistent with transgenic experiments in *Drosophila* suggesting that Dfz2 and Arrow initiate whereas Arrow amplifies Wg signaling (Baig-Lewis et al. 2007). Supporting this model, phosphorylation of an individual PPPSP motif depends profoundly on neighboring PPPSP motifs, and LRP6 requires minimally four PPPSPxS motifs to be quasi-competent for signaling (and LRP6 mutants with combinations of three PPPSPxS motifs are mostly inactive or even dominant negative) (MacDonald et al. 2008; Wolf et al. 2008). Therefore, five PPPSPxS motifs in LRP6 act like an in *cis* amplifier for signaling. Why then is a single PPPSPxS motif sufficient to transfer signaling to a heterologous receptor, LDLRΔN (Tamai et al. 2004; Zeng et al. 2005; MacDonald et al. 2008)? LDLRΔN, and therefore LDLRΔN-PPPSP, fortuitously form on membrane large aggregates

 Cite this article as *Cold Spring Harb Perspect Biol* doi: 10.1101/cshperspect.a007880

(Bilic et al. 2007; X-J Zhang and X He, unpubl.), which likely constitute in *trans* amplification.

The LRP6 "signalosome" model is based on observations that Wnt induces large-sized LRP6 aggregates in fractionation and immunostaining (Bilic et al. 2007), and that DVL is required for LRP6 phosphorylation in these aggregates in a manner that correlates with DIX-mediated polymerization (Schwarz-Romond et al. 2007a). When LRP6 is overexpressed together with FZD, DVL, Axin, and GSK3, LRP6 aggregates become particularly prominent and likely contain these other components (Fig. 4B) (Bilic et al. 2007; Schwarz-Romond et al. 2007a). Note that these results are consistent with those that led to the "initiation-amplification" model, suggesting that the two models may describe the same receptor activation events from different temporal/molecular and spatial/cellular perspectives, with DVL polymerization underlying and being emphasized in the signalosome model.

Another model, in which phosphorylated/activated LRP6 signals in an endosomal platform, posits that LRP6 and phosphorylation in lipid rafts (a cholesterol-rich microdomain in the plasma membrane) and its subsequent endocytosis through caveolin are both required for signaling (Yamamoto et al. 2006, 2008). Caveolin binds to LRP6 upon Wnt stimulation and is required for phosphorylated LRP6 to recruit Axin (Yamamoto et al. 2006). One may notice significant similarities between lipid raft/endosomal LRP6 versus signalosomes, because methods (gradient fractionation and immunostaining) used to define them are similar, and signalosomes indeed contain caveolin (Yamamoto et al. 2006; Bilic et al. 2007), suggesting that these two models overlap despite different emphases. PIP2 is suggested to promote LRP6 signalosome formation and thus phosphorylation (Pan et al. 2008), but its key roles in receptor endocytosis should be noted (Di Paolo and De Camilli 2006). A new transmembrane Wnt inhibitor, Waif1, binds to LRP6 and prevents LRP6 endocytosis but not its phosphorylation (Kagermeier-Schenk et al. 2011), lending support to this "endosomal signaling" model. Another transmembrane protein, PRR (prorenin receptor), which binds to FZD and LRP6 and also to the vacuolar H^+-adenosine triphosphatase (V-ATPase) complex that acidifies vesicles, is required for LRP6 phosphorylation by CK1 (Cruciat et al. 2010), consistent with the possibility of a vesicular LRP6 signalosome (or endosome) where LRP6 phosphorylation occurs.

An extension of the endosomal signaling model posits that LRP6 endocytosis via caveolin versus clathrin activates and inhibits signaling, respectively (Yamamoto et al. 2008). However, different results exist regarding this issue (Blitzer and Nusse 2006). Another obstacle is that caveolin1$^{-/-}$ mice show apparently higher β-catenin signaling in mammary and intestinal stem cells (Li et al. 2005; Sotgia et al. 2005). In addition, caveolin does not exist in *Drosophila*, and therefore a different mechanism has to be envisioned for Wg signaling. Thus, some cautions are warranted, particularly when perturbations of general endocytic pathways are performed and thus broad cellular events beyond Wnt signaling are affected (Gagliardi et al. 2008). Indeed, manipulating endocytic molecules in *Drosophila* has also yielded contradictory results regarding Wg signaling (Piddini et al. 2005; Rives et al. 2006; Seto and Bellen 2006). Finally, some of the cell biological studies of LRP6 and Wnt signaling have relied on overexpression of LRP6 and other components. Whether the endogenous LRP6 and other Wnt signaling proteins behave similarly is a major caveat that needs to be taken into account.

INHIBITION OF β-CATENIN PHOSPHORYLATION POST RECEPTOR ACTIVATION

How activation of the Wnt receptor complex, in particular phosphorylation of LRP6, leads to inhibition of β-catenin phosphorylation remains not fully understood and debated. An earlier model (Fig. 5A), prior to knowledge of LRP6 phosphorylation, suggests that Wnt signaling via DVL leads to disruption of the Axin complex (Kimelman and Xu 2006; MacDonald et al. 2009). Indeed, upon Wnt stimulation, reduced association of Axin with β-catenin and GSK3 is observed (Kishida et al. 1999a; Li et al.

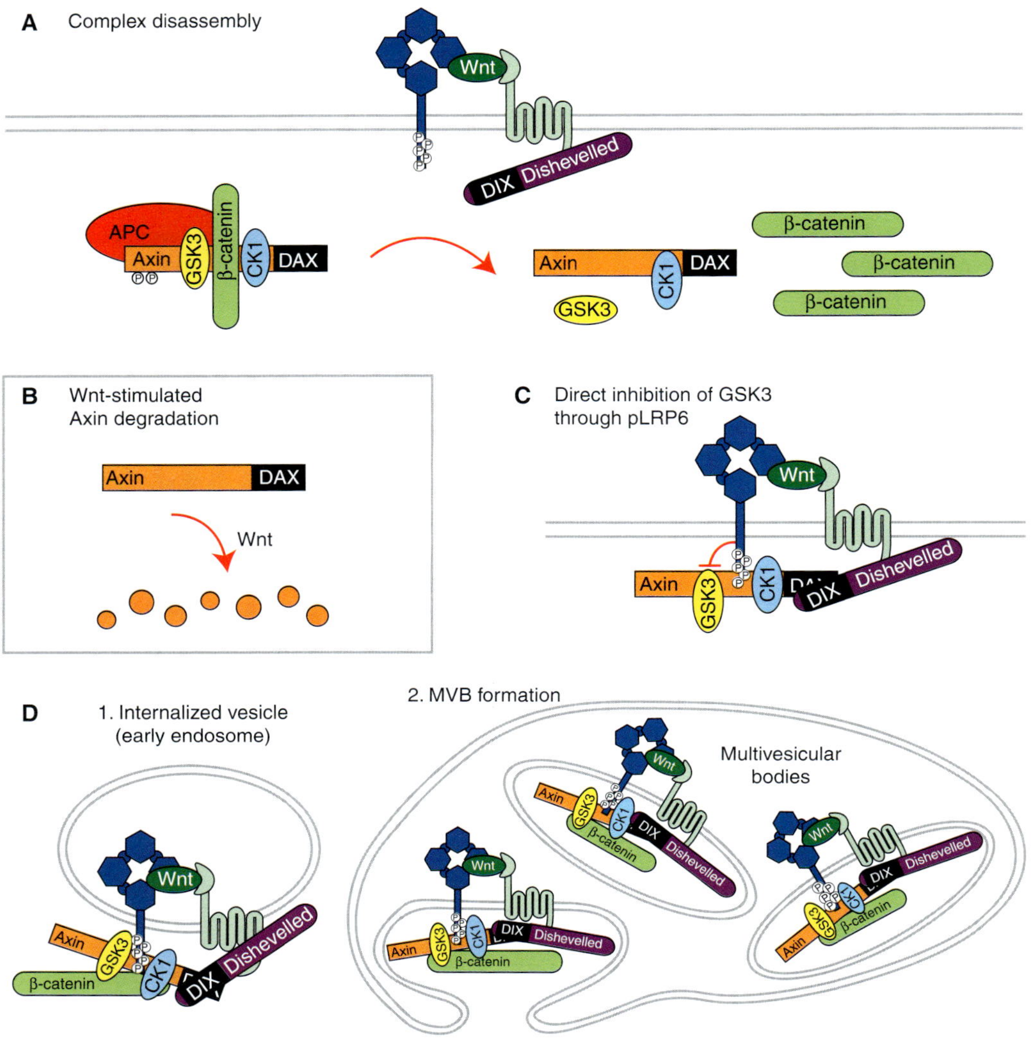

Figure 5. Models for Wnt-induced inhibition of β-catenin phosphorylation. (*A*) Axin complex disassembly. (*B*) Axin protein degradation. (*C*) GSK3 inhibition by phosphorylated PPPSPxS motifs. (*D*) GSK3 inclusion by multivesicular bodies.

1999; Itoh et al. 2000; Liu et al. 2005), and this is accompanied by hypophosphorylation of Axin (Willert et al. 1999; Yamamoto et al. 1999), which itself is a substrate for GSK3 (and CK1) (Ikeda et al. 1998; Jho et al. 1999). This model suggests that displacement from or inhibition of GSK3 in the Axin complex causes Axin hypophosphorylation, which further results in reduced β-catenin binding to Axin and thus β-catenin phosphorylation (Kimelman and Xu 2006). Some of these earlier studies included a role for FRAT/GBP (Frequently Rearranged in Advanced T-cell lymphoma/GSK3-Binding Protein), which is a family of GSK3-binding proteins that displace GSK3 from Axin and thus inhibit GSK3 (Yost et al. 1998; Ferkey and Kimelman 2002; Dajani et al. 2003). Because FRAT also binds to DVL, an extension of the model is that DVL through FRAT displaces GSK3 in the Axin complex, leading to inhibition

 Cite this article as *Cold Spring Harb Perspect Biol* doi: 10.1101/cshperspect.a007880

of β-catenin phosphorylation. However, triple knockouts of all three *Frat* genes produced viable and fertile mice, arguing that FRAT is dispensable for Wnt signaling (van Amerongen et al. 2005). This is also true for Wg signaling, because *Drosophila* does not have a FRAT homolog. Although a FRAT-based model is no longer favored, the core part of the "complex disruption" model and data leading to it remain valid and should not be discounted. A recent study showed that protein phosphatase 1 (PP1) is required for Wnt signaling through Axin dephosphorylation that disrupts Axin–GSK3 interaction (Luo et al. 2007), consistent with the suggestion that active Axin dephosphorylation together with inhibition of Axin phosphorylation by GSK3 are involved in Axin complex disruption (Willert et al. 1999).

A study suggests that APC, via an unknown mechanism, is broadly required for GSK3 activity and that Wnt stimulation results in dissociation of APC from the Axin–GSK3 complex, resulting in down-regulation of GSK3 (Valvezan et al. 2012), again consistent with the complex disruption model.

Earlier results showed that overexpression of constitutively activated LRP5 or Arrow resulted in decreased Axin protein levels in mammalian and *Drosophila* systems (Fig. 5B) (Mao et al. 2001b; Tolwinski et al. 2003). However, the Wnt-induced decrease in Axin protein levels lags behind the increase in β-catenin protein levels (Willert et al. 1999; Yamamoto et al. 1999), and thus is unlikely to be the primary mechanism for β-catenin stabilization. However, Axin degradation may be important for long-term or chronic Wnt stimulation. It is worth noting that Tankyrase has been shown to promote Axin PARsylation and degradation (Huang et al. 2009), but Tankyrase-induced Axin degradation does not seem to be under Wnt regulation.

To integrate recent findings of LRP6 phosphorylation and DVL acting primarily through LRP6 phosphorylation, a newer model has emerged in which phosphorylated PPPSPxS motifs directly inhibit GSK3 (Cselenyi et al. 2008; Piao et al. 2008; Wu et al. 2009). This model is based on findings that LRP6 binds to GSK3 (Zeng et al. 2005; Mi et al. 2006) and that phosphorylated recombinant LRP6 cytoplasmic domain and, in fact, phosphorylated PPPSPxS peptides bind to and inhibit β-catenin phosphorylation by GSK3 in vitro and activate β-catenin signaling in vivo (Cselenyi et al. 2008; Piao et al. 2008; Wu et al. 2009). This inhibition mechanism bears apparent similarity to GSK3 inhibition during insulin signaling by its own amino-terminal region, whose phosphorylation (at S21 in GSK3α or S9 in GSK3β by the AKT kinase) creates an inhibitory and binding pseudosubstrate for GSK3 inhibition (Dajani et al. 2001; Frame et al. 2001). This model fits well with the requirement of LRP6 PPPSPxS phosphorylation in signaling and the hypothesis of DVL-dependent signalosomes in recruitment of Axin–GSK3 to the proximity of LRP6 and can account for the specificity of inhibition of GSK3 phosphorylation of β-catenin (but not of other GSK3 substrates outside of the Axin complex). This model is also compatible with the "complex disruption" model, because LRP6 inhibition of GSK3 will also result in Axin hypophosphorylation and thus Axin complex disruption. Implicit in this model is that phosphorylated PPPSPxS motifs may have multiple roles in Wnt signaling: binding to Axin, enhancing mutual PPPSPxS phosphorylation, and binding to and inhibition of GSK3. In this context, it is noted that five PPPSPxS motifs in LRP5/6 are spaced tightly within a region of 120 residues (Fig. 3B), leading to speculation that an LRP6 may only interact with a single Axin molecule at any given time, that is, only one phosphorylated PPPSPxS motif is occupied by Axin, while the remaining ones provide a high local concentration to sustain LRP6–Axin interaction (Wu et al. 2009). In this scenario, LRP6 via its five phosphorylated PPPSPxS motifs can engage Axin and inhibit GSK3 simultaneously, a possibility that is also conceivable if LRP6 aggregates in trans in signalosomes. Related to this issue is whether Axin binds to LRP6 indirectly via GSK3 (Piao et al. 2008), but recent evidence does not support the view of indirect LRP6–Axin interaction (MacDonald et al. 2011). Further studies will be needed to clarify the mechanism of phosphorylated PPPSPxS motifs in Axin and GSK3 binding and GSK3 inhibition.

A drastically different model was proposed recently, in which multivesicular body (MVB) formation after endocytosis of the Wnt receptor wraps GSK3 into the deep interior of the MVB, thereby physically separating GSK3 from its cytoplasmic substrates including β-catenin (Fig. 5D) (Taelman et al. 2010). This "MVB inclusion" model has several features: (1) It requires MVBs to round up most of the cellular GSK3 proteins upon Wnt signaling, because gene deletions of Gsk3α and/or Gsk3β suggest that cells can perform Wnt signaling when two (or less than three) of the four Gsk3 alleles are deleted (Doble et al. 2007). (2) Because GSK3 levels do not overtly decrease during Wnt signaling, it requires a mechanism for GSK3 subsequently to escape MVBs, which normally shuttles the cargo to lysosomes for degradation. (3) It suggests that genes involved in MVB formation are required for Wnt signaling, for which supporting evidence exists in *Xenopus* but not in *Drosophila* (Piddini et al. 2005; Rives et al. 2006; Seto and Bellen 2006; Taelman et al. 2010). (4) It predicts that most GSK3 substrates (in fact, ~20% of cellular proteins) in addition to β-catenin are under Wnt regulation (Taelman et al. 2010), thus departing from the prevailing view that roles of GSK3 in Wnt and other signaling (such as insulin) pathways are insulated from one another. Another unexpected finding is that β-catenin itself promotes inclusion of GSK3 into MVBs (Taelman et al. 2010), implying a new β-catenin function downstream of the Wnt receptors, acting before its role as a transcription coactivator in the nucleus. How to integrate/reconcile this provocative MVB model with the existing data and models remains to be considered (Metcalfe and Bienz 2011), although the possibility of MVB actions in the longer time course of Wnt treatment has been discussed.

CONCLUDING REMARKS

Our understanding of the Wnt receptor complex has progressed tremendously but remains rudimentary. Many unanswered questions linger, including the specificity and stoichiometry of the receptors with Wnt ligands and their antagonists, newly and yet-to-be identified components of the receptor complex, molecular and dynamic interactions that assemble and disassemble the receptor complex and downstream signaling components, and ultimately the atomic structures of some or all of these components in Wnt-OFF and Wnt-ON states. These will represent huge challenges for generations of researchers to accomplish.

ACKNOWLEDGMENTS

We thank members of the Xi He laboratory for their critical reading and comments on the manuscript, and Claudia Janda and Chris Garcia for providing the Figure 1D template. X.H. acknowledges grant support from the National Institutes of Health (NIH) (RO1 GM057603, RO1 GM057603S1, RO1 GM074241, and RO1 AR060359) and Boston Children's Hospital Intellectual and Developmental Disabilities Research Center (P30 HD-18655).

REFERENCES

*Reference is also in this collection.

Adamska M, Larroux C, Adamski M, Green K, Lovas E, Koop D, Richards GS, Zwafink C, Degnan BM. 2010. Structure and expression of conserved Wnt pathway components in the demosponge *Amphimedon queenslandica*. *Evol Dev* **12:** 494–518.

Ahn VE, Chu ML, Choi HJ, Tran D, Abo A, Weis WI. 2011. Structural basis of Wnt signaling inhibition by Dickkopf binding to LRP5/6. *Dev Cell* **21:** 862–873.

Ahumada A, Slusarski DC, Liu X, Moon RT, Malbon CC, Wang HY. 2002. Signaling of rat Frizzled-2 through phosphodiesterase and cyclic GMP. *Science* **298:** 2006–2010.

Axelrod JD, Miller JR, Shulman JM, Moon RT, Perrimon N. 1998. Differential recruitment of Dishevelled provides signaling specificity in the planar cell polarity and Wingless signaling pathways. *Genes Dev* **12:** 2610–2622.

Baig-Lewis S, Peterson-Nedry W, Wehrli M. 2007. Wingless/Wnt signal transduction requires distinct initiation and amplification steps that both depend on Arrow/LRP. *Dev Biol* **306:** 94–111.

Bayly R, Axelrod JD. 2011. Pointing in the right direction: New developments in the field of planar cell polarity. *Nat Rev Genet* **12:** 385–391.

Behrens J, Jerchow BA, Wurtele M, Grimm J, Asbrand C, Wirtz R, Kuhl M, Wedlich D, Birchmeier W. 1998. Functional interaction of an axin homolog, conductin, with β-catenin, APC, and GSK3β. *Science* **280:** 596–599.

Bhanot P, Brink M, Samos CH, Hsieh JC, Wang Y, Macke JP, Andrew D, Nathans J, Nusse R. 1996. A new member of the frizzled family from *Drosophila* functions as a Wingless receptor. *Nature* **382:** 225–230.

Biechele TL, Kulikauskas RM, Toroni RA, Lucero OM, Swift RD, James RG, Robin NC, Dawson DW, Moon RT, Chien AJ. 2012. Wnt/β-catenin signaling and AXIN1 regulate apoptosis triggered by inhibition of the mutant kinase BRAFV600E in human melanoma. *Sci Signal* **5:** ra3.

Bilic J, Huang YL, Davidson G, Zimmermann T, Cruciat CM, Bienz M, Niehrs C. 2007. Wnt induces LRP6 signalosomes and promotes Dishevelled-dependent LRP6 phosphorylation. *Science* **316:** 1619–1622.

Blitzer JT, Nusse R. 2006. A critical role for endocytosis in Wnt signaling. *BMC Cell Biol* **7:** 28.

Bourhis E, Tam C, Franke Y, Bazan JF, Ernst J, Hwang J, Costa M, Cochran AG, Hannoush RN. 2010. Reconstitution of a Frizzled8–Wnt3a–LRP6 signaling complex reveals multiple Wnt and Dkk1 binding sites on LRP6. *J Biol Chem* **285:** 9172–9179.

Bourhis E, Wang W, Tam C, Hwang J, Zhang Y, Spittler D, Huang OW, Gong Y, Estevez A, Zilberleyb I, et al. 2011. Wnt antagonists bind through a short peptide to the first β-propeller domain of LRP5/6. *Structure* **19:** 1433–1442.

Boutros M, Paricio N, Strutt DI, Mlodzik M. 1998. Dishevelled activates JNK and discriminates between JNK pathways in planar polarity and wingless signaling. *Cell* **94:** 109–118.

Boutros M, Mihaly J, Bouwmeester T, Mlodzik M. 2000. Signaling specificity by Frizzled receptors in *Drosophila*. *Science* **288:** 1825–1828.

Bridges CB, Brehme KS. 1944. *The mutants of* Drosophila melanogaster. Carnegie Institution of Washington, Washington, DC.

Brown SD, Twells RC, Hey PJ, Cox RD, Levy ER, Soderman AR, Metzker ML, Caskey CT, Todd JA, Hess JF. 1998. Isolation and characterization of LRP6, a novel member of the low density lipoprotein receptor gene family. *Biochem Biophys Res Commun* **248:** 879–888.

Capelluto DG, Kutateladze TG, Habas R, Finkielstein CV, He X, Overduin M. 2002. The DIX domain targets Dishevelled to actin stress fibres and vesicular membranes. *Nature* **419:** 726–729.

Carroll CE, Marada S, Stewart DP, Ouyang JX, Ogden SK. 2012. The extracellular loops of Smoothened play a regulatory role in control of Hedgehog pathway activation. *Development* **139:** 612–621.

Cervenka I, Wolf J, Masek J, Krejci P, Wilcox WR, Kozubik A, Schulte G, Gutkind JS, Bryja V. 2011. Mitogen-activated protein kinases promote WNT/β-catenin signaling via phosphorylation of LRP6. *Mol Cell Biol* **31:** 179–189.

Chen CM, Struhl G. 1999. Wingless transduction by the Frizzled and Frizzled2 proteins of *Drosophila*. *Development* **126:** 5441–5452.

Chen M, Philipp M, Wang J, Premont RT, Garrison TR, Caron MG, Lefkowitz RJ, Chen W. 2009. G protein-coupled receptor kinases phosphorylate LRP6 in the Wnt pathway. *J Biol Chem* **284:** 35040–35048.

Chen S, Bubeck D, MacDonald BT, Liang WX, Mao JH, Malinauskas T, Llorca O, Aricescu AR, Siebold C, He X, et al. 2011. Structural and functional studies of LRP6 ectodomain reveal a platform for Wnt signaling. *Dev Cell* **21:** 848–861.

Cheng Z, Biechele T, Wei Z, Morrone S, Moon RT, Wang L, Xu W. 2011. Crystal structures of the extracellular domain of LRP6 and its complex with DKK1. *Nat Struct Mol Biol* **18:** 1204–1210.

Clevers H. 2006. Wnt/β-catenin signaling in development and disease. *Cell* **127:** 469–480.

Cong F, Schweizer L, Varmus H. 2004a. Casein kinase Iε modulates the signaling specificities of Dishevelled. *Mol Cell Biol* **24:** 2000–2011.

Cong F, Schweizer L, Varmus H. 2004b. Wnt signals across the plasma membrane to activate the β-catenin pathway by forming oligomers containing its receptors, Frizzled and LRP. *Development* **131:** 5103–5115.

Cruciat CM, Ohkawara B, Acebron SP, Karaulanov E, Reinhard C, Ingelfinger D, Boutros M, Niehrs C. 2010. Requirement of prorenin receptor and vacuolar H^+-ATPase-mediated acidification for Wnt signaling. *Science* **327:** 459–463.

Cselenyi CS, Jernigan KK, Tahinci E, Thorne CA, Lee LA, Lee E. 2008. LRP6 transduces a canonical Wnt signal independently of Axin degradation by inhibiting GSK3's phosphorylation of β-catenin. *Proc Natl Acad Sci* **105:** 8032–8037.

Culi J, Mann RS. 2003. Boca, an endoplasmic reticulum protein required for Wingless signaling and trafficking of LDL receptor family members in *Drosophila*. *Cell* **112:** 343–354.

Culi J, Springer TA, Mann RS. 2004. Boca-dependent maturation of β-propeller/EGF modules in low-density lipoprotein receptor proteins. *EMBO J* **23:** 1372–1380.

Dajani R, Fraser E, Roe SM, Young N, Good V, Dale TC, Pearl LH. 2001. Crystal structure of glycogen synthase kinase 3β: Structural basis for phosphate-primed substrate specificity and autoinhibition. *Cell* **105:** 721–732.

Dajani R, Fraser E, Roe SM, Yeo M, Good VM, Thompson V, Dale TC, Pearl LH. 2003. Structural basis for recruitment of glycogen synthase kinase 3β to the Axin-APC scaffold complex. *EMBO J* **22:** 494–501.

Dann CE, Hsieh JC, Rattner A, Sharma D, Nathans J, Leahy DJ. 2001. Insights into Wnt binding and signalling from the structures of two Frizzled cysteine-rich domains. *Nature* **412:** 86–90.

Davidson G, Wu W, Shen J, Bilic J, Fenger U, Stannek P, Glinka A, Niehrs C. 2005. Casein kinase 1γ couples Wnt receptor activation to cytoplasmic signal transduction. *Nature* **438:** 867–872.

Davidson G, Shen J, Huang YL, Su Y, Karaulanov E, Bartscherer K, Hassler C, Stannek P, Boutros M, Niehrs C. 2009. Cell cycle control of wnt receptor activation. *Dev Cell* **17:** 788–799.

Del Valle-Perez B, Arques O, Vinyoles M, de Herreros AG, Dunach M. 2011. Coordinated action of CK1 isoforms in canonical Wnt signaling. *Mol Cell Biol* **31:** 2877–2888.

Di Paolo G, De Camilli P. 2006. Phosphoinositides in cell regulation and membrane dynamics. *Nature* **443:** 651–657.

Djiane A, Yogev S, Mlodzik M. 2005. The apical determinants aPKC and dPatj regulate Frizzled-dependent planar cell polarity in the *Drosophila* eye. *Cell* **121:** 621–631.

Doble BW, Patel S, Wood GA, Kockeritz LK, Woodgett JR. 2007. Functional redundancy of GSK-3α and GSK-3β in Wnt/β-catenin signaling shown by using an allelic series of embryonic stem cell lines. *Dev Cell* **12:** 957–971.

Ettenberg SA, Charlat O, Daley MP, Liu S, Vincent KJ, Stuart DD, Schuller AG, Yuan J, Ospina B, Green J, et al. 2010. Inhibition of tumorigenesis driven by different Wnt proteins requires blockade of distinct ligand-binding regions by LRP6 antibodies. *Proc Natl Acad Sci* **107:** 15473–15478.

Fahmy OG, Fahmy M. 1959. New mutants report. *Drosoph Inf Serv* **33:** 82–94.

Ferkey DM, Kimelman D. 2002. Glycogen synthase kinase-3β mutagenesis identifies a common binding domain for GBP and Axin. *J Biol Chem* **277:** 16147–16152.

Fiedler M, Mendoza-Topaz C, Rutherford TJ, Mieszczanek J, Bienz M. 2011. Dishevelled interacts with the DIX domain polymerization interface of Axin to interfere with its function in down-regulating β-catenin. *Proc Natl Acad Sci* **108:** 1937–1942.

Frame S, Cohen P, Biondi RM. 2001. A common phosphate binding site explains the unique substrate specificity of GSK3 and its inactivation by phosphorylation. *Mol Cell* **7:** 1321–1327.

Freeman M, Bienz M. 2001. EGF receptor/Rolled MAP kinase signalling protects cells against activated Armadillo in the *Drosophila* eye. *EMBO Rep* **2:** 157–162.

Gagliardi M, Piddini E, Vincent JP. 2008. Endocytosis: A positive or a negative influence on Wnt signalling? *Traffic* **9:** 1–9.

Gong Y, Bourhis E, Chiu C, Stawicki S, DeAlmeida VI, Liu BY, Phamluong K, Cao TC, Carano RA, Ernst, et al. 2010. Wnt isoform-specific interactions with coreceptor specify inhibition or potentiation of signaling by LRP6 antibodies. *PLoS ONE* **5:** e12682.

Guder C, Philipp I, Lengfeld T, Watanabe H, Hobmayer B, Holstein TW. 2006. The Wnt code: Cnidarians signal the way. *Oncogene* **25:** 7450–7460.

Hadjihannas MV, Bernkopf DB, Bruckner M, Behrens J. 2012. Cell cycle control of Wnt/β-catenin signalling by conductin/axin2 through CDC20. *EMBO Rep* **13:** 347–354.

He X, Semenov M, Tamai K, Zeng X. 2004. LDL receptor-related proteins 5 and 6 in Wnt/β-catenin signaling: Arrows point the way. *Development* **131:** 1663–1677.

Hering H, Sheng M. 2002. Direct interaction of Frizzled-1, -2, -4, and -7 with PDZ domains of PSD-95. *FEBS Lett* **521:** 185–189.

Hey PJ, Twells RC, Phillips MS, Yusuke N, Brown SD, Kawaguchi Y, Cox R, Guochun X, Dugan V, Hammond H, et al. 1998. Cloning of a novel member of the low-density lipoprotein receptor family. *Gene* **216:** 103–111.

Hsieh JC, Rattner A, Smallwood PM, Nathans J. 1999. Biochemical characterization of Wnt–Frizzled interactions using a soluble, biologically active vertebrate Wnt protein. *Proc Natl Acad Sci* **96:** 3546–3551.

Hsieh JC, Lee L, Zhang L, Wefer S, Brown K, DeRossi C, Wines ME, Rosenquist T, Holdener BC. 2003. *mesd* encodes an LRP5/6 chaperone essential for specification of mouse embryonic polarity. *Cell* **112:** 355–367.

Huang SM, Mishina YM, Liu S, Cheung A, Stegmeier F, Michaud GA, Charlat O, Wiellette E, Zhang Y, Wiessner S, et al. 2009. Tankyrase inhibition stabilizes axin and antagonizes Wnt signalling. *Nature* **461:** 614–620.

Ikeda S, Kishida S, Yamamoto H, Murai H, Koyama S, Kikuchi A. 1998. Axin, a negative regulator of the Wnt signaling pathway, forms a complex with GSK-3β and β-catenin and promotes GSK-3β-dependent phosphorylation of β-catenin. *EMBO J* **17:** 1371–1384.

Itoh K, Antipova A, Ratcliffe MJ, Sokol S. 2000. Interaction of Dishevelled and *Xenopus*Axin-related protein is required for Wnt signal transduction. *Mol Cell Biol* **20:** 2228–2238.

* Jackson BM, Eisenmann DM. 2012. β-Catenin-dependent Wnt signaling in *C. elegans*: Teaching an old dog a new trick. *Cold Spring Harb Perspect Biol* doi: 10.1101/cshperspect.a007948.

Janda CY, Waghray D, Levin AM, Thomas C, Garcia KC. 2012. Structural basis of Wnt recognition by Frizzled. *Science* **337:** 59–64.

Jeon H, Meng W, Takagi J, Eck MJ, Springer TA, Blacklow SC. 2001. Implications for familial hypercholesterolemia from the structure of the LDL receptor YWTD-EGF domain pair. *Nat Struct Biol* **8:** 499–504.

Jernigan KK, Cselenyi CS, Thorne CA, Hanson AJ, Tahinci E, Hajicek N, Oldham WM, Lee LA, Hamm HE, Hepler, et al. 2010. Gβγ activates GSK3 to promote LRP6-mediated β-catenin transcriptional activity. *Sci Signal* **3:** ra37.

Jho E, Lomvardas S, Costantini F. 1999. A GSK3β phosphorylation site in axin modulates interaction with β-catenin and Tcf-mediated gene expression. *Biochem Biophys Res Commun* **266:** 28–35.

Jiang J, Hui CC. 2008. Hedgehog signaling in development and cancer. *Dev Cell* **15:** 801–812.

Jung H, Lee SK, Jho EH. 2011. Mest/Peg1 inhibits Wnt signalling through regulation of LRP6 glycosylation. *Biochem J* **436:** 263–269.

Kagermeier-Schenk B, Wehner D, Ozhan-Kizil G, Yamamoto H, Li J, Kirchner K, Hoffmann C, Stern P, Kikuchi A, Schambony A, et al. 2011. Waif1/5T4 inhibits Wnt/β-catenin signaling and activates noncanonical Wnt pathways by modifying LRP6 subcellular localization. *Dev Cell* **21:** 1129–1143.

Katanaev VL, Ponzielli R, Semeriva M, Tomlinson A. 2005. Trimeric G protein-dependent Frizzled signaling in *Drosophila*. *Cell* **120:** 111–122.

Kennerdell JR, Carthew RW. 1998. Use of dsRNA-mediated genetic interference to demonstrate that Frizzled and Frizzled 2 act in the Wingless pathway. *Cell* **95:** 1017–1026.

Kimelman D, Xu W. 2006. β-Catenin destruction complex: Insights and questions from a structural perspective. *Oncogene* **25:** 7482–7491.

Kishida S, Yamamoto H, Ikeda S, Kishida M, Sakamoto I, Koyama S, Kikuchi A. 1998. Axin, a negative regulator of the Wnt signaling pathway, directly interacts with adenomatous polyposis coli and regulates the stabilization of β-catenin. *J Biol Chem* **273:** 10823–10826.

Kishida M, Koyama S, Kishida S, Matsubara K, Nakashima S, Higano K, Takada R, Takada S, Kikuchi A. 1999a. Axin prevents Wnt-3a-induced accumulation of β-catenin. *Oncogene* **18:** 979–985.

Kishida S, Yamamoto H, Hino S, Ikeda S, Kishida M, Kikuchi A. 1999b. DIX domains of Dvl and axin are necessary for protein interactions and their ability to regulate β-catenin stability. *Mol Cell Biol* **19:** 4414–4422.

Kishida M, Hino S, Michiue T, Yamamoto H, Kishida S, Fukui A, Asashima M, Kikuchi A. 2001. Synergistic activation of the Wnt signaling pathway by Dvl and casein kinase Iε. *J Biol Chem* **276:** 33147–33155.

Klein TJ, Jenny A, Djiane A, Mlodzik M. 2006. CKIε/*discs overgrown* promotes both Wnt-Fz/β-catenin and Fz/PCP signaling in *Drosophila*. *Curr Biol* **16:** 1337–1343.

Klingensmith J, Nusse R. 1994. Signaling by Wingless in *Drosophila*. *Dev Biol* **166:** 396–414.

Krasnow RE, Wong LL, Adler PN. 1995. Dishevelled is a component of the Frizzled signaling pathway in *Drosophila*. *Development* **121:** 4095–4102.

Li L, Yuan H, Weaver CD, Mao J, Farr GH III, Sussman DJ, Jonkers J, Kimelman D, Wu D. 1999. Axin and Frat1 interact with Dvl and GSK, bridging Dvl to GSK in Wnt-mediated regulation of LEF-1. *EMBO J* **18:** 4233–4240.

Li J, Hassan GS, Williams TM, Minetti C, Pestell RG, Tanowitz HB, Frank PG, Sotgia F, Lisanti MP. 2005. Loss of caveolin-1 causes the hyper-proliferation of intestinal crypt stem cells, with increased sensitivity to whole body γ-radiation. *Cell Cycle* **4:** 1817–1825.

Liang J, Fu Y, Cruciat CM, Jia S, Wang Y, Tong Z, Tao Q, Ingelfinger D, Boutros M, Meng A, et al. 2011. Transmembrane protein 198 promotes LRP6 phosphorylation and Wnt signaling activation. *Mol Cell Biol* **31:** 2577–2590.

Lighthouse JK, Zhang L, Hsieh JC, Rosenquist T, Holdener BC. 2011. MESD is essential for apical localization of megalin/LRP2 in the visceral endoderm. *Dev Dyn* **240:** 577–588.

Liu C, Li Y, Semenov M, Han C, Baeg GH, Tan Y, Zhang Z, Lin X, He X. 2002. Control of β-catenin phosphorylation/degradation by a dual-kinase mechanism. *Cell* **108:** 837–847.

Liu X, Rubin JS, Kimmel AR. 2005. Rapid, Wnt-induced changes in GSK3β associations that regulate β-catenin stabilization are mediated by Gα proteins. *Curr Biol* **15:** 1989–1997.

Liu CC, Pearson C, Bu G. 2009. Cooperative folding and ligand-binding properties of LRP6 β-propeller domains. *J Biol Chem* **284:** 15299–15307.

Logan CY, Nusse R. 2004. The Wnt signaling pathway in development and disease. *Annu Rev Cell Dev Biol* **20:** 781–810.

Luo W, Peterson A, Garcia BA, Coombs G, Kofahl B, Heinrich R, Shabanowitz J, Hunt DF, Yost HJ, Virshup DM. 2007. Protein phosphatase 1 regulates assembly and function of the β-catenin degradation complex. *EMBO J* **26:** 1511–1521.

MacDonald BT, Yokota C, Tamai K, Zeng X, He X. 2008. Wnt signal amplification via activity, cooperativity, and regulation of multiple intracellular PPPSP motifs in the Wnt co-receptor LRP6. *J Biol Chem* **283:** 16115–16123.

MacDonald BT, Tamai K, He X. 2009. Wnt/β-catenin signaling: Components, mechanisms, and diseases. *Dev Cell* **17:** 9–26.

MacDonald BT, Semenov MV, Huang H, He X. 2011. Dissecting molecular differences between Wnt coreceptors LRP5 and LRP6. *PLoS ONE* **6:** e23537.

Major MB, Camp ND, Berndt JD, Yi X, Goldenberg SJ, Hubbert C, Biechele TL, Gingras AC, Zheng N, Maccoss MJ, et al. 2007. Wilms tumor suppressor WTX negatively regulates WNT/β-catenin signaling. *Science* **316:** 1043–1046.

Manning G, Whyte DB, Martinez R, Hunter T, Sudarsanam S. 2002. The protein kinase complement of the human genome. *Science* **298:** 1912–1934.

Mao B, Wu W, Li Y, Hoppe D, Stannek P, Glinka A, Niehrs C. 2001a. LDL-receptor-related protein 6 is a receptor for Dickkopf proteins. *Nature* **411:** 321–325.

Mao J, Wang J, Liu B, Pan W, Farr GH III, Flynn C, Yuan H, Takada S, Kimelman D, Li L, et al. 2001b. Low-density lipoprotein receptor-related protein-5 binds to Axin and regulates the canonical Wnt signaling pathway. *Mol Cell* **7:** 801–809.

Metcalfe C, Bienz M. 2011. Inhibition of GSK3 by Wnt signalling—Two contrasting models. *J Cell Sci* **124:** 3537–3544.

Metcalfe C, Mendoza-Topaz C, Mieszczanek J, Bienz M. 2010. Stability elements in the LRP6 cytoplasmic tail confer efficient signalling upon DIX-dependent polymerization. *J Cell Sci* **123:** 1588–1599.

Mi K, Dolan PJ, Johnson GV. 2006. The low density lipoprotein receptor-related protein 6 interacts with glycogen synthase kinase 3 and attenuates activity. *J Biol Chem* **281:** 4787–4794.

Mulligan KA, Fuerer C, Ching W, Fish M, Willert K, Nusse R. 2012. Secreted Wingless-interacting molecule (Swim) promotes long-range signaling by maintaining Wingless solubility. *Proc Natl Acad Sci* **109:** 370–377.

Nikopoulos K, Venselaar H, Collin RW, Riveiro-Alvarez R, Boonstra FN, Hooymans JM, Mukhopadhyay A, Shears D, van Bers M, de Wijs IJ, et al. 2010. Overview of the mutation spectrum in familial exudative vitreoretinopathy and Norrie disease with identification of 21 novel variants in FZD4, LRP5, and NDP. *Hum Mutat* **31:** 656–666.

Nusslein-Volhard C, Wieschaus E, Kluding H. 1984. Mutations affecting the pattern of the larval cuticle in *Drosophila melanogaster*. I. Zygotic loci on the second chromosome. *Roux's Arch Dev Biol* **193:** 267–282.

Pan W, Choi SC, Wang H, Qin Y, Volpicelli-Daley L, Swan L, Lucast L, Khoo C, Zhang X, Li L, et al. 2008. Wnt3a-mediated formation of phosphatidylinositol 4,5-bisphosphate regulates LRP6 phosphorylation. *Science* **321:** 1350–1353.

Peeters MC, van Westen GJ, Li Q, AP IJ. 2010. Importance of the extracellular loops in G protein-coupled receptors for ligand recognition and receptor activation. *Trends Pharmacol Sci* **32:** 35–42.

Perrimon N, Mahowald AP. 1987. Multiple functions of segment polarity genes in *Drosophila*. *Dev Biol* **119:** 587–600.

Peters JM, McKay RM, McKay JP, Graff JM. 1999. Casein kinase I transduces Wnt signals. *Nature* **401:** 345–350.

Phillips BT, Kimble J. 2009. A new look at TCF and β-catenin through the lens of a divergent *C. elegans* Wnt pathway. *Dev Cell* **17:** 27–34.

Piao S, Lee SH, Kim H, Yum S, Stamos JL, Xu Y, Lee SJ, Lee J, Oh S, Han JK, et al. 2008. Direct inhibition of GSK3β by the phosphorylated cytoplasmic domain of LRP6 in Wnt/β-catenin signaling. *PLoS ONE* **3:** e4046.

Piddini E, Marshall F, Dubois L, Hirst E, Vincent JP. 2005. Arrow (LRP6) and Frizzled2 cooperate to degrade Wingless in *Drosophila* imaginal discs. *Development* **132:** 5479–5489.

Pinson KI, Brennan J, Monkley S, Avery BJ, Skarnes WC. 2000. An LDL-receptor-related protein mediates Wnt signalling in mice. *Nature* **407:** 535–538.

Povelones M, Nusse R. 2005. The role of the cysteine-rich domain of Frizzled in Wingless–Armadillo signaling. *EMBO J* **24:** 3493–3503.

Povelones M, Howes R, Fish M, Nusse R. 2005. Genetic evidence that *Drosophila* Frizzled controls planar cell polarity and Armadillo signaling by a common mechanism. *Genetics* **171:** 1643–1654.

Price MA. 2006. CKI, there's more than one: Casein kinase I family members in Wnt and Hedgehog signaling. *Genes Dev* **20:** 399–410.

Punchihewa C, Ferreira AM, Cassell R, Rodrigues P, Fujii N. 2009. Sequence requirement and subtype specificity in the high-affinity interaction between human Frizzled and Dishevelled proteins. *Protein Sci* **18:** 994–1002.

Qin Y, Li L, Pan W, Wu D. 2009. Regulation of phosphatidylinositol kinases and metabolism by Wnt3a and Dvl. *J Biol Chem* **284:** 22544–22548.

Reed SC. 1937. The inheritance and expression of Fused, a new mutation in the house mouse. *Genetics* **22:** 1–13.

Regimbald-Dumas Y, He X. 2011. Wnt signalling: What the X@# is WTX? *EMBO J* **30:** 1415–1417.

Rives AF, Rochlin KM, Wehrli M, Schwartz SL, DiNardo S. 2006. Endocytic trafficking of Wingless and its receptors, Arrow and DFrizzled-2, in the *Drosophila* wing. *Dev Biol* **293:** 268–283.

Robitaille J, MacDonald ML, Kaykas A, Sheldahl LC, Zeisler J, Dube MP, Zhang LH, Singaraja RR, Guernsey DL, Zheng B, et al. 2002. Mutant frizzled-4 disrupts retinal angiogenesis in familial exudative vitreoretinopathy. *Nat Genet* **32:** 326–330.

Rulifson EJ, Wu CH, Nusse R. 2000. Pathway specificity by the bifunctional receptor Frizzled is determined by affinity for Wingless. *Mol Cell* **6:** 117–126.

Schulte G, Bryja V. 2007. The Frizzled family of unconventional G-protein-coupled receptors. *Trends Pharmacol Sci* **28:** 518–525.

Schwarz-Romond T, Merrifield C, Nichols BJ, Bienz M. 2005. The Wnt signalling effector Dishevelled forms dynamic protein assemblies rather than stable associations with cytoplasmic vesicles. *J Cell Sci* **118:** 5269–5277.

Schwarz-Romond T, Fiedler M, Shibata N, Butler PJ, Kikuchi A, Higuchi Y, Bienz M. 2007a. The DIX domain of Dishevelled confers Wnt signaling by dynamic polymerization. *Nat Struct Mol Biol* **14:** 484–492.

Schwarz-Romond T, Metcalfe C, Bienz M. 2007b. Dynamic recruitment of axin by Dishevelled protein assemblies. *J Cell Sci* **120:** 2402–2412.

Seto ES, Bellen HJ. 2006. Internalization is required for proper Wingless signaling in *Drosophila melanogaster*. *J Cell Biol* **173:** 95–106.

Shafer B, Onishi K, Lo C, Colakoglu G, Zou Y. 2011. Vangl2 promotes Wnt/planar cell polarity-like signaling by antagonizing Dvl1-mediated feedback inhibition in growth cone guidance. *Dev Cell* **20:** 177–191.

Simons M, Gault WJ, Gotthardt D, Rohatgi R, Klein TJ, Shao Y, Lee HJ, Wu AL, Fang Y, Satlin LM, et al. 2009. Electrochemical cues regulate assembly of the Frizzled/Dishevelled complex at the plasma membrane during planar epithelial polarization. *Nat Cell Biol* **11:** 286–294.

Sotgia F, Williams TM, Cohen AW, Minetti C, Pestell RG, Lisanti MP. 2005. Caveolin-1-deficient mice have an increased mammary stem cell population with upregulation of Wnt/β-catenin signaling. *Cell Cycle* **4:** 1808–1816.

Springer TA. 1998. An extracellular β-propeller module predicted in lipoprotein and scavenger receptors, tyrosine kinases, epidermal growth factor precursor, and extracellular matrix components. *J Mol Biol* **283:** 837–862.

* Stamos JL, Weis WI. 2012. The β-catenin destruction complex. *Cold Spring Harb Perspect Biol* doi: 10.1101/cshperspect.a007898.

Szuts D, Freeman M, Bienz M. 1997. Antagonism between EGFR and Wingless signalling in the larval cuticle of *Drosophila*. *Development* **124:** 3209–3219.

Taelman VF, Dobrowolski R, Plouhinec JL, Fuentealba LC, Vorwald PP, Gumper I, Sabatini DD, De Robertis EM. 2010. Wnt signaling requires sequestration of glycogen synthase kinase 3 inside multivesicular endosomes. *Cell* **143:** 1136–1148.

Takada R, Satomi Y, Kurata T, Ueno N, Norioka S, Kondoh H, Takao T, Takada S. 2006. Monounsaturated fatty acid modification of Wnt protein: Its role in Wnt secretion. *Dev Cell* **11:** 791–801.

Tamai K, Semenov M, Kato Y, Spokony R, Liu C, Katsuyama Y, Hess F, Saint-Jeannet JP, He X. 2000. LDL-receptor-related proteins in Wnt signal transduction. *Nature* **407:** 530–535.

Tamai K, Zeng X, Liu C, Zhang X, Harada Y, Chang Z, He X. 2004. A mechanism for Wnt coreceptor activation. *Mol Cell* **13:** 149–156.

Tanneberger K, Pfister AS, Brauburger K, Schneikert J, Hadjihannas MV, Kriz V, Schulte G, Bryja V, Behrens J. 2011. Amer1/WTX couples Wnt-induced formation of $PtdIns_{4,5}P2$ to LRP6 phosphorylation. *EMBO J* **30:** 1433–1443.

Tauriello DV, Jordens I, Kirchner K, Slootstra JW, Kruitwagen T, Bouwman BA, Noutsou M, Rudiger SG, Schwamborn K, Schambony A, et al. 2012. Wnt/β-catenin signaling requires interaction of the Dishevelled DEP domain and C terminus with a discontinuous motif in Frizzled. *Proc Natl Acad Sci* **109:** E812–E820.

Tebben AJ, Schnur DM. 2011. Beyond rhodopsin: G protein-coupled receptor structure and modeling incorporating the β2-adrenergic and adenosine A(2A) crystal structures. *Methods Mol Biol* **672:** 359–386.

Tolwinski NS, Wehrli M, Rives A, Erdeniz N, DiNardo S, Wieschaus E. 2003. Wg/Wnt signal can be transmitted through arrow/LRP5,6 and Axin independently of Zw3/Gsk3β activity. *Dev Cell* **4:** 407–418.

Umbhauer M, Djiane A, Goisset C, Penzo-Mendez A, Riou JF, Boucaut JC, Shi DL. 2000. The C-terminal cytoplasmic Lys-thr-X-X-X-Trp motif in Frizzled receptors mediates Wnt/β-catenin signalling. *EMBO J* **19:** 4944–4954.

Valvezan AJ, Zhang F, Diehl JA, Klein PS. 2012. Adenomatous polyposis coli (APC) regulates multiple signaling pathways by enhancing glycogen synthase kinase-3 (GSK-3) activity. *J Biol Chem* **287:** 3823–3832.

van Amerongen R, Nawijn M, Franca-Koh J, Zevenhoven J, van der Gulden H, Jonkers J, Berns A. 2005. Frat is dispensable for canonical Wnt signaling in mammals. *Genes Dev* **19:** 425–430.

van Amerongen R, Mikels A, Nusse R. 2008. Alternative wnt signaling is initiated by distinct receptors. *Sci Signal* **1:** re9. doi: 10.1126/scisignal.135re9.

Vinson CR, Conover S, Adler PN. 1989. A *Drosophila* tissue polarity locus encodes a protein containing seven potential transmembrane domains. *Nature* **338:** 263–264.

Wan M, Li J, Herbst K, Zhang J, Yu B, Wu X, Qiu T, Lei W, Lindvall C, Williams BO, et al. 2011. LRP6 mediates cAMP generation by G protein-coupled receptors through regulating the membrane targeting of $G\alpha_s$. *Sci Signal* **4:** ra15.

Wehrli M, Dougan ST, Caldwell K, O'Keefe L, Schwartz S, Vaizel-Ohayon D, Schejter E, Tomlinson A, DiNardo S. 2000. *arrow* encodes an LDL-receptor-related protein essential for Wingless signalling. *Nature* **407:** 527–530.

Wheatley M, Wootten D, Conner M, Simms J, Kendrick R, Logan R, Poyner D, Barwell J. 2012. Lifting the lid on GPCRs: The role of extracellular loops. *Br J Pharmacol* **165:** 1688–1703.

Willert K, Brink M, Wodarz A, Varmus H, Nusse R. 1997. Casein kinase 2 associates with and phosphorylates dishevelled. *EMBO J* **16:** 3089–3096.

Willert K, Shibamoto S, Nusse R. 1999. Wnt-induced dephosphorylation of Axin releases β-catenin from the Axin complex. *Genes Dev* **13:** 1768–1773.

Wolf J, Palmby TR, Gavard J, Williams BO, Gutkind JS. 2008. Multiple PPPS/TP motifs act in a combinatorial fashion to transduce Wnt signaling through LRP6. *FEBS Lett* **582:** 255–261.

Wong LL, Adler PN. 1993. Tissue polarity genes of *Drosophila* regulate the subcellular location for prehair initiation in pupal wing cells. *J Cell Biol* **123:** 209–221.

Wong HC, Mao J, Nguyen JT, Srinivas S, Zhang W, Liu B, Li L, Wu D, Zheng J. 2000. Structural basis of the recognition of the dishevelled DEP domain in the Wnt signaling pathway. *Nat Struct Biol* **7:** 1178–1184.

Wong HC, Bourdelas A, Krauss A, Lee HJ, Shao Y, Wu D, Mlodzik M, Shi DL, Zheng J. 2003. Direct binding of the PDZ domain of Dishevelled to a conserved internal sequence in the C-terminal region of Frizzled. *Mol Cell* **12:** 1251–1260.

Wu CH, Nusse R. 2002. Ligand receptor interactions in the Wnt signaling pathway in *Drosophila*. *J Biol Chem* **277:** 41762–41769.

Wu G, Huang H, Garcia Abreu J, He X. 2009. Inhibition of GSK3 phosphorylation of β-catenin via phosphorylated PPPSPXS motifs of Wnt coreceptor LRP6. *PLoS ONE* **4:** e4926. doi: 10.1371/journal.pone.0004926.

Yamamoto H, Kishida S, Uochi T, Ikeda S, Koyama S, Asashima M, Kikuchi A. 1998. Axil, a member of the Axin family, interacts with both glycogen synthase kinase 3β and β-catenin and inhibits axis formation of *Xenopus* embryos. *Mol Cell Biol* **18:** 2867–2875.

Yamamoto H, Kishida S, Kishida M, Ikeda S, Takada S, Kikuchi A. 1999. Phosphorylation of Axin, a Wnt signal negative regulator, by glycogen synthase kinase-3β regulates its stability. *J Biol Chem* **274:** 10681–10684.

Yamamoto A, Nagano T, Takehara S, Hibi M, Aizawa S. 2005. Shisa promotes head formation through the inhibition of receptor protein maturation for the caudalizing factors, Wnt and FGF. *Cell* **120:** 223–235.

Yamamoto H, Komekado H, Kikuchi A. 2006. Caveolin is necessary for Wnt-3a-dependent internalization of LRP6 and accumulation of β-catenin. *Dev Cell* **11:** 213–223.

Yamamoto H, Sakane H, Yamamoto H, Michiue T, Kikuchi A. 2008. Wnt3a and Dkk1 regulate distinct internalization pathways of LRP6 to tune the activation of β-catenin signaling. *Dev Cell* **15:** 37–48.

Yanfeng WA, Tan C, Fagan RJ, Klein PS. 2006. Phosphorylation of frizzled-3. *J Biol Chem* **281:** 11603–11609.

Yost C, Farr GH, Pierce SB III, Ferkey DM, Chen MM, Kimelman D. 1998. GBP, an inhibitor of GSK-3, is implicated in *Xenopus* development and oncogenesis. *Cell* **93:** 1031–1041.

Yu A, Xing Y, Harrison SC, Kirchhausen T. 2010. Structural analysis of the interaction between Dishevelled2 and clathrin AP-2 adaptor, a critical step in noncanonical Wnt signaling. *Structure* **18:** 1311–1320.

Zeng L, Fagotto F, Zhang T, Hsu W, Vasicek TJ, Perry WL, Lee JJ III, Tilghman SM, Gumbiner BM, Costantini F. 1997. The mouse *Fused* locus encodes Axin, an inhibitor of the Wnt signaling pathway that regulates embryonic axis formation. *Cell* **90:** 181–192.

Zeng X, Tamai K, Doble B, Li S, Huang H, Habas R, Okamura H, Woodgett J, He X. 2005. A dual-kinase mechanism for Wnt co-receptor phosphorylation and activation. *Nature* **438:** 873–877.

Zeng X, Huang H, Tamai K, Zhang X, Harada Y, Yokota C, Almeida K, Wang J, Doble B, Woodgett J, et al. 2008. Initiation of Wnt signaling: Control of Wnt coreceptor Lrp6 phosphorylation/activation via Frizzled, Dishevelled and Axin functions. *Development* **135:** 367–375.

The β-Catenin Destruction Complex

Jennifer L. Stamos and William I. Weis

Departments of Structural Biology and Molecular and Cellular Physiology, Stanford University School of Medicine, Stanford, California 94305

Correspondence: bill.weis@stanford.edu

The Wnt/β-catenin pathway is highly regulated to insure the correct temporal and spatial activation of its target genes. In the absence of a Wnt stimulus, the transcriptional coactivator β-catenin is degraded by a multiprotein "destruction complex" that includes the tumor suppressors Axin and adenomatous polyposis coli (APC), the Ser/Thr kinases GSK-3 and CK1, protein phosphatase 2A (PP2A), and the E3-ubiquitin ligase β-TrCP. The complex generates a β-TrCP recognition site by phosphorylation of a conserved Ser/Thr-rich sequence near the β-catenin amino terminus, a process that requires scaffolding of the kinases and β-catenin by Axin. Ubiquitinated β-catenin is degraded by the proteasome. The molecular mechanisms that underlie several aspects of destruction complex function are poorly understood, particularly the role of APC. Here we review the molecular mechanisms of destruction complex function and discuss several potential roles of APC in β-catenin destruction.

β-Catenin, first discovered as a part of the adherens junction complex with cadherin and α-catenin (Ozawa et al. 1989), also serves as a transcriptional coactivator of Wnt target gene expression. In the absence of an extracellular Wnt stimulus, the nonjunctional pool of cytoplasmic β-catenin is targeted for proteolysis by a large multiprotein assembly termed the β-catenin destruction complex (Fig. 1). Wnt binding to cell–surface receptors turns off β-catenin destruction, and the resulting stabilized β-catenin translocates to the nucleus and binds to TCF/Lef proteins to activate Wnt target gene transcription (Behrens et al. 1996; Molenaar et al. 1996). The importance of β-catenin destruction first came to light with the discovery that mutations of the adenomatous polyposis coli protein associated with familial and sporadic colon cancers produce accumulation of β-catenin and inappropriate activation of target genes (Munemitsu et al. 1995; Korinek et al. 1997; Morin et al. 1997; Rubinfeld et al. 1997a; Clevers 2006). The precise temporal activation or inhibition of Wnt target genes necessary during development may explain the need for the rapid and sensitive responses of β-catenin protein levels.

The destruction complex is likely a dynamic multiprotein assembly, but its core components include, in addition to β-catenin itself, the Ser/Thr kinases glycogen synthase kinase 3 (GSK-3) and casein kinase 1 (CK1), the scaffolding protein Axin, the adenomatous polyposis coli (APC) protein, and the E3-ubiquitin ligase

Cite this article as *Cold Spring Harb Perspect Biol* doi: 10.1101/cshperspect.a007898

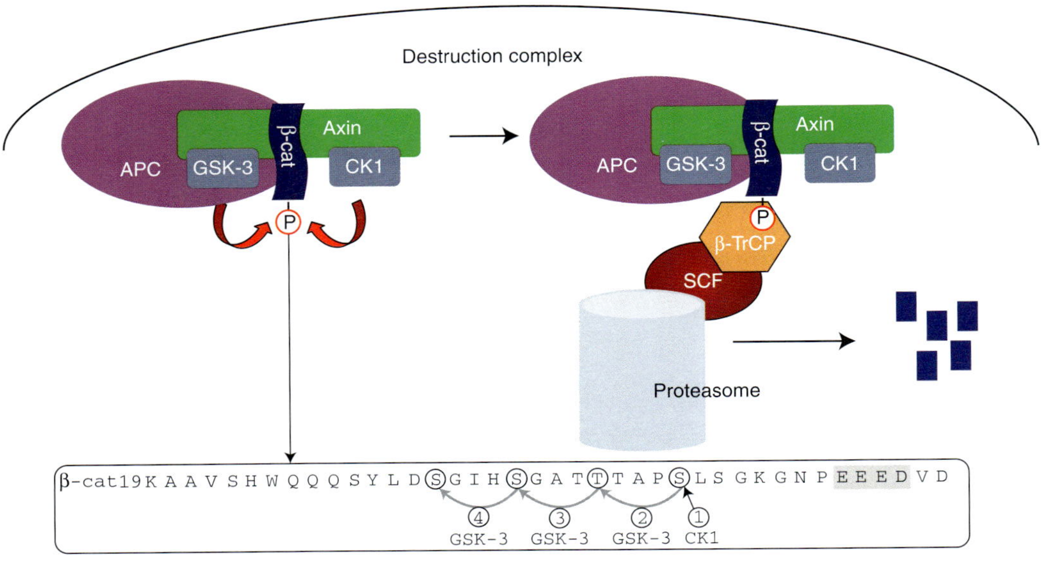

Figure 1. Overview of the destruction complex. The phosphorylated sequence in the amino terminus of β-catenin is shown, with CK1 and GSK-3 sites circled. The acidic cluster that primes CK1 phosphorylation is highlighted in gray. SCF, the Skp1/cullin/F-box complex.

β-TrCP (Fig. 1). Protein phosphatase 2A (PP2A) also associates with the complex (Hsu et al. 1999; Seeling et al. 1999; Ratcliffe et al. 2000; Yamamoto et al. 2001). Mutations in destruction complex components associated with various cancers result in inappropriate stabilization of β-catenin and Wnt target gene expression in the absence of a Wnt stimulus (Dominguez et al. 1995; Jiang and Struhl 1998; Marikawa and Elinson 1998; Peters et al. 1999; Liu et al. 2000; Satoh et al. 2000; Heisenberg et al. 2001). For example, in the absence of a Wnt signal, the half-life of β-catenin was found to be 50 min in AtT20 cells, but 3 h in the colon cancer cell line SW480, which bears a mutated APC protein (Munemitsu et al. 1996).

Although the core components of the β-catenin destruction complex and their binding interactions are understood in molecular detail, a complete molecular understanding of β-catenin destruction has been elusive. In particular, the essential role of APC in this process, as well as the relationship between the kinase/scaffold complexes and the ubiquitination machinery, remain unclear. Here we review what is known about these core destruction complex functions, focusing on potential molecular mechanisms.

β-CATENIN DESTRUCTION OCCURS VIA PROTEASOMAL DEGRADATION

The actual destruction of β-catenin is accomplished by the proteasome, which proteolytically degrades β-catenin (Aberle et al. 1997; Orford et al. 1997). β-Catenin is presented to the proteasome through its interaction with the F-box containing E3-ligase protein β-TrCP, an adaptor protein that forms a complex with the Skp1/Cullin machinery to attach ubiquitin to its binding partners (Hart et al. 1999; Kitagawa et al. 1999; Latres et al. 1999; Liu et al. 1999; Winston et al. 1999). The binding site for β-TrCP on β-catenin is a short peptide that encompasses two conserved serines, Ser33 and Ser37, which when phosphorylated interact with the β-propeller domain of β-TrCP (Orford et al. 1997; Wu et al. 2003). Cancer cells such as the colon cancer line HCT116 in which the key phosphorylated serines or threonines in β-catenin are mutated or deleted have inappropriately high levels of β-catenin that activate Wnt target genes (Morin et al. 1997).

PHOSPHORYLATION BY THE DESTRUCTION COMPLEX

The phosphorylation of β-catenin Ser33 and Ser37 by GSK-3 is the best-characterized function of the destruction complex. CK1 initially phosphorylates Ser45 of β-catenin (Amit et al. 2002; Liu et al. 2002), and is targeted to this site by a cluster of acidic residues located seven amino acids carboxy-terminally (Fig. 1) (Marin et al. 2003). GSK-3 phosphorylates residues in the sequence **S/T**-*X*-*X*-*X*-pS/pT, in which the bold residue represents the phosphorylation site, and *X* is any amino acid (Fiol et al. 1987; Frame et al. 2001). Phosphorylated Ser45 primes GSK-3-mediated phosphorylation of T41, which in turn primes successive phosphorylation of S37 and S33 by GSK-3 to generate the β-TrCP-binding site (Fig. 1) (Yost et al. 1996; Hagen et al. 2002; Liu et al. 2002; Wu and He 2006).

A number of CK1 isoforms can be expressed, including a membrane-tethered version (CK1γ). The α isoform consists almost solely of the catalytic kinase domain, whereas the δ and ε isoforms have carboxy-terminal extensions that can be autophosphorylated, which leads to autoinhibition (Cegielska et al. 1998). The cytoplasmic isoforms α, δ, and ε can bind to Axin, and purified CK1α, δ, or ε can phosphorylate β-catenin at S45 (Amit et al. 2002). RNAi experiments in both mammalian cells and *Drosophila* and genetic experiments in mice indicate that CK1α is the principal isoform responsible for phosphorylating Ser45 (Liu et al. 2002; Matsubayashi et al. 2004; Elyada et al. 2011).

Mammalian GSK-3 is expressed from two different genes, termed GSK-3α and GSK-3β, which differ principally by the presence of an extended, glycine-rich amino terminus in GSK-3α (Woodgett 1990). The single GSK-3 in *Drosophila* more closely resembles the shorter GSK-3β. Although isoform-specific roles have been observed in other pathways (e.g., McManus et al. 2005; Patel et al. 2008; Kaidanovich-Beilin et al. 2009), GSK-3α and β appear to function redundantly in β-catenin destruction (Doble et al. 2007). Many studies of the vertebrate Wnt pathway have focused on GSK-3β. Doble et al. (2007) noted that this is likely an historical bias that arose from apparent differences in the ability of the two isoforms to rescue GSK-3 mutants in *Drosophila* (Siegfried et al. 1992; Ruel et al. 1993), but expression of the two forms was not equal in those studies. GSK-3 can be phosphorylated at a tyrosine on its activation loop (Tyr279/Tyr216 in α/β isoforms), which increases enzymatic activity approximately five-fold, a modest increase compared to the activation of other kinases (Dajani et al. 2003). Mutating this residue to Phe appears to have little (Itoh et al. 1995; Dajani et al. 2003) or moderate (Doble et al. 2007) effects on β-catenin activity.

In addition to its role in the Wnt/β-catenin pathway, GSK-3 has regulatory roles in many other cellular processes, including glycogen biosynthesis, microtubule stability, cell-cycle control, and inflammatory pathways (Frame and Cohen 2001). GSK-3 activity is regulated in several of these pathways by phosphorylation of a serine (Ser21/Ser9, α/β isoforms) present in a conserved sequence near the amino terminus of the protein. When phosphorylated by other serine/threonine kinases, including Akt and p70S6K (Cross et al. 1995; Peyrollier et al. 2000), this phosphopeptide inhibits GSK-3 activity. Although autoinhibition by the phosphorylated amino-terminal peptide is important in non-Wnt GSK-3 functions, mutation of the Ser21/Ser9 does not affect Wnt signaling, and this residue does not become phosphorylated on Wnt stimulation (Ding et al. 2000; McManus et al. 2005). Furthermore, the binding of GSK-3 to Axin, discussed below, probably targets a fraction of cytoplasmic GSK-3 to the destruction complex and insulates GSK-3 from regulatory proteins not involved in Wnt signaling. For example, Ser21/Ser9 of Axin-bound GSK-3 does not become phosphorylated during insulin signaling and activation of Akt (Ding et al. 2000).

AXIN, THE CENTRAL PHOSPHORYLATION SCAFFOLD

Axin was discovered as the product of the mouse *Fused* locus, whose disruption caused the duplication of the embryonic body axis (Zeng et al. 1997). *Drosophila* possess only one Axin gene, but vertebrates and nematodes have two, Axin and Axin 2/Conductin. In the mouse, these two isoforms appear to be mechanistically interchangeable, and vary only in their patterns of transcription. Axin is more widely expressed, whereas Axin 2/Conductin has a more restricted expression pattern that is controlled in part by Wnt signaling (Jho et al. 2002; Leung et al. 2002; Chia and Costantini 2005). A study of hepatocellular carcinomas showed that Axin mutations lead to inappropriate β-catenin-mediated transcription (Satoh et al. 2000). Overexpression of Axin in colon cancer cells can rescue the accumulation of β-catenin that results from APC mutation (Behrens et al. 1998; Hart et al. 1998; Kishida et al. 1998; Sakanaka et al. 1998).

Axin is a largely unstructured, flexible protein (Spink et al. 2000; Noutsou et al. 2011) that contains CK1, GSK-3, and β-catenin-binding sites (Fig. 2) (Ikeda et al. 1998; Yamamoto et al. 1999; Rubinfeld et al. 2001; Liu et al. 2002). The region carboxy-terminal to the β-catenin and GSK-3-binding sites has been reported to bind to the catalytic domain of the PP2A or to the PR61 regulatory subunit of PP2A (Fig. 2) (Hsu et al. 1999; Ikeda et al. 2000; Yamamoto et al. 2001), as well as the phosphatase PP1 (Luo et al. 2007). The simultaneous binding of Axin to a kinase and β-catenin enforces close proximity and thereby increases the effective concentration of enzyme and substrate. This "scaffold effect" is seen with purified proteins; phosphorylation of β-catenin by CK1 and GSK-3β is greatly enhanced by the presence of Axin (Ikeda et al. 1998; Dajani et al. 2003; Ha et al. 2004). Crystal structures have detailed the binding interactions between Axin and β-catenin (Xing et al. 2003), and Axin and GSK-3 (Dajani et al. 2003). The structure of the GSK-3β–Axin complex shows that Axin binds to a region far removed from the catalytic site (Dajani et al. 2003), and we have found that Axin has no effect on the intrinsic catalytic capability of GSK-3 (unpubl.). Likewise, Axin binds to a region of CK1 far from the catalytic site, and a pure scaffolding effect without enhancement of intrinsic catalytic activity has been shown for this interaction (Sobrado et al. 2005).

In addition to the β-catenin, kinase, and PP2A binding sites, Axin also possesses two

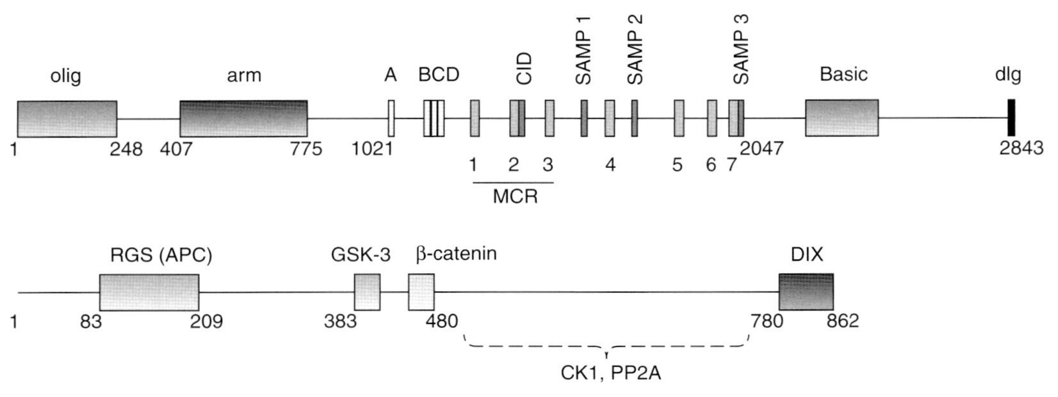

Figure 2. Primary structures of human Axin and APC. Key protein–protein interaction regions are indicated. Structured domains are RGS, regulator of G-protein signaling homology; DIX, domain common to Dishevelled and Axin; olig, coiled-coil dimerization domain of human APC; and arm, armadillo repeat domain. 15-mer repeats are marked A–D; 20-mer repeats are marked 1–7. The mutation cluster region (MCR) is the site of the majority of APC truncations found in colorectal cancers. Dashed lines indicate that mapping of the CK1 and PP2A sites on Axin are based on deletion studies and have not been verified with purified proteins.

 Cite this article as *Cold Spring Harb Perspect Biol* doi: 10.1101/cshperspect.a007898

folded, structured domains. Near the amino terminus, the RGS domain, named for its homology to the regulators of G-protein signaling protein family, binds to the APC protein (Behrens et al. 1998; Kishida et al. 1998; Spink et al. 2000). The interaction with APC is required for efficient phosphorylation of APC (see below) (Fagotto et al. 1999; Ikeda et al. 2000). At the carboxyl terminus, the DIX domain mediates homomeric complex formation, as well as binding to other DIX domain-containing proteins, most notably Dishevelled (Fig. 2) (Fagotto et al. 1999; Kishida et al. 1999; Julius et al. 2000; Capelluto et al. 2002; Schwarz-Romond et al. 2007a; Fiedler et al. 2011; Liu et al. 2011), an interaction crucial for transducing a Wnt signal to turn off β-catenin destruction (MacDonald and He 2012).

Axin is a key control point for β-catenin destruction by virtue of its ability to promote efficient phosphorylation of β-catenin and APC; indeed, the GSK-3 and β-catenin-binding sites are essential for Axin function (Fagotto et al. 1999). In addition, deletion of the Axin RGS domain in both vertebrates and *Drosophila* compromises β-catenin destruction in vivo (Fagotto et al. 1999; Peterson-Nedry et al. 2008), demonstrating the importance of the Axin–APC interaction. However, the APC interaction mediated by the RGS domain is dispensable when Axin is overexpressed in SW480 colon cancer cells (see below) (Kohler et al. 2009; Roberts et al. 2011). Measurements of destruction complex components in mammalian kidney epithelial cells and *Xenopus* oocytes gave different absolute and relative concentrations of APC and Axin (Salic et al. 2000; Lee et al. 2003; Tan et al. 2012), suggesting that Wnt signaling operates over a range of component concentrations that vary with cell type. Studies in *Drosophila* have suggested that precise APC levels are important in modulating Wnt responses; a twofold change in APC2 was shown to profoundly affect signaling in photoreceptor development, indicating that APC levels are likely tuned to optimize responses to morphogen gradients (Benchabane et al. 2008). Thus, although Axin is clearly a key component of the destruction complex, the relative levels of all components appear to be critical for both homeostasis and responsiveness to Wnt.

THE ROLE OF APC IN β-CATENIN DESTRUCTION

APC is an ~310-kDa protein whose central ~1000 amino acids contain short peptide motifs that bind to β-catenin or Axin (Fig. 2). The APC protein in most organisms, including *Drosophila*, expresses in two isoforms, designated APC and APC2. In addition to the central β-catenin- and Axin-binding region, the amino-terminal portion of APC contains a dimerization domain in vertebrates (Day and Alber 2000) and an armadillo repeat (arm) domain. The arm domain interacts with a number of cytoskeletal regulators (Kawasaki et al. 2000; Jimbo et al. 2002; Watanabe et al. 2004; Breitman et al. 2008), and the B56 regulatory subunit of PP2A (Seeling et al. 1999). The carboxy-terminal region of APC1 features a microtubule interaction region (McCartney and Nathke 2008) that is dispensable for β-catenin destruction (Smits et al. 1999). These activities are thought to be important for other aspects of APC biology, including roles in spindle formation, kinetochore attachment, and maintenance of microtubule stability (McCartney and Nathke 2008), but it is not known if interactions with cytoskeletal regulators are important to the function of APC in β-catenin destruction.

APC proteins contain multiple independent binding sites for β-catenin, comprising homologous repeats of either 15 or 20 amino acids (Fig. 2) (Eklof Spink et al. 2001). Human APC contains four 15-mers (designated 15RA-D) and seven 20-mers (20R1-7) (Fig. 3). The 20-mer repeats can be phosphorylated by GSK-3 and CK1, which enhances affinity for β-catenin up to 1500-fold (Fig. 3) (Rubinfeld et al. 1996; Rubinfeld et al. 1997b; Ha et al. 2004; Liu et al. 2006). Phosphorylation of the 20-mers in vivo likely requires Axin, as a scaffold effect can be shown using purified Axin and the kinases (Ikeda et al. 2000; Su et al. 2008). The level of phosphorylation heterogeneity is not known under physiological conditions, however. The stoichiometry of the APC–β-catenin interaction is

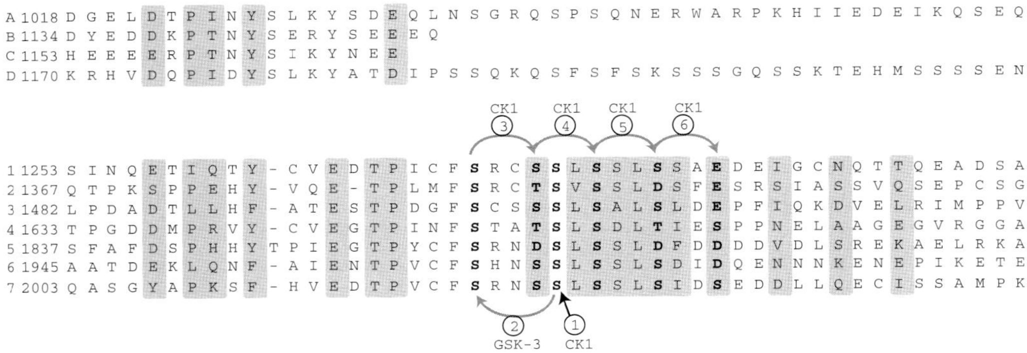

Figure 3. Alignment of β-catenin-binding sequences of human APC, showing the 15-mers (*top*) and 20-mers (*bottom*). Residues that directly contact β-catenin, as observed in crystal structures of β-catenin bound to 15RA (Spink et al. 2001) and 20R3 (Ha et al. 2004; Xing et al. 2004), are shaded. 20-mer residues phosphorylated by CK1 and GSK-3 are shown in bold, with the phosphorylation order determined by priming rules for each kinase. (From Ha et al. 2004; adapted, with express permission, from the authors.)

also unknown, and may vary depending on the status of Wnt signaling. An early experiment showed that APC protein extracted from cells appears to be bound to only one or two molecules of β-catenin (Rubinfeld et al. 1995).

Interspersed with the β-catenin-binding 20-mer repeats are Axin-binding sequences of ~16 amino acids, termed SAMP repeats for the conserved amino acid sequence present in the Axin-binding sites of mammalian APC (Fig. 2) (Behrens et al. 1998; Spink et al. 2000). APCs contain two or three SAMP repeats, depending on species and isoform. Mammalian cell culture studies showed that the APC–Axin interaction contributes to β-catenin destruction (Behrens et al. 1998; Hart et al. 1998), and at least one SAMP repeat in combination with at least two β-catenin-binding repeats were required (Rubinfeld et al. 1997a). In mice, at least one SAMP repeat is needed to regulate Wnt signaling (Smits et al. 1999). Truncations of APC that remove all SAMP repeats are oncogenic, although certain of these truncations may retain some level of β-catenin regulation (Smits et al. 1999; Kohler et al. 2009).

Experiments that rely on APC truncations to remove the SAMP repeats by necessity also remove many of the 20-mer β-catenin-binding sites, which confounds interpretations of the role of the APC–Axin interaction. Peifer and colleagues generated a *Drosophila* APC2 variant in which the SAMP repeats were deleted cleanly rather than by truncation, so that all β-catenin-binding sites were retained. This mutant APC was nonfunctional in animals, although β-catenin destruction and TCF signaling down-regulation were retained in SW480 cells (Roberts et al. 2011). Curiously, Roberts et al. 2011 found that APC2 lacking SAMP repeats colocalized with Axin, but the mutant APC2 did not coimmunoprecipitate with Axin, suggesting a weak and indirect interaction. It is possible that β-catenin itself bridged Axin and the mutant APC2, because it can interact with Axin and the APC 15-mer or nonphosphorylated 20-mer repeats simultaneously (Ha et al. 2004).

The molecular roles of APC in β-catenin destruction have not been determined decisively. Several distinct and not necessarily mutually exclusive functions for APC in regulating β-catenin have been proposed, which we consider in the following subsections.

Phosphorylation of β-Catenin

The ability of β-catenin, Axin, and APC to interact with one another might suggest a stabilizing role for APC in promoting β-catenin phosphorylation by Axin-bound kinases. Bioinformatic analyses, nuclear magnetic resonance (NMR) and circular dichroism (CD) spectroscopy, and proteolytic sensitivity experiments indicate that, like the central region of Axin (see above), the central portion of APC is intrinsically unstructured (Li and Nathke 2005; Liu et al. 2006). Moreover, biochemical and structural studies have shown that Axin and nonphosphorylated APC repeats can bind simultaneously to β-catenin (Ha et al. 2004). These properties imply that APC can scaffold the Axin–β-catenin interaction. However, addition of purified APC fragments that contain β-catenin and Axin-binding sites to purified β-catenin, Axin, GSK-3β, and CK1 does not enhance β-catenin phosphorylation (Su et al. 2008; JL Stamos and WI Weis, unpubl.).

Measurements of two mammalian kidney cell lines have shown that Axin concentrations are in the range 110–150 nM; total and cytosolic β-catenin concentrations are in the hundreds and tens of nM range, respectively (Tan et al. 2012). APC is found at 12–40× lower concentrations than Axin in these cells. In contrast, in *Xenopus* oocyte extracts Axin is present at sub-nM concentrations, roughly 2000× lower than β-catenin and 5000× lower than APC (Salic et al. 2000; Lee et al. 2003). Given these concentrations and the weak affinity (μM-range dissociation constants) of the Axin-APC, Axin-β-catenin, and nonphosphorylated APC-β-catenin interactions (Spink et al. 2000; Choi et al. 2006; Liu et al. 2006), it could be argued that the stabilizing effect of a three-way interaction network helps to insure efficient β-catenin phosphorylation in vivo. However, APC truncations that remove the Axin-binding SAMP repeats do not prevent β-catenin phosphorylation (Sadot et al. 2002; Yang et al. 2006), indicating that APC does not enhance Axin-mediated phosphorylation of β-catenin.

Klein and colleagues reported recently that addition of a large fragment of APC encompassing the β-catenin and Axin-binding region enhances phosphorylation of β-catenin in the presence or absence of Axin, and also promotes phosphorylation of other GSK-3 substrates (Valvezan et al. 2012). These findings would seem to contradict those of (Su et al. 2008), and there is at present no evidence for direct APC–GSK-3 interactions required for scaffolding or for allosteric activation. Further studies with purified proteins will be needed to assess the molecular basis of these findings.

Destruction Complex Cycling

An alternative role for APC in β-catenin destruction was suggested by the observation that, in contrast to the simultaneous binding of nonphosphorylated APC repeats and Axin to β-catenin, phosphorylation of the APC 20-mer repeats enhances their affinity by binding to a surface of β-catenin that overlaps the Axin-binding site (Fig. 4); the higher affinity of the phosphorylated 20-mer versus Axin enables displacement of Axin from β-catenin (Xing et al. 2003; Ha et al. 2004; Xing et al. 2004). Kimelman, Xu, and colleagues (Xing et al. 2003) proposed that CK1 and GSK-3 phosphorylate APC after phosphorylating β-catenin; the phosphorylated APC 20-mer would displace β-catenin from Axin, freeing Axin to interact with another β-catenin molecule. This model also postulates that PP2A, which has been reported to bind to Axin and APC (Hsu et al. 1999; Seeling et al. 1999; Yamamoto et al. 2001) and can modulate APC phosphorylation (Ikeda et al. 2000), dephosphorylates β-catenin-bound APC to reset the system for another round of β-catenin phosphorylation and release.

The Kimelman/Xu model is appealing in its explanation of the roles of conserved APC 20-mer phosphorylation sites, the competition between phosphorylated APC and Axin for β-catenin, and the presence of PP2A in the destruction complex, but there are several caveats. First, phosphorylation of APC would have to occur after β-catenin phosphorylation, because phosphorylated APC would prevent Axin from binding to β-catenin. Given the unstructured nature of Axin and APC, it is difficult to envision how such an ordered sequence could be achieved at a molecular level—it seems unlikely that random collisions of these flexible scaffold proteins could produce a strict order of interaction of β-catenin, then APC, with the Axin-bound kinases. A stochastic mechanism would be physically plausible but inefficient. Second, the postulated

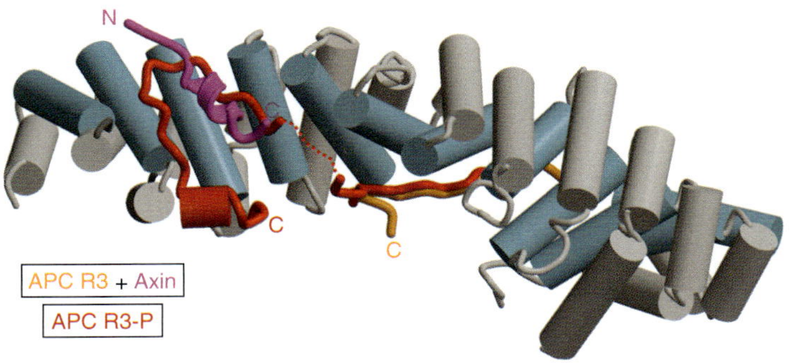

Figure 4. Competition between phosphorylated APC and Axin for β-catenin. Crystal structures of nonphosphorylated APC R3 (orange) (Ha et al. 2004), Axin (purple) (Xing et al. 2003), and phosphorylated APC R3-P (red) (Ha et al. 2004; Xing et al. 2004) bound to the armadillo repeat domain of β-catenin (blue and gray) were superimposed to show the relationship between the bound ligands. The portion of APC that becomes ordered on phosphorylation by CK1 and GSK-3 (compare red and orange structures) binds to an overlapping surface with Axin, and biochemical studies show that these two ligands compete for binding. This competition is the basis for models suggesting that Axin is displaced from β-catenin when the Axin-bound kinases phosphorylate APC.

role of PP2A is not supported by in vitro experiments with purified PP1, whose catalytic domain is very similar to that of PP2A, which showed that phosphorylated APC repeat 3 bound to β-catenin could not be dephosphorylated over 3 h (Ha et al. 2004). This would be inconsistent with the measured half-life of β-catenin (Munemitsu et al. 1996). Equivalent experiments have not been performed with heterotrimeric PP2A, however, and the effect of PP2A interactions with Axin and/or APC have not been assessed in this context. Finally, Peifer and colleagues showed that an APC2 construct in which only the highest affinity 20-mer repeat is retained is unable to promote β-catenin destruction (Roberts et al. 2011), indicating that high-affinity β-catenin binding by phosphorylated APC is not sufficient for destruction complex function.

It is important to note that there is no evidence that phosphorylated β-catenin must dissociate from the destruction complex to be degraded. Indeed, a ubiquitin ligase is part of the destruction complex (Hart et al. 1999; Liu et al. 1999; Major et al. 2007), and there is evidence that other components of the ubiquitination machinery physically associate with the proteasome in other systems (Crosas et al. 2006; Lee et al. 2011). It has been suggested that unfolding of β-catenin by proteasome-associated chaperones provides the energy to remove β-catenin from phosphorylated APC (Ha et al. 2004), which would then allow PP2A associated with the complex to dephosphorylate the APC 20-mers. Indeed, proteasomal activity has been shown to selectively remove and degrade individual components of multiprotein complexes (e.g., see Rape et al. 2001; Shi et al. 2008).

Interfacing to the Ubiquitin/Proteasome Pathway

APC truncations present in several colon cancer cell lines, including SW480, DLD-1, and HT-29, do not prevent β-catenin phosphorylation, but ablate or strongly reduce ubiquitination of β-catenin (Sadot et al. 2002; Yang et al. 2006; Su et al. 2008). Ubiquitination requires the presence of APC 20-mer repeat 2 (20R2), which, importantly, does not bind to β-catenin (Liu et al. 2006; Kohler et al. 2008). Furthermore, transfection of SW480 cells with a construct of APC truncated after 20R3 can still down-regulate β-catenin in a cell culture system (Yang et al. 2006; Kohler et al. 2008), whereas deletion of 20R2 and the strongly conserved sequence that follows it, designated the "catenin inhibitory domain" (CID; also called "sequence B" by Peifer and coworkers), prevents β-catenin destruction (Kohler et al. 2009; Roberts et al. 2011).

 Cite this article as *Cold Spring Harb Perspect Biol* doi: 10.1101/cshperspect.a007898

These data show that the 20R2-CID region of APC has an essential role in promoting β-catenin ubiquitination that is mechanistically downstream from β-catenin phosphorylation, and is independent of β-catenin-binding activity. Su et al. (2008) found that β-TrCP did not coimmunoprecipitate β-catenin from cell lines bearing APCs truncated before 20R2, but coprecipitation was restored by transfection with wild-type APC, suggesting that 20R2-CID mediates interactions with β-TrCP, either directly or through other proteins.

An alternative possibility for the essential role of 20R2-CID in β-catenin ubiquitination comes from the observation that in the absence of APC, Ser33/Ser37 phosphorylated β-catenin was rapidly dephosphorylated in vitro by PP2A, whereas full-length APC or APC fragments containing SAMP repeats could protect β-catenin from PP2A (Su et al. 2008). This suggests that APC prevents PP2A from accessing the amino-terminal phosphorylation sites on β-catenin (Su et al. 2008). Su et al. (2008) proposed that phosphorylated APC spatially orients phospho-β-catenin to prevent its dephosphorylation, although it is unclear how this could work given the unstructured, flexible nature of both APC and the amino-terminal region of β-catenin. The protection occurs only with phosphorylated APC, and the investigators argue APC must bind to β-catenin sufficiently tightly to protect the β-TrCP recognition sequence of β-catenin from PP2A. However, the protective effect of phosphorylated APC was observed even under conditions in which unphosphorylated and phosphorylated APC were equivalently occupied by β-catenin, so enhanced affinity would not appear to produce this result. Instead, binding of the phosphorylated APC 20-mers to β-catenin (Ha et al. 2004; Xing et al. 2004) might create a novel interaction surface for PP2A that inhibits its action on the β-catenin amino terminus.

Sequestration of β-Catenin

Comparison of β-catenin localization in wild-type and APC mutant cells suggested that APC and Axin can retain β-catenin in the cytoplasm (Tolwinski and Wieschaus 2001; Brocardo et al. 2005; Krieghoff et al. 2006; McCartney et al. 2006). Both Axin and APC contain CRM-1-dependent nuclear export signal sequences, and have been reported to mediate nuclear-cytoplasmic shuttling of β-catenin (Rosin-Arbesfeld et al. 2000; Henderson and Fagotto 2002; Cong and Varmus 2004). However, direct observation of intercompartmental movement using fluorescence recovery after photobleaching showed that APC and Axin enriched cytoplasmic β-catenin in a CRM-1 independent manner, and neither increased the rate of nuclear-cytoplasmic shuttling of β-catenin, suggesting that these proteins retain β-catenin in the cytosol (Krieghoff et al. 2006).

Cytoplasmic sequestration of β-catenin would insure that no signaling occurs before its actual destruction. Peifer and colleagues tested the ability of various *Drosophila* APC2 mutants to rescue β-catenin destruction in SW480 cells, and their effects in flies lacking APC2 only or both APC1 and APC2. Note that APC2 is the major form in fly embryos, but in single APC2 mutants, APC1 provides residual activity such that levels of β-catenin are only slightly elevated, producing effects on cell fate but not leading to complete nonviability (McCartney et al. 1999). By deleting R2 and/or the CID to prevent β-catenin destruction, Roberts et al. (2011) could assess the function of β-catenin-binding activity of the 15-mers and the other 20-mer repeats. An APC2 mutant lacking all 20-mers except for the high-affinity 20R3 was able to reduce nuclear β-catenin enrichment, consistent with a role in cytoplasmic retention. Likewise, Kohler et al. (2009) found that cell lines overexpressing a truncated human APC that included 20R3 displayed down-regulation of β-catenin, whereas removal of 20R3 (but retention of 20R2-CID) produced relatively poor down-regulation. Thus, a lack of high-affinity β-catenin-binding sites impairs down-regulation, consistent with the importance of cytoplasmic retention.

Roberts et al. (2011) found that an APC variant lacking all 20-mers displayed cytoplasmic β-catenin retention activity, presumably owing to the presence of the 15-mers, and could rescue cell-fate defects in single APC2 mutants, although it failed to rescue lethality of double APC1/APC2 mutant flies. Importantly, removal of all 15- and 20-mer repeats destroyed the ability to rescue the single APC2 mutant flies. These data suggest that cytoplasmic retention is an important function of APC, but that no one β-catenin-binding repeat is essential for this purpose. These experiments show that retention activity has at least an important modulatory role, but an APC variant containing only 20R2 and the CID, but none of the other β-catenin-binding repeats, would have to be tested to confirm an absolute requirement for β-catenin binding. The apparent need for cytoplasmic retention is consistent with the observation that colon tumors bearing mutant APCs carry at least one allele encoding a truncated protein. The "just right" model hypothesizes that these mutations are selected to provide sufficient destruction activity to avoid programmed cell death resulting from excessive β-catenin-mediated signaling (Albuquerque et al. 2002); buffering of free β-catenin levels by APC would provide the needed activity.

Ha et al. (2004) proposed that the presence of low- and high-affinity binding sites on APC provides a wide dynamic range for sequestering cytoplasmic β-catenin under different conditions. Binding of APC and Axin via the SAMP-RGS interaction would enable phosphorylation of 20-mer repeats regardless of the presence of β-catenin. High-affinity β-catenin-binding sites produced by APC phosphorylation would sequester β-catenin and prevent its nuclear localization when no Wnt signal is present and the concentration of free β-catenin is low. When β-catenin levels increase following a Wnt stimulus, binding to the lower affinity sites (both 15- and nonphosphorylated 20-mers) would help to reduce free β-catenin levels to turn off the signal. This model, however, does not explain how Axin would be able to access β-catenin bound to high-affinity APC sites to scaffold phosphorylation of β-catenin. It is possible that this is accomplished through the APC–Axin RGS interaction; alternatively, PP2A-mediated dephosphorylation suffices for this purpose, but, as noted above, no measurements of APC dephosphorylation kinetics have been reported.

Destruction Complex Localization

A pool of APC localizes to the basolateral cortex or cell–cell junctions of vertebrate epithelial cells and tissues, and *Drosophila* APC2 likewise localizes to the cell cortex through interactions with the microtubule and actin cytoskeletons (Näthke et al. 1996; McCartney et al. 1999, 2006; Yu et al. 1999; Rosin-Arbesfeld et al. 2001; Langford et al. 2006a,b; Grohmann et al. 2007; Zhou et al. 2011). In studies of nuclear export by APC, it was found that truncated APCs lacking cytoskeletal interaction regions shuttle between the nucleus and cytoplasm more efficiently than the full-length protein, suggesting that cytoskeletal interactions help to retain APC in the cytosol (Brocardo et al. 2005), consistent with a role in cytoplasmic retention of β-catenin.

Activation of the coreceptors LRP5/6 and Fzd results in recruitment of the destruction complex to the membrane and formation of membrane-associated puncta associated with β-catenin activation (Cliffe et al. 2003; Tolwinski et al. 2003; Bilic et al. 2007; Schwarz-Romond et al. 2007b; Hendriksen et al. 2008; MacDonald and He 2012). The recently described WTX/AMER1 protein binds to the APC arm domain and to PIP_2, and appears to be important for targeting the complex to the plasma membrane, although this protein appears to be specific to vertebrates (Grohmann et al. 2007; Tanneberger et al. 2011). In addition, the carboxy-terminal-binding protein (CtBP) binds to the 15-mer repeats of APC and promotes polymerization of truncated APC to form cytoplasmic puncta (Hamada and Bienz 2004; Schneikert et al. 2011), but the relationship between these puncta and the Wnt-stimulated assemblies is unclear.

Collectively, these observations suggest that APC localization might be important in its ability to modulate β-catenin destruction. However, Roberts et al. (2012) recently found that targeting fly APC2 to several different nonphysiological subcellular locations, including mitochondrial membranes and cytoplasmic vesicles, did not significantly alter down-regulation of β-catenin protein levels, TOPFLASH activation, or *Drosophila* embryo development. Moreover, these APC2 targeting variants do not appear to enter the nucleus at any time (Roberts et al. 2012), indicating that the ability of APC to enter the nucleus has either a relatively minor role in Wnt signaling, or is related to functions of APC apart from the Wnt pathway.

CONCLUDING REMARKS

A great deal is known about elements of the β-catenin destruction complex, but a complete molecular understanding of its function has remained elusive. This is due in part to the complexity of APC, the study of which is complicated by its other roles apart from Wnt signaling (McCartney and Nathke 2008). Outstanding problems include whether β-catenin binding by APC is an essential or modulatory aspect of the destruction complex, how the 20R2-CID region of APC functions in β-catenin ubiquitination, and the mechanism by which amino-terminally phosphorylated and ubiquitinated β-catenin is handed off to the proteasome. The function of PP2A in the complex is another open question. Finally, although the importance of cellular localization of destruction complex components has been challenged recently (Roberts et al. 2012), data from other systems would be useful in assessing potential roles of localization on Wnt signal transduction through LRP5/6, Fzd, and Dvl.

Many of the proteins that are necessary for β-catenin destruction, including GSK-3, CK1, Axin, and APC, also have roles in stabilizing β-catenin on Wnt stimulation (Liu et al. 2002; Cong et al. 2004; Zeng et al. 2005, 2008; Takacs et al. 2008; Tanneberger et al. 2011), thus confounding genetic analyses and cell culture overexpression studies in difficult-to-predict ways. Biochemical studies of purified protein complexes, as well as experiments that carefully control expression levels, will be important in resolving many of the mysteries surrounding the destruction complex.

ACKNOWLEDGMENTS

This work is supported by grant GM56169 from the U.S. National Institutes of Health.

REFERENCES

*Reference is also in this collection.

Aberle H, Bauer A, Stappert J, Kispert A, Kemler R. 1997. β-Catenin is a target for the ubiquitin-proteasome pathway. *EMBO J* **16:** 3797–3804.

Albuquerque C, Breukel C, van der Luijt R, Fidalgo P, Lage P, Slors FJ, Leitao CN, Fodde R, Smits R. 2002. The "just-right" signaling model: APC somatic mutations are selected based on a specific level of activation of the β-catenin signaling cascade. *Hum Mol Genet* **11:** 1549–1560.

Amit S, Hatzubai A, Birman Y, Andersen JS, Ben-Shushan E, Mann M, Ben-Neriah Y, Alkalay I. 2002. Axin-mediated CKI phosphorylation of β-catenin at Ser 45: A molecular switch for the Wnt pathway. *Genes Dev* **16:** 1066–1076.

Behrens J, von Kries JP, Kühl M, Bruhn L, Wedlich D, Grosschedl R, Birchmeier W. 1996. Functional interaction of β-catenin with the transcription factor LEF-1. *Nature* **382:** 638–642.

Behrens J, Jerchow B-A, Würtele M, Grimm J, Asbrand C, Wirtz R, Kühl M, Wedlich D, Birchmeier W. 1998. Functional interaction of an Axin homolog, conductin, with β-catenin, APC, and GSK3β. *Science* **280:** 596–599.

Benchabane H, Hughes EG, Takacs CM, Baird JR, Ahmed Y. 2008. Adenomatous polyposis coli is present near the minimal level required for accurate graded responses to the Wingless morphogen. *Development* **135:** 963–971.

Bilic J, Huang YL, Davidson G, Zimmermann T, Cruciat CM, Bienz M, Niehrs C. 2007. Wnt induces LRP6 signalosomes and promotes dishevelled-dependent LRP6 phosphorylation. *Science* **316:** 1619–1622.

Breitman M, Zilberberg A, Caspi M, Rosin-Arbesfeld R. 2008. The armadillo repeat domain of the APC tumor suppressor protein interacts with Striatin family members. *Biochim Biophys Acta* **1783:** 1792–1802.

Brocardo M, Nathke IS, Henderson BR. 2005. Redefining the subcellular location and transport of APC: New insights using a panel of antibodies. *EMBO Rep* **6:** 184–190.

Capelluto DG, Kutateladze TG, Habas R, Finkielstein CV, He X, Overduin M. 2002. The DIX domain targets dishevelled to actin stress fibres and vesicular membranes. *Nature* **419:** 726–729.

Cegielska A, Gietzen KF, Rivers A, Virshup DM. 1998. Autoinhibition of Casein Kinase Ie (CKIε) is relieved by protein phosphatases and limited proteolysis. *J Biol Chem* **273:** 1357–1364.

Chia IV, Costantini F. 2005. Mouse axin and axin2/conductin proteins are functionally equivalent in vivo. *Mol Cell Biol* **25:** 4371–4376.

Choi H-J, Huber AH, Weis WI. 2006. Thermodynamics of β-catenin–ligand interactions. The roles of the N- and C-terminal tails in modulating binding affinity. *J Biol Chem* **281:** 1027–1038.

Clevers H. 2006. Wnt/β-catenin signaling in development and disease. *Cell* **127:** 469–480.

Cliffe A, Hamada F, Bienz M. 2003. A role of Dishevelled in relocating Axin to the plasma membrane during wingless signaling. *Curr Biol* **13:** 960–966.

Cong F, Varmus H. 2004. Nuclear-cytoplasmic shuttling of Axin regulates subcellular localization of β-catenin. *Proc Natl Acad Sci* **101:** 2882–2887.

Cong F, Schweizer L, Varmus H. 2004. Casein kinase Iε modulates the signaling specificities of dishevelled. *Mol Cell Biol* **24:** 2000–2011.

Crosas B, Hanna J, Kirkpatrick DS, Zhang DP, Tone Y, Hathaway NA, Buecker C, Leggett DS, Schmidt M, King RW, et al. 2006. Ubiquitin chains are remodeled at the proteasome by opposing ubiquitin ligase and deubiquitinating activities. *Cell* **127:** 1401–1413.

Cross DA, Alessi DR, Cohen P, Andjelkovich M, Hemmings BA. 1995. Inhibition of glycogen synthase kinase-3 by insulin mediated by protein kinase B. *Nature* **378:** 785–789.

Dajani R, Faser E, Roe SM, Yeo M, Good VM, Thompson V, Dale TC, Pearl LH. 2003. Structural basis for recruitment of glycogen synthase kinase 3β to the axin-APC scaffold complex. *EMBO J* **22:** 494–501.

Day CL, Alber T. 2000. Crystal structure of the amino-terminal coiled-coil domain of the APC tumor suppressor. *J Mol Biol* **301:** 147–156.

Ding VW, Chen RH, McCormick F. 2000. Differential regulation of glycogen synthase kinase 3β by insulin and Wnt signaling. *J Biol Chem* **275:** 32475–32481.

Doble BW, Patel S, Wood GA, Kockeritz LK, Woodgett JR. 2007. Functional redundancy of GSK-3α and GSK-3β in Wnt/β-catenin signaling shown by using an allelic series of embryonic stem cell lines. *Dev Cell* **12:** 957–971.

Dominguez I, Itoh K, Sokol SY. 1995. Role of glycogen synthase kinase 3β as a negative regulator of dorsoventral axis formation in *Xenopus* embryos. *Proc Natl Acad Sci* **92:** 8498–8502.

Eklof Spink K, Fridman SG, Weis WI. 2001. Molecular mechanisms of β-catenin recognition by adenomatous polyposis coli revealed by the structure of an APC-β-catenin complex. *EMBO J* **20:** 6203–6212.

Elyada E, Pribluda A, Goldstein RE, Morgenstern Y, Brachya G, Cojocaru G, Snir-Alkalay I, Burstain I, Haffner-Krausz R, Jung S, et al. 2011. CKIα ablation highlights a critical role for p53 in invasiveness control. *Nature* **470:** 409–413.

Fagotto F, Jho E, Zeng L, Kurth T, Joos T, Kaufmann C, Costantini F. 1999. Domains of axin involved in protein-protein interactions, wnt pathway inhibition, and intracellular localization. *J Cell Biol* **145:** 741–756.

Fiedler M, Mendoza-Topaz C, Rutherford TJ, Mieszczanek J, Bienz M. 2011. Dishevelled interacts with the DIX domain polymerization interface of Axin to interfere with its function in down-regulating β-catenin. *Proc Natl Acad Sci* **108:** 1937–1942.

Fiol CJ, Mahrenholz AM, Wang Y, Roeske RW, Roach PJ. 1987. Formation of protein kinase recognition sites by covalent modification of the substrate. Molecular mechanism for the synergistic action of casein kinase II and glycogen synthase kinase 3. *J Biol Chem* **262:** 14042–14048.

Frame S, Cohen P. 2001. GSK3 takes centre stage more than 20 years after its discovery. *Biochem J* **359:** 1–16.

Frame S, Cohen P, Biondi RM. 2001. A common phosphate binding site explains the unique substrate specificity of GSK3 and its inactivation by phosphorylation. *Mol Cell* **7:** 1321–1327.

Grohmann A, Tanneberger K, Alzner A, Schneikert J, Behrens J. 2007. AMER1 regulates the distribution of the tumor suppressor APC between microtubules and the plasma membrane. *J Cell Sci* **120:** 3738–3747.

Ha N-C, Tonozuka T, Stamos JL, Choi H-J, Weis WI. 2004. Mechanism of phosphorylation-dependent binding of APC to β-catenin and its role in β-catenin degradation. *Mol Cell* **15:** 511–521.

Hagen T, Daniel ED, Culbert AA, Reith AD. 2002. Expression and characterization of GSK-3 mutants and their effect on β-catenin phosphorylation in intact cells. *J Biol Chem* **277:** 23330–23335.

Hamada F, Bienz M. 2004. The APC tumor suppressor binds to C-terminal binding protein to divert nuclear β-catenin from TCF. *Dev Cell* **7:** 677–685.

Hart MJ, de los Santos R, Albert IN, Rubinfeld B, Polakis P. 1998. Downregulation of β-catenin by human Axin and its association with the APC tumor suppressor, β-catenin, and GSK3β. *Curr Biol* **8:** 573–581.

Hart M, Concordet JP, Lassot I, Albert I, del los Santos R, Durand H, Perret C, Rubinfeld B, Margottin F, Benarous R, et al. 1999. The F-box protein β-TrCP associates with phosphorylated β-catenin and regulates its activity in the cell. *Curr Biol* **9:** 207–210.

Heisenberg CP, Houart C, Take-Uchi M, Rauch GJ, Young N, Coutinho P, Masai I, Caneparo L, Concha ML, Geisler R, et al. 2001. A mutation in the Gsk3-binding domain of zebrafish Masterblind/Axin1 leads to a fate transformation of telencephalon and eyes to diencephalon. *Genes Dev* **15:** 1427–1434.

Henderson BR, Fagotto F. 2002. The ins and outs of APC and β-catenin nuclear transport. *EMBO Rep* **3:** 834–839.

Hendriksen J, Jansen M, Brown CM, van der Velde H, van Ham M, Galjart N, Offerhaus GJ, Fagotto F, Fornerod M. 2008. Plasma membrane recruitment of dephosphorylated β-catenin upon activation of the Wnt pathway. *J Cell Sci* **121:** 1793–1802.

Hsu W, Zeng L, Costantini F. 1999. Identification of a domain of Axin that binds to the serine/threonine protein phosphatase 2A and a self-binding domain. *J Biol Chem* **274:** 3439–3445.

Ikeda S, Kishida S, Yamamoto H, Murai H, Koyama S, Kikuchi A. 1998. Axin, a negative regulator of the Wnt signaling pathway, forms a complex with GSK-3β and β-catenin and promotes GSK-3β-depedent phosphorylation of β-catenin. *EMBO J* **17:** 1371–1384.

Ikeda S, Kishida M, Matsuura Y, Usui H, Kikuchi A. 2000. GSK-3β-dependent phosphorylation of adenomatous polyposis coli gene product can be modulated by β-catenin and protein phosphatase 2A complexed with Axin. *Oncogene* **19:** 537–545.

Itoh K, Tang TL, Neel BG, Sokol SY. 1995. Specific modulation of ectodermal cell fates in *Xenopus* embryos by glycogen synthase kinase. *Development* **121:** 3979–3988.

Jho EH, Zhang T, Domon C, Joo CK, Freund JN, Costantini F. 2002. Wnt/β-catenin/Tcf signaling induces the transcription of Axin2, a negative regulator of the signaling pathway. *Mol Cell Biol* **22:** 1172–1183.

Jiang J, Struhl G. 1998. Regulation of the Hedgehog and Wingless signalling pathways by the F-box/WD40-repeat protein Slimb. *Nature* **391:** 493–496.

Jimbo T, Kawasaki Y, Koyama R, Sato R, Takada S, Haraguchi K, Akiyama T. 2002. Identification of a link between the tumour suppressor APC and the kinesin superfamily. *Nat Cell Biol* **4:** 323–327.

Julius MA, Schelbert B, Hsu W, Fitzpatrick E, Jho E, Fagotto F, Costantini F, Kitajewski J. 2000. Domains of axin and disheveled required for interaction and function in wnt signaling. *Biochem Biophys Res Commun* **276:** 1162–1169.

Kaidanovich-Beilin O, Lipina TV, Takao K, van Eede M, Hattori S, Laliberte C, Khan M, Okamoto K, Chambers JW, Fletcher PJ, et al. 2009. Abnormalities in brain structure and behavior in GSK-3α mutant mice. *Mol Brain* **2:** 35.

Kawasaki Y, Senda T, Ishidate T, Koyama R, Mirishita T, Iwayama Y, Higuchi O, Akiyama T. 2000. Asef, a link between the tumor suppressor APC and G-protein signaling. *Science* **289:** 1194–1197.

Kishida S, Yamamoto H, Ikeda S, Kishida M, Sakamoto I, Koyama S, Kikuchi A. 1998. Axin, a negative regulator of the Wnt signaling pathway, directly interacts with Adenomatous Polyposis Coli and regulates the stabilization of β-catenin. *J Biol Chem* **273:** 10823–10826.

Kishida S, Yamamoto H, Hino S, Ikeda S, Kishida M, Kikuchi A. 1999. DIX domains of Dvl and axin are necessary for protein interactions and their ability to regulate β-catenin stability. *Mol Cell Biol* **19:** 4414–4422.

Kitagawa M, Hatakeyama S, Shirane M, Matsumoto M, Ishida N, Hattori K, Nakmichi I, Kikuchi A, Nakayama K, Nakayama K. 1999. An F-box protein, FWD1, mediates ubiquitin-dependent proteolysis of β-catenin. *EMBO J* **18:** 2401–2410.

Kohler EM, Derungs A, Daum G, Behrens J, Schneikert J. 2008. Functional definition of the mutation cluster region of adenomatous polyposis coli in colorectal tumours. *Hum Mol Genet* **17:** 1978–1987.

Kohler EM, Chandra SH, Behrens J, Schneikert J. 2009. β-Catenin degradation mediated by the CID domain of APC provides a model for the selection of APC mutations in colorectal, desmoid and duodenal tumours. *Hum Mol Genet* **18:** 213–226.

Korinek V, Barker N, Morin PJ, van Wichen D, de Weger R, Kinzler KW, Vogelstein B, Clevers H. 1997. Constitutive transcriptional activation by a β-catenin–Tcf complex in $APC^{-/-}$ colon carcinoma. *Science* **275:** 1784–1787.

Krieghoff E, Behrens J, Mayr B. 2006. Nucleo-cytoplasmic distribution of β-catenin is regulated by retention. *J Cell Sci* **119:** 1453–1463.

Langford KJ, Askham JM, Lee T, Adams M, Morrison EE. 2006a. Examination of actin and microtubule dependent APC localisations in living mammalian cells. *BMC Cell Biol* **7:** 3.

Langford KJ, Lee T, Askham JM, Morrison EE. 2006b. Adenomatous polyposis coli localization is both cell type and cell context dependent. *Cell Motil Cytoskel* **63:** 483–492.

Latres E, Chiaur DS, Pagano M. 1999. Human F box protein β-Trcp associates with the Cul1/Skp1 complex and regulates the stability of β-catenin. *Oncogene* **18:** 849–854.

Lee E, Salic A, Krüger R, Heinrich R, Kirschner MW. 2003. The roles of APC and axin derived from experimental and theoretical analysis of the Wnt pathway. *PLoS Biol* **1:** 116–132.

Lee MJ, Lee BH, Hanna J, King RW, Finley D. 2011. Trimming of ubiquitin chains by proteasome-associated deubiquitinating enzymes. *Mol Cell Proteomics* **10:** R110. 003871.

Leung JY, Kolligs FT, Wu R, Zhai Y, Kuick R, Hanash S, Cho KR, Fearon ER. 2002. Activation of AXIN2 expression by β-catenin–T cell factor. A feedback repressor pathway regulating Wnt signaling. *J Biol Chem* **277:** 21657–21665.

Li Z, Nathke IS. 2005. Tumor-associated NH2-terminal fragments are the most stable part of the Adenomatous Polyposis Coli protein and can be regulated by interactions with COOH-terminal domains. *Cancer Res* **65:** 5195–5204.

Liu C, Kato Y, Zhang Z, Do VM, Yankner BA, He X. 1999. β-Trcp couples β-catenin phosphorylation-degradation and regulates *Xenopus* axis formation. *Proc Natl Acad Sci* **96:** 6273–6278.

Liu W, Dong X, Mai M, Seelan RS, Taniguchi K, Krishnadath KK, Halling KC, Cunningham JM, Boardman LA, Qian C, et al. 2000. Mutations in AXIN2 cause colorectal cancer with defective mismatch repair by activating β-catenin/TCF signalling. *Nat Genet* **26:** 146–147.

Liu C, Li Y, Semenov M, Han C, Baeg G-H, Tan Y, Zhang Z, Lin X, He X. 2002. Control of β-catenin phosphorylation/degradation by a dual-kinase mechanism. *Cell* **108:** 837–847.

Liu J, Xing Y, Hinds TR, Zheng J, Xu W. 2006. The third 20 amino acid repeat is the tightest binding site of APC for β-catenin. *J Mol Biol* **360:** 133–144.

Liu YT, Dan QJ, Wang J, Feng Y, Chen L, Liang J, Li Q, Lin SC, Wang ZX, Wu JW. 2011. Molecular basis of Wnt activation via the DIX domain protein Ccd1. *J Biol Chem* **286:** 8597–8608.

Luo W, Peterson A, Garcia BA, Coombs G, Kofahl B, Heinrich R, Shabanowitz J, Hunt DF, Yost HJ, Virshup DM. 2007. Protein phosphatase 1 regulates assembly and function of the β-catenin degradation complex. *EMBO J* **26:** 1511–1521.

* MacDonald BT, He X. 2012. Frizzled and LRP5/6 receptors for Wnt/β-catenin signalings. *Cold Spring Harb Perspect Biol* doi: 10.1101/cshperspect.a007880.

Major MB, Camp ND, Berndt JD, Yi X, Goldenberg SJ, Hubbert C, Biechele TL, Gingras AC, Zheng N, Maccoss MJ, et al. 2007. Wilms tumor suppressor WTX negatively regulates WNT/β-catenin signaling. *Science* **316:** 1043–1046.

Marikawa Y, Elinson RP. 1998. β-TrCP is a negative regulator of Wnt/β-catenin signaling pathway and dorsal axis formation in *Xenopus* embryos. *Mech Dev* **77:** 75–80.

Marin O, Bustos VH, Cesaro L, Meggio F, Pagano MA, Antonelli M, Allende CC, Pinna LA, Allende JE. 2003. A noncanonical sequence phosphorylated by casein kinase 1 in β-catenin may play a role in casein kinase 1 targeting of important signaling proteins. *Proc Natl Acad Sci* **100:** 10193–10200.

Matsubayashi H, Sese S, Lee J-S, Shirakawa T, Iwatsubo T, Tomita T, Yanagawa S. 2004. Biochemical characterization of the *Drosophila* Wingless signaling pathway based on RNA interference. *Mol Cell Biol* **24:** 2012–2024.

McCartney BM, Nathke IS. 2008. Cell regulation by the Apc protein Apc as master regulator of epithelia. *Curr Opin Cell Biol* **20:** 186–193.

McCartney BM, Dierick HA, Kirkpatrick C, Moline MM, Baas A, Peifer M, Bejsovec A. 1999. *Drosophila* APC2 is a cytoskeletally-associated protein that regulates wingless signaling in the embryonic epidermis. *J Cell Biol* **146:** 1303–1318.

McCartney BM, Price MH, Webb RL, Hayden MA, Holot LM, Zhou M, Bejsovec A, Peifer M. 2006. Testing hypotheses for the functions of APC family proteins using null and truncation alleles in *Drosophila*. *Development* **133:** 2407–2418.

McManus EJ, Sakamoto K, Armit LJ, Ronaldson L, Shpiro N, Marquez R, Alessi DR. 2005. Role that phosphorylation of GSK3 plays in insulin and Wnt signalling defined by knockin analysis. *EMBO J* **24:** 1571–1583.

Molenaar M, van de Wetering M, Oosterwegel M, Peterson-Maduro J, Godsave S, Korinek V, Roose J, Destrée O, Clevers H. 1996. XTcf-3 transcription factor mediates β-catenin-induced axis formation in *Xenopus* embryos. *Cell* **86:** 391–399.

Morin PJ, Sparks AB, Korinek V, Barker N, Clevers H, Vogelstein B, Kinzler KW. 1997. Activation of β-catenin–Tcf signaling in colon cancer by mutations in β-catenin or APC. *Science* **275:** 1787–1790.

Munemitsu S, Albert I, Souza B, Rubinfeld B, Polakis P. 1995. Regulation of intracellular β-catenin levels by the adenomatous polyposis coli (APC) tumor-suppressor protein. *Proc Natl Acad Sci* **92:** 3046–3050.

Munemitsu S, Albert I, Rubinfeld B, Polakis P. 1996. Deletion of an amino-terminal sequence stabilizes β-catenin in vivo and promotes hyperphosphorylation of the adenomatous polyposis coli tumor suppressor protein. *Mol Cell Biol* **16:** 4088–4094.

Näthke IS, Adams CL, Polakis P, Sellin JH, Nelson WJ. 1996. The adenomatous polyposis coli tumor suppressor protein localizes to plasma membrane sites involved in active cell migration. *J Cell Biol* **134:** 165–179.

Noutsou M, Duarte AM, Anvarian Z, Didenko T, Minde DP, Kuper I, de Ridder I, Oikonomou C, Friedler A, Boelens R, et al. 2011. Critical scaffolding regions of the tumor suppressor Axin1 are natively unfolded. *J Mol Biol* **405:** 773–786.

Orford K, Crockett C, Jensen JP, Weissman AM, Byers SW. 1997. Serine phosphorylation-regulated ubiquitination and degradation of β-catenin. *J Biol Chem* **272:** 24735–24738.

Ozawa M, Baribault H, Kemler R. 1989. The cytoplasmic domain of the cell adhesion molecule uvomorulin associates with three independent proteins structurally related in different species. *EMBO J* **8:** 1711–1717.

Patel S, Doble BW, MacAulay K, Sinclair EM, Drucker DJ, Woodgett JR. 2008. Tissue-specific role of glycosen synthase kinase 3β in glucose homeostatis and insulin action. *Mol Cell Biol* **28:** 6314–6328.

Peters JM, McKay RM, McKay JP, Graff JM. 1999. Casein kinase I transduces Wnt signals. *Nature* **401:** 345–350.

Peterson-Nedry W, Erdeniz N, Kremer S, Yu J, Baig-Lewis S, Wehrli M. 2008. Unexpectedly robust assembly of the Axin destruction complex regulates Wnt/Wg signaling in *Drosophila* as revealed by analysis in vivo. *Dev Biol* **320:** 226–241.

Peyrollier K, Hajduch E, Blair AS, Hyde R, Hundal HS. 2000. L-leucine availability regulates phosphatidylinositol 3-kinase, p70 S6 kinase and glycogen synthase kinase-3 activity in L6 muscle cells: Evidence for the involvement of the mammalian target of rapamycin (mTOR) pathway in the L-leucine-induced up-regulation of system A amino acid transport. *Biochem J* **350** (Pt 2): 361–368.

Rape M, Hoppe T, Gorr I, Kalocay M, Richly H, Jentsch S. 2001. Mobilization of processed, membrane-tethered SPT23 transcription factor by CDC48(UFD1/NPL4), a ubiquitin-selective chaperone. *Cell* **107:** 667–677.

Ratcliffe MJ, Itoh K, Sokol SY. 2000. A positive role for the PP2A catalytic subunit in Wnt signal transduction. *J Biol Chem* **275:** 35680–35683.

Roberts DM, Pronobis MI, Poulton JS, Waldmann JD, Stephenson EM, Hanna S, Peifer M. 2011. Deconstructing the β-catenin destruction complex: Mechanistic roles for the tumor suppressor APC in regulating Wnt signaling. *Mol Biol Cell* **22:** 1845–1863.

Roberts DM, Pronobis MI, Poulton JS, Kane EG, Peifer M. 2012. Regulation of Wnt signaling by the tumor suppressor APC does not require ability to enter the nucleus nor a particular cytoplasmic location. *Mol Biol Cell* doi: 10.1091/mbc.E11-11-0965.

Rosin-Arbesfeld R, Townsley F, Bienz M. 2000. The APC tumour suppressor has a nuclear export function. *Nature* **406:** 1009–1012.

Rosin-Arbesfeld R, Ihrke G, Bienz M. 2001. Actin-dependent membrane association of the APC tumour suppressor in polarized mammalian epithelial cells. *EMBO J* **20:** 5929–5939.

Rubinfeld B, Souza B, Albert I, Munemitsu S, Polakis P. 1995. The APC protein and E-cadherin form similar but independent complexes with α-catenin, β-catenin, and plakoglobin. *J Biol Chem* **270:** 5549–5555.

Rubinfeld B, Albert I, Porfiri E, Fiol C, Munemitsu S, Polakis P. 1996. Binding of GSK3β to the APC-β-catenin complex and regulation of complex assembly. *Science* **272:** 1023–1026.

Rubinfeld B, Albert I, Porfiri E, Munemitsu S, Polakis P. 1997a. Loss of β-catenin regulation by the APC tumor suppressor protein correlates with loss of structure due to common somatic mutations of the gene. *Cancer Res* **57:** 4624–4630.

Rubinfeld B, Robbins P, El-Gamil M, Albert I, Porfiri E, Polakis P. 1997b. Stabilization of β-catenin by genetic defects in melanoma cell lines. *Science* **275:** 1790–1792.

Rubinfeld B, Tice DA, Polakis P. 2001. Axin-dependent phosphorylation of the Adenomatous Polyposis Coli protein mediated by casein kinase 1ε. *J Biol Chem* **276:** 39037–39045.

Ruel L, Bourouis M, Heitzler P, Pantesco V, Simpson P. 1993. *Drosophila* shaggy kinase and rat glycogen synthase kinase-3 have conserved activities and act downstream of Notch. *Nature* **362:** 557–560.

Sadot E, Conacci-Sorrel M, Zhurinsky J, Shnizer D, Lando Z, Zharhary D, Kam Z, Ben-Ze'ev A, Geiger B. 2002. Regulation of S33/S37 phosphorylated β-catenin in normal and transformed cells. *J Cell Sci* **115:** 2771–2780.

Sakanaka C, Weiss JB, Williams LT. 1998. Bridging of β-catenin and glycogen synthase kinase-3β by axin and inhibition of β-catenin-mediated transcription. *Proc Natl Acad Sci* **95:** 3020–3023.

Salic A, Lee E, Mayer L, Kirschner MW. 2000. Control of β-catenin stability: Reconstitution of the cytoplasmic steps of the Wnt pathway in *Xenopus* egg extracts. *Mol Cell* **5:** 523–532.

Satoh S, Daigo Y, Furukawa Y, Kato T, Miwa N, Nishiwaki T, Kawasoe T, Ishiguro H, Fujita M, Tokino T, et al. 2000. *AXIN1* mutations in hepatocellular carcinomas, and growth suppression in cancer cells by virus-mediated transfer of *AXIN1*. *Nat Genet* **24:** 245–250.

Schneikert J, Brauburger K, Behrens J. 2011. APC mutations in colorectal tumours from FAP patients are selected for CtBP-mediated oligomerization of truncated APC. *Hum Mol Genet* **20:** 3554–3564.

Schwarz-Romond T, Fiedler M, Shibata N, Butler PJ, Kikuchi A, Higuchi Y, Bienz M. 2007a. The DIX domain of Dishevelled confers Wnt signaling by dynamic polymerization. *Nat Struct Mol Biol* **14:** 484–492.

Schwarz-Romond T, Metcalfe C, Bienz M. 2007b. Dynamic recruitment of axin by Dishevelled protein assemblies. *J Cell Sci* **120:** 2402–2412.

Seeling JM, Miller JR, Gil R, Moon RT, White R, Virshup DM. 1999. Regulation of β-catenin signaling by the B56 subunit of protein phosphatase 2A. *Science* **283:** 2089–2091.

Shi W, Ma Z, Willers H, Akhtar K, Scott SP, Zhang J, Powell S. 2008. Disassembly of MDC1 foci is controlled by ubiquitin-proteasome-dependent degradation. *J Biol Chem* **283:** 31608–31616.

Siegfried E, Chou TB, Perrimon N. 1992. Wingless signaling acts through zeste-white 3, the *Drosophila* homolog of glycogen synthase kinase-3, to regulate engrailed and establish cell fate. *Cell* **71:** 1167–1179.

Smits R, Kielman MF, Breukel C, Zurcher C, Neufeld K, Jagmohan-Changur S, Hofland N, van Dijk J, White R, Edelmann W, et al. 1999. Apc1638T: A mouse model delineating critical domains of the adenomatous polyposis coli protein involved in tumorigenesis and development. *Genes Dev* **13:** 1309–1321.

Sobrado P, Jedlicki A, Bustos VH, Allende CC, Allende JE. 2005. Basic region of residues 228–231 of protein kinase CK1α is involved in its interaction with axin: Binding to axin does not affect the kinase activity. *J Cell Biochem* **94:** 217–224.

Spink KE, Polakis P, Weis WI. 2000. Structural basis of the Axin-adenomatous polyposis coli interaction. *EMBO J* **19:** 2270–2279.

Spink KE, Fridman SG, Weis WI. 2001. Molecular mechanisms of β-catenin recognition by adenomatous polyposis coli revealed by the structure of an APC/β-catenin complex. *EMBO J* **20:** 6203–6212.

Su Y, Fu C, Ishikawa S, Stella A, Kojima M, Shitoh K, Schreiber EM, Day BW, Liu B. 2008. APC is essential for targeting phosphorylated β-catenin to the SCFβ-TrCP ubiquitin ligase. *Mol Cell* **32:** 652–661.

Takacs CM, Baird JR, Hughes EG, Kent SS, Benchabane H, Paik R, Ahmed Y. 2008. Dual positive and negative regulation of wingless signaling by adenomatous polyposis coli. *Science* **319:** 333–336.

Tan CW, Gardiner BS, Hirokawa Y, Layton MJ, Smith DW, Burgess AW. 2012. Wnt signalling pathway parameters for mammalian cells. *PLoS ONE* **7:** e31882.

Tanneberger K, Pfister AS, Brauburger K, Schneikert J, Hadjihannas MV, Kriz V, Schulte G, Bryja V, Behrens J. 2011. Amer1/WTX couples Wnt-induced formation of PtdIns(4,5)P2 to LRP6 phosphorylation. *EMBO J* **30:** 1433–1443.

Tolwinski NS, Wieschaus E. 2001. Armadillo nuclear import is regulated by cytoplasmic anchor Axin and nuclear anchor dTCF/Pan. *Development* **128:** 2107–2117.

Tolwinski NS, Wehrli M, Rives A, Erdeniz N, DiNardo S, Wieschaus E. 2003. Wg/Wnt signal can be transmitted through arrow/LRP5,6 and Axin independently of Zw3/Gsk3β activity. *Dev Cell* **4:** 407–418.

Valvezan AJ, Zhang F, Diehl JA, Klein PS. 2012. Adenomatous polyposis coli (APC) regulates multiple signaling pathways by enhancing glycogen synthase kinase-3 (GSK-3) activity. *J Biol Chem* **287:** 3823–3832.

Watanabe T, Wang S, Noritake J, Sato K, Fukata M, Takefuji M, Nakagawa M, Izumi N, Akiyama T, Kaibuchi K. 2004. Interaction with IQGAP1 links APC to Rac1, Cdc42, and actin filaments during cell polarization and migration. *Dev Cell* **7:** 871–883.

Winston JT, Strack P, Beer-Romero P, Chu CY, Elledge SJ, Harper JW. 1999. The $SCF^{\beta\text{-}TRCP}$-ubiquitin ligase complex associates specifically with phosphorylated destruction motifs in IκBα and β-catenin and stimulates IκBα ubiquitination in vitro. *Genes Dev* **13:** 270–283.

Woodgett JR. 1990. Molecular cloning and expression of glycogen synthase kinase-3/factor A. *EMBO J* **9:** 2431–2438.

Wu G, He X. 2006. Threonine 41 in β-catenin serves as a key phosphorylation relay residue in β-catenin degradation. *Biochemistry* **45:** 5319–5323.

Wu G, Xu G, Schulman BA, Jeffrey PD, Harper JW, Pavletich NP. 2003. Structure of a β-TrCP1-Skp1-β-catenin complex: Destruction motif binding and lysine specificity of the SCF(β-TrCP1) ubiquitin ligase. *Mol Cell* **11:** 1445–1456.

Xing Y, Clemens WK, Kimmelman D, Xu W. 2003. Crystal structure of a β-catenin/Axin complex suggests a mechanism for the β-catenin destruction complex. *Genes Dev* **17:** 2753–2764.

Xing Y, Clements WK, Le Trong I, Hinde TR, Stenkamp R, Kimelman D, Xu W. 2004. Crystal structure of a β-catenin/APC complex reveals a critical role for APC phosphorylation in APC function. *Mol Cell* **15:** 523–533.

Yamamoto H, Kishida S, Kishida M, Ikeda S, Takada S, Kikuchi A. 1999. Phosphorylation of Axin, A Wnt signal negative regulator, by glycogen synthase kinase-3β regulates its stability. *J Biol Chem* **274:** 10681–10684.

Yamamoto H, Hinoi T, Michiue T, Fukui A, Usui H, Janssens V, Van Hoof C, Goris J, Asashima M, Kikuchi A. 2001. Inhibition of the Wnt signaling pathway by the PR61 subunit of protein phosphatase 2A. *J Biol Chem* **276:** 26875–26882.

Yang J, Zhang W, Evans PM, Chen X, He X, Liu C. 2006. Adenomatous polyposis coli (APC) differentially regulates β-catenin phosphorylation and ubiquitination in colon cancer cells. *J Biol Chem* **281:** 17751–17757.

Yost C, Torres M, Miller JR, Huang E, Kimelman D, Moon RT. 1996. The axis-inducing activity, stability, and subcellular distribution of β-catenin is regulated in *Xenopus* embryos by glycogen synthase kinase 3. *Genes Dev* **10:** 1443–1454.

Yu X, Waltzer L, Bienz M. 1999. A new *Drosophila* APC homologue associated with adhesive zones of epithelial cells. *Nat Cell Biol* **1:** 144–151.

Zeng L, Fagotto F, Zhang T, Hsu W, Vasicek TJ, Perry WL 3rd, Lee JJ, Tilghman SM, Gumbiner BM, Costantini F. 1997. The mouse Fused locus encodes Axin, an inhibitor of the Wnt signaling pathway that regulates embryonic axis formation. *Cell* **90:** 181–192.

Zeng X, Tamai K, Doble B, Li S, Huang H, Habas R, Okamura H, Woodgett J, He X. 2005. A dual-kinase mechanism for Wnt co-receptor phosphorylation and activation. *Nature* **438:** 873–877.

Zeng X, Huang H, Tamai K, Zhang X, Harada Y, Yokota C, Almeida K, Wang J, Doble B, Woodgett J, et al. 2008. Initiation of Wnt signaling: Control of Wnt coreceptor Lrp6 phosphorylation/activation via frizzled, dishevelled and axin functions. *Development* **135:** 367–375.

Zhou MN, Kunttas-Tatli E, Zimmerman S, Zhouzheng F, McCartney BM. 2011. Cortical localization of APC2 plays a role in actin organization but not in Wnt signaling in *Drosophila*. *J Cell Sci* **124:** 1589–1600.

TCF/LEFs and Wnt Signaling in the Nucleus

Ken M. Cadigan[1] and Marian L. Waterman[2]

[1]Department of Molecular, Cellular and Developmental Biology, University of Michigan, Ann Arbor, Michigan 48109-1048

[2]Department of Microbiology and Molecular Genetics, University of California, Irvine, Irvine, California 92697-4025

Correspondence: mlwaterm@uci.edu

T-cell factor/lymphoid enhancer factor (TCF/LEF) transcription factors are the major end point mediators of Wnt/Wingless signaling throughout metazoans. TCF/LEFs are multifunctional proteins that use their sequence-specific DNA-binding and context-dependent interactions to specify which genes will be regulated by Wnts. Much of the work to define their actions has focused on their ability to repress target gene expression when Wnt signals are absent and to recruit β-catenin to target genes for activation when Wnts are present. Recent advances have highlighted how these on/off actions are regulated by Wnt signals and stabilized β-catenin. In contrast to invertebrates, which typically contain one TCF/LEF protein that can both activate and repress Wnt targets, gene duplication and isoform complexity of the family in vertebrates have led to specialization, in which individual TCF/LEF isoforms have distinct activities.

Wnt signals play important roles during animal development (Logan and Nusse 2004), as well as in adult tissues that are refreshed and repaired by stem cells (Haegebarth and Clevers 2009). It is the essential function of Wnt signaling in stem cell self-renewal and cell proliferation that links this pathway to problems of aging and disease such as cancer and diabetes (Polakis 2007; Laudes 2011). The term "Wnt signaling" does not imply a single-purpose signal transduction system. Rather, it refers to a diverse collection of signals triggered by Wnt ligand–receptor interactions that direct cell behavior in multiple ways: cell polarity, movement, proliferation, differentiation, survival and self-renewal. Diversity in Wnt signaling derives from the diversity of its components, its set of 19 ligands, ten receptors, alternative receptors, its signal transduction components, as well as the cell's particular developmental history. Despite this complexity, many Wnt signals act through a single mediator, β-catenin, to regulate gene expression. Wnt ligand–receptor interactions at the plasma membrane are communicated to target genes by the translocation of β-catenin into the nucleus where it partners with DNA-binding proteins that recognize specific sequence motifs in promoters and enhancers of target genes. The central 12-armadillo repeat array of β-catenin is the main mediator of transcription factor interactions, whereas domains in the amino and carboxy termini carry potent transcription-activating functions (Orsulic and Peifer 1996; van de Wetering et al. 1997; Hsu

Cite this article as *Cold Spring Harb Perspect Biol* doi: 10.1101/cshperspect.a007906

et al. 1998). Thus, once β-catenin is recruited to target genes, transcription is activated via the actions of these domains and an array of transcriptional coactivators (Mosimann et al. 2009; Cadigan 2012).

Multiple transcription factors that recruit β-catenin to Wnt targets have been identified and this review will summarize those that are best characterized. However, the nuclear mediators most closely associated with Wnt/β-catenin action are the TCF/LEFs, high-mobility group (HMG) DNA-binding proteins with multiple domains for protein interaction and regulation. This review will focus attention on this family of factors and discuss recent advances that shed light on how Wnt signaling works in stem cell niches and differentiation.

TCF/LEF TRANSCRIPTION FACTOR FAMILY

Almost all invertebrate genomes carry a single TCF/LEF ortholog, whereas vertebrates have expanded to a family of four members (five in the Zebrafish) (Fig. 1) (Arce et al. 2006; Archbold et al. 2012). The first discoveries of TCF/LEFs came from searches for transcription regulators of cell-fate markers in human T lymphocytes. "T cell factor 1" (HUGO gene name *TCF7* [van de Wetering et al. 1991]) and "lymphoid enhancer factor 1" (HUGO gene name *LEF1* [Travis et al. 1991; Waterman et al. 1991]) were discovered in T and B cells. T cell factor 3 and T cell factor 4 were found later through low-stringency hybridization screens with *TCF7* cDNAs (HUGO gene names, *TCF7L1*, *TCF7L2* [Korinek et al. 1998b]). The immunology-connected nomenclature is thus historical, and as knockout studies and expression studies attest, these factors play important regulatory roles in almost all tissues of the body (Oosterwegel et al. 1993; Brunner et al. 1997; van de Wetering et al. 1997; Lin et al. 1998; Archbold et al. 2012).

DNA-Binding Domain

All TCF/LEFs contain a highly conserved HMG box and a small peptide motif of basic residues ("basic tail"); together the HMG box and basic tail comprise the HMG DNA-binding domain (HMG DBD) (Fig. 1). The HMG DBD can recognize specific DNA sequences with nanomolar affinity (Giese et al. 1991; van de Wetering and Clevers 1992; Love et al. 1995). In addition to DNA sequence specificity, the HMG box has a DNA bending function. It recognizes its specific nucleotide sequence in the minor groove of the DNA and enforces a bend in the helix between 90° and 127° (Giese et al. 1995; Love et al. 1995). Both directed and random screen studies have identified a consensus recognition sequence for the HMG DBD; 5′-SCTTTGATS-3′ (Fig. 2) (van de Wetering et al. 1997; van Beest et al. 2000; Hallikas and Taipale 2006; Atcha et al. 2007). Recent chromatin immunoprecipitation experiments to define TCF/LEF-β-catenin-binding patterns genome wide identify this consensus as the most frequently occurring sequence in TCF4 and β-catenin-binding peaks (Hatzis et al. 2008; Blahnik et al. 2010; Bottomly et al. 2010; Norton et al. 2011). The small basic tail motif is located nine residues carboxy terminal to the HMG box and serves two purposes: elevating DNA-binding affinity through contact with the positively charged DNA backbone and functioning as a strong nuclear localization signal for interactions with importins (Giese et al. 1991; Prieve et al. 1998).

β-Catenin-Binding Domain

The connection between TCF/LEFs and Wnt signaling came from yeast two-hybrid screens suggesting that β-catenin could bind tightly to LEF1 and TCF1, an interaction subsequently delimited to a conserved motif in the amino terminus of TCF/LEFs (Fig. 1) (Behrens et al. 1996; Huber et al. 1996; Molenaar et al. 1996; van de Wetering et al. 1997; Graham et al. 2000; Poy et al. 2001). Deletion of this domain was an important step in establishing TCF/LEFs as downstream mediators of Wnt and β-catenin. Truncated "dominant negatives" could no longer bind to β-catenin, they suppressed the ability of overexpressed Wnt or β-catenin to induce secondary axes in frog embryos, and they phenocopied the segment polarity defect *wingless* mutants in *Drosophila* (Behrens et al. 1996; Huber et al. 1996; Molenaar et al. 1996; van de

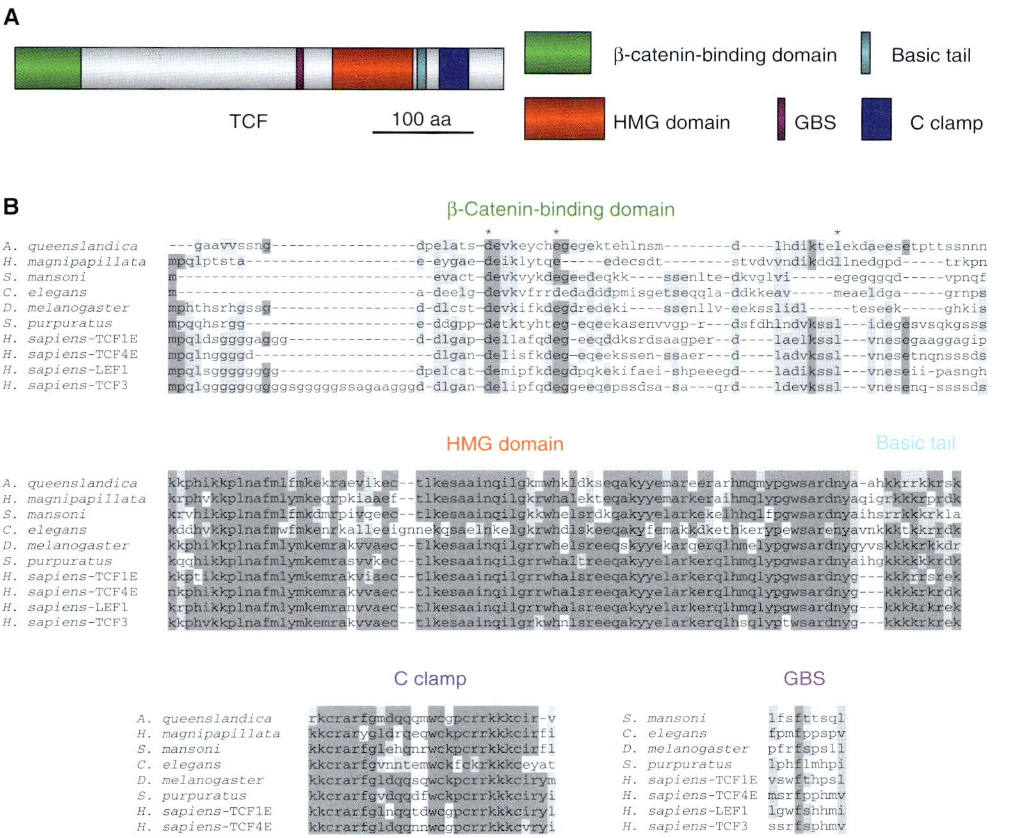

Figure 1. Hallmarks of the TCF/LEF family. (*A*) Schematic of TCF domains, showing the location of five conserved domains found in this family, the amino-terminal β-catenin-binding domain, the Groucho binding sequence (GBS), the high-mobility group (HMG) domain followed by a nuclear localization signal (basic tail), and the C clamp. The specific TCF shown is from the sea urchin *Strongylocentrotus purpuratus*. (*B*) Alignments of the five domains from six invertebrate TCFs and the four human family members. Consensus residues are highlighted in dark gray with conservative changes in light gray. The extent of conservation is far greater in the HMG, basic tail, and C-clamp domains compared with the β-catenin-binding domain and GBS. Three residues in the β-catenin-binding domain that contact β-catenin and are essential for interaction are marked with asterisks (Graham et al. 2000; Poy et al. 2001). A GBS could not be identified in TCF/LEFs from porifera and cnidarians and some flatworms, whereas a C clamp is found in nearly all invertebrate TCFs (it is not found in some flatworm TCF/LEFs) and in some vertebrate TCF1 and TCF4 isoforms. The sequences shown are from the sponge *Amphimedon queenslandica* (gene bank accession number ADO16566.1), the cnidarian *Hydra magnipapillata* (XP_002159974.1), the parasitic flatworm *Schistosoma mansoni* (XP_002572116.1), the nematode *Caenorhabditis elegans* (NP_491053.3), the dipteran *Drosophila melanogaster* (NP_726522), the sea urchin *Strongylocentrotus purpuratus* (NP_999640) and human TCF1E (EAW62279.1), TCF4E (CAB97213.1), and LEF1 (NP_001124185), and TCF3 (NP_112573.1).

Wetering et al. 1997). These experiments showed how powerfully suppressive amino-terminally truncated TCF/LEFs can be. As we will discuss below, these forms occur naturally in the vertebrate family (see "TCF/LEF heterogeneity").

C Clamp

Alignment of TCF sequences reveals a third region as a highly conserved domain (Fig. 1B) (van de Wetering et al. 1997). This domain is

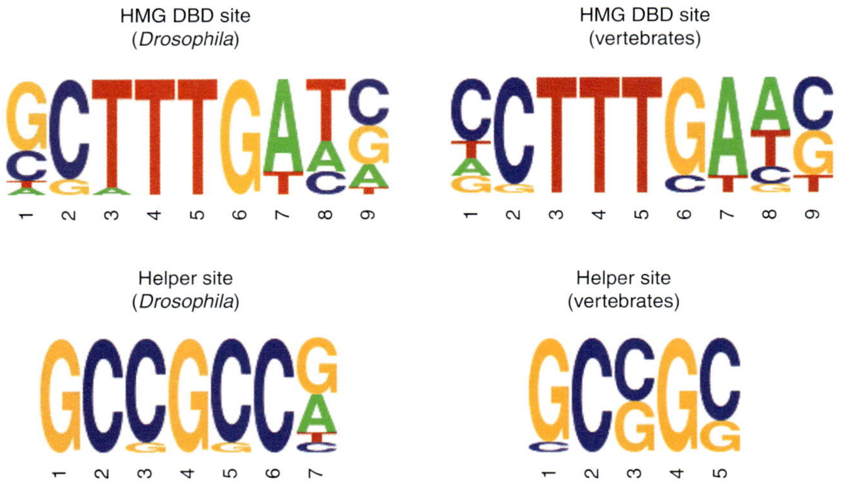

Figure 2. High-mobility group DNA-binding domain (HMG DBD) and Helper sites of *Drosophila* and vertebrate TCF/LEFs. Sequence logos were constructed from TCF/Pangolin sites described in Chang et al. (2008) and from a collection of vertebrate Wnt-regulated enhancers (N Hoverter, ML Waterman, and KM Cadigan, unpubl.).

small, enriched in basic, cysteine, and aromatic residues, and located carboxy terminal to the HMG DBD. The function of this domain was defined only recently when its requirement for gene regulation in human cancer cells (Atcha et al. 2007) and *Drosophila* (Chang et al. 2008) led to the discovery that it is a second DNA-binding domain. The domain is called the "C clamp" to highlight the absolute requirement for four cysteine residues in DNA binding (Atcha et al. 2007), and the first had been previously noted by the Clevers group as the "CRARF" domain (van de Wetering et al. 1997). Vertebrate TCF/LEFs containing the C clamp have been suggested to be important in regulating specific target genes (Wohrle et al. 2007; Weise et al. 2010). Although there is not yet an nuclear magnetic resonance (NMR) or crystal structure of the C clamp, experiments show that it has specific DNA-binding activities. The C clamp carries specificity for a secondary, GC-rich sequence called a "Helper site" (Fig. 2) that can occur with variable spacing and orientation relative to the Wnt response element (Atcha et al. 2007; Chang et al. 2008). It is unusual, but not unprecedented for transcription factors to carry two sequence-specific DNA-binding domains and, as will be discussed below, the C clamp is no longer present in every vertebrate TCF/LEF family member. However, one of the genome-wide chromatin immunoprecipitation studies for TCF4 binding noted enrichment of a Helper-like GC-rich sequence at a subset of targets (Hatzis et al. 2008). This suggests that despite the loss of the C clamp in a couple of family members, this domain is functionally relevant in vertebrates. Understanding the contribution of the C clamp to Wnt regulation in vertebrates and invertebrates requires further study.

Gro/TLE-Binding Domain

TCF/LEFs repress gene transcription in the absence of available β-catenin. Several corepressors participate in this process, such as myeloid translocation gene related-1 (Mtgr1) (Moore et al. 2008), corepressor of Pan (Coop) (Song et al. 2010), and hydrogen peroxide-inducible clone (HIC5) (Ghogomu et al. 2006; Li et al. 2011). The most intensively studied TCF/LEF co-repressors are the Groucho/transducin-like enhancer of split (Gro/TLE) repressor family. Genetic studies in *Drosophila* and *C. elegans* support a role for their Gro/TLE orthologs repressing Wnt targets (Cavallo et al. 1998; Calvo

 Cite this article as *Cold Spring Harb Perspect Biol* doi: 10.1101/cshperspect.a007906

et al. 2001). Gro/TLEs contain a conserved glutamine-rich Q domain at their amino termini that binds to the central portion of TCF/LEFs as well as the portion containing the HMG DBD (Daniels and Weis 2005; Arce et al. 2009). A short motif in the central domain has been identified that is essential for binding of LEF1 to TLE1, and similar regions can be found in most other TCF/LEFs (Fig. 1) (Arce et al. 2009). Whether this small motif is the core site of bona fide Gro/TLE interactions in TCF/LEF-assembled complexes at endogenous target genes remains to be determined. Other small sequence motifs have been discovered between the β-catenin-binding and HMG domains and they appear to impart additional modes of repression (see "TCF/LEF heterogeneity").

EVOLUTION OF TCF/LEFs

Examination of the predicted protein sequences from sequenced genomes provides a picture of the evolution of the TCF/LEF family. Defining proteins solely by sequence similarity has an important caveat, nicely illustrated by the SYS-1 protein, which clearly functions as a β-catenin in *C. elegans* and is structurally similar to other β-catenins despite having almost no conservation at the primary sequence level (Kidd et al. 2005; Liu et al. 2008a). Leaving this consideration aside, examination of predicted proteomes suggests that Wnts, β-catenin, and TCF/LEFs are metazoan inventions, because they are absent in Choanoflagellates (a single-cell eukaryote) and present in Sponges (King et al. 2008; Adamska et al. 2010; Srivastava et al. 2010). Nearly all invertebrate genomes examined contain one recognizable TCF gene that codes for a TCF/LEF containing an amino-terminal β-catenin-binding domain, an HMG domain followed by a basic tail, and a C-clamp domain (Fig. 1). The one exception is found in *Platyhelminthes*, in which the parasitic flatworm *Schistosoma mansoni* genome (Berriman et al. 2009) contains three TCF/LEF genes and the genome of Planaria (*Schmidtea mediterranea*) has five (C Petersen, pers. comm.). These flatworm TCF/LEFs all have well conserved HMG and basic tail motifs, but C clamps are found in only one TCF/LEF from *S. mansoni* and two TCF/LEFs from *S. mediterranea* (C Peterson, pers. comm.). It is not yet clear which TCF/LEFs in these organisms mediate Wnt signaling, but it is interesting to note that a similar situation exists in amphibians and mammals, which contain four TCF genes that produce a large number of isoforms, but only the TCF1E and TCF4E isoforms contain a C clamp (Atcha et al. 2007). These findings are consistent with a rule in which TCFs from organisms containing a single TCF gene will have a C clamp, whereas this domain is lost from some TCFs in organisms possessing multiple TCF genes. Evolution to multiple-TCF genomes may enable organisms to refine Wnt signals to specific subsets of target genes.

TCF/LEF HETEROGENEITY

Refinement of Wnt signaling through TCF specialization is evident in vertebrates in which the TCF/LEF family has expanded to four loci with alternative promoters and alternative messenger RNA (mRNA) splicing. These capabilities create heterogeneous patterns of activity. Alternative promoters produce truncated dominant–negative forms of LEF1, TCF1, and TCF4, similar to the Wnt-interfering, amino-terminal deleted TCF/LEF forms created for epistasis analysis (van de Wetering et al. 1996; Hovanes et al. 2001; Vacik and Lemke 2011). Alternative mRNA splicing produces TCF/LEFs missing functional domains such as the C clamp and other small regions amino terminal of the HMG DBD domain (van de Wetering et al. 1996; Duval et al. 2000; Hovanes et al. 2000; Prokunina-Olsson et al. 2009b; Weise et al. 2010). With one exception for TCF4 (Kennell et al. 2003), nearly all variations retain the HMG DBD suggesting that TCF/LEF isoforms exert their spectrum of activities via specific DNA binding.

The functional outcome of all this heterogeneity is that TCF/LEF isoforms enable Wnt signals to be interpreted differently. For example, although Wnt can activate gene transcription by directing the formation of β-catenin-TCF/LEF complexes with transcription-activating complexes (Mosimann et al. 2009;

Cadigan 2012), a recent example in *Xenopus* shows how it also uses β-catenin to disengage TCF3 and corepressors (Hikasa et al. 2010; Mahmoudi et al. 2010; Mohan et al. 2010). Heterogeneity also means that TCF/LEFs can specialize for unique gene targeting, or that they can engage distinct activating or repressing cofactors. To illustrate, TCF3 is generally known as a repressor of Wnt target genes (Kim et al. 2000; Merrill et al. 2004) and the full-length, β-catenin-binding form of TCF4 is also sometimes associated with target gene repression (Tang et al. 2008). In contrast, full-length TCF1 and LEF1 are more often linked to Wnt target gene activation (Reya et al. 2000; Kratochwil et al. 2002; Liu et al. 2005). The mechanistic basis for these differences is not completely known but they seem to track to small repressor-recruiting motifs in the central region and/or the carboxyl terminus of the proteins. For example, the five amino acid motifs LVPQ and SxxSS that are present obligatorily in TCF3 appear as alternatively spliced variations in TCF4 and not at all in TCF1 and LEF1 (Pukrop et al. 2001; Gradl et al. 2002; Liu et al. 2005). Original studies in *Xenopus* systems showed that forms with SxxSS behave like repressors, whereas forms missing this motif act like activators (Liu et al. 2005). This was recently confirmed in a study of human hepatocellular carcinoma in which of the 14 alternatively spliced TCF4 isoforms expressed in liver cancer cells, those that contained the SxxSS motif were growth suppressive, whereas those lacking the element were Wnt activating and growth promoting (Tsedensodnom et al. 2011). The fact that SxxSS forms of TCF4 are generated through alternative splicing means that the prevalence of these forms could differ among cell types and cell stages. Such variations could be one reason why TCF4 has been assigned either activating or repressing, oncogenic or tumor suppressor functions in intestine and colon cancer cells (Korinek et al. 1998a; Tang et al. 2008; Angus-Hill et al. 2011).

Other alternatively spliced regions of TCFs have been noted to increase or decrease target gene activation potential (Weise et al. 2010; Le Bacquer et al. 2011). Perhaps the most interesting of these are the CtBP motifs in the carboxyl terminus of invertebrate TCF/LEFs and mammalian TCF3 and TCF4 (PXDLS) (Brannon et al. 1999; Valenta et al. 2003). Direct CtBP:TCF-binding interactions are controversial and have been variously deemed irrelevant or functional depending on the study (Brannon et al. 1999; Valenta et al. 2003; Hamada and Bienz 2004; Fang et al. 2006). In *Drosophila* cells, CtBP represses Wnt targets in the absence of signaling, and CtBP associates with Wnt target gene chromatin independently of TCF/LEF (Fang et al. 2006). Interestingly, homo-oligomerization of CtBP is required for repression of some Wnt targets, but CtBP is also necessary for activation of other targets as a TCF/LEF-β-catenin-associated monomer (Bhambhani et al. 2011). These novel findings suggest that CtBP could be part of the heterogeneity and context-dependent actions of TCF/LEFs, including the activating/repressing functions of mammalian TCF3 and TCF4—a correlation that might refocus attention on the question of whether CtBP does or does not directly interact with the PXDLS motifs in the carboxyl terminus of these two TCF/LEFs.

THE TCF/LEF TRANSCRIPTIONAL SWITCH

For genes whose expression is induced by Wnt/β-catenin signaling, the standard model is that TCF/LEF represses transcription in the absence of signaling, and is converted to an activator by association with β-catenin. This model is well supported in *Drosophila* and *C. elegans*, organisms that contain a single TCF/LEF gene. However, the situation is more complex in vertebrates, in which individual TCF/LEF proteins have become more specialized in their ability to repress and/or activate Wnt target genes.

In *Drosophila* tissues and cell culture, loss or depletion of TCF/LEF (also known as Pangolin, or Pan) results in a strong reduction in activation of Wnt targets (Brunner et al. 1997; van de Wetering et al. 1997; Stadeli and Basler 2005; Fang et al. 2006; Parker et al. 2008). In addition, reduced TCF/Pan also causes derepression of several Wnt targets in the absence of signaling (Cavallo et al. 1998; Fang et al. 2006; Liu et al. 2008b). Studies with POP1, the nematode TCF/LEF, also support this dual role with both repression (Rocheleau et al. 1997; Thorpe et al. 1997; Maduro et al. 2002) and activation of Wnt targets being severely compromised in POP1 mutants (Herman 2001; Shetty et al. 2005; Lam et al. 2006). TCF/LEF loss-of-function studies are nicely complemented in both organisms by analysis of reporter genes whose expressions are driven by Wnt response elements (WREs). Mutation of the TCF/LEF binding sites in these WREs results in a reduction of activation by Wnt signaling, but also depression of reporter activity in cells where the WRE is normally not active (Yang et al. 2000; Knirr and Frasch 2001; Shetty et al. 2005). These results support the model depicted in Figure 3A,C, in which these invertebrate TCF/LEFs restrict WRE activity in cells receiving little or no Wnt signal, and activate the WRE in cells with sufficient Wnt signaling. It should be noted that the repression role for TCF/LEF is not observed at every WRE, presumably owing to the lack of elements that would activate gene expression in the absence of functional TCF/LEF sites (Lam et al. 2006; Chang et al. 2008).

In vertebrate systems, initial studies with simple reporters containing multimerized TCF/LEF sites indicated that all four TCFs can mediate activation of transcription by β-catenin (Molenaar et al. 1996; Korinek et al. 1997; van de Wetering et al. 1997; Hsu et al. 1998) and are bound by corepressors of the Gro/TLE family (Brantjes et al. 2001). However, loss-of-function analyses support the view that repression and activation capabilities of these factors have been divided among the family. In *Xenopus* ventral–lateral patterning, systematic knockdown of TCF/LEFs with morpholinos combined with a rescue by chimeric TCFs fused to repressor or activator domains, found that TCF3 and TCF4 acted as repressors, whereas LEF1 and TCF1

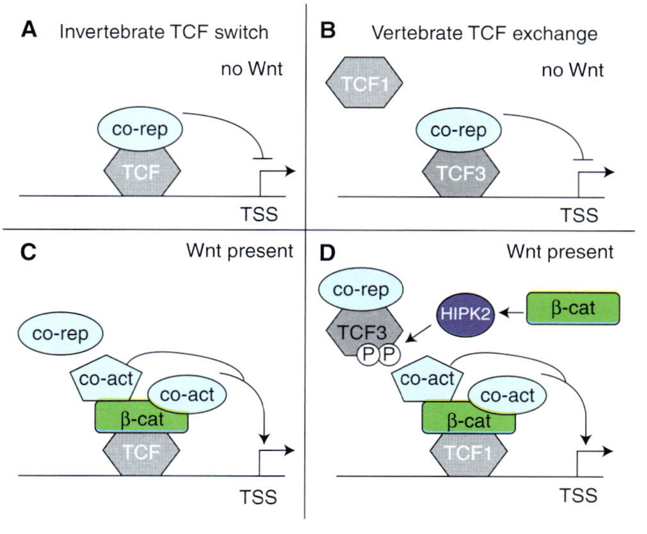

Figure 3. Regulation of TCF/LEFs in invertebrates and vertebrates. (*A*) Most invertebrates have one TCF/LEF gene producing a protein that can recruit corepressors to Wnt targets in the absence of signaling. (*B*) In vertebrates, TCF/LEFs are more specialized, with TCF3 often fulfilling this repressive role. (*C*) When high levels of β-catenin are found in the nucleus after Wnt signaling, it binds to invertebrate TCF, displacing (or inactivating) corepressors and recruiting coactivators through its amino- and carboxy-terminal transactivation domains. (*D*) For at least some vertebrate Wnt targets, a TCF/LEF exchange occurs, in which Wnt/β-catenin signaling promotes HIPK2-dependent phosphorylation of TCF3, causing it to leave Wnt target gene chromatin, allowing TCF1-β-catenin to occupy the WRE, facilitating transcriptional activation. See text for more information.

acted as activators (Liu et al. 2005). A similar separation of roles was observed at a different developmental stage during Wnt-dependent induction of the Spemann organizer. Here, TCF3 and TCF1 provided repressor roles and TCF4 an activator function (Houston et al. 2002; Standley et al. 2006). TCF3 seems to act most often as a repressor as evident in genetic studies in other systems (Kim et al. 2000; Merrill et al. 2004; Cole et al. 2008; Yi et al. 2011), whereas LEF1 appears restricted to activation of Wnt targets (Reya et al. 2000; Kratochwil et al. 2002). TCF1 and TCF4 have both repressor and activator functions, depending on the cells or tissues examined (Korinek et al. 1998a; Galceran et al. 1999; Roose et al. 1999; Tang et al. 2008; Nguyen et al. 2009). An important limitation of these studies is the heterogeneity of TCF isoforms produced in cells. As described above, some TCF1 and TCF4 isoforms lack the β-catenin-binding domain and act as dominant negatives, so they would behave genetically as repressors of the pathway in a mouse knockout (Roose et al. 1999).

In addition to a Wnt-directed exchange of corepressors for the coactivator β-catenin, differential activities of vertebrate TCFs raise the possibility that the transcriptional switch promoted by Wnt could also involve an exchange of TCF family members. Direct evidence for this type of mechanism has been found in *Xenopus* for the *Vent2* gene, in which TCF/LEF repression and activation occurs through a single binding site (Hikasa et al. 2010). Wnt stimulation promotes the phosphorylation of repressor-acting TCF3 by homeodomain-interacting protein kinase (HIPK2), which results in its dissociation from the WRE (Hikasa et al. 2010). TCF1 is not phosphorylated by HIPK2 and replaces TCF3 on the *Vent2* WRE for gene activation (Fig. 3B) (Hikasa and Sokol 2011). Wnt-HIPK2 signaling also promotes the dissociation of LEF1 and TCF4 from the *Vent2* WRE (Hikasa et al. 2010), suggesting that the HIPK2 pathway could play a prominent role in regulating TCF/LEF chromatin occupancy in many contexts. One potential example is embryonic stem cells, in which it has recently been shown that Wnt signaling promotes an exchange of TCF3 and TCF1 at several Wnt targets (Yi et al. 2011). Whether HIPK2 is involved in the exchange in stem cells is not known.

TCF/LEF-β-CATENIN REGULATION OF HISTONE MODIFICATIONS

The transcriptional switch model for regulation of TCF/LEF activity by Wnt/β-catenin signaling proposes that in the absence of signaling, TCF/LEF recruits corepressors to target gene chromatin. Upon β-catenin association with TCF/LEF, these corepressors are displaced or inactivated, and a variety of coactivators are recruited via interactions with β-catenin. The list of transcriptional coregulators involved in this switch is large (Willert and Jones 2006; Parker et al. 2007; Cadigan and Peifer 2009; Mosimann et al. 2009; Cadigan 2012) and a comprehensive list will not be presented here. Rather, the connection between these factors and histone modifications of Wnt target gene chromatin will be discussed given the wealth of evidence that the state of chromatin has a profound influence on transcription (Berger 2007; Suganuma and Workman 2011).

The most extensive chromatin modification to be examined in the context of Wnt target gene regulation is histone acetylation. Addition of acetyl groups to several conserved lysine residues on the amino termini of histone H3 and H4 subunits is highly correlated with transcriptional activation. This reaction is catalyzed by histone acetyltransferases (HATs) and counteracted by histone deacetylases (HDACs). The role of HDACs in repression of Wnt targets in the absence of signaling was first proposed using an artificial Wnt reporter (Billin et al. 2000), but it has been extended to endogenous targets as well (Kioussi et al. 2002; Sierra et al. 2006; Lyu et al. 2011). In addition, TCF/LEF1-binding corepressors such as members of the Groucho/transducin-like enhancer of split (Gro/TLE) and myeloid translocation gene (MTG) families, which contribute to Wnt target gene repression (Cavallo et al. 1998; Roose et al. 1998; Moore et al. 2008), are known to act in part by binding to HDACs (Chen et al. 1999; Brantjes et al. 2001; Linggi et al. 2005).

 Cite this article as *Cold Spring Harb Perspect Biol* doi: 10.1101/cshperspect.a007906

On the activation side of the transcriptional switch, there is abundant evidence that CREB-binding protein (CBP) and its close relative p300, both of which encode HATs, can bind to the carboxy-terminal transactivation domain of β-catenin and mediate activation of Wnt target gene expression (Mosimann et al. 2009; Teo and Kahn 2010). Recruitment of CBP/p300 to Wnt target gene chromatin has been correlated with an increase in H3 and H4 acetylation (Kioussi et al. 2002; Sierra et al. 2006; Parker et al. 2008; Badis et al. 2009; Lyu et al. 2011). In *Drosophila* cells, Wnt/β-catenin signaling causes a widespread increase in H3/H4 acetylation across the entire Wnt target locus (up to 40 kb), even though CBP recruitment is limited to the site of TCF binding (Parker et al. 2008). However, depletion of CBP resulted in loss of all the Wnt-dependent histone acetylation, supporting a model in which CBP directly or indirectly catalyzes the widespread modification. Although the evidence outlined above suggests a major role for histone acetylation in activation of some Wnt targets, other studies have found no evidence for a significant change in histone acetylation at other Wnt targets in response to pathway activation (Wohrle et al. 2007; Blythe et al. 2010).

Trimethylation of the fourth lysine of H3 (H3-K4me3) has also been linked to activation of Wnt targets and the function of the Wnt coactivator Pygopus. H3-K4me3 is a chromatin mark found at proximal promoters, a signature which is positively correlated with transcription (Berger 2007; Suganuma and Workman 2011). Wnt/β-catenin signaling causes an elevation of H3-K4me3 at various target promoters (Parker et al. 2008; Blythe et al. 2010; Chen et al. 2010) and the mixed-lineage-leukemia proteins (MLL1/2), the methyltransferases that catalyze this modification, are required for full activation of several Wnt targets (Sierra et al. 2006; Chen et al. 2010).

Pygopus proteins (Pygo in flies; Pygo1 and Pygo2 in mammals) contain PHD domains and are Wnt coactivators that bind to the amino-terminal transactivation domain of β-catenin indirectly through Legless (in flies) or BCL9 and BCL9-2 (in mammals) (Kramps et al. 2002; Jessen et al. 2008). The PHD domain of Pygo can bind H3-K4me3 (Fiedler et al. 2008; Gu et al. 2009; Kessler et al. 2009). Although the role of this interaction is controversial in *Drosophila* (Kessler et al. 2009), in mammalian systems Pygo2 also interacts with MLL2 (Chen et al. 2010), suggesting a model in which Pygopus proteins act to connect WREs bound by TCF and β-catenin with proximal promoters of Wnt targets (Cadigan and Peifer 2009). This idea is further strengthened by physical interactions between Pygo and Mediator subunits (Carrera et al. 2008) and TAF4 (a TFIID subunit) (Wright and Tjian 2009), both of which physically associate with the RNA polymerase II complex. Indeed, there is recent evidence that several WREs in the c-myc locus form loops with the proximal promoter to affect transcriptional activation (Wright et al. 2010; Yochum et al. 2010; Yochum 2011).

Other recent examples of histone modification complexes that participate in activation of Wnt targets are protein arginine methyltransferase 2 (PRMT2) in *Xenopus*, which promotes H3K8me signatures (Blythe et al. 2010), and the DOT-COM complex, responsible for H3K79me3, a chromatin mark associated with transcriptional elongation (Steger et al. 2008; Mahmoudi et al. 2010; Mohan et al. 2010). PRMT2 was biochemically associated with β-catenin in *Xenopus* embryonic extracts, and is required for Wnt/β-catenin signaling-dependent establishment of the dorsal (Spemann) organizer (Blythe et al. 2010). A more general role in the activation of other Wnt targets remains to be tested. Components of the DOT/COM complex, such as DOTL1 (the methyltransferase subunit; also called DOT1) are required for the pathway in *Drosophila* and mammalian cells (Mahmoudi et al. 2010; Mohan et al. 2010). In the latter system, transcription profiling suggested that DOTL1 functions predominately in the regulation of genes activated by Wnt/β-catenin signaling (Mohan et al. 2010). Further investigations will be needed to determine the role of the DOTL1 complex in general versus Wnt-specific gene activation, as well as exploring the mechanism by which DOT-COM and PRMT8 complexes are recruited to WRE chromatin.

In addition to the chromatin marks discussed above, there are others (e.g., H4K20me via SET8 and H3R17me via CARM1) that have been linked to activation of Wnt targets (Li et al. 2011; Ou et al. 2011). Although still to be investigated, it seems likely that not all Wnt targets will require all the aforementioned chromatin marks, and a more genome-wide systematic analysis may reveal the existence of classes of chromatin signatures that define subsets of target genes. Adding to this complexity in vertebrate systems is the likelihood that different target genes will require different TCFs for regulation. For example, loss of TCF3 in a neuroblast stem cell results in global elevation of H3/H4ac levels (Lluis et al. 2011), a change consistent with its role in transcriptional repression. Perhaps targets that are more dependent on TCF3 will also show larger differences in histone acetylation in response to pathway activation.

TCF/LEF EXPRESSION AND REGULATION

Because TCF/LEFs are potent transcription regulators, it is important to control their activity. New insights into regulation of transcription and protein localization show how Wnt signaling and other signals control TCF/LEFs beyond just making β-catenin and its chromatin modifiers available.

TCF/LEFs are expressed in distinct but broadly overlapping patterns such that it is common for more than one TCF/LEF to be coexpressed in any one cell. Double knockout studies in mice show that functional redundancy is common, as mice missing more than one TCF/LEF develop early and severe embryonic lethal phenotypes, much more than single knockouts (Galceran et al. 1999; Gregorieff et al. 2004). TCF3 is unique in that its knockout leads to an early embryonic lethal phenotype (Houston et al. 2002; Merrill et al. 2004; Liu et al. 2005). Given that TCF3 is the most abundant family member in embryonic stem cells, this phenotype makes sense. However, single knockout experiments also show that TCF/LEFs have specialized in their function and expression patterns. What are the pathways that control expression, and do these pathways regulate all TCF/LEFs or specific members?

First, TCF/LEF gene transcription can be regulated by the Wnt pathway. With the exception of *TCF3* (*TCF7L1*), all of the TCF/LEF promoters (including the promoters for dominant–negative isoforms) contain predicted WREs (Fig. 4) (K Pate and ML Waterman, unpubl.). These WREs have been experimentally validated in the *LEF1* and *TCF7* promoters and both genes often appear as Wnt-regulated transcripts in microarray studies (Roose et al. 1999; Hovanes et al. 2001; Li et al. 2006). In addition, genome-wide ChIP-ChIP experiments show that the promoters for full-length *LEF1*, *TCF7*, and *TCF7L2* are occupied by TCF4 (Hatzis et al. 2008; Bottomly et al. 2010). The *TC7L1* promoter (for TCF3) is the lone exception; it is not occupied by TCF4 and its promoter does not have identifiable TCF/LEF binding sites.

Second, regulation of the internal promoters for dominant–negative TCF/LEFs is equally important and could vary during development or disease progression. For example, dominant–negative forms of TCF1/*TCF7* are detected in normal colon crypt epithelial cells but they are absent in colon cancer cells (Najdi et al. 2009). A switch in isoform expression from a Wnt-opposing dominant negative in normal cells, to a Wnt-promoting full-length isoform in cancer cells could explain why knockout of *TCF7* in mice produces adenomas in the intestine, whereas knockout of *TCF7* in colon cancer cell lines slows their growth (Tang et al. 2008). Likewise, the *LEF1* locus is aberrantly activated in colon cancer, but only the promoter for full-length LEF1. The promoter for dnLEF1 is silent (Hovanes et al. 2001). Reexpression of dnTCF1 or dnTCF4 in colon cancer cells can shut down Wnt signaling and force a stall in the G1 phase of the cell cycle (van de Wetering et al. 2002). How is it that dnTCF/dnLEF promoters are silent in cancer? A partial explanation is available for the dnLEF1 promoter: Its promoter is silenced through active repression by an upstream distal repressor element that makes WREs in the promoter inaccessible (Li et al. 2006; Yokoyama et al. 2009). The YY1 transcription factor

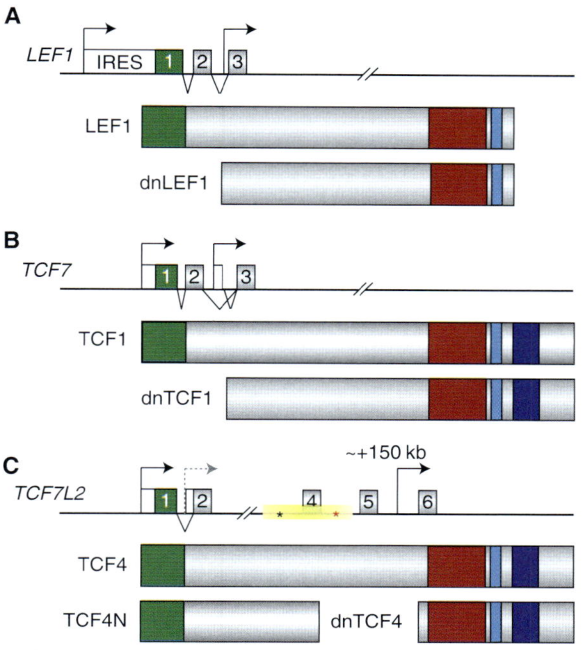

Figure 4. TCF/LEF isoform diversity. (*A*) Schematic of the *LEF1* locus with two promoters. The first promoter produces an mRNA with an internal ribosome entry site in the 5′ UTR (encoded by exon1) (Jimenez et al. 2005; Tsai et al. 2011). A second promoter in intron2 produces a truncated, dominant–negative form of LEF1 missing the β-catenin-binding domain (green rectangle) (Hovanes et al. 2001; Li et al. 2006). Red and blue domains are the HMG and basic domains for DNA binding (refer to Fig. 1). (*B*) Schematic of the *TCF7* locus (which codes for TCF1). Similar to *LEF1*, the *TCF7* locus codes for full-length and dominant–negative forms through the use of two promoters (van de Wetering et al. 1996). (*C*) The *TCF7L2* locus (codes for TCF4) produces multiple truncated isoforms in addition to the full-length form. A putative intron1 promoter has been proposed that produces a dnTCF4 similar to dnLEF1 and dnTCF1 (Duval et al. 2000). A third promoter in intron5 has been defined (Vacik and Lemke 2011). This promoter produces a dnTCF4 isoform because it is missing the β-catenin-binding domain. The promoter lies immediately downstream from a 92-kb genomic interval (yellow rectangle) that contains the strongest known risk alleles for type 2 diabetes (black asterisk; rs7903146 and rs12255372 [Grant et al. 2006]) as well as an alternative polyadenylation signal that might produce the previously discovered inhibitory form referred to as TCF4N (red asterisk) (Kennell et al. 2003; Locke et al. 2011).

is necessary for this repression as it binds to the core promoter and communicates with the distal repressor (Yokoyama et al. 2009).

Another example comes from the recent discovery of a promoter for dominant–negative TCF4 and an intronic enhancer connected to risk of diabetes. The strongest diabetes-associated polymorphism identified to date occurs in an intervening sequence of the *TCF4/TCF7L2* gene (Grant et al. 2006; Lyssenko et al. 2007; and reviewed in Doria et al. 2008). Single-nucleotide polymorphisms (SNP) in introns can change splicing efficiencies and influence the relative expression of spliced isoforms (Prokunina-Olsson et al. 2009a,b), but the most significant diabetes-associated SNP lies within a 92-kb genomic interval that has additional activities. A recent bacterial artificial chromosome (BAC) survey in transgenic mice determined that this region contains a strong regulatory enhancer (Savic et al. 2011). This region also contains a signal for alternative polyadenylation for production of a truncated, inhibitory TCF4 isoform missing the HMG DNA-binding domain (Fig. 4) (Kennell et al. 2003; Locke et al. 2011). The SNP-containing interval is clearly an important regulator of the levels of *TCF7L2* gene expression. But there is an even deeper consideration. A recent study of TCF4 expression during development showed that the fifth intron of the *TCF7L2* locus harbors a highly conserved promoter for a dominant–negative isoform of TCF4—a form missing the β-catenin-binding domain but retaining the HMG DNA-binding domain. This promoter is activated during embryogenesis by VAX1/VAX2, a homeobox regulator essential for negative feedback of Wnt signaling. Although the VAX-induced dnTCF4 form is most prevalent in the brain, it might be important to reassess how the diabetes-associated SNPs located upstream affect the production of full-length versus dominant–negative TCF4 in tissues such as the pancreas. Such changes would be mostly invisible in microarray and RNA-seq data unless one was specifically looking for it. Sorting out the connection between expression and activity of TCF4 isoforms is important because out of all the vertebrate TCF/LEFs, this family member shows the most polymorphisms and mutations in cancer and diabetes (Grant et al. 2006; Sjoblom et al. 2006; Cauchi et al. 2007; Bass et al. 2011; Mullighan et al. 2011).

As illustrated with the HIPK2-directed "TCF exchange," TCF/LEF localization appears to be dynamic on target gene chromatin. As mentioned above, Wnt signals direct a swap of TCF3 for TCF1 in a β-catenin- and HIPK2-dependent manner in *Xenopus* embryos and mouse embryonic stem cells (mESCs) (Hikasa et al. 2010; Hikasa and Sokol 2011; Yi et al. 2011). The swap results in activation of Wnt target genes because TCF3 is repressive, whereas TCF1 is activating. Multiple sites of phosphorylation in the central portion of xTCF3 are essential target residues for the switch, and these sites are also phosphorylated in the corresponding region of xTCF4 and mLEF1 (Fig. 5) (Hikasa and Sokol 2011). It is worth noting that these residues are conserved in human orthologs of TCF3, TCF4, and LEF1, but they are not present in any TCF1 homolog (Fig. 5). This suggests that Wnt-triggered HIPK2 signals may be especially effective at promoting TCF1 occupancy of WREs. It is also worth emphasizing that HIPK2 activation and the TCF3/TCF1 swap is β-catenin dependent.

The Wnt- and β-catenin-dependent feature of HIPK2 action has intriguing parallels to regulation of the *C. elegans* ortholog POP1. In this case, a Wnt signal that is delivered asymmetrically to dividing daughters of the EMS progenitor cell directs the alternative catenin WRM1 and a Nemo-like kinase (NLK) called LIT1, to bind the carboxyl terminus of POP1 and phosphorylate its context-dependent regulatory domain (CRD) (Lo et al. 2004; Yang et al. 2011). NLK phosphorylates CRD residues that are similar to those targeted by Wnt/β-catenin/HIPK2 (Fig. 5). Phosphorylated POP1 is recognized by 14-3-3 proteins for rapid nuclear export, an action that lowers

```
HIPK2
     LEF1  SLSPPIPRT-SNKVPVVQPSHAVHPLTPLITYSDEHFSPGSHPSHIPSDV  178
     TCF3  TRSPSPAHL-SNKVPVVQHPHHMHPLTPLITYSNDHFSPGSPPTHLSPEI  215
     TCF4  ARSPSPAHIVSNKVPVVQHPHHVHPLTPLITYSNEHFTPGNPPPHLPADV  201

LIT1/NLK
     LEF1  SLSPPIPRT-SNKVPVVQPSHAVHPLTPLITYSDEHFSPGSHPSHIPSDV  178
     TCF4  ARSPSPAHIVSNKVPVVQHPHHVHPLTPLITYSNEHFTPGNPPPHLPADV  201
     POP1                     ..PPYAAALRSPSLMFPMGAMSPTFPMFPPSPVY  130

TNIK
     TCF3  TRSPSPAHL-SNKVPVVQHPHHMHPLTPLITYSNDHFSPGSPPTHLSPEI  215
     TCF4  ARSPSPAHIVSNKVPVVQHPHHVHPLTPLITYSNEHFTPGNPPPHLPADV  201
```

Figure 5. TCF/LEF phosphorylation sites for HIPK2, LIT1/NLK, and TNIK. Amino acid alignment of human LEF1, TCF3, and TCF4 and the *C. elegans* ortholog POP1. The indicated regions are amino terminal of the HMG domain. Residues in red are validated, mapped phosphorylation sites by the kinase HIPK2 (Hikasa and Sokol 2011), the kinase NLK and its *C. elegans* ortholog LIT1 (Ishitani et al. 2003; Lo et al. 2004), and TNIK (Mahmoudi et al. 2009).

 Cite this article as *Cold Spring Harb Perspect Biol* doi: 10.1101/cshperspect.a007906

nuclear concentrations of POP1 and results in an increase in target gene activation (Thorpe et al. 1997; Maduro et al. 2002; Shetty et al. 2005). Activation of mammalian NLK also has negative actions on TCF/LEFs. In this case, NLK is activated by Wnt ligands such as Wnt5a to phosphorylate TCF4 and LEF1 and trigger their dissociation from DNA (Ishitani et al. 1999, 2003). Similar to LIT1 and HIPK2, Wnt directs the formation of a complex between NLK, β-catenin, and the TCF/LEF protein, and once again, the residues phosphorylated by NLK are a subset of the LIT1- and HIPK2-dependent phosphorylations (Fig. 5) (Sokol 2011).

There is one other kinase that phosphorylates TCFs in a Wnt- and β-catenin-dependent manner. TNIK (for Traf2- and Nck-interacting kinase) was discovered through proteomic analysis of TCF4 complexes in mouse intestine and human colon cancer (Mahmoudi et al. 2009). This kinase is recruited by β-catenin to WRE-bound TCFs for phosphorylation. TNIK is a Germinal Center/Ste20-type kinase, very different from NLK and HIPK2, yet it targets a residue in the CRD region of TCF3 and TCF4 that is identical to a HIPK2 site (Fig. 5) (Shitashige et al. 2010). TNIK actions enhance Wnt target gene activation much like the outcomes of HIPK2 and LIT1 action but how it does this is unknown (Mahmoudi et al. 2009; Satow et al. 2010; Shitashige et al. 2010). TNIK could certainly function in a manner that is completely different from HIPK2, but it is also possible that it triggers a transcriptional switch between TCF/LEFs. In fact, TCF4 has been recently proposed to function as a growth suppressor in colon cancer cells, whereas TCF1 has been observed to promote Wnt signals and proliferation (at least in colon cancer) (Tang et al. 2008; Angus-Hill et al. 2011). Because TCF1 is expressed as a full-length activating form in colon cancer, and because it is not targeted by any of these kinases, it would be interesting to know whether TNIK can activate Wnt target gene transcription by directing a transcription switch between TCF4 and TCF1.

TCF1 appears as a stand-alone in that it is not phosphorylated by HIPK2, NLK, or TNIK, but its TCF1 actions are controlled in other ways. First, normal colon crypts express a dominant–negative form of TCF1 that can oppose its full-length counterpart (Najdi et al. 2009). As mentioned above, this Wnt-suppressing form is missing in colon cancer. A second tier of negative regulation uses specific Wnt signals (Wnt1, Wnt5a) to trigger nuclear export of TCF1 (Najdi et al. 2009; R Najdi, unpubl.). This action shares similarities with the Wnt-directed inhibition of other TCF/LEFs because β-catenin appears to be involved, but whether TCF1 is phosphorylated by a different kinase for export is not known.

The heterogeneity inherent in the TCF/LEF family means that linking the molecular effects of β-catenin-dependent kinases (HIPK2, NLK, and TNIK) to their overall effect on Wnt signaling in any one cell type or tissue may not be straightforward because multiple TCF/LEFs with different activities are often coexpressed. For example, can we conclude that Wnt-HIPK2 always results in Wnt target gene activation and that Wnt-NLK-nuclear export of TCF1 always leads to Wnt target gene repression? The answer is clearly no. Because the vertebrate family has expanded and specialized, the functional outcome of HIPK2 or NLK kinases on Wnt signaling depends on how they alter the relative concentrations of activating and repressing TCF/LEFs versus the available pool of β-catenin (Sokol 2011). In fact, Wnt-NLK-TCF1 export predicts a very different outcome for normal colon epithelia in which dnTCF1 is expressed (export would lead to Wnt target gene activation), versus colon cancer cells that express full-length TCF1 (export would reduce Wnt target expression) (Najdi et al. 2009). Another important case in point is the observation in *C. elegans* that exporting the single TCF ortholog POP1 from the nucleus leads to overall Wnt target gene activation. It may seem paradoxical that reduction of TCF/POP1 concentrations in the nucleus leads to target gene activation, but a recently proposed model suggests that nuclear concentrations of TCFs are as important as nuclear β-catenin concentrations for establishing an optimal β-catenin:TCF ratio (Phillips and Kimble 2009). Too much TCF might translate into target gene repression rather than activation.

Consistent with this, quantitative modeling of Wnt signaling in *Xenopus* shows that sensing fold changes in Wnt signals works best by monitoring the relative amounts of two interacting factors (i.e., β-catenin and TCF) than by trying to measure absolute levels of stabilized β-catenin (Goentoro and Kirschner 2009). Thus, it could be that Wnt signals regulate gene expression by dynamically controlling the concentrations of TCF/LEFs in the nucleus as much as they do β-catenin.

NON-TCF RELAYS: WNT/β-CATENIN SIGNALING THROUGH NON-TCF/LEF PROTEINS

As outlined thus far in this review, TCFs are major nuclear recipients of Wnt/β-catenin signaling. However, there are other transcription factors that can bind β-catenin and activate transcription. These include type I and type II nuclear receptors (Mulholland et al. 2005; Beildeck et al. 2010), several members of the SOX family (Kormish et al. 2010), FOXO proteins (Essers et al. 2005; Almeida et al. 2007), the homeodomain proteins Prop1 and PitX2 (Kioussi et al. 2002; Olson et al. 2006), hypoxia-inducible factor 1α (HIF1α) (Kaidi et al. 2007), and the bHLH protein MyoD (Kim et al. 2008). Here we summarize some of this work, to highlight general questions regarding non-TCF proteins and their role in Wnt/β-catenin transcription and biology.

One well-studied example of β-catenin interacting with a nuclear receptor is the androgen receptor (AR). The ligand-binding domain of AR binds directly to the Arm repeats of β-catenin (Mulholland et al. 2002; Yang et al. 2002; Song et al. 2003). Androgen increases the strength of the AR-β-catenin interaction (Truica et al. 2000; Song et al. 2003) and promotes nuclear accumulation of the complex (Mulholland et al. 2002; Pawlowski et al. 2002; Song et al. 2003). Consistent with these studies, Wnt/β-catenin signaling promotes transcriptional activation by AR at the level of simple reporter constructs (Truica et al. 2000; Mulholland et al. 2002; Verras et al. 2004) and AR target genes (Masiello et al. 2004; Cronauer et al. 2005; Read et al. 2007). Both AR and β-catenin can bind to TIF2/GRIP1, a p160 steroid receptor coactivator and the three proteins may act together in a complex to activate transcription (Song and Gelmann 2005).

The AR-β-catenin interaction likely plays a crucial role in prostate cancer, in which AR is required for progress of this cancer at all stages (Culig et al. 2002; Niu et al. 2010). Mutations stabilizing β-catenin are found in 5%–7% of prostate cancers (Voeller et al. 1998; Chesire et al. 2000) and other proteins that elevate nuclear β-catenin levels are also found at high frequency in tumors (Rios-Doria et al. 2004; Chen et al. 2006). Most prostate cancers initially respond to a reduction in androgen levels, but eventually convert to an androgen-independent state (Culig et al. 2002; Niu et al. 2010). In the mouse, expression of a stabilized form of β-catenin can promote prostate hyperplasia pre- or post-castration (Bierie et al. 2003; Yu et al. 2009) and the physical association of AR and β-catenin is stronger post-castration (Wang et al. 2008) or with an AR mutant that is hormone insensitive (Masiello et al. 2004). Thus, β-catenin may play an essential role in enabling AR to operate independently in low androgen environments (Schweizer et al. 2008).

β-catenin-binding proteins have the potential to compete with TCF/LEF for limiting amounts of β-catenin. For example, overexpression of AR inhibits activation of the TCF/LEF reporter TOPFLASH by β-catenin (Chesire and Isaacs 2002; Pawlowski et al. 2002). The significance of this competition is not clear for AR, but has been proposed to be significant in the case of HIF-1α (Kaidi et al. 2007). Hypoxia (which increases HIF1α levels) reduces TCF/LEF-β-catenin transcriptional activity, which is correlated with cell-cycle exit of colorectal cancer cells (Kaidi et al. 2007), and it also regulates gene expression in a β-catenin-dependent manner (Kaidi et al. 2007; Zhao et al. 2011). However in some hypoxic stem cells, there appears to be sufficient β-catenin to interact with both TCFs and HIF1α (Mazumdar et al. 2010). In colorectal cancer cell lines, a positive-feedback loop occurs, in which HIF1α represses APC transcription, elevating β-catenin protein levels, which promotes high levels of HIF1α transcript and protein (Newton et al. 2010). These data indicate a complex interaction between HIF1α, TCF/LEFs, and β-catenin, which may differ between cell types.

Given the large number of transcription factors that can bind and recruit β-catenin to regulatory sequences, how much of Wnt/β-catenin is mediated through TCFs in vertebrate systems? To answer this question, mutations that specifically abolish β-catenin binding will have to be examined in vivo. For example, the vitamin D receptor (VDR) a type II nuclear receptor, can bind β-catenin independent of vitamin D to promote transcription (Palmer et al. 2001; Malloy et al. 2002; Shah et al. 2006). Recessive mutations of the human VDR gene result in hereditary vitamin D-dependent rickets (HVDDR) and are usually accompanied by epidermal defects such as alopecia, i.e., hair loss (Malloy and Feldman 2011). However, some VDR alleles display HVDDR without alopecia (Malloy et al. 2002; Malloy and Feldman 2011). Characterization of one such mutation showed that it lacked the ability to respond to vitamin D (Malloy et al. 2002) but still retained the ability to be activated by β-catenin (Shah et al. 2006; Palmer et al. 2008). Given recent data that β-catenin and VDR act together to promote hair follicle development (Palmer et al. 2008; Baker et al. 2010), these data suggest that VDR acts in bones primarily through vitamin D and through β-catenin in hair formation. Although this model requires further testing, it provides a blueprint for dissecting the importance of β-catenin interactions for other transcription factors.

CONCLUDING REMARKS

The transcription regulatory actions of β-catenin are much more varied and complex than originally assumed when TCF/LEFs were first identified as a β-catenin partner. This review has highlighted how β-catenin participates in the switch-in/switch-out of different TCFs at target genes, how it recruits varied and multiple histone modifying complexes, and how it interacts with other types of transcription factors. Likewise, we have reviewed how the transcription regulatory actions of TCF/LEFs are equally complex and powerful. Genomes with multiple TCF/LEFs have evolved individual family members with specialized functions, a feature that might explain how some TCFs can act as growth suppressors or tumor suppressors and how some TCFs access selective subsets of target genes. Multiple isoforms from alternative promoters and alternative splicing create many of these specialized functions. Defining the regulatory networks that control isoform expression is not only important for understanding how each TCF/LEF gene functions in stem cells and their differentiating progeny, but it is important for understanding how these isoforms participate (or oppose) disease states such as diabetes or cancer. Tools that detect specific TCF/LEF forms, and mutations that selectively remove one form or another will be critical for this understanding.

ACKNOWLEDGMENTS

We acknowledge members of the Cadigan and Waterman laboratories for helpful discussions. The work is supported by NIH grant GM082994 and NSF grant 0950348 to K.M.C., and CA096878, CA108697, and CIRM #RB2-01629 to M.L.W.

REFERENCES

Adamska M, Larroux C, Adamski M, Green K, Lovas E, Koop D, Richards GS, Zwafink C, Degnan BM. 2010. Structure and expression of conserved Wnt pathway components in the desmosponge Amphimedon queenslandica. *Evol Dev* **12:** 494–518.

Almeida M, Han L, Martin-Millan M, O'Brien CA, Manolagas SC. 2007. Oxidative stress antagonizes Wnt signaling in osteoblast precursors by diverting β-catenin from T cell factor- to forkhead box O-mediated transcription. *J Biol Chem* **282:** 27298–27305.

Angus-Hill ML, Elbert KM, Hidalgo J, Capecchi MR. 2011. T-cell factor 4 functions as a tumor suppressor whose disruption modulates colon cell proliferation and tumorigenesis. *Proc Natl Acad Sci* **108:** 4914–4919.

Arce L, Yokoyama NN, Waterman ML. 2006. Diversity of LEF/TCF action in development and disease. *Oncogene* **25:** 7492–7504.

Arce L, Pate KT, Waterman ML. 2009. Groucho binds two conserved regions of LEF-1 for HDAC-dependent repression. *BMC Cancer* **9:** 159.

Archbold HC, Yang YX, Chen L, Cadigan KM. 2012. How do they do Wnt they do?: Regulation of transcription by the Wnt/β-catenin pathway. *Acta Physiol (Oxford)* **204:** 74–109.

Atcha FA, Syed A, Wu B, Hoverter NP, Yokoyama NN, Ting JH, Munguia JE, Mangalam HJ, Marsh JL, Waterman ML. 2007. A unique DNA binding domain converts T-cell factors into strong Wnt effectors. *Mol Cell Biol* **27:** 8352–8363.

Badis G, Berger MF, Philippakis AA, Talukder S, Gehrke AR, Jaeger SA, Chan ET, Metzler G, Vedenko A, Chen X, et al. 2009. Diversity and complexity in DNA recognition by transcription factors. *Science* **324:** 1720–1723.

Baker CM, Verstuyf A, Jensen KB, Watt FM. 2010. Differential sensitivity of epidermal cell subpopulations to β-catenin-induced ectopic hair follicle formation. *Dev Biol* **343:** 40–50.

Bass AJ, Lawrence MS, Brace LE, Ramos AH, Drier Y, Cibulskis K, Sougnez C, Voet D, Saksena G, Sivachenko A, et al. 2011. Genomic sequencing of colorectal adenocarcinomas identifies a recurrent VTI1A-TCF7L2 fusion. *Nat Genet* **43:** 964–968.

Behrens J, von Kries JP, Kuhl M, Bruhn L, Wedlich D, Grosschedl R, Birchmeier W. 1996. Functional interaction of β-catenin with the transcription factor LEF-1. *Nature* **382:** 638–642.

Beildeck ME, Gelmann EP, Byers SW. 2010. Cross-regulation of signaling pathways: An example of nuclear hormone receptors and the canonical Wnt pathway. *Exp Cell Res* **316:** 1763–1772.

Berger SL. 2007. The complex language of chromatin regulation during transcription. *Nature* **447:** 407–412.

Berriman M, Haas BJ, LoVerde PT, Wilson RA, Dillon GP, Cerqueira GC, Mashiyama ST, Al-Lazikani B, Andrade LF, Ashton PD, et al. 2009. The genome of the blood fluke *Schistosoma mansoni*. *Nature* **460:** 352–358.

Bhambhani C, Chang JL, Akey DL, Cadigan KM. 2011. The oligomeric state of CtBP determines its role as a transcriptional co-activator and co-repressor of Wingless targets. *EMBO J* **30:** 2031–2043.

Bierie B, Nozawa M, Renou JP, Shillingford JM, Morgan F, Oka T, Taketo MM, Cardiff RD, Miyoshi K, Wagner KU, et al. 2003. Activation of β-catenin in prostate epithelium induces hyperplasias and squamous transdifferentiation. *Oncogene* **22:** 3875–3887.

Billin AN, Thirlwell H, Ayer DE. 2000. β-Catenin-histone deacetylase interactions regulate the transition of LEF1 from a transcriptional repressor to an activator. *Mol Cell Biol* **20:** 6882–6890.

Blahnik KR, Dou L, O'Geen H, McPhillips T, Xu X, Cao AR, Iyengar S, Nicolet CM, Ludäscher B, Korf I, et al. 2010. Sole-Search: An integrated analysis program for peak detection and functional annotation using ChIP-seq data. *Nucleic Acids Res* **38:** e13.

Blythe SA, Cha SW, Tadjuidje E, Heasman J, Klein PS. 2010. β-Catenin primes organizer gene expression by recruiting a histone H3 arginine 8 methyltransferase, Prmt2. *Dev Cell* **19:** 220–231.

Bottomly D, Kyler SL, McWeeney SK, Yochum GS. 2010. Identification of β-catenin binding regions in colon cancer cells using ChIP-Seq. *Nucleic Acids Res* **38:** 5735–5745.

Brannon M, Brown JD, Bates R, Kimelman D, Moon RT. 1999. XCtBP is a XTcf-3 co-repressor with roles throughout Xenopus development. *Development (Cambridge, England)* **126:** 3159–3170.

Brantjes H, Roose J, van De Wetering M, Clevers H. 2001. All Tcf HMG box transcription factors interact with Groucho-related co-repressors. *Nucleic Acids Res* **29:** 1410–1419.

Brunner E, Peter O, Schweizer L, Basler K. 1997. Pangolin encodes a Lef-1 homologue that acts downstream of Armadillo to transduce the Wingless signal in *Drosophila*. *Nature* **385:** 829–833.

Cadigan KM. 2012. TCFs and Wnt/β-catenin signaling more than one way to throw the switch. *Curr Top Dev Biol* **98:** 1–34.

Cadigan KM, Peifer M. 2009. Wnt signaling from development to disease: Insights from model systems. *Cold Spring Harb Perspect Biol* **1:** 002881.

Calvo D, Victor M, Gay F, Sui G, Luke MP, Dufourcq P, Wen G, Maduro M, Rothman J, Shi Y. 2001. A POP-1 repressor complex restricts inappropriate cell type-specific gene transcription during *Caenorhabditis elegans* embryogenesis. *EMBO J* **20:** 7197–7208.

Carrera I, Janody F, Leeds N, Duveau F, Treisman JE. 2008. Pygopus activates Wingless target gene transcription through the mediator complex subunits Med12 and Med13. *Proc Natl Acad Sci* **105:** 6644–6649.

Cauchi S, El Achhab Y, Choquet H, Dina C, Krempler F, Weitgasser R, Nejjari C, Patsch W, Chikri M, Meyre D, et al. 2007. TCF7L2 is reproducibly associated with type 2 diabetes in various ethnic groups: A global meta-analysis. *J Mol Med (Berl)* **85:** 777–782.

Cavallo RA, Cox RT, Moline MM, Roose J, Polevoy GA, Clevers H, Peifer M, Bejsovec A. 1998. *Drosophila* Tcf and Groucho interact to repress Wingless signalling activity. *Nature* **395:** 604–608.

Chang MV, Chang JL, Gangopadhyay A, Shearer A, Cadigan KM. 2008. Activation of wingless targets requires bipartite recognition of DNA by TCF. *Curr Biol* **18:** 1877–1881.

Chen G, Fernandez J, Mische S, Courey AJ. 1999. A functional interaction between the histone deacetylase Rpd3 and the corepressor groucho in *Drosophila* development. *Genes Dev* **13:** 2218–2230.

Chen SY, Wulf G, Zhou XZ, Rubin MA, Lu KP, Balk SP. 2006. Activation of β-catenin signaling in prostate cancer by peptidyl-prolyl isomerase Pin1-mediated abrogation of the androgen receptor-β-catenin interaction. *Mol Cell Biol* **26:** 929–939.

Chen J, Luo Q, Yuan Y, Huang X, Cai W, Li C, Wei T, Zhang L, Yang M, Liu Q, et al. 2010. Pygo2 associates with MLL2 histone methyltransferase and GCN5 histone acetyltransferase complexes to augment Wnt target gene expression and breast cancer stem-like cell expansion. *Mol Cell Biol* **30:** 5621–5635.

Chesire DR, Isaacs WB. 2002. Ligand-dependent inhibition of β-catenin/TCF signaling by androgen receptor. *Oncogene* **21:** 8453–8469.

Chesire DR, Ewing CM, Sauvageot J, Bova GS, Isaacs WB. 2000. Detection and analysis of β-catenin mutations in prostate cancer. *The Prostate* **45:** 323–334.

Cole MF, Johnstone SE, Newman JJ, Kagey MH, Young RA. 2008. Tcf3 is an integral component of the core regulatory circuitry of embryonic stem cells. *Genes Dev* **22:** 746–755.

Cronauer MV, Schulz WA, Ackermann R, Burchardt M. 2005. Effects of WNT/β-catenin pathway activation on signaling through T-cell factor and androgen receptor in prostate cancer cell lines. *Int J Oncol* **26:** 1033–1040.

Culig Z, Klocker H, Bartsch G, Hobisch A. 2002. Androgen receptors in prostate cancer. *Endocr Relat Cancer* **9:** 155–170.

Daniels DL, Weis WI. 2005. β-Catenin directly displaces Groucho/TLE repressors from Tcf/Lef in Wnt-mediated transcription activation. *Nat Struct Mol Biol* **12:** 364–371.

Doria A, Patti ME, Kahn CR. 2008. The emerging genetic architecture of type 2 diabetes. *Cell Metab* **8:** 186–200.

Duval A, Rolland S, Tubacher E, Bui H, Thomas G, Hamelin R. 2000. The human T-cell transcription factor-4 gene: Structure, extensive characterization of alternative splicings, and mutational analysis in colorectal cancer cell lines. *Cancer Res* **60:** 3872–3879.

Essers MA, de Vries-Smits LM, Barker N, Polderman PE, Burgering BM, Korswagen HC. 2005. Functional interaction between β-catenin and FOXO in oxidative stress signaling. *Science* **308:** 1181–1184.

Fang M, Li J, Blauwkamp T, Bhambhani C, Campbell N, Cadigan KM. 2006. C-Terminal-binding protein directly activates and represses Wnt transcriptional targets in *Drosophila*. *EMBO J* **25:** 2735–2745.

Fiedler M, Sanchez-Barrena MJ, Nekrasov M, Mieszczanek J, Rybin V, Muller J, Evans P, Bienz M. 2008. Decoding of methylated histone H3 tail by the Pygo-BCL9 Wnt signaling complex. *Mol Cell* **30:** 507–518.

Galceran J, Farinas I, Depew MJ, Clevers H, Grosschedl R. 1999. Wnt3a$^{-/-}$like phenotype and limb deficiency in Lef1$^{-/-}$Tcf1$^{-/-}$ mice. *Genes Dev* **13:** 709–717.

Ghogomu SM, van Venrooy S, Ritthaler M, Wedlich D, Gradl D. 2006. HIC-5 is a novel repressor of lymphoid enhancer factor/T-cell factor-driven transcription. *J Biol Chem* **281:** 1755–1764.

Giese K, Amsterdam A, Grosschedl R. 1991. DNA-binding properties of the HMG domain of the lymphoid-specific transcriptional regulator LEF-1. *Genes Dev* **5:** 2567–2578.

Giese K, Kingsley C, Kirshner JR, Grosschedl R. 1995. Assembly and function of a TCR α enhancer complex is dependent on LEF-1-induced DNA bending and multiple protein-protein interactions. *Genes Dev* **9:** 995–1008.

Goentoro L, Kirschner MW. 2009. Evidence that fold-change, and not absolute level, of β-catenin dictates Wnt signaling. *Mol Cell* **36:** 872–884.

Gradl D, Konig A, Wedlich D. 2002. Functional diversity of Xenopus lymphoid enhancer factor/T-cell factor transcription factors relies on combinations of activating and repressing elements. *J Biol Chem* **277:** 14159–14171.

Graham TA, Weaver C, Mao F, Kimelman D, Xu W. 2000. Crystal structure of a β-catenin/Tcf complex. *Cell* **103:** 885–896.

Grant SF, Thorleifsson G, Reynisdottir I, Benediktsson R, Manolescu A, Sainz J, Helgason A, Stefansson H, Emilsson V, Helgadottir A, et al. 2006. Variant of transcription factor 7-like 2 (TCF7L2) gene confers risk of type 2 diabetes. *Nat Genet* **38:** 320–323.

Gregorieff A, Grosschedl R, Clevers H. 2004. Hindgut defects and transformation of the gastro-intestinal tract in Tcf4$^{-/-}$/Tcf1$^{-/-}$ embryos. *EMBO J* **23:** 1825–1833.

Gu B, Sun P, Yuan Y, Moraes RC, Li A, Teng A, Agrawal A, Rheaume C, Bilanchone V, Veltmaat JM, et al. 2009. Pygo2 expands mammary progenitor cells by facilitating histone H3 K4 methylation. *J Cell Biol* **185:** 811–826.

Haegebarth A, Clevers H. 2009. Wnt signaling, lgr5, and stem cells in the intestine and skin. *Am J Pathol* **174:** 715–721.

Hallikas O, Taipale J. 2006. High-throughput assay for determining specificity and affinity of protein-DNA binding interactions. *Nat Protoc* **1:** 215–222.

Hamada F, Bienz M. 2004. The APC tumor suppressor binds to C-terminal binding protein to divert nuclear beta-catenin from TCF. *Dev Cell* **7:** 677–685.

Hatzis P, van der Flier LG, van Driel MA, Guryev V, Nielsen F, Denissov S, Nijman IJ, Koster J, Santo EE, Welboren W, et al. 2008. Genome-wide pattern of TCF7L2/TCF4 chromatin occupancy in colorectal cancer cells. *Mol Cell Biol* **28:** 2732–2744.

Herman M. 2001. *C. elegans* POP-1/TCF functions in a canonical Wnt pathway that controls cell migration and in a noncanonical Wnt pathway that controls cell polarity. *Development (Cambridge, England)* **128:** 581–590.

Hikasa H, Sokol SY. 2011. Phosphorylation of TCF proteins by homeodomain-interacting protein kinase 2. *J Biol Chem* **286:** 12093–12100.

Hikasa H, Ezan J, Itoh K, Li X, Klymkowsky MW, Sokol SY. 2010. Regulation of TCF3 by Wnt-dependent phosphorylation during vertebrate axis specification. *Dev Cell* **19:** 521–532.

Houston DW, Kofron M, Resnik E, Langland R, Destree O, Wylie C, Heasman J. 2002. Repression of organizer genes in dorsal and ventral Xenopus cells mediated by maternal XTcf3. *Development (Cambridge, England)* **129:** 4015–4025.

Hovanes K, Li TWH, Waterman ML. 2000. The human *LEF-1* gene contains a promoter preferentially active in lymphocytes and encodes multiple isoforms derived from alternative splicing. *Nucleic Acids Res* **28:** 1994–2003.

Hovanes K, Li TW, Munguia JE, Truong T, Milovanovic T, Lawrence Marsh J, Holcombe RF, Waterman ML. 2001. β-Catenin-sensitive isoforms of lymphoid enhancer factor-1 are selectively expressed in colon cancer. *Nat Genet* **28:** 53–57.

Hsu S-C, Galceran J, Grosschedl R. 1998. Modulation of transcriptional regulation by LEF-1 in response to Wnt-1 signaling and association with β-catenin. *Mol Cell Biol* **18:** 4807–4818.

Huber O, Korn R, McLaughlin J, Ohsugi M, Herrmann B, Kemler R. 1996. Nuclear localization of β-catenin by interaction with transcription factor LEF-1. *Mech Dev* **59:** 3–10.

Ishitani T, Ninomiya-Tsuji J, Nagai S, Nishita M, Meneghini M, Barker N, Waterman M, Bowerman B, Clevers H, Shibuya H, et al. 1999. The TAK1-NLK-MAPK-related pathway antagonizes signalling between β-catenin and transcription factor TCF. *Nature* **399:** 798–802.

Ishitani T, Kishida S, Hyodo-Miura J, Ueno N, Yasuda J, Waterman M, Shibuya H, Moon RT, Ninomiya-Tsuji J, Matsumoto K. 2003. The TAK1-NLK mitogen-activated protein kinase cascade functions in the Wnt-5a/Ca^{2+} pathway to antagonize Wnt/β-catenin signaling. *Mol Cell Biol* **23:** 131–139.

Jessen S, Gu B, Dai X. 2008. Pygopus and the Wnt signaling pathway: A diverse set of connections. *Bioessays* **30:** 448–456.

Jimenez J, Jang GM, Semler BL, Waterman ML. 2005. An internal ribosome entry site mediates translation of lymphoid enhancer factor-1. *RNA* **11:** 1385–1399.

Kaidi A, Williams AC, Paraskeva C. 2007. Interaction between β-catenin and HIF-1 promotes cellular adaptation to hypoxia. *Nat Cell Biol* **9:** 210–217.

Kennell JA, O'Leary EE, Gummow BM, Hammer GD, MacDougald OA. 2003. T-cell factor 4N (TCF-4N), a novel isoform of mouse TCF-4, synergizes with β-catenin to coactivate C/EBPα and steroidogenic factor 1 transcription factors. *Mol Cell Biol* **23:** 5366–5375.

Kessler R, Hausmann G, Basler K. 2009. The PHD domain is required to link *Drosophila* Pygopus to Legless/β-catenin and not to histone H3. *Mech Dev* **126:** 752–759.

Kidd AR III, Miskowski JA, Siegfried KR, Sawa H, Kimble J. 2005. A β-catenin identified by functional rather than sequence criteria and its role in Wnt/MAPK signaling. *Cell* **121:** 761–772.

Kim CH, Oda T, Itoh M, Jiang D, Artinger KB, Chandrasekharappa SC, Driever W, Chitnis AB. 2000. Repressor activity of Headless/Tcf3 is essential for vertebrate head formation. *Nature* **407:** 913–916.

Kim CH, Neiswender H, Baik EJ, Xiong WC, Mei L. 2008. β-Catenin interacts with MyoD and regulates its transcription activity. *Mol Cell Biol* **28:** 2941–2951.

King N, Westbrook MJ, Young SL, Kuo A, Abedin M, Chapman J, Fairclough S, Hellsten U, Isogai Y, Letunic I, et al. 2008. The genome of the choanoflagellate Monosiga brevicollis and the origin of metazoans. *Nature* **451:** 783–788.

Kioussi C, Briata P, Baek SH, Rose DW, Hamblet NS, Herman T, Ohgi KA, Lin C, Gleiberman A, Wang J, et al. 2002. Identification of a Wnt/Dvl/β-catenin →Pitx2 pathway mediating cell-type-specific proliferation during development. *Cell* **111:** 673–685.

Knirr S, Frasch M. 2001. Molecular integration of inductive and mesoderm-intrinsic inputs governs even-skipped enhancer activity in a subset of pericardial and dorsal muscle progenitors. *Dev Biol* **238:** 13–26.

Korinek V, Barker N, Morin PJ, van Wichen D, de Weger R, Kinzler KW, Vogelstein B, Clevers H. 1997. Constitutive transcriptional activation by a β-catenin-Tcf complex in $APC^{-/-}$ colon carcinoma. *Science* **275:** 1784–1787.

Korinek V, Barker N, Moerer P, van Donselaar E, Huls G, Peters PJ, Clevers H. 1998a. Depletion of epithelial stem-cell compartments in the small intestine of mice lacking Tcf-4. *Nat Genet* **19:** 379–383.

Korinek V, Barker N, Willert K, Molenaar M, Roose J, Wagenaar G, Markman M, Lamers W, Destree O, Clevers H. 1998b. Two members of the Tcf family implicated in Wnt/β-catenin signaling during embryogenesis in the mouse. *Mol Cell Biol* **18:** 1248–1256.

Kormish JD, Sinner D, Zorn AM. 2010. Interactions between SOX factors and Wnt/β-catenin signaling in development and disease. *Dev Dyn* **239:** 56–68.

Kramps T, Peter O, Brunner E, Nellen D, Froesch B, Chatterjee S, Murone M, Zullig S, Basler K. 2002. Wnt/wingless signaling requires BCL9/legless-mediated recruitment of pygopus to the nuclear β-catenin-TCF complex. *Cell* **109:** 47–60.

Kratochwil K, Galceran J, Tontsch S, Roth W, Grosschedl R. 2002. FGF4, a direct target of LEF1 and Wnt signaling, can rescue the arrest of tooth organogenesis in $Lef1^{-/-}$ mice. *Genes Dev* **16:** 3173–3185.

Lam N, Chesney MA, Kimble J. 2006. Wnt signaling and CEH-22/tinman/Nkx2.5 specify a stem cell niche in *C. elegans*. *Curr Biol* **16:** 287–295.

Laudes M. 2011. Role of WNT signalling in the determination of human mesenchymal stem cells into preadipocytes. *J Mol Endocrinol* **46:** R65–R72.

Le Bacquer O, Shu L, Marchand M, Neve B, Paroni F, Kerr Conte J, Pattou F, Froguel P, Maedler K. 2011. TCF7L2 splice variants have distinct effects on β-cell turnover and function. *Hum Mol Genet* **20:** 1906–1915.

Li TW, Ting JH, Yokoyama NN, Bernstein A, van de Wetering M, Waterman ML. 2006. Wnt activation and alternative promoter repression of LEF1 in colon cancer. *Mol Cell Biol* **26:** 5284–5299.

Li Z, Nie F, Wang S, Li L. 2011. Histone H4 Lys 20 monomethylation by histone methylase SET8 mediates Wnt target gene activation. *Proc Natl Acad Sci* **108:** 3116–3123.

Lin R, Hill RJ, Priess JR. 1998. POP-1 and anterior-posterior fate decisions in *C. elegans* embryos. *Cell* **92:** 229–239.

Linggi BE, Brandt SJ, Sun ZW, Hiebert SW. 2005. Translating the histone code into leukemia. *J Cell Biochem* **96:** 938–950.

Liu F, van den Broek O, Destree O, Hoppler S. 2005. Distinct roles for Xenopus Tcf/Lef genes in mediating specific responses to Wnt/β-catenin signalling in mesoderm development. *Development (Cambridge, England)* **132:** 5375–5385.

Liu J, Phillips BT, Amaya MF, Kimble J, Xu W. 2008a. The *C. elegans* SYS-1 protein is a bona fide β-catenin. *Dev Cell* **14:** 751–761.

Liu YI, Chang MV, Li HE, Barolo S, Chang JL, Blauwkamp TA, Cadigan KM. 2008b. The chromatin remodelers ISWI and ACF1 directly repress Wingless transcriptional targets. *Dev Biol* **323:** 41–52.

Lluis F, Ombrato L, Pedone E, Pepe S, Merrill BJ, Cosma MP. 2011. T-cell factor 3 (Tcf3) deletion increases somatic cell reprogramming by inducing epigenome modifications. *Proc Natl Acad Sci* **108:** 11912–11917.

Lo MC, Gay F, Odom R, Shi Y, Lin R. 2004. Phosphorylation by the β-catenin/MAPK complex promotes 14-3-3-mediated nuclear export of TCF/POP-1 in signal-responsive cells in *C. elegans*. *Cell* **117:** 95–106.

Locke JM, Da Silva Xavier G, Rutter GA, Harries LW. 2011. An alternative polyadenylation signal in TCF7L2 generates isoforms that inhibit T cell factor/lymphoid-enhancer factor (TCF/LEF)-dependent target genes. *Diabetologia* **54:** 3078–3082.

Logan CY, Nusse R. 2004. The Wnt signaling pathway in development and disease. *Annu Rev Cell Dev Biol* **20:** 781–810.

Love JJ, Li X, Case DA, Giese K, Grosschedl R, Wright PE. 1995. Structural basis for DNA bending by the architectural transcription factor LEF-1. *Nature* **376:** 791–795.

Lyssenko V, Lupi R, Marchetti P, Del Guerra S, Orho-Melander M, Almgren P, Sjogren M, Ling C, Eriksson KF, Lethagen AL, et al. 2007. Mechanisms by which common variants in the TCF7L2 gene increase risk of type 2 diabetes. *J Clin Invest* **117:** 2155–2163.

Lyu J, Jho EH, Lu W. 2011. Smek promotes histone deacetylation to suppress transcription of Wnt target gene brachyury in pluripotent embryonic stem cells. *Cell Res* **21:** 911–921.

Maduro MF, Lin R, Rothman JH. 2002. Dynamics of a developmental switch: Recursive intracellular and intranuclear redistribution of *Caenorhabditis elegans* POP-1 parallels Wnt-inhibited transcriptional repression. *Dev Biol* **248:** 128–142.

Mahmoudi T, Li VS, Ng SS, Taouatas N, Vries RG, Mohammed S, Heck AJ, Clevers H. 2009. The kinase TNIK is an essential activator of Wnt target genes. *EMBO J* **28:** 3329–3340.

Mahmoudi T, Boj SF, Hatzis P, Li VS, Taouatas N, Vries RG, Teunissen H, Begthel H, Korving J, Mohammed S, et al. 2010. The leukemia-associated Mllt10/Af10-Dot1l are Tcf4/β-catenin coactivators essential for intestinal homeostasis. *PLoS Biol* **8:** e1000539.

Malloy PJ, Feldman D. 2011. The role of vitamin D receptor mutations in the development of alopecia. *Mol Cell Endocrinol* **347:** 90–96.

Malloy PJ, Xu R, Peng L, Clark PA, Feldman D. 2002. A novel mutation in helix 12 of the vitamin D receptor impairs coactivator interaction and causes hereditary 1,25-dihydroxyvitamin D-resistant rickets without alopecia. *Mol Endocrinol (Baltimore, MD)* **16:** 2538–2546.

Masiello D, Chen SY, Xu Y, Verhoeven MC, Choi E, Hollenberg AN, Balk SP. 2004. Recruitment of β-catenin by wild-type or mutant androgen receptors correlates with ligand-stimulated growth of prostate cancer cells. *Mol Endocrinol (Baltimore, MD)* **18:** 2388–2401.

Mazumdar J, O'Brien WT, Johnson RS, LaManna JC, Chavez JC, Klein PS, Simon MC. 2010. O2 regulates stem cells through Wnt/β-catenin signalling. *Nat Cell Biol* **12:** 1007–1013.

Merrill BJ, Pasolli HA, Polak L, Rendl M, Garcia-Garcia MJ, Anderson KV, Fuchs E. 2004. Tcf3: A transcriptional regulator of axis induction in the early embryo. *Development (Cambridge, England)* **131:** 263–274.

Mohan M, Herz HM, Takahashi YH, Lin C, Lai KC, Zhang Y, Washburn MP, Florens L, Shilatifard A. 2010. Linking H3K79 trimethylation to Wnt signaling through a novel Dot1-containing complex (DotCom). *Genes Dev* **24:** 574–589.

Molenaar M, van de Wetering M, Oosterwegel M, Peterson-Maduro J, Godsave S, Korinek V, Roose J, Destree O, Clevers H. 1996. XTCF-3 transcription factor mediates β-catenin-induced axis formation in *Xenopus* embryos. *Cell* **86:** 391–399.

Moore AC, Amann JM, Williams CS, Tahinci E, Farmer TE, Martinez JA, Yang G, Luce KS, Lee E, Hiebert SW. 2008. Myeloid translocation gene family members associate with T-cell factors (TCFs) and influence TCF-dependent transcription. *Mol Cell Biol* **28:** 977–987.

Mosimann C, Hausmann G, Basler K. 2009. β-Catenin hits chromatin: Regulation of Wnt target gene activation. *Nat Rev* **10:** 276–286.

Mulholland DJ, Cheng H, Reid K, Rennie PS, Nelson CC. 2002. The androgen receptor can promote β-catenin nuclear translocation independently of adenomatous polyposis coli. *J Biol Chem* **277:** 17933–17943.

Mulholland DJ, Dedhar S, Coetzee GA, Nelson CC. 2005. Interaction of nuclear receptors with the Wnt/β-catenin/Tcf signaling axis: Wnt you like to know? *Endocr Rev* **26:** 898–915.

Mullighan CG, Zhang J, Kasper LH, Lerach S, Payne-Turner D, Phillips LA, Heatley SL, Holmfeldt L, Collins-Underwood JR, Ma J, et al. 2011. CREBBP mutations in relapsed acute lymphoblastic leukaemia. *Nature* **471:** 235–239.

Najdi R, Syed A, Arce L, Theisen H, Ting JH, Atcha F, Nguyen AV, Martinez M, Holcombe RF, Edwards RA, et al. 2009. AWnt kinase network alters nuclear localization of TCF-1 in colon cancer. *Oncogene* **28:** 4133–4146.

Newton IP, Kenneth NS, Appleton PL, Nathke I, Rocha S. 2010. Adenomatous polyposis coli and hypoxia-inducible factor-1α have an antagonistic connection. *Mol Cell Biol* **21:** 3630–3638.

Nguyen H, Merrill BJ, Polak L, Nikolova M, Rendl M, Shaver TM, Pasolli HA, Fuchs E. 2009. Tcf3 and Tcf4 are essential for long-term homeostasis of skin epithelia. *Nat Genet* **41:** 1068–1075.

Niu Y, Chang TM, Yeh S, Ma WL, Wang YZ, Chang C. 2010. Differential androgen receptor signals in different cells explain why androgen-deprivation therapy of prostate cancer fails. *Oncogene* **29:** 3593–3604.

Norton L, Fourcaudot M, Abdul-Ghani MA, Winnier D, Mehta FF, Jenkinson CP, Defronzo RA. 2011. Chromatin occupancy of transcription factor 7-like 2 (TCF7L2) and its role in hepatic glucose metabolism. *Diabetologia* **54:** 3132–3142.

Olson LE, Tollkuhn J, Scafoglio C, Krones A, Zhang J, Ohgi KA, Wu W, Taketo MM, Kemler R, Grosschedl R, et al. 2006. Homeodomain-mediated β-catenin-dependent switching events dictate cell-lineage determination. *Cell* **125:** 593–605.

Oosterwegel M, van de Wetering M, Timmerman J, Kruisbeek A, Destree O, Meijlink F, Clevers H. 1993. Differential expression of the HMG box factors TCF-1 and LEF-1 during murine embryogenesis. *Development (Cambridge, England)* **118:** 439–448.

Orsulic S, Peifer M. 1996. An in vivo structure-function study of armadillo, the β-catenin homologue, reveals both separate and overlapping regions of the protein required for cell adhesion and for wingless signaling. *J Cell Biol* **134:** 1283–1300.

Ou CY, LaBonte MJ, Manegold PC, So AY, Ianculescu I, Gerke DS, Yamamoto KR, Ladner RD, Kahn M, Kim JH, et al. 2011. A coactivator role of CARM1 in the dysregulation of β-catenin activity in colorectal cancer cell growth and gene expression. *Mol Cancer Res* **9:** 660–670.

Palmer HG, Gonzalez-Sancho JM, Espada J, Berciano MT, Puig I, Baulida J, Quintanilla M, Cano A, de Herreros AG, Lafarga M, et al. 2001. Vitamin D(3) promotes the differentiation of colon carcinoma cells by the induction of E-cadherin and the inhibition of β-catenin signaling. *J Cell Biol* **154:** 369–387.

Palmer HG, Anjos-Afonso F, Carmeliet G, Takeda H, Watt FM. 2008. The vitamin D receptor is a Wnt effector that controls hair follicle differentiation and specifies tumor type in adult epidermis. *PLoS ONE* **3:** e1483.

Parker DS, Blauwkamp T, Cadigan KM. 2007. Wnt/β-catenin-mediated transcriptional regulation. *Adv Devel Biol* **17:** 1–61.

Parker DS, Ni YY, Chang JL, Li J, Cadigan KM. 2008. Wingless signaling induces widespread chromatin remodeling of target loci. *Mol Cell Biol* **28:** 1815–1828.

Pawlowski JE, Ertel JR, Allen MP, Xu M, Butler C, Wilson EM, Wierman ME. 2002. Liganded androgen receptor interaction with β-catenin: Nuclear co-localization and modulation of transcriptional activity in neuronal cells. *J Biol Chem* **277:** 20702–20710.

Phillips BT, Kimble J. 2009. A new look at TCF and β-catenin through the lens of a divergent *C. elegans* Wnt pathway. *Dev Cell* **17:** 27–34.

Polakis P. 2007. The many ways of Wnt in cancer. *Curr Opin Genet Dev* **17:** 45–51.

Poy F, Lepourcelet M, Shivdasani RA, Eck MJ. 2001. Structure of a human Tcf4-β-catenin complex. *Nat Struct Biol* **8:** 1053–1057.

Prieve MG, Guttridge KL, Munguia JE, Waterman ML. 1998. Differential importin-α recognition and nuclear transport by nuclear localization signals within the high-mobility-group DNA binding domains of lymphoid enhancer factor 1 and T-cell factor 1. *Mol Cell Biol* **18:** 4819–4832.

Prokunina-Olsson L, Kaplan LM, Schadt EE, Collins FS. 2009a. Alternative splicing of TCF7L2 gene in omental and subcutaneous adipose tissue and risk of type 2 diabetes. *PLoS ONE* **4:** e7231.

Prokunina-Olsson L, Welch C, Hansson O, Adhikari N, Scott LJ, Usher N, Tong M, Sprau A, Swift A, Bonnycastle LL, et al. 2009b. Tissue-specific alternative splicing of TCF7L2. *Hum Mol Genet* **18:** 3795–3804.

Pukrop T, Gradl D, Henningfeld KA, Knochel W, Wedlich D, Kuhl M. 2001. Identification of two regulatory elements within the high mobility group box transcription factor XTCF-4. *J Biol Chem* **276:** 8968–8978.

Read JT, Rahmani M, Boroomand S, Allahverdian S, McManus BM, Rennie PS. 2007. Androgen receptor regulation of the versican gene through an androgen response element in the proximal promoter. *J Biol Chem* **282:** 31954–31963.

Reya T, O'Riordan M, Okamura R, Devaney E, Willert K, Nusse R, Grosschedl R. 2000. Wnt signaling regulates B lymphocyte proliferation through a LEF-1 dependent mechanism. *Immunity* **13:** 15–24.

Rios-Doria J, Kuefer R, Ethier SP, Day ML. 2004. Cleavage of β-catenin by calpain in prostate and mammary tumor cells. *Cancer Res* **64:** 7237–7240.

Rocheleau CE, Downs WD, Lin R, Wittmann C, Bei Y, Cha YH, Ali M, Priess JR, Mello CC. 1997. Wnt signaling and an APC-related gene specify endoderm in early *C. elegans* embryos. *Cell* **90:** 707–716.

Roose J, Molenaar M, Peterson J, Hurenkamp J, Brantjes H, Moerer P, van de Wetering M, Destree O, Clevers H. 1998. The Xenopus Wnt effector XTcf-3 interacts with Groucho-related transcriptional repressors. *Nature* **395:** 608–612.

Roose J, Huls G, van Beest M, Moerer P, van der Horn K, Goldschmeding R, Logtenberg T, Clevers H. 1999. Synergy between tumor suppressor APC and the β-catenin-Tcf4 target Tcf1. *Science* **285:** 1923–1926.

Satow R, Shitashige M, Jigami T, Honda K, Ono M, Hirohashi S, Yamada T. 2010. Traf2- and Nck-interacting kinase is essential for canonical Wnt signaling in Xenopus axis formation. *J Biol Chem* **285:** 26289–26294.

Savic D, Ye H, Aneas I, Park SY, Bell GI, Nobrega MA. 2011. Alterations in TCF7L2 expression define its role as a key regulator of glucose metabolism. *Genome Res* **21:** 1417–1425.

Schweizer L, Rizzo CA, Spires TE, Platero JS, Wu Q, Lin TA, Gottardis MM, Attar RM. 2008. The androgen receptor can signal through Wnt/β-catenin in prostate cancer cells as an adaptation mechanism to castration levels of androgens. *BMC Cell Biol* **9:** 4.

Shah S, Islam MN, Dakshanamurthy S, Rizvi I, Rao M, Herrell R, Zinser G, Valrance M, Aranda A, Moras D, et al. 2006. The molecular basis of vitamin D receptor and β-catenin crossregulation. *Mol Cell* **21:** 799–809.

Shetty P, Lo MC, Robertson SM, Lin R. 2005. *C. elegans* TCF protein, POP-1, converts from repressor to activator as a result of Wnt-induced lowering of nuclear levels. *Dev Biol* **285:** 584–592.

Shitashige M, Satow R, Jigami T, Aoki K, Honda K, Shibata T, Ono M, Hirohashi S, Yamada T. 2010. Traf2- and Nck-interacting kinase is essential for Wnt signaling and colorectal cancer growth. *Cancer Res* **70:** 5024–5033.

Sierra J, Yoshida T, Joazeiro CA, Jones KA. 2006. The APC tumor suppressor counteracts β-catenin activation and H3K4 methylation at Wnt target genes. *Genes Dev* **20:** 586–600.

Sjoblom T, Jones S, Wood LD, Parsons DW, Lin J, Barber TD, Mandelker D, Leary RJ, Ptak J, Silliman N, et al. 2006. The consensus coding sequences of human breast and colorectal cancers. *Science* **314:** 268–274.

Sokol SY. 2011. Maintaining embryonic stem cell pluripotency with Wnt signaling. *Development (Cambridge, England)* **138:** 4341–4350.

Song LN, Gelmann EP. 2005. Interaction of β-catenin and TIF2/GRIP1 in transcriptional activation by the androgen receptor. *J Biol Chem* **280:** 37853–37867.

Song LN, Herrell R, Byers S, Shah S, Wilson EM, Gelmann EP. 2003. β-catenin binds to the activation function 2 region of the androgen receptor and modulates the effects of the N-terminal domain and TIF2 on ligand-dependent transcription. *Mol Cell Biol* **23:** 1674–1687.

Song H, Goetze S, Bischof J, Spichiger-Haeusermann C, Kuster M, Brunner E, Basler K. 2010. Coop functions as a corepressor of Pangolin and antagonizes Wingless signaling. *Genes Dev* **24:** 881–886.

Srivastava M, Simakov O, Chapman J, Fahey B, Gauthier ME, Mitros T, Richards GS, Conaco C, Dacre M, Hellsten

U, et al. 2010. The Amphimedon queenlandica genome and the evolution of animal complexity. *Nature* **466:** 720–726.

Stadeli R, Basler K. 2005. Dissecting nuclear Wingless signalling: Recruitment of the transcriptional co-activator Pygopus by a chain of adaptor proteins. *Mech Dev* **122:** 1171–1182.

Standley HJ, Destree O, Kofron M, Wylie C, Heasman J. 2006. Maternal XTcf1 and XTcf4 have distinct roles in regulating Wnt target genes. *Dev Biol* **289:** 318–328.

Steger DJ, Lefterova MI, Ying L, Stonestrom AJ, Schupp M, Zhuo D, Vakoc AL, Kim JE, Chen J, Lazar MA, et al. 2008. DOT1L/KMT4 recruitment and H3K79 methylation are ubiquitously coupled with gene transcription in mammalian cells. *Mol Cell Biol* **28:** 2825–2839.

Suganuma T, Workman JL. 2011. Signals and combinatorial functions of histone modifications. *Annu Rev Biochem* **80:** 473–499.

Tang W, Dodge M, Gundapaneni D, Michnoff C, Roth M, Lum L. 2008. A genome-wide RNAi screen for Wnt/β-catenin pathway components identifies unexpected roles for TCF transcription factors in cancer. *Proc Natl Acad Sci* **105:** 9697–9702.

Teo JL, Kahn M. 2010. The Wnt signaling pathway in cellular proliferation and differentiation: A tale of two coactivators. *Adv Drug Deliv Rev* **62:** 1149–1155.

Thorpe CJ, Schlesinger A, Carter JC, Bowerman B. 1997. Wnt signaling polarizes an early *C. elegans* blastomere to distinguish endoderm from mesoderm. *Cell* **90:** 695–705.

Travis A, Amsterdam A, Belanger C, Grosschedl R. 1991. LEF-1, a gene encoding a lymphoid-specific protein with an HMG domain, regulates T-cell receptor α enhancer function. *Genes Dev* **5:** 880–894.

Truica CI, Byers S, Gelmann EP. 2000. β-catenin affects androgen receptor transcriptional activity and ligand specificity. *Cancer Res* **60:** 4709–4713.

Tsai BP, Wang X, Huang L, Waterman ML. 2011. Quantitative profiling of in vivo-assembled RNA-protein complexes using a novel integrated proteomic approach. *Mol Cell Proteomics* **10:** M110.007385.

Tsedensodnom O, Koga H, Rosenberg SA, Nambotin SB, Carroll JJ, Wands JR, Kim M. 2011. Identification of T-cell factor-4 isoforms that contribute to the malignant phenotype of hepatocellular carcinoma cells. *Exp Cell Res* **317:** 920–931.

Vacik T, Lemke G. 2011. Dominant-negative isoforms of Tcf/Lef proteins in development and disease. *Cell Cycle* **10:** 4199–4200.

Valenta T, Lukas J, Korinek V. 2003. HMG box transcription factor TCF-4's interaction with CtBP1 controls the expression of the Wnt target Axin2/Conductin in human embryonic kidney cells. *Nucleic Acids Res* **31:** 2369–2380.

van Beest M, Dooijes D, van De Wetering M, Kjaerulff S, Bonvin A, Nielsen O, Clevers H. 2000. Sequence-specific high mobility group box factors recognize 10-12-base pair minor groove motifs. *J Biol Chem* **275:** 27266–27273.

van de Wetering M, Clevers H. 1992. Sequence-specific interaction of the HMG box proteins TCF-1 and SRY occurs within the minor groove of a Watson-Crick double helix. *EMBO J* **11:** 3039–3044.

van de Wetering M, Oosterwegel M, Dooijes D, Clevers H. 1991. Identification and cloning of TCF-1, a T lymphocyte specific transcription factor containing a sequence-specific HMG box. *EMBO J* **10:** 123–132.

van de Wetering M, Castrop J, Korinek V, Clevers H. 1996. Extensive alternative splicing and dual promoter usage generate Tcf-protein isoforms with differential transcription control properties. *Mol Cell Biol* **16:** 745–752.

van de Wetering M, Cavallo R, Dooijes D, van Beest M, van Es J, Loureiro J, Ypma A, Hursh D, Jones T, Bejsovec A, et al. 1997. Armadillo coactivates transcription driven by the product of the *Drosophila* segment polarity gene *dTCF*. *Cell* **88:** 789–799.

van de Wetering M, Sancho E, Verweij C, de Lau W, Oving I, Hurlstone A, van der Horn K, Batlle E, Coudreuse D, Haramis AP, et al. 2002. The β-catenin/TCF-4 complex imposes a crypt progenitor phenotype on colorectal cancer cells. *Cell* **111:** 241–250.

Verras M, Brown J, Li X, Nusse R, Sun Z. 2004. Wnt3a growth factor induces androgen receptor-mediated transcription and enhances cell growth in human prostate cancer cells. *Cancer Res* **64:** 8860–8866.

Voeller HJ, Truica CI, Gelmann EP. 1998. β-catenin mutations in human prostate cancer. *Cancer Res* **58:** 2520–2523.

Wang G, Wang J, Sadar MD. 2008. Crosstalk between the androgen receptor and β-catenin in castrate-resistant prostate cancer. *Cancer Res* **68:** 9918–9927.

Waterman ML, Fischer WH, Jones KA. 1991. A thymus-specific member of the HMG protein family regulates the human T cell receptor C α enhancer. *Genes Dev* **5:** 656–669.

Weise A, Bruser K, Elfert S, Wallmen B, Wittel Y, Wohrle S, Hecht A. 2010. Alternative splicing of Tcf7l2 transcripts generates protein variants with differential promoter-binding and transcriptional activation properties at Wnt/β-catenin targets. *Nucleic Acids Res* **38:** 1964–1981.

Willert K, Jones KA. 2006. Wnt signaling: Is the party in the nucleus? *Genes Dev* **20:** 1394–1404.

Wohrle S, Wallmen B, Hecht A. 2007. Differential control of Wnt target genes involves epigenetic mechanisms and selective promoter occupancy by T-cell factors. *Mol Cell Biol* **27:** 8164–8177.

Wright KJ, Tjian R. 2009. Wnt signaling targets ETO coactivation domain of TAF4/TFIID in vivo. *Proc Natl Acad Sci* **106:** 55–60.

Wright JB, Brown SJ, Cole MD. 2010. Upregulation of c-MYC in cis through a large chromatin loop linked to a cancer risk-associated single-nucleotide polymorphism in colorectal cancer cells. *Mol Cell Biol* **30:** 1411–1420.

Yang X, van Beest M, Clevers H, Jones T, Hursh DA, Mortin MA. 2000. Decapentaplegic is a direct target of dTcf repression in the *Drosophila* visceral mesoderm. *Development (Cambridge, England)* **127:** 3695–3702.

Yang F, Li X, Sharma M, Sasaki CY, Longo DL, Lim B, Sun Z. 2002. Linking β-catenin to androgen-signaling pathway. *J Biol Chem* **277:** 11336–11344.

Yang XD, Huang S, Lo MC, Mizumoto K, Sawa H, Xu W, Robertson S, Lin R. 2011. Distinct and mutually inhibitory binding by two divergent β-catenins coordinates TCF levels and activity in *C. elegans*. *Development (Cambridge, England)* **138:** 4255–4265.

Yi F, Pereira L, Hoffman JA, Shy BR, Yuen CM, Liu DR, Merrill BJ. 2011. Opposing effects of Tcf3 and Tcf1 control Wnt stimulation of embryonic stem cell self-renewal. *Nat Cell Biol* **13:** 762–770.

Yochum GS. 2011. Multiple Wnt/ss-catenin responsive enhancers align with the MYC promoter through long-range chromatin loops. *PLoS ONE* **6:** e18966.

Yochum GS, Sherrick CM, Macpartlin M, Goodman RH. 2010. A β-catenin/TCF-coordinated chromatin loop at MYC integrates 5′ and 3′ Wnt responsive enhancers. *Proc Natl Acad Sci* **107:** 145–150.

Yokoyama NN, Pate KT, Waterman ML. 2009. A role for YY1 in repression of dominant negative LEF-1 expression. *Nucleic Acids Res* **38:** 6375–6388.

Yu X, Wang Y, Jiang M, Bierie B, Roy-Burman P, Shen MM, Taketo MM, Wills M, Matusik RJ. 2009. Activation of β-catenin in mouse prostate causes HGPIN and continuous prostate growth after castration. *The Prostate* **69:** 249–262.

Zhao JH, Luo Y, Jiang YG, He DL, Wu CT. 2011. Knockdown of β-catenin through shRNA cause a reversal of EMT and metastatic phenotypes induced by HIF-1α. *Cancer Invest* **29:** 377–382.

Alternative Wnt Pathways and Receptors

Renée van Amerongen

Department of Developmental Biology, Stanford University, Stanford, California 94305; Netherlands Cancer Institute, Division of Molecular Oncology, Plesmanlaan 121, 1066 CX Amsterdam, The Netherlands

Correspondence: r.v.amerongen@nki.nl

In addition to activating β-catenin/TCF transcriptional complexes, Wnt proteins can elicit a variety of other responses. These are often lumped together under the denominator "alternative" or "non-canonical" Wnt signaling, but they likely comprise distinct signaling events. In this article, I discuss how the use of different ligand and receptor combinations is thought to give rise to these alternative Wnt-signaling responses. Although many of the biochemical details remain to be resolved, it is evident that alternative Wnt signaling plays important roles in regulating tissue morphogenesis during embryonic development.

The very notion of an "alternative" mode of Wnt signaling implies the existence of a default response. That honor befalls the Wnt- and Frizzed/LRP-mediated stabilization of β-catenin and its concomitant association with LEF/TCF transcription factors (historically called "canonical" Wnt signaling and hereafter referred to as Wnt/β-catenin signaling), which remains the best understood and most studied response to Wnt ligand stimulation. There are multiple instances, however, in which Wnt proteins elicit a biological response that does not appear to involve the activation of β-catenin/TCF complexes. For lack of a better description, these responses are often jointly referred to as "alternative" Wnt signaling.

An obvious first question, then, might be to ask how many alternative responses exist. Perhaps less obvious is the answer to this question: We don't really know. The main reason for this is the fact that β-catenin-independent signaling responses remain relatively poorly characterized at the molecular level. As a result, we lack the robust readouts that exist to probe Wnt/β-catenin signaling. Fortunately, we are gaining an ever-increasing understanding of the molecular players involved in and the developmental events controlled by alternative Wnt signaling.

ROLES FOR β-CATENIN-INDEPENDENT WNT SIGNALING IN DEVELOPMENT

Before discussing the initiation and progression of alternative Wnt-signaling responses, I shall introduce some of the physiological processes that are controlled by β-catenin-independent Wnt signaling in vivo. In general, rather than affecting cell proliferation or cell fate specification, alternative Wnt signaling controls various aspects of cell migration and polarity, thereby having profound effects on tissue morphogenesis.

Cite this article as *Cold Spring Harb Perspect Biol* doi: 10.1101/cshperspect.a007914

Planar Cell Polarity

The process in which cells align themselves within the plane of a tissue is called planar cell (or tissue) polarity (PCP). Although it is observed in a wide variety of tissues, the two best-known examples are the uniform array of distally pointing hairs in the fly wing and the coordinated polarization of cochlear hair cells in the mammalian inner ear (Fig. 1A,B). However, PCP can also encompass the alignment of larger structures rather than individual cells, such as the multicellular ommatidia in the compound fly eye, and even entire hair follicles in the mammalian skin (Montcouquiol et al. 2003; Devenport and Fuchs 2008; Jenny 2010; Maung and Jenny 2011).

Although the Wnt-pathway components frizzled (fz) and dishevelled (dsh) are core PCP proteins (Fig. 1C), it should be stressed that to date there is no evidence for the involvement of a ligand in either the establishment or maintenance of PCP in the *Drosophila* wing and ommatidia (Chen et al. 2008). This issue has been long debated, however, and recent evidence suggests that this might not hold true for all tissues (Shimizu et al. 2011). Furthermore, Wnt proteins have been implicated in PCP in vertebrates and

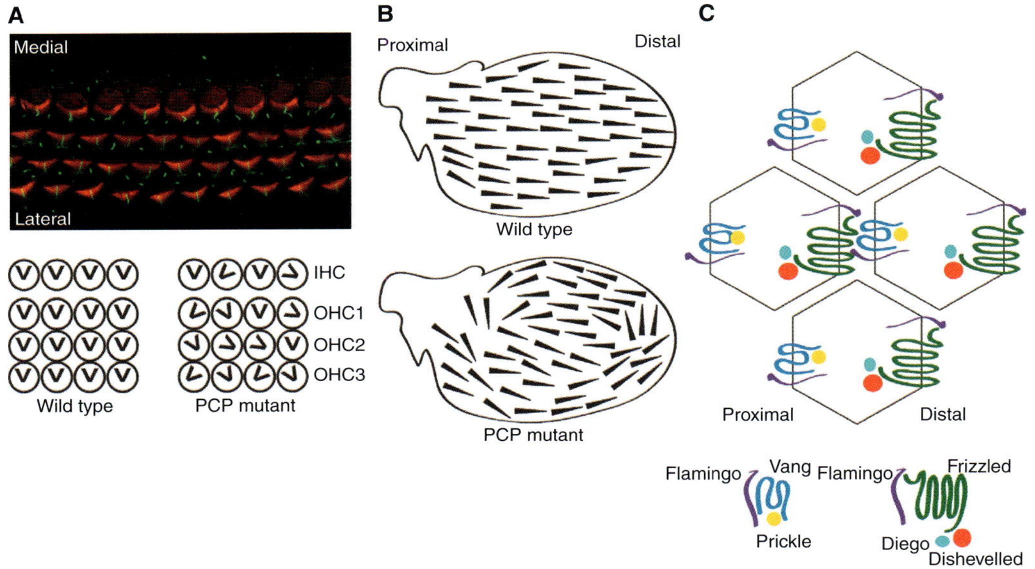

Figure 1. Planar cell polarity (PCP) as an example of alternative Wnt signaling. (*A*, *top*) PCP in the sensory epithelium of the mammalian cochlea manifests itself in the alignment of inner ear hair cells, which carry characteristic V-shaped stereociliary bundles of actin hairs on the apical cell surface (visualized here by phalloidin staining in red; kinocilia are stained green with an antibody against acetylated tubulin). (*Bottom*) PCP defects are scored based on the misorientation of these sterociliary bundles on inner (IHC) and outer (OHC) hair cells in whole-mount preparations of the organ of Corti. (*B*, *top*) In a textbook example of PCP, the surface of the fly wing is covered with a neatly organized array of hairs, all of which point toward the distal side. (*Bottom*) Defects in PCP are typically scored based on the misorientation of wing hairs. (*C*) Molecular control of PCP in the *Drosophila* wing. Complexes of the transmembrane proteins flamingo and vang, together with the cytoplasmic protein prickle, accumulate at the proximal end of the cell. At the distal border, flamingo and frizzled accumulate together with dishevelled and diego. Interactions between the extracellular domains of flamingo, frizzled, and vang are thought to play an important role in maintaining these complexes at opposite sides of the cell boundary (Chen et al. 2008; Wu and Mlodzik 2008). In addition, diego has been proposed to promote fz/dsh signaling, whereas vang/pk is thought to prevent recruitment of dsh to the cell membrane, something that is a prerequisite for PCP signaling (Tree et al. 2002; Das et al. 2004; Jenny et al. 2005).

Caenorhabditis elegans (Dabdoub et al. 2003; Dabdoub and Kelley 2005; Wu and Herman 2006; Qian et al. 2007; Yamamoto et al. 2008).

Convergent Extension

The same core players that control PCP have been shown to regulate a seemingly unrelated developmental process, called convergent extension (CE) (Keller and Danilchik 1988; Warga and Kimmel 1990; Wallingford et al. 2000). This morphogenetic event refers to the simultaneous lengthening and narrowing of a tissue as the result of directional cell intercalation. As such, CE movements control extension of the primary body axis and neural tube closure (Keller et al. 1985; Davidson and Keller 1999). For experimental reasons, CE is mostly studied in *Xenopus* and zebrafish, in which gastrulation defects are relatively easy to score. In addition, CE can be monitored and manipulated in *Xenopus* animal cap assays (Wilson et al. 1989; Hertzano et al. 2004). Both Wnts and Frizzleds (Fzds), as well as Dishevelled (Dvl) and other core PCP proteins, have been implicated in CE in amphibians, fish, and mammals (Djiane et al. 2000; Heisenberg et al. 2000; Carreira-Barbosa et al. 2003; Kilian et al. 2003; Torban et al. 2004; Moeller et al. 2006; Wang et al. 2006a; Kim et al. 2008). However, despite the fact that PCP and CE are often used interchangeably when discussing the role of alternative Wnt signaling, it is important to bear in mind that they are two distinct biological phenomena that are often studied in different species. Although they share the same core components, different downstream effector mechanisms are therefore prone to exist, and PCP and CE are unlikely to be controlled by completely identical signal transduction events (Fig. 2) (see also the section on Downstream Signaling Events). Future studies will have to determine how similar (or different) their underlying machineries really are.

Other Developmental Consequences of Defects in Alternative Wnt Signaling

Because the proteins previously implicated in PCP and CE are found to control multiple aspects of mammalian development, the possibility arises that these processes may underlie polarized cell movements in general. Indeed, the developmental consequences of deregulated alternative Wnt signaling are substantial and include deficits in left/right patterning (Zhang and Levin 2009; Antic et al. 2010; Song et al. 2010), branching morphogenesis (Shabani et al. 2008; Karner et al. 2009; Yates et al. 2010), cardiac outflow tract formation (Hamblet et al. 2002; Phillips et al. 2005; Etheridge et al. 2008), intestinal elongation (Cervantes et al. 2009; Yamada et al. 2010), limb outgrowth (Yamaguchi et al. 1999; Afzal et al. 2000; van Bokhoven et al. 2000; Schwarzer et al. 2009; Person et al. 2010; Gao et al. 2011), and neural tube closure (Kibar et al. 2001, 2011; Montcouquiol et al. 2003; Seo et al. 2011).

Of note, not all β-catenin-independent responses to Wnt-ligand stimulation have been definitively linked to PCP/CE. However, recent reports showing the hitherto unknown involvement of PCP proteins other than Fzd and Dvl in Wnt-mediated axon guidance and differentiation underscore that this may simply be a matter of time (Shafer et al. 2011; Shimizu et al. 2011).

ALTERNATIVE WNT-SIGNALING MECHANISMS

Initiation of the Wnt-signaling response is first and foremost decided at the level of ligand and receptor binding. In all animal species, however, both Wnts and Fzds belong to large, multigene families, complicating a comprehensive analysis of their signaling activities. Further diversification of the signaling response comes from the use of different (co-)receptors and cytoplasmic effectors. With this in mind, future studies have to take into account that the outcome of a given Wnt-signaling event is determined by the ligand and receptor context and the developmental history of the receiving cell.

At the tissue level, the response is ultimately controlled by a "Wnt code" or "Wnt landscape" that integrates the complex expression patterns and different signaling inputs of multiple Wnt ligands and (ant)agonists (Gleason et al. 2006; Guder et al. 2006; Green et al. 2008; Zinovyeva et al. 2008; van Amerongen and Nusse 2009; Janssen et al. 2010; Fossat et al. 2011).

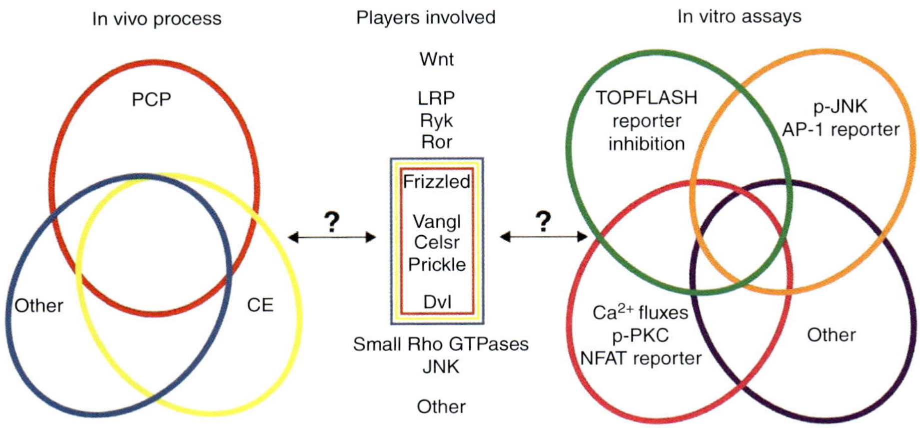

Figure 2. Knowledge gaps. In the absence of robust and quantitative readouts, it has remained difficult to dissect the biological processes (*left*) in which alternative Wnt-signaling components (*center*) play a role in distinct signal transduction events (*right*). For instance, despite the fact that multiple Wnt-pathway components show a genetic interaction with core PCP/CE proteins (boxed in the *center*), it remains unknown how Wnts and their (co-)receptors are tied to the core PCP/CE machinery at the biochemical level. As an example, the core PCP/CE protein *vang*/Vangl2 plays a central role in tissue morphogenesis. Animals deficient for *vang*/Vangl2 display severe defects in PCP (red ellipse; e.g., cochlear hair cell orientation) and CE (yellow ellipse; e.g., neural tube closure), as well as in developmental processes that have not (yet) been definitively linked to either PCP or CE (blue ellipse; e.g., lung branching morphogenesis). A genetic interaction with Vangl2 has been shown for both Wnt5a and Ror2, and Wnt5a/Ror2 has been reported to phosphorylate Vangl2 in the murine limb bud (Qian et al. 2007; Yamamoto et al. 2008; Gao et al. 2011). The most robust assay to monitor Wnt5a/Ror2 signaling in vitro is inhibition of the Wnt/β-catenin-responsive TOPFLASH luciferase reporter (green ellipse). However, Wnt5a and Ror2 have also been reported to activate JNK and AP-1 in mammalian cell culture experiments (orange ellipse). It is unknown how far these two responses overlap and whether they involve Vangl2. Conversely, it is unknown whether either Wnt5a or Ror2 plays a direct role in the diverse tissue-morphogenetic events in which Vangl2 has been implicated.

Use of Alternative Wnt Ligands

From the moment it became clear that Wnt proteins were members of an evolutionarily conserved family of signaling molecules (Baker 1987; Rijsewijk et al. 1987), attempts were made to classify them into groups with distinct properties. Two assays that were regularly used for this were the morphological transformation of C57MG cells and axis duplication in *Xenopus* embryos. The former scores the ability of individual Wnt proteins to change these cells from a cuboidal to a spindle-like cell shape (Jue et al. 1992), whereas the latter makes use of the fact that ectopic Wnt/β-catenin signaling on the ventral side of the early embryo will induce formation of a secondary body axis (McMahon and Moon 1989; Sokol et al. 1991). Although no one has ever reported an exhaustive side-by-side comparison of the behavior of all 19 mammalian Wnt genes in both assays, it was clear that the capacity of some Wnt proteins (e.g., Wnt3a), but not others (e.g., Wnt5a), to induce transformation and cause an increase in the cellular levels of β-catenin in C57MG cells (Wong et al. 1994; Shimizu et al. 1997) corresponded to their ability to induce axis duplication in *Xenopus* (Du et al. 1995). The transforming Wnts became known as "canonical" and were assumed to activate β-catenin/TCF signaling. In contrast, the non-transforming or "non-canonical" Wnts were thought to use alternative signaling mechanisms. Since then, Wnt3a and Wnt5a have become the prototypes for either class.

Although valuable, these assays do have some drawbacks. For instance, the C57MG morphological transformation assay is not quantifiable and therefore difficult to interpret, especially for weak transforming genes (e.g., Wnt6, Wnt7B), which might explain why they have an effect in some studies, but not others (Wong et al. 1994; Shimizu et al. 1997). More importantly, the strict subdivision into canonical and non-canonical Wnts is a vast oversimplification, and an important caveat of these early classification attempts is the fact that they overlooked the importance of receptor context.

Instead of inducing axis duplication upon injection into early *Xenopus* embryos, exogenous Wnt5a does not affect development until after the onset of gastrulation, ultimately resulting in complex head and tail malformations. In addition, Wnt5a blocks the elongation of blastula cap explants, first suggesting a role in controlling the CE movements associated with gastrulation (Moon et al. 1993). However, Wnt5a was subsequently shown to be able to induce an ectopic axis when it was coinjected with Fz5 (He et al. 1997). Likewise, Wnt5a was found to activate β-catenin/TCF signaling when it was added to 293 cells overexpressing a combination of Fzd4/LRP5 receptors (Mikels and Nusse 2006). Furthermore, when Wnt5a or Wnt11 was fused to Fzd5, both were able to induce a secondary axis in *Xenopus* embryos—something that is never observed for either Wnt alone (Holmen et al. 2002).

Importantly, in vivo studies support the idea that individual Wnt proteins can have different functions depending on the time and place of their activity. Formation of the primary body axis in *Xenopus* depends on Wnt/β-catenin signaling on the dorsal side of the fertilized oocyte. Quite surprisingly, maternal Wnt5a and Wnt11, at the time generally considered noncanonical Wnt proteins, were shown to be required for this process (Tao et al. 2005; Cha et al. 2008). Even more remarkable was the discovery that Wnt5a and Wnt11 function together in a multimeric complex to induce Wnt/β-catenin signaling during primary axis formation (Cha et al. 2009). Yet the same proteins clearly control β-catenin-independent gastrulation events at subsequent stages of development (Cha et al. 2008), as illustrated by the fact that both the Wnt11/*Silberblick* and Wnt5a/*Pipetail* zebrafish mutants display defects in convergent extension movements (Heisenberg et al. 2000; Kilian et al. 2003).

Thus, the same Wnt proteins can have both β-catenin-dependent and -independent functions depending on the cellular context, suggesting that any definitive conclusions regarding their signaling activity should be made on a case-by-case basis. That being said, it is a fact that some Wnts are less often associated with β-catenin/TCF signaling than others. Precisely what controls ligand and receptor affinity or intrinsic differences in signaling capacity remains to be determined.

Alternative Use of Frizzled and Dishevelled

Genetic screens in *Drosophila* provide a powerful tool for dissecting signaling pathways. Using epistasis experiments, it was shown that *wingless* (*wg*/Wnt) functioned upstream of a *frizzled* (*fz*/Fzd) transmembrane receptor and used *dishevelled* (*dsh*/Dvl) and *zeste-white* (*zw*/GSK3) to ultimately signal through *armadillo* (*arm*/β-catenin) and *pangolin* (*pan*/TCF), thereby providing a blueprint of the now well-known Wnt/β-catenin pathway (Siegfried et al. 1992, 1994; Noordermeer et al. 1994; Bhanot et al. 1996; van de Wetering et al. 1997). In addition to controlling segment polarity, however, both *fz* and *dsh*, but not *arm*, were found to be required for the establishment of PCP in collaboration with a completely distinct set of signaling partners (Vinson et al. 1989; Wong and Adler 1993; Axelrod et al. 1998). These include *flamingo*, *vang*, and *prickle* (Celsr1/2/3, Vangl1/2, and Prickle1/2/3 in vertebrates) (see also Fig. 1C) (see section on Core PCP/CE Proteins).

Frizzled

Similar to the situation observed for Wnts, not all Frizzled proteins are equal. The mammalian genome encodes ten Frizzled proteins. Of these only Fzd3 and Fzd6 have been implicated in PCP signaling so far (Guo et al. 2004; Montcouquiol et al. 2006; Wang et al. 2006b, 2010). Of the four *frizzled* receptors present in *Drosophila*, *fz* and *fz2* function redundantly in transmitting the *wg*/*arm* signal (Bhat 1998; Kennerdell and

Carthew 1998; Bhanot et al. 1999). However, *fz2* has a higher affinity for *wg*, and only *fz* is required for PCP signaling (Rulifson et al. 2000) or capable of recruiting *dsh* to the cell membrane (Axelrod et al. 1998).

An intrinsic difference between the two receptors was shown to reside in their cytoplasmic tails, which affected both their subcellular localization and their capacity to activate either the Wnt/β-catenin or the PCP pathway (Boutros et al. 2000; Wu et al. 2004). Despite the fact that no Wnt protein appears to be required for PCP signaling in the *Drosophila* wing, the *fz* cysteine-rich domain (CRD), which constitutes the *wg* binding site, is indispensable and might interact with *flamingo/vang* complexes on adjacent cells (Wu and Herman 2006; Wu and Mlodzik 2008).

Again, however, the involvement in either Wnt/β-catenin or alternative signaling responses is not black and white. Even though certain Fzds may be more prone to participate in one pathway than the other, a single Fzd receptor can have multiple signaling activities. For instance, Fzd7 has been well documented to control β-catenin-dependent as well as -independent developmental events in *Xenopus* (Medina et al. 2000; Sumanas et al. 2000; Abu-Elmagd et al. 2006). In addition, overexpression of Fzd7 was shown to inhibit convergent extension movements through Dvl (Djiane et al. 2000), whereas loss of Fzd7 was shown to disrupt the separation of mesodermal and ectodermal tissues in a Dvl-independent manner (Medina et al. 2000; Winklbauer et al. 2001). Similarly, Fzd2 has been linked to both β-catenin-dependent as well as -independent responses in mammalian cells (Liu et al. 1999; Ma and Wang 2007; Verkaar et al. 2009; Sato et al. 2010). This shows that, similar to Wnts, individual Fzds can signal through multiple independent branches (see below), only some of which may require Dvl.

Dishevelled

The first indication for a dual signaling role for Dishevelled (Dvl) came from the identification of the *Drosophila* dsh^1 allele, which causes defects in PCP, but not in Wnt/β-catenin signaling (Adler 1992). Since then, much effort has been made to map the domains that are required for the different signaling activities of Dvl. The protein contains three well-conserved domains: an amino-terminal DIX domain, a central PDZ domain, and a carboxy-terminal DEP domain. Of these, the DIX domain is critically required for inducing Wnt/β-catenin signaling. It has been shown to mediate dynamic polymerization, resulting in the recruitment of Fzd/LRP complexes into large signalosomes (Bilic et al. 2007; Schwarz-Romond et al. 2007a,b; Fiedler et al. 2011). Deletion of the DIX domain renders Dvl functionally inactive in a Wnt/β-catenin-responsive TOPFLASH reporter assay as well as in *Xenopus* axis duplication experiments. Conversely, the PDZ and DEP domains are largely dispensable for axis duplication (Rothbacher et al. 2000). Instead, deletion mutants lacking the DEP domain are defective in disrupting CE movements (Wallingford et al. 2000).

The observation that wild-type *dsh* was recruited to the plasma membrane by *fz* (the receptor involved in PCP signaling), but not by *fz2* and *wg* (the ligand/receptor combination that controls *wg/arm* signaling), in *Xenopus* animal cap cells suggested that membrane localization was critical for the role of *dsh* in establishing PCP/CE (Axelrod et al. 1998). In support of this, the mutant *dsh* protein encoded by the dsh^1 allele was found to contain a K → M transition mutation in the DEP domain and had lost its capacity to be recruited to the cell membrane.

These observations were elegantly confirmed in the mouse, using an allelic series of BAC transgenes (Wang et al. 2006a). The mammalian genome encodes three *Dvl* homologs, and mice deficient for both *Dvl1* and *Dvl2* display severe neural tube closure defects as well as cochlear hair cell orientation defects, similar to mice that carry a mutated version of the core PCP gene *Vangl2* (Kibar et al. 2001; Hamblet et al. 2002; Wang et al. 2005). The neural tube closure defects observed in the $Dvl1^{-/-}; Dvl2^{-/-}$ mutants could be rescued by either a *Dvl2* or a *Dvl2ΔDIX*, but not by a *Dvl2ΔDEP* BAC transgene or by a *Dvl2* BAC transgene carrying a single point mutation identical to that observed in the *Drosophila* dsh^1 allele (Wang et al. 2006a).

Summarizing, the amino-terminal DIX domain of Dvl is dispensable, whereas its carboxy-terminal DEP domain is required for PCP/CE signaling. The opposite holds true for Wnt/β-catenin signaling. In addition to differences in subcellular localization, the different pools of Dvl also appear to use different downstream effectors to activate the different branches of the Wnt pathway (see below).

Alternative Receptors

The interaction with Wnt ligands is mediated by the Fzd CRD domain. This same domain, which is also used by the extracellular antagonists of the secreted Frizzled-related protein (SFRP) family, was found to be present in the otherwise completely unrelated tyrosine kinase receptors Ror1 and Ror2 (Saldanha et al. 1998). A different Wnt binding domain, used by the extracellular Wnt-inhibitory factor (WIF), resides in the extracellular domain of the tyrosine kinase-related receptor Ryk (Yoshikawa et al. 2003; Inoue et al. 2004; Lu et al. 2004). Since then, both Ryk and Ror have been shown to be bona fide Wnt-pathway components that have been linked to both β-catenin-dependent and -independent Wnt-signaling events.

Ror

The single-pass transmembrane proteins of the Ror family are evolutionarily conserved across vertebrate and invertebrate species, including *Drosophila*, *C. elegans*, *Xenopus*, mice, and humans. The genomes of most species harbor two homologs. Studies performed in mammals to date have focused on the role of Ror2, but both Ror1 and Ror2 have been shown to mediate alternative Wnt signaling (Paganoni et al. 2010), although at present the experimental data support a role for Ror1 as a pseudokinase (Gentile et al. 2011). The fact that Ror1 as well as Ror2 can bind Wnt ligands directly (Oishi et al. 2003; Mikels and Nusse 2006; Fukuda et al. 2008; Liu et al. 2008) suggests that both can function as bona fide Wnt receptors. Indeed, Ror2 has been shown to interact with multiple Wnt proteins (Billiard et al. 2005; Mikels and Nusse 2006; Winkel et al. 2008).

Ror2-deficient mice present, among others, with craniofacial abnormalities and shortened limbs. Their external appearance somewhat resembles that of Wnt5a-null mice, suggesting that Wnt5a and Ror2 might have a similar function during embryonic development (Oishi et al. 1999; Yamaguchi et al. 1999; Schwabe et al. 2004). Indeed, Ror2 has since been shown to transmit a Wnt5a signal that results in inhibition of the Wnt/β-catenin signaling response (Mikels and Nusse 2006; Mikels et al. 2009). One of the earliest indications of the existence of such an inhibitory branch of the pathway was the observation that *Xenopus* axis duplication induced by *XWnt1* could be inhibited by *XWnt5a* (Torres et al. 1996). Since then, Wnt5a has been shown to counteract Wnt/β-catenin signaling in a broad range of cells and tissues (MacLeod et al. 2007; Nemeth et al. 2007; Roarty et al. 2009).

One possible mechanism by means of which Ror (or Fzd proteins for that matter) can affect Wnt signaling is by the competition for and sequestration of other Wnt ligands. This mechanism has been shown to exist for the Ror homolog *cam-1* in *C. elegans* (Green et al. 2007). However, the Wnt5a/Ror2-mediated inhibition of Wnt/β-catenin signaling was shown to require an intact tyrosine kinase domain (Liu et al. 2008; Mikels et al. 2009), suggesting an active signaling function. In contrast, Ror2 has also been reported to enhance Wnt/β-catenin signaling as a coreceptor together with Fzd, independent from its intracellular, carboxy-terminal domain (Li et al. 2008), suggesting that Ror may have both kinase-dependent and -independent functions. The final verdict is also still out on whether Ror2 mainly functions alone or in concert with other receptors (Mikels and Nusse 2006; Li et al. 2008; Yamamoto et al. 2008; Nishita et al. 2010; Sato et al. 2010). Studies that have analyzed ligand/receptor complex formation have mostly depended on protein overexpression, and as such, it is unknown to what extent these complexes exist at the physiological level. Furthermore, despite the binding partners and potential effectors that have been reported for Ror2, the precise molecular makeup of the Ror2 signaling cascade remains unknown (Matsuda et al. 2003; Sammar et al. 2004, 2009; Liu et al. 2007; Winkel et al. 2008; van Wijk et al. 2009; Feike et al. 2010; Witte et al. 2010). Wnt5a/Ror2-mediated inhibition of Wnt/β-catenin signaling does not appear to involve NFAT, CamKII, or calcium fluxes (Topol et al. 2003; Mikels and Nusse 2006). Although Wnt5a has been shown to induce the expression of the E3 ubiquitin ligase *Siah2* in a human colon adenocarcinoma cell line, resulting in the degradation of β-catenin (Topol et al. 2003), other data suggest that the Wnt5a/Ror2-mediated inhibition of Wnt/β-catenin signaling occurs downstream from β-catenin nuclear entry (Verkaar et al. 2010). In agreement with the latter, we found no evidence of a transcriptional response following stimulation of Ror2-expressing 293 cells with purified Wnt5a, even though the Wnt3a-mediated induction of *Axin2* expression was prevented (R van Amerongen and R Nusse, unpubl.). Of note, Wnt5a/Ror2 signaling has also been reported to inhibit convergent extension movements through the activation of JNK/AP-1 (Oishi et al. 2003). Similar to Fzd, therefore, Ror proteins might elicit multiple, independent signaling responses.

Interestingly, a novel Ror1/Ror2 double-knockout mouse model was recently reported to resemble the Wnt5a-null phenotype to a larger extent than existing models. Yet, in mouse embryo fibroblasts, the combined loss of Ror1 and Ror2 was shown to solely affect the phosphorylation of Dvl, but not the phosphorylation of c-Jun or the Wnt5a-mediated inhibition of Wnt/β-catenin signaling (Ho et al. 2012). Thus, to what extent all of the aforementioned signaling activities of Wnt5a and Ror2 play a role under physiological conditions remains to be determined. In either case, it appears that Wnt5a/Ror signaling controls complex cell movements during embryonic development.

Ryk

Ryk transmembrane receptors are also conserved across vertebrate and invertebrate species. Mammalian Ryk is critically required for axon guidance and neurite outgrowth in response to multiple Wnt ligands (Lu et al. 2004; Keeble et al. 2006). It was shown to bind both Fzd and Dvl and to be essential for Wnt/β-catenin signaling (Lu et al. 2004; Berndt et al. 2011). Similar to Fzd and Ror, Ryk can have different functions depending on the cellular context. For instance, the fly homologs *derailed1* and *derailed2* play opposing roles in Wnt5a signaling during development of the *Drosophila* olfactory system (Sakurai et al. 2009).

Mammalian Ryk, as well as its *C. elegans* homolog *lin-18*, are catalytically inactive, suggesting a coreceptor function. Indeed, although the extracellular WIF domain of *lin-18* and *derailed* was required, the intracellular atypical kinase domain was found to be largely dispensable for their function (Inoue et al. 2004; Taillebourg et al. 2005). Although Ryk has mostly been studied in a neurobiological context, it was also shown to regulate gastrulation movements. In *Xenopus*, it was found to cooperate with Fzd7 and Wnt11 during CE (Kim et al. 2008), whereas in zebrafish, it was shown to control directional cell movements in response to Wnt5b (Lin et al. 2010). Similar to Ror1 and Ror2, at present the signaling mechanism downstream from Ryk remains incompletely understood. Interestingly, Ryk was shown to undergo cleavage, after which the carboxy-terminal fragment translocated to the nucleus in response to Wnt3 stimulation. This processing appears to be essential for the function of Ryk in controlling the differentiation of neuronal progenitor cells (Lyu et al. 2008, 2009). Ryk signaling during axon guidance in *Drosophila* was shown to require Src kinases (Wouda et al. 2008). However, at present there is no evidence that the Src-mediated signaling events involve Ryk cleavage.

Taken together, multiple independent responses are likely to exist downstream from ligand and receptor binding for both Frizzled and non-Frizzled receptors. Future studies will have to be aimed at understanding how specificity is achieved in light of such dynamic interactions.

Downstream Signaling Events

Most alternative Wnt-signaling responses remain ill characterized at the molecular level. The main reason for this stems from the fact that it has proven difficult to find a reliable experimental system in which to study alternative Wnt-signal transduction events to begin with, because complex developmental and tissue morphogenetic processes such as PCP and CE are not easily translated to a tissue culture dish (Endo et al. 2012). Molecular analyses have been further complicated by the fact that alternative Wnt signaling often leads to cytoskeletal and migratory changes, rather than a transcriptional response, resulting in the lack of specific and easily quantifiable in vitro readouts, analogous to the TOPFLASH reporter assay for Wnt/β-catenin signaling.

Unlike β-catenin/TCF, many of the implicated downstream components (e.g., small Rho GTPases, JNK/AP-1, Ca^{2+}/NFAT) are not exclusive to alternative Wnt signaling and instead play a role in a vast array of cellular processes. Furthermore, in contrast to later-generation TOPFLASH constructs, existing luciferase reporters for monitoring AP-1 and NFAT signaling, for instance, show low-amplitude responses following stimulation with Wnt ligands, and it is probably safe to say that they have cost many a researcher a headache. The fact that many of these readouts are not as robust as the typical TOPFLASH reporter assay has surely contributed to some disagreement as to whether the measured responses are "real," and at present there is no consensus in the field that these reporters truly reflect alternative Wnt signaling.

Core PCP/CE Proteins

PCP is established by the asymmetric distribution of the so-called core PCP proteins at the cell membrane (Fig. 1C) (for review, see Bayly and Axelrod 2011; Goodrich and Strutt 2011; Gray et al. 2011). In cells of the fly wing, flamingo/vang/prickle complexes accumulate on the proximal side of the cell, whereas complexes of flamingo/fz/dsh/diego localize to the distal side. The asymmetric localization of Fzd, Vangl, and Prickle is also observed in cochlear hair and support cells, suggesting that a comparable, although not identical, mechanism operates in mammals (Montcouquiol et al. 2006; Deans et al. 2007).

These same proteins have also been shown to mediate CE movements in frogs and fish (Djiane et al. 2000; Wallingford and Harland 2001; Darken et al. 2002; Veeman et al. 2003) and/or to control neural tube closure in mice (Curtin et al. 2003; Wang et al. 2006a; Torban et al. 2008).

One interesting characteristic is that both gain- and loss-of-function mutants for core PCP/CE proteins display typical (albeit not identical) PCP/CE phenotypes, suggesting that it is the correct gradient and not the absence or presence of signaling per se that is required for proper PCP/CE (Krasnow and Adler 1994; Djiane et al. 2000; Darken et al. 2002).

PCP/CE Effectors

Precisely how the core PCP/CE machinery connects to its downstream effectors remains unknown (Fig. 2). The first insight into the PCP signaling cascade came from *Drosophila*. Flies deficient for *rhoA* (a small Rho GTPase) and *jun* (an AP-1 transcription factor) presented with typical PCP defects in the eye and the wing (Kockel et al. 1997; Strutt et al. 1997; Weber et al. 2000). These and other genes, namely, *hep* (MKK7), *bsk* (JNK), and the small Rho GTPase *rac*, but not *cdc42*, were also mapped into the PCP pathway in genetic epistasis experiments, based on the ability of their loss-of-function mutants to rescue the PCP defects induced by overexpression of either *fz* or *dsh* (Strutt et al. 1997; Boutros et al. 1998).

Other genes shown to be involved in *Drosophila* PCP signaling downstream from *fz* and *dsh* are *inturned*, *fuzzy*, and *multiple wing hairs*, which together appear to be responsible for restricting wing hair formation to the distal side of the cell (Adler 1992; Lee and Adler 2002). However, these effectors only control PCP in the wing and not in the ommatidia. Conversely, *nemo* was found to be required for PCP in the fly eye, but not the wing (Choi and Benzer 1994).

The vertebrate orthologs of *inturned* and *fuzzy* are required for ciliogenesis. As a result of aberrant cilia formation, both mice and frogs deficient for Inturned or Fuzzy have defects in Hedgehog signaling (Park et al. 2006; Gray et al. 2009; Zeng et al. 2010). However, unlike the loss of core PCP proteins such as Vangl2, loss of Inturned or Fuzzy does not result in clear CE defects. Moreover, unlike Wnt5a or Ror2, neither Inturned nor Fuzzy shows a genetic interaction with Vangl2 in neural tube closure (Heydeck and Liu 2011). In contrast, the vertebrate ortholog of *nemo*, Nemo like kinase (NLK), genetically interacts with Wnt11 in CE during zebrafish development (Thorpe and Moon 2004). Together, these findings not only underscore that important differences may exist between the signal transduction events that control PCP and those that regulate CE, but also that even within the same species, the effectors that function downstream from the core PCP/CE protein complexes are likely to be tissue specific.

Rho, Rac, and Cdc42 in Vertebrates

The small Rho GTPases Rho, Rac, and Cdc42 have all been implicated in alternative Wnt-signal transduction in vertebrates, but their functions appear to be divergent. Evidence for the involvement of Cdc42 comes from the observation that exogenous Cdc42 affects *Xenopus* CE movements and the demonstration that Cdc42 is activated downstream of Wnt, Fzd, and PKC (Djiane et al. 2000; Choi and Han 2002; Penzo-Mendez et al. 2003). Both endogenous Rho and Rac were shown to be required for CE movements during gastrulation in the frog. In addition, the overexpression of Dvl1 and Dvl2, as well as of Fzd1 and Fzd7 but not Fzd2 or Fzd5, resulted in the activation of Rho and Rac, but not Cdc42, in 293T cells (Habas et al. 2001, 2003).

Although both Rho and Rac can activate JNK (Li et al. 1999; Moriguchi et al. 1999; Habas et al. 2001, 2003; Marinissen et al. 2004; Kim and Han 2005; Rosso et al. 2005), it has been suggested that Dvl activates two separate branches in mammalian cells. The first results in the activation of Rho, but not JNK, and requires full-length Dvl, whereas the second results in the activation of Rac and JNK and requires only the Dvl DEP domain (Habas et al. 2003). The Dvl-mediated activation of Rho was, in fact, shown to actively repress the activation of Rac during the directional migration of neural crest cells in developing zebrafish embryos (Matthews et al. 2008).

JNK and AP-1 in Vertebrates

Many experiments in amphibians, mice, and mammalian cell lines have placed JNK and AP-1 signaling downstream from a Wnt-signaling cascade. For instance, in *Xenopus* animal cap explants, the expression of dominant-negative JNK or its activator MKK7 rescued CE movements that had been blocked by exogenous Wnt5a (Yamanaka et al. 2002). Likewise, in mice deficient for Wnt9B, which develop cystic kidneys, the levels of activated JNK2 were found to be significantly decreased (Karner et al. 2009).

Wnt as well as Dvl proteins have also been shown to activate JNK in mammalian NIH3T3 and 293T cells (Boutros et al. 1998; Li et al. 1999; Habas et al. 2003). In addition, both Wnt and Dvl can activate AP-1 transcriptional complexes, causing AP-1 luciferase reporters to be used as a surrogate readout to monitor alternative Wnt-signaling responses (Li et al. 1999; Tufan et al. 2002; Yamanaka et al. 2002; Le Floch et al. 2005). A more specific and robust assay was recently described in the form of a luciferase reporter based on an ATF2 response element (Ohkawara and Niehrs 2011). This reporter was used in *Xenopus* embryos to monitor the effects of both gains and losses of Wnts, their receptors, and various intracellular mediators, including JNK. At present it remains unknown, however, whether this reporter can be used to monitor alternative Wnt-signaling responses in other species as well.

Other Effectors: G-Proteins, Ca^{2+}, and NFAT

Not all alternative Wnt-signaling events have been linked to PCP and/or CE processes, and this article would not be complete without discussing them.

Frizzled proteins have officially been classified as a distinct class within the superfamily of G-protein-coupled receptors (GPCRs), although they lack certain features that are well conserved in other, conventional GPCRs (Schulte 2010).

It should be noted that, to date, controversy remains in the field regarding the requirement of G-proteins for Wnt-signal transduction, despite the fact that they have been implicated in both Wnt/β-catenin and alternative signaling responses. For instance, RNAi-mediated depletion of $G\alpha_o$ and $G\alpha_q$ or treatment with the G-protein inhibitor pertussis toxin (PTX) was reported to prevent the Wnt3a-mediated increase in cytosolic β-catenin levels in mammalian cells (Liu et al. 2005), whereas in *Drosophila*, G_o hypomorphic clones showed compromised expression of *wg* target genes (Katanaev et al. 2005). Until G-proteins are shown to be directly required for transmitting Wnt signals by using a genetic knockout approach in mice, the GPCR nature of Fzds will likely remain contested.

Signaling through GPCRs generally results in the activation of heterotrimeric G-proteins, phospholipase C, and Ca^{2+} fluxes. The first evidence for the existence of a Wnt/Ca^{2+} pathway came from experiments in zebrafish, where the injection of ectopic Wnt5a and Fzd2, but not Wnt8 or Fzd1, was shown to cause an increase in the frequency of intracellular Ca^{2+} transients (Slusarski et al. 1997a,b). Calcium fluxes control the activity of multiple intracellular proteins, including PKC, CamKII, and calcineurin, and, indeed, all of these have been implicated in alternative Wnt signaling. For instance, Wnt5a and Fzd2 were also shown to promote the translocation of PKC from the cytoplasm to the cell membrane (Sheldahl et al. 1999), whereas Wnt5a and Wnt11 were shown to activate CamKII (Kuhl et al. 2000). All of these activities could be inhibited by pertussis toxin (PTX), as could the aforementioned effect of Fzd7 on tissue separation in *Xenopus* (Winklbauer et al. 2001), suggesting the involvement of G-proteins.

In *Xenopus* animal cap explants, Wnt5a and Fzd2 were shown to promote the nuclear translocation of NFAT (Saneyoshi et al. 2002). This process is controlled by the calcium-dependent phosphatase calcineurin, thereby indicating NFAT as a downstream effector of the Wnt/Ca^{2+} pathway. Indeed, Wnt5a can induce slight activation of an NFAT luciferase reporter in a Dvl-dependent manner in mouse F9 teratocarcinoma cells overexpressing Fzd2 (Ma et al. 2010). However, the precise requirement for Dvl in Wnt/Ca^{2+} signaling remains to be determined (Winklbauer et al. 2001; Sheldahl et al. 2003).

CONCLUDING REMARKS

Probing any of the alternative Wnt-signaling responses discussed above will remain a challenge until their precise molecular makeup is unraveled. For this, more specific players need to be identified, and the exact temporal order of signaling events needs to be revealed. One of the major challenges for future research will be to grasp how the different Wnt-signaling responses interact during development. For instance, Wnt/β-catenin and Wnt/Ca^{2+} signaling were shown to have opposing activities in convergent extension, with PKC and CamKII impacting on the signaling activities of β-catenin (Kuhl et al. 2001). Likewise, distinct Wnt-signaling pathways were shown to have opposing roles in tail fin regeneration (Stoick-Cooper et al. 2007). In *Drosophila*, an attractive *wnt4/fz* and a repulsive *wnt5/ryk* signal together control salivary gland migration (Harris and Beckendorf 2007). As such, multiple Wnt pathways might collaborate in controlling complex morphogenetic events, by each regulating different aspects of cell polarization and directional migration. For instance, development of the organ of Corti in the cochlea, which has become the mammalian paradigm for studying PCP, displays aspects of both PCP and CE. Whereas disruption of Vangl2 function causes the typical misorientation of inner ear hair cells associated with the former, loss of Wnt5a results in a shorter and wider cochlear duct, with an extra row of hair cells, more reminiscent of the latter (Montcouquiol et al. 2003; Qian et al. 2007). It is evident, however, that few if any of the extracellular and transmembrane components are exclusively dedicated to a single branch of the pathway. This includes proteins that were previously thought to be solely required for Wnt/β-catenin signaling, such as Lrp6, showing that cross talk between the different branches might be more common than previously anticipated (Tahinci et al. 2007; Andersson et al. 2010).

A second aspect worthy of attention concerns the initiation of the different signaling events. At present, our understanding of how the cellular context determines the assembly of Wnt/(co-)receptor complexes, and how these complexes elicit specific responses remains ill understood. Recent data suggest that part of the answer may lie in the localization of receptor complexes to distinct microdomains within the cell membrane and corresponding different mechanisms of receptor internalization (Sato et al. 2010). As the literature on this topic is expanding, it is important to remember that at the same time many of the ligand and receptor combinations remain unexplored. As a result, we have only begun to uncover the tip of the iceberg when it comes to understanding the role of alternative Wnt signaling in development and disease.

ACKNOWLEDGMENTS

This work is supported by a KWF fellowship from the Dutch Cancer Society.

REFERENCES

Abu-Elmagd M, Garcia-Morales C, Wheeler GN. 2006. Frizzled7 mediates canonical Wnt signaling in neural crest induction. *Dev Biol* **298:** 285–298.

Adler PN. 1992. The genetic control of tissue polarity in *Drosophila*. *Bioessays* **14:** 735–741.

Afzal AR, Rajab A, Fenske CD, Oldridge M, Elanko N, Ternes-Pereira E, Tuysuz B, Murday VA, Patton MA, Wilkie AO, et al. 2000. Recessive Robinow syndrome, allelic to dominant brachydactyly type B, is caused by mutation of ROR2. *Nat Genet* **25:** 419–422.

Andersson ER, Bryjova L, Biris K, Yamaguchi TP, Arenas E, Bryja V. 2010. Genetic interaction between Lrp6 and Wnt5a during mouse development. *Dev Dyn* **239:** 237–245.

Antic D, Stubbs JL, Suyama K, Kintner C, Scott MP, Axelrod JD. 2010. Planar cell polarity enables posterior localization of nodal cilia and left-right axis determination during mouse and *Xenopus* embryogenesis. *PLoS ONE* **5:** e8999.

Axelrod JD, Miller JR, Shulman JM, Moon RT, Perrimon N. 1998. Differential recruitment of Dishevelled provides signaling specificity in the planar cell polarity and Wingless signaling pathways. *Genes Dev* **12:** 2610–2622.

Baker NE. 1987. Molecular cloning of sequences from wingless, a segment polarity gene in *Drosophila*: The spatial distribution of a transcript in embryos. *EMBO J* **6:** 1765–1773.

Bayly R, Axelrod JD. 2011. Pointing in the right direction: New developments in the field of planar cell polarity. *Nat Rev Genet* **12:** 385–391.

Berndt JD, Aoyagi A, Yang P, Anastas JN, Tang L, Moon RT. 2011. Mindbomb 1, an E3 ubiquitin ligase, forms a complex with RYK to activate Wnt/β-catenin signaling. *J Cell Biol* **194:** 737–750.

Bhanot P, Brink M, Samos CH, Hsieh JC, Wang Y, Macke JP, Andrew D, Nathans J, Nusse R. 1996. A new member of the *frizzled* family from *Drosophila* functions as a Wingless receptor. *Nature* **382:** 225–230.

Bhanot P, Fish M, Jemison JA, Nusse R, Nathans J, Cadigan KM. 1999. Frizzled and Dfrizzled-2 function as redundant receptors for Wingless during *Drosophila* embryonic development. *Development* **126:** 4175–4186.

Bhat KM. 1998. *frizzled* and *frizzled 2* play a partially redundant role in Wingless signaling and have similar requirements to Wingless in neurogenesis. *Cell* **95:** 1027–1036.

Bilic J, Huang YL, Davidson G, Zimmermann T, Cruciat CM, Bienz M, Niehrs C. 2007. Wnt induces LRP6 signalosomes and promotes Dishevelled-dependent LRP6 phosphorylation. *Science* **316:** 1619–1622.

Billiard J, Way DS, Seestaller-Wehr LM, Moran RA, Mangine A, Bodine PV. 2005. The orphan receptor tyrosine kinase Ror2 modulates canonical Wnt signaling in osteoblastic cells. *Mol Endocrinol* **19:** 90–101.

Boutros M, Paricio N, Strutt DI, Mlodzik M. 1998. Dishevelled activates JNK and discriminates between JNK pathways in planar polarity and *wingless* signaling. *Cell* **94:** 109–118.

Boutros M, Mihaly J, Bouwmeester T, Mlodzik M. 2000. Signaling specificity by Frizzled receptors in *Drosophila*. *Science* **288:** 1825–1828.

Carreira-Barbosa F, Concha ML, Takeuchi M, Ueno N, Wilson SW, Tada M. 2003. Prickle 1 regulates cell movements during gastrulation and neuronal migration in zebrafish. *Development* **130:** 4037–4046.

Cervantes S, Yamaguchi TP, Hebrok M. 2009. Wnt5a is essential for intestinal elongation in mice. *Dev Biol* **326:** 285–294.

Cha SW, Tadjuidje E, Tao Q, Wylie C, Heasman J. 2008. Wnt5a and Wnt11 interact in a maternal Dkk1-regulated fashion to activate both canonical and non-canonical signaling in *Xenopus* axis formation. *Development* **135:** 3719–3729.

Cha SW, Tadjuidje E, White J, Wells J, Mayhew C, Wylie C, Heasman J. 2009. Wnt11/5a complex formation caused by tyrosine sulfation increases canonical signaling activity. *Curr Biol* **19:** 1573–1580.

Chen WS, Antic D, Matis M, Logan CY, Povelones M, Anderson GA, Nusse R, Axelrod JD. 2008. Asymmetric homotypic interactions of the atypical cadherin flamingo mediate intercellular polarity signaling. *Cell* **133:** 1093–1105.

Choi KW, Benzer S. 1994. Rotation of photoreceptor clusters in the developing *Drosophila* eye requires the *nemo* gene. *Cell* **78:** 125–136.

Choi SC, Han JK. 2002. *Xenopus* Cdc42 regulates convergent extension movements during gastrulation through Wnt/Ca^{2+} signaling pathway. *Dev Biol* **244:** 342–357.

Curtin JA, Quint E, Tsipouri V, Arkell RM, Cattanach B, Copp AJ, Henderson DJ, Spurr N, Stanier P, Fisher EM, et al. 2003. Mutation of Celsr1 disrupts planar polarity of inner ear hair cells and causes severe neural tube defects in the mouse. *Curr Biol* **13:** 1129–1133.

Dabdoub A, Kelley MW. 2005. Planar cell polarity and a potential role for a Wnt morphogen gradient in stereociliary bundle orientation in the mammalian inner ear. *J Neurobiol* **64:** 446–457.

Dabdoub A, Donohue MJ, Brennan A, Wolf V, Montcouquiol M, Sassoon DA, Hseih JC, Rubin JS, Salinas PC, Kelley MW. 2003. Wnt signaling mediates reorientation of outer hair cell stereociliary bundles in the mammalian cochlea. *Development* **130:** 2375–2384.

Darken RS, Scola AM, Rakeman AS, Das G, Mlodzik M, Wilson PA. 2002. The planar polarity gene *strabismus* regulates convergent extension movements in *Xenopus*. *EMBO J* **21:** 976–985.

Das G, Jenny A, Klein TJ, Eaton S, Mlodzik M. 2004. Diego interacts with Prickle and Strabismus/Van Gogh to localize planar cell polarity complexes. *Development* **131:** 4467–4476.

Davidson LA, Keller RE. 1999. Neural tube closure in *Xenopus laevis* involves medial migration, directed protrusive activity, cell intercalation and convergent extension. *Development* **126:** 4547–4556.

Deans MR, Antic D, Suyama K, Scott MP, Axelrod JD, Goodrich LV. 2007. Asymmetric distribution of Prickle-like 2 reveals an early underlying polarization of vestibular sensory epithelia in the inner ear. *J Neurosci* **27:** 3139–3147.

Devenport D, Fuchs E. 2008. Planar polarization in embryonic epidermis orchestrates global asymmetric morphogenesis of hair follicles. *Nat Cell Biol* **10:** 1257–1268.

Djiane A, Riou J, Umbhauer M, Boucaut J, Shi D. 2000. Role of *frizzled 7* in the regulation of convergent extension movements during gastrulation in *Xenopus laevis*. *Development* **127:** 3091–3100.

Du SJ, Purcell SM, Christian JL, McGrew LL, Moon RT. 1995. Identification of distinct classes and functional domains of Wnts through expression of wild-type and chimeric proteins in *Xenopus* embryos. *Mol Cell Biol* **15:** 2625–2634.

Endo M, Nishita M, Minami Y. 2012. Analysis of Wnt/planar cell polarity pathway in cultured cells. *Methods Mol Biol* **839:** 201–214.

Etheridge SL, Ray S, Li S, Hamblet NS, Lijam N, Tsang M, Greer J, Kardos N, Wang J, Sussman DJ, et al. 2008. Murine Dishevelled 3 functions in redundant pathways with Dishevelled 1 and 2 in normal cardiac outflow tract, cochlea, and neural tube development. *PLoS Genet* **4:** e1000259.

Feike AC, Rachor K, Gentzel M, Schambony A. 2010. Wnt5a/Ror2-induced upregulation of xPAPC requires xShcA. *Biochem Biophys Res Commun* **400:** 500–506.

Fiedler M, Mendoza-Topaz C, Rutherford TJ, Mieszczanek J, Bienz M. 2011. Dishevelled interacts with the DIX domain polymerization interface of Axin to interfere with its function in down-regulating β-catenin. *Proc Natl Acad Sci* **108:** 1937–1942.

Fossat N, Jones V, Khoo PL, Bogani D, Hardy A, Steiner K, Mukhopadhyay M, Westphal H, Nolan PM, Arkell R, et al. 2011. Stringent requirement of a proper level of canonical WNT signalling activity for head formation in mouse embryo. *Development* **138:** 667–676.

Fukuda T, Chen L, Endo T, Tang L, Lu D, Castro JE, Widhopf GF II, Rassenti LZ, Cantwell MJ, Prussak CE, et al. 2008. Antisera induced by infusions of autologous Ad-CD154-leukemia B cells identify ROR1 as an oncofetal antigen and receptor for Wnt5a. *Proc Natl Acad Sci* **105:** 3047–3052.

Gao B, Song H, Bishop K, Elliot G, Garrett L, English MA, Andre P, Robinson J, Sood R, Minami Y, et al. 2011. Wnt signaling gradients establish planar cell polarity by inducing Vangl2 phosphorylation through Ror2. *Dev Cell* **20:** 163–176.

Gentile A, Lazzari L, Benvenuti S, Trusolino L, Comoglio PM. 2011. Ror1 is a pseudokinase that is crucial for Met-driven tumorigenesis. *Cancer Res* **71:** 3132–3141.

Gleason JE, Szyleyko EA, Eisenmann DM. 2006. Multiple redundant Wnt signaling components function in two processes during *C. elegans* vulval development. *Dev Biol* **298:** 442–457.

Goodrich LV, Strutt D. 2011. Principles of planar polarity in animal development. *Development* **138:** 1877–1892.

Gray RS, Abitua PB, Wlodarczyk BJ, Szabo-Rogers HL, Blanchard O, Lee I, Weiss GS, Liu KJ, Marcotte EM, Wallingford JB, et al. 2009. The planar cell polarity effector Fuz is essential for targeted membrane trafficking, ciliogenesis and mouse embryonic development. *Nat Cell Biol* **11:** 1225–1232.

Gray RS, Roszko I, Solnica-Krezel L. 2011. Planar cell polarity: Coordinating morphogenetic cell behaviors with embryonic polarity. *Dev Cell* **21:** 120–133.

Green JL, Inoue T, Sternberg PW. 2007. The *C. elegans* ROR receptor tyrosine kinase, CAM-1, non-autonomously inhibits the Wnt pathway. *Development* **134:** 4053–4062.

Green JL, Inoue T, Sternberg PW. 2008. Opposing Wnt pathways orient cell polarity during organogenesis. *Cell* **134:** 646–656.

Guder C, Philipp I, Lengfeld T, Watanabe H, Hobmayer B, Holstein TW. 2006. The Wnt code: Cnidarians signal the way. *Oncogene* **25:** 7450–7460.

Guo N, Hawkins C, Nathans J. 2004. Frizzled6 controls hair patterning in mice. *Proc Natl Acad Sci* **101:** 9277–9281.

Habas R, Kato Y, He X. 2001. Wnt/Frizzled activation of Rho regulates vertebrate gastrulation and requires a novel Formin homology protein Daam1. *Cell* **107:** 843–854.

Habas R, Dawid IB, He X. 2003. Coactivation of Rac and Rho by Wnt/Frizzled signaling is required for vertebrate gastrulation. *Genes Dev* **17:** 295–309.

Hamblet NS, Lijam N, Ruiz-Lozano P, Wang J, Yang Y, Luo Z, Mei L, Chien KR, Sussman DJ, Wynshaw-Boris A. 2002. Dishevelled 2 is essential for cardiac outflow tract development, somite segmentation and neural tube closure. *Development* **129:** 5827–5838.

Harris KE, Beckendorf SK. 2007. Different Wnt signals act through the Frizzled and RYK receptors during *Drosophila* salivary gland migration. *Development* **134:** 2017–2025.

He X, Saint-Jeannet JP, Wang Y, Nathans J, Dawid I, Varmus H. 1997. A member of the Frizzled protein family mediating axis induction by Wnt-5A. *Science* **275:** 1652–1654.

Heisenberg CP, Tada M, Rauch GJ, Saude L, Concha ML, Geisler R, Stemple DL, Smith JC, Wilson SW. 2000.

Silberblick/Wnt11 mediates convergent extension movements during zebrafish gastrulation. *Nature* **405:** 76–81.

Hertzano R, Montcouquiol M, Rashi-Elkeles S, Elkon R, Yucel R, Frankel WN, Rechavi G, Moroy T, Friedman TB, Kelley MW, et al. 2004. Transcription profiling of inner ears from Pou4f3(ddl/ddl) identifies Gfi1 as a target of the Pou4f3 deafness gene. *Hum Mol Genet* **13:** 2143–2153.

Heydeck W, Liu A. 2011. PCP effector proteins inturned and fuzzy play nonredundant roles in the patterning but not convergent extension of mammalian neural tube. *Dev Dyn* **240:** 1938–1948.

Ho HY, Susman MW, Bikoff JB, Ryu YK, Jonas AM, Hu L, Kuruvilla R, Greenberg ME. 2012. Wnt5a-Ror-Dishevelled signaling constitutes a core developmental pathway that controls tissue morphogenesis. *Proc Natl Acad Sci* **109:** 4044–4051.

Holmen SL, Salic A, Zylstra CR, Kirschner MW, Williams BO. 2002. A novel set of Wnt–Frizzled fusion proteins identifies receptor components that activate β-catenin-dependent signaling. *J Biol Chem* **277:** 34727–34735.

Inoue T, Oz HS, Wiland D, Gharib S, Deshpande R, Hill RJ, Katz WS, Sternberg PW. 2004. *C. elegans* LIN-18 is a Ryk ortholog and functions in parallel to LIN-17/Frizzled in Wnt signaling. *Cell* **118:** 795–806.

Janssen R, Le Gouar M, Pechmann M, Poulin F, Bolognesi R, Schwager EE, Hopfen C, Colbourne JK, Budd GE, Brown SJ, et al. 2010. Conservation, loss, and redeployment of Wnt ligands in protostomes: Implications for understanding the evolution of segment formation. *BMC Evol Biol* **10:** 374.

Jenny A. 2010. Planar cell polarity signaling in the *Drosophila* eye. *Curr Top Dev Biol* **93:** 189–227.

Jenny A, Reynolds-Kenneally J, Das G, Burnett M, Mlodzik M. 2005. Diego and Prickle regulate Frizzled planar cell polarity signalling by competing for Dishevelled binding. *Nat Cell Biol* **7:** 691–697.

Jue SF, Bradley RS, Rudnicki JA, Varmus HE, Brown AM. 1992. The mouse Wnt-1 gene can act via a paracrine mechanism in transformation of mammary epithelial cells. *Mol Cell Biol* **12:** 321–328.

Karner CM, Chirumamilla R, Aoki S, Igarashi P, Wallingford JB, Carroll TJ. 2009. Wnt9b signaling regulates planar cell polarity and kidney tubule morphogenesis. *Nat Genet* **41:** 793–799.

Katanaev VL, Ponzielli R, Semeriva M, Tomlinson A. 2005. Trimeric G protein-dependent frizzled signaling in *Drosophila*. *Cell* **120:** 111–122.

Keeble TR, Halford MM, Seaman C, Kee N, Macheda M, Anderson RB, Stacker SA, Cooper HM. 2006. The Wnt receptor Ryk is required for Wnt5a-mediated axon guidance on the contralateral side of the corpus callosum. *J Neurosci* **26:** 5840–5848.

Keller R, Danilchik M. 1988. Regional expression, pattern and timing of convergence and extension during gastrulation of *Xenopus laevis*. *Development* **103:** 193–209.

Keller RE, Danilchik M, Gimlich R, Shih J. 1985. The function and mechanism of convergent extension during gastrulation of *Xenopus laevis*. *J Embryol Exp Morphol* **89:** 185–209.

Kennerdell JR, Carthew RW. 1998. Use of dsRNA-mediated genetic interference to demonstrate that *frizzled* and *frizzled 2* act in the Wingless pathway. *Cell* **95:** 1017–1026.

Kibar Z, Vogan KJ, Groulx N, Justice MJ, Underhill DA, Gros P. 2001. Ltap, a mammalian homolog of *Drosophila* Strabismus/Van Gogh, is altered in the mouse neural tube mutant Loop-tail. *Nat Genet* **28:** 251–255.

Kibar Z, Salem S, Bosoi CM, Pauwels E, De Marco P, Merello E, Bassuk AG, Capra V, Gros P. 2011. Contribution of VANGL2 mutations to isolated neural tube defects. *Clin Genet* **80:** 76–82.

Kilian B, Mansukoski H, Barbosa FC, Ulrich F, Tada M, Heisenberg CP. 2003. The role of Ppt/Wnt5 in regulating cell shape and movement during zebrafish gastrulation. *Mech Dev* **120:** 467–476.

Kim GH, Han JK. 2005. JNK and ROKα function in the noncanonical Wnt/RhoA signaling pathway to regulate *Xenopus* convergent extension movements. *Dev Dyn* **232:** 958–968.

Kim GH, Her JH, Han JK. 2008. Ryk cooperates with Frizzled 7 to promote Wnt11-mediated endocytosis and is essential for *Xenopus laevis* convergent extension movements. *J Cell Biol* **182:** 1073–1082.

Kockel L, Zeitlinger J, Staszewski LM, Mlodzik M, Bohmann D. 1997. Jun in *Drosophila* development: Redundant and nonredundant functions and regulation by two MAPK signal transduction pathways. *Genes Dev* **11:** 1748–1758.

Krasnow RE, Adler PN. 1994. A single frizzled protein has a dual function in tissue polarity. *Development* **120:** 1883–1893.

Kuhl M, Sheldahl LC, Malbon CC, Moon RT. 2000. Ca^{2+}/calmodulin-dependent protein kinase II is stimulated by Wnt and Frizzled homologs and promotes ventral cell fates in *Xenopus*. *J Biol Chem* **275:** 12701–12711.

Kuhl M, Geis K, Sheldahl LC, Pukrop T, Moon RT, Wedlich D. 2001. Antagonistic regulation of convergent extension movements in *Xenopus* by Wnt/β-catenin and Wnt/Ca^{2+} signaling. *Mech Dev* **106:** 61–76.

Lee H, Adler PN. 2002. The function of the Frizzled pathway in the *Drosophila* wing is dependent on Inturned and Fuzzy. *Genetics* **160:** 1535–1547.

Le Floch N, Rivat C, De Wever O, Bruyneel E, Mareel M, Dale T, Gespach C. 2005. The proinvasive activity of Wnt-2 is mediated through a noncanonical Wnt pathway coupled to GSK-3β and c-Jun/AP-1 signaling. *FASEB J* **19:** 144–146.

Li L, Yuan H, Xie W, Mao J, Caruso AM, McMahon A, Sussman DJ, Wu D. 1999. Dishevelled proteins lead to two signaling pathways. Regulation of LEF-1 and c-Jun N-terminal kinase in mammalian cells. *J Biol Chem* **274:** 129–134.

Li C, Chen H, Hu L, Xing Y, Sasaki T, Villosis MF, Li J, Nishita M, Minami Y, Minoo P. 2008. Ror2 modulates the canonical Wnt signaling in lung epithelial cells through cooperation with Fzd2. *BMC Mol Biol* **9:** 11.

Lin S, Baye LM, Westfall TA, Slusarski DC. 2010. Wnt5b–Ryk pathway provides directional signals to regulate gastrulation movement. *J Cell Biol* **190:** 263–278.

Liu X, Liu T, Slusarski DC, Yang-Snyder J, Malbon CC, Moon RT, Wang H. 1999. Activation of a Frizzled-2/β-adrenergic receptor chimera promotes Wnt signaling and differentiation of mouse F9 teratocarcinoma cells via Gαo and Gαt. *Proc Natl Acad Sci* **96:** 14383–14388.

Liu Y, Shi J, Lu CC, Wang ZB, Lyuksyutova AI, Song XJ, Zou Y. 2005. Ryk-mediated Wnt repulsion regulates posterior-directed growth of corticospinal tract. *Nat Neurosci* **8:** 1151–1159.

Liu Y, Ross JF, Bodine PV, Billiard J. 2007. Homodimerization of Ror2 tyrosine kinase receptor induces 14-3-3β phosphorylation and promotes osteoblast differentiation and bone formation. *Mol Endocrinol* **21:** 3050–3061.

Liu Y, Rubin B, Bodine PV, Billiard J. 2008. Wnt5a induces homodimerization and activation of Ror2 receptor tyrosine kinase. *J Cell Biochem* **105:** 497–502.

Lu W, Yamamoto V, Ortega B, Baltimore D. 2004. Mammalian Ryk is a Wnt coreceptor required for stimulation of neurite outgrowth. *Cell* **119:** 97–108.

Lyu J, Yamamoto V, Lu W. 2008. Cleavage of the Wnt receptor Ryk regulates neuronal differentiation during cortical neurogenesis. *Dev Cell* **15:** 773–780.

Lyu J, Wesselschmidt RL, Lu W. 2009. Cdc37 regulates Ryk signaling by stabilizing the cleaved Ryk intracellular domain. *J Biol Chem* **284:** 12940–12948.

Ma L, Wang HY. 2007. Mitogen-activated protein kinase p38 regulates the Wnt/cyclic GMP/Ca^{2+} non-canonical pathway. *J Biol Chem* **282:** 28980–28990.

Ma L, Wang Y, Malbon CC, Wang HY. 2010. Dishevelled-3 C-terminal His single amino acid repeats are obligate for Wnt5a activation of non-canonical signaling. *J Mol Signal* **5:** 19.

MacLeod RJ, Hayes M, Pacheco I. 2007. Wnt5a secretion stimulated by the extracellular calcium-sensing receptor inhibits defective Wnt signaling in colon cancer cells. *Am J Physiol Gastrointest Liver Physiol* **293:** G403–G411.

Marinissen MJ, Chiariello M, Tanos T, Bernard O, Narumiya S, Gutkind JS. 2004. The small GTP-binding protein RhoA regulates c-jun by a ROCK–JNK signaling axis. *Mol Cell* **14:** 29–41.

Matsuda T, Suzuki H, Oishi I, Kani S, Kuroda Y, Komori T, Sasaki A, Watanabe K, Minami Y. 2003. The receptor tyrosine kinase Ror2 associates with the melanoma-associated antigen (MAGE) family protein Dlxin-1 and regulates its intracellular distribution. *J Biol Chem* **278:** 29057–29064.

Matthews HK, Marchant L, Carmona-Fontaine C, Kuriyama S, Larrain J, Holt MR, Parsons M, Mayor R. 2008. Directional migration of neural crest cells in vivo is regulated by Syndecan-4/Rac1 and non-canonical Wnt signaling/RhoA. *Development* **135:** 1771–1780.

Maung SM, Jenny A. 2011. Planar cell polarity in *Drosophila*. *Organogenesis* **7:** 165–179.

McMahon AP, Moon RT. 1989. Ectopic expression of the proto-oncogene int-1 in *Xenopus* embryos leads to duplication of the embryonic axis. *Cell* **58:** 1075–1084.

Medina A, Reintsch W, Steinbeisser H. 2000. *Xenopus frizzled 7* can act in canonical and non-canonical Wnt signaling pathways: Implications on early patterning and morphogenesis. *Mech Dev* **92:** 227–237.

Mikels AJ, Nusse R. 2006. Purified Wnt5a protein activates or inhibits β-catenin–TCF signaling depending on receptor context. *PLoS Biol* **4:** e115.

Mikels A, Minami Y, Nusse R. 2009. Ror2 receptor requires tyrosine kinase activity to mediate Wnt5A signaling. *J Biol Chem* **284:** 30167–30176.

Moeller H, Jenny A, Schaeffer HJ, Schwarz-Romond T, Mlodzik M, Hammerschmidt M, Birchmeier W. 2006. Diversin regulates heart formation and gastrulation movements in development. *Proc Natl Acad Sci* **103:** 15900–15905.

Montcouquiol M, Rachel RA, Lanford PJ, Copeland NG, Jenkins NA, Kelley MW. 2003. Identification of Vangl2 and Scrb1 as planar polarity genes in mammals. *Nature* **423:** 173–177.

Montcouquiol M, Sans N, Huss D, Kach J, Dickman JD, Forge A, Rachel RA, Copeland NG, Jenkins NA, Bogani D, et al. 2006. Asymmetric localization of Vangl2 and Fz3 indicate novel mechanisms for planar cell polarity in mammals. *J Neurosci* **26:** 5265–5275.

Moon RT, Campbell RM, Christian JL, McGrew LL, Shih J, Fraser S. 1993. Xwnt-5A: A maternal Wnt that affects morphogenetic movements after overexpression in embryos of *Xenopus laevis*. *Development* **119:** 97–111.

Moriguchi T, Kawachi K, Kamakura S, Masuyama N, Yamanaka H, Matsumoto K, Kikuchi A, Nishida E. 1999. Distinct domains of mouse dishevelled are responsible for the c-Jun N-terminal kinase/stress-activated protein kinase activation and the axis formation in vertebrates. *J Biol Chem* **274:** 30957–30962.

Nemeth MJ, Topol L, Anderson SM, Yang Y, Bodine DM. 2007. Wnt5a inhibits canonical Wnt signaling in hematopoietic stem cells and enhances repopulation. *Proc Natl Acad Sci* **104:** 15436–15441.

Nishita M, Itsukushima S, Nomachi A, Endo M, Wang Z, Inaba D, Qiao S, Takada S, Kikuchi A, Minami Y. 2010. Ror2/Frizzled complex mediates Wnt5a-induced AP-1 activation by regulating Dishevelled polymerization. *Mol Cell Biol* **30:** 3610–3619.

Noordermeer J, Klingensmith J, Perrimon N, Nusse R. 1994. *dishevelled* and *armadillo* act in the Wingless signalling pathway in *Drosophila*. *Nature* **367:** 80–83.

Ohkawara B, Niehrs C. 2011. An ATF2-based luciferase reporter to monitor non-canonical Wnt signaling in *Xenopus* embryos. *Dev Dyn* **240:** 188–194.

Oishi I, Takeuchi S, Hashimoto R, Nagabukuro A, Ueda T, Liu ZJ, Hatta T, Akira S, Matsuda Y, Yamamura H, et al. 1999. Spatio-temporally regulated expression of receptor tyrosine kinases, mRor1, mRor2, during mouse development: Implications in development and function of the nervous system. *Genes Cells* **4:** 41–56.

Oishi I, Suzuki H, Onishi N, Takada R, Kani S, Ohkawara B, Koshida I, Suzuki K, Yamada G, Schwabe GC, et al. 2003. The receptor tyrosine kinase Ror2 is involved in non-canonical Wnt5a/JNK signalling pathway. *Genes Cells* **8:** 645–654.

Paganoni S, Bernstein J, Ferreira A. 2010. Ror1–Ror2 complexes modulate synapse formation in hippocampal neurons. *Neuroscience* **165:** 1261–1274.

Park TJ, Haigo SL, Wallingford JB. 2006. Ciliogenesis defects in embryos lacking inturned or fuzzy function are associated with failure of planar cell polarity and Hedgehog signaling. *Nat Genet* **38:** 303–311.

Penzo-Mendez A, Umbhauer M, Djiane A, Boucaut JC, Riou JF. 2003. Activation of Gβγ signaling downstream of Wnt-11/Xfz7 regulates Cdc42 activity during *Xenopus* gastrulation. *Dev Biol* **257:** 302–314.

Person AD, Beiraghi S, Sieben CM, Hermanson S, Neumann AN, Robu ME, Schleiffarth JR, Billington CJ Jr, van Bokhoven H, Hoogeboom JM, et al. 2010. WNT5A mutations in patients with autosomal dominant Robinow syndrome. *Dev Dyn* **239:** 327–337.

Phillips HM, Murdoch JN, Chaudhry B, Copp AJ, Henderson DJ. 2005. Vangl2 acts via RhoA signaling to regulate polarized cell movements during development of the proximal outflow tract. *Circ Res* **96:** 292–299.

Qian D, Jones C, Rzadzinska A, Mark S, Zhang X, Steel KP, Dai X, Chen P. 2007. Wnt5a functions in planar cell polarity regulation in mice. *Dev Biol* **306:** 121–133.

Rijsewijk F, Schuermann M, Wagenaar E, Parren P, Weigel D, Nusse R. 1987. The *Drosophila* homolog of the mouse mammary oncogene *int-1* is identical to the segment polarity gene *wingless*. *Cell* **50:** 649–657.

Roarty K, Baxley SE, Crowley MR, Frost AR, Serra R. 2009. Loss of TGF-β or Wnt5a results in an increase in Wnt/β-catenin activity and redirects mammary tumour phenotype. *Breast Cancer Res* **11:** R19.

Rosso SB, Sussman D, Wynshaw-Boris A, Salinas PC. 2005. Wnt signaling through Dishevelled, Rac and JNK regulates dendritic development. *Nat Neurosci* **8:** 34–42.

Rothbacher U, Laurent MN, Deardorff MA, Klein PS, Cho KW, Fraser SE. 2000. Dishevelled phosphorylation, subcellular localization and multimerization regulate its role in early embryogenesis. *EMBO J* **19:** 1010–1022.

Rulifson EJ, Wu CH, Nusse R. 2000. Pathway specificity by the bifunctional receptor Frizzled is determined by affinity for Wingless. *Mol Cell* **6:** 117–126.

Sakurai M, Aoki T, Yoshikawa S, Santschi LA, Saito H, Endo K, Ishikawa K, Kimura K, Ito K, Thomas JB, et al. 2009. Differentially expressed Drl and Drl-2 play opposing roles in Wnt5 signaling during *Drosophila* olfactory system development. *J Neurosci* **29:** 4972–4980.

Saldanha J, Singh J, Mahadevan D. 1998. Identification of a Frizzled-like cysteine rich domain in the extracellular region of developmental receptor tyrosine kinases. *Protein Sci* **7:** 1632–1635.

Sammar M, Stricker S, Schwabe GC, Sieber C, Hartung A, Hanke M, Oishi I, Pohl J, Minami Y, Sebald W, et al. 2004. Modulation of GDF5/BRI-b signalling through interaction with the tyrosine kinase receptor Ror2. *Genes Cells* **9:** 1227–1238.

Sammar M, Sieber C, Knaus P. 2009. Biochemical and functional characterization of the Ror2/BRIb receptor complex. *Biochem Biophys Res Commun* **381:** 1–6.

Saneyoshi T, Kume S, Amasaki Y, Mikoshiba K. 2002. The Wnt/calcium pathway activates NF-AT and promotes ventral cell fate in *Xenopus* embryos. *Nature* **417:** 295–299.

Sato A, Yamamoto H, Sakane H, Koyama H, Kikuchi A. 2010. Wnt5a regulates distinct signalling pathways by binding to Frizzled2. *EMBO J* **29:** 41–54.

Schulte G. 2010. International Union of Basic and Clinical Pharmacology. LXXX. The class Frizzled receptors. *Pharmacol Rev* **62:** 632–667.

Schwabe GC, Trepczik B, Suring K, Brieske N, Tucker AS, Sharpe PT, Minami Y, Mundlos S. 2004. Ror2 knockout mouse as a model for the developmental pathology of autosomal recessive Robinow syndrome. *Dev Dyn* **229:** 400–410.

Schwarzer W, Witte F, Rajab A, Mundlos S, Stricker S. 2009. A gradient of ROR2 protein stability and membrane localization confers brachydactyly type B or Robinow syndrome phenotypes. *Hum Mol Genet* **18:** 4013–4021.

Schwarz-Romond T, Fiedler M, Shibata N, Butler PJ, Kikuchi A, Higuchi Y, Bienz M. 2007a. The DIX domain of Dishevelled confers Wnt signaling by dynamic polymerization. *Nat Struct Mol Biol* **14:** 484–492.

Schwarz-Romond T, Metcalfe C, Bienz M. 2007b. Dynamic recruitment of axin by Dishevelled protein assemblies. *J Cell Sci* **120:** 2402–2412.

Seo JH, Zilber Y, Babayeva S, Liu J, Kyriakopoulos P, De Marco P, Merello E, Capra V, Gros P, Torban E. 2011. Mutations in the planar cell polarity gene, Fuzzy, are associated with neural tube defects in humans. *Hum Mol Genet* **20:** 4324–4333.

Shabani M, Asgarian-Omran H, Vossough P, Sharifian RA, Faranoush M, Ghragozlou S, Khoshnoodi J, Roohi A, Jeddi-Tehrani M, Mellstedt H, et al. 2008. Expression profile of orphan receptor tyrosine kinase (ROR1) and Wilms' tumor gene 1 (WT1) in different subsets of B-cell acute lymphoblastic leukemia. *Leuk Lymphoma* **49:** 1360–1367.

Shafer B, Onishi K, Lo C, Colakoglu G, Zou Y. 2011. Vangl2 promotes Wnt/planar cell polarity-like signaling by antagonizing Dvl1-mediated feedback inhibition in growth cone guidance. *Dev Cell* **20:** 177–191.

Sheldahl LC, Park M, Malbon CC, Moon RT. 1999. Protein kinase C is differentially stimulated by Wnt and Frizzled homologs in a G-protein-dependent manner. *Curr Biol* **9:** 695–698.

Sheldahl LC, Slusarski DC, Pandur P, Miller JR, Kuhl M, Moon RT. 2003. Dishevelled activates Ca^{2+} flux, PKC, and CamKII in vertebrate embryos. *J Cell Biol* **161:** 769–777.

Shimizu H, Julius MA, Giarre M, Zheng Z, Brown AM, Kitajewski J. 1997. Transformation by Wnt family proteins correlates with regulation of β-catenin. *Cell Growth Differ* **8:** 1349–1358.

Shimizu K, Sato M, Tabata T. 2011. The Wnt5/planar cell polarity pathway regulates axonal development of the *Drosophila* mushroom body neuron. *J Neurosci* **31:** 4944–4954.

Siegfried E, Chou TB, Perrimon N. 1992. *wingless* signaling acts through *zeste-white 3*, the *Drosophila* homolog of *glycogen synthase kinase-3*, to regulate *engrailed* and establish cell fate. *Cell* **71:** 1167–1179.

Siegfried E, Wilder EL, Perrimon N. 1994. Components of Wingless signalling in *Drosophila*. *Nature* **367:** 76–80.

Slusarski DC, Corces VG, Moon RT. 1997a. Interaction of Wnt and a Frizzled homologue triggers G-protein-linked phosphatidylinositol signalling. *Nature* **390:** 410–413.

Slusarski DC, Yang-Snyder J, Busa WB, Moon RT. 1997b. Modulation of embryonic intracellular Ca^{2+} signaling by Wnt-5A. *Dev Biol* **182:** 114–120.

Sokol S, Christian JL, Moon RT, Melton DA. 1991. Injected Wnt RNA induces a complete body axis in *Xenopus* embryos. *Cell* **67:** 741–752.

Song H, Hu J, Chen W, Elliott G, Andre P, Gao B, Yang Y. 2010. Planar cell polarity breaks bilateral symmetry by controlling ciliary positioning. *Nature* **466:** 378–382.

Stoick-Cooper CL, Weidinger G, Riehle KJ, Hubbert C, Major MB, Fausto N, Moon RT. 2007. Distinct Wnt signaling pathways have opposing roles in appendage regeneration. *Development* **134:** 479–489.

Strutt DI, Weber U, Mlodzik M. 1997. The role of RhoA in tissue polarity and Frizzled signalling. *Nature* **387:** 292–295.

Sumanas S, Strege P, Heasman J, Ekker SC. 2000. The putative Wnt receptor *Xenopus frizzled-7* functions upstream of β-catenin in vertebrate dorsoventral mesoderm patterning. *Development* **127:** 1981–1990.

Tahinci E, Thorne CA, Franklin JL, Salic A, Christian KM, Lee LA, Coffey RJ, Lee E. 2007. Lrp6 is required for convergent extension during *Xenopus* gastrulation. *Development* **134:** 4095–4106.

Taillebourg E, Moreau-Fauvarque C, Delaval K, Dura JM. 2005. In vivo evidence for a regulatory role of the kinase activity of the linotte/derailed receptor tyrosine kinase, a *Drosophila* Ryk ortholog. *Dev Genes Evol* **215:** 158–163.

Tao Q, Yokota C, Puck H, Kofron M, Birsoy B, Yan D, Asashima M, Wylie CC, Lin X, Heasman J. 2005. Maternal Wnt11 activates the canonical Wnt signaling pathway required for axis formation in *Xenopus* embryos. *Cell* **120:** 857–871.

Thorpe CJ, Moon RT. 2004. *nemo-like kinase* is an essential co-activator of Wnt signaling during early zebrafish development. *Development* **131:** 2899–2909.

Topol L, Jiang X, Choi H, Garrett-Beal L, Carolan PJ, Yang Y. 2003. Wnt-5a inhibits the canonical Wnt pathway by promoting GSK-3-independent β-catenin degradation. *J Cell Biol* **162:** 899–908.

Torban E, Wang HJ, Groulx N, Gros P. 2004. Independent mutations in mouse Vangl2 that cause neural tube defects in looptail mice impair interaction with members of the Dishevelled family. *J Biol Chem* **279:** 52703–52713.

Torban E, Patenaude AM, Leclerc S, Rakowiecki S, Gauthier S, Andelfinger G, Epstein DJ, Gros P. 2008. Genetic interaction between members of the Vangl family causes neural tube defects in mice. *Proc Natl Acad Sci* **105:** 3449–3454.

Torres MA, Yang-Snyder JA, Purcell SM, DeMarais AA, McGrew LL, Moon RT. 1996. Activities of the Wnt-1 class of secreted signaling factors are antagonized by the Wnt-5A class and by a dominant negative cadherin in early *Xenopus* development. *J Cell Biol* **133:** 1123–1137.

Tree DR, Shulman JM, Rousset R, Scott MP, Gubb D, Axelrod JD. 2002. Prickle mediates feedback amplification to generate asymmetric planar cell polarity signaling. *Cell* **109:** 371–381.

Tufan AC, Daumer KM, DeLise AM, Tuan RS. 2002. AP-1 transcription factor complex is a target of signals from both WnT-7a and N-cadherin-dependent cell–cell adhesion complex during the regulation of limb mesenchymal chondrogenesis. *Exp Cell Res* **273:** 197–203.

van Amerongen R, Nusse R. 2009. Towards an integrated view of Wnt signaling in development. *Development* **136:** 3205–3214.

van Bokhoven H, Celli J, Kayserili H, van Beusekom E, Balci S, Brussel W, Skovby F, Kerr B, Percin EF, Akarsu N, et al. 2000. Mutation of the gene encoding the ROR2 tyrosine kinase causes autosomal recessive Robinow syndrome. *Nat Genet* **25:** 423–426.

van de Wetering M, Cavallo R, Dooijes D, van Beest M, van Es J, Loureiro J, Ypma A, Hursh D, Jones T, Bejsovec A, et al. 1997. Armadillo coactivates transcription driven by the product of the *Drosophila* segment polarity gene dTCF. *Cell* **88:** 789–799.

van Wijk NV, Witte F, Feike AC, Schambony A, Birchmeier W, Mundlos S, Stricker S. 2009. The LIM domain protein Wtip interacts with the receptor tyrosine kinase Ror2 and inhibits canonical Wnt signalling. *Biochem Biophys Res Commun* **390:** 211–216.

Veeman MT, Slusarski DC, Kaykas A, Louie SH, Moon RT. 2003. Zebrafish prickle, a modulator of noncanonical Wnt/Fz signaling, regulates gastrulation movements. *Curr Biol* **13:** 680–685.

Verkaar F, van Rosmalen JW, Smits JF, Blankesteijn WM, Zaman GJ. 2009. Stably overexpressed human Frizzled-2 signals through the β-catenin pathway and does not activate Ca^{2+}-mobilization in Human Embryonic Kidney 293 cells. *Cell Signal* **21:** 22–33.

Verkaar F, Blankesteijn WM, Smits JF, Zaman GJ. 2010. β-Galactosidase enzyme fragment complementation for the measurement of Wnt/β-catenin signaling. *FASEB J* **24:** 1205–1217.

Vinson CR, Conover S, Adler PN. 1989. A *Drosophila* tissue polarity locus encodes a protein containing seven potential transmembrane domains. *Nature* **338:** 263–264.

Wallingford JB, Harland RM. 2001. *Xenopus dishevelled* signaling regulates both neural and mesodermal convergent extension: Parallel forces elongating the body axis. *Development* **128:** 2581–2592.

Wallingford JB, Rowning BA, Vogeli KM, Rothbacher U, Fraser SE, Harland RM. 2000. Dishevelled controls cell polarity during *Xenopus* gastrulation. *Nature* **405:** 81–85.

Wang J, Mark S, Zhang X, Qian D, Yoo SJ, Radde-Gallwitz K, Zhang Y, Lin X, Collazo A, Wynshaw-Boris A, et al. 2005. Regulation of polarized extension and planar cell polarity in the cochlea by the vertebrate PCP pathway. *Nat Genet* **37:** 980–985.

Wang J, Hamblet NS, Mark S, Dickinson ME, Brinkman BC, Segil N, Fraser SE, Chen P, Wallingford JB, Wynshaw-Boris A. 2006a. Dishevelled genes mediate a conserved mammalian PCP pathway to regulate convergent extension during neurulation. *Development* **133:** 1767–1778.

Wang Y, Guo N, Nathans J. 2006b. The role of Frizzled3 and Frizzled6 in neural tube closure and in the planar polarity of inner-ear sensory hair cells. *J Neurosci* **26:** 2147–2156.

Wang Y, Chang H, Nathans J. 2010. When whorls collide: The development of hair patterns in *frizzled 6* mutant mice. *Development* **137:** 4091–4099.

Warga RM, Kimmel CB. 1990. Cell movements during epiboly and gastrulation in zebrafish. *Development* **108:** 569–580.

Weber U, Paricio N, Mlodzik M. 2000. Jun mediates Frizzled-induced R3/R4 cell fate distinction and planar polarity determination in the *Drosophila* eye. *Development* **127:** 3619–3629.

Wilson PA, Oster G, Keller R. 1989. Cell rearrangement and segmentation in *Xenopus*: Direct observation of cultured explants. *Development* **105:** 155–166.

Winkel A, Stricker S, Tylzanowski P, Seiffart V, Mundlos S, Gross G, Hoffmann A. 2008. Wnt-ligand-dependent interaction of TAK1 (TGF-β-activated kinase-1) with the receptor tyrosine kinase Ror2 modulates canonical Wnt-signalling. *Cell Signal* **20:** 2134–2144.

Winklbauer R, Medina A, Swain RK, Steinbeisser H. 2001. Frizzled-7 signalling controls tissue separation during *Xenopus* gastrulation. *Nature* **413:** 856–860.

Witte F, Bernatik O, Kirchner K, Masek J, Mahl A, Krejci P, Mundlos S, Schambony A, Bryja V, Stricker S. 2010. Negative regulation of Wnt signaling mediated by CK1-phosphorylated Dishevelled via Ror2. *FASEB J* **24:** 2417–2426.

Wong LL, Adler PN. 1993. Tissue polarity genes of *Drosophila* regulate the subcellular location for prehair initiation in pupal wing cells. *J Cell Biol* **123:** 209–221.

Wong GT, Gavin BJ, McMahon AP. 1994. Differential transformation of mammary epithelial cells by Wnt genes. *Mol Cell Biol* **14:** 6278–6286.

Wouda RR, Bansraj MR, de Jong AW, Noordermeer JN, Fradkin LG. 2008. Src family kinases are required for WNT5 signaling through the Derailed/RYK receptor in the *Drosophila* embryonic central nervous system. *Development* **135:** 2277–2287.

Wu M, Herman MA. 2006. A novel noncanonical Wnt pathway is involved in the regulation of the asymmetric B cell division in *C. elegans*. *Dev Biol* **293:** 316–329.

Wu J, Mlodzik M. 2008. The frizzled extracellular domain is a ligand for Van Gogh/Stbm during nonautonomous planar cell polarity signaling. *Dev Cell* **15:** 462–469.

Wu J, Klein TJ, Mlodzik M. 2004. Subcellular localization of frizzled receptors, mediated by their cytoplasmic tails, regulates signaling pathway specificity. *PLoS Biol* **2:** e158.

Yamada M, Udagawa J, Matsumoto A, Hashimoto R, Hatta T, Nishita M, Minami Y, Otani H. 2010. Ror2 is required for midgut elongation during mouse development. *Dev Dyn* **239:** 941–953.

Yamaguchi TP, Bradley A, McMahon AP, Jones S. 1999. A Wnt5a pathway underlies outgrowth of multiple structures in the vertebrate embryo. *Development* **126:** 1211–1223.

Yamamoto S, Nishimura O, Misaki K, Nishita M, Minami Y, Yonemura S, Tarui H, Sasaki H. 2008. Cthrc1 selectively activates the planar cell polarity pathway of Wnt signaling by stabilizing the Wnt–receptor complex. *Dev Cell* **15:** 23–36.

Yamanaka H, Moriguchi T, Masuyama N, Kusakabe M, Hanafusa H, Takada R, Takada S, Nishida E. 2002. JNK functions in the non-canonical Wnt pathway to regulate convergent extension movements in vertebrates. *EMBO Rep* **3:** 69–75.

Yates LL, Schnatwinkel C, Murdoch JN, Bogani D, Formstone CJ, Townsend S, Greenfield A, Niswander LA, Dean CH. 2010. The PCP genes Celsr1 and Vangl2 are required for normal lung branching morphogenesis. *Hum Mol Genet* **19:** 2251–2267.

Yoshikawa S, McKinnon RD, Kokel M, Thomas JB. 2003. Wnt-mediated axon guidance via the *Drosophila* Derailed receptor. *Nature* **422:** 583–588.

Zeng H, Hoover AN, Liu A. 2010. PCP effector gene *Inturned* is an important regulator of cilia formation and embryonic development in mammals. *Dev Biol* **339:** 418–428.

Zhang Y, Levin M. 2009. Left–right asymmetry in the chick embryo requires core planar cell polarity protein Vangl2. *Genesis* **47:** 719–728.

Zinovyeva AY, Yamamoto Y, Sawa H, Forrester WC. 2008. Complex network of Wnt signaling regulates neuronal migrations during *Caenorhabditis elegans* development. *Genetics* **179:** 1357–1371.

The Evolution of the Wnt Pathway

Thomas W. Holstein

Department of Molecular Evolution and Genomics, Centre for Organismal Studies, Heidelberg University, D-69120 Heidelberg, Germany

Correspondence: thomas.holstein@cos.uni-heidelberg.de

Wnt genes are important regulators of embryogenesis and cell differentiation in vertebrates and insects. New data revealed by comparative genomics have now shown that members of the Wnt signaling pathway can be found in all clades of metazoans, but not in fungi, plants, or unicellular eukaryotes. This article focuses on new data from recent genomic analyses of several basal metazoan organisms, providing evidence that the Wnt pathway was a primordial signaling pathway during evolution. The formation of a Wnt signaling center at the site of gastrulation was instrumental for the formation of a primary, anterior–posterior body axis, which can be traced throughout animal evolution.

Wnt genes are specific for metazoans, a monophyletic group of eukaryotes with a common origin in protozoans (Ruiz-Trillo et al. 2008; Schierwater et al. 2009). So far, no *Wnt* genes have been described from any unicellular eukaryotes, neither from choanoflagellates, which are presumed to represent the common ancestor of animals (King et al. 2008), nor from other protists or fungi (Sebe-Pedros et al. 2011).

Fossil records from 580-million-year (Myr)-old Ediacaran assemblages indicate that the first metazoans were diploblastic organisms similar to modern sponges and cnidarians (Xiao and Laflamme 2009). These organisms are primarily radially symmetric (Radiata) and can be distinguished from the rest of animals showing two body axes (Bilateria). Paleontological records of the Burgess Shale assemblages document that this bilaterian diversification occurred within a short period in the lower Cambrium (Conway Morris 2000). However, major questions concerning the origin of the major phyla remain unsettled, indicating that the deep nodes in metazoan evolution are difficult to resolve (Rokas et al. 2005).

Bilaterian animals are patterned along two major body axes (Niehrs 2010): the anterior–posterior (AP) axis, which is oriented parallel to the gut and, in a perpendicular orientation, the dorsal–ventral (DV) axis. Increasing evidence suggests that the primary body axis is the AP axis, whereas the secondary body axis is the DV axis (Niehrs 2010; Holstein et al. 2011). A localized Wnt/β-catenin signaling center determines the orientation of the primary body axis in both bilaterians and non-bilaterians, indicating that the conserved Wnt signaling pathway is at the base of polarized development in all metazoans (Petersen and Reddien 2009). The secondary DV axis, which is patterned by the antagonistic BMP-Chordin network, has been conserved between vertebrate and invertebrate phyla in

Cite this article as *Cold Spring Harb Perspect Biol* doi: 10.1101/cshperspect.a007922

patterning the DV axis (De Robertis 2010). Prebilaterian groups like sponges and cnidarians lack a DV axis. The antagonistic BMP-Chordin network seems absent in sponges, whereas in cnidarians it is initially expressed along the primary body axis (Rentzsch et al. 2008).

Bilaterians are further split into protostomes and deuterostomes based on the site of origin of their mouth (Brusca and Brusca 2002). The protostomes comprise two major phyla designated as Lophotrochozoa and Ecdysozoa, the latter including two prominent model organisms, *Drosophila melanogaster* and *Caenorhabditis elegans*. Deuterostomes include chordates (i.e., vertebrates, urochordates, and the cephalochordate *Amphioxus*), hemichordates, and echinoderms. The transition from radial to bilateral symmetry and the origin of the two major bilaterian groups are a matter of ongoing debate with controversial hypotheses (De Robertis and Sasai 1996; Arendt et al. 2001; Martindale and Hejnol 2009; Schierwater et al. 2009). Based primarily on the expression of *Wnt*s and anterior marker genes, even the existence of a common ur-bilaterian ancestor has been questioned, positing that protostome and deuterostome lineages evolved independently from a radially symmetric cnidarian-, gastrula-like organism (Meinhardt 2002, 2004, 2006). The evolution of *Wnt* genes has therefore not only an impact for our understanding of stem cell self-renewal, cell differentiation, and apoptosis, but also for the evolution and diversification of the metazoan body plans.

ROOTS OF WNT SIGNALING IN PRE-METAZOANS

The Wnt signaling pathway is an invention of the first multicellular animals (Metazoa), because in no single-cell organism (Protozoa) has a complete Wnt signaling pathway been discovered. Nevertheless, it is possible to identify modules of this signaling pathway in several protozoans. This suggests that a successful assembly of these modules was a driving force for the formation of tissues and a signaling center at the transition of single-cell to multicellular organisms. Figure 1 summarizes the evolutionary origin of genes involved in the Wnt signaling pathway(s).

Wnt Ligands

No genes encoding for Wnt ligands have been found in any protozoan. This is different from other metazoan-specific signaling pathways, for example, Hedgehog signaling, where proteins containing a Hedge and/or Hog domain of the Hedgehog ligand can be traced back to a variety of single-cell organisms, including red algae and dinoflagellates (Ingham et al. 2011). No members of the Wnt secretion machinery like Wntless (Banziger et al. 2006; Bartscherer et al. 2006) and Porcupine (Kadowaki et al. 1996) have been found outside of the Metazoa. The invention of Wnt ligands must therefore be tightly coupled with the secretory control of Wnt proteins.

Wnt Receptors and Mediators

Wnt ligands bind to the cysteine-rich domain (CRD) of Frizzled (Fzd) receptors (Dann et al. 2001) in a ternary complex with low-density lipoprotein receptor–related proteins (LRP5/6) (Tamai et al. 2000), recruiting Dishevelled (Dsh) to the membrane. GSK3 and CK1γ phosphorylate the cytoplasmic domain of LRP5/6 (Heuberger and Birchmeier 2010), which, in turn, recruits Axin to the membrane and stabilizes β-catenin.

Wnt-related receptors have been found in the cellular slime mold *Dictyostelium discoideum*. In these social amoebae, several *Fzd*-related genes have been described, but no LRP5/6 orthologs (Eichinger et al. 2005; Prabhu and Eichinger 2006). Sixteen *Fzd*-related genes encode for a putative cysteine-rich domain (CRD) and have been denoted *Frizzled/Smoothened-like* (*fsl*) (Harwood 2008). Two members (*fslJ* and *fslK*) contain an additional KTXXXW motif, which is required for activation of the canonical pathway (Harwood 2008).

Three main components of the Wnt signal transduction cascade, that is, GSK3, CK1, and an ortholog related to β-catenin called Aardvark, have also been found in *Dictyostelium*

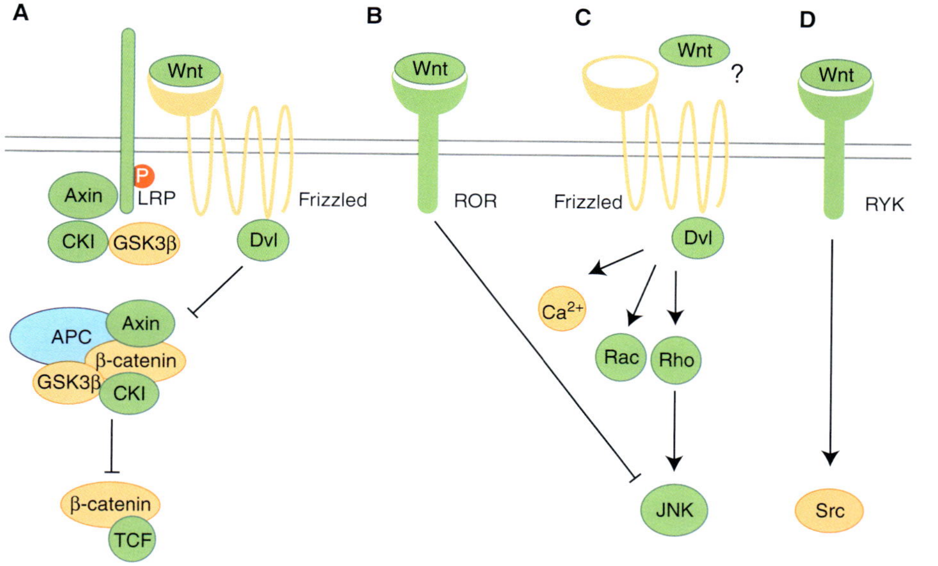

Figure 1. Evolution of Wnt pathways. Proteins involved in different Wnt pathways (*A–D*) are shown and labeled according to their occurrence in protozoans (yellow), pre-bilaterians (green), and bilaterians (blue). (*A*) Binding of Wnt to Frizzled receptor and LRP activates β-catenin/TCF signaling. The β-catenin destruction complex, including APC, Axin, GSK3β, and CKI, sequesters and phosphorylates β-catenin. Phosphorylation of LRP is crucial for Wnt/β-catenin signaling. (*B*) Binding of Wnt to the receptor tyrosine kinase (RTK) Ror2 activates Jnk and inhibits β-catenin/TCF signaling. (*C*) Frizzled receptors in planar cell polarity (PCP) of vertebrates act via downstream messengers including Dishevelled (Dvl), small Rho GTPases, Jnk, or Ca^{2+}. In flies, PCP can be activated independently of Wnt-Frizzled binding (?). (*D*) Binding of Wnt proteins to RYK RTKs results in the activation of Src proteins. (Adapted from van Amerongen and Nusse 2009; reprinted, with permission, from the authors.)

(Grimson et al. 2000; Coates et al. 2002), but other components of the Wnt signal transduction cascade, including Dsh, Axin, and APC proteins, have not been identified (Harwood 2008). A GSK3 ortholog (GSKA) acts in cell growth and differentiation processes of *Dictyostelium* (Harwood 2008), for example, in the homeostasis of prestalk and prespore cells (Harwood et al. 1995; Schilde et al. 2004). Although GSK3 was also identified in other eukaryotes (Srivastava et al. 2008), the Axin-binding site present in GSK3 and other non-metazoan GSK3-related kinases can be considered as a pre-adaptation for the function of GSK3 in Wnt signaling.

The most notable member of the Wnt pathway present in *Dictyostelium* is the β-catenin-related molecule Aardvark (Aar) (Grimson et al. 2000). Aardvark is essential for the cAMP-mediated induction of pre-spore cells, where it acts together with GSKA. However, in contrast to the negative regulation of canonical Wnt signaling by GSK3, that is, phosphorylation of β-catenin by GSK3, *Dictyostelium* GSKA acts positively on Aar to induce pspA expression (Harwood 2008). As pointed out by Harwood (2008), this mechanism is reminiscent to the regulation of noncanonical *Wnts* in *C. elegans* (Hardin and King 2008).

Recently, an α-catenin ortholog has been discovered that, together with the β-catenin ortholog Aar, plays a role in cell adhesion (Dickinson et al. 2011). Mutations of *Aar* result in a loss of cell junctions (Coates and Harwood 2001; Hardin and King 2008). This is remarkable because cadherin-like molecules have not been found in *Dictyostelium* and were only identified in the genome of the choanoflagellate

Monosiga brevicollis (Abedin and King 2008). This could indicate that β-catenin has an ancient function in cell adhesion that predates the evolution of its function in Wnt signaling (Dickinson et al. 2011).

Wnt Transcription Factors and Target Genes

No orthologs to the TCF/Lef transcription factors have been described outside metazoans, suggesting that these are metazoan-specific innovations (Srivastava et al. 2010). However, a *Brachyury*-related T-box gene was found in *Capsaspora owczarzaki*, a protist that is closely related to choanoflagellates and metazoans (Sebe-Pedros et al. 2011). Brachyury is a direct downstream target of Wnt signaling during posterior growth of vertebrates (Yamaguchi et al. 1999), where it directly regulates Wnt signaling (Martin and Kimelman 2008, 2009). At which point in animal evolution this positive Brachyury–Wnt loop was established is unknown so far (Martin and Kimelman 2008), but the coexpression of *Wnt* and *Brachyury* in basal metazoans (Technau et al. 2000) suggests that it was established during early metazoans evolution.

ORIGIN OF DIFFERENT WNT PATHWAYS IN PRE-BILATERIANS

Wnt signaling controls a variety of different cellular behaviors including cell proliferation, stem cell maintenance and differentiation, coordinated cell movement, and the establishment of tissue polarity (Croce and McClay 2008; van Amerongen and Nusse 2009). Based on their ability to induce an ectopic axis in *Xenopus* embryos (McMahon and Moon 1989) or planar cell polarity and convergent extension movements (Heisenberg et al. 2000), the different Wnts have frequently been classified as "canonical" or "noncanonical" Wnts. However, this is an oversimplification because how different Wnts signal is also determined by the receptors with which they interact. For example, the "noncanonical" Wnt5a can also signal in the canonical β-catenin pathway, if it is exposed to the right receptor on the cell, that is, Fz4 (van Amerongen et al. 2008; van Amerongen and Nusse 2009). As shown in Figure 1, in canonical β-catenin signaling, Wnts can bind to Frizzled and LRP and thereby activate β-catenin/TCF. Wnts can also interact in a β-catenin-independent manner with RYK receptors in Src signaling or with ROR in JNK signaling (via small Rho GTPases and c-Jun amino-terminal kinase) (van Amerongen and Nusse 2009). In some cases, multiple Wnts are required to induce a specific cellular response (van Amerongen and Nusse 2009).

It is difficult to judge which of the Wnt functions might represent the ancestral form—the β-catenin dependent or independent—or how this might be related to the specific structure of Wnt ligands. There are hints that canonical and noncanonical functions of Wnt proteins may be conserved during evolution (Rigo-Watermeier et al. 2011), but the distinction between both groups of ligands is not sharp (see above). It could therefore easily be that the first Wnt ligands had multiple functions in different pathways. Because the repertoire of Wnt receptors was smaller in basal metazoans compared with higher bilaterians, it is tempting to speculate that Wnt receptors in basal metazoans are more promiscuous than in vertebrates and that the evolution of Wnt receptors resulted in a larger and more refined Wnt-receptor repertoire.

Considering the core function of β-catenin in cell adhesion, an ancient role of β-catenin in cell adhesion might have further evolved to control local changes in cell signaling by Wnt ligands by creating a collective of cells with similar signaling properties (Fig. 2). These positional cues were probably the origin of the canonical pathway, fostered by the evolution of multidomain scaffolding proteins.

Porifera, Placozoa, and Ctenophora

Porifera (sponges), Placozoa, and Ctenophora (comb jellies) are simple metazoans for which a basal set of Wnt ligands, receptors, and cytoplasmic transducers has been identified. Their *Wnt* gene repertoire is less complex than that of any other metazoans. Although the phylogenetic position of placozoans and ctenophores is still under debate (Ruiz-Trillo et al. 2008), it is

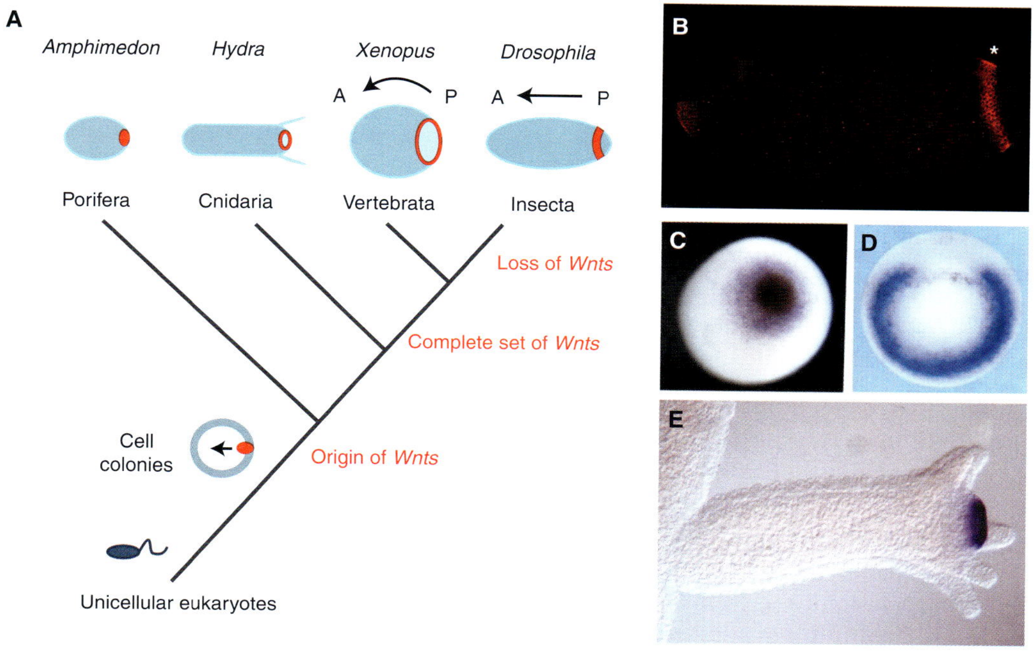

Figure 2. Evolution of Wnt signaling and axis formation during metazoan evolution. (*A*) Wnt signaling centers (red) evolved at the transition from unicellular (e.g., *Monosiga*) to multicellular eukaryotes. In all studied metazoans (sponges, cnidarians, deuterostomes, and protostomes), a posterior Wnt signaling center defines the posterior pole of the body axis. The number of *Wnt* gene subfamilies increased from the first metazoans to cnidarians, whose complete set of *Wnt* genes is only retained in basal deuterostomes and protostomes. Protostomes are characterized by frequent loss of *Wnt* gene subfamilies. (*B*) The posterior *Wg* (*Wnt1*) stripe in *Drosophila*. (Figure kindly provided by Drs. P. Vorwald and E. De Robertis, Howard Hughes Medical Institute, UC Los Angeles.) (*C*) *Wnt* expression at the blastopore of a *Amphimedon* gastrulae (*AmqWntA*), at the blastopore of an early *Xenopus* gastrulae (*Wnt8*) (figure kindly provided by Dr. Maja Adamska, SARS Center, Bergen), and (*D*) and (*E*) at the hypostome of *Hydra* polyps (*Wnt3*). (Fig. 2D from Steiner et al. 2006; reprinted, with permission, from The Company of Biologists © 2006; Fig. 2E from Hobmayer et al. 2000; reprinted, with permission, from the author.)

likely that they represent a metazoan clade that branched off early in metazoan evolution.

Placozoans, disc-shaped creeping creatures with unknown embryology, have two epithelial layers lacking any gastric cavity (Srivastava et al. 2008). The genome contains 11,514 protein-coding genes that are closely related to those of cnidarians (Srivastava et al. 2008). The only known species, *Trichoplax adhaerens*, has three unclassified *Wnt* genes and major components of canonical Wnt signaling (Dsh, Frz, GSK3, Axin, β-catenin TCF), but no Wnt antagonists like Dkk or secreted Frizzled-related proteins (Srivastava et al. 2008). The expression patterns of the *Wnt* genes and β-catenin are unknown.

Sponges have a simple body plan that develops from a blastula, producing a tube-shaped diploblastic larva that is similar to the planula larva of cnidarians. This diploblastic tube invaginates at multiple positions and develops a branched filtering channel system (Brusca and Brusca 2002). The filtering cells (choanocytes) are similar in morphology to the protozoan choanoflagellate *Monosiga* (King et al. 2008). The genome of the demosponge *Amphimedon queenslandica* shows significant conservation of gene families with those of cnidarians and bilaterians (Srivastava et al. 2008). Besides β-catenin, Fzd, and GSK3, the main components of the canonical Wnt/β-catenin pathway

(SFRP, Lrp5/6, Dvl, Axin, APC, TCF, and Groucho) are present, but those of the noncanonical pathways are missing (Adamska et al. 2007, 2010). *Amphimedon* contains three, and the homoscleromorph sponge *Oscarella* two, *Wnt* genes, which have proven difficult to classify (Lapebie et al. 2009; Adamska et al. 2010; Srivastava et al. 2010). *WntA* expression is restricted to the posterior pole of the *Amphimedon* larva, which is also marked by a pigment ring (Fig. 2C). In *Oscarella*, *Wnt-I* is also expressed at the canal openings (ostia), and β-catenin activation induces ectopic ostia (Lapebie et al. 2009). This is reminiscent of the blastoporal organizer in cnidarians (Windsor and Leys 2010), and future work must reveal to what extent these *Wnt* genes also contribute to noncanonical Wnt function.

In the ctenophore *Mnemiopsis leidyi*, orthologs of the main components of Wnt signaling have been found (β-catenin, Dishevelled, 2Fz, sFrp, TCF, Pygopus, Porcupine, LRP5/6, GSK3, APC, CK1, Groucho, Wntless), including four Wnt ligands (WntA, Wnt6, Wnt9, Wntx) (Pang et al. 2010). β-catenin is expressed at the oral pole during gastrulation, whereas the four Wnt ligands are expressed only late at the aboral side of the larva, when tentacles and sensory organs form (Pang et al. 2010).

Cnidaria

Cnidarians show an archetypal gastrula-shaped body plan. The genomes of the sea anemone *Nematostella* (Putnam et al. 2007) and the freshwater polyp *Hydra* (Chapman et al. 2010) reveal a gene repertoire with about 18,000 bona fide protein-coding genes, illustrating the high genomic complexity of the common bilaterian–cnidarian ancestor (Putnam et al. 2007; Chapman et al. 2010). All bilaterian *Wnt* gene subfamilies are present in these cnidarian genomes (Kusserow et al. 2005; Lee et al. 2006; Lengfeld et al. 2009). In *Nematostella*, *Wnt-9* has not been found (Kusserow et al. 2005; Lee et al. 2006), whereas in *Hydra*, *Wnt4*, *-6*, and *-A* are missing. Three *Hydra Wnt* genes were classified as *HyWnt9/10-a*, *-b*, and *-c* (Lengfeld et al. 2009). This completeness of *Wnt* gene subfamilies in cnidarians suggests that the common (eumetazoan) ancestor of cnidarians and bilaterians already possessed a complete repertoire of *Wnt* gene ligands (Kusserow et al. 2005).

Cnidarian Wnts act in the canonical and the PCP pathways. All core components of the Wnt receptor and the β-catenin destruction complex are present. The number of Frizzled (Fzd) receptors is lower than that of the ligands (four Fzd in *Hydra*, and six in *Nematostella*), suggesting that the radiation of *Wnt* genes was followed by diversification of the receptors. Similarly to sponges, *Nematostella* Axin is lacking a clear β-catenin interaction domain, and APC shows only clear Armadillo repeats and a PDZ-domain (Adamska et al. 2010), raising questions as to whether the assembly of the cnidarian β-catenin destruction complex is as complete as in bilaterians. Orthologs of PCP signaling like Strabismus/Van Gogh, RhoA, Rock-2, and RAC1 are present, and all secreted Wnt antagonists have been identified, that is, sFrp, Wnt inhibitory factor (WIF), Cerberus, and Dkk proteins (Augustin et al. 2006; Guder et al. 2006a; Lee et al. 2006; Holstein et al. 2011; Technau and Steele 2011). Components of Ca^{2+} signaling (PLC, PKC, CaMKII, and Calcineurin) are also present, but no functional data exist on the putative role of Wnt5 or any other cnidarian Wnt ligand in Ca^{2+} signaling.

Wnt/β-catenin signaling has a major function in gastrulation and anterior–posterior patterning throughout metazoan development, because most *Wnt* genes are expressed at the site of the blastopore (Fig. 2E). During gastrulation, β-catenin is activated at the site of the future blastopore, and blocking GSK3 causes extended gastrulation movements in *Nematostella* (Wikramanayake et al. 2003). Dsh is necessary for this blastoporal activation (Lee et al. 2007). In the hydrozoan *Clytia*, Fz1 and maternal Wnt3 are required for oral β-catenin activation (Momose and Houliston 2007; Momose et al. 2008). In *Nematostella* gastrulation and planula larvae, *Wnt*s are expressed in overlapping domains along the oral–aboral axis (Kusserow et al. 2005).

Most *Wnt* genes are also expressed in a similar cascade during head regeneration and budding of *Hydra* (Lengfeld et al. 2009). Here, canonical Wnt signaling acts upstream of noncanonical Wnt signaling, which is required for mass-tissue movements during tissue evaginations. Noncanonical processes are mediated by *Wnt5*, *Wnt8*, *Fzd2*, and *Dsh* together with JNK (Philipp et al. 2009). A role for *Nematostella* Wnts in *Xenopus* Wnt-5a/Ror2 and Wnt-11 PCP signaling was shown by different morphant phenotypes that differ in PAPC regulation, cell polarization, cell protrusion formation, and microtubule orientation (Rigo-Watermeier et al. 2011). NvWnt-5 rescued XWnt-11 and NvWnt-11 rescued XWnt-5a, suggesting that specific structures in Wnt ligands are important for receptor complex recognition in Wnt signaling.

PATTERNS OF GENE GAIN AND LOSS IN BILATERIANS

An intriguing feature in metazoan evolution is the frequent loss of *Wnt* gene subfamilies in protostomes (Table 1). All protostome lineages lack the *Wnt3* ortholog, whereas *Wnt1*, *Wnt5*, and *WntA* are well conserved in most protostomes. The other *Wnt* gene subfamilies have been significantly reduced in various groups (Table 1). Gene loss of *Wnt*s is also evident in some deuterostomes, but to a much lesser extent.

Lophotrochozoa

The lophotrochozoans were identified as a monophyletic group by molecular phylogenies (Halanych et al. 1995) and constitute forms with spiral cleavage and a trochophora larva (annelids and mollusks) and flatworms. Twelve Wnt subfamilies have been reported for annelids (*Capitella teleta* and *Platynereis dumerilii*), a major lophotrochozoan group. This indicates that basal protostomes also had an almost complete set of *Wnt* genes (Cho et al. 2010). Other annelids and mollusks have lost several *Wnt* gene subfamilies, that is, *Wnt8*, *-9*, and *-A* in *Helobdella robusta* or *Wnt4*, *-5*, *-6*, *-7*, *-8*, *-9*, *-11*, and *-16* in *Patella vulgata* (Prud'homme et al. 2002; Cho et al. 2010; Janssen et al. 2010; Riddiford and Olson 2011). The *Wnt* genes of *P. dumerilii* show blastoporal and posterior expression in the growth zone (segment addition zone) of trochophoran larvae, and at the posterior side of each segment (Janssen et al. 2010), reminiscent to arthropods (see below).

It has been suggested that ciliated acoel and turbellarian flatworms are the most ancient bilaterians (Ruiz-Trillo et al. 1999; Baguna and Riutort 2004). Their shape is similar to cnidarian planulae, but they have an elaborated gut with one pharyngeal opening derived from the blastopore. A survey of *Wnt* genes in platyhelminths, including *Schmidtea mediterranea* (turbellarian planarian), the parasitic *Schistosoma mansoni* (cestode), *Echinococcus granulosus* (trematode), and *Hymenolepis microstoma* (trematode) revealed a strong tendency toward gene loss (Riddiford and Olson 2011). Planarians possess nine *Wnt* genes, classified as members of the *Wnt1*, *-2*, *-5*, and *-11* subfamilies, with putative gene duplications in the *Wnt11* group (Gurley et al. 2010; Riddiford and Olson 2011). Because most platyhelminths are highly diverged, it will be essential for future studies to unravel the *Wnt* gene repertoire of acoel flatworms, the putative bilaterian stem group.

In *S. mediterranea*, all *Wnt* genes are predominantly expressed with sFrp genes in discrete, complex, overlapping domains along the A/P axis, in a manner that suggests a possible steady-state posterior-to-anterior gradient of β-catenin activity (Gurley et al. 2008, 2010; Petersen and Reddien 2008). This is reminiscent of the overlapping *Wnt* expression patterns observed in the radially symmetric cnidarian body plan, the main axis of which is determined by β-catenin signaling (Gurley et al. 2010). Wnt and sFrp genes play an important role in flatworm regeneration (Reddien and Sanchez Alvarado 2004; Gurley et al. 2008, 2010; Petersen and Reddien 2008; Iglesias et al. 2011). Down-regulation of β-catenin in regenerating planarians produces an enhancement of anterior head structures (Gurley et al. 2008; Iglesias et al. 2008; Petersen and Reddien 2008). Inhibition of APC or silencing of Axin results in the regeneration of tails from anterior-facing wounds (Gurley et al. 2008) or in two-tailed planarians without a brain (Iglesias et al. 2008). In line with these data, inhibition of the Wnt inhibitor

Table 1. Distribution of metazoan *Wnt* genes

		N	Wnt1	Wnt2	Wnt3	Wnt4	Wnt5	Wnt6	Wnt7	Wnt8	Wnt9	Wnt10	Wnt11	Wnt16	WntA	n.d.
	Choanoflagellata	0	0	0	0	0	0	0	0	0	0	0	0	0	0	0
Radiata	Placozoa	1	0	0	0	0	0	0	0	0	0	0	0	0	0	2
	Porifera	2	0	0	0	0	0	0	0	0	0	1(0)	1(2)	1(0)	0	0
	Ctenophora	1	0	0	0	0	0	1	0	0	1	1	0	0	1	0
	Cnidaria	2	1	1	1	1 (0)	1	1	1	1 (2)	0	1 (3)	1	1	1	0
Lophotrocho-zoa	Mollusca	2	1	2	0	1 (0)	1 (0)	1 (0)	1 (0)	0	1 (0)	1	1 (0)	1 (0)	1	0
	Annelida	3	1	1	0	1	1 (2)	1	1	1 (0)	1 (0)	1	1 (3)	1 (2)	1 (0)	0
	Platyhelminthes	4	1	1	0	1 (3)	1	0	0	0	0	0	1 (2)	0	0	0
Ecdysozoa	Nematoda	1	0	0	0	1	1	0	0	0	1	1	0	0	1	0
	Chelicerata	2	1	1	0	1	1	1	1 (2)	1	1	0	1 (2)	1	1 (0)	0
	Crustacea/Myriapoda	2	1	1	0	1	1	1	1	1	1	1	1	1	1	0
	Insecta	6	1	0	0	0	1	1 (0)	1 (0)	1 (0)	1 (0)	1 (0)	1 (0)	1 (0)	1 (0)	0
Deuterostomia	Echinodermata	1	1	0	1	1	1	1	1	1	1	1	0	1	1	0
	Hemichordata	1*	1	1	1	1	1	1	1	1	1	1	1	1	1	
	Urochordata	1	1	1	1	2	1	1	2	0	1	1	0	1	0	0
	Cephalochordata	1	1	0	1	1	1	1	1	1	0	1	1	0	0	0
	Vertebrata	>10	1 (2)	2 (5)	2 (3)	1 (2)	2	1 (2)	2 (3)	2 (3)	1 (3)	2 (3)	1 (3)	1	0	0

Notes: N indicates the number of species analyzed in a clade; numbers in parentheses indicate deviations of the number of Wnt orthologs found in species of a given clade. Colors in filled boxes indicate the abundance of *Wnt* gene subfamily members present in a given clade (gene duplication or gene loss). Note the loss of *Wnt3* in protostomes and the loss of *WntA* in chordates. Data are based on molecular Wnt phylogenies by Kusserow et al. (2005), Lee et al. (2006), Croce and McClay (2008), Srivastava et al. (2008), Lengfeld et al. (2009), Lapebie et al (2009), Adamska et al (2010), Gurley et al. (2010), Janssen et al (2010), and Riddiford and Olson (2011).

(*) The Wnts from the hemichordate *Saccoglossus kowalevskii* correspond to Gl: 259013321 (*Wnt1*), Gl: 259013323 (*Wnt2*), Gl: 259013327 (*Wnt3*), Gl: 291234645 (*Wnt4*), Gl: 269785077 (*Wnt5*), Gl: 268054405 (*Wnt6*), Gl: 269785051 (*Wnt7*), Gl: 269785053 (*Wnt8*), Gl: 269785055 (*Wnt9*), Gl: 291244845 (*Wnt10*), Gl: 269785075 (*Wnt11*), Gl: 197320537 (*Wnt16*), Gl: 284005183 (*WntA*).

notum induces the regeneration of anterior-facing tails instead of a head, and double-RNAi experiments indicate that *notum* inhibits Wnt signaling in a feedback inhibition to promote head regeneration (Petersen and Reddien 2011). Thus, Wnt/β-catenin signaling clearly controls the tail-versus-head (i.e., posterior–anterior) axis in intact and regenerating planarians.

Ecdysozoa

Gene loss has also occurred in various ecdysozoan groups (Table 1), the most prominent having occurred in *Caenorhabditis elegans* (nematodes), which is lacking eight *Wnt* genes subfamilies (*Wnt1–3, Wnt6–8, Wnt11*, and *Wnt16*). The other major ecdysozoan groups are arthropods. Basal arthropods show a rich *Wnt* gene repertoire, indicating that the common ancestor of arthropods only lacked *Wnt-3* (Janssen et al. 2010). Members of 12 *Wnt* gene subfamilies were characterized at the transcriptional level (Janssen et al. 2010). A relatively complete set of protostome *Wnt* genes was found in the genomes of the crustacean *Daphnia pulex* (12 *Wnt* subfamilies), and in chelicerates and myriapods, where *Wnt10* has been lost (Janssen et al. 2010). In comparison, insects show significant gene loss (Table 1). Only *Wnt1* (wingless) and *Wnt5* are present in all insects investigated so far. The other *Wnt* genes are absent in several insects, and there seems to be a tendency for increasing loss of *Wnt* genes, for example, *Wnt2* and *Wnt4* are lacking in all insects, *Wnt16* in all holometabolous insects, *Wnt11* in all dipterans, and *WntA* in *Drosophila* (Janssen et al. 2010).

The expression patterns of arthropod *Wnt* genes reveal that most overlap in a segment addition or growth zone, and in segmentally reiterated patterns defining the parasegments (Janssen et al. 2010).

These expression patterns are remarkably similar to annelids', suggesting similar Wnt regulation mechanisms during segment and parasegment formation in annelids and arthropods, respectively (Janssen et al. 2010). *Wnt1/Wg* is a prototypical *Wnt* gene for the posterior growth zone in annelids and arthropods. It is notable that even in *Drosophila* it shows an early posterior expression domain (Fig. 2B). In *Drosophila*, all segments are formed simultaneously (long germ band formation) and the posterior band of *Wg* is expressed in the early blastoderm before the segment-polarity stripes of *Wg* are formed (Vorwald-Denholtz and De Robertis 2011). This posterior *Wg* band might represent a remnant of a posterior signaling center (Vorwald-Denholtz and De Robertis 2011).

Of particular interest is the fate of *Wnt8* in arthropods. In the spider *Achaearanea* sp. Wnt8 is required for posterior development and the maintenance of the growth zone (McGregor et al. 2008). It is expressed in a solid circular domain at the posterior end of a radial symmetric embryo before its transition to axial symmetry (Fig. 1A in McGregor et al. 2008). In *Drosophila*, Wnt8 has a completely different function and acts as a feedback inhibitor of the NF-κB homolog Dorsal (the gene was therefore named *Wnt inhibitor of Dorsal, WntD*) in DV signaling and immunity (Gordon et al. 2005; Gordon and Nusse 2006). *WntD* expression is under the control of Toll/Dorsal signaling, and increased levels of WntD block the nuclear accumulation of Dorsal. WntD signals independently of the β-catenin homolog Armadillo (Ganguly et al. 2005; Gordon et al. 2005).

Deuterostomia

In deuterostomes, data from more than 10 completely sequenced vertebrate genomes show unambiguously that all *Wnt* gene subfamilies except *WntA* were present at the base of vertebrate radiation (Table 1). Because *WntA* was identified in the genome of echinoderms (Sodergren et al. 2006; Croce and McClay 2008), the common ancestor of deuterostomes must have had a complete set of *Wnt* genes. Genome data suggest some Wnt gene loss in the cephalochordate amphioxus (*Wnt2*, *-9*, *-16*, and *-A*) (Putnam et al. 2007) and the urochordate ascidian *Ciona* (*Wnt8* and *-11*) (Dehal et al. 2002). According to NCBI annotations, the only deuterostome with a complete set of *Wnt* gene subfamilies is the hemichordate *Saccoglossus kowalevskii* (Table 1). Although the *Saccoglossus* data must be confirmed by a specific phylogenetic study, it is likely that the deuterostome radiation started with a complete set of *Wnt* gene subfamilies.

The function and localization of Wnt and β-catenin signaling in deuterostome AP-axis formation has been recently reviewed in detail (De Robertis 2010; Niehrs 2010; Holstein et al. 2011). The general picture that emerges is that β-catenin plays an important function in the early polarization of the animal–vegetal axis of the embryo (Martin and Kimelman 2009). Similar to pre-bilaterian animals, β-catenin is vegetally localized in many basal deuterostomes. In sea urchins (echinoderms), development starts with clear polarity from the animal to the vegetal pole that resembles the oral–aboral body axis in cnidarians. β-catenin signaling marks the vegetal pole of the sea urchin embryo (Wikramanayake et al. 1998, 2004; Logan et al. 1999) and is required for the formation of the endoderm and mesoderm. Overactivation of β-catenin with LiCl results in vegetalized embryos with ectopic endoderm (Darras et al. 2011). Furthermore, in many chordates, a peak expression of β-catenin at the vegetal and blastoporal side defines a gradient along the AP axis. In both ascidians and amphioxus, nuclear-localized β-catenin marks the blastopore, which, in turn, defines the future posterior end of the embryo (Imai et al. 2000). Activation of β-catenin signaling by LiCl posteriorizes the amphioxus embryo, which is evident by loss of the neural plate (Yu et al. 2007; Onai et al. 2009). Similarly to the posterior growth zone in amphioxus tail buds (Schubert et al. 2001), nuclear β-catenin also forms a posterior-to-anterior gradient in vertebrates. Wnt signaling is involved in tail development in zebrafish (Agathon et al. 2003; Shimizu et al. 2005; Thorpe et al. 2005), where Wnt3a determines the anterior–posterior position of the somite determination front (Aulehla et al. 2003; Dunty et al. 2008; De Robertis 2010; Niehrs 2010).

SYNTENY AND EVOLUTION OF *WNT* GENES

Wnt genes show intriguing axial expression patterns along the anterior–posterior axis in cnidarians (e.g., in *Nematostella*). However, this is also true in many other bilaterians with complete sets of Wnts, where expression during embryogenesis or regeneration indicates an underlying order in transcriptional regulation. The pattern of overlapping Wnt ligands that is used in cell fate specification and patterning was characterized as the “Wnt code” (Guder et al. 2006a) or the “Wnt landscape” (Janssen et al. 2010).

Although *Wnt* genes do not show a proper Hox-type collinearity on the genomic level (Sullivan et al. 2007), there is a certain degree of *Wnt* clustering and synteny (see also Fig. 3). A primordial cluster of *Wnt* genes (*Wnt1*, *-6*, and *-10*) was first postulated by Nusse (2001) for the *Drosophila* and human genomes. Accordingly, the human genome contains duplicated copies of this cluster, but *Wnt6* was deleted from 12q13 and *Wnt1* was deleted from 2q35 (Fig. 3). The clustering of *Wnt1*, *-6*, and *-9/10* in a molecular phylogeny with *Wnts* from *Nematostella* (Kusserow et al. 2005) and *Hydra* (Lengfeld et al. 2009) supports this hypothesis. A second conserved cluster was found for *Wnt5* and *Wnt7* (Sullivan et al. 2007). New data on the arrangement of *Wnt* genes in the *Daphnia* genome revealed two syntenic clusters of these genes: *Wnt9*–*Wnt1*–*Wnt6*–*Wnt10* and *Wnt5*–*Wnt7* (Janssen et al. 2010). *Lottia gigantea* shows a similar organization of these *Wnt* genes and therefore reflects the ancient cluster of *Wnt* genes in eumetazoans (Nusse 2001; Sullivan et al. 2007; Cho et al. 2010; Janssen et al. 2010). These two *Wnt* clusters are also supported by recent molecular phylogenies on *Wnt* genes (Cho et al. 2010; Janssen et al. 2010). How the interacting Wnt ligands are regulated is rather unknown so far, but it is tempting to speculate that there is a regulative hierarchy of *Wnt* gene expression similar to that described for *Hydra* (Nakamura et al. 2011) or *Drosophila* (van de Wetering et al. 1997; Lessing and Nusse 1998).

THE ORGANIZER: EVOLUTION OF THE POSTERIOR WNT SIGNALING CENTER

Cnidarians are the only pre-bilaterian animals with a complete set of *Wnt* genes and a blastoporal signaling center that is similar to that in deuterostomes and protostomes. This signaling center has been investigated in detail in the freshwater polyp *Hydra*, where it was shown to exhibit inductive properties that are reminiscent of the classical vertebrate organizers.

The *Hydra* Organizer

The *Hydra* organizer corresponds to the oral end of an adult polyp, which has retained the signaling properties of the blastoporal signaling center. It is known as the *Hydra* "head" organizer. The naming is misleading because the *Hydra* "head," being a blastoporal region, represents the posterior pole of the body axis, and the *Hydra* "foot" is therefore the anterior pole (Meinhardt 2002; Niehrs 2010; Holstein et al. 2011). We therefore name it the "*Hydra* Organizer (HO)." A group of 10–50 ectodermal and endodermal epithelial cells constitutes this signaling center (Technau et al. 2000), and when transplanted to an ectopic site, the HO induces a secondary body axis by recruiting host tissue (Browne 1909; Broun and Bode 2002).

Short- and Long-Range Wnt Signaling

Activation of Wnt signaling stimulates the expression of all *Wnt* genes as well as of the Wnt target genes *Tcf* and *Brachyury*, resulting in numerous spot-like *Wnt* expression domains along the body column (Broun et al. 2005; Guder et al. 2006b; Nakamura et al. 2011). It is generally accepted that each of these Wnt signaling centers creates a gradient of diffusible Wnt, although thus far the distribution of none of the cnidarian Wnts has been visualized by using either antibodies or reporter constructs.

The parameters of the HO were determined experimentally under conditions of de novo pattern formation (Technau et al. 2000) in reaggregates from dissociated single-cell suspensions (Gierer et al. 1972). Clusters with a minimal size of five to 15 epithelial Wnt3-expressing cells are necessary and sufficient to form de novo a HO in an aggregate, generating a lateral inhibition with a diffusion range of up to 1000 μm (Technau et al. 2000).

These parameters fit with the reaction-diffusion model of *Hydra* pattern formation, assuming short-range autocatalytic activation and long-range inhibition (Gierer and Meinhardt 1972). In transgenic *Hydra*, an activator element of the *Hydra Wnt3* promoter was identified that maintains *Wnt3* transcription in the HO by recruiting TCF and β-catenin (Nakamura et al. 2011).

The molecular nature of long-range inhibition could also be explained by Wnt secretion.

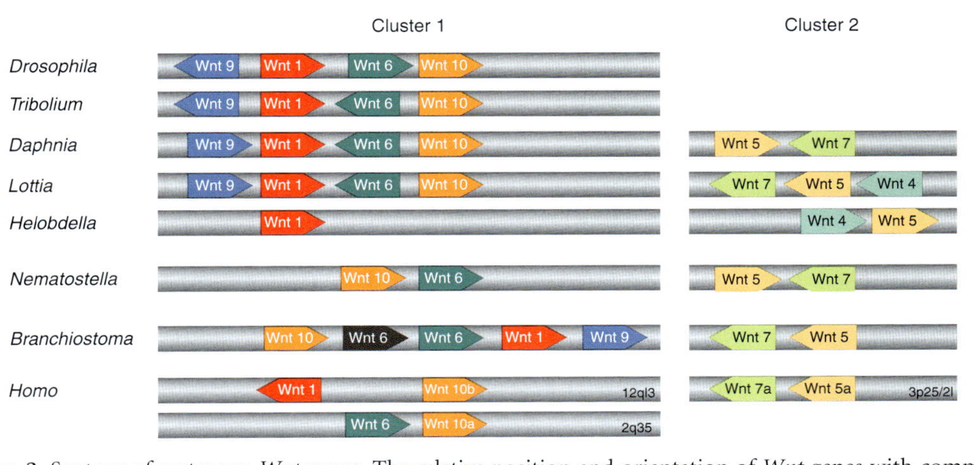

Figure 3. Synteny of metazoan *Wnt* genes. The relative position and orientation of *Wnt* genes with common scaffolds/chromosomes are shown for *D. melanogaster, Tribolium castaneum, D. pulex, L. gigantea, H. robusta, Nematostella vectensis, Branchiostoma floridae*, and *Homo sapiens* (not to scale). Numbers indicate specific chromosomes in the case of *H. sapiens*. (Figure created from data from Nusse [2001], Sullivan et al. [2007], Cho et al. [2010], and Janssen et al. [2010].) Note that there are more Wnts for a given species that are not clustered (see Table 1).

According to a model proposed by Bartscherer and Boutros (2008), Wnt ligands may have different diffusion ranges related to their mode of secretion into different vesicle populations (Bartscherer and Boutros 2008; Lorenowicz and Korswagen 2009). Wnt can be secreted with a short diffusion range at the apical side of an epithelial layer, whereas Wnt proteins that are associated with lipoprotein particles result in a longer range. We presume that both modes of Wnt secretion contribute to short- and long-range signaling in *Hydra* (Fig. 4B). Because β-catenin and Wnt3 act in an autoregulatory loop, a gradient of Wnt3 will result in a graded distribution of nuclear β-catenin along the body column (Guder et al. 2006a). A long-range Wnt3 signal can activate both a transcriptional repressor (see below) and the Wnt antagonist

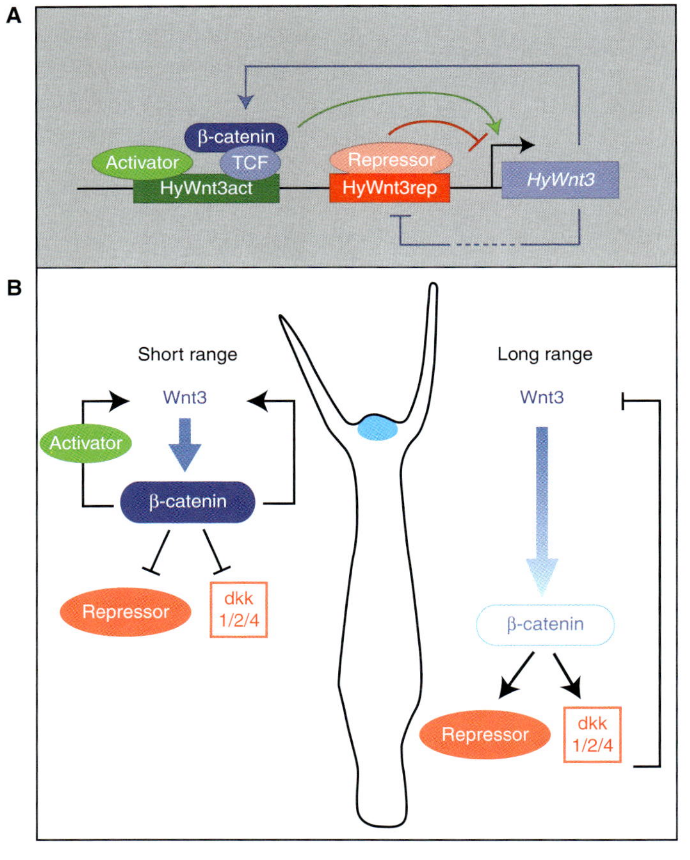

Figure 4. Transcriptional control of Wnt signaling in *Hydra*. (*A*) Model showing the transcriptional regulation of head organizer-specific *HyWnt3* expression. *HyWnt3* expression is controlled by two distinct *cis*-regulatory elements; the activator (HyWnt3act, green) and the repressor (HyWnt3rep, red) are positively and negatively regulated by Wnt/β-catenin signaling, respectively. The β-catenin/TCF complex and putative activators (light green) bind to HyWnt3act, and their combinatorial inputs act in *HyWnt3* transcription (green arrow). Potential repressors (red) bind to a repressor element (red) and inhibit *HyWnt3* expression. (Figure adapted from Nakamura et al. 2011; reprinted, with permission, from the author.) (*B*) Presumed distribution or activity of β-catenin depends on the diffusion range of Wnt3. Following short-range diffusion, the concentrations of Wnt3 and β-catenin are high, thereby repressing putative repressors and dkk1/2/4. Following long-range diffusion, the level of nuclear β-catenin is lower, inducing putative repressors including dkk1/2/4. Positive and negative regulation restricts *HyWnt3* expression (blue) to the head organizer region.

Dkk1/2/4 in the body column (Fig. 4B). Fields of lateral inhibition can be visualized by the suppression of Dkk1/2/4 expression in *Hydra* polyps after activation of Wnt signaling by using alsterpaullone treatment (Guder et al. 2006a,b). The double function of Wnt3 as a short-range activator and long-range inhibitor was recently proposed and elaborated in a model by Meinhardt (2012).

An alternative explanation for the long-range signaling could be that a cascade of sequential *Wnt* gene activation and expression could explain the "field" behavior of Wnt signaling, thereby extending its range. The cascade of consecutive *Wnt* activation during gastrulation in *Nematostella* (Kusserow et al. 2005) and during regeneration and bud formation in *Hydra* (Lengfeld et al. 2009) would support this alternative model of long-range Wnt signaling.

Local Restriction of the Organizer

In line with these new hypotheses, we found that Wnt signaling can also be effectively suppressed at the transcriptional level (Nakamura et al. 2011). The removal of a repressor element in the regulatory region of Wnt3 resulted in an expansion of the Wnt3 gene expression domain toward the gastric region (Fig. 5) (Nakamura et al. 2011). This shows that transcriptional regulation is essential to restrict Wnt expression to the site of the signaling center. Transcriptional regulation of *Wnt3* expression is probably at the core of Wnt/β-catenin signaling in *Hydra*, and only complemented by Dkk1/2/4. The creation of local sources for the secretion of Wnt ligands might have been an important step in the evolution of Wnt signaling centers. We propose that *cis*-regulatory control mechanisms combining short-range autoactivation and long-range repression resulted in a locally defined Wnt signaling center. This signaling center may be independent of the function of extracellular Wnt antagonists, which could explain why Wnt antagonists are absent from several metazoan lineages (Nakamura et al. 2011).

CONCLUDING REMARKS

The broad variety in animal forms appears to be founded on astonishingly few signaling pathways that are shared by all major metazoan phyla. The Wnt signaling pathways had a critical function in setting up an ancient blastoporal signaling center. Molecular components of this signaling network can only be traced in part to pre-metazoan origins, indicating that major novelties were linked to the emergence of the metazoan stem. A core set of *Wnt* genes was already present in sponges, but only in cnidarians is a complete functional repertoire of *Wnt* ligands present. All *Wnt* genes act in these archetypical animals during gastrulation, pattern formation, and regeneration, providing a basic tool set to understand the twist of canonical and noncanonical functions in Wnt signaling.

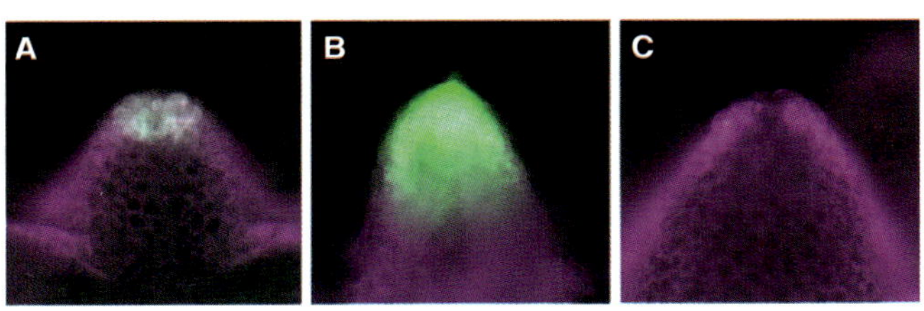

Figure 5. Expression of Wnt3 in the *Hydra* organizer. Reporter constructs for HyWnt3-EGFP (green) were used with an independent transformation marker (HyActin–RFP reporter gene, magenta) to ensure that all cells carry the EGFP reporter gene under examination. (*A*) The transgenic *Hydra* strain shows the localization of Wnt3 in epithelial cells at the tip of the hypostome with a complete Wnt3 promoter (HyWnt3FL). (*B*) Transgenic *Hydra* with a reporter construct lacking the HyWnt3 repressor element show dramatic expansion of expression. (*C*) Transgenic *Hydra* with a reporter construct lacking the HyWnt3 activator sequence including TCF binding sites show no expression. (From Nakamura et al. 2011; reprinted, with permission, from the author.)

Future work should focus on unraveling the *cis*-regulatory control mechanisms (Nakamura et al. 2011) that link the coordinated action of *Wnt* genes (van Amerongen and Nusse 2009).

ACKNOWLEDGMENTS

Thanks to Ildiko Somorjai (Heidelberg) and Suat Özbek (Heidelberg) for critically reading the manuscript, to Michael Boutros (Heidelberg) and our Wnt research group for discussions, to Hans Meinhardt (Tübingen) for sharing unpublished data, to Peggy Vorwald and Edward De Robertis (Los Angeles) for providing Fig. 2B, and to Maja Adamska (Bergen) for providing Fig. 2C. This work is supported by the DFG (FOR 1036/A1).

REFERENCES

Abedin M, King N. 2008. The premetazoan ancestry of cadherins. *Science* **319:** 946–948.

Adamska M, Degnan SM, Green KM, Adamski M, Craigie A, Larroux C, Degnan BM. 2007. Wnt and TGF-β expression in the sponge *Amphimedon queenslandica* and the origin of metazoan embryonic patterning. *PLoS ONE* **2:** e1031.

Adamska M, Larroux C, Adamski M, Green K, Lovas E, Koop D, Richards GS, Zwafink C, Degnan BM. 2010. Structure and expression of conserved Wnt pathway components in the demosponge *Amphimedon queenslandica*. *Evol Dev* **12:** 494–518.

Agathon A, Thisse C, Thisse B. 2003. The molecular nature of the zebrafish tail organizer. *Nature* **424:** 448–452.

Arendt D, Technau U, Wittbrodt J. 2001. Evolution of the bilaterian larval foregut. *Nature* **409:** 81–85.

Augustin R, Franke A, Khalturin K, Kiko R, Siebert S, Hemmrich G, Bosch TC. 2006. Dickkopf related genes are components of the positional value gradient in *Hydra*. *Dev Biol* **296:** 62–79.

Aulehla A, Wehrle C, Brand-Saberi B, Kemler R, Gossler A, Kanzler B, Herrmann BG. 2003. Wnt3a plays a major role in the segmentation clock controlling somitogenesis. *Dev Cell* **4:** 395–406.

Baguna J, Riutort M. 2004. The dawn of bilaterian animals: The case of acoelomorph flatworms. *Bioessays* **26:** 1046–1057.

Banziger C, Soldini D, Schutt C, Zipperlen P, Hausmann G, Basler K. 2006. Wntless, a conserved membrane protein dedicated to the secretion of Wnt proteins from signaling cells. *Cell* **125:** 509–522.

Bartscherer K, Boutros M. 2008. Regulation of Wnt protein secretion and its role in gradient formation. *EMBO Rep* **9:** 977–982.

Bartscherer K, Pelte N, Ingelfinger D, Boutros M. 2006. Secretion of Wnt ligands requires Evi, a conserved transmembrane protein. *Cell* **125:** 523–533.

Broun M, Bode HR. 2002. Characterization of the head organizer in hydra. *Development* **129:** 875–884.

Broun M, Gee L, Reinhardt B, Bode HR. 2005. Formation of the head organizer in *Hydra* involves the canonical Wnt pathway. *Development* **132:** 2907–2916.

Browne EN. 1909. The production of new hydranths in *Hydra* by the insertion of small grafts. *J Exp Zool* **7:** 1–37.

Brusca RC, Brusca GJ. 2002. *Invertebrates*. Sinauer, Sunderland, MA.

Chapman JA, Kirkness EF, Simakov O, Hampson SE, Mitros T, Weinmaier T, Rattei T, Balasubramanian PG, Borman J, Busam D, et al. 2010. The dynamic genome of *Hydra*. *Nature* **464:** 592–596.

Cho SJ, Valles Y, Giani VCJr, Seaver EC, Weisblat DA. 2010. Evolutionary dynamics of the *wnt* gene family: A lophotrochozoan perspective. *Mol Biol Evol* **27:** 1645–1658.

Coates JC, Harwood AJ. 2001. Cell–cell adhesion and signal transduction during *Dictyostelium* development. *J Cell Sci* **114:** 4349–4358.

Coates JC, Grimson MJ, Williams RS, Bergman W, Blanton RL, Harwood AJ. 2002. Loss of the β-catenin homologue Aardvark causes ectopic stalk formation in *Dictyostelium*. *Mech Dev* **116:** 117–127.

Conway Morris S. 2000. The Cambrian "explosion": Slow-fuse or megatonnage? *Proc Natl Acad Sci* **97:** 4426–4429.

Croce JC, McClay DR. 2008. Evolution of the Wnt pathways. *Methods Mol Biol* **469:** 3–18.

Dann CE, Hsieh JC, Rattner A, Sharma D, Nathans J, Leahy DJ. 2011. Insights into Wnt binding and signalling from the structures of two Frizzled cysteine-rich domains. *Nature* **412:** 86–90.

Darras S, Gerhart J, Terasaki M, Kirschner M, Lowe CJ. 2011. β-Catenin specifies the endomesoderm and defines the posterior organizer of the hemichordate *Saccoglossus kowalevskii*. *Development* **138:** 959–970.

Dehal P, Satou Y, Campbell RK, Chapman J, Degnan B, De Tomaso A, Davidson B, Di Gregorio A, Gelpke M, Goodstein DM, et al. 2002. The draft genome of *Ciona intestinalis*: Insights into chordate and vertebrate origins. *Science* **298:** 2157–2167.

De Robertis EM. 2010. Wnt signaling in axial patterning and regeneration: Lessons from planaria. *Sci Signal* **3:** pe21.

De Robertis EM, Sasai Y. 1996. A common plan for dorsoventral patterning in Bilateria. *Nature* **380:** 37–40.

Dickinson DJ, Nelson WJ, Weis WI. 2011. A polarized epithelium organized by β- and α-catenin predates cadherin and metazoan origins. *Science* **331:** 1336–1339.

Dunty WC Jr, Biris KK, Chalamalasetty RB, Taketo MM, Lewandoski M, Yamaguchi TP. 2008. Wnt3a/β-catenin signaling controls posterior body development by coordinating mesoderm formation and segmentation. *Development* **135:** 85–94.

Eichinger L, Pachebat JA, Glockner G, Rajandream MA, Sucgang R, Berriman M, Song J, Olsen R, Szafranski K, Xu Q, et al. 2005. The genome of the social amoeba *Dictyostelium discoideum*. *Nature* **435:** 43–57.

Ganguly A, Jiang J, Ip YT. 2005. *Drosophila* WntD is a target and an inhibitor of the Dorsal/Twist/Snail network in the gastrulating embryo. *Development* **132:** 3419–3429.

Gierer A, Meinhardt H. 1972. A theory of biological pattern formation. *Kybernetik* **12:** 30–39.

Gierer A, Berking S, Bode H, David CN, Flick K, Hansmann G, Schaller H, Trenkner E. 1972. Regeneration of *Hydra* from reaggregated cells. *Nat New Biol* **239:** 98–101.

Gordon MD, Nusse R. 2006. Wnt signaling: Multiple pathways, multiple receptors, and multiple transcription factors. *J Biol Chem* **281:** 22429–22433.

Gordon MD, Dionne MS, Schneider DS, Nusse R. 2005. WntD is a feedback inhibitor of Dorsal/NF-κB in *Drosophila* development and immunity. *Nature* **437:** 746–749.

Grimson MJ, Coates JC, Reynolds JP, Shipman M, Blanton RL, Harwood AJ. 2000. Adherens junctions and β-catenin-mediated cell signalling in a non-metazoan organism. *Nature* **408:** 727–731.

Guder C, Philipp I, Lengfeld T, Watanabe H, Hobmayer B, Holstein TW. 2006a. The Wnt code: Cnidarians signal the way. *Oncogene* **25:** 7450–7460.

Guder C, Pinho S, Nacak TG, Schmidt HA, Hobmayer B, Niehrs C, Holstein TW. 2006b. An ancient Wnt–Dickkopf antagonism in *Hydra*. *Development* **133:** 901–911.

Gurley KA, Rink JC, Sanchez Alvarado A. 2008. β-Catenin defines head versus tail identity during planarian regeneration and homeostasis. *Science* **319:** 323–327.

Gurley KA, Elliott SA, Simakov O, Schmidt HA, Holstein TW, Sanchez Alvarado A. 2010. Expression of secreted Wnt pathway components reveals unexpected complexity of the planarian amputation response. *Dev Biol* **347:** 24–39.

Halanych KM, Bacheller JD, Aguinaldo AM, Liva SM, Hillis DM, Lake JA. 1995. Evidence from 18S ribosomal DNA that the lophophorates are protostome animals. *Science* **267:** 1641–1643.

Hardin J, King RS. 2008. The long and the short of Wnt signaling in *C. elegans*. *Curr Opin Genet Dev* **18:** 362–367.

Harwood AJ. 2008. *Dictyostelium* development: A prototypic Wnt pathway? *Methods Mol Biol* **469:** 21–32.

Harwood AJ, Plyte SE, Woodgett J, Strutt H, Kay RR. 1995. Glycogen synthase kinase 3 regulates cell fate in *Dictyostelium*. *Cell* **80:** 139–148.

Heisenberg CP, Tada M, Rauch GJ, Saude L, Concha ML, Geisler R, Stemple DL, Smith JC, Wilson SW. 2000. Silberblick/Wnt11 mediates convergent extension movements during zebrafish gastrulation. *Nature* **405:** 76–81.

Heuberger J, Birchmeier W. 2010. Interplay of cadherin-mediated cell adhesion and canonical Wnt signaling. *Cold Spring Harb Perspect Biol* **2:** a002915.

Hobmayer B, Rentzsch F, Kuhn K, Happel CM, von Laue CC, Snyder P, Rothbächer U, Holstein TW. 2000. WNT signalling molecules act in axis formation in the diploblastic metazoan *Hydra*. *Nature* **407:** 186–189.

Holstein TW, Watanabe H, Ozbek S. 2011. Signaling pathways and axis formation in the lower metazoa. *Curr Top Dev Biol* **97:** 137–177.

Iglesias M, Gomez-Skarmeta JL, Salo E, Adell T. 2008. Silencing of Smed-β-catenin1 generates radial-like hypercephalized planarians. *Development* **135:** 1215–1221.

Iglesias M, Almuedo-Castillo M, Aboobaker AA, Salo E. 2011. Early planarian brain regeneration is independent of blastema polarity mediated by the Wnt/β-catenin pathway. *Dev Biol* **358:** 68–78.

Imai K, Takada N, Satoh N, Satou Y. 2000. β-Catenin mediates the specification of endoderm cells in ascidian embryos. *Development* **127:** 3009–3020.

Ingham PW, Nakano Y, Seger C. 2011. Mechanisms and functions of Hedgehog signalling across the metazoa. *Nat Rev Genet* **12:** 393–406.

Janssen R, Le Gouar M, Pechmann M, Poulin F, Bolognesi R, Schwager EE, Hopfen C, Colbourne JK, Budd GE, Brown SJ, et al. 2010. Conservation, loss, and redeployment of Wnt ligands in protostomes: Implications for understanding the evolution of segment formation. *BMC Evol Biol* **10:** 374.

Kadowaki T, Wilder E, Klingensmith J, Zachary K, Perrimon N. 1996. The segment polarity gene *porcupine* encodes a putative multitransmembrane protein involved in Wingless processing. *Genes Dev* **10:** 3116–3128.

King N, Westbrook MJ, Young SL, Kuo A, Abedin M, Chapman J, Fairclough S, Hellsten U, Isogai Y, Letunic I, et al. 2008. The genome of the choanoflagellate *Monosiga brevicollis* and the origin of metazoans. *Nature* **451:** 783–788.

Kusserow A, Pang K, Sturm C, Hrouda M, Lentfer J, Schmidt HA, Technau U, von Haeseler A, Hobmayer B, Martindale MQ, et al. 2005. Unexpected complexity of the *Wnt* gene family in a sea anemone. *Nature* **433:** 156–160.

Lapebie P, Gazave E, Ereskovsky A, Derelle R, Bezac C, Renard E, Houliston E, Borchiellini C. 2009. WNT/β-catenin signalling and epithelial patterning in the homoscleromorph sponge *Oscarella*. *PLoS ONE* **4:** e5823.

Lee PN, Pang K, Matus DQ, Martindale MQ. 2006. A WNT of things to come: Evolution of Wnt signaling and polarity in cnidarians. *Semin Cell Dev Biol* **17:** 157–167.

Lee PN, Kumburegama S, Marlow HQ, Martindale MQ, Wikramanayake AH. 2007. Asymmetric developmental potential along the animal–vegetal axis in the anthozoan cnidarian, *Nematostella vectensis*, is mediated by Dishevelled. *Dev Biol* **310:** 169–186.

Lengfeld T, Watanabe H, Simakov O, Lindgens D, Gee L, Law L, Schmidt HA, Ozbek S, Bode H, Holstein TW. 2009. Multiple Wnts are involved in *Hydra* organizer formation and regeneration. *Dev Biol* **330:** 186–199.

Lessing D, Nusse R. 1998. Expression of *wingless* in the *Drosophila* embryo: A conserved *cis*-acting element lacking conserved Ci-binding sites is required for *patched*-mediated repression. *Development* **125:** 1469–1476.

Logan CY, Miller JR, Ferkowicz MJ, McClay DR. 1999. Nuclear β-catenin is required to specify vegetal cell fates in the sea urchin embryo. *Development* **126:** 345–357.

Lorenowicz MJ, Korswagen HC. 2009. Sailing with the Wnt: Charting the Wnt processing and secretion route. *Exp Cell Res* **315:** 2683–2689.

Martin BL, Kimelman D. 2008. Regulation of canonical Wnt signaling by Brachyury is essential for posterior mesoderm formation. *Developmental cell* **15:** 121–133.

Martin BL, Kimelman D. 2009. Wnt signaling and the evolution of embryonic posterior development. *Curr Biol* **19:** R215–R219.

Martindale MQ, Hejnol A. 2009. A developmental perspective: Changes in the position of the blastopore during bilaterian evolution. *Dev Cell* **17:** 162–174.

McGregor AP, Pechmann M, Schwager EE, Feitosa NM, Kruck S, Aranda M, Damen WG. 2008. Wnt8 is required for growth-zone establishment and development of opisthosomal segments in a spider. *Curr Biol* **18:** 1619–1623.

McMahon AP, Moon RT. 1989. Ectopic expression of the proto-oncogene *int-1* in *Xenopus* embryos leads to duplication of the embryonic axis. *Cell* **58:** 1075–1084.

Meinhardt H. 2002. The radial-symmetric *Hydra* and the evolution of the bilateral body plan: An old body became a young brain. *Bioessays* **24:** 185–191.

Meinhardt H. 2004. Different strategies for midline formation in bilaterians. *Nat Rev Neurosci* **5:** 502–510.

Meinhardt H. 2006. Primary body axes of vertebrates: Generation of a near-Cartesian coordinate system and the role of Spemann-type organizer. *Dev Dyn* **235:** 2907–2919.

Meinhardt H. 2012. Modeling pattern formation in *Hydra*—a route to understand essential steps in development. *Int J Dev Biol* doi: 10.1387/ijdb.113483hm.

Momose T, Houliston E. 2007. Two oppositely localised Frizzled RNAs as axis determinants in a cnidarian embryo. *PLoS Biol* **5:** e70.

Momose T, Derelle R, Houliston E. 2008. A maternally localised Wnt ligand required for axial patterning in the cnidarian *Clytia hemisphaerica*. *Development* **135:** 2105–2113.

Nakamura Y, Tsiairis CD, Ozbek S, Holstein TW. 2011. Autoregulatory and repressive inputs localize *Hydra* Wnt3 to the head organizer. *Proc Natl Acad Sci* **108:** 9137–9142.

Niehrs C. 2010. On growth and form: A Cartesian coordinate system of Wnt and BMP signaling specifies bilaterian body axes. *Development* **137:** 845–857.

Nusse R. 2001. An ancient cluster of Wnt paralogues. *Trends Genet* **17:** 443.

Onai T, Lin HC, Schubert M, Koop D, Osborne PW, Alvarez S, Alvarez R, Holland ND, Holland LZ. 2009. Retinoic acid and Wnt/β-catenin have complementary roles in anterior/posterior patterning embryos of the basal chordate amphioxus. *Dev Biol* **332:** 223–233.

Pang K, Ryan JF, Mullikin JC, Baxevanis AD, Martindale MQ. 2010. Genomic insights into Wnt signaling in an early diverging metazoan, the ctenophore *Mnemiopsis leidyi*. *Evodevo* **1:** 10.

Petersen CP, Reddien PW. 2008. Smed-β-catenin-1 is required for anteroposterior blastema polarity in planarian regeneration. *Science* **319:** 327–330.

Petersen CP, Reddien PW. 2009. Wnt signaling and the polarity of the primary body axis. *Cell* **139:** 1056–1068.

Petersen CP, Reddien PW. 2011. Polarized notum activation at wounds inhibits Wnt function to promote planarian head regeneration. *Science* **332:** 852–855.

Philipp I, Aufschnaiter R, Ozbek S, Pontasch S, Jenewein M, Watanabe H, Rentzsch F, Holstein TW, Hobmayer B. 2009. Wnt/β-catenin and noncanonical Wnt signaling interact in tissue evagination in the simple eumetazoan *Hydra*. *Proc Natl Acad Sci* **106:** 4290–4295.

Prabhu Y, Eichinger L. 2006. The *Dictyostelium* repertoire of seven transmembrane domain receptors. *Eur J Cell Biol* **85:** 937–946.

Prud'homme B, Lartillot N, Balavoine G, Adoutte A, Vervoort M. 2002. Phylogenetic analysis of the *Wnt* gene family. Insights from lophotrochozoan members. *Curr Biol* **12:** 1395.

Putnam NH, Srivastava M, Hellsten U, Dirks B, Chapman J, Salamov A, Terry A, Shapiro H, Lindquist E, Kapitonov VV, et al. 2007. Sea anemone genome reveals ancestral eumetazoan gene repertoire and genomic organization. *Science* **317:** 86–94.

Reddien PW, Sanchez Alvarado A. 2004. Fundamentals of planarian regeneration. *Annu Rev Cell Dev Biol* **20:** 725–757.

Rentzsch F, Fritzenwanker JH, Scholz CB, Technau U. 2008. FGF signalling controls formation of the apical sensory organ in the cnidarian *Nematostella vectensis*. *Development* **135:** 1761–1769.

Riddiford N, Olson PD. 2011. *Wnt* gene loss in flatworms. *Dev Genes Evol* **221:** 187–197.

Rigo-Watermeier T, Kraft B, Ritthaler M, Wallkamm V, Holstein TW, Wedlich D. 2011. Functional conservation of *Nematostella* Wnts in canonical and noncanonical Wnt-signaling. *Biol Open* doi: 10.1242/bio.2011021.

Rokas A, Kruger D, Carroll SB. 2005. Animal evolution and the molecular signature of radiations compressed in time. *Science* **310:** 1933–1938.

Ruiz-Trillo I, Riutort M, Littlewood DT, Herniou EA, Baguna J. 1999. Acoel flatworms: Earliest extant bilaterian Metazoans, not members of Platyhelminthes. *Science* **283:** 1919–1923.

Ruiz-Trillo I, Roger AJ, Burger G, Gray MW, Lang BF. 2008. A phylogenomic investigation into the origin of metazoa. *Mol Biol Evol* **25:** 664–672.

Schierwater B, Eitel M, Jakob W, Osigus HJ, Hadrys H, Dellaporta SL, Kolokotronis SO, Desalle R. 2009. Concatenated analysis sheds light on early metazoan evolution and fuels a modern "urmetazoon" hypothesis. *PLoS Biol* **7:** e20.

Schilde C, Araki T, Williams H, Harwood A, Williams JG. 2004. GSK3 is a multifunctional regulator of *Dictyostelium* development. *Development* **131:** 4555–4565.

Schubert M, Holland LZ, Stokes MD, Holland ND. 2001. Three amphioxus *Wnt* genes (*AmphiWnt3*, *AmphiWnt5*, and *AmphiWnt6*) associated with the tail bud: The evolution of somitogenesis in chordates. *Dev Biol* **240:** 262–273.

Sebe-Pedros A, de Mendoza A, Lang BF, Degnan BM, Ruiz-Trillo I. 2011. Unexpected repertoire of Metazoan transcription factors in the unicellular Holozoan *Capsaspora owczarzaki*. *Mol Biol Evol* **28:** 1241–1254.

Shimizu T, Bae YK, Muraoka O, Hibi M. 2005. Interaction of *Wnt* and *caudal*-related genes in zebrafish posterior body formation. *Dev Biol* **279:** 125–141.

Sodergren E, Weinstock GM, Davidson EH, Cameron RA, Gibbs RA, Angerer RC, Angerer LM, Arnone MI, Burgess DR, Burke RD, et al. 2006. The genome of the sea urchin *Strongylocentrotus purpuratus*. *Science* **314:** 941–952.

Srivastava M, Begovic E, Chapman J, Putnam NH, Hellsten U, Kawashima T, Kuo A, Mitros T, Salamov A, Carpenter ML, et al. 2008. The *Trichoplax* genome and the nature of placozoans. *Nature* **454:** 955–960.

Srivastava M, Simakov O, Chapman J, Fahey B, Gauthier ME, Mitros T, Richards GS, Conaco C, Dacre M, Hellsten U, et al. 2010. The *Amphimedon queenslandica* genome and the evolution of animal complexity. *Nature* **466:** 720–726.

Steiner AB, Engleka MJ, Lu Q, Piwarzyk EC, Yaklichkin S, Lefebvre JL, Walters JW, Pineda-Salgado L, Labosky PA, Kessler DS. 2006. FoxD3 regulation of Nodal in the Spemann organizer is essential for *Xenopus* dorsal mesoderm development. *Development* **133:** 4827–4838.

Sullivan JC, Ryan JF, Mullikin JC, Finnerty JR. 2007. Conserved and novel Wnt clusters in the basal eumetazoan *Nematostella vectensis*. *Dev Genes Evol* **217:** 235–239.

Tamai K, Semenov M, Kato Y, Spokony R, Liu C, Katsuyama Y, Hess F, Saint-Jeannet JP, He X. 2000. LDL-receptor-related proteins in Wnt signal transduction. *Nature* **407:** 530–535.

Technau U, Steele RE. 2011. Evolutionary crossroads in developmental biology: Cnidaria. *Development* **138:** 1447–1458.

Technau U, Cramer von Laue C, Rentzsch F, Luft S, Hobmayer B, Bode HR, Holstein TW. 2000. Parameters of self-organization in *Hydra* aggregates. *Proc Natl Acad Sci* **97:** 12127–12131.

Thorpe CJ, Weidinger G, Moon RT. 2005. Wnt/β-catenin regulation of the Sp1-related transcription factor sp5l promotes tail development in zebrafish. *Development* **132:** 1763–1772.

van Amerongen R, Nusse R. 2009. Towards an integrated view of Wnt signaling in development. *Development* **136:** 3205–3214.

van Amerongen R, Mikels A, Nusse R. 2008. Alternative wnt singnaling is initiated by distinct receptors. *Sci Signal* **1:** re9.

van de Wetering M, Cavallo R, Dooijes D, van Beest M, van Es J, Loureiro J, Ypma A, Hursh D, Jones T, Bejsovec A, et al. 1997. Armadillo coactivates transcription driven by the product of the *Drosophila* segment polarity gene *dTCF*. *Cell* **88:** 789–799.

Vorwald-Denholtz PP, De Robertis EM. 2011. Temporal pattern of the posterior expression of Wingless in *Drosophila* blastoderm. *Gene Expr Patterns* **11:** 456–463.

Wikramanayake AH, Huang L, Klein WH. 1998. β-Catenin is essential for patterning the maternally specified animal–vegetal axis in the sea urchin embryo. *Proc Natl Acad Sci* **95:** 9343–9348.

Wikramanayake AH, Hong M, Lee PN, Pang K, Byrum CA, Bince JM, Xu R, Martindale MQ. 2003. An ancient role for nuclear β-catenin in the evolution of axial polarity and germ layer segregation. *Nature* **426:** 446–450.

Wikramanayake AH, Peterson R, Chen J, Huang L, Bince JM, McClay DR, Klein WH. 2004. Nuclear β-catenin-dependent Wnt8 signaling in vegetal cells of the early sea urchin embryo regulates gastrulation and differentiation of endoderm and mesodermal cell lineages. *Genesis* **39:** 194–205.

Windsor PJ, Leys SP. 2010. Wnt signaling and induction in the sponge aquiferous system: Evidence for an ancient origin of the organizer. *Evol Dev* **12:** 484–493.

Xiao S, Laflamme M. 2009. On the eve of animal radiation: Phylogeny, ecology and evolution of the Ediacara biota. *Trends Ecol Evol* **24:** 31–40.

Yamaguchi TP, Takada S, Yoshikawa Y, Wu N, McMahon AP. 1999. T (Brachyury) is a direct target of Wnt3a during paraxial mesoderm specification. *Genes Dev* **13:** 3185–3190.

Yu JK, Satou Y, Holland ND, Shin IT, Kohara Y, Satoh N, Bronner-Fraser M, Holland LZ. 2007. Axial patterning in cephalochordates and the evolution of the organizer. *Nature* **445:** 613–617.

Wnt/Wingless Signaling in *Drosophila*

Sharan Swarup and Esther M. Verheyen

Department of Molecular Biology and Biochemistry, Simon Fraser University, Burnaby, British Columbia V5A1S6, Canada

Correspondence: everheye@sfu.ca

The Wingless (Wg) pathway represents one of the best-characterized intercellular signaling networks. Studies performed in *Drosophila* over the last 30 years have contributed to our understanding of the role of Wg signaling in the regulation of tissue growth, polarity, and patterning. These studies have revealed mechanisms conserved in the vertebrate Wnt pathways and illustrate the elegance of using the *Drosophila* model to understand evolutionarily conserved modes of gene regulation. In this article, we describe the function of Wg signaling in patterning the *Drosophila* embryonic epidermis and wing imaginal disc. As well, we present an overview of the establishment of the Wg morphogen gradient and discuss the differential modes of Wg-regulated gene expression.

Evolutionarily conserved cell signaling pathways regulate the development of metazoans through their reiterative implementation, both spatially and temporally. Wnt signaling represents one such pathway that has multiple, essential roles during both embryogenesis and adult homeostasis to regulate cell proliferation, cell polarity, and the specification of cell fate (for review, see Wodarz and Nusse 1998). Wnt genes encode secreted glycoprotein ligands that can act both as short-range signaling molecules and long-range morphogens, depending on the developmental context. Members of the Wnt family are defined by sequence homology to Wnt-1 (Nusse and Varmus 1982; Nusse et al. 1984), the first identified Wnt protein, rather than by functional homology. As such, subsequent to the identification of Wnt-1, diverse Wnt-regulated processes have been identified that when aberrantly regulated result in myriad diseases, ranging from developmental disorders to cancers. Thus far, 19 vertebrate Wnt family members have been discovered, of which there are seven homologs in *Drosophila* (Table 1). Much of our understanding of the role of Wnt proteins during development has come as a result of genetic analyses of the *Drosophila wnt-1* (*Dwnt-1*) or *wingless* (*wg*) gene.

As the name suggests, the *wg* gene is required to pattern the *Drosophila* wings and other adult body structures. It was originally identified through a hypomorphic allele, wg^{1}, which harbors a deletion in a regulatory element of the gene and causes the variable transformation of the adult wing(s) to thoracic notum (Sharma and Chopra 1976; Babu 1977). Subsequent to characterization of the viable wg^{1} allele, large-scale genetic screens performed by Eric Wieschaus, Christiane Nusslein-Volhard, and colleagues yielded embryonic lethal,

Cite this article as *Cold Spring Harb Perspect Biol* doi: 10.1101/cshperspect.a007930

Table 1. Comparison of Wnt genes between *Drosophila* and vertebrates

Structural homology between *Drosophila* and vertebrate Wnt genes							
Drosophila genes	*Dwnt-1* or *wingless*	*Dwnt-2*	*Dwnt-5*	*Dwnt-4*	*Dwnt-6*	*Dwnt-8* or *Dwnt-D*	*Dwnt-10*
Vertebrate homologs	*Wnt-1*	*Wnt-7*	*Wnt-5*	*Wnt-9* *Wnt-14* *Wnt-15*	*Wnt-6*	*Wnt-8*	*Wnt-10*

loss-of-function alleles of *wg* (Nusslein-Volhard and Wieschaus 1980; Nusslein-Volhard et al. 1984). In the years that followed, the *wg* gene was cloned (Baker 1987; Cabrera et al. 1987; Rijsewijk et al. 1987), and through the use of conditional mutants, mosaics analyses, and ectopic expression, it was shown to have important roles at several stages of development in multiple tissues, including the embryonic ectoderm (Baker 1988a; Bejsovec and Martinez Arias 1991; Dougan and DiNardo 1992; Bejsovec and Wieschaus 1993), head (Schmidt-Ott and Technau 1992), midgut (Immerglück et al. 1990; Reuter et al. 1990; Thuringer and Bienz 1993; Bienz 1994), wing disc (Simcox et al. 1989; Cohen 1990; Cohen et al. 1993; Phillips and Whittle 1993; Williams et al. 1993), and leg disc (Baker 1988b; Campbell et al. 1993; Couso et al. 1993). Moreover, through genetic and biochemical analyses performed predominantly in *Drosophila* over the years, the molecular mechanism of canonical Wnt or Wg signaling has emerged. In the absence of the Wnt/Wg ligand, cytoplasmic levels of β-catenin/Armadillo (Arm), the transcriptional effector of the pathway, are kept low through its constitutive degradation by a protein destruction complex composed of Axin, APC, GSK3/Zw3, and CK1. As a result, Wnt/Wg-regulated genes are kept off by the DNA-binding transcription factor T-cell factor (Tcf) with the aid of other transcriptional corepressors. Binding of the Wnt/Wg ligand to its coreceptors, Frizzled2 (Fz2) and LRP/Arrow (Arr), initiates a sequence of cytoplasmic events that leads to the Dishevelled (Dsh)–mediated inactivation of the protein destruction complex, thereby allowing stabilized β-catenin/Arm to translocate to the nucleus, where it binds Tcf to direct the activation of Wnt/Wg-target genes (for review, see Bejsovec 2006).

In this article, we discuss the role of the Wg molecule as an organizing center during embryonic segmentation and patterning of the wing disc, because these are now considered the classic systems for demonstrating different aspects of Wg signaling.

FUNCTION OF WINGLESS SIGNALING IN THE EMBRYO

During *Drosophila* embryogenesis, a hierarchy of maternal and zygotic (gap, pair-rule, and segment polarity) genes progressively subdivides the embryonic syncytium into transverse regions that determine the anterior/posterior axis (for review, see Ingham and Martinez Arias 1992; St. Johnston and Nuesslein-Volhard 1992). The cellular blastoderm is formed during stage 14 of embryogenesis and coincides with the division of the anterior/posterior axis into segmental units as directed by the segment polarity genes *wg* and *hedgehog* (*hh*) (for review, see Perrimon 1994). These segment polarity genes interact with one another to define the segment boundaries and intrasegmental pattern of the embryo (Fig. 1). At the end of embryogenesis, the outcome of the segmentation and patterning events is a larva characterized on the ventral epidermis by an alternating pattern of protrusions called denticles that are separated by regions of naked cuticle (for review, see Martinez-Arias 1993). We here describe the mechanism through which Wg signaling establishes and patterns each segment to generate this stereotypical arrangement of denticles and naked epidermal cuticle. This process can be divided into four successive events: establishment of the organizer, asymmetric signaling from the organizer, subdivision of each segment into

Cite this article as *Cold Spring Harb Perspect Biol* doi: 10.1101/cshperspect.a007930

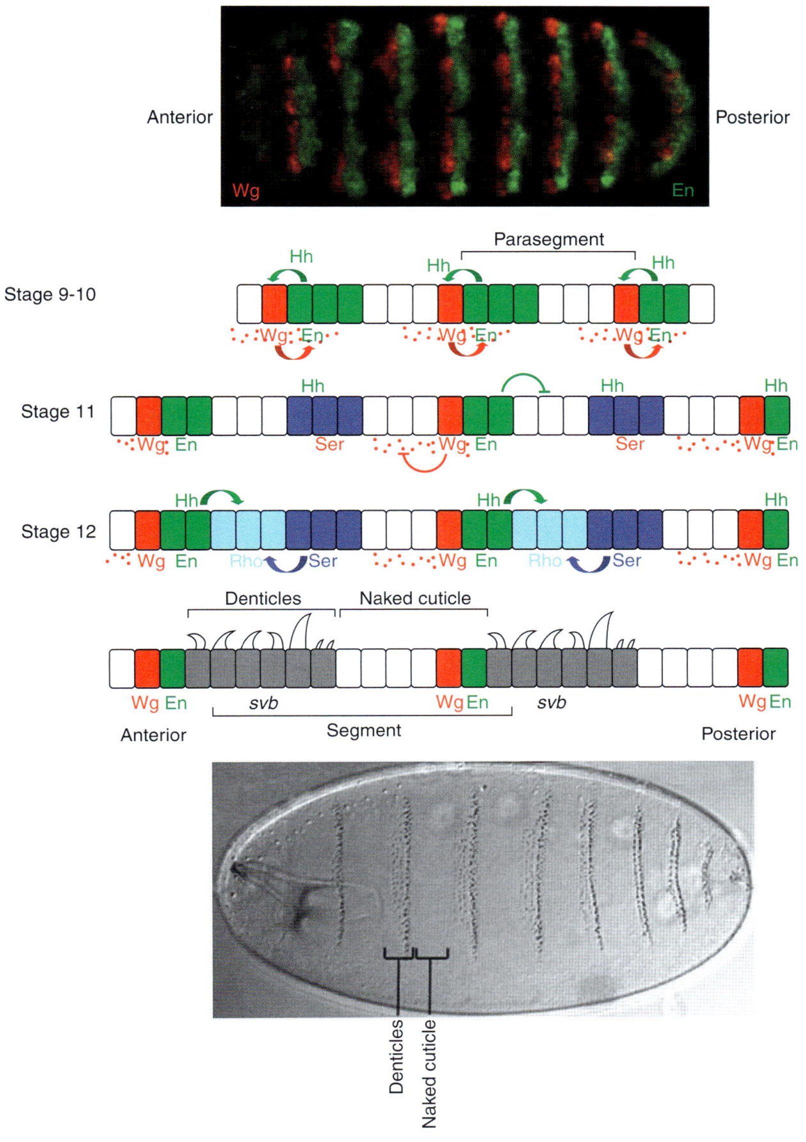

Figure 1. Wingless-regulated patterning of the *Drosophila* embryonic epidermis. The interplay between the Wg and Hh signaling pathways initially establishes the parasegment boundaries and subsequently directs the intrasegmental pattern to establish the stereotypical arrangement of denticles and naked cuticle at the end of embryogenesis (see text for details). The embryo is positioned with its anterior end to the left. (Top panel courtesy of L.R. Braid.)

signaling domains, and cell fate specification by the signaling domains (Fig. 1).

The expression of *wg* and *hh* is initiated by the pair-rule genes in adjacent, non-overlapping domains during stage 9–10 of embryogenesis, and subsequently, they reciprocally regulate each other to stabilize their expression (Fig. 1) (for review, see DiNardo et al. 1994). Wg protein that is transcribed and secreted from an anterior row of cells maintains the expression of a transcription factor, *engrailed* (*en*), in adjoining, posterior cells. The En-expressing cells, in turn, transcribe and secrete the Hh ligand, which reciprocally maintains Wg expression in the neighboring, anterior cells (DiNardo et al. 1988; Martinez Arias et al. 1988; Hidalgo and Ingham 1990; Bejsovec and Martinez Arias 1991). The interface between these two adjacent domains defines the parasegment boundary or organizer, with *en/hh* transcribed at the anterior and *wg* at the posterior end of each parasegment, respectively (Baker 1987; Lee et al. 1992; Mohler and Vani 1992).

Initially, after the parasegment boundary is established, the distribution of the Wg ligand is bidirectional and triggers a response through its signaling cascade at equivalent levels in both the anterior and posterior directions (Riggleman et al. 1990; Peifer et al. 1994). However, during stage 11, the distribution of extracellular Wg is only graded anterior to the Wg-expressing domain and is completely lost in the posterior direction (Fig. 1) (van den Heuvel et al. 1989; Gonzalez et al. 1991). This asymmetry in Wg distribution and correspondingly signaling activity is a result of its rapid endocytosis and degradation in cells posterior from which it is secreted, in a process that is promoted by Hh signaling (Sanson et al. 1999; Dubois et al. 2001). Interestingly, Wg signaling, in turn, attenuates Hh signaling anterior to the En/Hh-expressing cells, thereby allowing its activity in only the posterior direction (Gritzan et al. 1999). Thus, a pattern of polarized signaling of Wg and Hh at the parasegment boundary is formed that can direct cells on either side to follow distinct developmental programs.

The *wg*- and *en/hh*-expressing domains initially established successively specify two additional domains within each parasegment that are defined by the expression of *Serrate* (*Ser*) (a ligand for the Notch pathway) (Wiellette et al. 1999) and *rhomboid* (*rho*) (promotes EGFR signaling) (Golembo et al. 1996). Wg and Hh signaling antagonize the posterior and anterior boundaries, respectively, of *Ser* expression, thereby restricting its domain to the center of each parasegment (Alexandre et al. 1999; Gritzan et al. 1999; Sanson et al. 1999). *rho* expression is repressed by Wg signaling and is positively reinforced through a combination of Hh and Ser signaling, thereby establishing its domain immediately posterior to the *en/hh* domain (Alexandre et al. 1999). During stage 12, each parasegment on the ventral epidermis is divided into four domains that express specific genes that are responsible for the intraparasegmental patterning of the embryo. This period also coincides with the formation of a segmental groove at the posterior edge of each *en/hh* domain and defines the segment boundary (Fig. 1).

The four signaling domains established within each segment control the binary decision between specification of naked cuticle or denticle cell fates. The outcome between these two choices is dependent on the expression of a transcription factor encoded by the *shaven baby* (*svb*) gene, which is necessary and sufficient to direct denticle formation cell autonomously (Payre et al. 1999). The expression of *svb* is inhibited in cells that specify naked cuticle, whereas cells that make denticles express *svb*. Wg signaling specifies naked cuticle (Bejsovec and Martinez Arias 1991; Noordermeer et al. 1992; Lawrence et al. 1996) by repressing the expression of *svb* (Payre et al. 1999). Due to the asymmetric distribution of Wg, the repression of *svb* is asymmetric and results in one row of cells posterior to, and four rows of cells anterior to the Wg domain that produce naked cuticle (O'Keefe et al. 1997; Szuts et al. 1997), across the posterior half of each segment. The six rows toward the anterior half of each segment do not receive the Wg ligand but instead transduce the EGFR signal that promotes *svb* expression (Payre et al. 1999), thus resulting in the generation of denticles that vary in shape, size, and polarity. The outcome of the patterning events mediated by regulated cell signaling is a repeated mosaic of denticle belts and naked cuticle on the ventral epidermis of the embryo, composed of segments 11–12 cells wide along the anterior/posterior axis and 36–40 cells wide along the dorsal/ventral axis (Fig. 1). In *wg* loss-of-function mutants, *svb* expression is not repressed and naked cuticle is not specified, resulting in the excess specification of denticles. Conversely, when *wg* is ectopically expressed, excess naked cuticle is produced with

Cite this article as *Cold Spring Harb Perspect Biol* doi: 10.1101/cshperspect.a007930

a reduction or loss of denticles (Fig. 2) (Payre et al. 1999).

Components of the Wg signal transduction cascade were initially identified in large-scale genetic screens designed by Eric Wieschaus and Christiane Nusslein-Volhard to isolate zygotic mutations that disrupt the *Drosophila* embryonic cuticle. These screens yielded numerous genes that displayed segment polarity phenotypes when mutated, including both positive and negative regulators of the Wg pathway (Nusslein-Volhard and Wieschaus 1980; Jürgens et al. 1984; Nusslein-Volhard et al. 1984; Wieschaus et al. 1984). Mutations in *arm* and *arr* resemble the *wg* loss-of-function phenotype to specify excess denticles, whereas *naked*, an inhibitor of the pathway, was also isolated and displays a *wg* gain-of-function phenotype with an excess specification of naked cuticle. Although the aforementioned genetic screens led to the characterization of several downstream components, they did not yield the entire set of genes involved in the Wg pathway. This is because some genes are maternally contributed and zygotic mutants thus retain sufficient maternal product to pattern a relatively normal embryo. This problem was circumvented through the induction of mitotic recombination in the germline, thereby eliminating maternally contributed gene product, and this technique was used to genetically screen the X-chromosome for maternal effect mutations that disrupt patterning of the embryonic cuticle (Perrimon et al. 1989). *arm* was reisolated in this screen along with additional new components of the pathway—*dsh*, *porcupine*, and *zw3*.

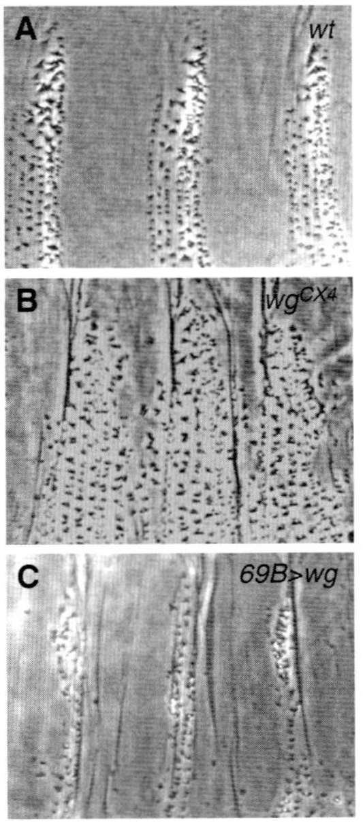

Figure 2. *wingless* mutant phenotypes of the *Drosophila* embryonic epidermis. (*A*) In a wild type (wt) embryo, *wg* expression in segmental stripes generates regions of naked cuticle that are intercalated by regions of denticles. (*B*) A wg^{CX4} mutant embryo results in an excess specification of denticles, whereas (*C*) the overexpression of *wg* (*69B-Gal4*>*UAS-wg*) results in the excess specification of naked cuticle.

Traditional loss-of-function epistasis experiments rely on evaluating phenotypes in double-mutant combinations, thus allowing one to determine which of the two distinct phenotypes "overrides" or is epistatic to the other. This information allows the ordering of genes within a pathway, because the gene that acts most downstream will generally be epistatic to those acting upstream. However, the similar phenotypes of many of the first identified patterning genes precluded such analyses. To overcome this limitation, numerous studies were performed using combinations of gain- and loss-of-function fly strains, which provided distinct phenotypes (e.g., excess naked cuticle due to gain of *wg* compared with excess denticles in an *arm* mutant). These studies allowed the relationships between the early pathway components to be defined and have continued to serve as a template for characterizing and positioning novel pathway members. Remarkably, already in 1994, the minimal components of Wg signaling were known and ordered into a rudimentary pathway leading from *wg* to *dsh*, *sgg*, *arm*, and into the nucleus to regulate *en* expression (Noordermeer et al. 1994; Siegfried et al. 1994). Thus, through the implementation of genetic screens, several components of the Wg pathway were first defined based on their ability to disrupt embryonic patterning, and the order of these genes in the pathway was deduced through genetic epistasis. The genetic and molecular characterization of these components formed the foundation of our current understanding of Wnt/Wg signaling.

FUNCTION OF WINGLESS SIGNALING IN THE WING IMAGINAL DISC

Like the embryonic ectoderm, the wing imaginal disc represents another tissue for which the function of Wg signaling has been well elucidated. The wing disc is an epithelial sac that is composed of 20 cells when it is formed during embryonic development and then proliferates during the larval stages to generate a disc of ~75,000 cells in the late third larval instar. At this stage, the cells comprising the major elements of the wing primordium can be identified through the use of molecular markers, including the notum, hinge, blade, and margin (Fig. 3). *wg* is expressed in two ring-like domains in the hinge region, along the dorsal/ventral compartment boundary dividing the wing blade, and in a broad band in the dorsal part of the disc. The inner ring-like domain frames the wing blade and gives rise to the hinge, whereas the dorsal/ventral boundary region forms the wing margin. The most dorsal part of the wing disc gives rise to the notum of the adult fly (Fig. 3).

As previously mentioned, the first wg^1 mutant displayed a transformation of wing structures into thoracic notal structures (Sharma and Chopra 1976; Morata and Lawrence 1977). Conversely, misexpression of *wg* at a specific time leads to the ectopic induction of wing structures in the notum (Ng et al. 1996; Klein and Arias 1998a). These phenotypic analyses suggest that the *wg* gene has a crucial function during the development of the wing disc to specify the region that eventually gives rise to the adult wing. During the second larval instar, *wg* is expressed in the ventral region of the wing disc and specifies the wing field, while the EGFR ligand, *vein*, is expressed in the dorsal region of the wing disc to specify the notum (Simcox et al. 1996; Wang et al. 2000). The loss of *vein* during the second larval instar results in the loss of all notal structures (Simcox et al. 1996; Wang et al. 2000), whereas the misexpression of *vein* in the wing field prevents wing formation and produces ectopic notum structures (Baonza et al. 2000; Wang et al. 2000), phenotypes that are complementary to those of *wg*. The expression of *vein* is restricted to the dorsal region of the wing disc by the suppressive influence of Wg signaling in the ventral region (Baonza et al. 2000). Conversely, *wg* expression is antagonized in the dorsal region of the disc by EGFR signaling to limit its expression to only the ventral region (Baonza et al. 2000; Wang et al. 2000). This mutual antagonism between the Wg and EGFR pathways segregates the early wing disc into notum and wing regions and provides an explanation for the wing-to-notum transformation that occurs in the absence of *wg* function.

Multiple pathways contribute to the refinement of the *wg* expression pattern as larval

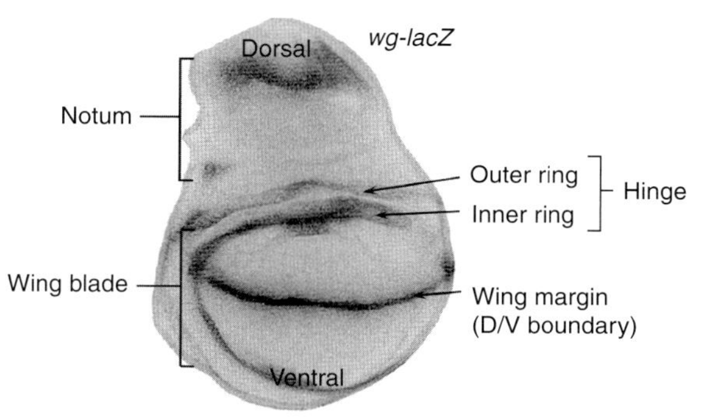

Figure 3. *wingless* expression in the wing imaginal disc. *wg* is expressed at the dorsal/ventral boundary, in two concentric rings and in a broad stripe in the dorsal region of a third instar wing disc. The expression of *wg* in the disc regulates the patterning of the wing margin, blade, hinge, and notum of the adult fly.

Cite this article as *Cold Spring Harb Perspect Biol* doi: 10.1101/cshperspect.a007930

development proceeds. During the late second instar, the expression of the selector gene *apterous* (*ap*) overlaps ventrally with that of *wg*, to divide the developing wing field into two regions (Williams et al. 1993; Ng et al. 1996). The Notch pathway ligands, *Serrate* and *Delta*, are transcribed at the Ap-expression boundary (dorsal/ventral boundary) and lead to the induction of downstream genes that are ultimately required for the establishment of the distal wing fates, including the wing blade and margin (Williams et al. 1994; Klein and Arias 1998b; Milan and Cohen 2000). Subdivision of the wing field is initiated by the Notch-induced expression of *vestigial* (*vg*) from its boundary enhancer (vg^{BE}). Vg specifies the wing blade and collaborates with Notch signaling to induce the expression of *wg* along the dorsal/ventral boundary (Kim et al. 1996; Klein and Arias 1998a, 1999). The Wg ligand is secreted from cells at the dorsal/ventral boundary and subsequently patterns the wing margin through activation of its target genes in a concentration-dependent manner (Couso et al. 1994; Neumann and Cohen 1997). After the establishment of the expression of *wg* and *vg* in the wing primordium, the wing blade begins to grow during the third instar. The expression of *vg* is induced in cells of the wing blade that lie outside of the domain of Notch signaling through its quadrant enhancer (vg^{QE}) (Kim et al. 1996). This enhancer requires input from Vg itself, Wg (produced at the dorsal/ventral boundary), and Dpp (produced at the anterior/posterior boundary) for its activity.

Flies deficient for *wg* do not develop wings, and mutant patches of cells that cannot respond to the Wg ligand are eliminated from the wing field in the wing disc (Chen and Struhl 1999). This suggests that there is an absolute requirement of Wg for growth of the wing. However, until recently it had been unclear whether cells of the wing disc required Wg signaling as an instructive signal or a permissive signal in order to proliferate. Early studies suggested that although the misexpression of *wg* has mitogenic effects in the hinge region, it does not induce cell proliferation within the wing pouch but rather respecifies these cells to assume a wing margin fate (Neumann and Cohen 1996). It was subsequently shown that high levels of Wg at the presumptive wing margin, in fact, arrest the cell cycle (O'Brochta and Bryant 1985; Johnston and Edgar 1998). The ability of uniform expression of moderate levels of Wg to indeed promote growth throughout the presumptive wing region was shown only recently (Baena-Lopez et al. 2009). In this study, the investigators propose a model in which Vg-expressing cells expand their numbers by inducing their neighboring, non-wing cells to express Vg, provided these cells also receive the Wg signal, and thereby be recruited into the wing field through a feed-forward mechanism. In this scenario, Wg is proposed to act as a permissive signal to enable the recruitment of cells into the wing field to promote growth of the wing disc (Zecca and Struhl 2007a,b).

DOES WINGLESS ALWAYS BEHAVE AS A MORPHOGEN?

As a minimal definition, a "morphogen" is a molecule that diffuses away from a localized source to directly instruct cell identities in a concentration-dependent manner. For a secreted signaling ligand such as Wg to qualify as a morphogen, it must form a graded distribution away from its source, directly act on cells at a distance rather than indirectly through a relay mechanism, and induce the commitment of cells in the field, as a function of their distance from the morphogen source, to distinct developmental fates through the expression of different sets of genes.

Experiments performed in the wing disc suggest that Wg does indeed behave as a morphogen in this tissue. During the third larval instar, Wg, secreted from cells at the dorsal/ventral organizing boundary of the wing disc, diffuses away from its source on either side and acts directly on cells at long range to induce the nested expression of its target genes *achaete* (*ac*), *distal-less* (*dll*), and *vestigial* (*vg*) (Fig. 4) (Zecca et al. 1996; Neumann and Cohen 1997). The modulation of Wg signaling through clonal analyses or ectopic expression of *arm* and *dsh*, positive regulators of the pathway, shows

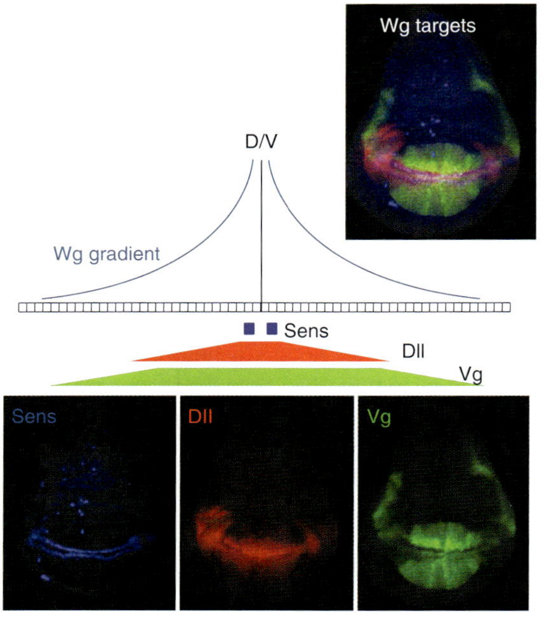

Figure 4. Morphogen gradient of Wingless. In the third instar wing disc, Wg is secreted from cells at the dorsal/ventral boundary and forms a concentration gradient on either side to regulate the expression of its target genes, *senseless* (Sens, blue, high threshold), *distalless* (Dll, red, medium threshold), and *vestigial* (Vg, green, low threshold). The expression of *distalless* is graded throughout its domain, whereas *vestigial* expression is graded only at its edges. *senseless*, like *achaete* (discussed in the text), requires high levels of Wg signaling and is expressed only in cells abutting the dorsal/ventral boundary.

a cell-autonomous change in Wg-responsive target gene expression, regardless of the position of the cell from the source of the Wg signal (Zecca et al. 1996; Neumann and Cohen 1997). These data suggest that the Wg ligand acts directly on cells distant from its source, and the expression of target genes is not activated via a secondary or relay signal. In addition, in contrast to the wild type secreted Wg, a membrane-tethered form of the ligand that is unable to diffuse away from its source, activates expression of target genes only in its immediate neighbors (Zecca et al. 1996; Neumann and Cohen 1997). Furthermore, through the use of a temperature-sensitive allele, wg^{ts}, it has been shown that the level of Wg activity minimally required to activate expression of *ac*, a high-threshold target gene; *dll*, a medium-threshold target gene; and *vg*, a low-threshold target gene; is progressively lower. Because the activity of Wg^{ts} is decreased (by raising the temperature), the expression of *ac* is lost at a temperature at which *dll* expression is retained, and as the temperature is further increased, the expression of *dll* is lost with no effect on *vg* expression. In both cases, there is a concomitant reduction of the expression domains of target genes toward the source of the ligand as the activity level of Wg is reduced. Consistent with this result, low levels of ectopically expressed Wg are able to activate *dll* but not *ac* (Neumann and Cohen 1997). These data suggest that Wg activates different target genes in a concentration-dependent manner and defines their expression domains through different activation thresholds. This does not exclude the possibility that activation of target genes requires other permissive signals, but it does argue that the level of the Wg ligand is the instructive signal. Lastly, the extracellular concentration gradient of the Wg protein can be detected up to 10 cell diameters away from the secreting cells at the dorsal/ventral boundary and represents direct evidence that Wg acts as a morphogen in the wing disc (Strigini and Cohen 2000).

Wingless has been proposed to behave as a morphogen additionally in the embryonic epidermis and midgut, but whether it actually does so in these developmental contexts is unclear. In the embryonic ectoderm, Wg secreted from the parasegment boundary adopts a graded distribution in the anterior direction up to four cells away to specify naked cuticle (Bejsovec and Martinez Arias 1991). Although in this context Wg fulfills the criteria of having a localized source, forming a gradient, and acting at long range, there is no evidence for a concentration-dependent induction of target genes. In fact, the epidermal phenotype of *wg* mutant embryos can be rescued through the ubiquitous expression of a *wg* transgene (Sampedro et al. 1993), suggesting that Wg acts in a gradient-independent manner in this tissue. Additionally, in a *wg* mutant background, the overexpression of a membrane-tethered form of Wg from cells that normally express the ligand can recapitulate the normal range of signaling, confirming that the restricted diffusion of Wg is dispensable for patterning of the embryonic cuticle (Pfeiffer et al. 2000). In the embryonic midgut, *wg* is expressed in the visceral mesoderm and directs the neighboring endoderm to differentiate into two different cell types: large, flat cells require high levels of signaling and develop immediately adjacent to the Wg source, whereas copper cells develop further away as their differentiation is repressed at high levels of Wg signaling. As the level of Wg is modified, there is a concomitant change in the domains of the large, flat cells and copper cells. Additionally, the *labial* gene is expressed in a graded manner in the region of copper cells, consistent with the proposed concentration-dependent effect of Wg. However, the effects of Wg have not been shown to be direct in this tissue (Hoppler and Bienz 1995). Thus, in the embryonic ectoderm and midgut, it is possible that Wg signaling does not define the pattern of the response but, rather, stabilizes patterns of gene expression that have been specified through other mechanisms.

ENDOCYTOSIS: SHAPING THE WINGLESS GRADIENT

Secreted Wg is theoretically capable of passive diffusion at a relatively fast rate in all directions over a long range. However, the graded distribution of Wg forms at a slower rate over a shorter range than that predicted through free diffusion and directionally along the epithelial surface. Moreover, the contrasting range of its activity during embryogenesis (4 cell diameters) (Bejsovec and Martinez Arias 1991; Pfeiffer et al. 2002) versus development of the wing disc (10 cell diameters) (Strigini and Cohen 2000) suggests that the movement of the Wg ligand in the extracellular environment is regulated. Recent experimental and theoretical studies favor a model in which the Wg gradient is formed and maintained through the combined effects of restricted diffusion and endocytosis. Once secreted, the Wg molecule undergoes restricted diffusion, as opposed to free diffusion, because of its interactions with lipid-based transport proteins, receptors on the surface of membranes, and heparan sulfate proteoglycans of the extracellular matrix (Lin and Perrimon 1999; Baeg et al. 2001, 2004; Lecourtois et al. 2001; Bornemann et al. 2004; Kirkpatrick et al. 2004; Han et al. 2005; Panakova et al. 2005; Piddini et al. 2005; Katanaev et al. 2008; Mulligan et al. 2012). In addition, the repeated vesicle-mediated endocytosis and resecretion of Wg, in a process called "planar transcytosis," and the endocytosis-mediated degradation of Wg also contribute to the formation of the Wg gradient, both extracellularly and intracellularly. We here review the evidence for the endocytosis-mediated regulation of the Wg gradient in the *Drosophila* embryo and wing disc.

At the subcellular level, Wg protein can be detected not only in the extracellular environment, but also within intracellular vesicles and multivesicular bodies in non-secreting cells. This clearly indicates that the Wg ligand is internalized and suggests that its gradient is regulated through endocytosis. The endocytic regulation of the Wg gradient has been investigated using mutations in components of the endocytic machinery. The formation of the Wg gradient in the embryo is a result of asymmetric endocytosis and trafficking to lysosomes, and perhaps transcytosis. As previously described, during embryogenesis, the initially symmetrical Wg gradient becomes asymmetrical due to increased Wg degradation in cells posterior to its source (Fig. 1) (Sanson et al. 1999; Dubois et al. 2001). Evidence for this model comes from experiments performed using an HRP (horse radish peroxidase)–Wg fusion protein expressed under the control of the endogenous *wg* promoter. Unlike the Wg portion of the fusion protein, the HRP moiety is stable throughout the endocytic pathway and thus serves as a tool to monitor Wg degradation in vivo. In cells anterior to the HRP-Wg source, both HRP and Wg can be detected in intracellular vesicles (presumed to be early endosomes), whereas only HRP can be detected in vesicles (presumably late endosomes) in cells posterior to the source. Moreover, posterior to the Wg source, many more HRP-positive intracellular compartments can be detected that extend beyond the Wg protein gradient, confirming that the degradation of Wg through the endocytic pathway limits the range of its gradient (Dubois et al. 2001). In *deep orange* (*dor*) mutants that have impaired trafficking to the lysosome, Wg accumulates in multivesicular bodies (Piddini et al. 2005). Lastly, in *clathrin heavy chain* (*chc*) mutants that cannot initiate clathrin-mediated endocytosis, the Wg gradient extends in the posterior direction, suggesting that clathrin is normally required for establishing and maintaining the asymmetric distribution of Wg (Desbordes et al. 2005). However, there is also contradictory evidence that suggests a role for endocytosis in Wg dispersal in the embryo. In temperature-sensitive *shibire*^{ts} (which encodes Dynamin, an essential component of both clathrin and caveolin-mediated endocytosis) mutant embryos that are shifted to the restrictive temperature, the Wg gradient is reduced to a narrow range around the Wg-expressing cells (Moline et al. 1999). This narrowed Wg gradient in *shibire*^{ts} embryos is not due to any effect on the secretion of the Wg ligand or its stability in non-secreting cells but, rather, can be attributed to an effect on the transport of the Wg ligand extracellularly (Moline et al. 1999). The visualization of the trafficking of a Wg-GFP fusion protein in live embryos reveals that Wg can be internalized and recycled back to the cell surface (Pfeiffer et al. 2000), providing additional evidence of a role for transcytosis in Wg distribution.

In the wing disc, the Wg protein can indeed traverse a *shibire* mutant clone and can be detected extracellularly both proximal and distal to the clone (Strigini and Cohen 2000). This suggests that in the wing disc, the Wg ligand once secreted is able to spread to non-expressing cells in the absence of endocytic trafficking. In fact, the inhibition of endocytosis leads to an extension of the Wg gradient, indicating that endocytosis serves to down-regulate the levels of Wg in the wing disc (Strigini and Cohen 2000; Piddini et al. 2005). In accordance with this hypothesis, more extracellular Wg is present within the *shibire* mutant clone than outside the clone (Strigini and Cohen 2000).

Notably, the majority of evidence from the embryonic epidermis and wing disc support different models of Wg transport and stability that contribute to the Wg gradient. Further studies need to be performed to resolve which mechanism of Wg distribution is predominant under physiological conditions.

WINGLESS-RESPONSIVE TARGET GENES: ACTIVATION VERSUS REPRESSION

Although the mechanistic details through which cell signaling pathways ultimately regulate gene expression to control the development of metazoans may differ, they all share at least three common, conserved features: default repression, activator insufficiency, and cooperative activation (for review, see Barolo et al. 2002; Affolter et al. 2008). Developmental signaling pathways regulate gene expression by a switch mechanism, whereby from an actively repressed state in the absence of the signal, genes are transcribed in the presence of the signal. The phenomenon of inhibition of gene transcription in the absence of signaling is referred to as "default repression." "Activator insufficiency" and "cooperative activation" refer to the inability of a signaling pathway to activate the same and complete set of target genes in all developmental contexts. The combination of these three features allows a signaling pathway to robustly activate specific target gene expression in response to the signal, in a context-dependent manner, while preventing target gene expression in the absence of the signal (for review, see Barolo et al. 2002; Affolter et al. 2008).

In the case of Wg signaling, default repression is exerted on the same pathway response element using the same signal-regulated transcription factor, Tcf, as is used in the presence of active signaling. In the absence of signaling, Tcf binds Wg-responsive elements within target genes along with other corepressors to suppress gene expression. When the Wg ligand is present, the same transcription factor Tcf binds Arm and recruits other coactivators to direct the expression of target genes (Brunner et al. 1997; van de Wetering et al. 1997; Lawrence et al. 2000; Schweizer et al. 2003). However, it has recently become apparent that target genes of the pathway are not only directly activated by the Wg signal, but can be directly repressed as well. Direct repression implies that following signaling, a gene's transcription is repressed without the intermediate transcriptional induction of a repressor. Signal-induced gene repression conflicts with the principle of default repression. If gene transcription is silenced through default repression before pathway activity, there is no opportunity to repress gene transcription following signaling. Indeed, for Wg signaling to repress transcription of a gene, default repression of the gene would have to be circumvented before signaling with a switch to a default activation state. Conceptually, the reversal of this feature would allow signaling pathways to robustly repress target gene expression in the presence of the signal, in a context-dependent manner, while activating the target gene in its absence.

Several target genes including *fz2*, *svb*, *rho*, *Ultrabithorax* (*Ubx*), *stripe* (*sr*), and *dpp* have been proposed to be directly repressed by Wg signaling. However, in the cases of *fz2*, *svb*, and *rho* (Cadigan et al. 1998; Payre et al. 1999; Sanson et al. 1999), the molecular studies addressing the *cis* and/or *trans* regulatory elements involved in transcriptional repression are lacking, while for *Ubx*, it is thought that repression of gene transcription occurs indirectly and is therefore a secondary effect of Wg signaling (Waltzer et al. 2001). Only in the cases of *sr* in the embryonic epidermis (Piepenburg et al. 2000) and *dpp* in the leg imaginal disc (Theisen et al. 2007), has it been convincingly shown that these genes are, in fact, repressed by Tcf in the presence of signaling. Both of these genes contain functional Tcf-binding sites in their response elements that are required for repression, and mutation of these sites results in a failure of Wg-mediated gene silencing.

In a recent study, several genes that are repressed by Wg signaling were identified from cultured *Drosophila* hemocytic cells. Surprisingly, the characterization of the *cis* regulatory element of one of these genes, Ugt36Bc, revealed that the Tcf recognition site is markedly different from a typical consensus Tcf binding site. Furthermore, the novel Tcf binding sites were not only required for Wg-induced repression, but were also essential for the default transcription of Ugt36Bc in the absence of signaling (Blauwkamp et al. 2008). This suggests that the nature of the Tcf binding site within the Wg-responsive element can determine the nature of the transcriptional output, both in the absence and presence of signaling, and is one possible mechanism through which Wg signaling can switch from activation to repression of certain genes.

CONCLUDING REMARKS

Over the last 30 years, the use of genetic analyses in *Drosophila* has elucidated both the function, in various contexts of development, and the molecular mechanism of the Wg pathway. In addition, the study of Wnt/Wg proteins first illustrated the relationship between normal development and oncogenesis. Owing to the vast array of genetic techniques available in *Drosophila* and its high degree of conservation, no doubt, novel components and Wg-regulated processes identified in future studies using this model system are likely to be directly applicable to vertebrate development.

ACKNOWLEDGMENTS

We gratefully acknowledge support from a Natural Sciences and Engineering Research Council of Canada Doctoral Research Award (to S.S.) and a Canadian Institutes for Health Research operating grant (to E.M.V.). We apologize to those colleagues whose work was not cited because of length restrictions.

REFERENCES

Affolter M, Pyrowolakis G, Weiss A, Basler K. 2008. Signal-induced repression: The exception or the rule in developmental signaling? *Dev Cell* **15:** 11–22.

Alexandre C, Lecourtois M, Vincent J. 1999. Wingless and Hedgehog pattern *Drosophila* denticle belts by regulating the production of short-range signals. *Development* **126:** 5689–5698.

Babu P. 1977. Early developmental subdivisions of the wing disk in *Drosophila*. *Mol Gen Genet* **151:** 289–294.

Baeg GH, Lin X, Khare N, Baumgartner S, Perrimon N. 2001. Heparan sulfate proteoglycans are critical for the organization of the extracellular distribution of Wingless. *Development* **128:** 87–94.

Baeg GH, Selva EM, Goodman RM, Dasgupta R, Perrimon N. 2004. The Wingless morphogen gradient is established by the cooperative action of Frizzled and Heparan Sulfate Proteoglycan receptors. *Dev Biol* **276:** 89–100.

Baena-Lopez LA, Franch-Marro X, Vincent JP. 2009. Wingless promotes proliferative growth in a gradient-independent manner. *Sci Signal* **2:** ra60.

Baker NE. 1987. Molecular cloning of sequences from *wingless*, a segment polarity gene in *Drosophila*: The spatial distribution of a transcript in embryos. *EMBO J* **6:** 1765–1773.

Baker NE. 1988a. Embryonic and imaginal requirements for *wingless*, a segment-polarity gene in *Drosophila*. *Dev Biol* **125:** 96–108.

Baker NE. 1988b. Transcription of the segment-polarity gene *wingless* in the imaginal discs of *Drosophila*, and the phenotype of a pupal-lethal *wg* mutation. *Development* **102:** 489–497.

Baonza A, Roch F, Martin-Blanco E. 2000. DER signaling restricts the boundaries of the wing field during *Drosophila* development. *Proc Natl Acad Sci* **97:** 7331–7335.

Barolo S, Stone T, Bang AG, Posakony JW. 2002. Default repression and Notch signaling: Hairless acts as an adaptor to recruit the corepressors Groucho and dCtBP to Suppressor of Hairless. *Genes Dev* **16:** 1964–1976.

Bejsovec A. 2006. Flying at the head of the pack: Wnt biology in *Drosophila*. *Oncogene* **25:** 7442–7449.

Bejsovec A, Martinez Arias A. 1991. Roles of *wingless* in patterning the larval epidermis of *Drosophila*. *Development* **113:** 471–485.

Bejsovec A, Wieschaus E. 1993. Segment polarity gene interactions modulate epidermal patterning in *Drosophila* embryos. *Development* **119:** 501–517.

Bienz M. 1994. Homeotic genes and positional signalling in the *Drosophila* viscera. *Trends Genet* **10:** 22–26.

Blauwkamp TA, Chang MV, Cadigan KM. 2008. Novel TCF-binding sites specify transcriptional repression by Wnt signalling. *EMBO J* **27:** 1436–1446.

Bornemann DJ, Duncan JE, Staatz W, Selleck S, Warrior R. 2004. Abrogation of heparan sulfate synthesis in *Drosophila* disrupts the Wingless, Hedgehog and Decapentaplegic signaling pathways. *Development* **131:** 1927–1938.

Brunner E, Peter O, Schweizer L, Basler K. 1997. *pangolin* encodes a Lef-1 homologue that acts downstream of Armadillo to transduce the Wingless signal in *Drosophila*. *Nature* **385:** 829–833.

Cabrera CV, Alonso MC, Johnston P, Phillips RG, Lawrence PA. 1987. Phenocopies induced with antisense RNA identify the *wingless* gene. *Cell* **50:** 659–663.

Cadigan KM, Fish MP, Rulifson EJ, Nusse R. 1998. Wingless repression of Drosophila frizzled 2 expression shapes the Wingless morphogen gradient in the wing. *Cell* **93:** 767–777.

Campbell G, Weaver T, Tomlinson A. 1993. Axis specification in the developing *Drosophila* appendage: The role of *wingless*, *decapentaplegic*, and the homeobox gene *aristaless*. *Cell* **74:** 1113–1123.

Chen CM, Struhl G. 1999. Wingless transduction by the Frizzled and Frizzled2 proteins of *Drosophila*. *Development* **126:** 5441–5452.

Cohen SM. 1990. Specification of limb development in the *Drosophila* embryo by positional cues from segmentation genes. *Nature* **343:** 173–177.

Cohen B, Simcox AA, Cohen SM. 1993. Allocation of the thoracic imaginal primordia in the *Drosophila* embryo. *Development* **117:** 597–608.

Couso JP, Bate M, Martinez-Arias A. 1993. A *wingless*-dependent polar coordinate system in *Drosophila* imaginal discs. *Science* **259:** 484–489.

Couso JP, Bishop SA, Martinez-Arias A. 1994. The *wingless* signalling pathway and the patterning of the wing margin in *Drosophila*. *Development* **120:** 621–636.

Desbordes SC, Chandraratna D, Sanson B. 2005. A screen for genes regulating the Wingless gradient in *Drosophila* embryos. *Genetics* **170:** 749–766.

DiNardo S, Sher E, Heemskerk-Jongens J, Kassis JA, O'Farrell PH. 1988. Two-tiered regulation of spatially patterned *engrailed* gene expression during *Drosophila* embryogenesis. *Nature* **332:** 604–609.

DiNardo S, Heemskerk J, Dougan S, O'Farrell PH. 1994. The making of a maggot: Patterning the *Drosophila* embryonic epidermis. *Curr Opin Genet Dev* **4:** 529–534.

Dougan S, DiNardo S. 1992. *Drosophila wingless* generates cell type diversity among *engrailed* expressing cells. *Nature* **360:** 347–350.

Dubois L, Lecourtois M, Alexandre C, Hirst E, Vincent JP. 2001. Regulated endocytic routing modulates Wingless signaling in *Drosophila* embryos. *Cell* **105:** 613–624.

Golembo M, Raz E, Shilo BZ. 1996. The *Drosophila* embryonic midline is the site of Spitz processing, and induces activation of the EGF receptor in the ventral ectoderm. *Development* **122:** 3363–3370.

Gonzalez F, Swales L, Bejsovec A, Skaer H, Martinez Arias A. 1991. Secretion and movement of Wingless protein in the epidermis of the *Drosophila* embryo. *Mech Dev* **35:** 43–54.

Gritzan U, Hatini V, DiNardo S. 1999. Mutual antagonism between signals secreted by adjacent Wingless and Engrailed cells leads to specification of complementary regions of the *Drosophila* parasegment. *Development* **126:** 4107–4115.

Han C, Yan D, Belenkaya TY, Lin X. 2005. *Drosophila* glypicans Dally and Dally-like shape the extracellular Wingless morphogen gradient in the wing disc. *Development* **132:** 667–679.

Hidalgo A, Ingham P. 1990. Cell patterning in the *Drosophila* segment: Spatial regulation of the segment polarity gene *patched*. *Development* **110:** 291–301.

Hoppler S, Bienz M. 1995. Two different thresholds of *wingless* signalling with distinct developmental consequences in the *Drosophila* midgut. *EMBO J* **14:** 5016–5026.

Immerglück K, Lawrenc PA, Bienz M. 1990. Induction across germ layers in *Drosophila* mediated by a genetic cascade. *Cell* **62:** 261–268.

Ingham PW, Martinez Arias A. 1992. Boundaries and fields in early embryos. *Cell* **68:** 221–235.

Johnston LA, Edgar BA. 1998. Wingless and Notch regulate cell-cycle arrest in the developing *Drosophila* wing. *Nature* **394:** 82–84.

Jürgens G, Wieschaus E, Nüsslein-Volhard C, Kluding H. 1984. Mutations affecting the pattern of the larval cuticle in *Drosophila melanogaster* ctII: Zygotic loci on the third chromosome. *Rouxs Arch Dev Biol* **193:** 283–295.

Katanaev VL, Solis GP, Hausmann G, Buestorf S, Katanayeva N, Schrock Y, Stuermer CA, Basler K. 2008. Reggie-1/flotillin-2 promotes secretion of the long-range signalling forms of Wingless and Hedgehog in *Drosophila*. *EMBO J* **27:** 509–521.

Kim J, Sebring A, Esch J, Kraus M, Vorwerk K, Magee J, Carroll S. 1996. Integration of positional signals and regulation of wing formation and identity by *Drosophila vestigial* gene. *Nature* **382:** 133–138.

Kirkpatrick CA, Dimitroff BD, Rawson JM, Selleck SB. 2004. Spatial regulation of Wingless morphogen distribution and signaling by Dally-like protein. *Dev Cell* **7:** 513–523.

Klein T, Arias AM. 1998a. Different spatial and temporal interactions between *Notch*, *wingless*, and *vestigial* specify proximal and distal pattern elements of the wing in *Drosophila*. *Dev Biol* **194:** 196–212.

Klein T, Arias AM. 1998b. Interactions among Delta, Serrate and Fringe modulate Notch activity during Drosophila wing development. *Development* **125:** 2951–2962.

Klein T, Arias AM. 1999. The *vestigial* gene product provides a molecular context for the interpretation of signals during the development of the wing in *Drosophila*. *Development* **126:** 913–925.

Lawrence PA, Sanson B, Vincent JP. 1996. Compartments, *wingless* and *engrailed*: Patterning the ventral epidermis of *Drosophila* embryos. *Development* **122:** 4095–4103.

Lawrence N, Dearden P, Hartley D, Roose J, Clevers H, Arias AM. 2000. dTcf antagonises Wingless signalling during the development and patterning of the wing in *Drosophila*. *Int J Dev Biol* **44:** 749–756.

Lecourtois M, Alexandre C, Dubois L, Vincent JP. 2001. Wingless capture by Frizzled and Frizzled2 in *Drosophila* embryos. *Dev Biol* **235:** 467–475.

Lee JJ, von Kessler DP, Parks S, Beachy PA. 1992. Secretion and localized transcription suggest a role in positional signaling for products of the segmentation gene *hedgehog*. *Cell* **71:** 33–50.

Lin X, Perrimon N. 1999. Dally cooperates with *Drosophila* Frizzled 2 to transduce Wingless signalling. *Nature* **400:** 281–284.

Martinez-Arias A. 1993. Development and patterning of the larval epidermis of *Drosophila*. In *The development of* Drosophila melanogaster (ed. M Bate, A Martinez-Arias), pp. 517–608. Cold Spring Harbor Laboratory Press, Cold Spring Harbor, NY.

Martinez Arias A, Baker NE, Ingham PW. 1988. Role of segment polarity genes in the definition and maintenance of cell states in the *Drosophila* embryo. *Development* **103:** 157–170.

Milan M, Cohen SM. 2000. Temporal regulation of apterous activity during development of the *Drosophila* wing. *Development* **127:** 3069–3078.

Mohler J, Vani K. 1992. Molecular organization and embryonic expression of the *hedgehog* gene involved in cell–cell communication in segmental patterning of *Drosophila*. *Development* **115:** 957–971.

Moline MM, Southern C, Bejsovec A. 1999. Directionality of Wingless protein transport influences epidermal patterning in the *Drosophila* embryo. *Development* **126:** 4375–4384.

Morata G, Lawrence PA. 1977. The development of *wingless*, a homeotic mutation of *Drosophila*. *Dev Biol* **56:** 227–240.

Mulligan K, Fuerer C, Ching W, Fish M, Willert K, Nusse R. 2012. Secreted Wingless-interacting molecule (Swim) promotes long-range signaling by maintaining Wingless solubility. *Proc Natl Acad Sci* **109:** 370–377.

Neumann CJ, Cohen SM. 1996. Distinct mitogenic and cell fate specification functions of *wingless* in different regions of the wing. *Development* **122:** 1781–1789.

Neumann CJ, Cohen SM. 1997. Long-range action of Wingless organizes the dorsal–ventral axis of the *Drosophila* wing. *Development* **124:** 871–880.

Ng M, Diaz-Benjumea FJ, Vincent JP, Wu J, Cohen SM. 1996. Specification of the wing by localized expression of Wingless protein. *Nature* **381:** 316–318.

Noordermeer J, Johnston P, Rijsewijk F, Nusse R, Lawrence P. 1992. The consequences of ubiquitous expression of the *wingless* gene in the *Drosophila* embryo. *Development* **116:** 711–719.

Noordermeer J, Klingensmith J, Perrimon N, Nusse R. 1994. *dishevelled* and *armadillo* act in the Wingless signalling pathway in *Drosophila*. *Nature* **367:** 80–83.

Nusse R, Varmus HE. 1982. Many tumors induced by the mouse mammary tumor virus contain a provirus integrated in the same region of the host genome. *Cell* **31:** 99–109.

Nusse R, van Ooyen A, Cox D, Fung YK, Varmus H. 1984. Mode of proviral activation of a putative mammary oncogene (*int-1*) on mouse chromosome 15. *Nature* **307:** 131–136.

Nusslein-Volhard C, Wieschaus E. 1980. Mutations affecting segment number and polarity in *Drosophila*. *Nature* **287:** 795–801.

Nusslein-Volhard C, Wiechaus E, Kluding H. 1984. Mutations affecting the pattern of the larval cuticle in *Drosophila melanogaster*. I. Zygotic loci on the second chromosome. *Roux's Arch Dev Biol* **193:** 267–282.

O'Brochta DA, Bryant PJ. 1985. A zone of non-proliferating cells at a lineage restriction boundary in *Drosophila*. *Nature* **313** 138–141.

O'Keefe L, Dougan ST, Gabay L, Raz E, Shilo BZ, DiNardo S. 1997. Spitz and Wingless, emanating from distinct borders, cooperate to establish cell fate across the Engrailed domain in the *Drosophila* epidermis. *Development* **124:** 4837–4845.

Panakova D, Sprong H, Marois E, Thiele C, Eaton S. 2005. Lipoprotein particles are required for Hedgehog and Wingless signalling. *Nature* **435:** 58–65.

Payre F, Vincent A, Carreno S. 1999. *ovo/svb* integrates Wingless and DER pathways to control epidermis differentiation. *Nature* **400:** 271–275.

Peifer M, Sweeton D, Casey M, Wieschaus E. 1994. *wingless* signal and Zeste-white 3 kinase trigger opposing changes in the intracellular distribution of Armadillo. *Development* **120:** 369–380.

Perrimon N. 1994. The genetic basis of patterned baldness in *Drosophila*. *Cell* **76:** 781–784.

Perrimon N, Engstrom L, Mahowald A. 1989. Zygotic lethals with specific maternal effect phenotypes in *Drosophila melanogaster* I loci on the X chromosome. *Genetics* **121:** 333–352.

Pfeiffer S, Alexandre C, Calleja M, Vincent JP. 2000. The progeny of *wingless*-expressing cells deliver the signal at a distance in *Drosophila* embryos. *Curr Biol* **10:** 321–324.

Pfeiffer S, Ricardo S, Manneville JB, Alexandre C, Vincent JP. 2002. Producing cells retain and recycle Wingless in *Drosophila* embryos. *Curr Biol* **12:** 957–962.

Phillips RG, Whittle JR. 1993. *wingless* expression mediates determination of peripheral nervous system elements in late stages of *Drosophila* wing disc development. *Development* **118:** 427–438.

Piddini E, Marshall F, Dubois L, Hirst E, Vincent JP. 2005. Arrow (LRP6) and Frizzled2 cooperate to degrade Wingless in *Drosophila* imaginal discs. *Development* **132:** 5479–5489.

Piepenburg O, Vorbruggen G, Jackle H. 2000. *Drosophila* segment borders result from unilateral repression of Hedgehog activity by Wingless signaling. *Mol Cell* **6:** 203–209.

Reuter R, Panganiban GE, Hoffmann FM, Scott MP. 1990. Homeotic genes regulate the spatial expression of putative growth factors in the visceral mesoderm of *Drosophila* embryos. *Development* **110:** 1031–1040.

Riggleman B, Schedl P, Wieschaus E. 1990. Spatial expression of the *Drosophila* segment polarity gene *armadillo* is posttranscriptionally regulated by *wingless*. *Cell* **63:** 549–560.

Rijsewijk F, Schuermann M, Wagenaar E, Parren P, Weigel D, Nusse R. 1987. The *Drosophila* homolog of the mouse mammary oncogene *int-1* is identical to the segment polarity gene *wingless*. *Cell* **50:** 649–657.

Sampedro J, Johnston P, Lawrence PA. 1993. A role for *wingless* in the segmental gradient of *Drosophila*? *Development* **117:** 677–687.

Sanson B, Alexandre C, Fascetti N, Vincent JP. 1999. Engrailed and Hedgehog make the range of Wingless asymmetric in *Drosophila* embryos. *Cell* **98:** 207–216.

Schmidt-Ott U, Technau GM. 1992. Expression of *en* and *wg* in the embryonic head and brain of *Drosophila* indicates a refolded band of seven segment remnants. *Development* **116:** 111–125.

Schweizer L, Nellen D, Basler K. 2003. Requirement for Pangolin/dTCF in *Drosophila* Wingless signaling. *Proc Natl Acad Sci* **100:** 5846–5851.

Sharma RP, Chopra VL. 1976. Effect of the *wingless* (wg^1) mutation on wing and haltere development in *Drosophila melanogaster*. *Dev Biol* **48:** 461–465.

Siegfried E, Wilder EL, Perrimon N. 1994. Components of *wingless* signalling in *Drosophila*. *Nature* **367:** 76–80.

Simcox AA, Roberts IJH, Hersperger H, Gribbin MC, Shearn A, Whittle JRS. 1989. Imaginal discs can be recovered from cultured embryos mutant for the segment-polarity genes *engrailed*, *naked* and *patched* but not from *wingless*. *Development* **107:** 715–722.

Simcox AA, Grumbling G, Schnepp B, Bennington-Mathias C, Hersperger E, Shearn A. 1996. Molecular, phenotypic, and expression analysis of *vein*, a gene required for growth of the *Drosophila* wing disc. *Dev Biol* **177:** 475–489.

St. Johnston D, Nuesslein-Volhard C. 1992. The origin of pattern and polarity in the *Drosophila* embryo. *Cell* **68:** 201–219.

Strigini M, Cohen SM. 2000. Wingless gradient formation in the *Drosophila* wing. *Curr Biol* **10:** 293–300.

Szuts D, Freeman M, Bienz M. 1997. Antagonism between EGFR and Wingless signalling in the larval cuticle of *Drosophila*. *Development* **124:** 3209–3219.

Theisen H, Syed A, Nguyen BT, Lukacsovich T, Purcell J, Srivastava GP, Iron D, Gaudenz K, Nie Q, Wan FY, et al. 2007. Wingless directly represses DPP morphogen expression via an Armadillo/TCF/Brinker complex. *PLoS ONE* **2:** e142.

Thuringer F, Bienz M. 1993. Indirect autoregulation of a homeotic *Drosophila* gene mediated by extracellular signaling. *Proc Natl Acad Sci* **90:** 3899–3903.

van den Heuvel M, Nusse R, Johnston P, Lawrence PA. 1989. Distribution of the *wingless* gene product in *Drosophila* embryos: A protein involved in cell–cell communication. *Cell* **59:** 739–749.

van de Wetering M, Cavallo R, Dooijes D, van Beest M, van Es J, Loureiro J, Ypma A, Hursh D, Jones T, Bejsovec A, et al. 1997. Armadillo coactivates transcription driven by the product of the *Drosophila* segment polarity gene *dTCF*. *Cell* **88:** 789–799.

Waltzer L, Vandel L, Bienz M. 2001. Teashirt is required for transcriptional repression mediated by high Wingless levels. *EMBO J* **20:** 137–145.

Wang SH, Simcox A, Campbell G. 2000. Dual role for *Drosophila* epidermal growth factor receptor signaling in early wing disc development. *Genes Dev* **14:** 2271–2276.

Wiellette EL, Harding KW, Mace KA, Ronshaugen MR, Wang FY, McGinnis W. 1999. *spen* encodes an RNP motif protein that interacts with Hox pathways to repress the development of head-like sclerites in the *Drosophila* trunk. *Development* **126:** 5373–5385.

Wieschaus E, Nüsslein-Volhard C, Jürgens G. 1984. Mutations affecting the pattern of larval cuticle in *Drosophila melanogaster* III: Zygotic loci on the X-chromosome and fourth chromosome. *Rouxs Arch Dev Biol* **193:** 296–307.

Williams JA, Paddock SW, Carroll SB. 1993. Pattern formation in a secondary field: A hierarchy of regulatory genes subdivides the developing *Drosophila* wing disc into discrete subregions. *Development* **117:** 571–584.

Williams JA, Paddock SW, Vorwerk K, Carroll SB. 1994. Organization of wing formation and induction of a wing-patterning gene at the dorsal/ventral compartment boundary. *Nature* **368:** 299–305.

Wodarz A, Nusse R. 1998. Mechanisms of Wnt signaling in development. *Annu Rev Cell Dev Biol* **14:** 59–88.

Zecca M, Struhl G. 2007a. Control of *Drosophila* wing growth by the vestigial quadrant enhancer. *Development* **134:** 3011–3020.

Zecca M, Struhl G. 2007b. Recruitment of cells into the *Drosophila* wing primordium by a feed-forward circuit of vestigial autoregulation. *Development* **134:** 3001–3010.

Zecca M, Basler K, Struhl G. 1996. Direct and long-range action of a wingless morphogen gradient. *Cell* **87:** 833–844.

β-Catenin-Dependent Wnt Signaling in *C. elegans*: Teaching an Old Dog a New Trick

Belinda M. Jackson[1] and David M. Eisenmann[2]

[1]Department of Medicine, Uniformed Services University of the Health Sciences, Bethesda, Maryland 20814

[2]Department of Biological Sciences, University of Maryland Baltimore County, Baltimore, Maryland 21250

Correspondence: eisenman@umbc.edu

Wnt signaling is an evolutionarily ancient pathway used to regulate many events during metazoan development. Genetic results from *Caenorhabditis elegans* more than a dozen years ago suggested that Wnt signaling in this nematode worm might be different than in vertebrates and *Drosophila*: the worm had a small number of Wnts, too many β-catenins, and some Wnt pathway components functioned in an opposite manner than in other species. Work over the ensuing years has clarified that *C. elegans* does possess a canonical Wnt/β-catenin signaling pathway similar to that in other metazoans, but that the majority of Wnt signaling in this species may proceed via a variant Wnt/β-catenin signaling pathway that uses some new components (mitogen-activated protein kinase signaling enzymes), and in which some conserved pathway components (β-catenin, T-cell factor [TCF]) are used in new and interesting ways. This review summarizes our current understanding of the canonical and novel TCF/β-catenin-dependent signaling pathways in *C. elegans*.

Wnts are secreted glycoproteins that function as extracellular signals in a wide group of metazoans, including cnidarians, nematodes, insects, and vertebrates. This evolutionary conservation suggests that cellular signaling mechanisms functioning in response to Wnts were part of the "developmental toolkit" in place at least 500 million years ago in the common ancestor to modern metazoans (Guder et al. 2006). Signaling regulated by Wnt ligand binding plays an important and often essential role in an astounding number of processes during invertebrate and vertebrate development, as well as functioning in the maintenance of adult stem cell populations in vertebrates and *Drosophila* (reviewed in Clevers 2006; Cadigan and Peifer 2009; MacDonald et al. 2009). In the nematode *Caenorhabditis elegans*, Wnt signaling is necessary for many events during both embryonic and larval development, and regulates many processes including cell migration, cell polarity, fate specification, axon guidance, synaptogenesis, and stem cell-like renewal (see Eisenmann 2005; Mizumoto and Sawa 2007b; Van Hoffelen and Herman 2008). Intriguingly, although *C. elegans* uses a Wnt/β-catenin or "canonical" signaling pathway similar to that in other metazoans, it also has a second, novel Wnt signaling pathway that uses additional β-catenins to effect transcriptional up-regulation of target genes by a mechanism distinct from that in the common Wnt pathway (Mizumoto and Sawa 2007b;

Cite this article as *Cold Spring Harb Perspect Biol* doi: 10.1101/cshperspect.a007948

Phillips and Kimble 2009). Here we summarize our current understanding of these two *C. elegans* signaling pathways and the ways they are similar to, and diverge from, Wnt signaling pathways in other species. Although recent work has extended our knowledge of Wnt signaling to other nematode species (Zheng et al. 2005; Rudel et al. 2008; Tian et al. 2008; Lin et al. 2009; Owraghi et al. 2010; Seetharaman et al. 2010; Penigault and Felix 2011a; Wang and Sommer 2011), in this review we will concern ourselves with a survey of Wnt signaling in the established model system *C. elegans*.

THE CONSERVED METAZOAN WNT/β-CATENIN SIGNALING PATHWAY

The best-characterized signaling pathways acting downstream from Wnts in vertebrates and *Drosophila* are the Wnt/β-catenin (i.e., canonical) pathway (MacDonald et al. 2009) and the β-catenin-independent Wnt/PCP (planar cell polarity) pathway (Axelrod 2009). As the vast majority of Wnt signaling during *C. elegans* development involves β-catenin-dependent signaling, in this review we will consider only the Wnt/β-catenin signaling pathway elucidated from vertebrate and *Drosophila* studies (Fig. 1; pathway function reviewed in Cadigan and Peifer 2009; MacDonald et al. 2009).

The main downstream effector of this pathway is a heterodimeric factor composed of the transcription activator β-catenin and a member of the T-cell factor/lymphoid enhancer factor (TCF/LEF) family of DNA-binding proteins. In the absence of ligand, the TCF/LEF protein is bound at target genes, where it interacts with transcriptional repressor proteins to keep gene expression low (Fig. 1A). Meanwhile, any β-catenin not located in the cadherin/catenin cell adhesion complex at the plasma membrane is acted on by the "destruction complex," which contains the core scaffolding components adenomatous polyposis coli protein (APC) and Axin and the kinases glycogen synthase kinase 3β (GSK3β) and casein kinase 1α (CK1α). The destruction complex phosphorylates β-catenin at conserved amino-terminal sites, leading to its ubiquitination and degradation by the proteasome.

Activation of the Wnt/β-catenin pathway is initiated when lipid-modified, secreted Wnt ligand binds to a coreceptor consisting of a member of the seven-transmembrane Frizzled receptor family and one of the single-transmembrane low-density lipoprotein receptor-related proteins, LRP5 or LRP6 (Fig. 1B). Ligand binding leads to phosphorylation of the LRP intracellular domain, the recruitment of Axin to the coreceptor, and the inactivation of the destruction complex. Hypophosphorylated cytoplasmic β-catenin is not degraded, and it translocates into the nucleus, where it heterodimerizes with TCF/LEF proteins to activate expression of Wnt-responsive target genes.

WNT PATHWAY COMPONENTS IN *C. elegans*

The *C. elegans* genome contains a complement of core Wnt pathway components similar to those in flies and vertebrates, with some interesting differences (see Table 1) (Ruvkun and Hobert 1998; Eisenmann 2005; Hardin and King 2008).

Table 1. *C. elegans* Wnt pathway components

Component	*C. elegans* homolog
Porcupine	MOM-1
Wntless	MIG-14
Retromer complex	VPS-26, VPS-29, VPS-35, others
Wnt	EGL-20, LIN-44, MOM-2, CWN-1, CWN-2
SFRP	SFRP-1
Frizzled	LIN-17, MOM-5, MIG-1, CFZ-2
LRP5/6	??
Ryk/Derailed	LIN-18
Ror	CAM-1
Dsh	MIG-5, DSH-1, DSH-2
GSK3β	GSK-3
CK1	KIN-19
Axin	PRY-1, AXL-1
APC	APR-1
β-TrCP/F-box protein	LIN-23, others
β-catenin	HMP-2, BAR-1, SYS-1, WRM-1
TCF	POP-1
Groucho/TLE	UNC-37

 Cite this article as *Cold Spring Harb Perspect Biol* doi: 10.1101/cshperspect.a007948

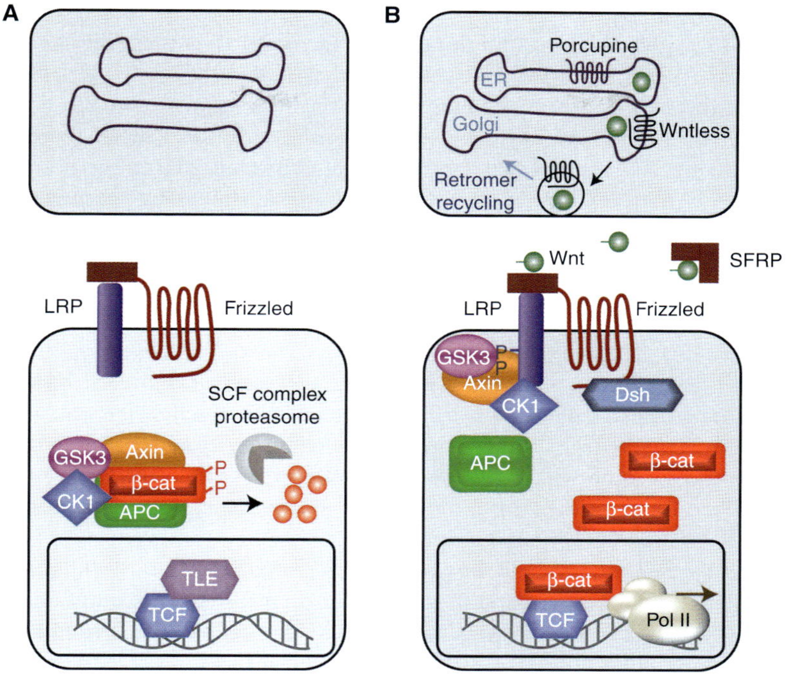

Figure 1. The conserved Wnt/β-catenin-dependent signaling pathway. Wnt-sending cells are shown above, Wnt-receiving cells below. (*A*) In the absence of signal, the destruction complex (including Axin, APC, GSK3β, and CK1α) phosphorylates β-catenin, targeting it for ubiquitination and destruction by the proteasome. The transcription factor TCF interacts with repressors (including transducing-like enhancers [TLE]) to keep target gene expression low. (*B*) After processing in the sending cell, some secreted Wnt binds to a Frizzled/LRP coreceptor, which leads to phosphorylation of the intracellular domain of LRP, inactivation of the destruction complex, and stabilization of β-catenin. After nuclear translocation β-catenin binds to TCF at Wnt target genes and up-regulates their expression. Some secreted Wnt ligand is sequestered by extracellular inhibitors such as SFRPs.

Wnt Ligands and Inhibitors: Reduced Complexity and Redundancy

The *C. elegans* genome contains only five genes encoding Wnt ligands: *lin-44*, *egl-20*, *mom-2*, *cwn-1*, and *cwn-2* (Shackleford et al. 1993; Herman et al. 1995; Rocheleau et al. 1997; Thorpe et al. 1997; Maloof et al. 1999), a lower number than in other animals. It may have once been assumed this small number was due to some "lower" evolutionary status of nematodes, but genome sequencing of diverse animal species now suggests that the early metazoan common ancestor had a large complement of Wnt genes and that nematode lineages have lost many of these genes (Guder et al. 2006; van Amerongen and Nusse 2009; Janssen et al. 2010; Pang et al. 2010). Surprisingly, only one of the *C. elegans* Wnts, MOM-2, appears to play a major role during embryogenesis: ~70% of *mom*-2 zygotic mutant animals die during embryogenesis, whereas <5% of *lin-44; cwn-1; cwn-2* or *lin-44; cwn-1; egl-20* mutants display embryonic lethality (Thorpe et al. 1997; Gleason et al. 2006). This result is surprising given the apparent reiterated use of Wnt signaling in a large number of asymmetric cell divisions during embryonic development (Park and Priess 2003; Huang et al. 2007). However, both *cwn-1* and *cwn-2* mutations enhance the *mom*-2 lethal phenotype, and the *cwn-1; cwn-2; mom*-2 triple mutant shows 100% lethality, indicating that there is some functional redundancy of Wnts in embryonic development (Gleason et al. 2006).

All five Wnts are also expressed during larval life in discrete locations along the anterior–posterior axis: The Wnts LIN-44, EGL-20, and CWN-1 are expressed in a nested pattern in the posterior, whereas CWN-2 is expressed in more anterior cells (Fig. 2) (Herman et al. 1995; Whangbo and Kenyon 1999; Coudreuse et al. 2006; Gleason et al. 2006; Pan et al. 2006; Kennerdell et al. 2009; Song et al. 2010; Harterink et al. 2011; Yamamoto et al. 2011). Wnts function in several developmental processes along the anterior–posterior axis during larval development, functioning in cell-fate specification, neuronal migration, and neuronal cell polarity (Eisenmann 2005). For several of these larval Wnt-dependent processes, the Wnt genes also display functional redundancy, such that strong phenotypes are only apparent when multiple Wnt genes are mutated (Zinovyeva and Forrester 2005; Gleason et al. 2006; Hilliard and Bargmann 2006; Pan et al. 2006; Prasad and Clark 2006; Green et al. 2008; Yamamoto et al. 2011). This indicates that while in some *C. elegans* processes, distinct Wnt and Wnt receptor pairings may be used; in other processes, the identity of individual Wnts may not be as important as the reception of a threshold amount of Wnt signal. This redundancy may provide a mechanism for developmental robustness in these processes.

This nested pattern of Wnt expression in the larva (which is set up in the late embryo [Harterink et al. 2011]) is intriguing because Wnt signaling is known to regulate the expression of three *C. elegans* Hox genes along the larval anterior–posterior axis. It has been proposed that Wnt signaling may have been used as an ancestral axis specification mechanism with patterning by Hox genes elaborated on top of this more ancestral mechanism (Guder et al. 2006; Ryan and Baxevanis 2007; Martin and Kimelman 2009; Petersen and Reddien 2009; Niehrs 2010). Further experiments will need to be performed to determine if specific Hox gene expression along the axis in *C. elegans* is patterned by the anterior–posterior Wnt gradient.

C. elegans expresses homologs of the Porcupine (MOM-1) and Wntless (MIG-14) proteins, which are required for modification and secretion of Wnts (Rocheleau et al. 1997; Banziger et al. 2006), and of the retromer complex, which acts in the recycling of Wntless in Wnt-sending cells (Belenkaya et al. 2008; Pan et al. 2008; Yang et al. 2008). Mutations affecting the *C. elegans* genes encoding Porcupine, Wntless, and retromer complex proteins lead to phenotypes in several processes that are consistent with a reduction in Wnt signaling (Rocheleau et al. 1997; Thorpe et al. 1997; Nishiwaki 1999; Eisenmann

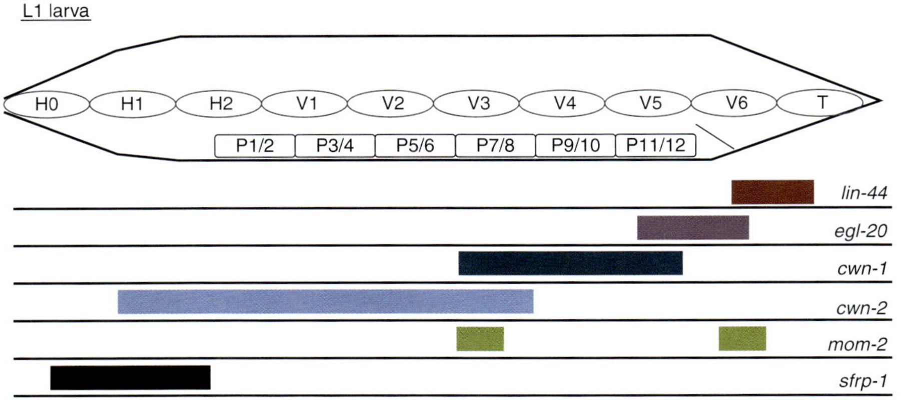

Figure 2. Expression of *C. elegans* Wnt ligands and a Wnt inhibitor along the anterior–posterior axis. (*Top*) Diagram of a newly hatched L1 larva; anterior is to the left, dorsal is up. The lateral hypodermal seam cells (H0–H2, V1–V6, T) and the ventral hypodermal cells (P1/2–P11/12) on the left side are shown. (*Bottom*) Domains of transcript accumulation along the anterior–posterior axis for each of the five Wnt genes and the Wnt inhibitor *sfrp-1* at this time in development. (Data from Harterink et al. 2011.)

and Kim 2000; Coudreuse et al. 2006; Prasad and Clark 2006; Pan et al. 2008; Yang et al. 2008).

In vertebrates a number of secreted Wnt inhibitors are known, which function to modulate response to Wnts (MacDonald et al. 2009). Until recently there was no strong genetic or genome sequence evidence for the presence of such factors in *C. elegans*, and the only known inhibitor of Wnt ligands was the receptor CAM-1 (see below). Recently, a single gene encoding a secreted Frizzled-related protein (SFRP-1) has been identified and shown to be expressed in anterior regions of the worm early in larval development (Fig. 2) (Harterink et al. 2011). In *sfrp-1* null mutants (which are viable), several neurons show overmigration phenotypes in which they migrate farther into the anterior than in wild type, and reduction of function of Wntless, the retromer complex, or Wnts themselves can suppress these phenotypes (e.g., Fig. 3E,F). These data suggest that this single secreted Wnt inhibitor likely diffuses from the anterior to bind Wnt ligands and sharpen the Wnt anterior–posterior gradient that controls these processes.

Wnt Receptors: No Coreceptor?

C. elegans has four genes encoding Frizzled receptors, *lin-17*, *mom-5*, *mig-1*, and *cfz-2* (Sawa et al. 1996; Rocheleau et al. 1997; Ruvkun and Hobert 1998; Zhao et al. 2003), and a single Ryk/Derailed receptor homolog encoded by *lin-18* (Inoue et al. 2004), which appear to function

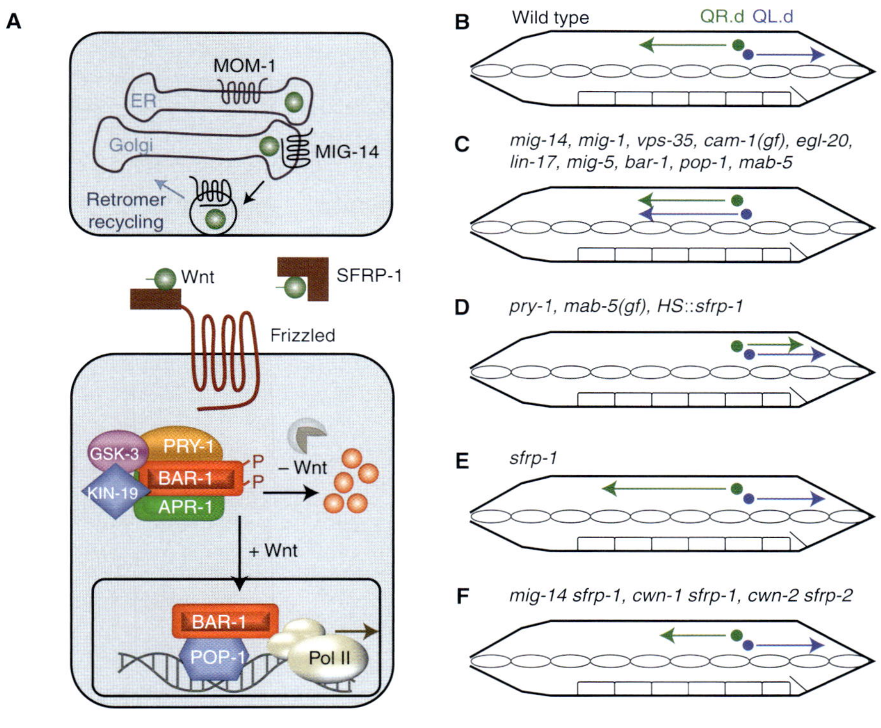

Figure 3. The *C. elegans* Wnt/BAR-1 canonical (WBC) signaling pathway. (*A*) The canonical Wnt signaling pathway in *C. elegans* that uses the β-catenin BAR-1. Components are drawn similar to those in Figure 1, with *C. elegans* protein names given. (*B–F*) Diagrams show an L1 larval stage animal as in Figure 2. The direction of migration of the descendants of the QR (on the right side) and QL (on the left side) neuroblasts are shown in wild type (*B*) and various mutant animals (*C–F*). *cam-1(gf)* refers to extra copies of the wild-type *cam-1* gene; *mab-5(gf)* indicates a mutation causing ectopic expression of *mab-5*; *HS::sfrp-1* indicates expression of *sfrp-1* from the heat shock promoter.

as Wnt receptors. As with the Wnt ligands, for some processes the receptors display genetic redundancy such that Wnt reduction-of-function phenotypes are only seen when multiple genes are mutated (Gleason et al. 2006; Pan et al. 2006; Green et al. 2008; Zinovyeva et al. 2008).

One of the most curious differences between Wnt pathway function in *C. elegans* and other species is the lack of evidence for a required LRP family coreceptor. Although there are several low-density lipoprotein receptor–related proteins encoded in the worm genome with the domain architecture found in LRP5/6 and Arrow, none have been identified as functioning in a Wnt-mediated process in forward genetic screens, or by directed testing via RNA interference (RNAi) (D Eisenmann, unpubl.). This negative result must be tempered by the caveats of possible functional redundancy and embryonic lethality for genetic analysis, and the apparent rapid divergence of worm homolog sequences in the case of genomic analysis. However, should it hold true that the Wnt binding in *C. elegans* uses a single receptor, this would raise interesting questions about the mechanism of pathway activation in this species, given the role of the LRP cytoplasmic domain in pathway activation in other species.

In addition to the five receptors mentioned above, the *cam-1* gene encodes a Ror tyrosine kinase receptor that functions as a negative Wnt ligand "sink" in several postembryonic processes; only the extracellular cysteine-rich domain is necessary for this activity (Kim and Forrester 2003; Forrester et al. 2004; Green et al. 2007, 2008). Recent results, however, suggest it may also function in a positive manner as a Wnt receptor in some processes (Hayashi et al. 2009; Kennerdell et al. 2009; Song et al. 2010; Yamamoto et al. 2011).

Disheveled

The worm genome encodes three Disheveled family members, MIG-5, DSH-1, and DSH-2, which function positively in Wnt signaling and which display genetic redundancy in several cases (Ruvkun and Hobert 1998; Walston et al. 2004; Eisenmann 2005; King et al. 2009).

Destruction Complex: Both Negative and Positive Roles in Wnt Signaling

Genes encoding the core β-catenin destruction complex components CK1α (*kin-19*) (Peters et al. 1999), GSK3β (*gsk-3*) (Schlesinger et al. 1999), Axin (*pry-1* and *axl-1*) (Maloof et al. 1999; Korswagen et al. 2002; Oosterveen et al. 2007), and APC (*apr-1*) (Rocheleau et al. 1997) are found in *C. elegans*. In the case of Axin and APC, the encoded proteins are fairly diverged from family members in other species, with several conserved domains absent. In the case of Axin, this low-level homology led to an initial declaration that the *C. elegans* genome did not encode an Axin homolog (Ruvkun and Hobert 1998). A newly discovered second *C. elegans* Axin homolog, AXL-1, is even less conserved than PRY-1, yet appears to be a functional Axin protein (Oosterveen et al. 2007). Interestingly, the function of *apr-1*, *gsk-3*, and *kin-19* appear to differ from process to process, acting positively in some cases (Rocheleau et al. 1997; Peters et al. 1999; Schlesinger et al. 1999). The early observation that the APC homolog APR-1 acts positively in the Wnt signaling pathway in the P2/EMS asymmetric cell division (see below) led to the idea that Wnt signaling in worms did not proceed via the canonical pathway (Han 1997). However, it is now clear that all of the destruction complex components do function as negative regulators of a canonical worm Wnt pathway in at least one process, vulval cell-fate specification (Gleason et al. 2002, 2006). The observation that the same proteins can function in opposite ways in different Wnt processes raises interesting questions about their substrates and function in those cases in which they act in a positive manner.

β-Catenins: The More the Merrier

Unique among sequenced metazoan genomes, *C. elegans* has four genes encoding β-catenins, *bar-1*, *hmp-2*, *sys-1*, and *wrm-1*; all but *wrm-1* were identified in forward genetic screens (Rocheleau et al. 1997; Costa et al. 1998; Eisenmann et al. 1998; Kidd et al. 2005). All four encoded proteins are more poorly conserved with β-catenins from other species than is commonly observed (Phillips and Kimble 2009); HMP-2 shows the highest identity to human β-catenin at ~30%. BAR-1, with slightly less identity, is the family member with most of the hallmarks of a stereotypical β-catenin: 12 Armadillo repeats, transcription activation domains, amino-terminal TCF-binding domain, conserved GSK-3 and CK1 phosphorylation sites in the amino terminus (Eisenmann et al. 1998; Natarajan et al. 2001), and an interaction with a ubiquitin ligase complex that regulates its ubiquitination and accumulation (Dreier et al. 2005). The last *C. elegans* β-catenin identified, SYS-1, is only 10% identical to human β-catenin, and was predicted to have only three Armadillo repeats, but was suggested to be a bona fide β-catenin based on genetic data (Siegfried et al. 2004; Kidd et al. 2005). Amazingly, the crystal structure of SYS-1 shows a protein with 12 Armadillo repeats and a structure very similar to that of vertebrate β-catenin (Liu et al. 2008). This example shows the strength of the genetic approach in model systems for identification of novel Wnt pathway components and regulators, and again illustrates the danger of relying solely on genome sequence analysis to identify homologous proteins.

One factor that may explain the lower conservation in the nematode β-catenins is that the dual functions of β-catenin in Wnt signaling and cell adhesion have been segregated among the four worm proteins. BAR-1 and SYS-1 interact with the TCF family member POP-1 as coactivators of gene expression in two different *C. elegans* Wnt pathways (Korswagen et al. 2000; Natarajan et al. 2001; Kidd et al. 2005). WRM-1 functions in a variant Wnt signaling pathway with SYS-1 and has acquired the novel function of regulation of TCF nuclear localization (described below). HMP-2 is the only protein with clear function in cell adhesion in a cadherin/catenin complex (Costa et al. 1998; Korswagen et al. 2000; Natarajan et al. 2001); however, recent work suggests it may also regulate Wnt signaling indirectly (Putzke and Rothman 2010; Sumiyoshi et al. 2011). Thus, by having to maintain only one of the several functions of β-catenin in other species, *C. elegans* β-catenin proteins may have been more free to diverge in sequence after gene duplication events occurred.

TCF: There Can Be Only One

In contrast to vertebrates, the *C. elegans* genome encodes only one TCF/LEF homolog, POP-1 (Lin et al. 1995), which appears to function in all Wnt-mediated processes in the worm, except those that do not involve a transcriptional response (e.g., mitotic spindle rotation) (Eisenmann 2005). Like other TCF proteins, POP-1 has an amino-terminal catenin-binding domain through which it interacts with β-catenins BAR-1 and SYS-1, and an HMG box DNA-binding domain (Korswagen et al. 2000; Natarajan et al. 2001; Kidd et al. 2005). It also possesses an extended carboxy-terminal tail (E tail) similar to that in alternatively spliced isoforms of human TCF1 and TCF4, which functions in auxiliary DNA-binding and protein–protein interactions (Atcha et al. 2003, 2007). Recent work suggests that this carboxy-terminal tail in POP-1 mediates the interaction with the β-catenin WRM-1 and results in phosphorylation of POP-1 and its nuclear export (Yang et al. 2011) (see below). As with other TCFs, POP-1 is known to function in both the repression of target genes in unsignaled cells and the up-regulation of gene expression upon pathway activation (Owraghi et al. 2010). One curious difference between POP-1 and other TCF family members is its DNA-binding site preference. Several in vitro and in vivo analyses of TCF family members indicate a preference for the core DNA site CTTTGWW (Arce et al. 2006; Hatzis et al. 2008; MacDonald et al. 2009). Although few bona fide POP-1 targets are known in *C. elegans*, it is clear that POP-1 also binds to sites with a T instead of a C nucleotide in the first position (e.g., TTTTGAT in *ceh-13* and *psa-3* [Streit et al. 2002; Arata et al. 2006] and TTTTGAA in *ceh-22* [Lam et al. 2006]). Based on the structure of LEF-1 bound to DNA, POP-1 differs from other TCF family members in amino acid sequence where contact is made at the first nucleotide of the core site (Love et al. 1995). Specifically, POP-1 has a glutamine residue where glutamic acid is found in other TCF family members. However, the substitution of these residues between TCF and POP-1 has no obvious effect on DNA-binding affinity (K Thompson and D Eisenmann, unpubl.). It is currently unclear what the structural basis for the more promiscuous DNA-binding preference of POP-1 is, or if other TCF family members may have a similar specificity.

TWO β-CATENIN-DEPENDENT WNT SIGNALING PATHWAYS IN *C. elegans*

The majority of Wnt-mediated events in *C. elegans* development use one of two signaling pathways. However unlike the two major Wnt pathways in vertebrates, the Wnt/β-catenin-dependent pathway and the Wnt/PCP pathway, both of the *C. elegans* pathways involve transcriptional regulation of target genes via a β-catenin/TCF complex (Eisenmann 2005; Phillips and Kimble 2009). One pathway functions mechanistically similarly to the evolutionarily ancient Wnt/β-catenin-dependent (canonical) pathway characterized in vertebrates and *Drosophila* (Figs. 1 and 3). The other pathway, which appears to play a more widespread role in *C. elegans* development, may be a nematode-specific variation on the canonical Wnt signaling pathway theme (Fig. 4).

THE WNT/BAR-1 CANONICAL (WBC) PATHWAY: OLD DOG

As mentioned above, homologs of most Wnt/β-catenin-dependent pathway components have been identified in *C. elegans* (Table 1), and genetic and molecular analysis indicates that in several Wnt-mediated processes these proteins function in a manner similar to their vertebrate and fly counterparts (Eisenmann 2005). In all of these processes, the β-catenin BAR-1 is uniquely used, and for this reason we term this pathway the Wnt/BAR-1 canonical (WBC) pathway (Fig. 3A). The *C. elegans* WBC pathway uses all of the positively acting core components of a canonical Wnt pathway (Wnt, Fz, Dsh, β-catenin, and TCF) except for LRP, and the homologs of the β-catenin destruction complex components (PRY-1/Axin, GSK-3/GSK3β, APR-1/APC, and KIN-19/CK1α) function as negative regulators of this pathway (Eisenmann 2005; Gleason et al. 2006). For example, in *pry-1/Axin* mutants, several Wnt-responsive Hox genes (*lin-39*, *mab-1*, and *egl-5*) are expressed outside their normal domains and mutant animals display phenotypes consistent with over- or ectopic activation of the Wnt pathway (Harris et al. 1996; Eisenmann et al. 1998; Maloof et al. 1999; Gleason et al. 2002, 2006; Howard and Sundaram 2002; Korswagen et al. 2002; Oosterveen et al. 2007). PRY-1 physically interacts with APR-1, GSK-3, and BAR-1 (Korswagen et al. 2002), presumably to phosphorylate BAR-1, which contains eight GSK-3 consensus phosphorylation sites (Eisenmann et al. 1998). Deletion of these sites results in a hyperactive form of the protein that causes Wnt pathway gain-of-function phenotypes when overexpressed (Eisenmann et al. 1998; Gleason et al. 2002, 2006; Korswagen et al. 2002). Therefore, these data suggest that a Wnt/β-catenin-dependent pathway similar to that characterized in other species also functions in *C. elegans* during larval development.

In three different larval developmental processes, WBC pathway activity is necessary to maintain Hox gene expression in Wnt-receiving cells: WBC activity maintains expression of the midbody Hox gene *lin-39* during vulval precursor cell-fate specification (Eisenmann et al. 1998; Penigault and Felix 2011b), and is believed to be required for expression of the posterior Hox gene *egl-5* during P12 fate specification (Jiang and Sternberg 1998; Eisenmann and Kim 2000; Li et al. 2009) and for expression of the mid/posterior Hox gene *mab-5* in the daughters of the QL neuroblast to mediate their proper cell migration (Harris et al. 1996; Maloof et al. 1999; Eisenmann and Kim 2000; Herman 2001; Korswagen et al. 2002). As an example of WBC pathway function in *C. elegans*, we consider this last process.

CANONICAL WNT SIGNALING REGULATES MIGRATION OF THE QL NEUROBLAST PROGENY

During the early L1 larval stage, a pair of neuroblasts are born in the posterior in equivalent positions on the left (QL) and on the right (QR) (Sulston and Horvitz 1977). Both cells begin with an anterior migration; however, the three QL neuroblast descendants (QL.d)

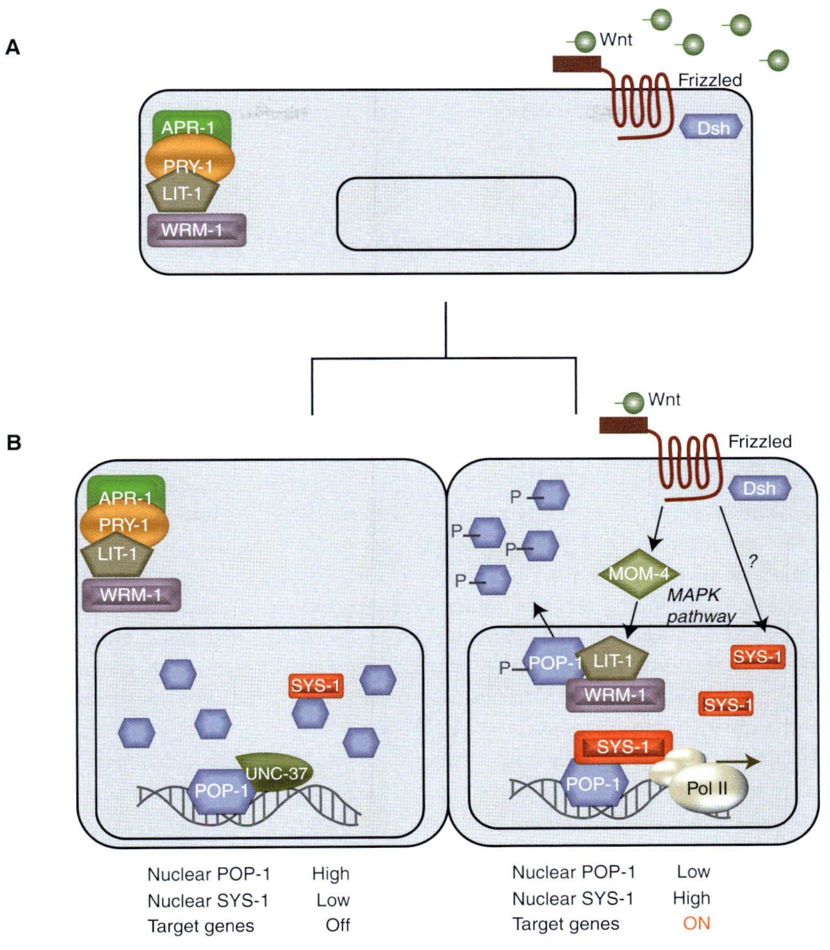

Figure 4. The *C. elegans* Wnt/β-catenin asymmetry (WBA) signaling pathway. (*A*) Asymmetric subcellular localization of Wnt pathway components in a mother cell before cell division. Frizzled and Disheveled proteins localize to the cortex on the side from which the Wnt signal is received, whereas PRY-1/Axin, APR-1/APC, WRM-1/β-catenin, and LIT-1/NLK proteins are found on the opposite side. Events in the nucleus are not shown. (*B*) Asymmetric pathway activation in the two daughters after cell division. In the daughter on the left, the Wnt and MAPK pathways are inactive, nuclear levels of POP-1/TCF are high, while nuclear levels of β-catenin SYS-1 are low. POP-1/TCF interacts with repressors to keep target gene expression low. In the daughter cell on the right, Wnt ligand activates WBA and MAPK pathways. The β-catenin WRM-1 interacts with LIT-1/NLK to phosphorylate POP-1/TCF, targeting it for nuclear export, thereby lowering the levels of nuclear POP-1. An unknown mechanism leads to an increase in nuclear levels of β-catenin SYS-1 in this daughter cell (question mark). POP-1 and SYS-1 interact to up-regulate expression of Wnt target genes (red rectangle). Unlabeled blue hexagons indicate high levels of nuclear POP-1 (*left*) or phosphorylated, cytoplasmic POP-1 (*right*).

migrate to the posterior in response to the Wnt EGL-20 (Herman 2002; Korswagen et al. 2002; Silhankova and Korswagen 2007), which is expressed in a posterior-to-anterior gradient (Coudreuse et al. 2006). Although both the QL and QR cells are exposed to the ligand, they have different intrinsic sensitivities. In the more sensitive QL neuroblast, exposure to EGL-20 leads to Wnt/β-catenin pathway activation, expression of the Hox gene *mab-5*, and posterior migration of progeny, whereas the less sensitive QR neuroblast and its progeny do not express *mab-5* and continue to migrate toward the anterior (Fig. 3B) (Salser and Kenyon 1992; Whangbo and Kenyon 1999). Loss of function of the positively acting WBC pathway genes *mig-14/Wntless*, *vps-35/retromer*, *egl-20/Wnt*, *lin-17/Fz*, *mig-1/Fz*, *mig-5/Dsh*, *bar-1/β-cat*, and *pop-1/TCF*, or overexpression of the negative regulators *gsk-3/GSK3β* and *pry-1/Axin*, leads to loss of *mab-5* expression and the anterior migration of the QL.d cells (Fig. 3C) (Harris et al. 1996; Maloof et al. 1999; Eisenmann and Kim 2000; Korswagen et al. 2000; Herman 2001; Zinovyeva et al. 2008). Conversely, loss of negative regulation by Axin in a *pry-1/axin* mutant or *pry-1; axl-1* double mutant causes ectopic expression of *mab-5* in the QR.d cells, which then migrate to the posterior (Fig. 3D) (Korswagen et al. 2002; Oosterveen et al. 2007).

The products of two genes appear to regulate the levels of EGL-20/Wnt along the anterior–posterior axis and modulate WBC pathway activation in the migrating neuroblast daughters. First, *cam-1* encodes a Ror receptor tyrosine kinase, and overexpression of *cam-1/Ror* reduces *mab-5* expression and results in the anterior migration of the QL.d (Kim and Forrester 2003; Forrester et al. 2004). This result is also observed upon overexpression of just the CAM-1 extracellular cysteine-rich domain and transmembrane domain, which suggests that CAM-1 may serve as a sink for the EGL-20 signal along the anterior–posterior axis. Second, the *sfrp-1* gene was recently shown to encode an SFRP family member that is expressed in the anterior (Fig. 2) (Harterink et al. 2011). Loss of *sfrp-1* function can suppress the QL.d anterior migration phenotype caused by loss of *mig-14/Wntless* (Fig. 3F) but not that caused by loss of *mab-5/Hox*, whereas overexpression of SFRP-1 leads to the QL.d migrating to even more posterior positions (Fig. 3D). These data suggest that SFRP-1 may regulate the activation of Wnt signaling in these and other cells migrating along the anterior–posterior axis by binding to Wnt proteins expressed from the posterior (EGL-20, LIN-44, and CWN-1) and creating a sharper Wnt gradient along this axis.

Thus, in this process, initial intrinsic differences in sensitivity to Wnt signals between two related cells leads to activation of a canonical Wnt/β-catenin-dependent pathway and target gene expression, which controls the final position of migrating cells along an axial Wnt gradient.

THE WNT/β-CATENIN ASYMMETRY (WBA) PATHWAY: NEW TRICK

The Wnt/β-catenin asymmetry (WBA) pathway was the first Wnt pathway characterized in *C. elegans* (Rocheleau et al. 1997; Thorpe et al. 1997), and it is known to regulate many asymmetric cell divisions during *C. elegans* embryonic and larval development (Fig. 4) (for review, see Eisenmann 2005; Mizumoto and Sawa 2007b; Phillips and Kimble 2009). The WBA pathway uses the diverged β-catenins SYS-1 and WRM-1, but not BAR-1, suggesting a strong segregation of β-catenin function between the two *C. elegans* pathways. Interestingly, this pathway may be a nematode-specific variation on the canonical Wnt/TCF-dependent signaling pathway theme.

For some cells whose fate is regulated by the WBA pathway, Frizzled and Disheveled proteins have been found to asymmetrically localize at the cortex of the mother cell on the side at which the Wnt signal is received, whereas negative regulators APC and Axin localize to the opposite side of the mother cell cortex (Park et al. 2004; Walston et al. 2004; Takeshita and Sawa 2005; Goldstein et al. 2006; Mizumoto and Sawa 2007a). Then, upon cell division and the asymmetrical distribution of these cytoplasmic components, Wnt pathway activation occurs in only one of the two daughter cells. Therefore, the hallmark of the WBA pathway is one or more asymmetries between two daughter cells in (1) the localization of upstream pathway components to the cell cortex, (2) the levels of downstream pathway components in the daughter cell nuclei, and (3) the activation of the Wnt signaling pathway (Fig. 4A,B) (Mizumoto and Sawa 2007b; Phillips and Kimble 2009).

In this pathway, Wnt-dependent activation of target genes by TCF and β-catenin occurs, but proceeds through a novel mechanism distinct from that seen in the Wnt/β-catenin canonical

 Cite this article as *Cold Spring Harb Perspect Biol* doi: 10.1101/cshperspect.a007948

signaling pathway (Fig. 4B). A major mechanistic difference with the canonical pathway is that in the WBA pathway, target gene activation requires a β-catenin (WRM-1) and mitogen-activated protein kinase/Nemo-like kinase (MAPK/NLK)-dependent nuclear export of phosphorylated POP-1/TCF from the nucleus of the cell in which the pathway is active. In this pathway, in addition to activating the Wnt pathway in one daughter, Wnt ligand binding activates the kinases MOM-4/TAK1 and LIT-1/NLK (Rocheleau et al. 1997; Thorpe et al. 1997; Meneghini et al. 1999). The diverged β-catenin WRM-1 binds to nuclear POP-1/TCF using a carboxy-terminal domain distinct from the canonical amino-terminal β-catenin-binding domain, and recruits the activated LIT-1/NLK to POP-1/TCF (Korswagen et al. 2000; Natarajan et al. 2001; Yang et al. 2011). The subsequent phosphorylation of POP-1 marks it for nuclear export by PAR-5/14-3-3 and CRM-1/Exportin (Rocheleau et al. 1997; Meneghini et al. 1999; Shin et al. 1999; Lo et al. 2004), leading to a lowering of POP-1/TCF levels in the nucleus of the signaled cell (Lin et al. 1998; Meneghini et al. 1999; Rocheleau et al. 1999; Herman 2001; Maduro et al. 2002; Park and Priess 2003; Siegfried et al. 2004; Huang et al. 2007; Phillips et al. 2007). Although the mechanism is unclear, in this daughter cell there is also a Wnt-induced increase in the levels of the β-catenin SYS-1 in the nucleus (Huang et al. 2007; Phillips et al. 2007). The end result of these Wnt-induced events is an asymmetric distribution of TCF and β-catenin between sister cell nuclei (Fig. 4B).

A decrease in nuclear levels of the pathway transcriptional effector may seem like an unusual outcome for the cell in which the pathway is activated and target genes are expressed. The current model (Phillips and Kimble 2009) suggests that in the nonactivated daughter cell, high POP-1/TCF levels and low SYS-1/β-catenin levels result in the formation of transcriptional repressor complexes between POP-1/TCF and corepressors such as UNC-37/Groucho and HDA-1/histone deacetylase, which act to repress target genes that regulate cell fate or behavior (Fig. 4B, left). Conversely, in the Wnt-activated daughter cell, the lower POP-1/TCF levels and higher SYS-1/β-catenin levels result in formation of a typical TCF/β-catenin bipartite transcription activator, which activates expression of target genes (Fig. 4B, right) (Calvo et al. 2001; Zhao et al. 2002; Maduro et al. 2005; Shetty et al. 2005; Arata et al. 2006; Lam et al. 2006; Huang et al. 2007; Owraghi et al. 2010). Molecular genetic manipulations that artificially alter the ratio of TCF to β-catenin between daughter cells lead to changes in gene expression or cell fate that are consistent with this model of Wnt pathway activation, switching the transcriptional behavior of TCF in the nucleus (Kidd et al. 2005; Lam et al. 2006; Huang et al. 2007; Bertrand and Hobert 2009).

The WBA pathway may be the major pathway mediating Wnt signals during development in the worm, as it controls numerous asymmetric cell divisions in the embryo, as well as the asymmetric division of the somatic gonad primordia and the stem cell-like divisions of the hypodermal seam cells in the larva (Eisenmann 2005; Mizumoto and Sawa 2007b; Phillips and Kimble 2009). As an example of the use of this pathway, we consider one of the first- and best-studied developmental processes regulated by the WBA pathway, the asymmetric division of the blastomere EMS in the early embryo to generate endodermal and mesodermal precursors.

P2/EMS SIGNALING IN EMBYONIC ENDODERM INDUCTION

At the four-cell stage in early embryogenesis, the EMS blastomere divides to generate an anterior MS daughter that generates mesodermal tissue and a posterior E daughter that generates all the worm's endoderm (Fig. 5A) (Sulston 1983). Blastomere isolation experiments showed that the posterior blastomere P2 is required for production of endoderm by EMS, indicating that a signal from P2 polarizes the asymmetric EMS division to generate a daughter with an E fate (Goldstein 1992, 1993; Goldstein et al. 2006). Forward and reverse genetic screens identified components of Wnt, MAPK, and Src signaling pathways as being required for endoderm formation (Lin et al. 1995; Kaletta et al. 1997;

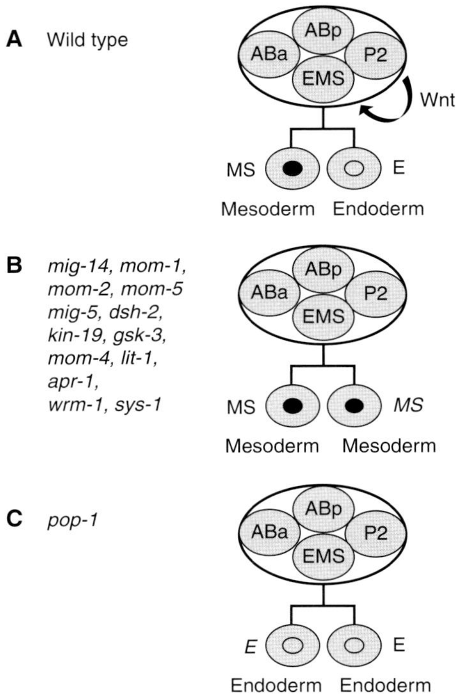

Figure 5. Wnt-dependent endoderm induction at the four-cell stage. (*A*) At the four-cell stage of embryogenesis, a Wnt signal from the posterior blastomere P2 instructs the EMS blastomere to divide asymmetrically, generating an anterior daughter (MS) that makes mesodermal cells and a posterior daughter (E) that is the endodermal founder cell. Nuclear levels of POP-1/TCF are high in MS (filled circle) and low in E (open circle). (*B*) Mutations in components of the WBA and MAPK signaling pathways lead to a symmetric division in which both daughters have high nuclear levels of POP-1/TCF and adopt an MS cell fate, resulting in more mesoderm (Mom phenotype). (*C*) Loss of POP-1/TCF leads to both daughters adopting the E cell fate and generating excess endodermal cells. Anterior is to the left, dorsal is up.

Rocheleau et al. 1997, 1999; Meneghini et al. 1999; Peters et al. 1999; Schlesinger et al. 1999; Shin et al. 1999; Bei et al. 2002; Thorpe and Moon 2004; Walston et al. 2004; Huang et al. 2007; Phillips et al. 2007). Wnt pathway components functioning in endoderm induction include MOM-1/Porcupine, MOM-2/Wnt, MIG-14/Wntless (a.k.a. MOM-3), MOM-5/Fz, DSH-2/Dsh, MIG-5/Dsh, KIN-19/CK1, GSK-3/GSK3β, APR-1/APC, WRM-1/β-catenin, SYS-1/β-catenin, and POP-1/TCF, while the MAPK signaling factors are TAP-1/TabAB1, MOM-4/TAK, and LIT-1/NLK (Meneghini et al. 1999; Rocheleau et al. 1999; Shin et al. 1999). The Tyr kinase receptor MES-1 and the Src homolog SRC-1 also participate in endoderm induction, but their role is less clear and they are not considered here (Bei et al. 2002; Zhang et al. 2008; Sugioka and Sawa 2010; Sumiyoshi et al. 2011). Mutations or RNAi reduction of function for these Wnt and MAPK pathway genes, with the exception of *pop-1/TCF*, leads to a Mom phenotype (more mesoderm), in which EMS divides symmetrically to give two MS-like daughters (Fig. 5B). *pop-1/TCF* mutations cause the opposite phenotype, the generation of two E-like daughters from EMS (Fig. 5C) (Lin et al. 1995). Notice that in this pathway, loss of the GSK3β, CK1α, and APC homologs causes the same phenotype as loss of ligand or receptor, indicating they function as positive regulators of this pathway. This initially led to the idea that this pathway was "noncanonical," but as described below, the outcome of WBA pathway activation is still a TCF/β-catenin-dependent activation of gene expression, suggesting the pathway represents a nematode variation on the canonical pathway.

Based on the current model of WBA function derived from studies on this and other processes (Fig. 4) (Eisenmann 2005; Mizumoto and Sawa 2007b; Van Hoffelen and Herman 2008; Phillips and Kimble 2009), Wnt signals (MOM-2) received on the posterior side of the EMS blastomere, where it contacts P2, result in the asymmetric localization of several pathway components to the anterior and posterior cortex of EMS (Fig. 4A). Upon EMS division this asymmetric localization continues, with WRM-1/β-catenin and LIT-1/NLK found primarily in the nucleus of the posterior daughter E, but not in the nucleus of the anterior daughter MS (Fig. 4B). In the anterior MS daughter nucleus, levels of POP-1/TCF are high, levels of the β-catenin SYS-1 are lower, and POP-1/TCF acts with transcriptional repressors such as histone deacetylase to silence expression of the endoderm-specific Wnt target genes *end-1* and *end-3* (Calvo et al. 2001; Maduro et al. 2002; Shetty et al. 2005; Owraghi et al. 2010). Conversely, in the posterior E daughter, the levels of POP-1/TCF are lower and the levels of SYS-1 are high relative to their levels in the MS cell, and together these factors activate expression of the target genes *end-1* and *end-3*, inducing that cell to adopt the endoderm fate (Huang et al. 2007). The asymmetry in POP-1 levels between the daughter nuclei is an easily observed marker for the differential activation of Wnt signaling between the anterior and posterior daughters (Fig. 5A). Consistent with this model, reduced gene expression of the WBA pathway positive regulators leads to both daughter cells having high levels of POP-1 and adopting the nonactivated MS cell fate (Fig. 5B), whereas manipulations that increase the SYS-1/POP-1 nuclear ratio (e.g., *pop-1* RNAi) lead to both daughter cells having low POP-1 levels and adopting the activated cell fate (Fig. 5C).

Thus, in this process the asymmetric distribution of a Wnt signal to a mother cell leads to an asymmetric localization of Wnt pathway components in the cortex of that cell, and the activation of a signaling pathway in only one of the daughters after cell division. Ligand binding results in activation of a MAPK signaling pathway and a variant Wnt signaling pathway in which several core components that act negatively in the canonical pathway appear to act positively to transduce signal. This variant pathway uses two diverged β-catenins to regulate the formation of a bipartite TCF/β-catenin activator protein in one daughter cell nucleus via a directed export of TCF from the signaled cell. Thus it appears that the worm has evolved a variant of the canonical Wnt/β-catenin pathway that uses ancient pathway components in novel ways to activate gene expression in response to Wnt ligand.

CONCLUDING REMARKS

Although a "canonical" Wnt/β-catenin pathway exists in *C. elegans* and controls several different types of processes (cell-fate specification, cell migration), a large number of Wnt-regulated developmental events in the worm use the variant Wnt pathway described above to regulate fate specification after asymmetric cell divisions. Although we have uncovered much about the function of these two pathways, a number of interesting questions remain about Wnt signaling in *C. elegans*.

Biochemical questions: First, as mentioned above, it remains to be determined if Wnt ligand binding in *C. elegans* uses an LRP5/6/Arrow homolog, as in other species. If it does not, it will show that Wnt signaling can proceed without the necessity of this component, which may have implications for Wnt signaling in other species. Second, it is unclear how *C. elegans* core components like APC and GSK-3 function in a negative manner in the WBC pathway and in some WBA-mediated processes, but appear to function in a positive manner in the P2/EMS signaling pathway (see Mizumoto and Sawa 2007b). Third, the fact that there is a fair amount of functional/genetic redundancy observed in *C. elegans* for both Wnt and Wnt receptors is interesting. It remains to be determined whether in these cases there are specific Wnt/receptor pairings mediating specific events, or whether there is true biochemical redundancy for these factors and all that matters is the total amount of Wnt ligand received and the number of receptors available to perceive it. In fact, for some WBA-mediated processes, it is unclear whether any Wnt proteins are involved at all (e.g., some early embryonic asymmetric divisions, somatic gonad division, seam cell divisions). For example, in the case of the larval seam cells, which divide in a stem cell-like asymmetric manner (Altun and Hall 2005), combining mutations in multiple Wnt receptors leads to symmetric divisions and defects in seam cell number, whereas animals lacking four or even all five Wnt genes do not display this phenotype (Gleason and Eisenmann 2010; Yamamoto et al. 2011). How the WBA pathway might function in a Wnt-independent but Wnt receptor-dependent manner is an intriguing question for future analysis. Finally, the exact mechanism by which cortical asymmetric localization of pathway components in the mother cell leads to asymmetric pathway activation in the daughters is only beginning to be understood (Mizumoto and Sawa 2007b).

Evolutionary questions: The major question here is whether the variant WBA signaling pathway exists and is used in other metazoans or whether it was a nematode-specific "invention" during evolution. Although hints of functions like those seen in the WBA pathway have been reported (see Phillips and Kimble 2009), there is currently no strong evidence that this pathway is used in other nonnematode species. This lack of evidence must be tempered by results from *C. elegans* that show that Wnt pathway components (such as SYS-1 and AXL-1) may lurk in genomes, but be overlooked by purely bioinformatic approaches. A final question is whether *C. elegans* uses a "noncanonical" β-catenin-independent Wnt/PCP pathway like that characterized in vertebrates and *Drosophila*. Although homologs of Wnt/PCP pathway components exist in the *C. elegans* genome, and mutations lead to weak phenotypes in some Wnt-mediated processes (Wu and Herman 2006, 2007; King et al. 2009), the evidence for a Wnt/PCP pathway in *C. elegans* is currently weak.

Therefore, in summary, it is possible that in terms of Wnt signaling, nematodes are "the same, but different" compared with other metazoans. Although still clearly relying on the major metazoan Wnt signaling pathway (WBC), they have also created a novel variant on the Wnt/β-catenin pathway (WBA), which mediates a major portion of Wnt signaling in this species.

ACKNOWLEDGMENTS

We thank members of the Eisenmann laboratory for critical reading of the manuscript, and thank *C. elegans* colleagues for discussions and sharing of unpublished results. The authors are supported by grants from National Institutes of Health (NIH)/National Institute of General Medical Sciences and National Science Foundation (D.M.E.) and an NIH Ruth L. Kirschstein National Research Service Award (B.M.J.).

REFERENCES

Altun ZF, Hall DH. 2009. Epithelial system, seam cells. *WormAtlas* doi: 10.3908/wormatlas.1.14.

Arata Y, Kouike H, Zhang Y, Herman MA, Okano H, Sawa H. 2006. Wnt signaling and a Hox protein cooperatively regulate *psa-3/Meis* to determine daughter cell fate after asymmetric cell division in *C. elegans*. *Dev Cell* **11:** 105–115.

Arce L, Yokoyama NN, Waterman ML. 2006. Diversity of LEF/TCF action in development and disease. *Oncogene* **25:** 7492–7504.

Atcha FA, Munguia JE, Li TW, Hovanes K, Waterman ML. 2003. A new β-catenin-dependent activation domain in T cell factor. *J Biol Chem* **278:** 16169–16175.

Atcha FA, Syed A, Wu B, Hoverter NP, Yokoyama NN, Ting JH, Munguia JE, Mangalam HJ, Marsh JL, Waterman ML. 2007. A unique DNA binding domain converts T-cell factors into strong Wnt effectors. *Mol Cell Biol* **27:** 8352–8363.

Axelrod JD. 2009. Progress and challenges in understanding planar cell polarity signaling. *Semin Cell Dev Biol* **20:** 964–971.

Banziger C, Soldini D, Schutt C, Zipperlen P, Hausmann G, Basler K. 2006. Wntless, a conserved membrane protein dedicated to the secretion of Wnt proteins from signaling cells. *Cell* **125:** 509–522.

Bei Y, Hogan J, Berkowitz LA, Soto M, Rocheleau CE, Pang KM, Collins J, Mello CC. 2002. SRC-1 and Wnt signaling act together to specify endoderm and to control cleavage orientation in early *C. elegans* embryos. *Dev Cell* **3:** 113–125.

Belenkaya TY, Wu Y, Tang X, Zhou B, Cheng L, Sharma YV, Yan D, Selva EM, Lin X. 2008. The retromer complex influences Wnt secretion by recycling Wntless from endosomes to the trans-Golgi network. *Dev Cell* **14:** 120–131.

Bertrand V, Hobert O. 2009. Linking asymmetric cell division to the terminal differentiation program of postmitotic neurons in *C. elegans*. *Dev Cell* **16:** 563–575.

Cadigan KM, Peifer M. 2009. Wnt signaling from development to disease: Insights from model systems. *Cold Spring Harb Perspect Biol* **1:** a002881.

Calvo D, Victor M, Gay F, Sui G, Luke MP, Dufourcq P, Wen G, Maduro M, Rothman J, Shi Y. 2001. A POP-1 repressor complex restricts inappropriate cell type-specific gene transcription during *Caenorhabditis elegans* embryogenesis. *EMBO J* **20:** 7197–7208.

Clevers H. 2006. Wnt/β-catenin signaling in development and disease. *Cell* **127:** 469–480.

Costa M, Raich W, Agbunag C, Leung B, Hardin J, Priess JR. 1998. A putative catenin–cadherin system mediates morphogenesis of the *Caenorhabditis elegans* embryo. *J Cell Biol* **141:** 297–308.

Coudreuse DY, Roel G, Betist MC, Destree O, Korswagen HC. 2006. Wnt gradient formation requires retromer function in Wnt-producing cells. *Science* **312:** 921–924.

Dreier L, Burbea M, Kaplan JM. 2005. LIN-23-mediated degradation of β-catenin regulates the abundance of GLR-1 glutamate receptors in the ventral nerve cord of *C. elegans*. *Neuron* **46:** 51–64.

Eisenmann DM. 2005. Wnt signaling. In *WormBook: The online review of* C. elegans *biology* (ed. The *C. elegans* Research Community). WormBook, doi: 10.1895/wormbook.1.7.1.

Eisenmann DM, Kim SK. 2000. Protruding vulva mutants identify novel loci and Wnt signaling factors that function during *Caenorhabditis elegans* vulva development. *Genetics* **156:** 1097–1116.

Eisenmann DM, Maloof JN, Simske JS, Kenyon C, Kim SK. 1998. The β-catenin homolog BAR-1 and LET-60 Ras coordinately regulate the Hox gene *lin-39* during *Caenorhabditis elegans* vulval development. *Development* **125:** 3667–3680.

Forrester WC, Kim C, Garriga G. 2004. The *Caenorhabditis elegans* Ror RTK CAM-1 inhibits EGL-20/Wnt signaling in cell migration. *Genetics* **168:** 1951–1962.

Gleason JE, Eisenmann DM. 2010. Wnt signaling controls the stem cell-like asymmetric division of the epithelial seam cells during *C. elegans* larval development. *Dev Biol* **348:** 58–66.

Gleason JE, Korswagen HC, Eisenmann DM. 2002. Activation of Wnt signaling bypasses the requirement for RTK/Ras signaling during *C. elegans* vulval induction. *Genes Dev* **16:** 1281–1290.

Gleason JE, Szyleyko EA, Eisenmann DM. 2006. Multiple redundant Wnt signaling components function in two processes during *C. elegans* vulval development. *Dev Biol* **298:** 442–457.

Goldstein B. 1992. Induction of gut in *Caenorhabditis elegans* embryos. *Nature* **357:** 255–257.

Goldstein B. 1993. Establishment of gut fate in the E lineage of *C. elegans*: The roles of lineage-dependent mechanisms and cell interactions. *Development* **118:** 1267–1277.

Goldstein B, Takeshita H, Mizumoto K, Sawa H. 2006. Wnt signals can function as positional cues in establishing cell polarity. *Dev Cell* **10:** 391–396.

Green JL, Inoue T, Sternberg PW. 2007. The *C. elegans* ROR receptor tyrosine kinase, CAM-1, non-autonomously inhibits the Wnt pathway. *Development* **134:** 4053–4062.

Green JL, Inoue T, Sternberg PW. 2008. Opposing Wnt pathways orient cell polarity during organogenesis. *Cell* **134:** 646–656.

Guder C, Philipp I, Lengfeld T, Watanabe H, Hobmayer B, Holstein TW. 2006. The Wnt code: Cnidarians signal the way. *Oncogene* **25:** 7450–7460.

Hall DH, Altun ZF. 2008. Epithelial system: Seam cells. In C. elegans *atlas*, pp. 35–41. Cold Spring Harbor Laboratory Press, Cold Spring Harbor, NY.

Han M. 1997. Gut reaction to Wnt signaling in worms. *Cell* **90:** 581–584.

Hardin J, King RS. 2008. The long and the short of Wnt signaling in *C. elegans*. *Curr Opin Genet Dev* **18:** 362–367.

Harris J, Honigberg L, Robinson N, Kenyon C. 1996. Neuronal cell migration in *C. elegans*: Regulation of Hox gene expression and cell position. *Development* **122:** 3117–3131.

Harterink M, Kim DH, Middelkoop TC, Doan TD, van Oudenaarden A, Korswagen HC. 2011. Neuroblast migration along the anteroposterior axis of *C. elegans* is controlled by opposing gradients of Wnts and a secreted Frizzled-related protein. *Development* **138:** 2915–2924.

Hatzis P, van der Flier LG, van Driel MA, Guryev V, Nielsen F, Denissov S, Nijman IJ, Koster J, Santo EE, Welboren W, et al. 2008. Genome-wide pattern of TCF7L2/TCF4 chromatin occupancy in colorectal cancer cells. *Mol Cell Biol* **28:** 2732–2744.

Hayashi Y, Hirotsu T, Iwata R, Kage-Nakadai E, Kunitomo H, Ishihara T, Iino Y, Kubo T. 2009. A trophic role for Wnt-Ror kinase signaling during developmental pruning in *Caenorhabditis elegans*. *Nat Neurosci* **12:** 981–987.

Herman MA. 2001. *C. elegans* POP-1/TCF functions in a canonical Wnt pathway that controls cell migration and in a noncanonical Wnt pathway that controls cell polarity. *Development* **128:** 581–590.

Herman MA. 2002. Control of cell polarity by noncanonical Wnt signaling in *C. elegans*. *Semin Cell Dev Biol* **13:** 233–241.

Herman MA, Vassilieva LL, Horvitz HR, Shaw JE, Herman RK. 1995. The *C. elegans* gene *lin-44*, which controls the polarity of certain asymmetric cell divisions, encodes a Wnt protein and acts cell nonautonomously. *Cell* **83:** 101–110.

Hilliard MA, Bargmann CI. 2006. Wnt signals and Frizzled activity orient anterior-posterior axon outgrowth in *C. elegans*. *Dev Cell* **10:** 379–390.

Howard RM, Sundaram MV. 2002. *C. elegans* EOR-1/PLZF and EOR-2 positively regulate Ras and Wnt signaling and function redundantly with LIN-25 and the SUR-2 Mediator component. *Genes Dev* **16:** 1815–1827.

Huang S, Shetty P, Robertson SM, Lin R. 2007. Binary cell fate specification during *C. elegans* embryogenesis driven by reiterated reciprocal asymmetry of TCF POP-1 and its coactivator β-catenin SYS-1. *Development* **134:** 2685–2695.

Inoue T, Oz HS, Wiland D, Gharib S, Deshpande R, Hill RJ, Katz WS, Sternberg PW. 2004. *C. elegans* LIN-18 is a Ryk ortholog and functions in parallel to LIN-17/Frizzled in Wnt signaling. *Cell* **118:** 795–806.

Janssen R, Le Gouar M, Pechmann M, Poulin F, Bolognesi R, Schwager EE, Hopfen C, Colbourne JK, Budd GE, Brown SJ, et al. 2010. Conservation, loss, and redeployment of Wnt ligands in protostomes: Implications for understanding the evolution of segment formation. *BMC Evol Biol* **10:** 374.

Jiang LI, Sternberg PW. 1998. Interactions of EGF, Wnt and HOM-C genes specify the P12 neuroectoblast fate in *C. elegans*. *Development* **125:** 2337–2347.

Kaletta T, Schnabel H, Schnabel R. 1997. Binary specification of the embryonic lineage in *Caenorhabditis elegans*. *Nature* **390:** 294–298.

Kennerdell JR, Fetter RD, Bargmann CI. 2009. Wnt-Ror signaling to SIA and SIB neurons directs anterior axon guidance and nerve ring placement in *C. elegans*. *Development* **136:** 3801–3810.

Kidd AR 3rd, Miskowski JA, Siegfried KR, Sawa H, Kimble J. 2005. A β-catenin identified by functional rather than sequence criteria and its role in Wnt/MAPK signaling. *Cell* **121:** 761–772.

Kim C, Forrester WC. 2003. Functional analysis of the domains of the *C. elegans* Ror receptor tyrosine kinase CAM-1. *Dev Biol* **264:** 376–390.

King RS, Maiden SL, Hawkins NC, Kidd AR III, Kimble J, Hardin J, Walston TD. 2009. The N- or C-terminal domains of DSH-2 can activate the *C. elegans* Wnt/β-catenin asymmetry pathway. *Dev Biol* **328:** 234–244.

Korswagen HC, Herman MA, Clevers HC. 2000. Distinct β-catenins mediate adhesion and signalling functions in *C. elegans*. *Nature* **406:** 527–532.

Korswagen HC, Coudreuse DY, Betist MC, van de Water S, Zivkovic D, Clevers HC. 2002. The Axin-like protein PRY-1 is a negative regulator of a canonical Wnt pathway in *C. elegans*. *Genes Dev* **16:** 1291–1302.

Lam N, Chesney MA, Kimble J. 2006. Wnt signaling and CEH-22/tinman/Nkx2.5 specify a stem cell niche in *C. elegans*. *Curr Biol* **16:** 287–295.

Li X, Kulkarni RP, Hill RJ, Chamberlin HM. 2009. HOM-C genes, Wnt signaling and axial patterning in the *C. elegans* posterior ventral epidermis. *Dev Biol* **332:** 156–165.

Lin R, Thompson S, Priess JR. 1995. *pop-1* encodes an HMG box protein required for the specification of a mesoderm precursor in early *C. elegans* embryos. *Cell* **83:** 599–609.

Lin R, Hill RJ, Priess JR. 1998. POP-1 and anterior–posterior fate decisions in *C. elegans* embryos. *Cell* **92:** 229–239.

Lin KT, Broitman-Maduro G, Hung WW, Cervantes S, Maduro MF. 2009. Knockdown of SKN-1 and the Wnt effector TCF/POP-1 reveals differences in endomesoderm specification in *C. briggsae* as compared with *C. elegans*. *Dev Biol* **325:** 296–306.

Liu J, Phillips BT, Amaya MF, Kimble J, Xu W. 2008. The *C. elegans* SYS-1 protein is a bona fide β-catenin. *Dev Cell* **14:** 751–761.

Lo MC, Gay F, Odom R, Shi Y, Lin R. 2004. Phosphorylation by the β-catenin/MAPK complex promotes 14-3-3-mediated nuclear export of TCF/POP-1 in signal-responsive cells in *C. elegans*. *Cell* **117:** 95–106.

Love JJ, Li X, Case DA, Giese K, Grosschedl R, Wright PE. 1995. Structural basis for DNA bending by the architectural transcription factor LEF-1. *Nature* **376:** 791–795.

MacDonald BT, Tamai K, He X. 2009. Wnt/β-catenin signaling: Components, mechanisms, and diseases. *Dev Cell* **17:** 9–26.

Maduro MF, Lin R, Rothman JH. 2002. Dynamics of a developmental switch: Recursive intracellular and intranuclear redistribution of *Caenorhabditis elegans* POP-1 parallels Wnt-inhibited transcriptional repression. *Dev Biol* **248:** 128–142.

Maduro MF, Kasmir JJ, Zhu J, Rothman JH. 2005. The Wnt effector POP-1 and the PAL-1/Caudal homeoprotein collaborate with SKN-1 to activate *C. elegans* endoderm development. *Dev Biol* **285:** 510–523.

Maloof JN, Whangbo J, Harris JM, Jongeward GD, Kenyon C. 1999. A Wnt signaling pathway controls Hox gene expression and neuroblast migration in *C. elegans*. *Development* **126:** 37–49.

Martin BL, Kimelman D. 2009. Wnt signaling and the evolution of embryonic posterior development. *Curr Biol* **19:** R215–R219.

Meneghini MD, Ishitani T, Carter JC, Hisamoto N, Ninomiya-Tsuji J, Thorpe CJ, Hamill DR, Matsumoto K, Bowerman B. 1999. MAP kinase and Wnt pathways converge to downregulate an HMG-domain repressor in *Caenorhabditis elegans*. *Nature* **399:** 793–797.

Mizumoto K, Sawa H. 2007a. Cortical β-catenin and APC regulate asymmetric nuclear β-catenin localization during asymmetric cell division in *C. elegans*. *Dev Cell* **12:** 287–299.

Mizumoto K, Sawa H. 2007b. Two βs or not two βs: Regulation of asymmetric division by β-catenin. *Trends Cell Biol* **17:** 465–473.

Natarajan L, Witwer NE, Eisenmann DM. 2001. The divergent *Caenorhabditis elegans* β-catenin proteins BAR-1, WRM-1 and HMP-2 make distinct protein interactions but retain functional redundancy in vivo. *Genetics* **159:** 159–172.

Niehrs C. 2010. On growth and form: A Cartesian coordinate system of Wnt and BMP signaling specifies bilaterian body axes. *Development* **137:** 845–857.

Nishiwaki K. 1999. Mutations affecting symmetrical migration of distal tip cells in *Caenorhabditis elegans*. *Genetics* **152:** 985–997.

Oosterveen T, Coudreuse DY, Yang PT, Fraser E, Bergsma J, Dale TC, Korswagen HC. 2007. Two functionally distinct Axin-like proteins regulate canonical Wnt signaling in *C. elegans*. *Dev Biol* **308:** 438–448.

Owraghi M, Broitman-Maduro G, Luu T, Roberson H, Maduro MF. 2010. Roles of the Wnt effector POP-1/TCF in the *C. elegans* endomesoderm specification gene network. *Dev Biol* **340:** 209–221.

Pan CL, Howell JE, Clark SG, Hilliard M, Cordes S, Bargmann CI, Garriga G. 2006. Multiple Wnts and Frizzled receptors regulate anteriorly directed cell and growth cone migrations in *Caenorhabditis elegans*. *Dev Cell* **10:** 367–377.

Pan CL, Baum PD, Gu M, Jorgensen EM, Clark SG, Garriga G. 2008. *C. elegans* AP-2 and retromer control Wnt signaling by regulating *mig-14/Wntless*. *Dev Cell* **14:** 132–139.

Pang K, Ryan JF, Mullikin JC, Baxevanis AD, Martindale MQ. 2010. Genomic insights into Wnt signaling in an early diverging metazoan, the ctenophore *Mnemiopsis leidyi*. *EvoDevo* **1:** 10.

Park FD, Priess JR. 2003. Establishment of POP-1 asymmetry in early *C. elegans* embryos. *Development* **130:** 3547–3556.

Park FD, Tenlen JR, Priess JR. 2004. *C. elegans* MOM-5/Frizzled functions in MOM-2/Wnt-independent cell polarity and is localized asymmetrically prior to cell division. *Curr Biol* **14:** 2252–2258.

Penigault JB, Felix MA. 2011a. Evolution of a system sensitive to stochastic noise: P3.p cell fate in *Caenorhabditis*. *Dev Biol* **357:** 419–427.

Penigault JB, Felix MA. 2011b. High sensitivity of *C. elegans* vulval precursor cells to the dose of posterior Wnts. *Dev Biol* **357:** 428–438.

Peters JM, McKay RM, McKay JP, Graff JM. 1999. Casein kinase I transduces Wnt signals. *Nature* **401:** 345–350.

Petersen CP, Reddien PW. 2009. Wnt signaling and the polarity of the primary body axis. *Cell* **139:** 1056–1068.

Phillips BT, Kimble J. 2009. A new look at TCF and β-catenin through the lens of a divergent *C. elegans* Wnt pathway. *Dev Cell* **17:** 27–34.

Phillips BT, Kidd AR III, King R, Hardin J, Kimble J. 2007. Reciprocal asymmetry of SYS-1/β-catenin and POP-1/TCF controls asymmetric divisions in *Caenorhabditis elegans*. *Proc Natl Acad Sci* **104:** 3231–3236.

Prasad BC, Clark SG. 2006. Wnt signaling establishes anteroposterior neuronal polarity and requires retromer in *C. elegans*. *Development* **133:** 1757–1766.

Putzke AP, Rothman JH. 2010. Repression of Wnt signaling by a Fer-type nonreceptor tyrosine kinase. *Proc Natl Acad Sci* **107:** 16154–16159.

Rocheleau CE, Downs WD, Lin R, Wittmann C, Bei Y, Cha YH, Ali M, Priess JR, Mello CC. 1997. Wnt signaling and an APC-related gene specify endoderm in early *C. elegans* embryos. *Cell* **90:** 707–716.

Rocheleau CE, Yasuda J, Shin TH, Lin R, Sawa H, Okano H, Priess JR, Davis RJ, Mello CC. 1999. WRM-1 activates the LIT-1 protein kinase to transduce anterior/posterior polarity signals in *C. elegans*. *Cell* **97:** 717–726.

Rudel D, Tian H, Sommer RJ. 2008. Wnt signaling in *Pristionchus pacificus* gonadal arm extension and the evolution of organ shape. *Proc Natl Acad Sci* **105:** 10826–10831.

Ruvkun G, Hobert O. 1998. The taxonomy of developmental control in *Caenorhabditis elegans*. *Science* **282:** 2033–2041.

Ryan JF, Baxevanis AD. 2007. Hox, Wnt, and the evolution of the primary body axis: Insights from the early-divergent phyla. *Biol. Direct* **2:** 37.

Salser SJ, Kenyon C. 1992. Activation of a *C. elegans Antennapedia* homologue in migrating cells controls their direction of migration. *Nature* **355:** 255–258.

Sawa H, Lobel L, Horvitz H. 1996. The *Caenorhabditis elegans* gene *lin-17*, which is required for certain asymmetric cell divisions, encodes a putative seven-transmembrane protein similar to the *Drosophila* Frizzled protein. *Genes Dev* **10:** 2189–2197.

Schlesinger A, Shelton CA, Maloof JN, Meneghini M, Bowerman B. 1999. Wnt pathway components orient a mitotic spindle in the early *Caenorhabditis elegans* embryo without requiring gene transcription in the responding cell. *Genes Dev* **13:** 2028–2038.

Seetharaman A, Cumbo P, Bojanala N, Gupta BP. 2010. Conserved mechanism of Wnt signaling function in the specification of vulval precursor fates in *C. elegans* and *C. briggsae*. *Dev Biol* **346:** 128–139.

Shackleford GM, Shivakumar S, Shiue L, Mason J, Kenyon C, Varmus HE. 1993. Two *wnt* genes in *Caenorhabditis elegans*. *Oncogene* **8:** 1857–1864.

Shetty P, Lo MC, Robertson SM, Lin R. 2005. *C. elegans* TCF protein, POP-1, converts from repressor to activator as a result of Wnt-induced lowering of nuclear levels. *Dev Biol* **285:** 584–592.

Shin TH, Yasuda J, Rocheleau CE, Lin R, Soto M, Bei Y, Davis RJ, Mello CC. 1999. MOM-4, a MAP kinase kinase kinase-related protein, activates WRM-1/LIT-1 kinase to transduce anterior/posterior polarity signals in *C. elegans*. *Mol Cell* **4:** 275–280.

Siegfried KR, Kidd AR III, Chesney MA, Kimble J. 2004. The *sys-1* and *sys-3* genes cooperate with Wnt signaling to establish the proximal-distal axis of the *Caenorhabditis elegans* gonad. *Genetics* **166:** 171–186.

Silhankova M, Korswagen HC. 2007. Migration of neuronal cells along the anterior–posterior body axis of *C. elegans*: Wnts are in control. *Curr Opin Genet Dev* **17:** 320–325.

Song S, Zhang B, Sun H, Li X, Xiang Y, Liu Z, Huang X, Ding M. 2010. A Wnt-Frz/Ror-Dsh pathway regulates neurite outgrowth in *Caenorhabditis elegans*. *PLoS Genet* **6:** e1001056.

Streit A, Kohler R, Marty T, Belfiore M, Takacs-Vellai K, Vigano MA, Schnabel R, Affolter M, Muller F. 2002. Conserved regulation of the *Caenorhabditis elegans labial/Hox1* gene *ceh-13*. *Dev Biol* **242:** 96–108.

Sugioka K, Sawa H. 2010. Regulation of asymmetric positioning of nuclei by Wnt and Src signaling and its roles in POP-1/TCF nuclear asymmetry in *Caenorhabditis elegans*. *Genes Cells* **15:** 397–407.

Sulston JE. 1983. Neuronal cell lineages in the nematode *Caenorhabditis elegans*. *Cold Spring Harb Symp Quant Biol* **48** (Pt 2): 443–452.

Sulston J, Horvitz H. 1977. Post-embryonic cell lineages of the nematode, *Caenorhabditis elegans*. *Dev Biol* **56:** 110–156.

Sumiyoshi E, Takahashi S, Obata H, Sugimoto A, Kohara Y. 2011. The β-catenin HMP-2 functions downstream of Src in parallel with the Wnt pathway in early embryogenesis of *C. elegans*. *Dev Biol* **355:** 302–312.

Takeshita H, Sawa H. 2005. Asymmetric cortical and nuclear localizations of WRM-1/β-catenin during asymmetric cell division in *C. elegans*. *Genes Dev* **19:** 1743–1748.

Thorpe CJ, Moon RT. 2004. *nemo-like kinase* is an essential co-activator of Wnt signaling during early zebrafish development. *Development* **131:** 2899–2909.

Thorpe CJ, Schlesinger A, Carter JC, Bowerman B. 1997. Wnt signaling polarizes an early *C. elegans* blastomere to distinguish endoderm from mesoderm. *Cell* **90:** 695–705.

Tian H, Schlager B, Xiao H, Sommer RJ. 2008. Wnt signaling induces vulva development in the nematode *Pristionchus pacificus*. *Curr Biol* **18:** 142–146.

van Amerongen R, Nusse R. 2009. Towards an integrated view of Wnt signaling in development. *Development* **136:** 3205–3214.

Van Hoffelen S, Herman MA. 2008. Analysis of Wnt signaling during *Caenorhabditis elegans* postembryonic development. *Methods Mol Biol* **469:** 87–102.

Walston T, Tuskey C, Edgar L, Hawkins N, Ellis G, Bowerman B, Wood W, Hardin J. 2004. Multiple Wnt signaling pathways converge to orient the mitotic spindle in early *C. elegans* embryos. *Dev Cell* **7:** 831–841.

Wang X, Sommer RJ. 2011. Antagonism of LIN-17/Frizzled and LIN-18/Ryk in nematode vulva induction reveals evolutionary alterations in core developmental pathways. *PLoS Biol* **9:** e1001110.

Whangbo J, Kenyon C. 1999. A Wnt signaling system that specifies two patterns of cell migration in *C. elegans*. *Mol Cell* **4:** 851–858.

Wu M, Herman MA. 2006. A novel noncanonical Wnt pathway is involved in the regulation of the asymmetric B cell division in *C. elegans*. *Dev Biol* **293:** 316–329.

Wu M, Herman MA. 2007. Asymmetric localizations of LIN-17/Fz and MIG-5/Dsh are involved in the asymmetric B cell division in *C. elegans*. *Dev Biol* **303:** 650–662.

Yamamoto Y, Takeshita H, Sawa H. 2011. Multiple Wnts redundantly control polarity orientation in *Caenorhabditis elegans* epithelial stem cells. *PLoS Genet* **7:** e1002308.

Yang PT, Lorenowicz MJ, Silhankova M, Coudreuse DY, Betist MC, Korswagen HC. 2008. Wnt signaling requires retromer-dependent recycling of MIG-14/Wntless in Wnt-producing cells. *Dev Cell* **14:** 140–147.

Yang XD, Huang S, Lo MC, Mizumoto K, Sawa H, Xu W, Robertson S, Lin R. 2011. Distinct and mutually inhibitory binding by two divergent β-catenins coordinates TCF levels and activity in *C. elegans*. *Development* **138:** 4255–4265.

Zhang H, Skop AR, White JG. 2008. Src and Wnt signaling regulate dynactin accumulation to the P2-EMS cell border in *C. elegans* embryos. *J Cell Sci* **121:** 155–161.

Zhao X, Yang Y, Fitch DH, Herman MA. 2002. TLP-1 is an asymmetric cell fate determinant that responds to Wnt signals and controls male tail tip morphogenesis in *C. elegans*. *Development* **129:** 1497–1508.

Zhao X, Sawa H, Herman MA. 2003. *tcl-2* encodes a novel protein that acts synergistically with Wnt signaling pathways in *C. elegans*. *Dev Biol* **256:** 276–289.

Zheng M, Messerschmidt D, Jungblut B, Sommer RJ. 2005. Conservation and diversification of Wnt signaling function during the evolution of nematode vulva development. *Nat Genet* **37:** 300–304.

Zinovyeva AY, Forrester WC. 2005. The *C. elegans* Frizzled CFZ-2 is required for cell migration and interacts with multiple Wnt signaling pathways. *Dev Biol* **285:** 447–461.

Zinovyeva AY, Yamamoto Y, Sawa H, Forrester WC. 2008. Complex network of Wnt signaling regulates neuronal migrations during *Caenorhabditis elegans* development. *Genetics* **179:** 1357–1371.

Wnt Signaling in Vertebrate Axis Specification

Hiroki Hikasa[1] and Sergei Y. Sokol[2]

[1]Division of Cancer Genetics, Medical Institute of Bioregulation, Kyushu University, Fukuoka 812-8582, Japan
[2]Mount Sinai School of Medicine, New York, New York 10029

Correspondence: sergei.sokol@mssm.edu

The Wnt pathway is a major embryonic signaling pathway that controls cell proliferation, cell fate, and body-axis determination in vertebrate embryos. Soon after egg fertilization, Wnt pathway components play a role in microtubule-dependent dorsoventral axis specification. Later in embryogenesis, another conserved function of the pathway is to specify the anteroposterior axis. The dual role of Wnt signaling in *Xenopus* and zebrafish embryos is regulated at different developmental stages by distinct sets of Wnt target genes. This review highlights recent progress in the discrimination of different signaling branches and the identification of specific pathway targets during vertebrate axial development.

Wnt pathways play major roles in cell-fate specification, proliferation and differentiation, cell polarity, and morphogenesis (Clevers 2006; van Amerongen and Nusse 2009). Signaling is initiated in the responding cell by the interaction of Wnt ligands with different receptors and coreceptors, including Frizzled, LRP5/6, ROR1/2, RYK, PTK7, and proteoglycans (Angers and Moon 2009; Kikuchi et al. 2009; MacDonald et al. 2009). Receptor activation is accompanied by the phosphorylation of Dishevelled (Yanagawa et al. 1995), which appears to transduce the signal to both the cell membrane and the nucleus (Cliffe et al. 2003; Itoh et al. 2005; Bilic et al. 2007). Another common pathway component is β-catenin, an abundant component of adherens junctions (Nelson and Nusse 2004; Grigoryan et al. 2008). In response to signaling, β-catenin associates with T-cell factors (TCFs) and translocates to the nucleus to stimulate Wnt target gene expression (Behrens et al. 1996; Huber et al. 1996; Molenaar et al. 1996).

This β-catenin-dependent activation of specific genes is often referred to as the "canonical" pathway. In the absence of Wnt signaling, β-catenin is destroyed by the protein complex that includes Axin, GSK3, and the tumor suppressor APC (Clevers 2006; MacDonald et al. 2009). Wnt proteins, such as Wnt1, Wnt3, and Wnt8, stimulate Frizzled and LRP5/6 receptors to inactivate this β-catenin destruction complex, and, at the same time, trigger the phosphorylation of TCF proteins by homeodomain-interacting protein kinase 2 (HIPK2) (Hikasa et al. 2010; Hikasa and Sokol 2011). Both β-catenin stabilization and the regulation of TCF protein function by phosphorylation appear to represent general strategies that are conserved in multiple systems (Sokol 2011). Thus, the signaling

Cite this article as *Cold Spring Harb Perspect Biol* doi: 10.1101/cshperspect.a007955

pathway consists of two branches that together regulate target gene expression (Fig. 1).

Other Wnt proteins, such as Wnt5a or Wnt11, strongly affect the cytoskeletal organization and morphogenesis without stabilizing β-catenin (Torres et al. 1996; Angers and Moon 2009; Wu and Mlodzik 2009). These "noncanonical" ligands do not influence TCF3 phosphorylation (Hikasa and Sokol 2011), but may use distinct receptors such as ROR1/2 and RYK instead of or in addition to Frizzled (Hikasa et al. 2002; Lu et al. 2004; Mikels and Nusse 2006; Nishita et al. 2006, 2010; Schambony and Wedlich 2007; Grumolato et al. 2010; Lin et al. 2010; Gao et al. 2011). In such cases, signaling mechanisms are likely to include planar cell polarity (PCP) components, such as Vangl2, Flamingo, Prickle, Diversin, Rho GTPases, and c-Jun amino-terminal kinases (JNKs), which do not directly affect β-catenin stability (Fig. 1) (Sokol 2000; Schwarz-Romond et al. 2002; Schambony and Wedlich 2007; Komiya and Habas 2008; Axelrod 2009; Itoh et al. 2009; Tada and Kai 2009; Sato et al. 2010; Gao et al. 2011). This simplistic dichotomy of the Wnt pathway does not preclude some Wnt ligands from using both β-catenin-dependent and -independent routes in a context-specific manner.

Despite the existence of many pathway branches, only the β-catenin-dependent branch

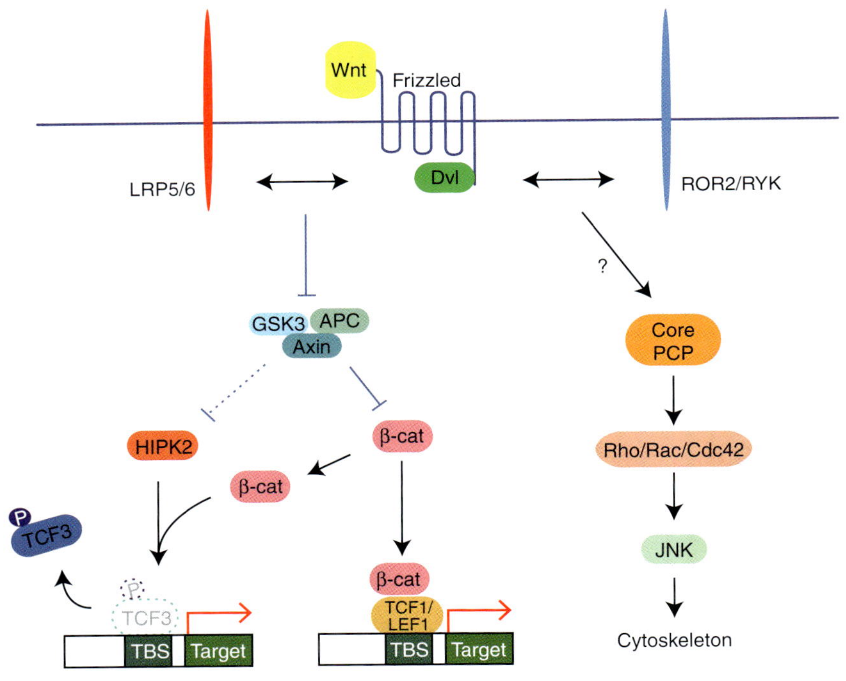

Figure 1. Conserved Wnt pathway branches and components. In the absence of Wnt signals, glycogen synthase kinase 3 (GSK3) binds Axin and APC to form the β-catenin destruction complex. Some Wnt proteins, such as Wnt8 and Wnt3a, stimulate Frizzled and LRP5/6 receptors to inhibit GSK3 activity and stabilize β-catenin (β-cat). Stabilized β-cat forms a complex with T-cell factors (e.g., TCF1/LEF1) to activate target genes. Moreover, GSK3 inhibition leads to target gene derepression by promoting TCF3 phosphorylation by homeodomain-interacting protein kinase 2 (HIPK2) through an unknown mechanism, for which β-catenin is required as a scaffold. This phosphorylation results in TCF3 removal from target promoters and gene activation. Other Wnt proteins, such as Wnt5a and Wnt11, use distinct receptors such as ROR2 and RYK, in addition to Frizzled, to control the the cytoskeletal organization through core planar cell polarity (PCP) proteins, small GTPases (Rho/Rac/Cdc42), and c-Jun amino-terminal kinase (JNK).

 Cite this article as *Cold Spring Harb Perspect Biol* doi: 10.1101/cshperspect.a007955

has been implicated in body-axis specification. Recent experiments in lower vertebrates have identified additional pathway components and targets and provided new insights into the underlying mechanisms.

ROLE OF CYTOPLASMIC DETERMINANTS AND WNT SIGNALING IN DORSOVENTRAL AXIS SPECIFICATION

The specification of the dorsoventral axis coincides with the formation of an essential embryonic signaling center known as the Spemann organizer in amphibians, the shield in zebrafish, and the node in the mouse (Harland and Gerhart 1997; Schier and Talbot 2005; Marlow 2010). During egg fertilization, sperm entry triggers microtubule-dependent rearrangement of the cortical cytoplasm from the vegetal pole toward the future dorsal side, a process known as the cortical rotation (Houliston and Elinson 1992). Experiments with cytoplasm microinjections or removal revealed vegetally localized dorsalizing activity that moves to the dorsal region before the first cleavage in *Xenopus* or zebrafish eggs (Fujisue et al. 1993; Holowacz and Elinson 1993; Mizuno et al. 1999; Ober and Schulte-Merker 1999). Treatment of fertilized *Xenopus* or zebrafish eggs with microtubule-depolymerizing agents, such as UV irradiation or nocodazole, causes embryo ventralization (Chung and Malacinski 1980; Scharf and Gerhart 1980; Jesuthasan and Stahle 1997). These observations indicated that the movement of the dorsalizing activity is essential for the establishment of the primary dorsoventral axis in both *Xenopus* and zebrafish.

One of the important outcomes of the cortical rotation is the accumulation of β-catenin in dorsal nuclei of early blastulae, whereas β-catenin remains largely cytoplasmic and cortical in ventral cells (Schneider et al. 1996; Schohl and Fagotto 2002). This difference in β-catenin localization can be detected as early as the 2- to 4-cell stage (Larabell et al. 1997). Supporting the critical role of this localization for dorsal development, several maternal effect mutants in zebrafish such as *hecate*, *ichabod*, and *tokkaebi* result in the reduction of β-catenin dorsal accumulation and defects in dorsal development (Kelly et al. 2000; Nojima et al. 2004; Lyman Gingerich et al. 2005; Bellipanni et al. 2006). The *ichabod* mutation has been mapped to the proximity of the gene encoding β-catenin-2, a second zebrafish β-catenin that is essential for maternal Wnt signaling (Bellipanni et al. 2006). Consistent with the ventralized phenotype of the *ichabod* embryos, maternal β-catenin-2 RNA is down-regulated. *Hecate* is another zebrafish mutation that is manifested by deficient dorsal development and enhanced intracellular calcium release; however, the mutated gene has not yet been identified (Lyman Gingerich et al. 2005). Positional cloning revealed that the *tokkaebi* gene encodes syntabulin, a microtubule-binding protein connecting kinesin I to syntaxin-containing cargo vesicles (Nojima et al. 2010). This finding further implicates microtubule-dependent trafficking in dorsoventral axis specification. Reminiscent of the key function of β-catenin in *Xenopus* and zebrafish, mouse embryos lacking β-catenin reveal abnormal visceral endoderm patterning before gastrulation and do not form the primitive streak (Haegel et al. 1995; Huelsken et al. 2000).

An essential role for Wnt/β-catenin signaling in dorsoventral axis determination has been initially suggested in gain-of-function experiments, in which Wnt and Dishevelled messenger RNAs (mRNAs) triggered ectopic organizer formation (McMahon and Moon 1989; Smith and Harland 1991; Sokol et al. 1991, 1995; Rothbacher et al. 1995). The requirement for β-catenin has been established by antisense oligonucleotide-mediated depletion of β-catenin in *Xenopus* and zebrafish embryos (Heasman et al. 1994; Bellipanni et al. 2006). β-catenin activates several major targets in *Xenopus* embryos that are responsible for Spemann organizer formation. Activation of many organizer-specific genes depends on functional TCF sites in their promoters (Brannon et al. 1997; Laurent et al. 1997; McKendry et al. 1997; Fan et al. 1998; Ryu et al. 2001; Leung et al. 2003). These early β-catenin targets include *Nodal* group genes that are essential for dorsal mesoderm formation in frogs, zebrafish, and mammals (Conlon et al. 1994; Feldman et al. 1998; Osada and Wright 1999; Takahashi et al. 2000; Yang et al. 2002; Hilton et al. 2003; Zhang et al. 2003). Besides *Nodals*, other direct targets of β-catenin are *Siamois* and *Twin* in *Xenopus* (Lemaire et al. 1995; Brannon et al. 1997; Laurent et al. 1997; Fan et al. 1998), and *Bozozok/Dharma/Nieuwkoid* in zebrafish (Ryu et al. 2001), which are required for organizer formation (Fig. 2; Table 1) (Fan and Sokol 1997; Kessler 1997; Koos and Ho 1998; Yamanaka et al. 1998; Fekany et al. 1999). Taken together, available data establish a key role for Wnt signaling and its early targets in early dorsoventral axis specification (Moon and Kimelman 1998; Schroeder et al. 1999; Sokol 1999; Tao et al. 2005).

Although lack of effect of a dominant–negative form of Dishevelled (Dvl) on axis specification does not eliminate Dvl participation; it suggests that β-catenin is stabilized by a downstream Wnt pathway component that is translocated dorsally during the cortical rotation (Sokol 1996; Harland and Gerhart 1997). One of the key factors is GSK3-binding protein or Frat, which inhibits GSK3 and is essential for dorsal development in *Xenopus* embryos (Yost et al. 1998). However, the triple knockout of the existing Frat homologs in the mouse does not produce a morphological phenotype, suggesting that this mechanism is not conserved in mammals (van Amerongen et al. 2005). More recently, antisense oligonucleotide-mediated depletion of maternal Wnt5a, Wnt11, and Fz7 indicated their roles in dorsoventral axis specification (Sumanas et al. 2000; Tao et al. 2005; Cha et al. 2008). These observations suggest a different model, in which Wnt5a and Wnt11 influence β-catenin levels in the oocyte, raising the question how β-catenin becomes localized to the dorsal side of the embryo. Because maternal Wnt11 RNA is vegetally localized (Ku and Melton 1993), one possibility is that Wnt11 is delivered to its future site of action by the cortical rotation (Tao et al. 2005). So far, dorsoventral patterning defects have not been described in embryos with genetically reduced Wnt5a or Wnt11 function (Rauch et al. 1997; Yamaguchi et al. 1999; Heisenberg et al. 2000; Kilian et al. 2003; Majumdar et al. 2003; Ulrich

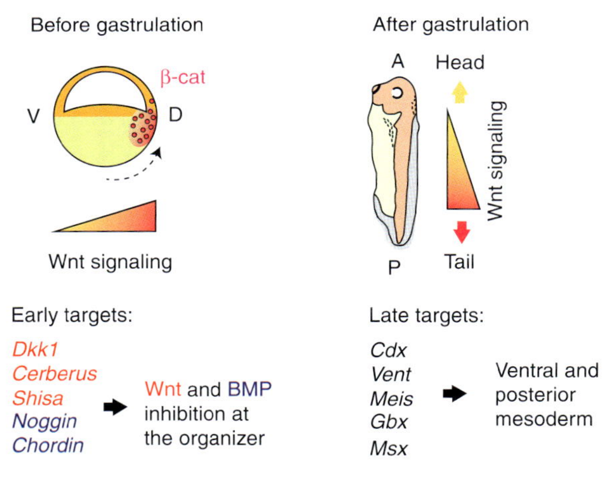

Figure 2. Axis specification by early and late Wnt signaling involves distinct targets. After cortical rotation (black dotted arrow), early β-catenin accumulation in the dorsal equatorial region activates gene targets to generate the Spemann organizer. β-Catenin up-regulates several Wnt and bone morphogenetic protein (BMP) antagonists, including *Dkk1, Cerberus, Shisa, Noggin*, and *Chordin*. The pathway is inhibited by Wnt antagonists in the anterior tissues, but the zygotic activation of Wnt8 causes ventral and posterior accumulation of β-catenin during gastrulation. The target genes of this late Wnt signaling, including *Cdx*, *Vent*, *Meis*, *Gbx*, and *Msx*, are critical for the specification of ventroposterior mesodermal fates and lead to tail formation.

 Cite this article as *Cold Spring Harb Perspect Biol* doi: 10.1101/cshperspect.a007955

Table 1. Putative Wnt targets during anteroposterior axis specification

Genes activated by Wnt3a	Genes inhibited by Dickkopf-1	References
Axin1/2	*Axin1/2*	Zeng et al. 1997; Jho et al. 2002
Kremen2	*Kremen2*	Mao et al. 2002
Cdx2, 4	*Cdx2, 4*	Isaacs et al. 1998
Gbx2	*Gbx2*	von Bubnoff et al. 1996
HoxA1/D1	*HoxA1/D1*	Kolm and Sive 1995
Esr9/10	*Esr9/10*	Li et al. 2003
Fz10	*Fz10*	Wheeler and Hoppler 1999; Garcia-Morales et al. 2009
Foxi1		Suri et al. 2005; Mir et al. 2007
Meis3	*Meis-like*	Elkouby et al. 2010
Msx1/2	*Msx1/2*	Maeda et al. 1997; Marazzi et al. 1997; Mathers et al. 1997; Suzuki et al. 1997; Willert et al. 2002
Nzl1	*Nzl1*	Andreazzoli et al. 2001; Nakamura et al. 2008
Nrh1		Bromley et al. 2004; Sasai et al. 2004
Riddle2		Shibata et al. 2008
Spr2		Weidinger et al. 2005
Vent/PV1	*Vent/PV1*	Onichtchouk et al. 1996
XARP	*XARP*	Itoh et al. 2000
Xmc		Frazzetto et al. 2002
Xpo	*Xpo*	Sato and Sargent 1991; Amaya et al. 1993; Hoppler et al. 1996

Only some of these genes have been shown to contain TCF/LEF sites in their promoters.

et al. 2003; Fossat et al. 2011). Potentially, these observations could be explained by functional redundancy or by the presence of maternally encoded proteins. On the other hand, overexpression of Wnt5a or Wnt11 does not lead to β-catenin stabilization (Torres et al. 1996; Mikels and Nusse 2006). Owing to these contradictory observations, the final conclusion regarding the involvement of these Wnt proteins in β-catenin stabilization and body-axis formation awaits additional experimental insights. Despite the variability in the existing models of vertebrate axis specification, there is a common outcome: the establishment of a signaling center that dorsalizes all three germ layers and mediates anteroposterior patterning.

ZYGOTIC WNT SIGNALING DURING ANTEROPOSTERIOR AXIS SPECIFICATION

Early studies revealed that the effect of Wnt proteins on embryogenesis is stage specific. When injected as mRNA, Wnt8 triggers secondary axis formation (Smith and Harland 1991; Sokol et al. 1991). However, when supplied as a plasmid, Wnt8 expression is initiated only at midblastula transition and the delayed stimulation leads to posteriorization, characterized by lack of head structures (Christian and Moon 1993). These observations correlated well with the time-sensitive effect of lithium chloride, a GSK3 inhibitor (Klein and Melton 1996), on anteroposterior patterning (Yamaguchi and Shinagawa 1989). Of note, Dkk1 RNA does not interfere with the early dorsal signaling, although being fully capable of inhibiting ectopic Wnt expression (Glinka et al. 1998; Brott and Sokol 2002), consistent with the idea that Wnt ligands are not accessible to Dkk-dependent inhibition at this early stage. These observations indicate that Wnt signaling plays different roles in axis specification at different developmental stages.

Several Wnt ligands, including Wnt3a, Wnt5a, Wnt8, and Wnt11, are expressed in the ventral or posterior region of the embryo (Christian et al. 1991a; Krauss et al. 1992; Moon et al. 1993; Kelly et al. 1995; Hong et al. 2008). On the other hand, multiple Wnt antagonists, such as Frzb/Sfrp3, Crescent, Shisa, and Dkk1, are expressed in the head region (Leyns et al. 1997; Wang et al. 1997; Glinka et al. 1998; Shibata et al. 2005; Yamamoto et al. 2005). This complementary expression pattern indicates a function for the Wnt pathway in anteroposterior axis specification. The involvement of Wnt signaling in anteroposterior patterning was first revealed in studies with overexpressed Wnt pathway modulators. Twofold titration of *Xenopus*

Dishevelled RNA caused graded appearance of region-specific positional markers, indicating a morphogenlike dose response with the highest point at the posterior of the embryo (Itoh and Sokol 1997). Wnt pathway antagonists, such as GSK3, Dkk1, and Shisa, cause anteriorization, in agreement with the proposed posteriorizing activity of Wnt ligands (Itoh et al. 1995; Glinka et al. 1998; Yamamoto et al. 2005). Further supporting this hypothesis, TCF-dependent reporter activity and the nuclear localization of endogenous β-catenin are elevated in the anterior-to-posterior gradient (Kiecker and Niehrs 2001; Dorsky et al. 2002).

The critical role of Wnt signaling in anteroposterior axis specification has been corroborated by depletion experiments for pathway components (Heasman et al. 2000; Erter et al. 2001; Lekven et al. 2001; Shimizu et al. 2005; Bellipanni et al. 2006; Hikasa et al. 2010) and the analysis of the corresponding zebrafish mutations. *Headless* (*hdl*) was identified as a point mutation in a *tcf3* gene homolog, which represses Wnt target genes in zebrafish embryos (Kim et al. 2000). Loss of forebrain and eyes and the modest expansion of ventral mesoderm in *hdl* embryos are consistent with up-regulated Wnt signaling. When both Headless and a closely related protein Tcf3b are down-regulated, this posteriorization is more severe (Dorsky et al. 2003). Similarly, *mastermind* zebrafish embryos with a mutation in *Axin1* are posteriorized with telencephalon and eyes transformed to diencephalon (Heisenberg et al. 2001). Thus, interference with Wnt pathway antagonists mimics Wnt protein overexpression and results in posteriorization. Phenotypes of mice lacking the genes for Dkk1, APC, TCF3, Wnt3, ICAT, and β-catenin also support the role of the Wnt pathway in posterior development (Takada et al. 1994; Liu et al. 1999b; Mukhopadhyay et al. 2001; Ishikawa et al. 2003; Merrill et al. 2004; Satoh et al. 2004). Combined with more recent studies of Cnidaria and Planaria (Holstein 2008; Martin and Kimelman 2009; Petersen and Reddien 2009), these findings establish a conserved key role for Wnt signaling in anteroposterior axis specification (Fig. 2).

PUTATIVE TRANSCRIPTIONAL TARGETS OF THE WNT PATHWAY IN AXIS SPECIFICATION

Molecular screens have identified transcriptional Wnt targets in early embryos and embryonic stem cells (Willert et al. 2002; Li et al. 2004; Hufton et al. 2006). A major obstacle to identifying Wnt gene targets is the stage and context dependence of cellular responses to Wnt ligands. Therefore, a special effort is needed to elucidate the Wnt targets that are relevant to anteroposterior axis specification.

Hybridization of RNA probes prepared from the control and Wnt3a-stimulated ectoderm cells with *Xenopus* copy DNA (cDNA) microarrays (Affymetrix) has identified a number of candidate Wnt targets (Table 1) (H Hikasa and SY Sokol, unpubl.). Among these are Wnt pathway components and modulators, such as Fz9, Fz10, Kremen 2, and the Axin family, highlighting the need for feedback regulation at this developmental stage. A second class contains many of the previously identified Wnt targets that are expressed ventrally or posteriorly. These genes are *Gbx2* (von Bubnoff et al. 1996), the *Cdx* group (Epstein et al. 1997; Haremaki et al. 2003; Shimizu et al. 2005; Keenan et al. 2006; Pilon et al. 2006), the *Vent*/*Vox*/*Ved*/*Xom* group (Gawantka et al. 1995; Ladher et al. 1996; Onichtchouk et al. 1996; Schmidt et al. 1996; Imai et al. 2001; Ramel and Lekven 2004; Thorpe and Moon 2004), and the *Meis* group (*Meis3*) (Choe et al. 2009; Elkouby et al. 2010). The remaining genes include some known bone morphogenetic protein (BMP)- and fibroblast growth factor (FGF)-inducible genes, reiterating the importance of the cross talk between these pathways, and novel putative Wnt targets (Table 1). Whereas some of these genes may be activated indirectly, *Meis3, Gbx2, Cdx4*, and *Vent1*/*2* are known to contain functional TCF-binding sites in their promoters that are required for Wnt responsiveness (Haremaki et al. 2003; Li et al. 2009; Elkouby et al. 2010; Hikasa et al. 2010). Supporting these findings, a complementary analysis revealed a similar list of genes that are inhibited in ventral mesoderm by Dkk1 (Table 1) (Hufton et al. 2006).

Cite this article as *Cold Spring Harb Perspect Biol* doi: 10.1101/cshperspect.a007955

The *Cdx* genes, a group of four conserved vertebrate homologes of *Drosophila caudal*, have been widely implicated in posterior development (Northrop and Kimelman 1994; Isaacs et al. 1998; Pillemer et al. 1998; van den Akker et al. 2002; Shimizu et al. 2005, 2006; Faas and Isaacs 2009; Young and Deschamps 2009; Young et al. 2009; Beck and Stringer 2010). Loss of the *Cdx4* gene function in the zebrafish mutant *kugelig* shows its essential role in posterior development (Davidson et al. 2003). The promoter of the *Xenopus* and mammalian *Cdx4* gene is Wnt inducible and contains multiple TCF/LEF binding sites (Haremaki et al. 2003; Shimizu et al. 2005; Pilon et al. 2006). Up-regulation of *Cdx* genes causes the anterior shift of posterior markers in *Xenopus* and zebrafish (Isaacs et al. 1998; Shimizu et al. 2005; Faas and Isaacs 2009). Similar to *Cdx* genes, *Vent* family proteins (*Vent/Vox/Ved/Xom*) in *Xenopus* and zebrafish are homeodomain-containing transcriptional repressors that promote ventroposterior fates by suppressing organizer genes such as *bozozok* (in zebrafish), *chordin*, and *goosecoid* (Schmidt et al. 1996; Melby et al. 1999, 2000; Imai et al. 2001; Ramel and Lekven 2004; Sander et al. 2007). Inactivation of *Vent* genes results in a severe loss of ventroposterior structures and enhanced dorsoanterior fate, similar to the Wnt8 knockdown (Ramel and Lekven 2004; Sander et al. 2007). Together, these findings show that the Wnt pathway controls posterior development through multiple gene targets.

Putative Wnt target genes that are involved in anteroposterior patterning differ from the early β-catenin targets that mark the Spemann organizer region. These organizer-specific genes include *Xenopus Siamois* (Lemaire et al. 1995; Brannon et al. 1997; Fan et al. 1998) and *Twin* (Laurent et al. 1997), zebrafish *Bozozok/dharma/nieuwkoid* (Koos and Ho 1998; Yamanaka et al. 1998; Fekany et al. 1999), *Xnr3* (Smith et al. 1995), *Xnr5/6* (Takahashi et al. 2000; Yang et al. 2002) and several Wnt antagonists: *Frzb/Sfrp3* (Leyns et al. 1997), *Dkk1* (Glinka et al. 1998), *Crescent* (Shibata et al. 2000), *Cerberus* (Bouwmeester et al. 1996), *Shisa* (Yamamoto et al. 2005), as well as other genes listed in Table 2. The difference between the early and late targets for the Wnt pathway corresponds to the switch between stage-specific Wnt signaling mechanisms that are critical for vertebrate axis specification. For additional information about Wnt targets in different experimental systems, the reader is referred to Roel Nusse's Wnt Window (http://www.stanford.edu/group/nusselab/cgi-bin/wnt/target_genes).

Table 2. Putative Wnt targets during dorsoventral axis specification

Genes activated by maternal β-catenin	References
Siamois	Lemaire et al. 1995; Brannon et al. 1997; Fan et al. 1998
Twin	Laurent et al. 1997
Xnr3	Smith et al. 1995
Goosecoid	Blumberg et al. 1991; Cho et al. 1991
Cerberus	Bouwmeester et al. 1996
Chordin	Sasai et al. 1994
Crescent	Shibata et al. 2000; Shibata et al. 2001
Noggin	Smith and Harland 1992
Xlim-1	Taira et al. 1994
Xnr5/6	Takahashi et al. 2000; Yang et al. 2002
Xnot1	von Dassow et al. 1993
Shisa	Yamamoto et al. 2005
Dkk1	Glinka et al. 1998
Frzb	Leyns et al. 1997
Frizzled 8	Deardorff et al. 1998; Itoh et al. 1998a
Bozozok	Koos and Ho 1998; Yamanaka et al. 1998; Fekany et al. 1999 (zebrafish-specific, also known as *Dharma* and *Nieuwkoid*)

Known antagonists of Wnt and BMP signaling are shown in bold.

REGULATION OF TCF PROTEIN FUNCTION DURING AXIS SPECIFICATION

How does Wnt signaling lead to target gene activation during anteroposterior patterning? A critical event seems to be TCF3 phosphorylation by HIPK2 allowing one to derepress the target genes (Hikasa et al. 2010; Sokol 2011), although in some cases TCF3 is likely to function in a Wnt-independent manner (Cole et al. 2008; Gribble et al. 2009). Homeodomain-interacting protein kinases (HIPKs) have been shown to function in a context-dependent manner in response to Wnt signaling (Wei et al. 2007; Lee et al. 2009; Louie et al. 2009; Kim et al. 2010). Loss-of-function studies reveal that HIPK2 is required in vivo for Wnt8-dependent ventroposterior development and antagonizes the function of TCF3 in axis specification (Hikasa et al. 2010). Besides HIPK2, TCF3 phosphorylation requires β-catenin. TCF3 with a mutation that disrupts β-catenin binding is not phosphorylated (Hikasa et al. 2010). Thus, β-catenin appears to act by providing a scaffold for TCF3 phosphorylation by HIPK2. This is a novel function of β-catenin that differs from its commonly proposed role as transcriptional coactivator (Behrens et al. 1996; van de Wetering et al. 1997; Cavallo et al. 1998; Daniels and Weis 2005). Although TCF3 phosphorylation has been shown to require LRP6 activity and GSK3 inhibition (Hikasa and Sokol 2011), the mechanistic role of GSK3 in HIPK2 activation remains a mystery.

Nemolike kinase (Nlk) has been previously reported to phosphorylate TCF proteins (Ishitani et al. 1999; Meneghini et al. 1999) and is a positive regulator of Wnt8 signaling in anteroposterior patterning in zebrafish embryos (Thorpe and Moon 2004). In mammalian cells, Nlk and HIPK2 have been shown to regulate the stability of Myb oncoprotein in response to Wnt1 (Kanei-Ishii et al. 2004), indicating that these kinases might also function together during axis specification. The pathway leading to TCF regulation in vertebrates is reminiscent of Wnt signaling in *C. elegans*, because LIT-1, a homolog of Nlk, and WRM-1, a β-catenin paralog, phosphorylates POP-1/TCF to enhance its nuclear export, leading to transcriptional derepression (Rocheleau et al. 1997, 1999; Meneghini et al. 1999; Lo et al. 2004). Because POP-1 plays a dual role in gene target regulation, the reader is referred to other reviews for additional information (Phillips and Kimble 2009; Sokol 2011).

Chromatin precipitation analysis has shown that phosphorylated TCF3 is removed from target promoters, resulting in gene activation (Hikasa et al. 2010). Subsequent experiments showed that *Vent2* gene activation is accompanied by the removal of TCF3 from the promoter and its replacement by TCF1, an activator form of TCF that is resistant to HIPK2-mediated phosphorylation (Hikasa and Sokol 2011). This finding strongly supports the view that the four vertebrate homologs of TCF, expressed in a tissue-specific manner (Molenaar et al. 1998; Dorsky et al. 1999, 2003; Konig et al. 2000; Roel et al. 2003; Veien et al. 2005), play diverse roles in Wnt signaling and axis specification (Hamilton et al. 2001; Houston et al. 2002; Roel et al. 2002; Liu et al. 2005; Standley et al. 2006). Although both TCF3 and TCF1 can bind β-catenin and Groucho corepressors (Roose et al. 1998; Brantjes et al. 2001), the two proteins appear to function very differently. TCF3 serves exclusively as a transcriptional repressor, whereas TCF1 is a transcriptional activator (Kim et al. 2000; Gradl et al. 2002; Houston et al. 2002; Merrill et al. 2004; Liu et al. 2005; Nguyen et al. 2006; Gribble et al. 2009; Hikasa and Sokol 2011). Additional studies are needed to explain the difference in the activity of TCF1 and TCF3, which is likely caused by a distinct set of interacting partners.

PATHWAY SWITCH MECHANISMS

Whereas the early dorsal accumulation of β-catenin is essential for the activation of organizer genes, it is later followed by the stabilization of β-catenin ventrally (Kiecker and Niehrs 2001; Schohl and Fagotto 2002; Hikasa et al. 2010). This reversed distribution of β-catenin correlates with the ventroposterior expression of Wnt8 (Christian et al. 1991a,b; Kelly et al. 1995; Lekven et al. 2001) and dorsoanterior expression of several Wnt antagonists in both *Xenopus* and zebrafish (Fig. 2). Thus, a likely explanation for this change is that dorsal β-catenin stimulates the transcription of pathway antagonists, such as *Dkk1, Shisa, FrzB*, and, likely, *Axin* family proteins (Leyns et al. 1997; Glinka et al. 1998; Yamamoto et al. 2005). The negative feedback from Wnt antagonists decreases β-catenin in the dorsoanterior region, whereas ventroposterior activity of Wnt8 up-regulates β-catenin ventrally. In agreement with this model, the increased level of β-catenin in ventroposterior tissues relies mostly on Wnt8 activity, rather than dorsally derived signals (Hikasa et al. 2010).

Although the molecular cause for the switch from early dorsal Wnt signaling to late ventroposterior Wnt signaling has not yet been identified, the gene targets for each round of signaling appear to be very different. A number of questions remain to be answered. (1) What prevents late Wnt targets from being activated early at the dorsal side? (2) Why are the early targets silent ventrally at gastrulation when β-catenin is stabilized in response to Wnt8 signals at later stages? A possible explanation of this context dependence is related to epigenetic changes in chromatin that underlie the developmental control of transcription (Li 2002; Kouzarides 2007; Ng and Gurdon 2008; Akkers et al. 2009; Xu et al. 2010). A plasmid reporter for the *Siamois* gene failed to respond to a constitutive TCF activator after gastrulation (Darken and Wilson 2001), suggesting that the stage-dependent regulation of target genes depends on some downstream components of the pathway. Moreover, β-catenin can modulate histone methylation of poised genes, indicating local epigenetic changes in chromatin (Blythe et al. 2010). Thus, ventroposterior Wnt targets, such as *Vent* or *Cdx*, may be poised for activation at a later time, predicting the existence of epigenetic differences in the chromatin state that are critical for axis specification.

Another explanation is that functionally diverse vertebrate TCF proteins are responsible for the early to late transition in Wnt target specificity. Besides the distinct roles of TCF proteins as discussed above, the switch from early to late Wnt signaling can be linked to the phosphorylation of TCF by Wnt8/HIPK2, attenuating target DNA binding. This takes place only after the beginning of gastrulation, likely depending on a factor that becomes available after the onset of zygotic transcription. Notably, Wnt8/HIPK2 signal can trigger the phosphorylation of TCF3, TCF4, and Lef1 (but not TCF1), owing to the conservation of the HIPK2 phosphorylation sites. Thus, context-dependent activation of early and late Wnt target genes may be explained by the opposing action of tissue-specific TCF proteins and their distinct response to Wnt8-mediated phosphorylation (Sokol 2011). Whether this explanation is correct remains to be experimentally tested.

CROSS TALK WITH OTHER PATHWAYS

Organizer-derived posteriorizing signals reach neuroectoderm during gastrulation to specify the anteroposterior axis (Nieuwkoop 1952; Doniach et al. 1992; Holowacz and Sokol 1999). Besides Wnt signaling, this process requires additional pathways, such as BMP and FGF signaling. The region- and stage-specific integration of distinct signals contributes to the differential selection of pathway targets that is crucial for the establishment of embryonic axes. The best known examples are the cross talk between the Wnt pathway and the BMP pathway in the ventroposterior region (Itasaki and Hoppler 2010) or FGF/IGF (insulinlike growth factor) receptor tyrosine kinase signaling dorsally and posteriorly (Schohl and Fagotto 2002; Pera et al. 2003; Kudoh et al. 2004; Marchal et al. 2009).

Both BMP and Wnt signaling are involved in setting up ventroposterior gene expression in vertebrate embryos (Hoppler and Moon 1998; De Robertis and Kuroda 2004; Schier and Talbot 2005). The synergistic function of Wnt8 and BMP4 in ventral mesoderm patterning has been first shown in *Xenopus* embryos (Hoppler and Moon 1998) and confirmed by genetic studies in zebrafish. Reduced *Vent* expression in zebrafish embryos depleted of β-catenin or containing a mutation in the *Wnt8* gene (Ramel and Lekven 2004; Bellipanni et al. 2006) indicates a requirement for the Wnt pathway in *Vent* activation. The *Wnt8* and *Swr* (*Bmp2*) double mutants display a progressive reduction of *Vent* gene expression and a concomitant expansion of the dorsal domain (Ramel and Lekven 2004; Ramel et al. 2005). The synergy between BMP and Wnt signaling has been documented in many other developmental processes, including bone, kidney, limb, and tooth development (Perantoni 2003; Tucker and Sharpe 2004;

Hartmann 2006; Yokoyama 2008; Itasaki and Hoppler 2010).

In contrast to the BMP pathway that is active in the ventral region of the blastula embryo, FGF signaling seems to be essential for dorsal posterior patterning (Amaya et al. 1991; Cox and Hemmati-Brivanlou 1995; Lamb and Harland 1995; Holowacz and Sokol 1999; Schohl and Fagotto 2002; Kudoh et al. 2004; Marchal et al. 2009; Martin and Kimelman 2009). Many Wnt-inducible genes, including *Cdx4, Xpo, Meis*, and *Marginal coil (Xmc)*, have been previously identified as FGF-responsive genes (Table 1) (Amaya et al. 1993; Frazzetto et al. 2002; Aamar and Frank 2004; Chung et al. 2004; Keenan et al. 2006). In studies using dominant–negative FGF receptor constructs and specific morpholino oligonucleotides, FGF signaling has been reported to be essential for the Wnt pathway to activate its targets during anteroposterior axis development (McGrew et al. 1997; Maegawa et al. 2006; Burks et al. 2009). The cross talk between the Wnt and the FGF pathway has been preserved in many other models, including brain patterning, limb formation, tail regeneration, and lung and kidney morphogenesis (Martinez et al. 1999; Schedl and Hastie 2000; Kawakami et al. 2001; Lupo et al. 2002; Villanueva et al. 2002; Lin and Slack 2008).

The mechanisms underlying these examples of pathway cross talk are being revealed. In some cases two pathways operate sequentially, with one pathway inducing the other. For example, FGF8a signaling transcriptionally activates Wnt8 to promote neural crest development (Fig. 3A) (Hong et al. 2008). Alternatively, two pathways may function in parallel, when a target promoter contains transcription factor binding sites that are specific for each pathway (Fig. 3B). Both Wnt and BMP signals promote ventral fate and directly activate the expression of *Vent* genes. The *Vent2* promoter contains a unique Wnt-responsive TCF-binding site and the Smad and OAZ sites needed for the BMP response (Candia et al. 1997; Hata et al. 2000; Karaulanov et al. 2004; Hikasa et al. 2010). Mutagenesis of individual promoter elements indicates that the Wnt and the BMP pathways separately control the *Vent2* gene (Hikasa et al. 2010).

Similarly, a combination of Wnt and FGF signals leads to a robust expression of *Cdx4* in *Xenopus* and zebrafish (Shimizu et al. 2005; Keenan et al. 2006). The regulatory region of the *Cdx4 gene* contains multiple TCF and Ets

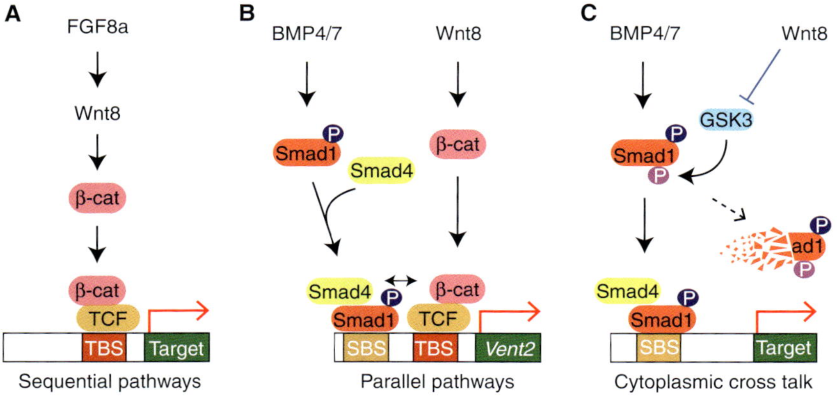

Figure 3. Cross talk between Wnt, BMP, and FGF pathways during axis specification. (*A*) Sequential interaction of two pathways. FGF8a induces the expression of Wnt8 transcripts to promote neural crest formation in *Xenopus* neurulae. (*B*) Parallel interaction of two pathways at a target promoter (*Vent2*). *Vent2* is up-regulated by BMP proteins acting through Smad1 interacting with the Smad binding site (SBS), whereas Wnt proteins act through TCF and the TCF-binding site (TBS). Both signals are responsible for the maximal activation of *Vent2*. (*C*) The cytoplasmic integration of two pathways. GSK3 phosphorylates the activated form of Smad1, marking it for degradation. Wnt proteins signal to inhibit or sequester GSK3, thereby promoting BMP target gene activation.

 Cite this article as *Cold Spring Harb Perspect Biol* doi: 10.1101/cshperspect.a007955

DNA-binding elements that are located in close proximity (Haremaki et al. 2003). This structure of the *Cdx4* regulatory region explains the observation that *Cdx4* is a direct target of both Wnt and FGF pathways, because Ets factors are known to regulate gene expression downstream from FGF (Nentwich et al. 2009; Znosko et al. 2010). This transcriptional cross talk does not preclude a more close interaction, in which two transcription factors interact physically, resulting in a synergistic response (Fig. 3B). The physical interaction of LEF1 with Smad3 and Smad4 may help explain the cross talk between Wnt and TGFβ signaling, relevant to the specification of the primary dorsoventral axis by early Wnt signaling (Sokol and Melton 1992; Sokol 1993; Labbe et al. 2000; Nishita et al. 2000).

A direct cytoplasmic mechanism has been proposed for Wnt-BMP and FGF-BMP pathway cross talk. Smad1 proteins are phosphorylated by MAPK in response to FGF and IGF, targeting Smad1 for degradation by the proteasome (Massague 2003; Pera et al. 2003; Sapkota et al. 2007). Similarly, GSK3 phosphorylates Smad1 and this phosphorylation can be inhibited by Wnt signaling in tissue culture cells (Fig. 3C) (Fuentealba et al. 2007; Sapkota et al. 2007). These phosphorylation events may provide specific interfaces for pathway cross talk. Nevertheless, this mode of regulation may not be the main mechanism for Wnt8-mediated anteroposterior axis determination, because no significant change in the amount of active Smad1 (phosphorylated at the carboxyl terminus) is observed in *Xenopus* embryos with altered Wnt8 activity (Hikasa et al. 2010). On the contrary, the requirement for the TCF-binding site in *Vent2* reporter activation and the regulation of TCF by Wnt signaling reiterates the importance of TCF regulation for Wnt signaling (Hikasa et al. 2010).

Besides FGF and BMP signaling, there are examples of Wnt pathway cross talk with other signaling pathways, especially those linked to proteasome-mediated protein degradation. The β-transducin repeat-containing protein (βTrCP) is an E3 ubiquitin ligase that represents a nodal point for down-regulation of β-catenin, Gli, and NF-κB levels (Liu et al. 1999a; Maniatis 1999). As Wnt signaling has been shown to up-regulate βTrCP activity in a negative-feedback loop (Spiegelman et al. 2000), the effect on the Hh and NF-κB signaling may be expected. Also, the Wnt pathway negatively regulates the activity of GSK3 (Cook et al. 1996; Itoh et al. 1998b; Taelman et al. 2010), an enzyme that has been implicated in the regulation (e.g., degradation) of a large number of proteins, in addition to β-catenin (Zhou et al. 2004; Xu et al. 2009). These findings suggest the interaction of the Wnt pathway with many signaling pathways, which is likely to be better understood in the course of future studies.

CONCLUDING REMARKS

Recent studies support the view that Wnt signaling plays a dual role in vertebrate axis specification. Initially, the maternally encoded components of the pathway help to establish the dorsoventral axis, whereas at a later stage the zygotic Wnt pathway is involved in anteroposterior axis specification. These two roles are mediated by two distinct sets of specific targets, yet the underlying mechanisms for target selection are unclear. Further work is needed to understand the role of maternally produced Wnt ligands in early axis determination in the context of the abundant evidence for the involvement of microtubule-dependent trafficking in the dorsal accumulation of β-catenin. Although HIPK2-mediated phosphorylation of TCF3 is a major route for anteroposterior target activation, this phosphorylation is undetectable before gastrulation, suggesting an alternative mechanism for early target derepression (Hikasa et al. 2010). Other possibilities should also be explored, given that β-catenin can use other cofactors, besides TCFs, for transcriptional control (Olson et al. 2006; Takao et al. 2007; Abu-Remaileh et al. 2010). Future research will be driven by the need to better understand pathway target selection and the contribution of Wnt5 and Wnt11 signaling to body-axis specification.

ACKNOWLEDGMENTS

We thank Stefan Hoppler, Keiji Itoh, and Jenya Grinblat for comments on the manuscript. We apologize to those investigators, whose work has not been cited here because of limited space. The work in the Sokol laboratory is supported by National Institutes of Health grants.

REFERENCES

Aamar E, Frank D. 2004. Xenopus Meis3 protein forms a hindbrain-inducing center by activating FGF/MAP kinase and PCP pathways. *Development* **131:** 153–163.

Abu-Remaileh M, Gerson A, Farago M, Nathan G, Alkalay I, Zins Rousso S, Gur M, Fainsod A, Bergman Y. 2010. Oct-3/4 regulates stem cell identity and cell fate decisions by modulating Wnt/β-catenin signalling. *EMBO J* **29:** 3236–3248.

Akkers RC, van Heeringen SJ, Jacobi UG, Janssen-Megens EM, Francoijs KJ, Stunnenberg HG, Veenstra GJ. 2009. A hierarchy of H3K4me3 and H3K27me3 acquisition in spatial gene regulation in *Xenopus* embryos. *Dev Cell* **17:** 425–434.

Amaya E, Musci TJ, Kirschner MW. 1991. Expression of a dominant negative mutant of the FGF receptor disrupts mesoderm formation in *Xenopus* embryos. *Cell* **66:** 257–270.

Amaya E, Stein PA, Musci TJ, Kirschner MW. 1993. FGF signalling in the early specification of mesoderm in *Xenopus*. *Development* **118:** 477–487.

Andreazzoli M, Broccoli V, Dawid IB. 2001. Cloning and expression of *noz1*, a zebrafish zinc finger gene related to *Drosophila* nocA. *Mech Dev* **104:** 117–120.

Angers S, Moon RT. 2009. Proximal events in Wnt signal transduction. *Nat Rev Mol Cell Biol* **10:** 468–477.

Axelrod JD. 2009. Progress and challenges in understanding planar cell polarity signaling. *Semin Cell Dev Biol* **20:** 964–971.

Beck F, Stringer EJ. 2010. The role of Cdx genes in the gut and in axial development. *Biochem Soc Trans* **38:** 353–357.

Behrens J, von Kries JP, Kuhl M, Bruhn L, Wedlich D, Grosschedl R, Birchmeier W. 1996. Functional interaction of β-catenin with the transcription factor LEF-1. *Nature* **382:** 638–642.

Bellipanni G, Varga M, Maegawa S, Imai Y, Kelly C, Myers AP, Chu F, Talbot WS, Weinberg ES. 2006. Essential and opposing roles of zebrafish β-catenins in the formation of dorsal axial structures and neurectoderm. *Development* **133:** 1299–1309.

Bilic J, Huang YL, Davidson G, Zimmermann T, Cruciat CM, Bienz M, Niehrs C. 2007. Wnt induces LRP6 signalosomes and promotes dishevelled-dependent LRP6 phosphorylation. *Science* **316:** 1619–1622.

Blumberg B, Wright CV, De Robertis EM, Cho KW. 1991. Organizer-specific homeobox genes in *Xenopus laevis* embryos. *Science* **253:** 194–196.

Blythe SA, Cha SW, Tadjuidje E, Heasman J, Klein PS. 2010. β-Catenin primes organizer gene expression by recruiting a histone H3 arginine 8 methyltransferase, Prmt2. *Dev Cell* **19:** 220–231.

Bouwmeester T, Kim S, Sasai Y, Lu B, De Robertis EM. 1996. Cerberus is a head-inducing secreted factor expressed in the anterior endoderm of Spemann's organizer. *Nature* **382:** 595–601.

Brannon M, Gomperts M, Sumoy L, Moon RT, Kimelman D. 1997. A β-catenin/XTcf-3 complex binds to the siamois promoter to regulate dorsal axis specification in *Xenopus*. *Genes Dev* **11:** 2359–2370.

Brantjes H, Roose J, van De Wetering M, Clevers H. 2001. All Tcf HMG box transcription factors interact with Groucho-related co-repressors. *Nucleic Acids Res* **29:** 1410–1419.

Bromley E, Knapp D, Wardle FC, Sun BI, Collins-Racie L, LaVallie E, Smith JC, Sive HL. 2004. Identification and characterisation of the posteriorly-expressed *Xenopus* neurotrophin receptor homolog genes fullback and fullback-like. *Gene Expr Patterns* **5:** 135–140.

Brott BK, Sokol SY. 2002. Regulation of Wnt/LRP signaling by distinct domains of Dickkopf proteins. *Mol Cell Biol* **22:** 6100–6110.

Burks PJ, Isaacs HV, Pownall ME. 2009. FGF signalling modulates transcriptional repression by *Xenopus* groucho-related-4. *Biol Cell* **101:** 301–308.

Candia AF, Watabe T, Hawley SH, Onichtchouk D, Zhang Y, Derynck R, Niehrs C, Cho KW. 1997. Cellular interpretation of multiple TGF-β signals: Intracellular antagonism between activin/BVg1 and BMP-2/4 signaling mediated by Smads. *Development* **124:** 4467–4480.

Cavallo RA, Cox RT, Moline MM, Roose J, Polevoy GA, Clevers H, Peifer M, Bejsovec A. 1998. *Drosophila* Tcf and Groucho interact to repress Wingless signalling activity. *Nature* **395:** 604–608.

Cha SW, Tadjuidje E, Tao Q, Wylie C, Heasman J. 2008. Wnt5a and Wnt11 interact in a maternal Dkk1-regulated fashion to activate both canonical and non-canonical signaling in *Xenopus* axis formation. *Development* **135:** 3719–3729.

Cho KW, Blumberg B, Steinbeisser H, De Robertis EM. 1991. Molecular nature of Spemann's organizer: The role of the *Xenopus* homeobox gene goosecoid. *Cell* **67:** 1111–1120.

Choe SK, Lu P, Nakamura M, Lee J, Sagerstrom CG. 2009. Meis cofactors control HDAC and CBP accessibility at Hox-regulated promoters during zebrafish embryogenesis. *Dev Cell* **17:** 561–567.

Christian JL, Moon RT. 1993. Interactions between Xwnt-8 and Spemann organizer signaling pathways generate dorsoventral pattern in the embryonic mesoderm of *Xenopus*. *Genes Dev* **7:** 13–28.

Christian JL, Gavin BJ, McMahon AP, Moon RT. 1991a. Isolation of cDNAs partially encoding four *Xenopus* Wnt-1/int-1-related proteins and characterization of their transient expression during embryonic development. *Dev Biol* **143:** 230–234.

Christian JL, McMahon JA, McMahon AP, Moon RT. 1991b. Xwnt-8, a *Xenopus* Wnt-1/int-1-related gene responsive to mesoderm-inducing growth factors, may play a role in ventral mesodermal patterning during embryogenesis. *Development* **111:** 1045–1055.

Chung HM, Malacinski GM. 1980. Establishment of the dorsal/ventral polarity of the amphibian embryo: Use of ultraviolet irradiation and egg rotation as probes. *Dev Biol* **80:** 120–133.

Chung HA, Hyodo-Miura J, Kitayama A, Terasaka C, Nagamune T, Ueno N. 2004. Screening of FGF target genes in *Xenopus* by microarray: Temporal dissection of the signalling pathway using a chemical inhibitor. *Genes Cells* **9:** 749–761.

Clevers H. 2006. Wnt/β-catenin signaling in development and disease. *Cell* **127:** 469–480.

Cliffe A, Hamada F, Bienz M. 2003. A role of Dishevelled in relocating Axin to the plasma membrane during wingless signaling. *Curr Biol* **13:** 960–966.

Cole MF, Johnstone SE, Newman JJ, Kagey MH, Young RA. 2008. Tcf3 is an integral component of the core regulatory circuitry of embryonic stem cells. *Genes Dev* **22:** 746–755.

Conlon FL, Lyons KM, Takaesu N, Barth KS, Kispert A, Herrmann B, Robertson EJ. 1994. A primary requirement for nodal in the formation and maintenance of the primitive streak in the mouse. *Development* **120:** 1919–1928.

Cook D, Fry MJ, Hughes K, Sumathipala R, Woodgett JR, Dale TC. 1996. Wingless inactivates glycogen synthase kinase-3 via an intracellular signalling pathway which involves a protein kinase C. *EMBO J* **15:** 4526–4536.

Cox WG, Hemmati-Brivanlou A. 1995. Caudalization of neural fate by tissue recombination and bFGF. *Development* **121:** 4349–4358.

Daniels DL, Weis WI. 2005. β-Catenin directly displaces Groucho/TLE repressors from Tcf/Lef in Wnt-mediated transcription activation. *Nat Struct Mol Biol* **12:** 364–371.

Darken RS, Wilson PA. 2001. Axis induction by wnt signaling: Target promoter responsiveness regulates competence. *Dev Biol* **234:** 42–54.

Davidson AJ, Ernst P, Wang Y, Dekens MP, Kingsley PD, Palis J, Korsmeyer SJ, Daley GQ, Zon LI. 2003. cdx4 mutants fail to specify blood progenitors and can be rescued by multiple hox genes. *Nature* **425:** 300–306.

Deardorff MA, Tan C, Conrad LJ, Klein PS. 1998. Frizzled-8 is expressed in the Spemann organizer and plays a role in early morphogenesis. *Development* **125:** 2687–2700.

De Robertis EM, Kuroda H. 2004. Dorsal-ventral patterning and neural induction in *Xenopus* embryos. *Annu Rev Cell Dev Biol* **20:** 285–308.

Doniach T, Phillips CR, Gerhart JC. 1992. Planar induction of anteroposterior pattern in the developing central nervous system of *Xenopus laevis*. *Science* **257:** 542–545.

Dorsky RI, Snyder A, Cretekos CJ, Grunwald DJ, Geisler R, Haffter P, Moon RT, Raible DW. 1999. Maternal and embryonic expression of zebrafish lef1. *Mech Dev* **86:** 147–150.

Dorsky RI, Sheldahl LC, Moon RT. 2002. A transgenic Lef1/β-catenin-dependent reporter is expressed in spatially restricted domains throughout zebrafish development. *Dev Biol* **241:** 229–237.

Dorsky RI, Itoh M, Moon RT, Chitnis A. 2003. Two tcf3 genes cooperate to pattern the zebrafish brain. *Development* **130:** 1937–1947.

Elkouby YM, Elias S, Casey ES, Blythe SA, Tsabar N, Klein PS, Root H, Liu KJ, Frank D. 2010. Mesodermal Wnt signaling organizes the neural plate via Meis3. *Development* **137:** 1531–1541.

Epstein M, Pillemer G, Yelin R, Yisraeli JK, Fainsod A. 1997. Patterning of the embryo along the anterior-posterior axis: The role of the caudal genes. *Development* **124:** 3805–3814.

Erter CE, Wilm TP, Basler N, Wright CV, Solnica-Krezel L. 2001. Wnt8 is required in lateral mesendodermal precursors for neural posteriorization in vivo. *Development* **128:** 3571–3583.

Faas L, Isaacs HV. 2009. Overlapping functions of Cdx1, Cdx2, and Cdx4 in the development of the amphibian *Xenopus tropicalis*. *Dev Dyn* **238:** 835–852.

Fan MJ, Sokol SY. 1997. A role for Siamois in Spemann organizer formation. *Development* **124:** 2581–2589.

Fan MJ, Gruning W, Walz G, Sokol SY. 1998. Wnt signaling and transcriptional control of Siamois in *Xenopus* embryos. *Proc Natl Acad Sci* **95:** 5626–5631.

Fekany K, Yamanaka Y, Leung T, Sirotkin HI, Topczewski J, Gates MA, Hibi M, Renucci A, Stemple D, Radbill A, et al. 1999. The zebrafish bozozok locus encodes Dharma, a homeodomain protein essential for induction of gastrula organizer and dorsoanterior embryonic structures. *Development* **126:** 1427–1438.

Feldman B, Gates MA, Egan ES, Dougan ST, Rennebeck G, Sirotkin HI, Schier AF, Talbot WS. 1998. Zebrafish organizer development and germ-layer formation require nodal-related signals. *Nature* **395:** 181–185.

Fossat N, Jones V, Khoo PL, Bogani D, Hardy A, Steiner K, Mukhopadhyay M, Westphal H, Nolan PM, Arkell R, et al. 2011. Stringent requirement of a proper level of canonical WNT signalling activity for head formation in mouse embryo. *Development* **138:** 667–676.

Frazzetto G, Klingbeil P, Bouwmeester T. 2002. *Xenopus* marginal coil (Xmc), a novel FGF inducible cytosolic coiled-coil protein regulating gastrulation movements. *Mech Dev* **113:** 3–14.

Fuentealba LC, Eivers E, Ikeda A, Hurtado C, Kuroda H, Pera EM, De Robertis EM. 2007. Integrating patterning signals: Wnt/GSK3 regulates the duration of the BMP/Smad1 signal. *Cell* **131:** 980–993.

Fujisue M, Kobayakawa Y, Yamana K. 1993. Occurrence of dorsal axis-inducing activity around the vegetal pole of an uncleaved *Xenopus* egg and displacement to the equatorial region by cortical rotation. *Development* **118:** 163–170.

Gao B, Song H, Bishop K, Elliot G, Garrett L, English MA, Andre P, Robinson J, Sood R, Minami Y, et al. 2011. Wnt signaling gradients establish planar cell polarity by inducing Vangl2 phosphorylation through Ror2. *Dev Cell* **20:** 163–176.

Garcia-Morales C, Liu CH, Abu-Elmagd M, Hajihosseini MK, Wheeler GN. 2009. Frizzled-10 promotes sensory neuron development in *Xenopus* embryos. *Dev Biol* **335:** 143–155.

Gawantka V, Delius H, Hirschfeld K, Blumenstock C, Niehrs C. 1995. Antagonizing the Spemann organizer: Role of the homeobox gene Xvent-1. *EMBO J* **14:** 6268–6279.

Glinka A, Wu W, Delius H, Monaghan AP, Blumenstock C, Niehrs C. 1998. Dickkopf-1 is a member of a new family of secreted proteins and functions in head induction. *Nature* **391:** 357–362.

Gradl D, Konig A, Wedlich D. 2002. Functional diversity of *Xenopus* lymphoid enhancer factor/T-cell factor transcription factors relies on combinations of activating and repressing elements. *J Biol Chem* **277:** 14159–14171.

Gribble SL, Kim HS, Bonner J, Wang X, Dorsky RI. 2009. Tcf3 inhibits spinal cord neurogenesis by regulating sox4a expression. *Development* **136:** 781–789.

Grigoryan T, Wend P, Klaus A, Birchmeier W. 2008. Deciphering the function of canonical Wnt signals in development and disease: Conditional loss- and gain-of-function mutations of β-catenin in mice. *Genes Dev* **22:** 2308–2341.

Grumolato L, Liu G, Mong P, Mudbhary R, Biswas R, Arroyave R, Vijayakumar S, Economides AN, Aaronson SA. 2010. Canonical and noncanonical Wnts use a common mechanism to activate completely unrelated coreceptors. *Genes Dev* **24:** 2517–2530.

Haegel H, Larue L, Ohsugi M, Fedorov L, Herrenknecht K, Kemler R. 1995. Lack of β-catenin affects mouse development at gastrulation. *Development* **121:** 3529–3537.

Hamilton FS, Wheeler GN, Hoppler S. 2001. Difference in XTcf-3 dependency accounts for change in response to β-catenin-mediated Wnt signalling in *Xenopus* blastula. *Development* **128:** 2063–2073.

Haremaki T, Tanaka Y, Hongo I, Yuge M, Okamoto H. 2003. Integration of multiple signal transducing pathways on Fgf response elements of the *Xenopus* caudal homologue Xcad3. *Development* **130:** 4907–4917.

Harland R, Gerhart J. 1997. Formation and function of Spemann's organizer. *Annu Rev Cell Dev Biol* **13:** 611–667.

Hartmann C. 2006. A Wnt canon orchestrating osteoblastogenesis. *Trends Cell Biol* **16:** 151–158.

Hata A, Seoane J, Lagna G, Montalvo E, Hemmati-Brivanlou A, Massague J. 2000. OAZ uses distinct DNA- and protein-binding zinc fingers in separate BMP-Smad and Olf signaling pathways. *Cell* **100:** 229–240.

Heasman J, Crawford A, Goldstone K, Garner-Hamrick P, Gumbiner B, McCrea P, Kintner C, Noro CY, Wylie C. 1994. Overexpression of cadherins and underexpression of β-catenin inhibit dorsal mesoderm induction in early *Xenopus* embryos. *Cell* **79:** 791–803.

Heasman J, Kofron M, Wylie C. 2000. β-Catenin signaling activity dissected in the early *Xenopus* embryo: A novel antisense approach. *Dev Biol* **222:** 124–134.

Heisenberg CP, Tada M, Rauch GJ, Saude L, Concha ML, Geisler R, Stemple DL, Smith JC, Wilson SW. 2000. Silberblick/Wnt11 mediates convergent extension movements during zebrafish gastrulation. *Nature* **405:** 76–81.

Heisenberg CP, Houart C, Take-Uchi M, Rauch GJ, Young N, Coutinho P, Masai I, Caneparo L, Concha ML, Geisler R, et al. 2001. A mutation in the Gsk3-binding domain of zebrafish Masterblind/Axin1 leads to a fate transformation of telencephalon and eyes to diencephalon. *Genes Dev* **15:** 1427–1434.

Hikasa H, Sokol SY. 2011. Phosphorylation of TCF proteins by homeodomain-interacting protein kinase 2. *J Biol Chem* **286:** 12093–12100.

Hikasa H, Shibata M, Hiratani I, Taira M. 2002. The *Xenopus* receptor tyrosine kinase Xror2 modulates morphogenetic movements of the axial mesoderm and neuroectoderm via Wnt signaling. *Development* **129:** 5227–5239.

Hikasa H, Ezan J, Itoh K, Li X, Klymkowsky MW, Sokol SY. 2010. Regulation of TCF3 by Wnt-dependent phosphorylation during vertebrate axis specification. *Dev Cell* **19:** 521–532.

Hilton E, Rex M, Old R. 2003. VegT activation of the early zygotic gene Xnr5 requires lifting of Tcf-mediated repression in the *Xenopus* blastula. *Mech Dev* **120:** 1127–1138.

Holowacz T, Elinson RP. 1993. Cortical cytoplasm, which induces dorsal axis formation in *Xenopus*, is inactivated by UV irradiation of the oocyte. *Development* **119:** 277–285.

Holowacz T, Sokol S. 1999. FGF is required for posterior neural patterning but not for neural induction. *Dev Biol* **205:** 296–308.

Holstein TW. 2008. Wnt signaling in cnidarians. *Methods Mol Biol* **469:** 47–54.

Hong CS, Park BY, Saint-Jeannet JP. 2008. Fgf8a induces neural crest indirectly through the activation of Wnt8 in the paraxial mesoderm. *Development* **135:** 3903–3910.

Hoppler S, Moon RT. 1998. BMP-2/-4 and Wnt-8 cooperatively pattern the *Xenopus* mesoderm. *Mech Dev* **71:** 119–129.

Hoppler S, Brown JD, Moon RT. 1996. Expression of a dominant-negative Wnt blocks induction of MyoD in *Xenopus* embryos. *Genes Dev* **10:** 2805–2817.

Houliston E, Elinson RP. 1992. Microtubules and cytoplasmic reorganization in the frog egg. *Curr Top Dev Biol* **26:** 53–70.

Houston DW, Kofron M, Resnik E, Langland R, Destree O, Wylie C, Heasman J. 2002. Repression of organizer genes in dorsal and ventral *Xenopus* cells mediated by maternal XTcf3. *Development* **129:** 4015–4025.

Huber O, Korn R, McLaughlin J, Ohsugi M, Herrmann BG, Kemler R. 1996. Nuclear localization of β-catenin by interaction with transcription factor LEF-1. *Mech Dev* **59:** 3–10.

Huelsken J, Vogel R, Brinkmann V, Erdmann B, Birchmeier C, Birchmeier W. 2000. Requirement for β-catenin in anterior-posterior axis formation in mice. *J Cell Biol* **148:** 567–578.

Hufton AL, Vinayagam A, Suhai S, Baker JC. 2006. Genomic analysis of *Xenopus* organizer function. *BMC Dev Biol* **6:** 27.

Imai Y, Gates MA, Melby AE, Kimelman D, Schier AF, Talbot WS. 2001. The homeobox genes vox and vent are redundant repressors of dorsal fates in zebrafish. *Development* **128:** 2407–2420.

Isaacs HV, Pownall ME, Slack JM. 1998. Regulation of Hox gene expression and posterior development by the *Xenopus* caudal homologue Xcad3. *EMBO J* **17:** 3413–3427.

Ishikawa TO, Tamai Y, Li Q, Oshima M, Taketo MM. 2003. Requirement for tumor suppressor Apc in the morphogenesis of anterior and ventral mouse embryo. *Dev Biol* **253:** 230–246.

Ishitani T, Ninomiya-Tsuji J, Nagai S, Nishita M, Meneghini M, Barker N, Waterman M, Bowerman B, Clevers H,

Shibuya H, et al. 1999. The TAK1-NLK-MAPK-related pathway antagonizes signalling between β-catenin and transcription factor TCF. *Nature* **399:** 798–802.

Itasaki N, Hoppler S. 2010. Crosstalk between Wnt and bone morphogenic protein signaling: A turbulent relationship. *Dev Dyn* **239:** 16–33.

Itoh K, Sokol SY. 1997. Graded amounts of *Xenopus* dishevelled specify discrete anteroposterior cell fates in prospective ectoderm. *Mech Dev* **61:** 113–125.

Itoh K, Tang TL, Neel BG, Sokol SY. 1995. Specific modulation of ectodermal cell fates in *Xenopus* embryos by glycogen synthase kinase. *Development* **121:** 3979–3988.

Itoh K, Jacob J, Sokol YS. 1998a. A role for *Xenopus* Frizzled 8 in dorsal development. *Mech Dev* **74:** 145–157.

Itoh K, Krupnik VE, Sokol SY. 1998b. Axis determination in *Xenopus* involves biochemical interactions of axin, glycogen synthase kinase 3 and β-catenin. *Curr Biol* **8:** 591–594.

Itoh K, Antipova A, Ratcliffe MJ, Sokol S. 2000. Interaction of dishevelled and *Xenopus* axin-related protein is required for wnt signal transduction. *Mol Cell Biol* **20:** 2228–2238.

Itoh K, Brott BK, Bae GU, Ratcliffe MJ, Sokol SY. 2005. Nuclear localization is required for Dishevelled function in Wnt/β-catenin signaling. *J Biol* **4:** 3.

Itoh K, Jenny A, Mlodzik M, Sokol SY. 2009. Centrosomal localization of Diversin and its relevance to Wnt signaling. *J Cell Sci* **122:** 3791–3798.

Jesuthasan S, Stahle U. 1997. Dynamic microtubules and specification of the zebrafish embryonic axis. *Curr Biol* **7:** 31–42.

Jho EH, Zhang T, Domon C, Joo CK, Freund JN, Costantini F. 2002. Wnt/β-catenin/Tcf signaling induces the transcription of Axin2, a negative regulator of the signaling pathway. *Mol Cell Biol* **22:** 1172–1183.

Kanei-Ishii C, Ninomiya-Tsuji J, Tanikawa J, Nomura T, Ishitani T, Kishida S, Kokura K, Kurahashi T, Ichikawa-Iwata E, Kim Y, et al. 2004. Wnt-1 signal induces phosphorylation and degradation of c-Myb protein via TAK1, HIPK2, and NLK. *Genes Dev* **18:** 816–829.

Karaulanov E, Knochel W, Niehrs C. 2004. Transcriptional regulation of BMP4 synexpression in transgenic *Xenopus*. *EMBO J* **23:** 844–856.

Kawakami Y, Capdevila J, Buscher D, Itoh T, Rodriguez Esteban C, Izpisua Belmonte JC. 2001. WNT signals control FGF-dependent limb initiation and AER induction in the chick embryo. *Cell* **104:** 891–900.

Keenan ID, Sharrard RM, Isaacs HV. 2006. FGF signal transduction and the regulation of Cdx gene expression. *Dev Biol* **299:** 478–488.

Kelly GM, Greenstein P, Erezyilmaz DF, Moon RT. 1995. Zebrafish wnt8 and wnt8b share a common activity but are involved in distinct developmental pathways. *Development* **121:** 1787–1799.

Kelly C, Chin AJ, Leatherman JL, Kozlowski DJ, Weinberg ES. 2000. Maternally controlled β-catenin-mediated signaling is required for organizer formation in the zebrafish. *Development* **127:** 3899–3911.

Kessler DS. 1997. Siamois is required for formation of Spemann's organizer. *Proc Natl Acad Sci* **94:** 13017–13022.

Kiecker C, Niehrs C. 2001. A morphogen gradient of Wnt/β-catenin signalling regulates anteroposterior neural patterning in *Xenopus*. *Development* **128:** 4189–4201.

Kikuchi A, Yamamoto H, Sato A. 2009. Selective activation mechanisms of Wnt signaling pathways. *Trends Cell Biol* **19:** 119–129.

Kilian B, Mansukoski H, Barbosa FC, Ulrich F, Tada M, Heisenberg CP. 2003. The role of Ppt/Wnt5 in regulating cell shape and movement during zebrafish gastrulation. *Mech Dev* **120:** 467–476.

Kim CH, Oda T, Itoh M, Jiang D, Artinger KB, Chandrasekharappa SC, Driever W, Chitnis AB. 2000. Repressor activity of Headless/Tcf3 is essential for vertebrate head formation. *Nature* **407:** 913–916.

Kim EA, Kim JE, Sung KS, Choi DW, Lee BJ, Choi CY. 2010. Homeodomain-interacting protein kinase 2 (HIPK2) targets β-catenin for phosphorylation and proteasomal degradation. *Biochem Biophys Res Commun* **394:** 966–971.

Klein PS, Melton DA. 1996. A molecular mechanism for the effect of lithium on development. *Proc Natl Acad Sci* **93:** 8455–8459.

Kolm PJ, Sive HL. 1995. Regulation of the *Xenopus* labial homeodomain genes, HoxA1 and HoxD1: Activation by retinoids and peptide growth factors. *Dev Biol* **167:** 34–49.

Komiya Y, Habas R. 2008. Wnt signal transduction pathways. *Organogenesis* **4:** 68–75.

Konig A, Gradl D, Kuhl M, Wedlich D. 2000. The HMG-box transcription factor XTcf-4 demarcates the forebrain-midbrain boundary. *Mech Dev* **93:** 211–214.

Koos DS, Ho RK. 1998. The nieuwkoid gene characterizes and mediates a Nieuwkoop-center-like activity in the zebrafish. *Curr Biol* **8:** 1199–1206.

Kouzarides T. 2007. Chromatin modifications and their function. *Cell* **128:** 693–705.

Krauss S, Korzh V, Fjose A, Johansen T. 1992. Expression of four zebrafish wnt-related genes during embryogenesis. *Development* **116:** 249–259.

Ku M, Melton DA. 1993. Xwnt-11: A maternally expressed *Xenopus* wnt gene. *Development* **119:** 1161–1173.

Kudoh T, Concha ML, Houart C, Dawid IB, Wilson SW. 2004. Combinatorial Fgf and Bmp signalling patterns the gastrula ectoderm into prospective neural and epidermal domains. *Development* **131:** 3581–3592.

Labbe E, Letamendia A, Attisano L. 2000. Association of Smads with lymphoid enhancer binding factor 1/T cell-specific factor mediates cooperative signaling by the transforming growth factor-β and wnt pathways. *Proc Natl Acad Sci* **97:** 8358–8363.

Ladher R, Mohun TJ, Smith JC, Snape AM. 1996. Xom: A *Xenopus* homeobox gene that mediates the early effects of BMP-4. *Development* **122:** 2385–2394.

Lamb TM, Harland RM. 1995. Fibroblast growth factor is a direct neural inducer, which combined with noggin generates anterior-posterior neural pattern. *Development* **121:** 3627–3636.

Larabell CA, Torres M, Rowning BA, Yost C, Miller JR, Wu M, Kimelman D, Moon RT. 1997. Establishment of the dorso-ventral axis in *Xenopus* embryos is presaged by early asymmetries in β-catenin that are modulated by the Wnt signaling pathway. *J Cell Biol* **136:** 1123–1136.

Laurent MN, Blitz IL, Hashimoto C, Rothbacher U, Cho KW. 1997. The *Xenopus* homeobox gene twin mediates

Wnt induction of goosecoid in establishment of Spemann's organizer. *Development* **124:** 4905–4916.

Lee W, Swarup S, Chen J, Ishitani T, Verheyen EM. 2009. Homeodomain-interacting protein kinases (Hipks) promote Wnt/Wg signaling through stabilization of β-catenin/Arm and stimulation of target gene expression. *Development* **136:** 241–251.

Lekven AC, Thorpe CJ, Waxman JS, Moon RT. 2001. Zebrafish wnt8 encodes two wnt8 proteins on a bicistronic transcript and is required for mesoderm and neurectoderm patterning. *Dev Cell* **1:** 103–114.

Lemaire P, Garrett N, Gurdon JB. 1995. Expression cloning of Siamois, a *Xenopus* homeobox gene expressed in dorsal-vegetal cells of blastulae and able to induce a complete secondary axis. *Cell* **81:** 85–94.

Leung T, Soll I, Arnold SJ, Kemler R, Driever W. 2003. Direct binding of Lef1 to sites in the boz promoter may mediate pre-midblastula-transition activation of boz expression. *Dev Dyn* **228:** 424–432.

Leyns L, Bouwmeester T, Kim SH, Piccolo S, De Robertis EM. 1997. Frzb-1 is a secreted antagonist of Wnt signaling expressed in the Spemann organizer. *Cell* **88:** 747–756.

Li E. 2002. Chromatin modification and epigenetic reprogramming in mammalian development. *Nat Rev Genet* **3:** 662–673.

Li Y, Fenger U, Niehrs C, Pollet N. 2003. Cyclic expression of esr9 gene in *Xenopus* presomitic mesoderm. *Differentiation* **71:** 83–89.

Li H, Malbon CC, Wang HY. 2004. Gene profiling of Frizzled-1 and Frizzled-2 signaling: Expression of G-protein-coupled receptor chimeras in mouse F9 teratocarcinoma embryonal cells. *Mol Pharmacol* **65:** 45–55.

Li B, Kuriyama S, Moreno M, Mayor R. 2009. The posteriorizing gene Gbx2 is a direct target of Wnt signalling and the earliest factor in neural crest induction. *Development* **136:** 3267–3278.

Lin G, Slack JM. 2008. Requirement for Wnt and FGF signaling in *Xenopus* tadpole tail regeneration. *Dev Biol* **316:** 323–335.

Lin S, Baye LM, Westfall TA, Slusarski DC. 2010. Wnt5b-Ryk pathway provides directional signals to regulate gastrulation movement. *J Cell Biol* **190:** 263–278.

Liu C, Kato Y, Zhang Z, Do VM, Yankner BA, He X. 1999a. β-Trcp couples β-catenin phosphorylation-degradation and regulates *Xenopus* axis formation. *Proc Natl Acad Sci* **96:** 6273–6278.

Liu P, Wakamiya M, Shea MJ, Albrecht U, Behringer RR, Bradley A. 1999b. Requirement for Wnt3 in vertebrate axis formation. *Nat Genet* **22:** 361–365.

Liu F, van den Broek O, Destree O, Hoppler S. 2005. Distinct roles for *Xenopus* Tcf/Lef genes in mediating specific responses to Wnt/β-catenin signalling in mesoderm development. *Development* **132:** 5375–5385.

Lo MC, Gay F, Odom R, Shi Y, Lin R. 2004. Phosphorylation by the β-catenin/MAPK complex promotes 14-3-3-mediated nuclear export of TCF/POP-1 in signal-responsive cells in *C. elegans*. *Cell* **117:** 95–106.

Louie SH, Yang XY, Conrad WH, Muster J, Angers S, Moon RT, Cheyette BN. 2009. Modulation of the β-catenin signaling pathway by the dishevelled-associated protein Hipk1. *PLoS ONE* **4:** e4310.

Lu W, Yamamoto V, Ortega B, Baltimore D. 2004. Mammalian Ryk is a Wnt coreceptor required for stimulation of neurite outgrowth. *Cell* **119:** 97–108.

Lupo G, Harris WA, Barsacchi G, Vignali R. 2002. Induction and patterning of the telencephalon in *Xenopus* laevis. *Development* **129:** 5421–5436.

Lyman Gingerich J, Westfall TA, Slusarski DC, Pelegri F. 2005. Hecate, a zebrafish maternal effect gene, affects dorsal organizer induction and intracellular calcium transient frequency. *Dev Biol* **286:** 427–439.

MacDonald BT, Tamai K, He X. 2009. Wnt/β-catenin signaling: Components, mechanisms, and diseases. *Dev Cell* **17:** 9–26.

Maeda R, Kobayashi A, Sekine R, Lin JJ, Kung H, Maeno M. 1997. Xmsx-1 modifies mesodermal tissue pattern along dorsoventral axis in *Xenopus laevis* embryo. *Development* **124:** 2553–2560.

Maegawa S, Varga M, Weinberg ES. 2006. FGF signaling is required for β-catenin-mediated induction of the zebrafish organizer. *Development* **133:** 3265–3276.

Majumdar A, Vainio S, Kispert A, McMahon J, McMahon AP. 2003. Wnt11 and Ret/Gdnf pathways cooperate in regulating ureteric branching during metanephric kidney development. *Development* **130:** 3175–3185.

Maniatis T. 1999. A ubiquitin ligase complex essential for the NF-κB, Wnt/Wingless, and Hedgehog signaling pathways. *Genes Dev* **13:** 505–510.

Mao B, Wu W, Davidson G, Marhold J, Li M, Mechler BM, Delius H, Hoppe D, Stannek P, Walter C, et al. 2002. Kremen proteins are Dickkopf receptors that regulate Wnt/β-catenin signalling. *Nature* **417:** 664–667.

Marazzi G, Wang Y, Sassoon D. 1997. Msx2 is a transcriptional regulator in the BMP4-mediated programmed cell death pathway. *Dev Biol* **186:** 127–138.

Marchal L, Luxardi G, Thome V, Kodjabachian L. 2009. BMP inhibition initiates neural induction via FGF signaling and Zic genes. *Proc Natl Acad Sci* **106:** 17437–17442.

Marlow FL. 2010. *Maternal control of development in vertebrates: My mother made me do it!* Morgan & Claypool Life Sciences, San Rafel, CA.

Martin BL, Kimelman D. 2009. Wnt signaling and the evolution of embryonic posterior development. *Curr Biol* **19:** R215–R219.

Martinez S, Crossley PH, Cobos I, Rubenstein JL, Martin GR. 1999. FGF8 induces formation of an ectopic isthmic organizer and isthmocerebellar development via a repressive effect on Otx2 expression. *Development* **126:** 1189–1200.

Massague J. 2003. Integration of Smad and MAPK pathways: A link and a linker revisited. *Genes Dev* **17:** 2993–2997.

Mathers PH, Grinberg A, Mahon KA, Jamrich M. 1997. The Rx homeobox gene is essential for vertebrate eye development. *Nature* **387:** 603–607.

McGrew LL, Hoppler S, Moon RT. 1997. Wnt and FGF pathways cooperatively pattern anteroposterior neural ectoderm in *Xenopus*. *Mech Dev* **69:** 105–114.

McKendry R, Hsu SC, Harland RM, Grosschedl R. 1997. LEF-1/TCF proteins mediate wnt-inducible transcription from the *Xenopus* nodal-related 3 promoter. *Dev Biol* **192:** 420–431.

McMahon AP, Moon RT. 1989. Ectopic expression of the proto-oncogene int-1 in *Xenopus* embryos leads to duplication of the embryonic axis. *Cell* **58:** 1075–1084.

Melby AE, Clements WK, Kimelman D. 1999. Regulation of dorsal gene expression in *Xenopus* by the ventralizing homeodomain gene Vox. *Dev Biol* **211:** 293–305.

Melby AE, Beach C, Mullins M, Kimelman D. 2000. Patterning the early zebrafish by the opposing actions of bozozok and vox/vent. *Dev Biol* **224:** 275–285.

Meneghini MD, Ishitani T, Carter JC, Hisamoto N, Ninomiya-Tsuji J, Thorpe CJ, Hamill DR, Matsumoto K, Bowerman B. 1999. MAP kinase and Wnt pathways converge to downregulate an HMG-domain repressor in *Caenorhabditis elegans*. *Nature* **399:** 793–797.

Merrill BJ, Pasolli HA, Polak L, Rendl M, Garcia-Garcia MJ, Anderson KV, Fuchs E. 2004. Tcf3: A transcriptional regulator of axis induction in the early embryo. *Development* **131:** 263–274.

Mikels AJ, Nusse R. 2006. Purified Wnt5a protein activates or inhibits β-catenin-TCF signaling depending on receptor context. *PLoS Biol* **4:** e115.

Mir A, Kofron M, Zorn AM, Bajzer M, Haque M, Heasman J, Wylie CC. 2007. FoxI1e activates ectoderm formation and controls cell position in the *Xenopus* blastula. *Development* **134:** 779–788.

Mizuno T, Yamaha E, Kuroiwa A, Takeda H. 1999. Removal of vegetal yolk causes dorsal deficencies and impairs dorsal-inducing ability of the yolk cell in zebrafish. *Mech Dev* **81:** 51–63.

Molenaar M, van de Wetering M, Oosterwegel M, Peterson-Maduro J, Godsave S, Korinek V, Roose J, Destree O, Clevers H. 1996. XTcf-3 transcription factor mediates β-catenin-induced axis formation in *Xenopus* embryos. *Cell* **86:** 391–399.

Molenaar M, Roose J, Peterson J, Venanzi S, Clevers H, Destree O. 1998. Differential expression of the HMG box transcription factors XTcf-3 and XLef-1 during early *Xenopus* development. *Mech Dev* **75:** 151–154.

Moon RT, Kimelman D. 1998. From cortical rotation to organizer gene expression: Toward a molecular explanation of axis specification in *Xenopus*. *Bioessays* **20:** 536–545.

Moon RT, Campbell RM, Christian JL, McGrew LL, Shih J, Fraser S. 1993. Xwnt-5A: A maternal Wnt that affects morphogenetic movements after overexpression in embryos of *Xenopus laevis*. *Development* **119:** 97–111.

Mukhopadhyay M, Shtrom S, Rodriguez-Esteban C, Chen L, Tsukui T, Gomer L, Dorward DW, Glinka A, Grinberg A, Huang SP, et al. 2001. Dickkopf1 is required for embryonic head induction and limb morphogenesis in the mouse. *Dev Cell* **1:** 423–434.

Nakamura M, Choe SK, Runko AP, Gardner PD, Sagerstrom CG. 2008. Nlz1/Znf703 acts as a repressor of transcription. *BMC Dev Biol* **8:** 108.

Nelson WJ, Nusse R. 2004. Convergence of Wnt, β-catenin, and cadherin pathways. *Science* **303:** 1483–1487.

Nentwich O, Dingwell KS, Nordheim A, Smith JC. 2009. Downstream of FGF during mesoderm formation in *Xenopus*: The roles of Elk-1 and Egr-1. *Dev Biol* **336:** 313–326.

Ng RK, Gurdon JB. 2008. Epigenetic memory of an active gene state depends on histone H3.3 incorporation into chromatin in the absence of transcription. *Nat Cell Biol* **10:** 102–109.

Nguyen H, Rendl M, Fuchs E. 2006. Tcf3 governs stem cell features and represses cell fate determination in skin. *Cell* **127:** 171–183.

Nieuwkoop PD. 1952. Activation and organization of the central nervous system in amphibians. *J Exp Zool* **120:** 83–108.

Nishita M, Hashimoto MK, Ogata S, Laurent MN, Ueno N, Shibuya H, Cho KW. 2000. Interaction between Wnt and TGF-β signalling pathways during formation of Spemann's organizer. *Nature* **403:** 781–785.

Nishita M, Yoo SK, Nomachi A, Kani S, Sougawa N, Ohta Y, Takada S, Kikuchi A, Minami Y. 2006. Filopodia formation mediated by receptor tyrosine kinase Ror2 is required for Wnt5a-induced cell migration. *J Cell Biol* **175:** 555–562.

Nishita M, Itsukushima S, Nomachi A, Endo M, Wang Z, Inaba D, Qiao S, Takada S, Kikuchi A, Minami Y. 2010. Ror2/Frizzled complex mediates Wnt5a-induced AP-1 activation by regulating Dishevelled polymerization. *Mol Cell Biol* **30:** 3610–3619.

Nojima H, Shimizu T, Kim CH, Yabe T, Bae YK, Muraoka O, Hirata T, Chitnis A, Hirano T, Hibi M. 2004. Genetic evidence for involvement of maternally derived Wnt canonical signaling in dorsal determination in zebrafish. *Mech Dev* **121:** 371–386.

Nojima H, Rothhamel S, Shimizu T, Kim CH, Yonemura S, Marlow FL, Hibi M. 2010. Syntabulin, a motor protein linker, controls dorsal determination. *Development* **137:** 923–933.

Northrop JL, Kimelman D. 1994. Dorsal-ventral differences in Xcad-3 expression in response to FGF-mediated induction in *Xenopus*. *Dev Biol* **161:** 490–503.

Ober EA, Schulte-Merker S. 1999. Signals from the yolk cell induce mesoderm, neuroectoderm, the trunk organizer, and the notochord in zebrafish. *Dev Biol* **215:** 167–181.

Olson LE, Tollkuhn J, Scafoglio C, Krones A, Zhang J, Ohgi KA, Wu W, Taketo MM, Kemler R, Grosschedl R, et al. 2006. Homeodomain-mediated β-catenin-dependent switching events dictate cell-lineage determination. *Cell* **125:** 593–605.

Onichtchouk D, Gawantka V, Dosch R, Delius H, Hirschfeld K, Blumenstock C, Niehrs C. 1996. The Xvent-2 homeobox gene is part of the BMP-4 signalling pathway controlling dorsoventral patterning of *Xenopus* mesoderm. *Development* **122:** 3045–3053.

Osada SI, Wright CV. 1999. *Xenopus* nodal-related signaling is essential for mesendodermal patterning during early embryogenesis. *Development* **126:** 3229–3240.

Pera EM, Ikeda A, Eivers E, De Robertis EM. 2003. Integration of IGF, FGF, and anti-BMP signals via Smad1 phosphorylation in neural induction. *Genes Dev* **17:** 3023–3028.

Perantoni AO. 2003. Renal development: Perspectives on a Wnt-dependent process. *Semin Cell Dev Biol* **14:** 201–208.

Petersen CP, Reddien PW. 2009. Wnt signaling and the polarity of the primary body axis. *Cell* **139:** 1056–1068.

Phillips BT, Kimble J. 2009. A new look at TCF and β-catenin through the lens of a divergent *C. elegans* Wnt pathway. *Dev Cell* **17:** 27–34.

Pillemer G, Yelin R, Epstein M, Gont L, Frumkin Y, Yisraeli JK, Steinbeisser H, Fainsod A. 1998. The Xcad-2 gene can provide a ventral signal independent of BMP-4. *Mech Dev* **74:** 133–143.

Pilon N, Oh K, Sylvestre JR, Bouchard N, Savory J, Lohnes D. 2006. Cdx4 is a direct target of the canonical Wnt pathway. *Dev Biol* **289:** 55–63.

Ramel MC, Lekven AC. 2004. Repression of the vertebrate organizer by Wnt8 is mediated by Vent and Vox. *Development* **131:** 3991–4000.

Ramel MC, Buckles GR, Baker KD, Lekven AC. 2005. WNT8 and BMP2B co-regulate non-axial mesoderm patterning during zebrafish gastrulation. *Dev Biol* **287:** 237–248.

Rauch GJ, Hammerschmidt M, Blader P, Schauerte HE, Strahle U, Ingham PW, McMahon AP, Haffter P. 1997. Wnt5 is required for tail formation in the zebrafish embryo. *Cold Spring Harb Symp Quant Biol* **62:** 227–234.

Rocheleau CE, Downs WD, Lin R, Wittmann C, Bei Y, Cha YH, Ali M, Priess JR, Mello CC. 1997. Wnt signaling and an APC-related gene specify endoderm in early *C. elegans* embryos. *Cell* **90:** 707–716.

Rocheleau CE, Yasuda J, Shin TH, Lin R, Sawa H, Okano H, Priess JR, Davis RJ, Mello CC. 1999. WRM-1 activates the LIT-1 protein kinase to transduce anterior/posterior polarity signals in *C. elegans*. *Cell* **97:** 717–726.

Roel G, Hamilton FS, Gent Y, Bain AA, Destree O, Hoppler S. 2002. Lef-1 and Tcf-3 transcription factors mediate tissue-specific Wnt signaling during *Xenopus* development. *Curr Biol* **12:** 1941–1945.

Roel G, van den Broek O, Spieker N, Peterson-Maduro J, Destree O. 2003. Tcf-1 expression during *Xenopus* development. *Gene Expr Patterns* **3:** 123–126.

Roose J, Molenaar M, Peterson J, Hurenkamp J, Brantjes H, Moerer P, van de Wetering M, Destree O, Clevers H. 1998. The *Xenopus* Wnt effector XTcf-3 interacts with Groucho-related transcriptional repressors. *Nature* **395:** 608–612.

Rothbacher U, Laurent MN, Blitz IL, Watabe T, Marsh JL, Cho KWY. 1995. Functional conservation of the Wnt signaling pathway revealed by ectopic expression of *Drosophila* dishevelled in *Xenopus*. *Dev Biol* **170:** 717–721.

Ryu SL, Fujii R, Yamanaka Y, Shimizu T, Yabe T, Hirata T, Hibi M, Hirano T. 2001. Regulation of dharma/bozozok by the Wnt pathway. *Dev Biol* **231:** 397–409.

Sander V, Reversade B, De Robertis EM. 2007. The opposing homeobox genes Goosecoid and Vent1/2 self-regulate *Xenopus* patterning. *EMBO J* **26:** 2955–2965.

Sapkota G, Alarcon C, Spagnoli FM, Brivanlou AH, Massague J. 2007. Balancing BMP signaling through integrated inputs into the Smad1 linker. *Mol Cell* **25:** 441–454.

Sasai Y, Lu B, Steinbeisser H, Geissert D, Gont LK, De Robertis EM. 1994. *Xenopus* chordin: A novel dorsalizing factor activated by organizer-specific homeobox genes. *Cell* **79:** 779–790.

Sasai N, Nakazawa Y, Haraguchi T, Sasai Y. 2004. The neurotrophin-receptor-related protein NRH1 is essential for convergent extension movements. *Nat Cell Biol* **6:** 741–748.

Sato SM, Sargent TD. 1991. Localized and inducible expression of *Xenopus-posterior* (Xpo), a novel gene active in early frog embryos, encoding a protein with a "CCHC" finger domain. *Development* **112:** 747–754.

Sato A, Yamamoto H, Sakane H, Koyama H, Kikuchi A. 2010. Wnt5a regulates distinct signalling pathways by binding to Frizzled2. *EMBO J* **29:** 41–54.

Satoh K, Kasai M, Ishidao T, Tago K, Ohwada S, Hasegawa Y, Senda T, Takada S, Nada S, Nakamura T, et al. 2004. Anteriorization of neural fate by inhibitor of β-catenin and T cell factor (ICAT), a negative regulator of Wnt signaling. *Proc Natl Acad Sci* **101:** 8017–8021.

Schambony A, Wedlich D. 2007. Wnt-5A/Ror2 regulate expression of XPAPC through an alternative noncanonical signaling pathway. *Dev Cell* **12:** 779–792.

Scharf SR, Gerhart JC. 1980. Determination of the dorsal-ventral axis in eggs of *Xenopus laevis*: Complete rescue of uv-impaired eggs by oblique orientation before first cleavage. *Dev Biol* **79:** 181–198.

Schedl A, Hastie ND. 2000. Cross-talk in kidney development. *Curr Opin Genet Dev* **10:** 543–549.

Schier AF, Talbot WS. 2005. Molecular genetics of axis formation in zebrafish. *Annu Rev Genet* **39:** 561–613.

Schmidt JE, von Dassow G, Kimelman D. 1996. Regulation of dorsal-ventral patterning: The ventralizing effects of the novel *Xenopus* homeobox gene Vox. *Development* **122:** 1711–1721.

Schneider S, Steinbeisser H, Warga RM, Hausen P. 1996. β-Catenin translocation into nuclei demarcates the dorsalizing centers in frog and fish embryos. *Mech Dev* **57:** 191–198.

Schohl A, Fagotto F. 2002. β-Catenin, MAPK and Smad signaling during early *Xenopus* development. *Development* **129:** 37–52.

Schroeder KE, Condic ML, Eisenberg LM, Yost HJ. 1999. Spatially regulated translation in embryos: Asymmetric expression of maternal Wnt-11 along the dorsal-ventral axis in *Xenopus*. *Dev Biol* **214:** 288–297.

Schwarz-Romond T, Asbrand C, Bakkers J, Kuhl M, Schaeffer HJ, Huelsken J, Behrens J, Hammerschmidt M, Birchmeier W. 2002. The ankyrin repeat protein Diversin recruits Casein kinase Iε to the ε-catenin degradation complex and acts in both canonical Wnt and Wnt/JNK signaling. *Genes Dev* **16:** 2073–2084.

Shibata M, Ono H, Hikasa H, Shinga J, Taira M. 2000. *Xenopus* crescent encoding a Frizzled-like domain is expressed in the Spemann organizer and pronephros. *Mech Dev* **96:** 243–246.

Shibata M, Itoh M, Ohmori SY, Shinga J, Taira M. 2001. Systematic screening and expression analysis of the head organizer genes in *Xenopus* embryos. *Dev Biol* **239:** 241–256.

Shibata M, Itoh M, Hikasa H, Taira S, Taira M. 2005. Role of crescent in convergent extension movements by modulating Wnt signaling in early *Xenopus* embryogenesis. *Mech Dev* **122:** 1322–1339.

Shibata T, Takahashi Y, Tasaki J, Saito Y, Izutsu Y, Maeno M. 2008. A role of D domain-related proteins in differentiation and migration of embryonic cells in *Xenopus laevis*. *Mech Dev* **125:** 284–298.

Shimizu T, Bae YK, Muraoka O, Hibi M. 2005. Interaction of Wnt and caudal-related genes in zebrafish posterior body formation. *Dev Biol* **279:** 125–141.

Shimizu T, Bae YK, Hibi M. 2006. Cdx-Hox code controls competence for responding to Fgfs and retinoic acid in zebrafish neural tissue. *Development* **133:** 4709–4719.

Smith WC, Harland RM. 1991. Injected Xwnt-8 RNA acts early in *Xenopus* embryos to promote formation of a vegetal dorsalizing center. *Cell* **67:** 753–765.

Smith WC, Harland RM. 1992. Expression cloning of noggin, a new dorsalizing factor localized to the Spemann organizer in *Xenopus* embryos. *Cell* **70:** 829–840.

Smith WC, McKendry R, Ribisi S Jr, Harland RM. 1995. A nodal-related gene defines a physical and functional domain within the Spemann organizer. *Cell* **82:** 37–46.

Sokol SY. 1993. Mesoderm formation in *Xenopus* ectodermal explants overexpressing Xwnt8: Evidence for a cooperating signal reaching the animal pole by gastrulation. *Development* **118:** 1335–1342.

Sokol SY. 1996. Analysis of *Dishevelled* signalling pathways during *Xenopus* development. *Curr Biol* **6:** 1456–1467.

Sokol SY. 1999. Wnt signaling and dorso-ventral axis specification in vertebrates. *Curr Opin Genet Dev* **9:** 405–410.

Sokol SY. 2000. A role for Wnts in morpho-genesis and tissue polarity. *Nat Cell Biol* **2:** E124–E125.

Sokol SY. 2011. Wnt signaling through T-cell factor phosphorylation. *Cell Res* **21:** 1002–1012.

Sokol SY, Melton DA. 1992. Interaction of Wnt and activin in dorsal mesoderm induction in *Xenopus*. *Dev Biol* **154:** 348–355.

Sokol S, Christian JL, Moon RT, Melton DA. 1991. Injected Wnt RNA induces a complete body axis in *Xenopus* embryos. *Cell* **67:** 741–752.

Sokol SY, Klingensmith J, Perrimon N, Itoh K. 1995. Dorsalizing and neuralizing properties of Xdsh, a maternally expressed *Xenopus* homolog of dishevelled. *Development* **121:** 3487.

Spiegelman VS, Slaga TJ, Pagano M, Minamoto T, Ronai Z, Fuchs SY. 2000. Wnt/β-catenin signaling induces the expression and activity of βTrCP ubiquitin ligase receptor. *Mol Cell* **5:** 877–882.

Standley HJ, Destree O, Kofron M, Wylie C, Heasman J. 2006. Maternal XTcf1 and XTcf4 have distinct roles in regulating Wnt target genes. *Dev Biol* **289:** 318–328.

Sumanas S, Strege P, Heasman J, Ekker SC. 2000. The putative wnt receptor *Xenopus* frizzled-7 functions upstream of β-catenin in vertebrate dorsoventral mesoderm patterning. *Development* **127:** 1981–1990.

Suri C, Haremaki T, Weinstein DC. 2005. Xema, a foxi-class gene expressed in the gastrula stage *Xenopus* ectoderm, is required for the suppression of mesendoderm. *Development* **132:** 2733–2742.

Suzuki A, Ueno N, Hemmati-Brivanlou A. 1997. *Xenopus* msx1 mediates epidermal induction and neural inhibition by BMP4. *Development* **124:** 3037–3044.

Tada M, Kai M. 2009. Noncanonical Wnt/PCP signaling during vertebrate gastrulation. *Zebrafish* **6:** 29–40.

Taelman VF, Dobrowolski R, Plouhinec JL, Fuentealba LC, Vorwald PP, Gumper I, Sabatini DD, De Robertis EM. 2010. Wnt signaling requires sequestration of glycogen synthase kinase 3 inside multivesicular endosomes. *Cell* **143:** 1136–1148.

Taira M, Otani H, Saint-Jeannet JP, Dawid IB. 1994. Role of the LIM class homeodomain protein Xlim-1 in neural and muscle induction by the Spemann organizer in *Xenopus*. *Nature* **372:** 677–679.

Takada S, Stark KL, Shea MJ, Vassileva G, McMahon JA, McMahon AP. 1994. Wnt-3a regulates somite and tailbud formation in the mouse embryo. *Genes Dev* **8:** 174–189.

Takahashi S, Yokota C, Takano K, Tanegashima K, Onuma Y, Goto J, Asashima M. 2000. Two novel nodal-related genes initiate early inductive events in *Xenopus* Nieuwkoop center. *Development* **127:** 5319–5329.

Takao Y, Yokota T, Koide H. 2007. β-Catenin up-regulates Nanog expression through interaction with Oct-3/4 in embryonic stem cells. *Biochem Biophys Res Commun* **353:** 699–705.

Tao Q, Yokota C, Puck H, Kofron M, Birsoy B, Yan D, Asashima M, Wylie CC, Lin X, Heasman J. 2005. Maternal wnt11 activates the canonical wnt signaling pathway required for axis formation in *Xenopus* embryos. *Cell* **120:** 857–871.

Thorpe CJ, Moon RT. 2004. Nemo-like kinase is an essential co-activator of Wnt signaling during early zebrafish development. *Development* **131:** 2899–2909.

Torres MA, Yang-Snyder JA, Purcell SM, DeMarais AA, McGrew LL, Moon RT. 1996. Activities of the Wnt-1 class of secreted signaling factors are antagonized by the Wnt-5A class and by a dominant negative cadherin in early *Xenopus* development. *J Cell Biol* **133:** 1123–1137.

Tucker A, Sharpe P. 2004. The cutting-edge of mammalian development; how the embryo makes teeth. *Nat Rev Genet* **5:** 499–508.

Ulrich F, Concha ML, Heid PJ, Voss E, Witzel S, Roehl H, Tada M, Wilson SW, Adams RJ, Soll DR, et al. 2003. Slb/Wnt11 controls hypoblast cell migration and morphogenesis at the onset of zebrafish gastrulation. *Development* **130:** 5375–5384.

van Amerongen R, Nusse R. 2009. Towards an integrated view of Wnt signaling in development. *Development* **136:** 3205–3214.

van Amerongen R, Nawijn M, Franca-Koh J, Zevenhoven J, van der Gulden H, Jonkers J, Berns A. 2005. Frat is dispensable for canonical Wnt signaling in mammals. *Genes Dev* **19:** 425–430.

van den Akker E, Forlani S, Chawengsaksophak K, de Graaff W, Beck F, Meyer BI, Deschamps J. 2002. Cdx1 and Cdx2 have overlapping functions in anteroposterior patterning and posterior axis elongation. *Development* **129:** 2181–2193.

van de Wetering M, Cavallo R, Dooijes D, van Beest M, van Es J, Loureiro J, Ypma A, Hursh D, Jones T, Bejsovec A, et al. 1997. Armadillo coactivates transcription driven by the product of the *Drosophila* segment polarity gene dTCF. *Cell* **88:** 789–799.

Veien ES, Grierson MJ, Saund RS, Dorsky RI. 2005. Expression pattern of zebrafish tcf7 suggests unexplored domains of Wnt/β-catenin activity. *Dev Dyn* **233:** 233–239.

Villanueva S, Glavic A, Ruiz P, Mayor R. 2002. Posteriorization by FGF, Wnt, and retinoic acid is required for neural crest induction. *Dev Biol* **241:** 289–301.

von Bubnoff A, Schmidt JE, Kimelman D. 1996. The *Xenopus laevis* homeobox gene Xgbx-2 is an early marker of anteroposterior patterning in the ectoderm. *Mech Dev* **54:** 149–160.

von Dassow G, Schmidt JE, Kimelman D. 1993. Induction of the *Xenopus* organizer: Expression and regulation of Xnot, a novel FGF and activin-regulated homeo box gene. *Genes Dev* **7:** 355–366.

Wang S, Krinks M, Lin K, Luyten FP, Moos M Jr. 1997. Frzb, a secreted protein expressed in the Spemann organizer, binds and inhibits Wnt-8. *Cell* **88:** 757–766.

Wei G, Ku S, Ma GK, Saito S, Tang AA, Zhang J, Mao JH, Appella E, Balmain A, Huang EJ. 2007. HIPK2 represses β-catenin-mediated transcription, epidermal stem cell expansion, and skin tumorigenesis. *Proc Natl Acad Sci* **104:** 13040–13045.

Weidinger G, Thorpe CJ, Wuennenberg-Stapleton K, Ngai J, Moon RT. 2005. The Sp1-related transcription factors sp5 and sp5-like act downstream of Wnt/β-catenin signaling in mesoderm and neuroectoderm patterning. *Curr Biol* **15:** 489–500.

Wheeler GN, Hoppler S. 1999. Two novel *Xenopus frizzled* genes expressed in developing heart and brain. *Mech Dev* **86:** 203–207.

Willert J, Epping M, Pollack JR, Brown PO, Nusse R. 2002. A transcriptional response to Wnt protein in human embryonic carcinoma cells. *BMC Dev Biol* **2:** 8.

Wu J, Mlodzik M. 2009. A quest for the mechanism regulating global planar cell polarity of tissues. *Trends Cell Biol* **19:** 295–305.

Xu C, Kim NG, Gumbiner BM. 2009. Regulation of protein stability by GSK3 mediated phosphorylation. *Cell Cycle* **8:** 4032–4039.

Xu PF, Zhu KY, Jin Y, Chen Y, Sun XJ, Deng M, Chen SJ, Chen Z, Liu TX. 2010. Setdb2 restricts dorsal organizer territory and regulates left-right asymmetry through suppressing fgf8 activity. *Proc Natl Acad Sci* **107:** 2521–2526.

Yamaguchi Y, Shinagawa A. 1989. Marked alteration at midblastula transition in the effect of lithium on formation of the larval body pattern of *Xenopus laevis*. *Dev Growth Differ* **31:** 531–541.

Yamaguchi TP, Bradley A, McMahon AP, Jones S. 1999. A Wnt5a pathway underlies outgrowth of multiple structures in the vertebrate embryo. *Development* **126:** 1211–1223.

Yamamoto A, Nagano T, Takehara S, Hibi M, Aizawa S. 2005. Shisa promotes head formation through the inhibition of receptor protein maturation for the caudalizing factors, Wnt and FGF. *Cell* **120:** 223–235.

Yamanaka Y, Mizuno T, Sasai Y, Kishi M, Takeda H, Kim CH, Hibi M, Hirano T. 1998. A novel homeobox gene, dharma, can induce the organizer in a non-cell-autonomous manner. *Genes Dev* **12:** 2345–2353.

Yanagawa S, van Leeuwen F, Wodarz A, Klingensmith J, Nusse R. 1995. The dishevelled protein is modified by wingless signaling in *Drosophila*. *Genes Dev* **9:** 1087–1097.

Yang J, Tan C, Darken RS, Wilson PA, Klein PS. 2002. β-Catenin/Tcf-regulated transcription prior to the midblastula transition. *Development* **129:** 5743–5752.

Yokoyama H. 2008. Initiation of limb regeneration: The critical steps for regenerative capacity. *Dev Growth Differ* **50:** 13–22.

Yost C, Farr GH III, Pierce SB, Ferkey DM, Chen MM, Kimelman D. 1998. GBP, an inhibitor of GSK-3, is implicated in *Xenopus* development and oncogenesis. *Cell* **93:** 1031–1041.

Young T, Deschamps J. 2009. Hox, Cdx, and anteroposterior patterning in the mouse embryo. *Curr Top Dev Biol* **88:** 235–255.

Young T, Rowland JE, van de Ven C, Bialecka M, Novoa A, Carapuco M, van Nes J, de Graaff W, Duluc I, Freund JN, et al. 2009. Cdx and Hox genes differentially regulate posterior axial growth in mammalian embryos. *Dev Cell* **17:** 516–526.

Zeng L, Fagotto F, Zhang T, Hsu W, Vasicek TJ, Perry WL III, Lee JJ, Tilghman SM, Gumbiner BM, Costantini F. 1997. The mouse Fused locus encodes Axin, an inhibitor of the Wnt signaling pathway that regulates embryonic axis formation. *Cell* **90:** 181–192.

Zhang C, Basta T, Jensen ED, Klymkowsky MW. 2003. The β-catenin/VegT-regulated early zygotic gene Xnr5 is a direct target of SOX3 regulation. *Development* **130:** 5609–5624.

Zhou BP, Deng J, Xia W, Xu J, Li YM, Gunduz M, Hung MC. 2004. Dual regulation of Snail by GSK-3β-mediated phosphorylation in control of epithelial-mesenchymal transition. *Nat Cell Biol* **6:** 931–940.

Znosko WA, Yu S, Thomas K, Molina GA, Li C, Tsang W, Dawid IB, Moon AM, Tsang M. 2010. Overlapping functions of Pea3 ETS transcription factors in FGF signaling during zebrafish development. *Dev Biol* **342:** 11–25.

Wnt Signaling in Mammalian Development: Lessons from Mouse Genetics

Jianbo Wang[1], Tanvi Sinha[1], and Anthony Wynshaw-Boris[2]

[1]Department of Cell Biology, School of Medicine, University of Alabama at Birmingham, Birmingham, Alabama 35294

[2]Department of Pediatrics and Institute for Human Genetics, School of Medicine, University of California San Francisco, San Francisco, California 94143-0794

Correspondence: j18wang@uab.edu and wynshawborist@peds.ucsf.edu

Wnts are evolutionarily conserved signaling ligands critical for animal development. Genetic engineering in the mouse has enabled investigators to acquire a detailed activation profile of the β-catenin-dependent canonical Wnt pathway during mouse development, and to manipulate Wnt pathway activities with great spatial and temporal precision. Together, these studies have not only revealed important functions of Wnt signaling at multiple stages of early mouse development, but also elucidated how the Wnt pathway interacts with other pathways to form signaling networks that confer the unique features of mammalian embryogenesis. Additionally, the planar cell polarity pathway has emerged as an essential β-catenin independent noncanonical Wnt pathway that coordinates cell polarity and regulates tissue morphogenesis in various mammalian developmental processes. Importantly, studies of Wnt signaling in mouse development have also revealed important pathogenic mechanisms of several congenital disorders in humans.

Among the model organisms used for studying Wnt signaling, the mouse shares the greatest genetic, structural, and physiological similarities with humans. Genetic studies in mice, therefore, provide the most direct insight in the role of Wnt signaling in human biology and diseases. Both mice and humans have 19 Wnt ligands and 10 Frizzled (Fz) receptors. Based on studies in *Drosophila*, *Xenopus*, and zebrafish, Wnt signaling can be broadly divided into the canonical Wnt pathway that involves β-catenin-mediated transcriptional activation, and the noncanonical Wnt pathway that is β-catenin independent but shares some components with the canonical Wnt pathway. In this review, we will discuss how both canonical and noncanonical Wnt signaling contributes to the unique developmental processes in mammals.

THE ROLE OF CANONICAL WNT SIGNALING IN EARLY MOUSE DEVELOPMENT

In mice, as in flies and *Xenopus*, canonical Wnt signaling is initiated by binding of Wnt ligands to Fz receptors and coreceptors Lrp5/6, which recruits Dishevelled (Dvl) to inhibit the destruction complex from phosphorylating and earmarking β-catenin for degradation. Consequently, β-catenin is stabilized and translocated

Cite this article as *Cold Spring Harb Perspect Biol* doi: 10.1101/cshperspect.a007963

to the nucleus, where it binds to transcription factors Tcf/Lef (T-cell transcription factor/lymphocyte enhancer factor) to activate target gene transcription. Based on this paradigm, several very useful mouse lines were created. First, because there is only one *β-catenin* in the mouse, canonical Wnt signaling can be completely abolished by deleting *β-catenin*. Consequently, a set of *β-catenin* null and *LoxP* flanked (*floxed*) conditional null alleles were established (reviewed by Grigoryan et al. 2008). In many cases, these *β-catenin* loss-of-function (LOF) mutants have allowed investigators to bypass the genetic redundancy issues associated with virtually all other components of the pathway and determine how complete loss of Wnt signaling affects development. Second, because exon3 of *β-catenin* encodes all the residues responsible for its phosphorylation and degradation, deleting exon3 leads to stabilized β-catenin that constitutively activates canonical Wnt signaling. An exon3-floxed, conditional *β-catenin* gain-of-function (GOF) allele was established to study how proper control of Wnt signaling activity contributes to development (Harada et al. 1999; Grigoryan et al. 2008). Third, because β-catenin activates gene transcription through Tcf/Lef, a number of Wnt reporter lines were established by placing *LacZ* or *EGFP* cassette behind promoters that contain multiple TCF/LEF binding sites, allowing visualization of canonical Wnt signaling activity by β-galactosidase (β-gal) staining or GFP fluorescence (Jho et al. 2002; Lustig et al. 2002; Maretto et al. 2003; Mohamed et al. 2004; Currier et al. 2010). The use of these reporters in conjunction with *β-catenin* mutants and an array of other mutants have helped us understand the role of the Wnt pathway in early mouse development.

Implantation

In mice, the fertilized egg undergoes several rounds of cleavages and by embryonic day (E) 3.5, forms the blastocyst that contains an inner cell mass (ICM) surrounded by a layer of trophoblast epithelium (trophectoderm) (Fig. 1A). The trophoblast attaches and invades the maternal uterine epithelium during implantation at E4.5, while the ICM rapidly proliferates to form the epiblast where the entire embryo originates, and the primitive endoderm that encloses the epiblast (Fig. 1B).

In preimplantation blastocysts, a number of Wnts are expressed in the ICM or the trophectoderm (Kemp et al. 2005). Early analyses of Wnt reporters or immunostaining for stabilized, nonphosphorylated β-catenin failed to detect active canonical Wnt signaling in the embryos at this stage (Mohamed et al. 2004). However, a recent study, using a different Wnt reporter, revealed a transient activation of Wnt signaling in the ICM, which is extinguished on implantation (ten Berge et al. 2011). Consistent with this observation, this recent study also found that Wnt signaling activation is required to maintain the undifferentiated state of ICM-derived embryonic stem cells in vitro (ten Berge et al. 2011). Whether Wnt signaling is similarly required within the embryo for preimplantation development, however, remains unclear because both *β-catenin* and *Lrp5/6* null mutants display no overt defects prior to implantation (Huelsken et al. 2000; Kelly et al. 2004). Furthermore, it is interesting to note that even when the stabilized *β-catenin* GOF mutant is ectopically expressed from the oocyte stage, no aberrant accumulation of β-catenin or Wnt signaling activation was detected before implantation (Kemler et al. 2004), suggesting that mechanisms in addition to the destruction complex must be restricting β-catenin accumulation/nuclear translocation during preimplantation.

The Wnt ligands secreted by the blastocyst may also trigger signaling activation in the maternal uterine epithelium during implantation. Analysis of Wnt reporters showed Wnt signaling activation in the uterine epithelium adjacent to the implanting blastocyst. Luminal injection of Wnt7a medium or coated beads could induce Wnt reporter activation, whereas injection of a Wnt antagonist, Sfrp2 (secreted Frizzled-related protein), significantly reduced implantation efficiency (Mohamed et al. 2005). Therefore, Wnt signaling activation in the uterine epithelium, likely induced by the blastocyst, may be functionally important for the implantation process.

 Cite this article as *Cold Spring Harb Perspect Biol* doi: 10.1101/cshperspect.a007963

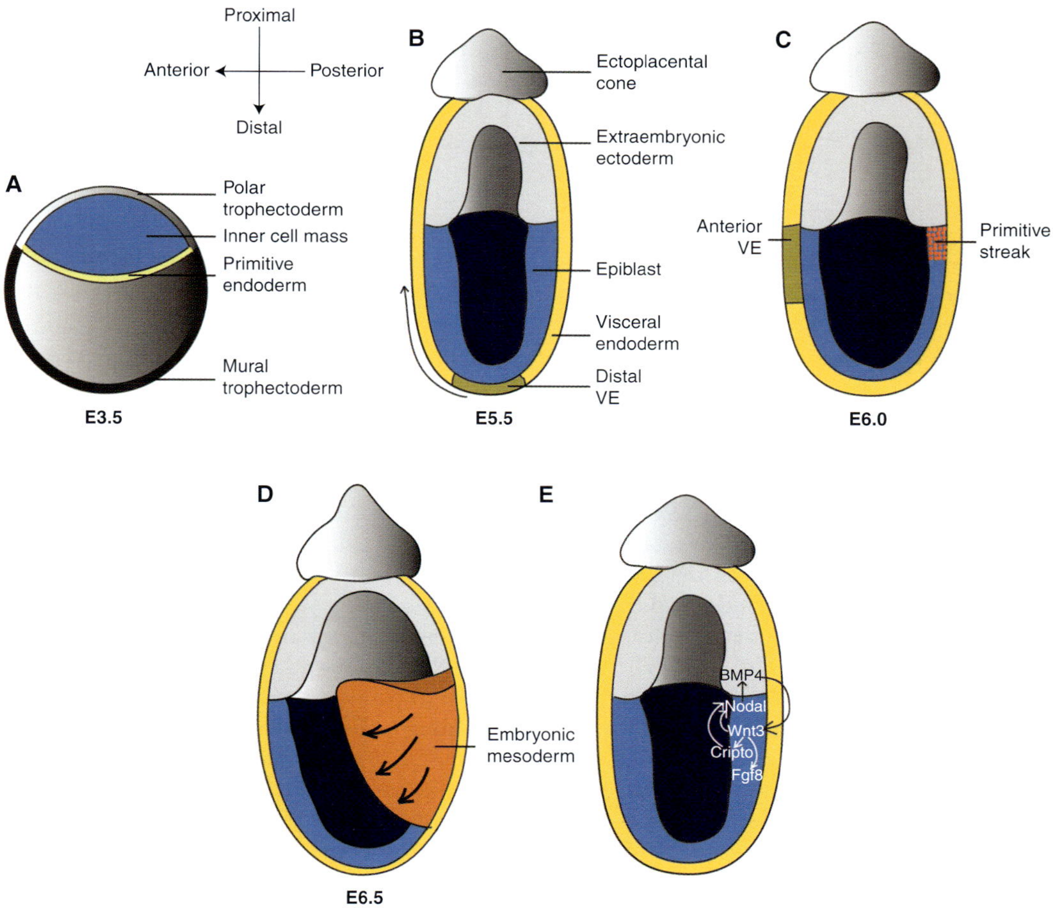

Figure 1. Early mouse development and the role of Wnt3 in mammalian gastrulation. (*A*) Preimplantation blastocyst stage mouse embryo. (*B*) After implantation, the inner cell mass (ICM) expands into a radially symmetric cup of epithelium known as the epiblast that gives rise to the entire embryo proper. The polar trophectoderm, initially overlying the ICM, gives rise to the ectoplacental cone and the extraembryonic ectoderm (ExE) that do not contribute to the embryo proper, but form the placenta and trophoblast giant cells to support embryonic development. The ExE and epiblast is also covered by a layer of visceral endoderm (VE) derived from the primitive endoerm in the blastocyst. (*C*) Between E5.5 and 6.0, the distal VE undergo proximally oriented unidirectional movement, leading to the formation of the anterior VE (AVE) that marks the future anterior side of the embryo. The AVE inhibits Nodal signaling anteriorly to restrict Nodal-induced primitive streak formation in the posterior epiblast adjacent to the ExE. (*D*) During gastrulation, epiblast cells ingress through the primitive streak to give rise to the mesoderm and definitive endoderm. (*E*) Wnt3-induced canonical Wnt signaling is an integral part of a global signaling network that regulates gastrulation. Epiblast-derived Nodal activates the expression of *Bmp4* in the neighboring ExE. Bmp4 signals back to activate *Wnt3* expression in the epiblast. In turn, Wnt3 potentiates Nodal signaling by promoting the expression of *Nodal* and its coreceptor *Cripto* (important for Nodal auto-induction), thereby establishing the Nodal-Bmp4-Wnt3 positive feedback loop to sustain gastrulation. In addition, Wnt3 is also required for Fgf8 expression, which promotes ingressing epiblast cells to migrate out of the primitive streak during gastrulation.

Postimplantation Development: Anterior-Posterior (A-P) Axis Formation and Gastrulation

By E5.5, shortly after implantation, the ICM-derived epiblast expands into a radially symmetric cup of epithelium (Fig. 1B). Proximally, the epiblast is adjacent to the extraembryonic ectoderm (ExE) derived from the polar trophectoderm, a subset of the trophectoderm apposing the ICM in the blastocyst (Fig. 1A). Both the epiblast and the ExE are covered by the visceral endoderm (VE), a single layer epithelium derived from the primitive endoderm.

E5.5 embryos display polarity only along the proximal-distal (P-D) axis from the ExE to the epiblast (Fig. 1B). Coordinated cell movement then rotates the P-D axis to establish the A-P axis: between E5.5 and 6.0, the VE cells at the distal tip of the embryo undergo proximally oriented unidirectional movement, leading to the formation of the anterior VE (AVE, Fig. 1B and C). The AVE marks the future anterior side of the embryo and patterns the underlying epiblast to form the head. The AVE also represses Nodal signaling to restrict the formation of the primitive streak to the proximal-posterior epiblast adjacent to the ExE between E6.0 to 6.5 (Fig. 1C) (Perea-Gomez et al. 2002; Stuckey et al. 2011). During gastrulation from E6.5–7.5, posterior epiblast cells undergo epithelial to mesenchymal transition and ingress through the primitive streak to give rise to the mesoderm and definitive endoderm (Fig. 1D) while the anterior epiblast forms the neural and surface ectoderm. Therefore, gastrulation establishes the three germ layers and a basic body plan.

Analyses of several Wnt reporters showed that their expression can be first detected consistently at E6.0 in the proximal-posterior epiblast in which the primitive streak will form (Fig. 1C); and as gastrulation proceeds, their expression extends distally in the epiblast, marking the forming primitive streak (Maretto et al. 2003; Mohamed et al. 2004; Currier et al. 2010). These studies suggest that canonical Wnt signaling is activated in the primitive streak from the onset of gastrulation. Indeed, several presumptive canonical Wnt genes, including *Wnt2b, Wnt3*, and *Wnt8a*, are expressed in a pattern very similar to the Wnt reporter activities prior to and during gastrulation (Liu et al. 1999; Yamaguchi 2008).

Genetic studies in the mouse showed that canonical Wnt signaling is also functionally required for gastrulation. In *Wnt3*$^{-/-}$ mutants, the primitive streak does not form and gastrulation does not occur (Liu et al. 1999; Barrow et al. 2007). Almost identical defects are observed when the function of both Lrp5 and -6 are disrupted in *Lrp5*$^{-/-}$; *Lrp6*$^{-/-}$ mutants or *mesd* (*mesoderm development*) mutants, which lacks a chaperone specifically required for trafficking Lrp5/6 to the plasma membrane (Hsieh et al. 2003; Kelly et al. 2004). Intriguingly, loss of *Wnt2b* or *Wnt8a* causes no embryonic defects (Yamaguchi 2008). Therefore, Wnt3 per se is both necessary and sufficient for activating Wnt signaling to initiate primitive streak formation and gastrulation.

More recent studies revealed that the canonical Wnt pathway is an integral part of a global signaling network that regulates gastrulation (Fig. 1E). *Wnt3* expression relies on Nodal, a Tgfβ ligand whose expression in the epiblast regulates P-D and A-P axes formation as well as gastrulation (Robertson et al. 2003). Nodal, however, regulates *Wnt3* expression indirectly through activating the expression of *Bmp4*, another Tgfβ family ligand, in the neighboring ExE (Ben-Haim et al. 2006). Bmp4 signals back to the adjacent epiblast to activate *Wnt3* expression (Ben-Haim et al. 2006; Miura et al. 2010). In turn, Wnt3 potentiates Nodal signaling by promoting the expression of *Nodal* and its coreceptor *Cripto* (Morkel et al. 2003; Ben-Haim et al. 2006), thereby establishing a Nodal-Bmp4-Wnt3 positive feedback loop to sustain gastrulation. In addition, Wnt3 is also required for Fgf8 expression in the primitive streak, which activates Fgf signaling to promote ingressing epiblast cells to migrate out of the streak during gastrulation (Sun et al. 1999; Morkel et al. 2003).

Consistent with β-catenin playing a central role in canonical Wnt signaling, *β-catenin* null mutants recapitulate the gastrulation defect in *Lrp5/6* and *Wnt3* mutants (Haegel et al. 1995; Huelsken et al. 2000). Unlike *Lrp5/6* and *Wnt3* mutants, though, *β-catenin*$^{-/-}$ mutants display an additional defect in A-P axis formation, because the distal VE fails to move proximally to form the AVE (Huelsken et al. 2000). Earlier studies attributed this defect to a more severe disruption of *Cripto* expression in *β-catenin*$^{-/-}$ mutants, because Cripto is a presumptive transcriptional target of β-catenin and a Nodal coreceptor essential for AVE migration (Ding et al. 1998; Morkel et al. 2003). However, *Cripto* expression is earlier and much broader than Wnt reporter expression that marks β-catenin induced transcriptional activity (Mohamed et al. 2004; ten Berge et al. 2008). Furthermore, in *β-catenin* GOF mutants, *Cripto* is highly expressed yet the AVE fails to form (Morkel et al. 2003). Therefore, whether the AVE defect in *β-catenin*$^{-/-}$ mutants is solely caused by the disruption of *Cripto* expression needs further investigation. Another study proposed that AVE migration might be propelled by a gradient of canonical Wnt signaling activity in the VE, established by Wnt3 and its extracellular inhibitors, such as Dkk1 (Dickkopf1), expressed in the prospective posterior and anterior side, respectively (Kimura-Yoshida et al. 2005). However, this hypothesis is not consistent with the chimera analyses of *β-catenin*$^{-/-}$ mutants, which concluded that β-catenin is required only in the epiblast, but not in the VE, for proper A-P axis formation (Huelsken et al. 2000). A third possibility is that the AVE defect in *β-catenin*$^{-/-}$ mutants may originate from cell proliferation defects. A recent study revealed that during AVE movement in wild-type embryos, the cell proliferation rate in the epiblast is 11-fold higher than that in the VE, and that this growth disparity is required to drive AVE migration (Stuckey et al. 2011). Therefore, the reduced cell proliferation manifested by *β-catenin*$^{-/-}$ mutant at pregastrulation stage (Huelsken et al. 2000) may also contribute to the failure of AVE migration.

Considering the indispensable role of Lrp5/6 and its homolog Arrow in Wnt signal transduction, it is intriguing that *Lrp5/6* and *mesd* mutants do not display the AVE defect manifested by *β-catenin* mutants (Hsieh et al. 2003; Kelly et al. 2004). It was proposed that this phenotypic difference may arise from a ligand-independent transcriptional activity of β-catenin in the mouse (Morkel et al. 2003). However, it is important to note that in addition to activating transcription in the Wnt pathway, β-catenin also functions as a structural protein to maintain adherens junctions (Nelson and Nusse 2004). Although *β-catenin*$^{-/-}$ mutants do not display overt disruption of cell adhesion owing to the compensatory substitution by plakoglobin (Huelsken et al. 2000; Lickert et al. 2002), loss of β-catenin may interfere with rapid cell proliferation that may not be sustained by plakoglobin alone, or morphogenetic movements in which plakoglobin may not be sufficient to confer the required dynamic remodeling of adherens junctions.

Wnt Signaling Suppression in Anterior Patterning

Although Wnt signaling activation is required for primitive streak formation and posterior patterning, it must be suppressed anteriorly for proper patterning of the anterior epiblast. Dkk1, which antagonizes Wnt signaling by binding to Lrp5/6, is initially expressed in a crescent-shaped domain around the AVE, and later in the primitive streak-derived anterior mesendoderm (AME) (Mukhopadhyay et al. 2001; Kimura-Yoshida et al. 2005). In *Dkk1*$^{-/-}$ null mutants, Wnt reporter expression expands anteriorly from E7.0 and head truncation occurs by E9.5 (Mukhopadhyay et al. 2001; Lewis et al. 2008). Chimera analysis indicated that AVE-derived Dkk1 is dispensable, whereas AME-derived Dkk1 is sufficient for anterior head patterning (Mukhopadhyay et al. 2001). Interestingly, the head truncation defect in *Dkk1*$^{-/-}$ mutants can be partially suppressed by halving the dosage of *Wnt3* (Lewis et al. 2008), indicating that a major function of Dkk1 is to modulate Wnt3 induced Wnt signaling. Furthermore, proper anterior patterning also requires inhibition of β-catenin mediated transactivation by ICAT (Inhibitor of β-catenin and T cell factor), an 81 amino acid protein that disrupts the interaction between β-catenin and TCF. *ICAT*$^{-/-}$ mice display head truncation defect similar to that in *Dkk1*$^{-/-}$ mutants (Satoh et al. 2004).

Tail Bud Formation, Somitogenesis, and Left-Right Determination

After gastrulation, Wnt3 expression is downregulated and eventually terminated by E8.5 (Liu et al. 1999). However, the primitive streak and its derivative, the tail bud, continue to produce mesoderm for another 5 days to form the paraxial presomitic mesoderm (PSM) and lateral plate mesoderm (LPM) that give rise to the trunk and tail. Analyses of Wnt reporters indicate that canonical Wnt signaling remains active in the primitive streak/tail bud, and this is due in part to Wnt3a, another canonical Wnt that starts to be expressed in the primitive streak from E7.5 (Nakaya et al. 2005). *Wnt3a*$^{-/-}$ mutants fail to form posterior mesoderm and the tail bud but instead develop ectopic neural tissue, indicating that Wnt3a is required for the maintenance and the function of the primitive streak (Takada et al. 1994; Yoshikawa et al. 1997). One of the most critical targets of Wnt3a is *T* (*Brachyury*), a transcription factor required for mesoderm formation (Fig. 2) (Yamaguchi et al. 1999b). *T* contains multiple Lef1 binding sites that are required for maintaining, but not initiating, *T* expression in the primitive streak and nascent mesoderm (Galceran et al. 2001). Finally, similar to *Wnt3a* and *T* mutants, *Lef1*/*Tcf1* double null mutants lack posterior mesoderm and develop ectopic neural tubes, whereas expression of a constitutively active Lef1-β-catenin fusion can suppress the defects in *Wnt3a* mutants, proving definitively that Wnt3a signals through the canonical Wnt pathway to regulate mesoderm formation (Galceran et al. 1999, 2001; Yamaguchi et al. 1999b).

Wnt3a secreted from the primitive streak also activates Wnt signaling in the flanking PSM, which undergoes periodic boundary formation that cleaves off a segment of the anterior PSM to form a somite every 2 hours along the A-P axis (Yamaguchi 2008). Rhythmic formation of somites requires a "segmentation clock," manifested as oscillating activation of the Notch signaling pathway in the PSM (Pourquie 2011). Wnt3a-induced Wnt signaling regulates the segmentation clock, at least in part, by activating Notch ligand Dll1 (*Delta-like1*) transcription in the PSM (Fig. 2) (Galceran et al. 2004; Hofmann et al. 2004; Nakaya et al. 2005). Secondly, a gradient of Wnt signaling activity, likely established by the graded expression of *Wnt3a* along the A-P axis, is thought to constitute part of the "determination wavefront" that positions the segmentation boundary during somitogenesis (Pourquie 2011). By manipulating Wnt activity gradient using β-catenin LOF and GOF mutants, studies show that high level

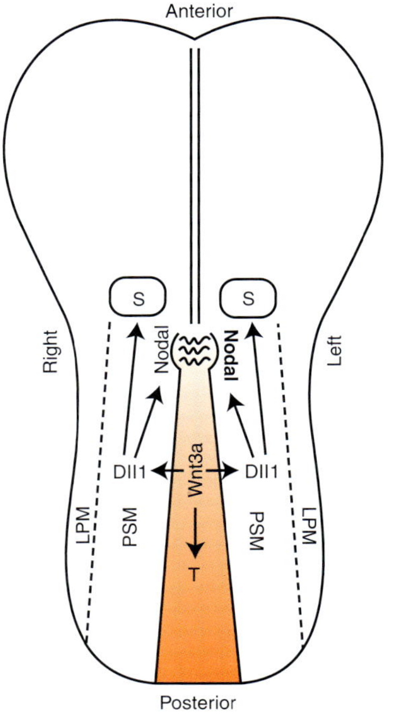

Figure 2. The role of Wnt3a in early postgastrulation mouse development. A schematic diagram of the ventral view of an early postgastrulation (E8.0) mouse embryo showing that expression of Wnt3a in the primitive streak activates canonical Wnt signaling to maintain *T* (*Brachyury*) expression, thereby sustaining mesoderm production. Primitive streak-derived Wnt3a also signals to the adjacent paraxial presomatic mesoderm (PSM) to active Notch ligand Dll1 expression to regulate somitogenesis. Wnt3a-activated Dll1 expression in the PSM also regulates *Nodal* expression at the periphery of the node. Left-ward flow generated by rotating node cilia then enhances Nodal signaling and eventually, *Nodal* expression, on the left side of the node, leading to the L-R axis determination. Dll1, Delta 1; PSM, presomitic mesoderm; LPM, lateral plate mesoderm; S, somite; T, Brachyury.

Wnt activity maintains posterior PSM in an undifferentiated state permissive of Notch signaling oscillation, while a gradual reduction of Wnt activity in the anterior PSM sets the boundary to arrest Notch signaling oscillation and to activate genes required for segment formation (Aulehla et al. 2008; Dunty et al. 2008).

Wnt3a is also required for left-right (L-R) axis determination because *Wnt3a* null mutants display laterality defects in multiple visceral organs (Nakaya et al. 2005). In mice, L-R determination relies on *Nodal* expression at the periphery of the node, a specialized structure located immediately rostral to the primitive streak (Fig. 2) (Brennan et al. 2002). *Nodal* is initially expressed symmetrically along both the right and left side of the node, but a leftward flow generated by node cilia rotation (see below) causes enhanced Nodal signaling on the left node, eventually breaking the L-R symmetry by up-regulating *Nodal* expression in the left node and left LPM. In *Wnt3a* mutants, *Nodal* expression in the node is diminished, leading to delayed and symmetric *Nodal* expression in the LPM and defective expression of several other genes essential for L-R determination, including *Lefty1/2* and *Pitx2* (Nakaya et al. 2005). Because *Nodal* expression in the node depends on Notch signaling, Wnt3a is thought to regulate L-R determination indirectly by activating *Dll1* expression in the adjacent PSM (Fig. 2) (Nakaya et al. 2005). In light of the finding that Wnt3 may regulate *Nodal* expression directly during gastrulation (Ben-Haim et al. 2006), it is possible that primitive streak-derived Wnt3a may also directly activate *Nodal* expression in the node.

NONCANONICAL WNT/PLANAR CELL POLARITY PATHWAY SIGNALING IN MOUSE DEVELOPMENT

Studies in zebrafish, *Xenopus*, and tissue culture have found that some Wnt ligands, such as Wnt5a and Wnt11, or overexpression of some Fz receptors, do not lead to β-catenin stabilization, but instead lead to divergent downstream events collectively termed the noncanonical Wnt signaling, which include JNK, Ca^{2+}, and small GTPase Rho/Rac/Cdc42 signaling activation (Veeman et al. 2003; Wallingford and Habas 2005). Because these downstream signaling events are diverse and may also serve as signaling outputs of many other pathways, currently we do not have an assay that can specifically determine noncanonical Wnt signaling activation. Therefore, our current understanding of noncanonical Wnt signaling in mouse development has been mostly limited to one branch known as the planar cell polarity (PCP) pathway, in which pioneering studies in *Drosophila* have identified most of the genetic components and provided critical mechanistic insights (Seifert and Mlodzik 2007; Zallen 2007; Strutt and Strutt 2009; Bayly and Axelrod 2011).

In flies, the PCP pathway coordinates cellular polarity on the plane of the epithelium, perpendicular to the apical-basal (A-B) polarity of the cell. Examples of PCP include uniformly oriented trichomes on the wing epithelium and bristles on the abdomen (Lawrence et al. 2004; Seifert and Mlodzik 2007). The PCP pathway shares Fz and Dvl with the canonical Wnt pathway, but also requires a distinct set of membrane or cytoplasmic proteins, including Flamingo, Diego, Prickle, and Van Gogh (Vang). Studies of these "core" PCP gene orthologs in *Xenopus* and zebrafish revealed an essential role of PCP signaling in convergent extension (CE), a morphogenetic process during which tissues elongate anterior-posteriorly and concomitantly narrow medial-laterally. During CE, PCP signaling has been shown to regulate polarized cell intercalation in mesodermal cells in *Xenopus*, and also to promote directional cell migration and oriented cell division in zebrafish (Heisenberg et al. 2000; Wallingford et al. 2000; Gong et al. 2004; Roszko et al. 2009). In light of these findings, mouse genetic studies have uncovered many important roles of PCP signaling in coordinating cell polarity and tissue morphogenesis in mammals.

Cell Polarity Establishment

Epiblast Cell Polarity

The canonical Wnt pathway is not the only Wnt related pathway important for gastrulation. Recently, Tao et al. found that the deletion of the mouse Prickle homolog, *mpk1*, resulted in early embryonic lethality, between E5.5 and E6.5, associated with failure of distal VE migration and primitive streak formation (Tao et al. 2009). Mutant epiblast tissue was disorganized, and epiblast cells displayed abnormal cell shapes, mislocalized extracellular matrix (ECM) proteins, and disrupted orientation of mitotic spindles with loss of A-B polarity of epiblast cells. They showed that *mpk1* genetically interacts with another core PCP gene *Vangl2* (an ortholog of *Vang*) in epiblast formation, suggesting that PCP components are commonly required for the establishment and/or the maintenance of epiblast A-B polarity. Thus, the PCP pathway may play a very early role in postimplantation establishment of A-P polarity and gastrulation.

Nodal Cilia

It was recently found that the three *Dvl* genes participate in the planar polarization of node cells to determine the rotational axis of nodal cilia (Hashimoto et al. 2010). Rotational movement of the node cilia generates a leftward fluid flow (see Fig. 2) in the mouse embryo because the cilia are posteriorly tilted (Nonaka et al. 2005; Okada et al. 2005), but it was not known how A-P information is translated into the posterior tilt of the node cilia. We found that the basal body of node cilia is initially positioned centrally but then gradually shifts toward the posterior side of the node cells, a shift that was impaired along with malpositioning of the basal body and unidirectional nodal flow in compound mutant mice lacking *Dvl* genes. The basal body was normally positioned in the node cells of $Wnt3a^{-/-}$ embryos (coding for a canonical Wnt involved in gastrulation), but inhibition of Rac1, a downstream effector of the noncanonical Wnt signaling pathway, impaired the polarized localization of the basal body in wild-type embryos. Dvl2-EGFP and Dvl3-EYFP proteins were localized to the apical side of the node cells, and their location was polarized to the posterior side of the cells before the posterior positioning of the basal body. Thus, posterior positioning of the basal body to provide posterior tilt to node cilia is determined by planar polarization mediated by noncanonical Wnt/PCP signaling. This interpretation was definitively proven when it was found that *Vangl1/2* double mutants displayed identical nodal cilia placement defects, resulting in disruption of leftward nodal flow and L-R asymmetry defects (Song et al. 2010). Vangl1 and -2 are thought to be exclusively part of the PCP pathway, unlike Dvls, which participate in canonical Wnt and noncanonical PCP pathways.

Stereociliary Bundle Orientation

The mammalian auditory sensory organ, the organ of Corti, consists of four rows of sensory hair cells along the lateral side of the cochlea duct. On the apical surface of each hair cell, there are a set of actin-filled microvilli called stereocilia that are organized into a V-shaped staircase. The vertex of each V-shaped stereociliary bundle is always pointed toward the lateral side of the cochlea, thereby displaying a distinct polarity reminiscent of fly PCP (Rida and Chen 2009). Indeed, in *Looptail* mutant mice ($Vangl2^{Lp/Lp}$) that harbor a spontaneous loss-of-function mutation in *Vangl2*, the stereociliary orientation becomes randomized, suggesting that a conserved mammalian PCP pathway is required to coordinate stereociliary orientation among hair cells (Montcouquiol et al. 2003). This hypothesis was validated by the findings of similar stereociliary orientation defects in mouse mutants of other PCP orthologs, including *Celsr1* (an ortholog of *flamingo* [Curtin et al. 2003]), *Fz3 and -6* (Wang et al. 2006b) and *Dvl1, -2, and -3* (Wang et al. 2005; Etheridge et al. 2008). Furthermore, mouse PCP proteins display asymmetric plasma membrane distribution in the organ of Corti, and mutation of one PCP gene perturbs the membrane localization of other PCP proteins (Wang et al. 2005, 2006b; Montcouquiol et al. 2006; Jones et al. 2008). The asymmetric membrane localization of PCP proteins is a unique feature first described in flies and is thought to provide a mechanism for PCP proteins to propagate polarity information across the entire field of epithelium (Amonlirdviman et al. 2005; Wu and Mlodzik 2008). Lastly, the asymmetric distribution of PCP proteins is observed prior to overt morphological polarization of stereociliary bundles (Wang et al. 2006a; Jones et al. 2008), consistent with PCP signaling playing a role in establishing stereocilia orientation in mice.

Although the above findings showed unequivocally that a mechanism highly conserved with the fly PCP pathway is involved in coordinating stereocilia orientation in mice, studies have also identified several genes that are indispensable for mouse PCP signaling, but are not known for fly PCP signaling. These include Scribble1 (Scrb1, homologous to fly scribble) known to be involved in apical-basal polarity determination (Bilder and Perrimon 2000; Montcouquiol et al. 2003), and two transmembrane tyrosine kinases Ptk7 and Ror2 (Lu et al. 2004; Yamamoto et al. 2008; Gao et al. 2011). In vertebrates, Ptk7 may function by recruiting Dvl to the plasma membrane (Wehner et al. 2011), whereas Ror2 may function together with Fz as a coreceptor for Wnt5a (Oishi et al. 2003; Yamamoto et al. 2008; Gao et al. 2011). Mutations in *Wnt5a* cause minor stereocilia mis-orientation and *Wnt5a* genetically interacts with *Vangl2* (Qian et al. 2007). Therefore, noncanonical Wnt ligands seem to be involved in PCP signaling during stereocilia orientation in mice. In contrast, studies in flies concluded that Wnt ligand is not required for PCP signaling (Povelones et al. 2005).

Another interesting finding in mice is that proper stereocilia orientation requires intact kinocilium, a single microtubule-filled primary cilium (similar to nodal cilium) located at the vertex of each V-shaped stereociliary bundle. Mutations that abolish kinocilium formation cause randomized stereocilia orientation despite proper asymmetric membrane localization of PCP proteins (Jones et al. 2008). These observations have led to the proposal that PCP signaling functions by providing cell–cell communication to establish coordinated polarity among adjacent cells, but not by specifying the intrinsic polarity within each cell in an instructive fashion (Rida and Chen 2009).

Hair Follicle Orientation

Mouse PCP signaling coordinates not only the orientation of stereocilia and nodal cilia that arise on single cells, but also the alignment of body hairs originating from hair follicles consisting of hundreds of cells. A detailed study showed that before basal epidermal cells invaginate to form hair follicles, PCP proteins are already enriched along their anterior and posterior plasma membrane. During invagination, only cells on the anterior side constrict basally, causing each hair follicle to form an anterior tilt that aligns with the global A-P axis of the body. In *Vangl2* and *Celsr1* embryos, anterior cells fail to undergo basal constriction, resulting in hair follicles that initially lack any tilt and later orient randomly (Devenport and Fuchs 2008). In *Fz6*$^{-/-}$ mice that survive postnatally, there are abnormal hair whorls, tufts, and waves throughout the body (Guo et al. 2004). Collectively, these results suggest that PCP signaling can orchestrate global hair follicle morphogenesis to control the directionality of body hairs.

Morphogenesis

Morphogenesis is the critical process through which tissues and organs gain their characteristic shapes. The anterior tilting of hair follicles is just one example of an increasing number of morphogenetic processes that have been linked to PCP signaling in mice. Consistent with *Xenopus* and zebrafish studies that have showed a central role of PCP signaling in CE, disruption of PCP signaling in mice leads to morphogenetic defects that manifest as shortening and widening of many tissues and organs, including the cochlea tube (Wang et al. 2005), stomach and intestine (Matsuyama et al. 2009). In some cases, aberrant tissue shapes in mouse PCP mutants also have profound effects that result in severe structural defects mimicking certain congenital disorders in humans. In this respect, investigation of mouse PCP mutants has been particularly instrumental in linking the mechanistic studies of PCP signaling in other model organisms with the understanding of the pathogenic mechanisms of several human birth defects.

Neural Tube (NT) Closure

One of the first and most characteristic morphogenetic defects found in mouse PCP mutants is a unique form of NT closure defect (NTD) called craniorachischisis, in which the entire NT from the mid-brain to the tail fails to close. Mutations in *Ptk7* as well as almost all mouse core PCP genes, including *Celsr1, Fz3/6, Dvl1/2/3* and *Vangl1/2*, cause this defect (Kibar et al. 2001; Murdoch et al. 2001; Curtin et al. 2003; Lu et al. 2004; Wang et al. 2006a,b; Etheridge et al. 2008; Torban et al. 2008). Furthermore, although *Wnt5a* and *Ror2* null mutants per se do not display NTDs, reducing *Vangl2* dosage by 50% in either mutant is sufficient to causes craniorachischisis (Qian et al. 2007; Gao et al. 2011). Therefore, Wnt5a may function together with at least one additional noncanonical Wnt ligand to regulate NT closure by activating Ror2 and Fz3/6 mediated PCP pathway.

During NT closure, the originally flat neuroepithelial sheet rolls up to form opposing neural folds, which appose and fuse along the dorsal midline to give rise to a hollow tube. Studies in *Xenopus* revealed that PCP signaling is required within the neuroepithelium to narrow the distance between the elevating neural folds, thereby facilitating their apposition and fusion at the dorsal midline (Wallingford and Harland 2002). In mice, we and others have found that during neurulation, the neural plate undergoes CE-like concomitant lengthening and narrowing, which can be measured as gradual increase of the length-to-width ratio (LWR) (Wang et al. 2006a; Savory et al. 2011). Disrupting PCP signaling significantly reduces LWR by 4 somite stage (ss), several hours before NT closure normally occurs (Wang et al. 2006a; Savory et al. 2011). Therefore, similar to *Xenopus*, PCP signaling in mice may also promote NT closure by regulating CE-like morphogenesis to reduce the width of the neural plate and shorten the distance between the opposing neural folds. Finally, in *Vangl2*$^{Lp/+}$ heterozygous mutants, a moderate reduction of the LWR is associated with delayed NT closure and partially penetrant, less severe NTDs like spinal bifida (Lu et al. 2004; Wang et al. 2006a), suggesting that perturbation of PCP-regulated CE may underlie a broad spectrum of NTDs.

These findings motivated several searches for PCP gene mutations in human NTDs that occur at a frequency of 1 in 1000 live births (Juriloff and Harris 2000). Indeed, missense mutations in *VANGL2* and its homolog *VANGL1* have been identified in patients with spinal bifida (Kibar et al. 2007, 2009, 2011) and miscarried fetuses with cranial NT defects such as anencephaly (Lei et al. 2010). These studies showed conclusively that disruption of PCP signaling is an important causative factor of NTDs in humans. Conversely, biochemical and functional studies of these NTD-associated *VANGL1/2* mutations in model organisms may also provide novel mechanistic insights into PCP signaling.

Limb Skeletal Development

Another example in which studies of mouse PCP mutants revealed important pathogenic mechanisms in human disorders is the recent finding that Wnt5a-initiated PCP signaling regulates limb morphogenesis and skeletal formation. In humans, *WNT5A* and *ROR2* mutations are associated with two congenital skeletal disorders: brachydactyly type B (BDB1), characterized by loss of phalanges of the digits, and Robinow syndrome (RRS), a broader skeletal dysplasia with features such as mesomelic limb shortening (Afzal et al. 2000; Oldridge et al. 2000; Person et al. 2010). These skeletal defects are recapitulated in *Wnt5a* and *Ror2* mutant mice, but how Wnt5a/Ror2 regulate limb skeletal formation was unclear (Yamaguchi et al. 1999a; Takeuchi et al. 2000; Topol et al. 2003; Raz et al. 2008).

Analysis of *Vangl2*$^{Lp/Lp}$ mouse mutants provided the first insight that Wnt5a/Ror2 signal through the PCP pathway to regulate limb skeletal formation. In *Vangl2*$^{Lp/Lp}$ mutants, most digits lack one phalanx, resembling the digit defects in *Ror2* mutant mice and human BDB1 patients. Halving *Wnt5a* dosage in *Vangl2*$^{Lp/Lp}$ mutants enhances the severity and penetrance of the digit defects and causes shortening of limb long bones similar to that in RRS patients,

 Cite this article as *Cold Spring Harb Perspect Biol* doi: 10.1101/cshperspect.a007963

suggesting that Wnt5a and Vangl2 function in the same pathway and disruption of PCP signaling underlies both BDB1 and RRS (Wang et al. 2011). Consistent with this idea, *Vangl2* also genetically interacts with *Ror2* during limb skeletal formation and a Wnt5a→ Ror2→ Vangl2 signaling cascade was proposed to be active in the limb, leading to Vangl2 phosphorylation at conserved serine/ threonine residues that are also important for its function in zebrafish CE (Gao et al. 2011).

Morphometric analysis revealed that by E11, before overt skeletal formation, the limb buds in *Wnt5a* and *Vangl2* mutants already possess aberrant shape and dimensions: their length along the P-D (proximal-distal) axis is reduced, but their width along the A-P axis and their thickness along the D-V (dorsal-ventral) axis are increased (Gros et al. 2010; Wang et al. 2011). Imaging studies indicate that mesenchymal cells in wild-type limb buds display elongated morphology, directional migration, and oriented cell division, all of which align preferentially with the P-D axis. Limb mesenchymal cell in *Wnt5a* mutants, however, are rounded and migrate and divide without a P-D bias (Gros et al. 2010). Collectively, these results suggest that Wnt5a-initiated PCP signaling may regulate polarized cell morphology and behavior to promote limb bud P-D lengthening while simultaneous limiting its widening and thickening along the A-P and D-V axes (Gros et al. 2010).

The altered limb bud shape and dimensions in *Wnt5a* and *Vangl2* mutants may affect limb skeletal formation in two ways (Fig. 3). First, corresponding to the shortening, widening and thickening of mutant limb buds, the limb mesenchyme-derived prechondrogenic condensates, which serve as templates for the future skeletal elements, also become shorter, wider and thicker (Yamaguchi et al. 1999a; Wang et al. 2011). Secondly, the aberrant limb bud shape and dimensions perturb the cross-talk among spatially distinct signaling centers important for limb growth, thereby disrupting distal phalanx formation. During limb development, the AER (apical ectodermal ridge), a distal ectodermal structure, is known to secret Fgf ligands (AER-Fgf) to promote cell proliferation and survival (Dudley et al. 2002; Sun et al. 2002; Yu and Ornitz 2008). AER-FGFs also stimulate limb mesenchymal cells located in the posterior ZPA (zone of polarizing activity) to produce Shh, which in turn maintains the expression of Gremlin 1 (Grem1) in the dorsal and ventral margin of the limb. Grem1 encodes an extracellular BMP inhibitor that counteracts the negative effect of Bmp on Fgf/ Shh signaling and AER maintenance (Fig. 3) (Khokha et al. 2003; Bastida et al. 2009; Benazet et al. 2009), thereby establishing the Fgf-Shh-Grem1 signaling loop to sustain limb growth. In *Vangl2* and *Wnt5a* mutants, the abnormal limb bud thickening and widening alter the distance among the three signaling centers (the distal AER, posterior ZPA, and dorsal-ventral margin of the limb), leading to reduced robustness of the Fgf-Shh-Grem1

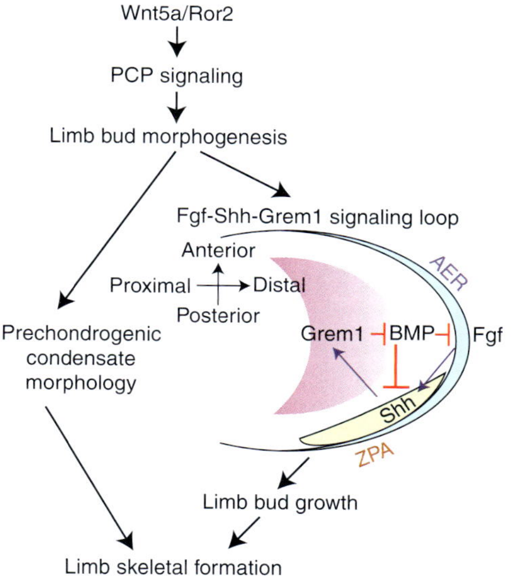

Figure 3. Wnt5a/Ror2 initiated PCP signaling regulates limb morphogenesis to promote limb skeletal formation. Wnt5a/Ror2-initiated PCP signaling regulates limb bud morphogenesis to control its shape and dimensions, which in turn ensure proper morphology of the prechondrogenic condensates in the limb and promote limb bud growth by maintaining cross-talk between spatially distinct signaling centers in the Fgf-Shh-Grem1 loop. AER: apical ectodermal ridge; Grem1: Gremlin 1; ZPA: zone of polarizing activity.

signaling loop and aberrant increase in Bmp signaling. Consequently, Fgf signaling is attenuated, resulting in limb growth defect and loss of distal phalanges. Importantly, halving *Bmp4* dosage in *Vangl2*$^{Lp/Lp}$ mutants partially rescued the loss of distal phalanges (Wang et al. 2011), providing definitive evidence that elevated Bmp signaling is causative to the digit defects in PCP mutant mice and presumably, human BDB1 patients. Collectively, the recent studies on *Wnt5a* and PCP mutants (Gros et al. 2010; Gao et al. 2011; Wang et al. 2011) provided not only an integrated view on how limb growth and morphogenesis coordinate to regulate skeletal formation, but also novel and critical insight into the pathogenic mechanisms of RRS and BDB1.

CONCLUDING REMARKS

Over the last decade, mouse genetic studies have provided a fundamental understanding of how the evolutionarily conserved canonical Wnt pathway, through regulating β-catenin stability, controls gene expression, cell proliferation, cell fate, and embryonic patterning during various stages of mammalian development (also see reviews (van Amerongen and Berns 2006; Grigoryan et al. 2008). Wnt signaling, however, is not simply “on” or “off,” and a major question remaining to be resolved is how the magnitude and duration of Wnt signaling activity impact mammalian development. Secondly, genetic studies have revealed that Wnt signaling functionally interacts with other pathways to form signaling networks that orchestrate mammalian embryogenesis. In most cases, however, we do not know the cellular and molecular mechanisms of these interactions. Answering these questions will provide great insight into stem cell biology, tissue engineering and regenerative medicine.

The rapidly expanding investigations on the mouse PCP pathway have established this noncanonical Wnt signaling branch as a central player in regulating tissue morphogenesis and cell polarity. These studies have also revealed how disrupting PCP-mediated morphogenesis can have profound effects that lead to structural defects, and provided critical insight into the etiology of human NTDs, RRS, and BDB1. The cellular basis behind PCP-mediated morphogenesis, however, remains largely unknown in mammals. At the molecular level, what are the effectors that function downstream from core PCP proteins to modulate cell behavior and polarity during mammalian development is another major question needs to be resolved in the future.

ACKNOWLEDGMENTS

JW is supported by AHA grants 0635262N and 11GRNT6980004, NIH grant R01HL109130, a pilot project from NIH grant 2P30AR046031 and start-up funds from the University of Alabama at Birmingham. AWB is supported by NIH grants from NICHD (HD43173) and NINDS (NS73159).

REFERENCES

Afzal AR, Rajab A, Fenske CD, Oldridge M, Elanko N, Ternes-Pereira E, Tuysuz B, Murday VA, Patton MA, Wilkie AO, et al. 2000. Recessive Robinow syndrome, allelic to dominant brachydactyly type B, is caused by mutation of ROR2. *Nat Genet* **25:** 419–422.

Amonlirdviman K, Khare NA, Tree DR, Chen WS, Axelrod JD, Tomlin CC. 2005. Mathematical modeling of planar cell polarity to understand domineering nonautonomy. *Science* **307:** 423–426.

Aulehla A, Wiegraebe W, Baubet V, Wahl MB, Deng C, Taketo M, Lewandoski M, Pourquie O. 2008. A β-catenin gradient links the clock and wavefront systems in mouse embryo segmentation. *Nat Cell Biol* **10:** 186–193.

Barrow JR, Howell WD, Rule M, Hayashi S, Thomas KR, Capecchi MR, McMahon AA. 2007. Wnt3 signaling in the epiblast is required for proper orientation of the anteroposterior axis. *Dev Biol* **312:** 312–320.

Bastida MF, Sheth R, Ros MM. 2009. A BMP-Shh negative-feedback loop restricts Shh expression during limb development. *Development* **136:** 3779–3789.

Bayly R, Axelrod JD. 2011. Pointing in the right direction: New developments in the field of planar cell polarity. *Nat Rev Genet* **12:** 385–391.

Ben-Haim N, Lu C, Guzman-Ayala M, Pescatore L, Mesnard D, Bischofberger M, Naef F, Robertson EJ, Constam DD. 2006. The nodal precursor acting via activin receptors induces mesoderm by maintaining a source of its convertases and BMP4. *Dev Cell* **11:** 313–323.

Benazet JD, Bischofberger M, Tiecke E, Goncalves A, Martin JF, Zuniga A, Naef F, Zeller R. 2009. A self-regulatory system of interlinked signaling feedback loops controls mouse limb patterning. *Science* **323:** 1050–1053.

Bilder D, Perrimon N. 2000. Localization of apical epithelial determinants by the basolateral PDZ protein Scribble. *Nature* **403:** 676–680.

Brennan J, Norris DP, Robertson EE. 2002. Nodal activity in the node governs left-right asymmetry. *Genes Dev* **16:** 2339–2344.

Currier N, Chea K, Hlavacova M, Sussman DJ, Seldin DC, Dominguez I. 2010. Dynamic expression of a LEF-EGFP Wnt reporter in mouse development and cancer. *Genesis* **48:** 183–194.

Curtin JA, Quint E, Tsipouri V, Arkell RM, Cattanach B, Copp AJ, Henderson DJ, Spurr N, Stanier P, Fisher EM, et al. 2003. Mutation of Celsr1 disrupts planar polarity of inner ear hair cells and causes severe neural tube defects in the mouse. *Curr Biol* **13:** 1129–1133.

Devenport D, Fuchs E. 2008. Planar polarization in embryonic epidermis orchestrates global asymmetric morphogenesis of hair follicles. *Nat Cell Biol* **10:** 1257–1268.

Ding J, Yang L, Yan YT, Chen A, Desai N, Wynshaw-Boris A, Shen MM. 1998. Cripto is required for correct orientation of the anterior-posterior axis in the mouse embryo. *Nature* **395:** 702–707.

Dudley AT, Ros MA, Tabin CC. 2002. A re-examination of proximodistal patterning during vertebrate limb development. *Nature* **418:** 539–544.

Dunty WC Jr, Biris KK, Chalamalasetty RB, Taketo MM, Lewandoski M, Yamaguchi TT. 2008. Wnt3a/β-catenin signaling controls posterior body development by coordinating mesoderm formation and segmentation. *Development* **135:** 85–94.

Etheridge SL, Ray S, Li S, Hamblet NS, Lijam N, Tsang M, Greer J, Kardos N, Wang J, Sussman DJ, et al. 2008. Murine dishevelled 3 functions in redundant pathways with dishevelled 1 and 2 in normal cardiac outflow tract, cochlea, and neural tube development. *PLoS Genet* **4.**

Galceran J, Farinas I, Depew MJ, Clevers H, Grosschedl R. 1999. Wnt3a$^{-/-}$–like phenotype and limb deficiency in Lef1$^{-/-}$Tcf1$^{-/-}$ mice. *Genes Dev* **13:** 709–717.

Galceran J, Hsu SC, Grosschedl R. 2001. Rescue of a Wnt mutation by an activated form of LEF-1: Regulation of maintenance but not initiation of Brachyury expression. *Proc Natl Acad Sci* **98:** 8668–8673.

Galceran J, Sustmann C, Hsu SC, Folberth S, Grosschedl R. 2004. LEF1-mediated regulation of Delta-like1 links Wnt and Notch signaling in somitogenesis. *Genes Dev* **18:** 2718–2723.

Gao B, Song H, Bishop K, Elliot G, Garrett L, English MA, Andre P, Robinson J, Sood R, Minami Y, et al. 2011. Wnt signaling gradients establish planar cell polarity by inducing Vangl2 phosphorylation through Ror2. *Dev Cell* **20:** 163–176.

Gong Y, Mo C, Fraser SS. 2004. Planar cell polarity signalling controls cell division orientation during zebrafish gastrulation. *Nature* **430:** 689–693.

Grigoryan T, Wend P, Klaus A, Birchmeier W. 2008. Deciphering the function of canonical Wnt signals in development and disease: Conditional loss- and gain-of-function mutations of β-catenin in mice. *Genes Dev* **22:** 2308–2341.

Gros J, Hu JK, Vinegoni C, Feruglio PF, Weissleder R, Tabin CC. 2010. WNT5A/JNK and FGF/MAPK pathways regulate the cellular events shaping the vertebrate limb bud. *Curr Biol* **20:** 1993–2002.

Guo N, Hawkins C, Nathans J. 2004. Frizzled6 controls hair patterning in mice. *Proc Natl Acad Sci* **101:** 9277–9281.

Haegel H, Larue L, Ohsugi M, Fedorov L, Herrenknecht K, Kemler R. 1995. Lack of β-catenin affects mouse development at gastrulation. *Development* **121:** 3529–3537.

Harada N, Tamai Y, Ishikawa T, Sauer B, Takaku K, Oshima M, Taketo MM. 1999. Intestinal polyposis in mice with a dominant stable mutation of the β-catenin gene. *Embo J* **18:** 5931–5942.

Hashimoto M, Shinohara K, Wang J, Ikeuchi S, Yoshiba S, Meno C, Nonaka S, Takada S, Hatta K, Wynshaw-Boris A, et al. 2010. Planar polarization of node cells determines the rotational axis of node cilia. *Nat Cell Biol* **12:** 170–176.

Heisenberg CP, Tada M, Rauch GJ, Saude L, Concha ML, Geisler R, Stemple DL, Smith JC, Wilson SS. 2000. Silberblick/Wnt11 mediates convergent extension movements during zebrafish gastrulation. *Nature* **405:** 76–81.

Hofmann M, Schuster-Gossler K, Watabe-Rudolph M, Aulehla A, Herrmann BG, Gossler A. 2004. WNT signaling, in synergy with T/TBX6, controls Notch signaling by regulating Dll1 expression in the presomitic mesoderm of mouse embryos. *Genes Dev* **18:** 2712–2717.

Hsieh JC, Lee L, Zhang L, Wefer S, Brown K, DeRossi C, Wines ME, Rosenquist T, Holdener BB. 2003. Mesd encodes an LRP5/6 chaperone essential for specification of mouse embryonic polarity. *Cell* **112:** 355–367.

Huelsken J, Vogel R, Brinkmann V, Erdmann B, Birchmeier C, Birchmeier W. 2000. Requirement for β-catenin in anterior-posterior axis formation in mice. *J Cell Biol* **148:** 567–578.

Jho EH, Zhang T, Domon C, Joo CK, Freund JN, Costantini F. 2002. Wnt/β-catenin/Tcf signaling induces the transcription of Axin2, a negative regulator of the signaling pathway. *Mol Cell Biol* **22:** 1172–1183.

Jones C, Roper VC, Foucher I, Qian D, Banizs B, Petit C, Yoder BK, Chen P. 2008. Ciliary proteins link basal body polarization to planar cell polarity regulation. *Nat Genet* **40:** 69–77.

Juriloff DM, Harris MJ. 2000. Mouse models for neural tube closure defects. *Hum Mol Genet* **9:** 993–1000.

Kelly OG, Pinson KI, Skarnes WW. 2004. The Wnt co-receptors Lrp5 and Lrp6 are essential for gastrulation in mice. *Development* **131:** 2803–2815.

Kemler R, Hierholzer A, Kanzler B, Kuppig S, Hansen K, Taketo MM, de Vries WN, Knowles BB, Solter D. 2004. Stabilization of β-catenin in the mouse zygote leads to premature epithelial-mesenchymal transition in the epiblast. *Development* **131:** 5817–5824.

Kemp C, Willems E, Abdo S, Lambiv L, Leyns L. 2005. Expression of all Wnt genes and their secreted antagonists during mouse blastocyst and postimplantation development. *Dev Dyn* **233:** 1064–1075.

Khokha MK, Hsu D, Brunet LJ, Dionne MS, Harland RR. 2003. Gremlin is the BMP antagonist required for maintenance of Shh and Fgf signals during limb patterning. *Nat Genet* **34:** 303–307.

Kibar Z, Vogan KJ, Groulx N, Justice MJ, Underhill DA, Gros P. 2001. Ltap, a mammalian homolog of Drosophila Strabismus/Van Gogh, is altered in the mouse neural tube mutant Loop-tail. *Nat Genet* **28:** 251–255.

Kibar Z, Torban E, McDearmid JR, Reynolds A, Berghout J, Mathieu M, Kirillova I, De Marco P, Merello E, Hayes JM, et al. 2007. Mutations in VANGL1 associated with neural-tube defects. *N Engl J Med* **356:** 1432–1437.

Kibar Z, Bosoi CM, Kooistra M, Salem S, Finnell RH, De Marco P, Merello E, Bassuk AG, Capra V, Gros P. 2009. Novel mutations in VANGL1 in neural tube defects. *Hum Mutat* **30.**

Kibar Z, Salem S, Bosoi C, Pauwels E, De Marco P, Merello E, Bassuk A, Capra V, Gros P. 2011. Contribution of VANGL2 mutations to isolated neural tube defects. *Clin Genet* **80:** 76–82.

Kimura-Yoshida C, Nakano H, Okamura D, Nakao K, Yonemura S, Belo JA, Aizawa S, Matsui Y, Matsuo I. 2005. Canonical Wnt signaling and its antagonist regulate anterior-posterior axis polarization by guiding cell migration in mouse visceral endoderm. *Dev Cell* **9:** 639–650.

Lawrence PA, Casal J, Struhl G. 2004. Cell interactions and planar polarity in the abdominal epidermis of Drosophila. *Development* **131:** 4651–4664.

Lei YP, Zhang T, Li H, Wu BL, Jin L, Wang HH. 2010. VANGL2 mutations in human cranial neural-tube defects. *N Engl J Med* **362:** 2232–2235.

Lewis SL, Khoo PL, De Young RA, Steiner K, Wilcock C, Mukhopadhyay M, Westphal H, Jamieson RV, Robb L, Tam PP. 2008. Dkk1 and Wnt3 interact to control head morphogenesis in the mouse. *Development* **135:** 1791–1801.

Lickert H, Kutsch S, Kanzler B, Tamai Y, Taketo MM, Kemler R. 2002. Formation of multiple hearts in mice following deletion of β-catenin in the embryonic endoderm. *Dev Cell* **3:** 171–181.

Liu P, Wakamiya M, Shea MJ, Albrecht U, Behringer RR, Bradley A. 1999. Requirement for Wnt3 in vertebrate axis formation. *Nat Genet* **22:** 361–365.

Lu X, Borchers AG, Jolicoeur C, Rayburn H, Baker JC, Tessier-Lavigne M. 2004. PTK7/CCK-4 is a novel regulator of planar cell polarity in vertebrates. *Nature* **430:** 93–98.

Lustig B, Jerchow B, Sachs M, Weiler S, Pietsch T, Karsten U, van de Wetering M, Clevers H, Schlag PM, Birchmeier W, et al. 2002. Negative feedback loop of Wnt signaling through upregulation of conductin/axin2 in colorectal and liver tumors. *Mol Cell Biol* **22:** 1184–1193.

Maretto S, Cordenonsi M, Dupont S, Braghetta P, Broccoli V, Hassan AB, Volpin D, Bressan GM, Piccolo S. 2003. Mapping Wnt/β-catenin signaling during mouse development and in colorectal tumors. *Proc Natl Acad Sci* **100:** 3299–3304.

Matsuyama M, Aizawa S, Shimono A. 2009. Sfrp controls apicobasal polarity and oriented cell division in developing gut epithelium. *PLoS Genet* **5:** e1000427.

Miura S, Singh AP, Mishina Y. 2010. Bmpr1a is required for proper migration of the AVE through regulation of Dkk1 expression in the pre-streak mouse embryo. *Dev Biol* **341:** 246–254.

Mohamed OA, Clarke HJ, Dufort D. 2004. β-catenin signaling marks the prospective site of primitive streak formation in the mouse embryo. *Dev Dyn* **231:** 416–424.

Mohamed OA, Jonnaert M, Labelle-Dumais C, Kuroda K, Clarke HJ, Dufort D. 2005. Uterine Wnt/β-catenin signaling is required for implantation. *Proc Natl Acad Sci* **102:** 8579–8584.

Montcouquiol M, Rachel RA, Lanford PJ, Copeland NG, Jenkins NA, Kelley MM. 2003. Identification of Vangl2 and Scrb1 as planar polarity genes in mammals. *Nature* **423:** 173–177.

Montcouquiol M, Sans N, Huss D, Kach J, Dickman JD, Forge A, Rachel RA, Copeland NG, Jenkins NA, Bogani D, et al. 2006. Asymmetric localization of Vangl2 and Fz3 indicate novel mechanisms for planar cell polarity in mammals. *J Neurosci* **26:** 5265–5275.

Morkel M, Huelsken J, Wakamiya M, Ding J, van de Wetering M, Clevers H, Taketo MM, Behringer RR, Shen MM, Birchmeier W. 2003. β-catenin regulates Cripto- and Wnt3-dependent gene expression programs in mouse axis and mesoderm formation. *Development* **130:** 6283–6294.

Mukhopadhyay M, Shtrom S, Rodriguez-Esteban C, Chen L, Tsukui T, Gomer L, Dorward DW, Glinka A, Grinberg A, Huang SP, et al. 2001. Dickkopf1 is required for embryonic head induction and limb morphogenesis in the mouse. *Dev Cell* **1:** 423–434.

Murdoch JN, Doudney K, Paternotte C, Copp AJ, Stanier P. 2001. Severe neural tube defects in the loop-tail mouse result from mutation of Lpp1, a novel gene involved in floor plate specification. *Hum Mol Genet* **10:** 2593–2601.

Nakaya MA, Biris K, Tsukiyama T, Jaime S, Rawls JA, Yamaguchi TT. 2005. Wnt3alinks left-right determination with segmentation and anteroposterior axis elongation. *Development* **132:** 5425–5436.

Nelson WJ, Nusse R. 2004. Convergence of Wnt, β-catenin, and cadherin pathways. *Science* **303:** 1483–1487.

Nonaka S, Yoshiba S, Watanabe D, Ikeuchi S, Goto T, Marshall WF, Hamada H. 2005. De novo formation of left-right asymmetry by posterior tilt of nodal cilia. *PLoS Biol* **3.**

Oishi I, Suzuki H, Onishi N, Takada R, Kani S, Ohkawara B, Koshida I, Suzuki K, Yamada G, Schwabe GC, et al. 2003. The receptor tyrosine kinase Ror2 is involved in non-canonical Wnt5a/JNK signalling pathway. *Genes Cells* **8:** 645–654.

Okada Y, Takeda S, Tanaka Y, Belmonte JC, Hirokawa N. 2005. Mechanism of nodal flow: A conserved symmetry breaking event in left-right axis determination. *Cell* **121:** 633–644.

Oldridge M, Fortuna AM, Maringa M, Propping P, Mansour S, Pollitt C, DeChiara TM, Kimble RB, Valenzuela DM, Yancopoulos GD, et al. 2000. Dominant mutations in ROR2, encoding an orphan receptor tyrosine kinase, cause brachydactyly type B. *Nat Genet* **24:** 275–278.

Perea-Gomez A, Vella FD, Shawlot W, Oulad-Abdelghani M, Chazaud C, Meno C, Pfister V, Chen L, Robertson E, Hamada H, et al. 2002. Nodal antagonists in the anterior visceral endoderm prevent the formation of multiple primitive streaks. *Dev Cell* **3:** 745–756.

Person AD, Beiraghi S, Sieben CM, Hermanson S, Neumann AN, Robu ME, Schleiffarth JR, Billington CJ, van Bokhoven H, Hoogeboom JM, et al. 2010. WNT5A mutations in patients with autosomal dominant Robinow syndrome. *Dev Dyn* **239:** 327–337.

Pourquie O. 2011. Vertebrate segmentation: From cyclic gene networks to scoliosis. *Cell* **145:** 650–663.

Povelones M, Howes R, Fish M, Nusse R. 2005. Genetic evidence that Drosophila frizzled controls planar cell polarity and Armadillo signaling by a common mechanism. *Genetics* **171:** 1643–1654.

Qian D, Jones C, Rzadzinska A, Mark S, Zhang X, Steel KP, Dai X, Chen P. 2007. Wnt5a functions in planar cell polarity regulation in mice. *Dev Biol* **306:** 121–133.

Raz R, Stricker S, Gazzerro E, Clor JL, Witte F, Nistala H, Zabski S, Pereira RC, Stadmeyer L, Wang X, et al. 2008. The mutation ROR2W749X, linked to human BDB, is a recessive mutation in the mouse, causing brachydactyly, mediating patterning of joints and modeling recessive Robinow syndrome. *Development* **135:** 1713–1723.

Rida PC, Chen P. 2009. Line up and listen: Planar cell polarity regulation in the mammalian inner ear. *Semin Cell Dev Biol* **20:** 978–985.

Robertson EJ, Norris DP, Brennan J, Bikoff EE. 2003. Control of early anterior-posterior patterning in the mouse embryo by TGF-β signalling. *Philos Trans R Soc Lond B Biol Sci* **358.**

Roszko I, Sawada A, Solnica-Krezel L. 2009. Regulation of convergence and extension movements during vertebrate gastrulation by the Wnt/PCP pathway. *Semin Cell Dev Biol* **20:** 986–997.

Satoh K, Kasai M, Ishidao T, Tago K, Ohwada S, Hasegawa Y, Senda T, Takada S, Nada S, Nakamura T, et al. 2004. Anteriorization of neural fate by inhibitor of β-catenin and T cell factor (ICAT), a negative regulator of Wnt signaling. *Proc Natl Acad Sci* **101:** 8017–8021.

Savory JG, Mansfield M, Rijli FM, Lohnes D. 2011. Cdx mediates neural tube closure through transcriptional regulation of the planar cell polarity gene Ptk7. *Development* **138:** 1361–1370.

Seifert JR, Mlodzik M. 2007. Frizzled/PCP signalling: A conserved mechanism regulating cell polarity and directed motility. *Nat Rev Genet* **8:** 126–138.

Song H, Hu J, Chen W, Elliott G, Andre P, Gao B, Yang Y. 2010. Planar cell polarity breaks bilateral symmetry by controlling ciliary positioning. *Nature* **466:** 378–382.

Strutt H, Strutt D. 2009. Asymmetric localisation of planar polarity proteins: Mechanisms and consequences. *Semin Cell Dev Biol* **20:** 957–963.

Stuckey DW, Clements M, Di-Gregorio A, Senner CE, Le Tissier P, Srinivas S, Rodriguez TT. 2011. Coordination of cell proliferation and anterior-posterior axis establishment in the mouse embryo. *Development* **138:** 1521–1530.

Sun X, Meyers EN, Lewandoski M, Martin GG. 1999. Targeted disruption of Fgf8 causes failure of cell migration in the gastrulating mouse embryo. *Genes Dev* **13:** 1834–1846.

Sun X, Mariani FV, Martin GG. 2002. Functions of FGF signalling from the apical ectodermal ridge in limb development. *Nature* **418:** 501–508.

Takada S, Stark KL, Shea MJ, Vassileva G, McMahon JA, McMahon AA. 1994. Wnt-3a regulates somite and tailbud formation in the mouse embryo. *Genes Dev* **8:** 174–189.

Takeuchi S, Takeda K, Oishi I, Nomi M, Ikeya M, Itoh K, Tamura S, Ueda T, Hatta T, Otani H, et al. 2000. Mouse Ror2 receptor tyrosine kinase is required for the heart development and limb formation. *Genes Cells* **5:** 71–78.

Tao H, Suzuki M, Kiyonari H, Abe T, Sasaoka T, Ueno N. 2009. Mouse prickle1, the homolog of a PCP gene, is essential for epiblast apical-basal polarity. *Proc Natl Acad Sci* **106:** 14426–14431.

ten Berge D, Koole W, Fuerer C, Fish M, Eroglu E, Nusse R. 2008. Wnt signaling mediates self-organization and axis formation in embryoid bodies. *Cell Stem Cell* **3:** 508–518.

ten Berge D, Kurek D, Blauwkamp T, Koole W, Maas A, Eroglu E, Siu RK, Nusse R. 2011. Embryonic stem cells require Wnt proteins to prevent differentiation to epiblast stem cells. *Nat Cell Biol* **13:** 1070–1075.

Topol L, Jiang X, Choi H, Garrett-Beal L, Carolan PJ, Yang Y. 2003. Wnt-5a inhibits the canonical Wnt pathway by promoting GSK-3-independent β-catenin degradation. *J Cell Biol* **162:** 899–908.

Torban E, Patenaude AM, Leclerc S, Rakowiecki S, Gauthier S, Andelfinger G, Epstein DJ, Gros P. 2008. Genetic interaction between members of the Vangl family causes neural tube defects in mice. *Proc Natl Acad Sci* **105:** 3449–3454.

van Amerongen R, Berns A. 2006. Knockout mouse models to study Wnt signal transduction. *Trends Genet* **22:** 678–689.

Veeman MT, Axelrod JD, Moon RR. 2003. A second canon. Functions and mechanisms of β-catenin-independent Wnt signaling. *Dev Cell* **5:** 367–377.

Wallingford JB, Habas R. 2005. The developmental biology of Dishevelled: An enigmatic protein governing cell fate and cell polarity. *Development* **132:** 4421–4436.

Wallingford JB, Harland RM. 2002. Neural tube closure requires Dishevelled-dependent convergent extension of the midline. *Development* **129:** 5815–5825.

Wallingford JB, Rowning BA, Vogeli KM, Rothbacher U, Fraser SE, Harland RR. 2000. Dishevelled controls cell polarity during Xenopus gastrulation. *Nature* **405:** 81–85.

Wang J, Mark S, Zhang X, Qian D, Yoo SJ, Radde-Gallwitz K, Zhang Y, Lin X, Collazo A, Wynshaw-Boris A, et al. 2005. Regulation of polarized extension and planar cell polarity in the cochlea by the vertebrate PCP pathway. *Nat Genet* **37:** 980–985.

Wang J, Hamblet NS, Mark S, Dickinson ME, Brinkman BC, Segil N, Fraser SE, Chen P, Wallingford JB, Wynshaw-Boris A. 2006a. Dishevelled genes mediate a conserved mammalian PCP pathway to regulate convergent extension during neurulation. *Development* **133:** 1767–1778.

Wang Y, Guo N, Nathans J. 2006b. The role of Frizzled3 and Frizzled6 in neural tube closure and in the planar polarity of inner-ear sensory hair cells. *J Neurosci* **26:** 2147–2156.

Wang B, Sinha T, Jiao K, Serra R, Wang J. 2011. Disruption of PCP signaling causes limb morphogenesis and skeletal defects and may underlie Robinow syndrome and brachydactyly type B. *Hum Mol Genet* **20:** 271–285.

Wehner P, Shnitsar I, Urlaub H, Borchers A. 2011. RACK1 is a novel interaction partner of PTK7 that is required for neural tube closure. *Development* **138:** 1321–1327.

Wu J, Mlodzik M. 2008. The frizzled extracellular domain is a ligand for Van Gogh/Stbm during nonautonomous planar cell polarity signaling. *Dev Cell* **15:** 462–469.

Yamaguchi TP. 2008. Genetics of Wnt signaling during early mammalian development. *Methods Mol Biol* **468:** 287–305.

Yamaguchi TP, Bradley A, McMahon AP, Jones S. 1999a. A Wnt5a pathway underlies outgrowth of multiple structures in the vertebrate embryo. *Development* **126:** 1211–1223.

Yamaguchi TP, Takada S, Yoshikawa Y, Wu N, McMahon AA. 1999b. T (Brachyury) is a direct target of Wnt3a during paraxial mesoderm specification. *Genes Dev* **13:** 3185–3190.

Yamamoto S, Nishimura O, Misaki K, Nishita M, Minami Y, Yonemura S, Tarui H, Sasaki H. 2008. Cthrc1 selectively activates the planar cell polarity pathway of Wnt signaling by stabilizing the Wnt-receptor complex. *Dev Cell* **15:** 23–36.

Yoshikawa Y, Fujimori T, McMahon AP, Takada S. 1997. Evidence that absence of Wnt-3a signaling promotes neuralization instead of paraxial mesoderm development in the mouse. *Dev Biol* **183:** 234–242.

Yu K, Ornitz DM. 2008. FGF signaling regulates mesenchymal differentiation and skeletal patterning along the limb bud proximodistal axis. *Development* **135:** 483–491.

Zallen JA. 2007. Planar polarity and tissue morphogenesis. *Cell* **129:** 1051–1063.

Wnt Pathway Regulation of Embryonic Stem Cell Self-Renewal

Bradley J. Merrill

Department of Biochemistry and Molecular Genetics, University of Illinois at Chicago, Chicago, Illinois 60607
Correspondence: merrillb@uic.edu

Embryonic stem cells (ESCs) can generate all of the cell types found in the adult organism. Remarkably, they retain this ability even after many cell divisions in vitro, as long as the culture conditions prevent differentiation of the cells. Wnt signaling and β-catenin have been shown to cause strong effects on ESCs both in terms of stimulating the expansion of stem cells and stimulating differentiation toward lineage committed cell types. The varied effects of Wnt signaling in ESCs, alongside the sometimes unconventional mechanisms underlying the effects, have generated a fair amount of controversy and intrigue regarding the role of Wnt signaling in pluripotent stem cells. Insights into the mechanisms of Wnt function in stem cells can be gained by examination of the causes for seemingly opposing effects of Wnt signaling on self-renewal versus differentiation.

For a single-cell embryo to eventually form an adult organism of trillions of cells, some cells in the early mammalian embryo must be able to generate all cell lineages in the animal. The potential to make all adult cell types defines the property of pluripotency, and it is maintained in proliferating cells through a process called self-renewal. As cells become specified to contribute to particular lineages, they typically lose the ability to make cell types from distinct lineages (Waddington 1957; Hochedlinger and Plath 2009). As such, pluripotency is lost during the initial steps of lineage commitment that occur during gastrulation (Beddington 1982, 1983; Lawson and Pedersen 1987; Lawson et al. 1991), which is a process that coordinates the generation of adult cell lineages with the elaboration of a basic three-dimensional body structure (Heisenberg and Solnica-Krezel 2008). In the mouse, pluripotency can be tested with various experiments; the gold standard is the injection of cells into a blastocyst-staged embryo followed by contribution to a diversity of cell types in the chimeric animal or chimeric embryo after gastrulation. Cells are typically considered to have been pluripotent only if they contributed to all three germ layers (endoderm, mesoderm, and ectoderm).

Embryonic stem cells (ESCs) are generated in vitro by outgrowths from a preimplantation-staged embryo, frequently a blastocyst. Pluripotent cells from the inner cell mass (ICM) of the blastocyst proliferate to form colonies, which can be expanded into ESC cultures. When culture conditions for in vitro propagation of mouse ESCs (mESCs) were first discovered more than 30 years ago (Evans and Kaufman 1981; Martin 1981), the critical achievement was finding

Cite this article as *Cold Spring Harb Perspect Biol* doi: 10.1101/cshperspect.a007971

conditions supporting indefinite ESC self-renewal, that is, maintenance of pluripotency following cell division. Compared with the other cell systems discussed below in this article, mESCs ostensibly display the greatest capacity for self-renewal and the highest ability to maintain pluripotency. As such, mESCs are typically thought to represent a primitive, or "naive," cellular state in the early embryo.

Several culture conditions can support self-renewal of mESCs. Initially, ESCs were grown in serum containing media atop a layer of mitotically inactivated fibroblasts, called feeder cells (Evans and Kaufman 1981). Feeder cells secrete the LIF cytokine, which binds a transmembrane receptor complex consisting of LIFR and gp130 proteins (Gearing et al. 1991; Gearing and Bruce 1992; Davis et al. 1993). LIF binding activates Jak/Stat signaling and Stat3 phosphorylation, which promotes ESC self-renewal (Niwa et al. 1998; Matsuda et al. 1999). Convincing proof of LIF's importance for self-renewal in vitro was shown when recombinant LIF protein was shown to be sufficient to replace feeder cells in ESC cultures (Smith et al. 1988; Williams et al. 1988; Nichols et al. 1990).

Essentially the same feeder cells can be used for both mESCs and human ESCs (hESCs); however, discrete activities of the feeders in terms of the cytokines they release are needed to effect optimal self-renewal for each cell. The LIF cytokine important for mESC self-renewal did not stimulate hESC self-renewal (Thomson et al. 1998). Instead, ERK signaling downstream from Fgf2 must accompany a feeder layer in serum-containing media for optimal hESC self-renewal (Xu et al. 2005). Interestingly, recombinant Fgf2 by itself could not replace feeders, and Fgf2 has been suggested to work in part by stimulating feeders to produce Activin/Nodal ligands; the combination of Fgf2 and Nodal/Activin is sufficient to support hESC self-renewal in serum-free chemically defined culture conditions (Vallier et al. 2004, 2009; James et al. 2005).

Clear differences exist between mESCs and hESCs. The colonies adopt different morphologies, they require distinct culture conditions for self-renewal, and they have significantly different gene expression signatures (Table 1). These differences make it interesting to compare hESCs with a different type of pluripotent mouse stem cell, called EpiSCs (for epiblast stem cells) (Brons et al. 2007; Tesar et al. 2007). Mouse EpiSCs are made from the epiblast of postimplantation-staged embryos between embryonic days 5.5 (E5.5) and E6.5 of embryogenesis (Brons et al. 2007; Tesar et al. 2007; Han et al.

Table 1. Pluripotent stem cell states: Naive and primed

	Mouse ESC	Human ESC	Mouse EpiSC
Effects of culture conditions			
Serum + Lif	Self-renewal	Differentiation	Differentiation
Wnt3a/GSK3inhibitor	Self-renewal	Differentiation	Differentiation
Fgf2 + Activin A	EpiSC	Self-renewal	Self-renewal
Gene expression profiles			
Oct4, Sox2, Nanog	High	High	High
Sox17, Eomes, Fgf5	Low	High	High
Klf4, Rex1, Stella	High	Low	Low
Activity in pluripotency tests			
Embryoid body	Pass	Pass	Pass
Teratoma formation	Pass	Pass	Pass
Blastocyst injection	Pass	Not determined	Poor
Tetraploid complementation	Pass	Not determined	Not determined
Epigenetic state	Naive	Primed	Primed

Three pluripotent cell systems are compared with respect to characteristics that describe their epigenetic state of pluripotency. See text for details.

Cite this article as *Cold Spring Harb Perspect Biol* doi: 10.1101/cshperspect.a007971

2010). Lineage specification of pluripotent epiblast cells begins soon after formation of a cup-like structure, and at E6.5, the cells in the epiblast begin to be specified to primary cell lineages during gastrulation. The in vivo cellular environment for ICM cells and postimplantation epiblast cells is considerably different, and it is not surprising that EpiSCs and mESCs display many different characteristics (Xu et al. 2010). However, it was somewhat surprising that EpiSCs share many characteristics with hESCs, including a common colony morphology, Fgf2 + Activin A culture conditions, and gene expression signatures (Table 1) (Brons et al. 2007; Tesar et al. 2007). Like mESCs and hESCs, EpiSCs pass pluripotency tests for in vitro differentiation and teratoma formation. Whereas mESC can efficiently convert (i.e., differentiate) into EpiSC-like cells when switched to Fgf2/Activin A media (Hanna et al. 2009; Greber et al. 2010), EpiSCs required genetic manipulation or reprogramming for efficient conversion to mES-like cells (Guo et al. 2009; Hanna et al. 2009; Greber et al. 2010; Guo and Smith 2010). Many investigators consider hESCs and mouse EpiSCs to be *primed* for differentiation as they reside in a less primitive differentiation state relative to the naive state of pluripotency in mESCs.

INTRINSIC TRANSCRIPTION FACTOR NETWORKS SUPPORT PLURIPOTENT CELL SELF-RENEWAL

All pluripotent cells are thought to use essentially the same core set of transcription factors in a conserved self-renewal network. Several reviews cover the topic comprehensively (Ng and Surani 2011; Young 2011), and here I cover only the highlights important for understanding the role of Wnt signaling in pluripotent cells. Three transcription factors—Oct4 (also called Pou5f1), Sox2, and Nanog—are considered to constitute a core self-renewal network for several reasons. Each is genetically necessary for self-renewal of pluripotent cells when the ICM cells divide rapidly and give rise to the epiblast (Nichols et al. 1998; Avilion et al. 2003; Mitsui et al. 2003). Gene ablation and RNAi-knockdown experiments showed that Oct4 and Sox2 are absolutely necessary for mESC self-renewal (Niwa et al. 2000; Avilion et al. 2003). Nanog$^{-/-}$ ESCs can be propagated in vitro, albeit at a reduced efficiency relative to wild-type mESCs (Chambers et al. 2007). Importantly, overexpression of Nanog has a potent and unique ability to prevent mESC differentiation (Chambers et al. 2003; Mitsui et al. 2003). Genome-wide chromatin immunoprecipitation (ChIP) experiments showed that Nanog, Oct4, and Sox2 all bind to a highly overlapping set of downstream target genes in each mESC and hESC (Boyer et al. 2005; Loh et al. 2006; Chen et al. 2008; Cole et al. 2008; Kim et al. 2008; Marson et al. 2008); notably, each protein occupied one another's gene regulatory region. RNAi knockdown of each Oct4, Sox2, or Nanog each caused largely overlapping effects on gene expression, consistent with each gene product regulating a common set of target genes (Loh et al. 2006). Many genes regulated by Oct4, Sox2, and Nanog include factors that play a downstream, yet important role in self-renewal, including Klf4, c-myc, Esrrb, Tbx3, and Nr5a2 (Gu et al. 2005; Ivanova et al. 2006; Jiang et al. 2008; Kim et al. 2008; Hall et al. 2009; Wagner et al. 2010). Together, these data are consistent with Oct4, Sox2, and Nanog being the central components of a feed-forward network of transcription factors that drives self-renewal of pluripotent cells (see Fig. 1).

EFFECTS OF WNTs ON PLURIPOTENT CELLS

Wnt/β-Catenin Signaling Inhibits Differentiation of Naive mESCs

For decades, mESCs indirectly contributed greatly to understanding Wnt signaling in mammals through the generation of knockout mice; however, direct examination of the effects of Wnt signaling in mESCs began in earnest only after the derivation of hESCs. Two groups that generated β-catenin knockout mESCs used them to generate mouse strains, which convincingly showed the requirement for β-catenin for gastrulation in mice (Haegel et al. 1995; Huelsken et al. 2000). This early work also

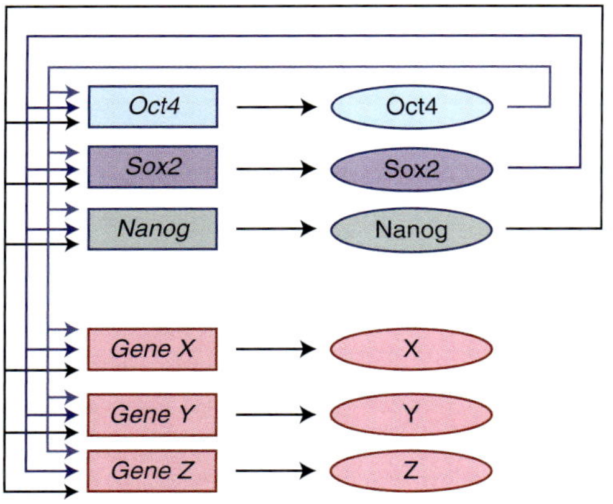

Figure 1. Feed-forward network of stem cell transcription factors. Three transcription factors—Oct4, Sox2, and Nanog—bind to one another's gene regulatory regions and control expression of one another's gene products. Downstream targets (genes X, Y, Z) stimulate pluripotent cell self-renewal through the functions of X, Y, and Z gene products. (The figure is modeled after those by Cole et al. 2008 and Young 2011.)

directly examined some effects in mESC. Kemler and colleagues used outgrowths from β-catenin knockout morulae and found that initial formation of ESC-like cells did not require β-catenin; however, β-catenin$^{-/-}$ cells displayed an adhesion defect causing colonies to appear as loosely associated cells (Haegel et al. 1995). Birchmeier and colleagues generated β-catenin$^{-/-}$ mESCs by performing two rounds of homologous recombination and found that the cells self-renewed sufficiently well enough to be used for blastocyst injection experiments (Huelsken et al. 2000). When injected into blastocysts, β-catenin$^{-/-}$ mESCs could contribute to ectoderm, mesoderm, and endoderm cell types in the embryo (Huelsken et al. 2000). Although they were performed before elucidation of the core pluripotency circuitry, these early observations indicated that β-catenin was not absolutely necessary for pluripotent cell self-renewal in classical mESC culture conditions (LIF^{+}, serum, feeders).

A positive effect of Wnt signaling in stimulating self-renewal was shown by a combination of experiments using Wnt-conditioned media, recombinant Wnt, and chemical inhibitors. When cultured without a feeder layer, the withdrawal of recombinant LIF caused mESCs to differentiate (Smith et al. 1988; Williams et al. 1988). This response is measured quite simply by scoring colony morphologies as differentiated (alkaline phosphatase (AP) negative, flattened colony) versus ESC-like (AP^{+}, rounded birefringent colony) after a period of 4–8 d of colony growth. Sato et al. (2004) showed that by adding an inhibitor of Gsk3, called BIO, or by adding Wnt3a-conditioned media to cells, exogenous LIF was no longer necessary for the formation of ESC-like colonies or for activity of the Rex1 promoter, which is specific for mESC. The conditioned media used in these early experiments probably also contained low levels of LIF or a similar cytokine, because unlike conditioned media, recombinant Wnt3a did not completely replace LIF; instead, recombinant Wnt3a sensitized the cells such that 100-fold lower-than-normal levels of exogenous LIF were able to support self-renewal (Ogawa et al. 2006). Similarly, the expression of Wnt6 and Wnt5a supplemented the capability of an ineffective feeder cell layer, called C-STO, into an efficient feeder cell layer that stimulated self-renewal of mESCs (Hao et al. 2006). These effects of Wnts appeared to function through canonical β-catenin signaling, because forced expression of a stable β-catenin in mESC was sufficient to allow low levels of exogenous LIF to stimulate self-renewal (Ogawa et al. 2006; Takao et al. 2007), and to allow C-STO feeders cells to function as an effective feeder cell layer (Hao et al. 2006). Each of these findings was important for showing a positive role for Wnt stimulation; however, they also indicated a potentially limited role for Wnts in stimulating self-renewal in the presence of serum-containing media.

Wnt signaling also regulates differentiation of mouse embryonic stem cells. When injected into a permissive site of a host mouse, normal mESCs form teratoma tumors consisting of diverse cell types derived from all three germ layers. Mouse ESCs harboring two mutant alleles of APC either failed to generate teratomas or formed tumors that were particularly deficient for neurectoderm differentiation (Kielman et al.

2002). The effects of APC mutation were attributed to increased β-catenin levels because stabilization of β-catenin through deletion of exon 3 (Harada et al. 1999) resulted in ostensibly the same effects in teratoma and colony-based assays (Kielman et al. 2002). Doble et al. (2007) genetically ablated both alleles of each GSK3α and GSK3β gene in a single mESC line, named GSK3 DKO for "double knockout." The analysis of GSK3 DKO cells elegantly proved that both GSK3α and β isoforms contributed essentially equally to the regulation of β-catenin in stem cells (Doble et al. 2007). Tcf-β-catenin target genes (Axin2, Brachyury, Cdx1) and TOPFlash activity were increased in GSK3 DKO cells without exogenous Wnt, indicating that endogenous GSK3 prevented activation of Wnt/Tcf/β-catenin signaling (Doble et al. 2007). Much like APC mutants, GSK3 DKO cells were unable to differentiate normally in either embryoid body (EB) or teratoma differentiation assays; they retained a significant population of cells expressing ESC marker genes, including Nanog, Oct4, and Rex1, whereas wild-type cells generated teratomas expressing diverse lineage markers (Doble et al. 2007; Kelly et al. 2011).

Wnt Signaling Stimulates Long-Term Self-Renewal of Naive mESCs

In addition to the short-term self-renewal and differentiation effects discussed above, several studies examined Wnt/β-catenin effects with more stringent tests of self-renewal, such as derivation of ESCs from embryos and long-term self-renewal in vitro. Austin Smith and colleagues noted that signals present in commonly used mESC cultures (serum-containing media) actually included significant differentiation-inducing signals (Ying et al. 2003, 2008; Kunath et al. 2007; Nichols et al. 2009). They posited that mESCs could exist in a ground state of self-renewal if extracellular signals were blocked. This model was supported by the observations that activation of ERK signaling promoted differentiation of mESCs and that mESC differentiation was constrained by adding an Fgf receptor (Fgfr) inhibitor and an ERK inhibitor (PD0325901) to a chemically defined media (Ying et al. 2008; Nichols et al. 2009). Using only Fgfr and ERK inhibitors did not sustain mESC cultures, because cells died without undergoing self-renewal divisions. Adding a GSK3 inhibitor alone, such as BIO or CHIR99021, sustained a short burst of self-renewal for only one or two passages (Ying et al. 2008). Remarkably, the combination of the three inhibitors, which was subsequently reduced to only two (PD0325901 + CHIR99021), was quite effective; these so-called two-inhibitor (2i) conditions supported the derivation of new mESC lines from blastocyst outgrowths and expansion into large cultures from a single cell of an already established mESC line (Ying et al. 2008; Nichols et al. 2009). Mouse ESCs grown in 2i conditions were cultured for several weeks with no apparent defects, and they maintained the ability to contribute to chimeric mice when injected into blastocysts (Ying et al. 2008). Recombinant Wnt3a could essentially replace the CHIR99021 to support self-renewal of mESC colonies + ERK inhibitors (Ying et al. 2008). Other, structurally dissimilar GSK3 inhibitors, RNAi knockdown of GSK3 enzymes, or genetic ablation of GSK3 in DKO cells all could replace CHIR99021, supporting the importance of GSK3 inhibition rather than potential off-target effects of inhibitors (Ying et al. 2008; Wray et al. 2011).

Activation of Wnt signaling or inhibition of GSK3 activity powerfully stimulated self-renewal of mESCs; however, exogenously activating Wnt/β-catenin does not appear to be absolutely necessary for pluripotency. The combination of LIF + ERK inhibitor without GSK3 inhibitors was able to support self-renewal of cells, albeit with a significantly lower efficiency than 2i conditions (Ying et al. 2008; Wray et al. 2011). Similarly, ERK inhibitors could be omitted when LIF was added together with GSK3 inhibition or recombinant Wnt3a (Ying et al. 2008; Berge et al. 2011; Wray et al. 2011). Indeed, some reports show that the positive effect of Wnt/β-catenin signaling on self-renewal is substantially greater than that of ERK inhibition (Berge et al. 2011). Consistent with this observation, the combination of exogenous LIF, Wnt3a, or GSK3 inhibitor, and ERK inhibitor together stimulated the most robust self-renewal of mESCs (Silva et al. 2008; Ying et al. 2008; Berge et al. 2011; Wray et al. 2011). These findings indicate that pluripotency does not absolutely require any one of the three components, that is, activation of LIF/Jak/Stat3 signaling, inhibition of ERK activity, or exogenous Wnt-stimulation/GSK3-inhibition.

It is important to note that mESCs secrete active Wnts (Berge et al. 2011). Blocking Wnt secretion from mESCs inhibited mESC self-renewal in a naive state (Berge et al. 2011). This discovery showed that endogenous Wnt/β-catenin contributes an underlying and necessary function for mESC self-renewal. The function is stimulated by exogenously supplied Wnt3a or GSK3 inhibitors to stimulate efficient self-renewal. Although the optimal self-renewal in 2i + LIF conditions indicates unique effects of each component, the effectiveness of any combination of two components indicates overlapping effects. The identity of the most important downstream effects for self-renewal remains an open topic of research for each component, including Wnt stimulation/GSK3 inhibition.

Effects of Wnt Stimulation and GSK3 Inhibition Depend on the State of Pluripotency

In contrast to mESCs, long-term self-renewal of hESCs did not benefit significantly from activating the Wnt signaling pathway (Dravid et al. 2005). This difference may be reflective of the differences between a primed versus naive state of pluripotency. Compared with the naive state of mESCs, hESCs more closely resemble the primed state of mEpiSCs (Fig. 1) (Brons et al. 2007; Tesar et al. 2007). Instead of stimulating self-renewal, Wnt signaling effectively stimulated differentiation of mESCs after they have been primed either by growth at a high density (Otero et al. 2004) or by initial commitment in EB assays (Lindsley et al. 2006; ten Berge et al. 2008). Similar to hESCs, EpiSCs responded to inhibiting GSK3 by differentiating (Greber et al. 2010). Interestingly, forced expression of Klf4, c-myc, or Nr5a2 enabled EpiSCs to be converted to a naive state when switched from Fgf2 + Activin A conditions to 2i conditions (Guo et al. 2009; Hanna et al. 2009; Guo and Smith 2010). This conversion required GSK3 inhibition. It was proposed that Klf4 and Nr5a2 expression effectively reprogrammed cells between metastable states of pluripotency (i.e., from a primed to a naive state), which could then be sustained by 2i conditions. Together, these observations indicated that GSK3 inhibition stimulated differentiation of the pluripotent cells primed for differentiation, but it stimulated self-renewal of naive pluripotent cells.

The importance of even subtle changes in the cellular state was neatly shown through examination of differences among individual cells in EpiSC cultures. Hans Scholer and colleagues isolated a culture of EpiSCs from an epiblast containing an Oct4-GFP promoter–reporter transgene (Han et al. 2010). Curiously, despite the EpiSC line expressing normal levels of Oct4 mRNA and protein, the Oct4-GFP activity was detectable in only ~0.5% of the EpiSCs. Although the mechanism underlying the effect on the promoter activity remains unclear, the activation or silencing of Oct4-GFP was dynamic. Sorted GFP-EpiSC cells generated cultures containing ~0.5% GFP^+ cells, and GFP^+ EpiSC cells generated cultures of predominantly GFP^- cells (Han et al. 2010). Thus, EpiSCs moved between the two cellular epigenetic states identified by Oct4-GFP activity. Gene expression signatures and blastocyst injection experiments indicated that GFP^+ EpiSCs were similar to naive cells from E5.5 embryos, and GFP^- EpiSCs were similar primed cells from E6.5 embryos (Han et al. 2010). When the two populations of cells were subjected to ERK and GSK3 inhibitor conditions, the GFP^- cells behaved as described for other EpiSC experiments, and GSK3 inhibition stimulated differentiation. In contrast, GSK3 inhibition plus ERK inhibition caused GFP^+ cells to revert to an overt ESC-like state. Thus, GSK3 inhibition stimulated self-renewal of cells in the naive state, and it stimulated differentiation of cells in the primed state.

Given the observation that Wnt/β-catenin signaling causes different effects on naive cells compared with primed cells, it is interesting to

 Cite this article as *Cold Spring Harb Perspect Biol* doi: 10.1101/cshperspect.a007971

note that Wnt/β-catenin controls the transition from naive to primed state. Support for this conclusion comes from work by ten Berge, Nusse, and colleagues, who showed that blocking Wnt ligands with Fz8CRD, a soluble Frizzled domain that sequesters Wnt proteins, caused mESCs to adopt EpiSC-like characteristics (ten Berge et al. 2011). Similarly, treating mESCs with IWP2, a porcupine inhibitor that blocks secretion of active Wnt proteins, inhibited AP^+ colony formation and mESC colony morphology while promoting a switch from naive (Rex1, Pecam1, Stella) to primed (Fgf5, Otx2, Dnmt3b) gene expression. The mechanism of action for IWP2 and Fz8CRD indicated that mESC cells secreted their own Wnt proteins that acted in a paracrine/autocrine manner. The colonies formed in the presence of IWP2 displayed a flattened EpiSC morphology and could be propagated using EpiSC culture conditions (Brons et al. 2007; Tesar et al. 2007; Berge et al. 2011). Together, these observations support a required function for Wnt signaling in preventing the transition from a naive to a primed state of pluripotency.

MECHANISMS OF WNT PATHWAY EFFECTS ON PLURIPOTENT CELLS

GSK3 Inhibition Functions through β-Catenin Stability

As is evident from the breadth of areas covered in this collective work, binding of a Wnt ligand to its receptor complex can cause a variety of downstream effects. Although GSK3 activity is necessary for regulating β-catenin levels, it is also important for cellular effects not normally associated with Wnt signaling, such as glycogen metabolism and translation control (Doble and Woodgett 2003). Moreover, the suggestion that β-catenin was not genetically necessary for mESCs to proliferate in LIF^+ serum conditions (Lyashenko et al. 2011; Wray et al. 2011) presented the possibility that β-catenin-independent effects of GSK3 inhibition and Wnt stimulation may affect pluripotent cell self-renewal. Therefore, it was important to determine if Wnt stimulation/Gsk3 inhibition affects function through β-catenin.

Several studies have addressed the role of β-catenin in mediating GSK3-inhibitor effects by using independent methods of increasing β-catenin levels in mESCs. When compared side-by-side with either mutating APC (Kielman et al. 2002) or ablating GSK3 in DKO cells (Doble et al. 2007), increasing β-catenin levels similarly inhibited the appearance of differentiated cell types in teratoma- and EB-based differentiation assays. Overexpression of β-catenin stimulated LIF-independent self-renewal of mESCs in serum-containing media (Ogawa et al. 2006; Takao et al. 2007). In addition, increasing β-catenin levels, inhibiting GSK3, or treating cells with Wnt3a all caused similar activation of endogenous target genes (Axin2, Cdx1, Brachyury) and transfected reporter plasmids (TOPFlash, Rex1, Nanog) (Kielman et al. 2002; Sato et al. 2004; Doble et al. 2007; Cole et al. 2008; Yi et al. 2008; Kelly et al. 2011). Thus, increasing β-catenin levels, inhibiting GSK3 activity, and adding exogenous Wnt3a all were sufficient to cause the similar pro-self-renewal effects on mESCs.

Until recently, surprisingly few studies genetically tested whether β-catenin was necessary for the effects of Wnts or GSK3 inhibition on pluripotent cells. Bradley Doble and colleagues used RNAi-knockdown techniques to determine whether β-catenin was necessary for differentiation defects in GSK3 DKO cells (Kelly et al. 2011). Teratomas formed from GSK3 DKO + control shRNA cells, but they failed to generate neurectoderm and retained significant numbers of Nanog-expressing cells. In contrast, GSK3 DKO + β-catenin-specific shRNA cells generated teratomas containing neurectoderm and lacking Nanog expression, similar to those formed from wild-type mESCs. Thus, β-catenin was necessary for differentiation defects in GSK3 DKO cells. Another approach independently taken by two groups (Austin Smith's group and Christine Hartmann's group) found that after conditionally ablating β-catenin from established mESC lines, cells formed abnormally scattered colonies consisting of poorly attached cells (Lyashenko et al.

2011; Wray et al. 2011), which was similar to the colony morphology of outgrowths from β-catenin-null blastocysts (Haegel et al. 1995). Interesting, the abnormal colony morphology returned to normal after several passages despite the continued absence of β-catenin, most likely because plakoglobin, a homolog of β-catenin that also binds to E-cadherin, replaced β-catenin at adherence junctions (Haegel et al. 1995; Huelsken et al. 2000; Lyashenko et al. 2011). Ablating β-catenin failed to block pluripotent cell proliferation in either 2i + LIF or LIF + ERK inhibitor conditions (Lyashenko et al. 2011; Wray et al. 2011). β-Catenin-null cells expressed markers of mESCs, including Oct4, Sox2, Nanog, Rex1, and Pecam1 (Lyashenko et al. 2011; Wray et al. 2011), which was consistent with β-catenin not being absolutely necessary for pluripotency. In contrast, EpiSC differentiation following IWP2 and Fz8CRD treatments indicated that Wnt signaling was necessary for self-renewal in a naive state (Berge et al. 2011). Thus, either a β-catenin-independent effect of Wnt proteins contributes to the naive state and/or the β-catenin-null cells exist in a primed state. Regardless of the state of pluripotency of β-catenin-deficient cells, it is important to note that the β-catenin was necessary for self-renewal under either 2i or LIF + GSK3-inhibitor conditions (Lyashenko et al. 2011; Wray et al. 2011), showing that β-catenin is the necessary downstream effector of GSK3 inhibition. Thus, the ability of GSK3 inhibition to support long-term self-renewal functions through β-catenin.

Exactly how β-catenin exerts pro-self-renewal effects in ESCs has been a somewhat controversial question, and it remains an emerging topic of research at the time of this article's preparation. All four Tcf/Lef proteins are expressed in mESCs (Pereira et al. 2006; Kelly et al. 2011); comparing Tcf/Lef mRNA levels with quantitative RT-PCR assays revealed that Tcf3 was by far the most abundant (Pereira et al. 2006). A wealth of data shows that Tcf3 plays an important role in regulating the self-renewal and differentiation of mESCs; however, some data reveal β-catenin activities that are independent of a Tcf3–β-catenin transactivator complex. Interestingly, research from other systems indicated that Tcf3 appears to be unique among Tcf/Lef proteins in terms of effects on target genes and the biochemical interactions that affect activities (Hikasa et al. 2010; Hikasa and Sokol 2011), supporting the potential for an unconventional effect of mammalian Tcf3 proteins in response to Wnt/β-catenin signaling. Below, a current view is presented regarding the underlying molecular mechanisms by which β-catenin stimulates self-renewal in pluripotent cells.

Potential Tcf-Independent Mechanisms: The β-Catenin–Oct4 Interaction

A physical interaction between β-catenin and Oct4 was first suggested by Takao et al. (2007) to be important for mESC self-renewal, and it was mechanistically expanded in subsequent work by several groups. Yehudit Bergman and colleagues found that Oct4 inhibited Tcf–β-catenin transcriptional activity in mESCs (Abu-Remaileh et al. 2010). The β-catenin–Oct4 interaction appears to be evolutionarily conserved, and *Xenopus* Oct4 homologs (Oct-60 and Oct-25) inhibited transcription of β-catenin-regulated genes during germ layer specification (Cao et al. 2007; Abu-Remaileh et al. 2010). An essentially identical effect of Oct4 inhibiting β-catenin stimulation of TOPFlash activity also occurred in the P19 embryoid carcinoma cell line (Marikawa et al. 2011). This Oct4 effect in P19 cells was shown to be cell-autonomous; however, overexpression of an Oct4-engrailed repressor fusion protein stimulated TOPFlash, whereas an Oct4-VP16 transactivator fusion protein inhibited TOPFlash activity. Although the heterologous nature of the Oct4-fusion proteins makes this experiment somewhat difficult to interpret, it indicates that the Oct4 regulation of Tcf–β-catenin transcriptional activity works through some mechanism other than Oct4 directly recruiting corepressors or coactivators to Tcf–β-catenin complexes.

The Oct4–β-catenin interaction was suggested to regulate β-catenin degradation. When Oct4 levels were genetically reduced in mESCs, total cellular levels of β-catenin increased; when Oct4 was increased, β-catenin levels decreased (Abu-Remaileh et al. 2010). The Oct4-mediated decrease in β-catenin was blocked by mutating GSK3-phosphorylation sites in β-catenin, by inhibiting the proteasome with MG132, or by mutating APC (Abu-Remaileh et al. 2010). In P19 cells, knocking down Oct4 stimulated several genes (Sp5, Wnt3, and Eomes), and this effect was significantly reduced by β-catenin shRNA (Marikawa et al. 2011). Taken together, these results were interpreted to suggest that β-catenin stimulates differentiation of mESC, and Oct4 promotes self-renewal by inhibiting β-catenin's pro-differentiation effect. Although this model draws appeal from consistency with some embryonic data showing that Wnt/β-catenin stimulates differentiation of primed cells toward mesoderm and endoderm lineages, it conflicts with observations that increasing β-catenin levels stimulated self-renewal and inhibited differentiation of naive mESCs. As such, it appears to be most relevant when considering the pro-differentiation effect of β-catenin specifically for cells in the primed state of pluripotency.

Tcf-independent mechanisms for β-catenin were also examined in GSK3 DKO cells. Stable expression of ΔNTcf4 in GSK3 DKO cells strikingly reduced the activation of well-characterized Tcf–β-catenin targets (TOPFlash, Axin2, Cdx1, Brachyury), indicating that ΔNTcf4 expression effectively blocked transactivation by Tcf–β-catenin complexes (Kelly et al. 2011). Importantly, the expression of ΔNTcf4 did not restore differentiation competence to GSK3 DKO cells, because teratomas and EBs made from GSK3 DKO + ΔNTcf4 mESC lacked neurectoderm and maintained Oct4 and Nanog expression. In contrast, β-catenin shRNA restored differentiation competence to DKO + Tcf4DN cells. In addition, forced expression of a β-catenin mutant (βcat33AΔC) failed to substantially activate Tcf–β-catenin targets (TOPFlash, Axin2, Cdx1, Brachyury), but it inhibited neurectoderm differentiation similar to forced expression of a fully functional, stable β-catenin (Kelly et al. 2011). Thus, there was a clear genetic separation of function between the activation of classical Tcf–β-catenin targets and the inhibition of differentiation. These observations strongly indicate an unconventional mechanism whereby β-catenin stimulates self-renewal, that is, one that does not require Tcf–β-catenin activator complexes. Doble and colleagues suggested that Oct4 may recruit β-catenin to promoters and Oct4–β-catenin complexes transcriptionally activate Oct4-bound self-renewal genes (Kelly et al. 2011). Although this model provides a possible explanation for pro-differentiation and pro-self-renewal effects of Wnt-stabilized β-catenin, it does not account for the role and activity of endogenously expressed Tcf/Lefs, including Tcf3.

Tcf Regulation of Wnt/β-Catenin Stimulation of Self-Renewal: Integral Role of Tcf3 in the Core Stem Cell Self-Renewal Network

Loss-of-function experiments showed that Tcf3 mediates important effects on the self-renewal of mESCs. Initial experiments showed that genetic ablation of Tcf3 allowed $Tcf3^{-/-}$ mESCs to self-renew in serum-containing media without LIF/Jak/Stat3 signaling (Pereira et al. 2006; Yi et al. 2008; Yi and Merrill 2010; Wray et al. 2011). $Tcf3^{-/-}$ mESCs also displayed delayed differentiation in EB assays and colony-based differentiation assays (Pereira et al. 2006; Yi et al. 2008; Wray et al. 2011). Thus, genetic ablation of endogenous Tcf3 caused effects that were similar to ectopic activation of Wnt–β-catenin signaling. Similarly, RNAi knockdown of total Tcf3 in mESCs stimulated the formation of ESC-like colonies in the absence of exogenous LIF (Cole et al. 2008), inhibited differentiation in monolayers treated with retinoic acid (Tam et al. 2008), and prevented EBs from developing a differentiated morphology (Tam et al. 2008). Conklin and colleagues found that mESCs express two Tcf3 isoforms at essentially equal levels (Salomonis et al. 2010); knockdown of either Tcf3 isoform was sufficient to inhibit differentiation in colony-based and EB-based differentiation assays, indicating that halving Tcf3 levels was sufficient to stimulate self-renewal (Salomonis et al. 2010).

An important role for Tcf3 was confirmed with unbiased forward genetic screens performed using mESCs. Patrick Paddison and colleagues used a library of shRNAs enriched for gene targets involved in chromatin regulation, signal transduction, and transcription (Schaniel et al. 2009). They transfected shRNAs into mESCs harboring a Nanog-GFP reporter, and cells were treated with retinoic acid, which induces differentiation and reduces Nanog-GFP activity. Tcf3 was one of the 12 targets whose knockdown allowed the maintenance of Nanog-GFP levels after retinoic acid. In a separate phenotype-driven genetic screen, the Blm^{-} mutator mESC system (Guo et al. 2004) was combined with *PiggyBAC* mutagenesis (Ding et al. 2005; Wang et al. 2008) to identify recessive mutations that inhibited neural differentiation in a colony-based assay (Guo et al. 2011). This pilot screen of 2000 random mutant mESC lines recovered five lines that failed to differentiate when stimulated; all five harbored a mutation in the Tcf3 gene and were likely identical by descent (Guo et al. 2011). The Tcf3 mutations were proven to be the cause of neurectoderm differentiation defects, because the defect was complemented with expression of wild-type Tcf3 protein. Interestingly, this same study reported that $Tcf3^{+/-}$ heterozygotes also displayed differentiation defects, consistent with the importance of the amount of Tcf3 in ESCs. Combined with results from the approaches that specifically targeted Tcf3, these unbiased screens show a central role for Tcf3 in mESC self-renewal and differentiation.

A breakthrough in understanding the mechanism of Wnt/β-catenin regulation of self-renewal came when Richard Young and colleagues examined the chromatin occupancy of Tcf3 in mESCs (Cole et al. 2008). They performed ChIP-on-chip and ChIP-seq experiments using an anti-Tcf3 antibody side-by-side with antibodies directed against Oct4, Sox2, and Nanog. Their results confirmed the co-occupancy of Oct4/Sox2/Nanog at common genes (Boyer et al. 2005; Loh et al. 2006), and they also observed an impressive amount of overlap with the occupancy of Tcf3 at the same genes (Cole et al. 2008; Marson et al. 2008). Independently performed ChIP-on-chip experiments by Bing Lim and colleagues also reported overlap in the genes bound by Tcf3 and those bound by Oct4 and Nanog (Tam et al. 2008). Perhaps the best illustration of co-occupancy came from ChIP-seq experiments, which showed that of the 1369 known genes identified as Tcf3-bound, 1173 were also bound by Oct4, and 942 were bound by each, Tcf3, Oct4, Sox2, and Nanog (Marson et al. 2008). Other studies showed Oct4 and Nanog mRNA and protein levels were increased when Tcf3 was knocked down or ablated (Pereira et al. 2006; Cole et al. 2008; Tam et al. 2008; Salomonis et al. 2010; Wagner et al. 2010). Transcriptional profiling of $Tcf3^{-/-}$ ESCs showed that Tcf3 ablation caused strongly opposite effects on gene expression compared with Oct4 knockdown and Nanog knockdown (Loh et al. 2006; Yi et al. 2008). It is notable that the effect on Oct4-regulated genes was measured in $Tcf3^{-/-}$ cells, which did not display elevated Oct4 protein levels (Pereira et al. 2006; Yi et al. 2008), and that knockdown of Nanog in $Tcf3^{-/-}$ cells did not reduce the levels of stem cell genes examined (Yi et al. 2011). Thus, the effects of Tcf3 ablation went beyond just Nanog and Oct4 stimulation and appeared to directly affect the genes co-occupied by Tcf3/Oct4/Sox2/Nanog. These results effectively integrated Tcf3 within the core self-renewal network and suggested that Tcf3 functioned as a limiter that restrains activity of the Oct4/Sox2/Nanog network (Yi et al. 2008).

Wnt-Stabilized β-Catenin Functions through Tcf-Dependent Mechanisms in mESCs

An early model suggested that Tcf3–β-catenin transactivator complexes directly stimulate expression of Tcf3/Oct4/Sox2/Nanog-bound stem cell genes in response to Wnt signaling (Cole et al. 2008). Although this model was consistent with the observation that Wnt3a stimulated genes that were bound by Tcf3 more frequently than those not bound by Tcf3 in mESCs (Cole et al. 2008), it was not consistent with molecular activities reported for Tcf3 in mESCs. For example, ChIP experiments showed that Tcf3 bound directly to enhancer regions of Nanog and Oct4 genes, which were increased by Tcf3 ablation (Pereira et al. 2006; Cole et al. 2008; Marson et al. 2008; Tam et al. 2008). Luciferase reporter experiments showed that Tcf3 directly repressed Nanog and Oct4 promoter activity (Pereira et al. 2006; Tam et al. 2008). Repression of Oct4 and Nanog promoters required DNA binding and the context regulator domain of Tcf3 but did not require β-catenin interaction (Roose et al. 1998; Arce et al. 2006; Pereira et al. 2006; Tam et al. 2008). Tam et al. (2008) reported that a protein recognized by a Tle2 antibody co-occupied Tcf3 sites by ChIP analysis and suggested that Tcf3–Grg complexes repress target genes in mESCs. Thus, there is strong evidence supporting Tcf3-repressor activity in mESCs and very little support for a Tcf3-activator activity in mESCs. Moreover, the separation of function observed between activation of Tcf–β-catenin targets and differentiation defects (Kelly et al. 2011) indicated that even if Tcf3–β-catenin complexes activated target genes, this was unlikely to be important for stimulating self-renewal.

An updated model explaining Tcf3's role in mESCs suggests that endogenous Tcf3 actually inhibits the self-renewal network via its repressor activity, and Wnt-stabilized β-catenin suppresses Tcf3-mediated repression. This model is directly supported by several recent observations: Overexpression of Tcf3 inhibited stem cell gene expression (Nishiyama et al. 2009) and stimulated rapid differentiation of mESCs (Yi et al. 2011). Treating Tcf3-overexpressing mESCs with Wnt3a effectively blocked differentiation and restored self-renewal (Yi et al. 2011). $Tcf3^{-/-}$ mESCs self-renewed in serum-free conditions without the addition of LIF or a GSK3 inhibitor, that is, in the presence of ERK inhibitor alone (Wray et al. 2011; Yi et al. 2011), indicating that ablation of Tcf3 effectively replaced the GSK3 inhibitor. Thus, the pro-self-renewal effects of Wnt/β-catenin signaling in mESCs appear to function by inhibiting Tcf3-repressor activity.

Austin Smith and colleagues suggested overexpressed β-cateninΔC proteins, such as those used in several studies (Kelly et al. 2011; Lyashenko et al. 2011; Wray et al. 2011), compete with Grg corepressor proteins for binding to Tcf3 (Daniels and Weis 2005). This causes β-cateninΔC to inhibit Tcf3 repressor activity, and the self-renewal stimulating ability of β-cateninΔC proteins supports a model wherein the inhibition of Tcf3 repression is the important activity necessary for self-renewal (Wray et al. 2011). However, Wnt/β-catenin clearly has additional activities in mESCs, as shown by stimulation of TOPFlash, Axin2, Cdx1, and Brachyury expression (Kielman et al. 2002; Doble and Woodgett 2003; Kelly et al. 2011; Yi et al. 2011). These observations bring forward a question of what is the role for β-catenin-mediated transactivation of those target genes not repressed by Tcf3 in mESCs. Transcriptional profiling of $Tcf3^{+/+}$ and $Tcf3^{-/-}$ mESC treated with Wnt3a addressed this possibility by revealing two distinct groups of Wnt3a-regulated genes: (1) those that were similarly increased by Tcf3 ablation and Wnt3a treatment and (2) those that were similarly increased by Wnt3a in $Tcf3^{-/-}$ and $Tcf3^{+/+}$ cells (Fig. 2) (Yi et al. 2011). Wnt3a effects on Group 1 genes correlated strongly with effects caused by Oct4, indicating that the Wnt3a suppression of Tcf3-repressed genes was important for self-renewal. Wnt3a effects on Group 2 genes did not correlate with effects caused by Oct4, indicating that activation by Wnt3a was unlikely to stimulate self-renewal. Moreover, the Group 2 genes were expressed at much lower levels in mESCs compared with Group 1 genes (Yi et al. 2011). Given the role of Wnt/β-catenin signaling in stimulating mesoderm formation in embryos and embryoid bodies, it is tempting to speculate that genes following Group 2–like responses would be important for the differentiation effects of Wnt signaling (see Fig. 3).

Precisely how Wnt3a derepresses Tcf3 targets in mESCs involves multiple molecular mechanisms and is not yet fully elucidated. Several experiments showed that the Tcf3–β-catenin interaction was important (Wray et al. 2011; Yi et al. 2011); however, the ability of Wnt3a to elevate the already high levels of stem cell gene expression in $Tcf3^{-/-}$ mESCs revealed an additional effect that did not require Tcf3 (Yi et al. 2011). Tcf1 protein is expressed in

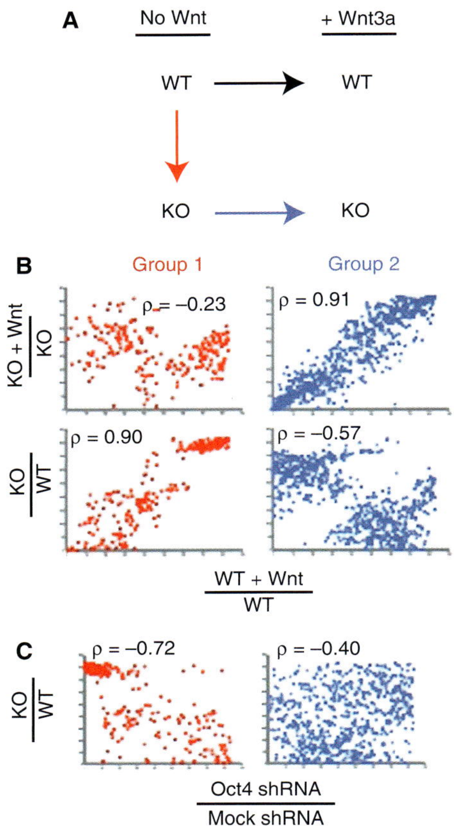

Figure 2. Relationship between Tcf3 and Wnt3a stimulation of gene expression. (*A*) Design of analysis shown in *B*. Genechip microarray data were acquired for $Tcf3^{+/+}$ mESCs (WT, wild type) and $Tcf3^{-/-}$ mESCs (KO, knockout) both without Wnt treatment or 24 h after Wnt3a treatment. Genes stimulated by Wnt3a or ablation of Tcf3 were categorized into two groups by comparing the effect of Wnt3a on WT (black arrow) with the effect of ablating Tcf3 (red arrow) and the effect of Wnt3a on KO (blue arrow). When the magnitude of the effect of Wnt3a (black arrow) was more similar to ablating Tcf3 (red arrow) than to Wnt3a on KO (blue arrow), the gene was assigned to Group 1 (red dot). When the magnitude of the effect of Wnt3a (black arrow) was more similar to Wnt3a on KO (blue arrow) than to ablating Tcf3 (red arrow), the gene was assigned to Group 2 (blue dot). (*B*) Each point represents a single gene and is plotted according to its rank-fold-effect for the comparisons shown on each axis. Pearson correlation coefficients indicate the relative strength of each relationship. Wnt3a stimulation of Group 1 genes in WT correlated well ($\rho = 0.90$) with ablation of Tcf3 but correlated poorly ($\rho = -0.23$) with Wnt3a stimulation in KO. Thus, Wnt3a effects on Group 1 genes were most similar to suppression of Tcf3-mediated repression. Conversely, Wnt3a stimulation of Group 2 genes in WT correlated well ($\rho = 0.91$) with Wnt3a stimulation in KO, indicating that Group 2 genes were activated by Wnt3a in a Tcf3-indpendent manner. (*C*) The effect of Oct4 shRNA (*x*-axis) on Group 1 (red dot) and Group 2 (blue dot) genes is compared with the effect of Tcf3 ablation (*y*-axis). Oct4 shRNA and Tcf3 ablation displayed a strong anti-correlation ($\rho = -0.72$) for Group 1 genes. A similar relationship was reported for Wnt3a on WT compared with Oct4 shRNA (data not shown). The relationships are consistent with Wnt3a suppression of Tcf3 repression being important for expression of Oct4-regulated, stem-cell genes. (Analyses and data are from Yi et al. 2011.)

mESCs (Kelly et al. 2011; Yi et al. 2011); unlike Tcf3, Tcf1 functioned as a transcriptional activator (Yi et al. 2011). The highly conserved HMG domain shared by Tcf1 and Tcf3 enables direct competition for regulation of target genes. Indeed, Wnt3a stimulated β-catenin recruitment to sites of Tcf3/Oct4/Sox2/Nanog DNA occupancy, and significant β-catenin recruitment was partially dependent on each Tcf3 and Tcf1 (Yi et al. 2011). In contrast, Oct4 was not able to recruit β-catenin to mESC chromatin without Tcf3/Tcf1; notably, this finding is inconsistent with the hypothesis that β-catenin stimulates the pluripotency network through Oct4–β-catenin coactivator complexes (Kelly et al. 2011). Combined with gene expression experiments, the β-catenin ChIP experiments showed that Tcf3 recruited β-catenin to target genes, yet Tcf3–β-catenin complexes were apparently ineffective at activating target genes in mESCs, whereas Tcf1–β-catenin complexes were sufficient to activate genes.

When considering the relative contribution of Tcf3–β-catenin and Tcf1–β-catenin interactions for maintenance of naive cell self-renewal, the effects of Tcf3 are more substantial. For example, Tcf1 shRNA reduced Wnt3a's stimulation of AP^+ colonies in $Tcf3^{+/+}$ mESCs, but that effect was significantly smaller than the stimulation of AP^+ colony formation caused by

Cite this article as *Cold Spring Harb Perspect Biol* doi: 10.1101/cshperspect.a007971

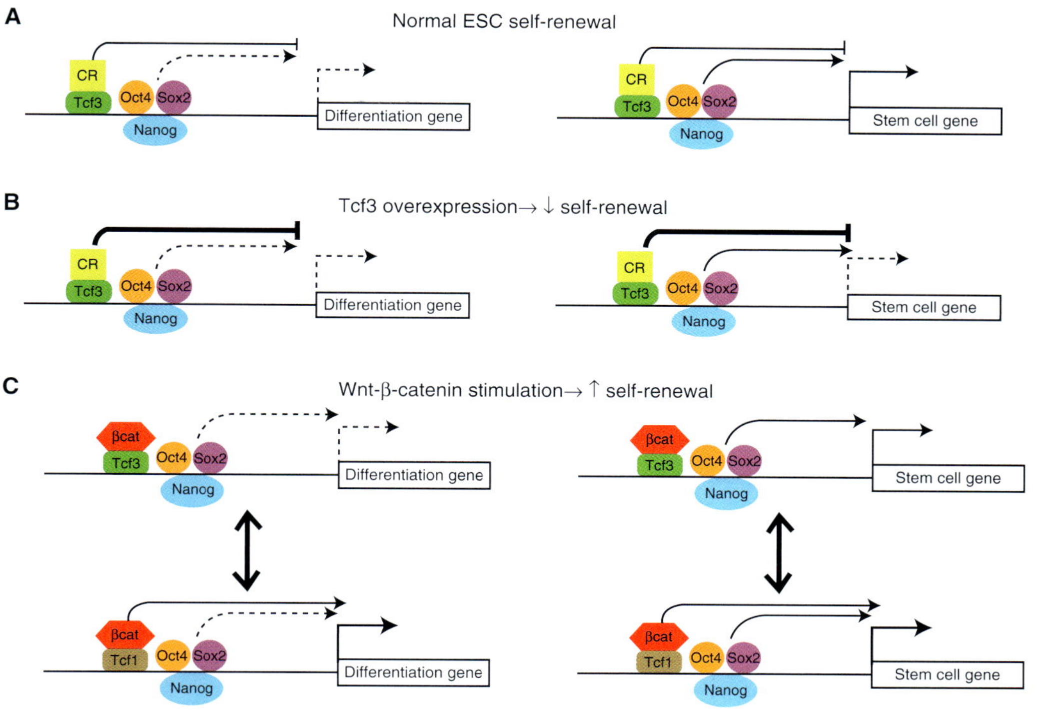

Figure 3. Model depicting effects of Wnt-stabilized β-catenin through Tcf proteins in mESCs. The two types of Wnt-regulated genes model the regulatory effects of Wnt3a and Tcf3 shown in Figure 2. Group 1 genes are modeled as "Stem cell gene," and Group 2 genes are modeled as "Differentiation gene." (*A*) The stem cell gene is expressed in self-renewing mESCs as a consequence of Oct4/Sox2/Nanog binding. Tcf3–corepressor (CR) interactions limit stem cell gene expression, but the effects of Tcf3–CR are counteracted by Oct4/Sox2/Nanog. For simplicity, the Tcf complexes and Oct4/Sox2/Nanog complexes occupy the same DNA strand; however, mutual exclusive and competitive binding of factors is also possible. The differentiation gene is unaffected by Tcf3–corepresor effects because it is not transcriptionally activated by other factors. Note that the differential effect of Oct4/Sox2/Nanog on stem cell versus differentiation genes has been observed; however, the mechanism underlying the different effects remains unclear. (*B*) Overexpression of Tcf3 reduces expression of stem cell genes by recruiting corepressor activity. The repression of essential stem cell genes is sufficient to promote differentiation, because other differentiation-inducing genes (i.e., those not regulated by Tcf) direct cell fate. (*C*) Wnt3a stimulates β-catenin stability, which binds to endogenous Tcf1 and Tcf3 proteins. Tcf3–β-catenin complexes bind to DNA and block the effects of Tcf3–CR complexes. Note that the same downstream effect could also be caused by Wnt-mediated reduction of Tcf3–DNA occupancy via posttranslational modification, such as phosphorylation by HIPK kinases described for XTCF-3 (Hikasa et al. 2010; Hikasa and Sokol 2011). Tcf1–β-catenin stimulates expression of both stem cell genes and differentiation genes; however, the high level of stem cell gene expression is sufficient to stimulate self-renewal.

Tcf3 ablation (Yi et al. 2008, 2011; Wray et al. 2011). It may be helpful to consider that Tcf3 binding to an already silent gene is relatively inconsequential, whereas the greatest effect of Tcf3 binding occurs on those target genes that are being activated through another transcriptional mechanism (for simplicity, Oct4/Sox2/Nanog complexes are shown as putative activators in Fig. 3). Therefore, Wnt/β-catenin inhibition of Tcf3 can be thought of as a stabilizer of the pluripotent state. In contrast, Wnt/β-catenin activation of silent genes in mESCs occurs through Tcf1 and can be thought of as a destabilizer of the pluripotent state or driver to a new cellular state. This model predicts that in naive cells, the effects of inhibiting Tcf3 dominate over the activating Tcf1. In primed cells, the effects of activating Tcf1 dominate over inhibiting Tcf3. A barrier to testing this model comes from the poor understanding of the biochemical mechanisms underlying the opposite activities of Tcf1 and Tcf3. Determining what makes Tcf3 different from the other Tcf/Lef proteins in mESCs should provide important insights into the self-renewal and differentiation of pluripotent cells.

CONCLUDING REMARKS

Recently, it has become clear that the so-called pluripotent cellular state described for early embryonic cells exists in several different variations. In the naive state of mESCs, Wnt/β-catenin signaling stimulates self-renewal by inhibiting the repressor activity of endogenously expressed Tcf3. The effect of Wnt/β-catenin signaling on naive cells effectively prevents them from transitioning into a primed state (Berge et al. 2011). Interestingly, once a cell has overcome Wnt/β-catenin effects and entered a primed state of pluripotency, Wnt/β-catenin stimulates differentiation through activation of classical Tcf–β-catenin target genes, such as Brachyury (Arnold et al. 2000; Lickert et al. 2005). Future research must elucidate more incisive differences between naive and primed pluripotent cells and how those differences cause Wnt/β-catenin signaling to stimulate remarkably opposite effects on the two cellular states.

ACKNOWLEDGMENTS

I apologize to the investigators whose contributions were not covered in this work because of space constraints. The author's research is supported by grants from the NIH (CA128571) and ACS (RSG GGC 112994).

REFERENCES

Abu-Remaileh M, Gerson A, Farago M, Nathan G, Alkalay I, Zins Rousso S, Gur M, Fainsod A, Bergman Y. 2010. Oct-3/4 regulates stem cell identity and cell fate decisions by modulating Wnt/β-catenin signalling. *EMBO J* **29:** 3236–3248.

Arce L, Yokoyama NN, Waterman ML. 2006. Diversity of LEF/TCF action in development and disease. *Oncogene* **25:** 7492–7504.

Arnold SJ, Stappert J, Bauer A, Kispert A, Herrmann BG, Kemler R. 2000. Brachyury is a target gene of the Wnt/β-catenin signaling pathway. *Mech Dev* **91:** 249–258.

Avilion AA, Nicolis SK, Pevny LH, Perez L, Vivian N, Lovell-Badge R. 2003. Multipotent cell lineages in early mouse development depend on SOX2 function. *Genes Dev* **17:** 126–140.

Beddington RS. 1982. An autoradiographic analysis of tissue potency in different regions of the embryonic ectoderm during gastrulation in the mouse. *J Embryol Exp Morphol* **69:** 265–285.

Beddington RS. 1983. Histogenetic and neoplastic potential of different regions of the mouse embryonic egg cylinder. *J Embryol Exp Morphol* **75:** 189–204.

Berge DT, Kurek D, Blauwkamp T, Koole W, Maas A, Eroglu E, Siu RK, Nusse R. 2011. Embryonic stem cells require Wnt proteins to prevent differentiation to epiblast stem cells. *Nat Cell Biol* **13:** 1070–1075.

Boyer LA, Lee TI, Cole MF, Johnstone SE, Levine SS, Zucker JP, Guenther MG, Kumar RM, Murray HL, Jenner RG, et al. 2005. Core transcriptional regulatory circuitry in human embryonic stem cells. *Cell* **122:** 947–956.

Brons IG, Smithers LE, Trotter MW, Rugg-Gunn P, Sun B, Chuva de Sousa Lopes SM, Howlett SK, Clarkson A, Ahrlund-Richter L, Pedersen RA, et al. 2007. Derivation of pluripotent epiblast stem cells from mammalian embryos. *Nature* **448:** 191–195.

Cao Y, Siegel D, Donow C, Knochel S, Yuan L, Knochel W. 2007. POU-V factors antagonize maternal VegT activity and β-catenin signaling in *Xenopus* embryos. *EMBO J* **26:** 2942–2954.

Chambers I, Colby D, Robertson M, Nichols J, Lee S, Tweedie S, Smith A. 2003. Functional expression cloning of Nanog, a pluripotency sustaining factor in embryonic stem cells. *Cell* **113:** 643–655.

Chambers I, Silva J, Colby D, Nichols J, Nijmeijer B, Robertson M, Vrana J, Jones K, Grotewold L, Smith A. 2007. Nanog safeguards pluripotency and mediates germline development. *Nature* **450:** 1230–1234.

Chen X, Xu H, Yuan P, Fang F, Huss M, Vega VB, Wong E, Orlov YL, Zhang W, Jiang J, et al. 2008. Integration of external signaling pathways with the core transcriptional network in embryonic stem cells. *Cell* **133:** 1106–1117.

Cole MF, Johnstone SE, Newman JJ, Kagey MH, Young RA. 2008. Tcf3 is an integral component of the core regulatory circuitry of embryonic stem cells. *Genes Dev* **22:** 746–755.

Daniels DL, Weis WI. 2005. β-Catenin directly displaces Groucho/TLE repressors from Tcf/Lef in Wnt-mediated transcription activation. *Nat Struct Mol Biol* **12:** 364–371.

Davis S, Aldrich TH, Stahl N, Pan L, Taga T, Kishimoto T, Ip NY, Yancopoulos GD. 1993. LIFRβ and gp130 as heterodimerizing signal transducers of the tripartite CNTF receptor. *Science* **260:** 1805–1808.

Ding S, Wu X, Li G, Han M, Zhuang Y, Xu T. 2005. Efficient transposition of the *piggyBac* (PB) transposon in mammalian cells and mice. *Cell* **122:** 473–483.

Doble BW, Woodgett JR. 2003. GSK-3: Tricks of the trade for a multi-tasking kinase. *J Cell Sci* **116:** 1175–1186.

Doble BW, Patel S, Wood GA, Kockeritz LK, Woodgett JR. 2007. Functional redundancy of GSK-3α and GSK-3β in Wnt/β-catenin signaling shown by using an allelic series of embryonic stem cell lines. *Dev Cell* **12:** 957–971.

Dravid G, Ye Z, Hammond H, Chen G, Pyle A, Donovan P, Yu X, Cheng L. 2005. Defining the role of Wnt/beta-catenin signaling in the survival, proliferation, and self-renewal of human embryonic stem cells. *Stem Cells* **23:** 1489–1501.

Evans MJ, Kaufman MH. 1981. Establishment in culture of pluripotential cells from mouse embryos. *Nature* **292:** 154–156.

Gearing DP, Bruce AG. 1992. Oncostatin M binds the high-affinity leukemia inhibitory factor receptor. *New Biol* **4:** 61–65.

Gearing DP, Thut CJ, VandeBos T, Gimpel SD, Delaney PB, King J, Price V, Cosman D, Beckmann MP. 1991. Leukemia inhibitory factor receptor is structurally related to the IL-6 signal transducer, gp130. *EMBO J* **10:** 2839–2848.

Greber B, Wu G, Bernemann C, Joo JY, Han DW, Ko K, Tapia N, Sabour D, Sterneckert J, Tesar P, et al. 2010. Conserved and divergent roles of FGF signaling in mouse epiblast stem cells and human embryonic stem cells. *Cell Stem Cell* **6:** 215–226.

Gu P, Goodwin B, Chung AC, Xu X, Wheeler DA, Price RR, Galardi C, Peng L, Latour AM, Koller BH, et al. 2005. Orphan nuclear receptor LRH-1 is required to maintain Oct4 expression at the epiblast stage of embryonic development. *Mol Cell Biol* **25:** 3492–3505.

Guo G, Smith A. 2010. A genome-wide screen in EpiSCs identifies Nr5a nuclear receptors as potent inducers of ground state pluripotency. *Development* **137:** 3185–3192.

Guo G, Wang W, Bradley A. 2004. Mismatch repair genes identified using genetic screens in Blm-deficient embryonic stem cells. *Nature* **429:** 891–895.

Guo G, Yang J, Nichols J, Hall JS, Eyres I, Mansfield W, Smith A. 2009. Klf4 reverts developmentally programmed restriction of ground state pluripotency. *Development* **136:** 1063–1069.

Guo G, Huang Y, Humphreys P, Wang X, Smith A. 2011. A *PiggyBac*-based recessive screening method to identify pluripotency regulators. *PLoS ONE* **6:** e18189. doi: 10.1371/journal.pone.0018189.

Haegel H, Larue L, Ohsugi M, Fedorov L, Herrenknecht K, Kemler R. 1995. Lack of β-catenin affects mouse development at gastrulation. *Development* **121:** 3529–3537.

Hall J, Guo G, Wray J, Eyres I, Nichols J, Grotewold L, Morfopoulou S, Humphreys P, Mansfield W, Walker R, et al. 2009. Oct4 and LIF/Stat3 additively induce Kruppel factors to sustain embryonic stem cell self-renewal. *Cell Stem Cell* **5:** 597–609.

Han DW, Tapia N, Joo JY, Greber B, Arauzo-Bravo MJ, Bernemann C, Ko K, Wu G, Stehling M, Do JT, et al. 2010. Epiblast stem cell subpopulations represent mouse embryos of distinct pregastrulation stages. *Cell* **143:** 617–627.

Hanna J, Markoulaki S, Mitalipova M, Cheng AW, Cassady JP, Staerk J, Carey BW, Lengner CJ, Foreman R, Love J, et al. 2009. Metastable pluripotent states in NOD-mouse-derived ESCs. *Cell Stem Cell* **4:** 513–524.

Hao J, Li TG, Qi X, Zhao DF, Zhao GQ. 2006. WNT/β-catenin pathway up-regulates Stat3 and converges on LIF to prevent differentiation of mouse embryonic stem cells. *Dev Biol* **290:** 81–91.

Harada N, Tamai Y, Ishikawa T, Sauer B, Takaku K, Oshima M, Taketo MM. 1999. Intestinal polyposis in mice with a dominant stable mutation of the β-catenin gene. *EMBO J* **18:** 5931–5942.

Heisenberg CP, Solnica-Krezel L. 2008. Back and forth between cell fate specification and movement during vertebrate gastrulation. *Curr Opin Genet Dev* **18:** 311–316.

Hikasa H, Sokol SY. 2011. Phosphorylation of TCF proteins by homeodomain-interacting protein kinase 2. *J Biol Chem* **286:** 12093–12100.

Hikasa H, Ezan J, Itoh K, Li X, Klymkowsky MW, Sokol SY. 2010. Regulation of TCF3 by Wnt-dependent phosphorylation during vertebrate axis specification. *Dev Cell* **19:** 521–532.

Hochedlinger K, Plath K. 2009. Epigenetic reprogramming and induced pluripotency. *Development* **136:** 509–523.

Huelsken J, Vogel R, Brinkmann V, Erdmann B, Birchmeier C, Birchmeier W. 2000. Requirement for β-catenin in anterior–posterior axis formation in mice. *J Cell Biol* **148:** 567–578.

Ivanova N, Dobrin R, Lu R, Kotenko I, Levorse J, DeCoste C, Schafer X, Lun Y, Lemischka IR. 2006. Dissecting self-renewal in stem cells with RNA interference. *Nature* **442:** 533–538.

James D, Levine AJ, Besser D, Hemmati-Brivanlou A. 2005. TGFβ/activin/nodal signaling is necessary for the maintenance of pluripotency in human embryonic stem cells. *Development* **132:** 1273–1282.

Jiang J, Chan YS, Loh YH, Cai J, Tong GQ, Lim CA, Robson P, Zhong S, Ng HH. 2008. A core Klf circuitry regulates self-renewal of embryonic stem cells. *Nat Cell Biol* **10:** 353–360.

Kelly KF, Ng DY, Jayakumaran G, Wood GA, Koide H, Doble BW. 2011. β-Catenin enhances Oct-4 activity and reinforces pluripotency through a TCF-independent mechanism. *Cell Stem Cell* **8:** 214–227.

Kielman MF, Rindapaa M, Gaspar C, van Poppel N, Breukel C, van Leeuwen S, Taketo MM, Roberts S, Smits R, Fodde R. 2002. Apc modulates embryonic stem-cell differentiation by controlling the dosage of β-catenin signaling. *Nat Genet* **32:** 594–605.

Kim J, Chu J, Shen X, Wang J, Orkin SH. 2008. An extended transcriptional network for pluripotency of embryonic stem cells. *Cell* **132:** 1049–1061.

Kunath T, Saba-El-Leil MK, Almousailleakh M, Wray J, Meloche S, Smith A. 2007. FGF stimulation of the Erk1/2 signalling cascade triggers transition of pluripotent embryonic stem cells from self-renewal to lineage commitment. *Development* **134:** 2895–2902.

Lawson KA, Pedersen RA. 1987. Cell fate, morphogenetic movement and population kinetics of embryonic endoderm at the time of germ layer formation in the mouse. *Development* **101:** 627–652.

Lawson KA, Meneses JJ, Pedersen RA. 1991. Clonal analysis of epiblast fate during germ layer formation in the mouse embryo. *Development* **113:** 891–911.

Lickert H, Cox B, Wehrle C, Taketo MM, Kemler R, Rossant J. 2005. Dissecting Wnt/β-catenin signaling during gastrulation using RNA interference in mouse embryos. *Development* **132:** 2599–2609.

Lindsley RC, Gill JG, Kyba M, Murphy TL, Murphy KM. 2006. Canonical Wnt signaling is required for development of embryonic stem cell-derived mesoderm. *Development* **133:** 3787–3796.

Loh YH, Wu Q, Chew JL, Vega VB, Zhang W, Chen X, Bourque G, George J, Leong B, Liu J, et al. 2006. The Oct4 and Nanog transcription network regulates pluripotency in mouse embryonic stem cells. *Nat Genet* **38:** 431–440.

Lyashenko N, Winter M, Migliorini D, Biechele T, Moon RT, Hartmann C. 2011. Differential requirement for the dual functions of β-catenin in embryonic stem cell self-renewal and germ layer formation. *Nat Cell Biol* **13:** 753–761.

Marikawa Y, Tamashiro DA, Fujita TC, Alarcon VB. 2011. Dual roles of Oct4 in the maintenance of mouse P19 embryonal carcinoma cells: As negative regulator of Wnt/β-catenin signaling and competence provider for Brachyury induction. *Stem Cells Dev* **20:** 621–633.

Marson A, Levine SS, Cole MF, Frampton GM, Brambrink T, Johnstone S, Guenther MG, Johnston WK, Wernig M, Newman J, et al. 2008. Connecting microRNA genes to the core transcriptional regulatory circuitry of embryonic stem cells. *Cell* **134:** 521–533.

Martin GR. 1981. Isolation of a pluripotent cell line from early mouse embryos cultured in medium conditioned by teratocarcinoma stem cells. *Proc Natl Acad Sci* **78:** 7634–7638.

Matsuda T, Nakamura T, Nakao K, Arai T, Katsuki M, Heike T, Yokota T. 1999. STAT3 activation is sufficient to maintain an undifferentiated state of mouse embryonic stem cells. *EMBO J* **18:** 4261–4269.

Mitsui K, Tokuzawa Y, Itoh H, Segawa K, Murakami M, Takahashi K, Maruyama M, Maeda M, Yamanaka S. 2003. The homeoprotein Nanog is required for maintenance of pluripotency in mouse epiblast and ES cells. *Cell* **113:** 631–642.

Ng HH, Surani MA. 2011. The transcriptional and signalling networks of pluripotency. *Nat Cell Biol* **13:** 490–496.

Nichols J, Evans EP, Smith AG. 1990. Establishment of germ-line-competent embryonic stem (ES) cells using differentiation inhibiting activity. *Development* **110:** 1341–1348.

Nichols J, Zevnik B, Anastassiadis K, Niwa H, Klewe-Nebenius D, Chambers I, Scholer H, Smith A. 1998. Formation of pluripotent stem cells in the mammalian embryo depends on the POU transcription factor Oct4. *Cell* **95:** 379–391.

Nichols J, Silva J, Roode M, Smith A. 2009. Suppression of Erk signalling promotes ground state pluripotency in the mouse embryo. *Development* **136:** 3215–3222.

Nishiyama A, Xin L, Sharov AA, Thomas M, Mowrer G, Meyers E, Piao Y, Mehta S, Yee S, Nakatake Y, et al. 2009. Uncovering early response of gene regulatory networks in ESCs by systematic induction of transcription factors. *Cell Stem Cell* **5:** 420–433.

Niwa H, Burdon T, Chambers I, Smith A. 1998. Self-renewal of pluripotent embryonic stem cells is mediated via activation of STAT3. *Genes Dev* **12:** 2048–2060.

Niwa H, Miyazaki J, Smith AG. 2000. Quantitative expression of Oct-3/4 defines differentiation, dedifferentiation or self-renewal of ES cells. *Nat Genet* **24:** 372–376.

Ogawa K, Nishinakamura R, Iwamatsu Y, Shimosato D, Niwa H. 2006. Synergistic action of Wnt and LIF in maintaining pluripotency of mouse ES cells. *Biochem Biophys Res Commun* **343:** 159–166.

Otero JJ, Fu W, Kan L, Cuadra AE, Kessler JA. 2004. β-Catenin signaling is required for neural differentiation of embryonic stem cells. *Development* **131:** 3545–3557.

Pereira L, Yi F, Merrill BJ. 2006. Repression of *nanog* gene transcription by Tcf3 limits embryonic stem cell self-renewal. *Mol Cell Biol* **26:** 7479–7491.

Roose J, Molenaar M, Peterson J, Hurenkamp J, Brantjes H, Moerer P, van de Wetering M, Destree O, Clevers H. 1998. The *Xenopus* Wnt effector XTcf-3 interacts with Groucho-related transcriptional repressors. *Nature* **395:** 608–612.

Salomonis N, Schlieve CR, Pereira L, Wahlquist C, Colas A, Zambon AC, Vranizan K, Spindler MJ, Pico AR, Cline MS, et al. 2010. Alternative splicing regulates mouse embryonic stem cell pluripotency and differentiation. *Proc Natl Acad Sci* **107:** 10514–10519.

Sato N, Meijer L, Skaltsounis L, Greengard P, Brivanlou AH. 2004. Maintenance of pluripotency in human and mouse embryonic stem cells through activation of Wnt signaling by a pharmacological GSK-3-specific inhibitor. *Nat Med* **10:** 55–63.

Schaniel C, Ang YS, Ratnakumar K, Cormier C, James T, Bernstein E, Lemischka IR, Paddison PJ. 2009. Smarcc1/Baf155 couples self-renewal gene repression with changes in chromatin structure in mouse embryonic stem cells. *Stem Cells* **27:** 2979–2991.

Silva J, Barrandon O, Nichols J, Kawaguchi J, Theunissen TW, Smith A. 2008. Promotion of reprogramming to ground state pluripotency by signal inhibition. *PLoS Biol* **6:** e253.

Smith AG, Heath JK, Donaldson DD, Wong GG, Moreau J, Stahl M, Rogers D. 1988. Inhibition of pluripotential embryonic stem cell differentiation by purified polypeptides. *Nature* **336:** 688–690.

Takao Y, Yokota T, Koide H. 2007. β-Catenin up-regulates Nanog expression through interaction with Oct-3/4 in embryonic stem cells. *Biochem Biophys Res Commun* **353:** 699–705.

Tam WL, Lim CY, Han J, Zhang J, Ang YS, Ng HH, Yang H, Lim B. 2008. T-cell factor 3 regulates embryonic stem cell pluripotency and self-renewal by the transcriptional control of multiple lineage pathways. *Stem Cells* **26:** 2019–2031.

ten Berge D, Koole W, Fuerer C, Fish M, Eroglu E, Nusse R. 2008. Wnt signaling mediates self-organization and axis formation in embryoid bodies. *Cell Stem Cell* **3:** 508–518.

Tesar PJ, Chenoweth JG, Brook FA, Davies TJ, Evans EP, Mack DL, Gardner RL, McKay RD. 2007. New cell lines from mouse epiblast share defining features with human embryonic stem cells. *Nature* **448:** 196–199.

Thomson JA, Itskovitz-Eldor J, Shapiro SS, Waknitz MA, Swiergiel JJ, Marshall VS, Jones JM. 1998. Embryonic stem cell lines derived from human blastocysts. *Science* **282:** 1145–1147.

Vallier L, Reynolds D, Pedersen RA. 2004. Nodal inhibits differentiation of human embryonic stem cells along the neuroectodermal default pathway. *Dev Biol* **275:** 403–421.

Vallier L, Mendjan S, Brown S, Chng Z, Teo A, Smithers LE, Trotter MW, Cho CH, Martinez A, Rugg-Gunn P, et al. 2009. Activin/Nodal signalling maintains pluripotency by controlling Nanog expression. *Development* **136:** 1339–1349.

Waddington CH. 1957. *The strategy of genes*. Geo Allen and Unwin, London.

Wagner RT, Xu X, Yi F, Merrill BJ, Cooney AJ. 2010. Canonical Wnt/β-catenin regulation of liver receptor homolog-1 mediates pluripotency gene expression. *Stem Cells* **28:** 1794–1804.

Wang W, Lin C, Lu D, Ning Z, Cox T, Melvin D, Wang X, Bradley A, Liu P. 2008. Chromosomal transposition of *PiggyBac* in mouse embryonic stem cells. *Proc Natl Acad Sci* **105:** 9290–9295.

Williams RL, Hilton DJ, Pease S, Willson TA, Stewart CL, Gearing DP, Wagner EF, Metcalf D, Nicola NA, Gough NM. 1988. Myeloid leukaemia inhibitory factor maintains the developmental potential of embryonic stem cells. *Nature* **336:** 684–687.

Wray J, Kalkan T, Gomez-Lopez S, Eckardt D, Cook A, Kemler R, Smith A. 2011. Inhibition of glycogen synthase kinase-3 alleviates Tcf3 repression of the pluripotency network and increases embryonic stem cell resistance to differentiation. *Nat Cell Biol* **13:** 838–845.

Xu RH, Peck RM, Li DS, Feng X, Ludwig T, Thomson JA. 2005. Basic FGF and suppression of BMP signaling sustain undifferentiated proliferation of human ES cells. *Nat Methods* **2:** 185–190.

Xu Y, Zhu X, Hahm HS, Wei W, Hao E, Hayek A, Ding S. 2010. Revealing a core signaling regulatory mechanism for pluripotent stem cell survival and self-renewal by small molecules. *Proc Natl Acad Sci* **107:** 8129–8134.

Yi F, Merrill BJ. 2010. Non-cell-autonomous stimulation of stem cell proliferation following ablation of Tcf3. *Exp Cell Res* **316:** 1050–1060.

Yi F, Pereira L, Merrill BJ. 2008. Tcf3 functions as a steady state limiter of transcriptional programs of mouse embryonic stem cell self renewal. *Stem Cells* **26:** 1951–1960.

Yi F, Pereira L, Hoffman JA, Shy BR, Yuen CM, Liu DR, Merrill BJ. 2011. Opposing effects of Tcf3 and Tcf1 control Wnt stimulation of embryonic stem cell self-renewal. *Nat Cell Biol* **13:** 762–770.

Ying QL, Nichols J, Chambers I, Smith A. 2003. BMP induction of Id proteins suppresses differentiation and sustains embryonic stem cell self-renewal in collaboration with STAT3. *Cell* **115:** 281–292.

Ying QL, Wray J, Nichols J, Batlle-Morera L, Doble B, Woodgett J, Cohen P, Smith A. 2008. The ground state of embryonic stem cell self-renewal. *Nature* **453:** 519–523.

Young RA. 2011. Control of the embryonic stem cell state. *Cell* **144:** 940–954.

Wnt Signaling, Stem Cells, and Cancer of the Gastrointestinal Tract

Arnout Schepers and Hans Clevers

Hubrecht Institute, KNAW and University Medical Centre Utrecht, 3584CT Utrecht, The Netherlands

Correspondence: h.clevers@hubrecht.eu

The Wnt signaling pathway was originally uncovered as one of the prototype developmental signaling cascades in invertebrates as well as in vertebrates. The first indication that Wnt signaling also plays a role in the adult animal came from the study of the intestine of Tcf-4 (*Tcf7L2*) knockout mice. The gastrointestinal epithelium continuously self-renews over the lifetime of an organism and is, in fact, the most rapidly self-renewing tissue of the mammalian body. Recent studies indicate that Wnt signaling plays a central role in the biology of gastrointestinal stem cells. Furthermore, mutational activation of the Wnt cascade is the principle cause of colon cancer.

The human stomach consists of the corpus and the pyloric antrum (Fig. 1A), segments with discrete functional properties (Karam 1999). Hydrochloric acid and zymogens are secreted in the corpus, whereas cells in the pyloric antrum mainly secrete mucus and gastric hormones. Like the intestine, the gastric epithelium has a glandular structure and is constantly renewed by progeny of stem cells. Tubular-shaped mucosal invaginations called gastric units consist of a pit and gland region, where the gland can be further subdivided into the isthmus, neck, and base. Each pit branches into a number of glands. The composition of the gastric units differs between the corpus and the pyloric antrum. The corpus has short pits and long glands (Fig. 1D), which contain a variety of surface mucous cells, parietal cells secreting acid and intrinsic factor, chief cells with zymogen granules, and hormone-secreting endocrine cells (Karam and Leblond 1992). Glands of the pyloric antrum are short and feed into a long pit (Fig. 1C). Pyloric units mainly consist of mucous cells, enteroendocrine cells, and occasional acid-secreting cells (Lee and Leblond 1985). In the corpus, mucus-secreting cells migrate from the gland to the surface epithelium, where they live for ~3 days. Parietal cells live for ~2 months and migrate bidirectionally toward the surface and the base of the gland. Chief cells migrate to the base of the gland and live for ~6 months (Karam 1993; Karam and Leblond 1993a,b).

There is one major difference between the human and mouse stomach, i.e., the presence of a forestomach in the latter. The rodent forestomach epithelium resembles that of the esophagus or the epidermis of the skin, i.e., it is a keratinizing stratified epithelium.

Cite this article as *Cold Spring Harb Perspect Biol* doi: 10.1101/cshperspect.a007989

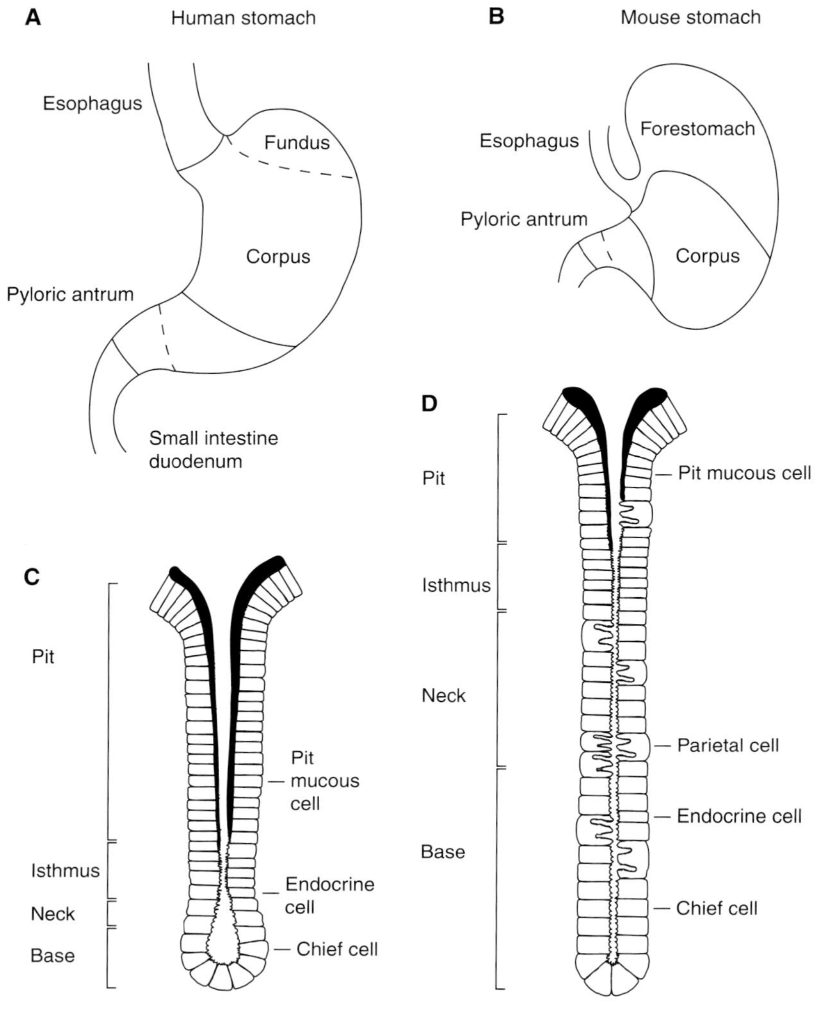

Figure 1. Anatomy of the human and mouse stomach. (*A*) Human stomach. (*B*) Mouse stomach. (*C*) Gastric unit of the pyloric antrum. (*D*) Gastric unit of the corpus.

THE INTESTINE

The intestinal wall consists of three main layers. The outer layer contains smooth muscle cells responsible for peristaltic movement of the stool toward the anus (Leedham et al. 2005). The middle layer of connective tissue, the stroma, holds many nerves and lymphatic vessels, whereas the inner layer is made up of a simple epithelium, i.e., a one-cell-layer epithelial sheet. Anatomically and functionally, the intestinal tract consists of the small intestine (duodenum, jejunum, and ileum) and the colon. The small intestine is mainly dedicated to nutrient absorption, whereas the colon mainly functions to resorb water. This functional difference is reflected by the structure and composition of the epithelium. The small intestinal epithelium is organized in a glandular structure with crypts and fingerlike structures called villi that maximize the exchange surface. Villi are composed of a mixture of differentiated cells: absorptive enterocytes, mucus-secreting goblet cells, hormone-producing enteroendocrine cells, secretory tuft cells (Gerbe et al. 2011), and cup cells with unknown function (Madara 1982). The colon contains crypts, but has a flat surface epithelium rather than villi. In accordance with their function, absorptive enterocytes are most abundant in the proximal small intestine, whereas mucus-secreting goblet cells are most abundant in the distal part of the small intestine and colon, reflecting the need for lubrication after water resorption from the stool. All epithelial cells derive from progenitor cells in the proliferative crypts. A mouse intestine contains about a million crypts, each of which generates approximately 300 new cells per crypt per day (Hagemann et al. 1970). The driving force behind this renewal are the intestinal stem cells that are located at the crypt base and divide on average once per day (Potten et al. 1990; Barker et al. 2007). They give rise to a pool of transit-amplifying cells that divide every 12–16 h, gradually migrate upward, and differentiate to fuel the villi with mature functional cells. Being pushed up by the newly generated cells, these cells keep migrating to the tops of the villi, where they eventually undergo apoptosis and are shed into the lumen. Paneth cells of the small intestine do not follow this migration pattern. They remain at the bottom of the crypt where they reside for 6–8 weeks and classically are believed to regulate the local microbial environment by secreting antimicrobial peptides and lysozyme (Bjerknes and Cheng 1981a,b; Ireland et al. 2005).

Wnt SIGNALING CONSTITUTES THE MAJOR DRIVING FORCE BEHIND HOMEOSTATIC SELF-RENEWAL OF THE CRYPT

The Wnt signaling pathway was originally uncovered as one of the prototype developmental signaling cascades in invertebrates as well as vertebrates (Cadigan and Nusse 1997). The first indication that Wnt signaling also plays a role in the adult came from the study of Tcf-4 (*Tcf7L2*) knockout mice (Korinek et al. 1998). At birth, *Tcf7L2*$^{-/-}$ mice lack proliferative stem cells in the prospective crypt regions ("intervillus pockets") between villi. Subsequent studies documented similar roles for Wnt signaling in other adult stem cell compartments, such as the skin and bone marrow (Reya and Clevers 2005). Later it was shown that loss of Wnt signaling by genetic deletion of β-catenin (Fevr et al. 2007) or through inhibition of Wnt receptors by recombinant Dkk1 (Pinto et al. 2003; Kuhnert et al. 2004) resulted in complete ablation of intestinal crypts in the adult mouse. A complete inventory of the expression of Wnt receptors and ligands in the murine intestine is given in (Gregorieff et al. 2005).

Wnt signals affect cell-fate determination in the small intestine. In addition to proliferative defects, *Tcf7L2*$^{-/-}$ mice lack enteroendocrine cells (Korinek et al. 1998), whereas inhibition of Wnt by Dkk1 results in loss of secretory lineages (Pinto et al. 2003). A more recent study found that Wnt is necessary for the development of early secretory progenitors, but not required for the differentiation and maturation of enteroendocrine precursors (Wang et al. 2007). Paneth cells at the base of the crypt receive abundant Wnt signals, have high levels of nuclear β-catenin, but do not proliferate. Rather, these postmitotic cells use Wnt signals for their terminal differentiation (Andreu et al. 2005; van Es et al. 2005). This is at least in part mediated through the Wnt target gene Sox9. Conditional knockout of this transcription factor resulted in a loss of Paneth cells, increased cell proliferation in the crypts, and an increased number of cells expressing c-Myc (Bastide et al. 2007; Mori-Akiyama et al. 2007).

Nuclear β-catenin is highest in cells at the base of the crypt, implying that a Wnt gradient exists along the crypt axis (Kongkanuntn et al. 1999; van de Wetering et al. 2002). Indeed, gene-expression analysis of horizontally dissected human colon samples showed the highest expression of the Wnt target program in the base of the crypts and a gradual decrease toward the top (Kosinski et al. 2007). The Wnt gradient is believed to control the spatial organization of the epithelium. For example, EphB2 and B3, well-known sorting receptors, are Wnt target genes. Their expression follows the Wnt gradient, i.e., they are highest at the crypt bottom. Their ligands, e.g., EphrinB1, are expressed in an opposite fashion, i.e., highest on the villi. Repulsive forces between EphB receptors and Ephrin ligands position cells along the crypt-villus axis. This is particularly evident for Paneth cells that uniquely express EphB3. In EphB3 mutant mice, Paneth cells do not occupy the crypt bottom but are scattered along the crypt–villus axis (Batlle et al. 2002).

ACTIVATING MUTATIONS IN THE WNT PATHWAY CAUSE INTESTINAL CANCER

Abnormal Wnt signaling has been associated with cancer in many tissues (Reya and Clevers 2005). In the majority of sporadic colorectal cancers the adenomatous polyposis coli (*APC*) gene is mutated (Kinzler and Vogelstein 1996). Colorectal cancers without APC mutations often have Axin2 mutations (Liu et al. 2000) or activating point mutations in the destruction motif of β-catenin (Morin et al. 1997). The *APC* gene was identified to be mutated in familial adenomatous polyposis (FAP) patients (Kinzler et al. 1991; Nishisho et al. 1991). These patients have one defective *APC* allele and develop colon polyps by loss of heterozygosity (LOH). Polyps are benign but may progress into malignant adenocarcinoma over time by sustaining additional mutations. Loss of APC leads to inappropriate stabilization of β-catenin (Rubinfeld et al. 1996; Korinek et al. 1997). Overactive Wnt signaling is one of the key events in polyp initiation (Oshima et al. 1997), but additional mutations are required to progress into adenocarcinoma. Later stages of tumor progression mainly depend on activation of RAS or inactivation of P53, SMAD4, and PTEN (Fearon and Vogelstein 1990; Janssen et al. 2006; Marsh et al. 2008; Luo et al. 2009).

Several mouse models exist for FAP. Of note, mice develop polyps predominantly in the small intestine, whereas FAP patients mainly have polyps in the colon. The first mouse model for FAP (multiple intestinal neoplasia or min) was isolated after random mutagenesis (Moser et al. 1990). These mice harbor a mutation in the *Apc* gene; therefore the strain was called Apc^{min} (Su et al. 1992). Since then, several Apc mutations have been generated by gene-targeting techniques, e.g., $Apc^{\Delta 716}$, $Apc^{\Delta 1638}$, and the hypomorphic alleles Apc^{neoR} and Apc^{neoF} (Fodde et al. 1994; Oshima et al. 1995; Li et al.). These mice all develop adenomas after LOH, but polyp numbers and the location of eventual tumors are different. Conditional deletion of both copies of a conditional floxed *Apc* allele in adult animals converted the entire epithelium into "crypt-progenitor-like" cells within days (Sansom et al. 2004), a phenomenon dependent on an intact *cMyc* allele (Sansom et al. 2007). In mice with truncated Apc proteins, β-catenin/T-cell factor (TCF) transcriptional activity inversely correlated with Apc levels. The extent of Wnt activation turned out to be the major factor determining the number of polyps.

Analysis of LOH and point mutations in adenomas from colorectal cancer patients provided insight in the diversity of APC mutations (Albuquerque et al. 2002). The APC protein has seven 20-amino-acid repeats that are involved in the down-regulation of β-catenin. The analysis showed that there is selection for APC genotypes that have one or two repeats left. Based on these findings it was proposed that there is a level of Wnt activity which is "just right" for tumor formation. Such a quantitative nature of Wnt signaling was also described in embryonic development (Kielman et al. 2002).

From these studies, it was suggested that benign transformation of APC-mutant epithelial cells involves the constitutive expression of a genetic program that is normally involved in self-renewal of crypt (stem) cells. Initial studies pointed to cMyc and CyclinD1 as pivotal players in the transforming Wnt target gene program in colorectal cancer (He et al. 1998; Tetsu and McCormick 1999). When microarraying became available, the "global" genetic program that is controlled by β-catenin/TCF in colorectal cancer cells was indeed found to resemble the genetic program of the proliferative compartment of the intestinal crypt (van de Wetering et al. 2002; Sabates-Bellver et al. 2007; van der Flier et al. 2007).

INTESTINAL STEM CELLS

The existence of a self-renewing, multipotent stem cell population in adult crypts was first shown by tracking randomly introduced, somatic

 Cite this article as *Cold Spring Harb Perspect Biol* doi: 10.1101/cshperspect.a007989

mutations (Winton and Ponder 1990; Bjerknes and Cheng 2002). Clones of cells marked by these sporadic mutations were long-lived and contained all of the different epithelial cell lineages, indicating a common progenitor. Tracking of spontaneous mitochondrial mutations revealed the presence of multiple multipotent cells in adult crypts (Taylor et al. 2003). Today, it is generally accepted that stem cells reside at the base of the crypt; however, there has been much debate about the exact location of the stem cells. Cells located at the so-called +4 position between the Paneth cells and the transit-amplifying cells were first claimed to be the stem cells. Support for the +4 cell came from studies based on radiosensitivity and label retention of these cells (Potten et al. 1974). Several markers, including Bmi1, mTert, and Hopx, have been proposed to mark the +4 cell (Sangiorgi and Capecchi 2008; Montgomery et al. 2011; Takeda et al. 2011). Small cycling cells intermingled between the Paneth cells, so-called crypt base columnar (CBC) cells, were proposed as alternative stem cells (Cheng and Leblond 1974). Later marking studies also hinted at the CBC cells as being the stem cells (Bjerknes and Cheng 1999, 2002).

By definition, a stem cell can renew itself and give rise to all the specialized progeny within a given tissue. Two approaches can test this: transplantation of sorted, candidate stem cells into recipient animals, or lineage tracing by genetic marking. The latter approach has been successfully used in identifying Wnt-dependent stem cells in the intestinal tract. To perform lineage tracing, a candidate gene must be identified that—on its own—marks a putative stem cell. By transgenesis or gene-targeting technology, mice can be generated that express an inducible recombinase such as the tamoxifen-inducible CreERT2 (Hayashi and McMahon 2002) under the control of the regulatory elements of the candidate gene.

The first study addressing the existence of intestinal stem cells by lineage tracing used the gene encoding the leucine-rich repeat containing G-protein receptor Lgr5 (Gpr49). The global elucidation of the Wnt-driven genomic program was key to the identification of Lgr5 and the subsequent confirmation of CBC cells being ISCs (van de Wetering et al. 2002; Barker et al. 2007). Because Wnt is the driving force behind proliferation in the crypt, multiple individual Wnt target genes were analyzed to identify potential stem cell-specific genes. Whereas most target genes were expressed throughout the crypt, a small set of genes had a more restricted expression (van der Flier et al. 2007). One of these genes, *Lgr5*, was exclusively expressed in the small cycling CBC cells. With the first molecular marker for CBC cells in hand, several studies were performed to test the stem cell potential of these cells. First, a LacZ ($Lgr5^{LacZ}$) and an enhanced green fluorescent protein (EGFP) ($Lgr5^{EGFP\text{-}IRES\text{-}CreERT2}$; IRES, internal ribosome entry site) knockin were generated to confirm Lgr5 expression in the CBC cells. Then, $Lgr5^{EGFP\text{-}IRES\text{-}CreERT2}$ mice were crossed with a cAMP response element (CRE) -activatable $Rosa26^{LacZ}$ reporter mouse strain to be able to mark Lgr5-positive cells and follow their offspring (Fig. 2A–C). Lgr5-positive cells were shown to generate all different cell types (Fig. 2D,E) of the intestinal epithelium for the lifetime of the mouse. The tracings confirmed the CBC stem cell zone model originally proposed by Bjerknes and Cheng (1981c). Sorting and gene-expression analysis of Lgr5-EGFP cells revealed that these stem cells express several additional Wnt target genes such as *Olfm4* and *Ascl2* (van der Flier et al. 2009a,b). The transcription factor Ascl2 is of particular interest as it appears to function as the master regulator of the intestinal stem cell (van der Flier et al. 2009b). Culture conditions were subsequently developed that allow the outgrowth of a single small intestinal Lgr5 stem cell to form ever-expanding, self-organizing "miniguts," closely resembling the normal gut epithelium (Sato et al. 2009). This three-dimensional organoid culture system is based on EGF and on Rspondin1, a secreted Wnt agonist previously shown to induce crypt hyperplasia in vivo (Kim et al. 2005). Using this organoid technology it was shown that Paneth cells, which are intermingled with Lgr5 stem cells at crypt bottoms, constitute at least part of the crypt stem cell niche: They produce the essential stem cell growth factors EGF and

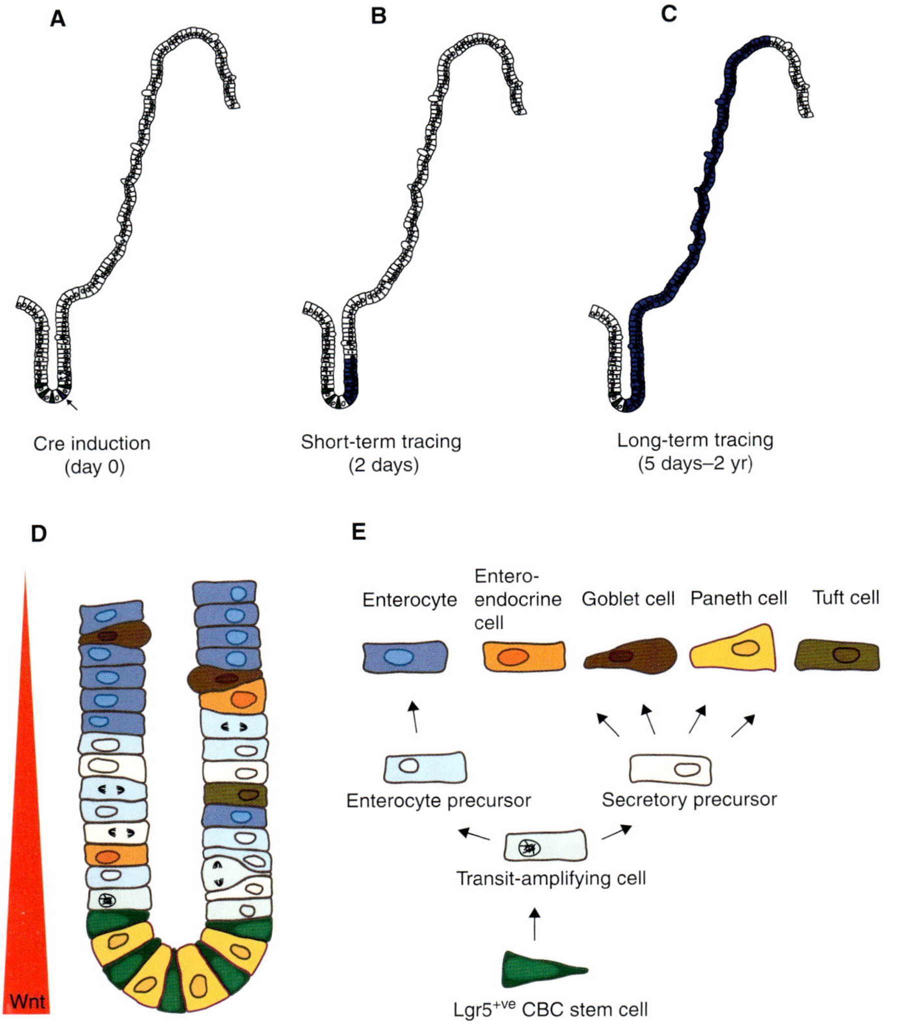

Figure 2. Lgr5-positive CBC cells give rise to all epithelial lineages. (*A*) Low dose of tamoxifen activates Cre recombinase sporadically in CBC cells. (*B*) Multiple daughter cells are labeled 2 days after induction. (*C*) Induced Lgr5-positive cells generate labeled "ribbons" containing all different cell types of the intestinal epithelium for the lifetime of the mouse. (*D*) Architecture of a small intestinal crypt with the Wnt gradient depicted in red. (*E*) Schematic overview of the generation of different cell lineages from Lgr5^{+ve} CBC stem cells.

Wnt3 and carry the Notch ligands Dll1 and Dll4 (Sato et al. 2011). Recently, it was reported that hepatocyte growth factor, secreted by stromal myofibroblasts, can enhance β-catenin-dependent transcription and maintain stem cell properties (Vermeulen et al. 2010).

The Lgr5 stem cells display some unexpected characteristics. Lineage tracing of stem cells using a multicolor reporter showed that they divide symmetrically and compete for essential niche signals (i.e., direct contact with Paneth cells) to maintain stemness (Snippert et al. 2010; Sato et al. 2011). ISCs that lose contact with Paneth cells lose stemness and differentiate into transit-amplifying cells. Paneth cell numbers remain constant under normal homeostasis, but it is currently unknown how this is regulated. Lgr5 stem cells are never quiescent: they have telomerase activity, which allows them to complete ~1000 cell divisions in the lifetime of a mouse (Schepers et al. 2011).

Many other potential stem cell markers have been reported, although most are only supported by positional information (reviewed by Barker et al. 2010a). Yet, lineage tracing studies involving CD133 (Zhu et al. 2009) and Sox9 (Furuyama et al. 2011), two genes that mark broader populations in the crypt including CBC cells, have confirmed the Lgr5-based findings. In addition, Bmi1 (Sangiorgi and Capecchi 2008; Tian et al. 2011) and Hopx (Takeda et al. 2011), both proposed +4 markers, have been used to trace all cell types in the crypt. Interestingly, $Bmi1^{+}/Lgr5^{-}$ cells were shown to be able to rescue the intestinal epithelium after deletion of $Lgr5^{+}$ stem cells (Tian et al. 2011). Single-molecule transcript counting has revealed that Bmi1 expression is coexpressed in CBC cells at lower crypt positions and is nearly constant throughout the crypt axis (Itzkovitz et al. 2011). Until now the nature of the $Bmi1^{+}/Lgr5^{-}$ cells remains unclear.

GASTRIC STEM CELLS

At birth, gastric units are polyclonal but they become monoclonal during adulthood (Nomura et al. 1998). Gastric self-renewal is driven by gastric stem cells (Thompson et al. 1990) and as in the small intestine there is some discrepancy about the exact location of the stem cells.

In 1948, studies using incorporation of labeled nucleotides suggested that renewal of gastric cells was driven by one or a few cells in the isthmus of both the corpus and the pyloric antrum (Leblond et al. 1948). The first evidence for multipotent stem cells in stomach epithelium came from "inverse tracing" with transgenic $Rosa26^{LacZ}$ mice. Chemical random mutagenesis resulted in loss of ubiquitous transgene expression at low frequency. LacZ-negative clones were found containing all four gastric cell lineages (Bjerknes and Cheng 2002). This indicated the existence of a multipotent progenitor/stem cell. Clones consisting of only one lineage were still present 48 weeks after mutagenesis, indicating the presence of long-lived monopotent progenitors. Because the mutations occurred in random cells, the identity of the stem cell could not be determined. Study of inheritance patterns of random mitochondrial mutations revealed that each gastric unit contains more than one multipotent adult stem cell (McDonald et al. 2008).

These findings position proliferating stem cells in the isthmus, where their direct progeny differentiates and migrates up and down toward the pit and the gland. Although this model is still considered valid for the corpus, new evidence using Lgr5 lineage tracing places multipotent stem cells at the bottom of the gastric units of the pyloric antrum (Barker et al. 2010b). Using the same strategies as for the intestine, it was shown that Lgr5 marks three to four cells at the bottom of pyloric gastric units. No Lgr5 expression was found in the isthmus. Lineage tracing with $Lgr5^{EGFP\text{-}IRES\text{-}CreERT2}/Rosa26^{LacZ}$ mice showed that Lgr5 cells in the stomach are cycling and give rise to all the gastric cell lineages. Examination of the ultrastructure of Lgr5 cells by electron microscopy showed a large centrally located nucleus, limited basal rough endoplasmic reticulum (ER), apical microvilli, and no secretory granules. Expression profile analysis comparing cells with high Lgr5 expression ($Lgr5^{hi}$) with their assumed daughter cells ($Lgr5^{lo}$) showed an enrichment of Wnt target genes. Culture conditions were developed under which the Lgr5 cells were able to form and maintain in vitro gastric organoids. The "ministomachs" were grown under similar conditions as the previously described intestinal organoids (Sato et al. 2009), but had an additional dependence on Wnt3a for growth and required FGF10 for budding into multiunit structures. The gastric organoids expressed Lgr5 and several gastric epithelial markers, however, no pit or enteroendocrine cells were observed. Differentiation toward these cell lineages did occur after reduction of the Wnt3a concentration (Barker et al. 2010b).

Work with a Villin-Cre transgenic mouse has revealed an independent stem cell population: rare progenitor cells at or just below the isthmus and in the base of antral glands (Qiao et al. 2007). These cells were shown to be quiescent under normal homeostatic conditions, but were able to divide on stimulation with the cytokine interferon γ. The reactivation subsequently gave rise to all lineages of antral glands. The fact that all gastric cells, including the quiescent stem cells in isthmus, are derived from Lgr5-positive cells (Barker et al. 2010b) suggests that stem cells may switch between a cycling Lgr5-positive state and a quiescent Lgr5-negative state.

THE CELL OF ORIGIN OF INTESTINAL CANCER

Tumors are thought to originate from a single cell. Two concepts were proposed for the origin and migration of tumor cells in the intestine: the "top-down" and "bottom-up" models. The initiation of cancer could be caused by reprogramming of early progenitors or even more differentiated cells, which migrate top down. The "late progenitor" cell of origin of cancer was suggested based on the presence of dysplastic cells at the luminal surface that carried mutations in Apc, whereas crypt cells were normal (Shih et al. 2001). Owing to the rapid turnover of the intestine, a mutated transit-amplifying or villus cell would have little time to form a tumor. However, positional cues like the expression of EphB2 are lost in many adenocarcinomas (Batlle et al. 2005), and it has been shown that mutations in Apc can block progression along the crypt–villus axis (Sansom et al. 2004). It was shown that cancer can originate from mutations in short-lived progenitors, but that this happens very infrequently and such adenomas develop slowly (Barker et al. 2009).

The bottom-up model defines stem cells as the cell of origin. Apc deletion in Lgr5+, CD133+, or Bmi1+ cells in the base of the crypt indeed resulted in efficient and fast generation of adenomas (Sangiorgi and Capecchi 2008; Barker et al. 2009; Zhu et al. 2009). The notion that stem cells may play a role in carcinogenesis is supported by the properties attributed to stem cells and cancer (initiating) cells. Both are thought to be able to self-renew and give rise to more differentiated daughters, have active telomerase, and be resistant to apoptosis (Wicha et al. 2006).

Within established adenomas or colorectal carcinomas, cells may reside that display stem cell-like properties, so-called cancer stem cells (CSCs). Several markers have been used to isolate potential CSCs and test their tumorigenic potential in xenograft assays (Dalerba et al. 2007; O'Brien et al. 2007; Ricci-Vitiani et al. 2007). Cells sorted based on CD133 or a combination of epithelial cell adhesion molecule (EpCAM) and CD44 were able to reproduce tumors in immunodeficient mice on serial transplantation. Interestingly, CD133 and CD44 are also expressed in the proliferative compartment of crypts (Zeilstra et al. 2008; Snippert et al. 2009). Genetic profiling of Ephb2+ and Lgr5+ expressing cells in the crypt revealed a gene signature specific for ISCs (Barker et al. 2007; Jung et al. 2011). A large part of the ISC-specific program was overexpressed in colorectal cancer samples. Furthermore, the ISC program was enriched in aggressive, poorly differentiated colorectal cancers compared to more benign well-differentiated cancers. Although proliferative genes were enriched in all tumor cells, ISC-specific genes marked only a subset of tumor cells (Jung et al. 2011). Heterogeneity in tumors is also reflected by the fact that not all tumor cells have high nuclear β-catenin (Fodde and Brabletz 2007). Mainly cells migrating into neighboring stromal tissues and cells located at the invasive front show high nuclear β-catenin staining. Because the tumors were derived from Apc-deficient cells with constitutively high Wnt signaling, Apc-independent β-catenin down-regulating mechanisms may be present within tumors.

Cells with the highest Wnt activity were found to define the colon CSCs (Vermeulen et al. 2010). In addition, it was shown that certain CSC properties can be gained or lost depending on the microenvironment. High Wnt activity was preferentially found in cells close to myofibroblasts in xenografts, and coculture of colorectal cancers with colon myofibroblasts or addition of conditioned medium from these cells markedly increased clonogenicity. These results suggest that stemness of colorectal cancer cells is a dynamic process that can be influenced by external factors. Genetic instability in tumors can further modify CSCs and make it difficult to find markers of a targetable pool of cells with stem cell potential (Fodde et al. 2001; Soreide et al. 2006; Odoux et al. 2008).

Mutations in APC, Axin, and β-catenin are also present in gastric cancer (Ebert et al. 2002; Ushijima and Sasako 2004; Pan et al. 2008). Gastric cancer patients show accumulation of nuclear β-catenin and abnormal methylation of Wnt pathway genes (Nakatsuru et al. 1992; Park et al. 1999; Clements et al. 2002). Apc^{min} mice develop tumors in the stomach (Tomita et al. 2007) and APC germline mutations in humans give a 10 times higher risk of gastric cancer (Offerhaus et al. 1992). High Wnt signaling cooperates with inflammatory signaling pathways to contribute to gastric tumor development (Oshima et al. 2006). For example, Helicobacter infection would lead to tumor necrosis factor-α (TNF-α) expression in macrophages, which in turn stimulates surrounding cells to promote Wnt signaling (Oguma et al. 2008). Apc deletion in Lgr5+ cells in the stomach leads to efficient tumor formation, suggesting gastric stem cells to be the cells of origin of gastric cancer (Barker et al. 2010b).

Lgr RECEPTORS MEDIATE R-SPONDIN SIGNALING

The fact that Lgr5 marks stem cells in multiple tissues (Barker et al. 2010a) raises interest in its function. The study of the role of the Lgr5 receptor has been complicated by the lack of a known natural ligand. Lgr5 belongs to a family of serpentine receptors that are defined by the presence of a large extracellular, ligand-binding domain. Lgr1, −2, and −3 represent the G-protein-coupled receptors for the hormones luteinizing hormone (LH), follicle-stimulating factor (FSH), and thyroid-stimulating hormone (TSH). Lgr7 and -8 are receptors for insulinlike molecules such as Relaxin (reviewed by Barker and Clevers 2010). Lgr4, −5, and −6 comprise a subfamily within this group of receptors. The proposed fly ortholog of this subfamily, dLgr2, is the receptor for Bursicon (Baker and Truman 2002). Because the two cysteine-knot-containing subunits of Bursicon are orthologous to vertebrate bone morphogenetic protein antagonists, it was long thought that the ligand for the mammalian Lgr4/5/6 receptors could be found within this family of secreted molecules.

Lgr5 null mice die within 1 day after birth owing to defects caused by fusion of the tongue to the floor of the mouth (Morita et al. 2004). It has been reported that Lgr5 null neonatal mice show premature Paneth cell differentiation (Garcia et al. 2009). However, conditional deletion of Lgr5 in the intestine does not result in a clear crypt phenotype (de Lau et al. 2011). In the intestinal crypt, Lgr5 is coexpressed with Lgr4, which has a broader expression pattern throughout the proliferative compartment. Conditional deletion of Lgr4 results in loss of proliferating cells in the crypt and eventual disappearance of many crypts (de Lau et al. 2011). Additional deletion of Lgr5 enhances this phenotype. Microarrying revealed the loss of a Wnt target gene program, implying that Lgr4 and Lgr5 act agonistically in the Wnt pathway. In a separate study of a hypomorphic Lgr4 mutant mouse, crypt proliferation was moderately affected, but intestinal organoids could not be grown unless the Wnt pathway was activated by the GSK3-inhibitor lithium (Mustata et al. 2011).

R-spondins are small secreted proteins that have been associated with β-catenin signaling previously. Yet, there has not been a consensus on their mechanism of action. Functionally, R-spondins were shown to activate β-catenin signaling (Kazanskaya et al. 2004, 2008; Kim et al. 2005) and transgenic overexpression of R-spondin1 or injection of recombinant R-spondin protein results in dramatic crypt expansion (Kim et al. 2005). In disease models, R-spondin1 accelerates regeneration of the intestine after irradiation (Bhanja et al. 2009; Takashima et al. 2011). Reduction of R-spondin2 protein did not affect lithium-induced β-catenin activity in *Xenopus* (Kazanskaya et al. 2004), indicating that R-spondin functions upstream of the destruction complex.

Two independent studies have shown that all four R-spondin proteins can bind to the leucine-rich domains of Lgr4, Lgr5, and Lgr6 (Carmon et al. 2011; de Lau et al. 2011). Moreover, de Lau et al. (2011) found that the Lgr proteins

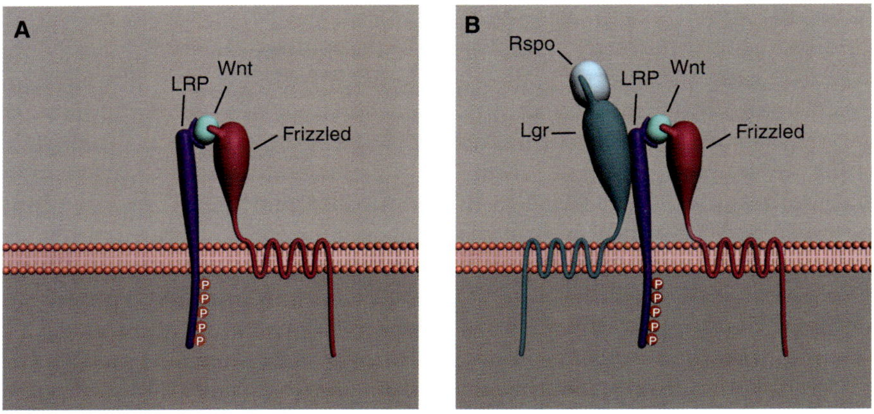

Figure 3. Schematic representation of the Wnt signaling complex. (*A*) The classical complex comprised of Wnt, Frizzled, and low-density lipoprotein receptor-related protein (LRP). (*B*) New proposed model including the Lgr receptor and R-spondin.

reside as facultative components within Frizzled/Lrp5–6 complexes. Removal of endogenous Lgr4 was found to abolish the enhancing effect of R-spondin1 but did not inhibit the stimulating effect of Wnt3a (Carmon et al. 2011; de Lau et al. 2011). Furthermore, the enhancing effect could be rescued with an Lgr5 construct with a mutated "DRY" domain (de Lau et al. 2011), which is supposed to be essential for G-protein coupling (Flanagan 2005). Carmon et al. (2011) also reported that G-protein signaling was not activated on Lgr5 engagement by R-spondin1.

Taken together, the Lgr5-related stem cell markers act as receptors for R-spondins. On R-spondin/Lgr5 interaction, the association of Lgr5 with the Frizzled complex subsequently allows the activating signal to be funneled into the canonical Wnt cascade (Fig. 3). The identification of the Lgr proteins as receptors for the R-spondin Wnt agonists fortifies the tight link between Wnt signaling and adult stem cell biology. The R-spondin proteins as well as the Lgr5 subfamily of receptors appear to be a vertebrate invention. The system functions to augment localized Wnt signals. Thus, these molecules allow expansion of small localized Wnt-driven stem cell niches into larger structures that include transit-amplifying compartments. This vertebrate adaptation of the Wnt pathway may have allowed the appearance of large body sizes within the animal kingdom.

ACKNOWLEDGMENTS

We thank members of the laboratory for critically reading the manuscript, and M. Isselman for the Wnt complex illustrations.

REFERENCES

Albuquerque C, Breukel C, van der Luijt R, Fidalgo P, Lage P, Slors FJ, Leitao CN, Fodde R, Smits R. 2002. The "just-right" signaling model: APC somatic mutations are selected based on a specific level of activation of the β-catenin signaling cascade. *Hum Mol Genet* **11:** 1549–1560.

Andreu P, Colnot S, Godard C, Gad S, Chafey P, Niwa-Kawakita M, Laurent-Puig P, Kahn A, Robine S, Perret C, et al. 2005. Crypt-restricted proliferation and commitment to the Paneth cell lineage following Apc loss in the mouse intestine. *Development* **132:** 1443–1451.

Baker JD, Truman JW. 2002. Mutations in the *Drosophila* glycoprotein hormone receptor, rickets, eliminate neuropeptide-induced tanning and selectively block a stereotyped behavioral program. *J Exp Biol* **205:** 2555–2565.

Barker N, Clevers H. 2010. Leucine-rich repeat-containing G-protein-coupled receptors as markers of adult stem cells. *Gastroenterology* **138:** 1681–1696.

Barker N, van Es JH, Kuipers J, Kujala P, van den Born M, Cozijnsen M, Haegebarth A, Korving J, Begthel H, Peters PJ, et al. 2007. Identification of stem cells in small intestine and colon by marker gene Lgr5. *Nature* **449:** 1003–1007.

Barker N, Ridgway RA, van Es JH, van de Wetering M, Begthel H, van den Born M, Danenberg E, Clarke AR, Sansom OJ, Clevers H. 2009. Crypt stem cells as the cells-of-origin of intestinal cancer. *Nature* **457:** 608–611.

Barker N, Bartfeld S, Clevers H. 2010a. Tissue-resident adult stem cell populations of rapidly self-renewing organs. *Cell Stem Cell* **7:** 656–670.

Barker N, Huch M, Kujala P, van de Wetering M, Snippert HJ, van Es JH, Sato T, Stange DE, Begthel H, van den Born M, et al. 2010b. Lgr5$^{(+ve)}$ stem cells drive self-renewal in the stomach and build long-lived gastric units in vitro. *Cell Stem Cell* **6:** 25–36.

Bastide P, Darido C, Pannequin J, Kist R, Robine S, Marty-Double C, Bibeau F, Scherer G, Joubert D, Hollande F, et al. 2007. Sox9 regulates cell proliferation and is required for Paneth cell differentiation in the intestinal epithelium. *J Cell Biol* **178:** 635–648.

Batlle E, Henderson JT, Beghtel H, van den Born MM, Sancho E, Huls G, Meeldijk J, Robertson J, van de Wetering M, Pawson T, et al. 2002. β-Catenin and TCF mediate cell positioning in the intestinal epithelium by controlling the expression of EphB/ephrinB. *Cell* **111:** 251–263.

Batlle E, Bacani J, Begthel H, Jonkheer S, Gregorieff A, van de Born M, Malats N, Sancho E, Boon E, Pawson T, et al. 2005. EphB receptor activity suppresses colorectal cancer progression. *Nature* **435:** 1126–1130.

Bhanja P, Saha S, Kabarriti R, Liu L, Roy-Chowdhury N, Roy-Chowdhury J, Sellers RS, Alfieri AA, Guha C. 2009. Protective role of R-spondin1, an intestinal stem cell growth factor, against radiation-induced gastrointestinal syndrome in mice. *PLoS ONE* **4:** e8014.

Bjerknes M, Cheng H. 1981a. The stem-cell zone of the small intestinal epithelium. I. Evidence from paneth cells in the adult mouse. *Am J Anat* **160:** 51–63.

Bjerknes M, Cheng H. 1981b. The stem-cell zone of the small intestinal epithelium. II. Evidence from paneth cells in the newborn mouse. *Am J Anat* **160:** 65–75.

Bjerknes M, Cheng H. 1981c. The stem-cell zone of the small intestinal epithelium. III. Evidence from columnar, enteroendocrine, and mucous cells in the adult mouse. *Am J Anat* **160:** 77–91.

Bjerknes M, Cheng H. 1999. Clonal analysis of mouse intestinal epithelial progenitors. *Gastroenterology* **116:** 7–14.

Bjerknes M, Cheng H. 2002. Multipotential stem cells in adult mouse gastric epithelium. *Am J Physiol Gastrointest Liver Physiol* **283:** G767–G777.

Cadigan KM, Nusse R. 1997. Wnt signaling: A common theme in animal development. *Genes Dev* **11:** 3286–3305.

Carmon KS, Gong X, Lin Q, Thomas A, Liu Q. 2011. R-spondins function as ligands of the orphan receptors LGR4 and LGR5 to regulate Wnt/β-catenin signaling. *Proc Natl Acad Sci* **108:** 11452–11457.

Cheng H, Leblond CP. 1974. Origin, differentiation and renewal of the four main epithelial cell types in the mouse small intestine. V. Unitarian theory of the origin of the four epithelial cell types. *Am J Anat* **141:** 537–561.

Clements WM, Wang J, Sarnaik A, Kim OJ, MacDonald J, Fenoglio-Preiser C, Groden J, Lowy AM. 2002. β-Catenin mutation is a frequent cause of Wnt pathway activation in gastric cancer. *Cancer Res* **62:** 3503–3506.

Dalerba P, Dylla SJ, Park IK, Liu R, Wang X, Cho RW, Hoey T, Gurney A, Huang EH, Simeone DM, et al. 2007. Phenotypic characterization of human colorectal cancer stem cells. *Proc Natl Acad Sci* **104:** 10158–10163.

De Lau W, Barker N, Low TY, Koo BK, Li VS, Teunissen H, Kujala P, Haegebarth A, Peters PJ, van de Wetering M, et al. 2011. Lgr5 homologues associate with Wnt receptors and mediate R-spondin signalling. *Nature* **476:** 293–297.

Ebert MP, Fei G, Kahmann S, Muller O, Yu J, Sung JJ, Malfertheiner P. 2002. Increased β-catenin mRNA levels and mutational alterations of the APC and β-catenin gene are present in intestinal-type gastric cancer. *Carcinogenesis* **23:** 87–91.

Fearon ER, Vogelstein B. 1990. A genetic model for colorectal tumorigenesis. *Cell* **61:** 759–767.

Fevr T, Robine S, Louvard D, Huelsken J. 2007. Wnt/β-catenin is essential for intestinal homeostasis and maintenance of intestinal stem cells. *Mol Cell Biol* **27:** 7551–7559.

Flanagan CA. 2005. A GPCR that is not "DRY." *Mol Pharmacol* **68:** 1–3.

Fodde R, Brabletz T. 2007. Wnt/β-catenin signaling in cancer stemness and malignant behavior. *Curr Opin Cell Biol* **19:** 150–158.

Fodde R, Edelmann W, Yang K, van Leeuwen C, Carlson C, Renault B, Breukel C, Alt E, Lipkin M, Khan PM, et al. 1994. A targeted chain-termination mutation in the mouse Apc gene results in multiple intestinal tumors. *Proc Natl Acad Sci* **91:** 8969–8973.

Fodde R, Kuipers J, Rosenberg C, Smits R, Kielman M, Gaspar C, van Es JH, Breukel C, Wiegant J, Giles RH, et al. 2001. Mutations in the APC tumour suppressor gene cause chromosomal instability. *Nat Cell Biol* **3:** 433–438.

Furuyama K, Kawaguchi Y, Akiyama H, Horiguchi M, Kodama S, Kuhara T, Hosokawa S, Elbahrawy A, Soeda T, Koizumi M, et al. 2011. Continuous cell supply from a Sox9-expressing progenitor zone in adult liver, exocrine pancreas and intestine. *Nat Genet* **43:** 34–41.

Garcia MI, Ghiani M, Lefort A, Libert F, Strollo S, Vassart G. 2009. LGR5 deficiency deregulates Wnt signaling and leads to precocious Paneth cell differentiation in the fetal intestine. *Dev Biol* **331:** 58–67.

Gerbe F, van Es JH, Makrini L, Brulin B, Mellitzer G, Robine S, Romagnolo B, Shroyer NF, Bourgaux JF, Pignodel C, et al. 2011. Distinct ATOH1 and Neurog3 requirements define tuft cells as a new secretory cell type in the intestinal epithelium. *J Cell Biol* **192:** 767–780.

Gregorieff A, Pinto D, Begthel H, Destree O, Kielman M, Clevers H. 2005. Expression pattern of Wnt signaling components in the adult intestine. *Gastroenterology* **129:** 626–638.

Hagemann RF, Sigdestad CP, Lesher S. 1970. A quantitative description of the intestinal epithelium of the mouse. *Am J Anat* **129:** 41–51.

Hayashi S, McMahon AP. 2002. Efficient recombination in diverse tissues by a tamoxifen-inducible form of Cre: A tool for temporally regulated gene activation/inactivation in the mouse. *Dev Biol* **244:** 305–318.

He TC, Sparks AB, Rago C, Hermeking H, Zawel L, da Costa LT, Morin PJ, Vogelstein B, Kinzler KW. 1998. Identification of c-MYC as a target of the APC pathway. *Science* **281:** 1509–1512.

Ireland H, Houghton C, Howard L, Winton DJ. 2005. Cellular inheritance of a Cre-activated reporter gene to determine Paneth cell longevity in the murine small intestine. *Dev Dyn* **233:** 1332–1336.

Itzkovitz S, Lyubimova A, Blat IC, Maynard M, van Es J, Lees J, Jacks T, Clevers H, van Oudenaarden A. 2011. Single-molecule transcript counting of stem-cell markers in the mouse intestine. *Nat Cell Biol* **14:** 106–114.

Janssen KP, Alberici P, Fsihi H, Gaspar C, Breukel C, Franken P, Rosty C, Abal M, El Marjou F, Smits R, et al. 2006. APC and oncogenic KRAS are synergistic in enhancing Wnt signaling in intestinal tumor formation and progression. *Gastroenterology* **131:** 1096–1109.

Jung P, Sato T, Merlos-Suarez A, Barriga FM, Iglesias M, Rossell D, Auer H, Gallardo M, Blasco MA, Sancho E, et al. 2011. Isolation and in vitro expansion of human colonic stem cells. *Nat Med* **17:** 1225–1227.

Karam SM. 1993. Dynamics of epithelial cells in the corpus of the mouse stomach. IV. Bidirectional migration of parietal cells ending in their gradual degeneration and loss. *Anat Rec* **236:** 314–332.

Karam SM. 1999. Lineage commitment and maturation of epithelial cells in the gut. *Front Biosci* **4:** D286–D298.

Karam SM, Leblond CP. 1992. Identifying and counting epithelial cell types in the "corpus" of the mouse stomach. *Anat Rec* **232:** 231–246.

Karam SM, Leblond CP. 1993a. Dynamics of epithelial cells in the corpus of the mouse stomach. II. Outward migration of pit cells. *Anat Rec* **236:** 280–296.

Karam SM, Leblond CP. 1993b. Dynamics of epithelial cells in the corpus of the mouse stomach. III. Inward migration of neck cells followed by progressive transformation into zymogenic cells. *Anat Rec* **236:** 297–313.

Kazanskaya O, Glinka A, del Barco Barrantes I, Stannek P, Niehrs C, Wu W. 2004. R-Spondin2 is a secreted activator of Wnt/β-catenin signaling and is required for Xenopus myogenesis. *Dev Cell* **7:** 525–534.

Kazanskaya O, Ohkawara B, Heroult M, Wu W, Maltry N, Augustin HG, Niehrs C. 2008. The Wnt signaling regulator R-spondin 3 promotes angioblast and vascular development. *Development* **135:** 3655–3664.

Kielman MF, Rindapaa M, Gaspar C, van Poppel N, Breukel C, van Leeuwen S, Taketo MM, Roberts S, Smits R, Fodde R. 2002. Apc modulates embryonic stem-cell differentiation by controlling the dosage of β-catenin signaling. *Nat Genet* **32:** 594–605.

Kim KA, Kakitani M, Zhao J, Oshima T, Tang T, Binnerts M, Liu Y, Boyle B, Park E, Emtage P, et al. 2005. Mitogenic influence of human R-spondin1 on the intestinal epithelium. *Science* **309:** 1256–1259.

Kinzler KW, Vogelstein B. 1996. Lessons from hereditary colorectal cancer. *Cell* **87:** 159–170.

Kinzler KW, Nilbert MC, Vogelstein B, Bryan TM, Levy DB, Smith KJ, Preisinger AC, Hamilton SR, Hedge P, Markham A, et al. 1991. Identification of a gene located at chromosome 5q21 that is mutated in colorectal cancers. *Science* **251:** 1366–1370.

Kongkanuntn R, Bubb VJ, Sansom OJ, Wyllie AH, Harrison DJ, Clarke AR. 1999. Dysregulated expression of β-catenin marks early neoplastic change in Apc mutant mice, but not all lesions arising in Msh2 deficient mice. *Oncogene* **18:** 7219–7225.

Korinek V, Barker N, Morin PJ, van Wichen D, de Weger R, Kinzler KW, Vogelstein B, Clevers H. 1997. Constitutive transcriptional activation by a β-catenin-Tcf complex in $APC^{-/-}$ colon carcinoma. *Science* **275:** 1784–1787.

Korinek V, Barker N, Moerer P, van Donselaar E, Huls G, Peters PJ, Clevers H. 1998. Depletion of epithelial stem-cell compartments in the small intestine of mice lacking Tcf-4. *Nat Genet* **19:** 379–383.

Kosinski C, Li VS, Chan AS, Zhang J, Ho C, Tsui WY, Chan TL, Mifflin RC, Powell DW, Yuen ST, et al. 2007. Gene expression patterns of human colon tops and basal crypts and BMP antagonists as intestinal stem cell niche factors. *Proc Natl Acad Sci* **104:** 15418–15423.

Kuhnert F, Davis CR, Wang HT, Chu P, Lee M, Yuan J, Nusse R, Kuo CJ. 2004. Essential requirement for Wnt signaling in proliferation of adult small intestine and colon revealed by adenoviral expression of Dickkopf-1. *Proc Natl Acad Sci* **101:** 266–271.

Leblond CP, Stevens CE, Bogoroch R. 1948. Histological localization of newly-formed desoxyribonucleic acid. *Science* **108:** 531–533.

Lee ER, Leblond CP. 1985. Dynamic histology of the antral epithelium in the mouse stomach: II. Ultrastructure and renewal of isthmal cells. *Am J Anat* **172:** 205–224.

Leedham SJ, Brittan M, McDonald SA, Wright NA. 2005. Intestinal stem cells. *J Cell Mol Med* **9:** 11–24.

Li Q, Ishikawa TO, Oshima M, Taketo MM. 2005. The threshold level of adenomatous polyposis coli protein for mouse intestinal tumorigenesis. *Cancer Res* **65:** 8622–8627.

Liu W, Dong X, Mai M, Seelan RS, Taniguchi K, Krishnadath KK, Halling KC, Cunningham JM, Boardman LA, Qian C, et al. 2000. Mutations in AXIN2 cause colorectal cancer with defective mismatch repair by activating β-catenin/TCF signalling. *Nat Genet* **26:** 146–147.

Luo F, Brooks DG, Ye H, Hamoudi R, Poulogiannis G, Patek CE, Winton DJ, Arends MJ. 2009. Mutated K-ras(Asp12) promotes tumourigenesis in Apc(Min) mice more in the large than the small intestines, with synergistic effects between K-ras and Wnt pathways. *Int J Exp Pathol* **90:** 558–574.

Madara JL. 1982. Cup cells: Structure and distribution of a unique class of epithelial cells in guinea pig, rabbit, and monkey small intestine. *Gastroenterology* **83:** 981–994.

Marsh V, Winton DJ, Williams GT, Dubois N, Trumpp A, Sansom OJ, Clarke AR. 2008. Epithelial Pten is dispensable for intestinal homeostasis but suppresses adenoma development and progression after Apc mutation. *Nat Genet* **40:** 1436–1444.

McDonald SA, Greaves LC, Gutierrez-Gonzalez L, Rodriguez-Justo M, Deheragoda M, Leedham SJ, Taylor RW, Lee CY, Preston SL, Lovell M, et al. 2008. Mechanisms of field cancerization in the human stomach: The expansion and spread of mutated gastric stem cells. *Gastroenterology* **134:** 500–510.

Montgomery RK, Carlone DL, Richmond CA, Farilla L, Kranendonk ME, Henderson DE, Baffour-Awuah NY, Ambruzs DM, Fogli LK, Algra S, et al. 2011. Mouse telomerase reverse transcriptase (mTert) expression marks slowly cycling intestinal stem cells. *Proc Natl Acad Sci* **108:** 179–184.

Mori-Akiyama Y, van den Born M, van Es JH, Hamilton SR, Adams HP, Zhang J, Clevers H, de Crombrugghe B. 2007. SOX9 is required for the differentiation of paneth cells in the intestinal epithelium. *Gastroenterology* **133:** 539–546.

Morin PJ, Sparks AB, Korinek V, Barker N, Clevers H, Vogelstein B, Kinzler KW. 1997. Activation of β-catenin-Tcf signaling in colon cancer by mutations in β-catenin or APC. *Science* **275:** 1787–1790.

Morita H, Mazerbourg S, Bouley DM, Luo CW, Kawamura K, Kuwabara Y, Baribault H, Tian H, Hsueh AJ. 2004. Neonatal lethality of LGR5 null mice is associated with ankyloglossia and gastrointestinal distension. *Mol Cell Biol* **24:** 9736–9743.

Moser AR, Pitot HC, Dove WF. 1990. A dominant mutation that predisposes to multiple intestinal neoplasia in the mouse. *Science* **247:** 322–324.

Mustata RC, Van Loy T, Lefort A, Libert F, Strollo S, Vassart G, Garcia MI. 2011. Lgr4 is required for Paneth cell differentiation and maintenance of intestinal stem cells ex vivo. *EMBO Rep* **12:** 558–564.

Nakatsuru S, Yanagisawa A, Ichii S, Tahara E, Kato Y, Nakamura Y, Horii A. 1992. Somatic mutation of the APC gene in gastric cancer: Frequent mutations in very well differentiated adenocarcinoma and signet-ring cell carcinoma. *Hum Mol Genet* **1:** 559–563.

Nishisho I, Nakamura Y, Miyoshi Y, Miki Y, Ando H, Horii A, Koyama K, Utsunomiya J, Baba S, Hedge P. 1991. Mutations of chromosome 5q21 genes in FAP and colorectal cancer patients. *Science* **253:** 665–669.

Nomura S, Esumi H, Job C, Tan SS. 1998. Lineage and clonal development of gastric glands. *Dev Biol* **204:** 124–135.

O'Brien CA, Pollett A, Gallinger S, Dick JE. 2007. A human colon cancer cell capable of initiating tumour growth in immunodeficient mice. *Nature* **445:** 106–110.

Odoux C, Fohrer H, Hoppo T, Guzik L, Stolz DB, Lewis DW, Gollin SM, Gamblin TC, Geller DA, Lagasse E. 2008. A stochastic model for cancer stem cell origin in metastatic colon cancer. *Cancer Res* **68:** 6932–6941.

Offerhaus GJ, Giardiello FM, Krush AJ, Booker SV, Tersmette AC, Kelley NC, Hamilton SR. 1992. The risk of upper gastrointestinal cancer in familial adenomatous polyposis. *Gastroenterology* **102:** 1980–1982.

Oguma K, Oshima H, Aoki M, Uchio R, Naka K, Nakamura S, Hirao A, Saya H, Taketo MM, Oshima M. 2008. Activated macrophages promote Wnt signalling through tumour necrosis factor-α in gastric tumour cells. *EMBO J* **27:** 1671–1681.

Oshima M, Oshima H, Kitagawa K, Kobayashi M, Itakura C, Taketo M. 1995. Loss of Apc heterozygosity and abnormal tissue building in nascent intestinal polyps in mice carrying a truncated Apc gene. *Proc Natl Acad Sci* **92:** 4482–4486.

Oshima H, Oshima M, Kobayashi M, Tsutsumi M, Taketo MM. 1997. Morphological and molecular processes of polyp formation in Apc(δ716) knockout mice. *Cancer Res* **57:** 1644–1649.

Oshima H, Matsunaga A, Fujimura T, Tsukamoto T, Taketo MM, Oshima M. 2006. Carcinogenesis in mouse stomach by simultaneous activation of the Wnt signaling and prostaglandin E2 pathway. *Gastroenterology* **131:** 1086–1095.

Pan KF, Liu WG, Zhang L, You WC, Lu YY. 2008. Mutations in components of the Wnt signaling pathway in gastric cancer. *World J Gastroenterol* **14:** 1570–1574.

Park WS, Oh RR, Park JY, Lee SH, Shin MS, Kim YS, Kim SY, Lee HK, Kim PJ, Oh ST, et al. 1999. Frequent somatic mutations of the β-catenin gene in intestinal-type gastric cancer. *Cancer Res* **59:** 4257–4260.

Pinto D, Gregorieff A, Begthel H, Clevers H. 2003. Canonical Wnt signals are essential for homeostasis of the intestinal epithelium. *Genes Dev* **17:** 1709–1713.

Potten CS, Kovacs L, Hamilton E. 1974. Continuous labelling studies on mouse skin and intestine. *Cell Tissue Kinet* **7:** 271–283.

Potten CS, Owen G, Roberts SA. 1990. The temporal and spatial changes in cell proliferation within the irradiated crypts of the murine small intestine. *Int J Radiat Biol* **57:** 185–199.

Qiao XT, Ziel JW, McKimpson W, Madison BB, Todisco A, Merchant JL, Samuelson LC, Gumucio DL. 2007. Prospective identification of a multilineage progenitor in murine stomach epithelium. *Gastroenterology* **133:** 1989–1998.

Reya T, Clevers H. 2005. Wnt signalling in stem cells and cancer. *Nature* **434:** 843–850.

Ricci-Vitiani L, Lombardi DG, Pilozzi E, Biffoni M, Todaro M, Peschle C, De Maria R. 2007. Identification and expansion of human colon-cancer-initiating cells. *Nature* **445:** 111–115.

Rubinfeld B, Albert I, Porfiri E, Fiol C, Munemitsu S, Polakis P. 1996. Binding of GSK3β to the APC-β-catenin complex and regulation of complex assembly. *Science* **272:** 1023–1026.

Sabates-Bellver J, Van der Flier LG, de Palo M, Cattaneo E, Maake C, Rehrauer H, Laczko E, Kurowski MA, Bujnicki JM, Menigatti M, et al. 2007. Transcriptome profile of human colorectal adenomas. *Mol Cancer Res* **5:** 1263–1275.

Sangiorgi E, Capecchi MR. 2008. Bmi1 is expressed in vivo in intestinal stem cells. *Nat Genet* **40:** 915–920.

Sansom OJ, Reed KR, Hayes AJ, Ireland H, Brinkmann H, Newton IP, Batlle E, Simon-Assmann P, Clevers H, Nathke IS, et al. 2004. Loss of Apc in vivo immediately perturbs Wnt signaling, differentiation, and migration. *Genes Dev* **18:** 1385–1390.

Sansom OJ, Meniel VS, Muncan V, Phesse TJ, Wilkins JA, Reed KR, Vass JK, Athineos D, Clevers H, Clarke AR. 2007. Myc deletion rescues Apc deficiency in the small intestine. *Nature* **446:** 676–679.

Sato T, Vries RG, Snippert HJ, van de Wetering M, Barker N, Stange DE, van Es JH, Abo A, Kujala P, Peters PJ, et al. 2009. Single Lgr5 stem cells build crypt-villus structures in vitro without a mesenchymal niche. *Nature* **459:** 262–265.

Sato T, van Es JH, Snippert HJ, Stange DE, Vries RG, van den Born M, Barker N, Shroyer NF, van de Wetering M, Clevers H. 2011. Paneth cells constitute the niche for Lgr5 stem cells in intestinal crypts. *Nature* **469:** 415–418.

Schepers AG, Vries R, van den Born M, van de Wetering M, Clevers H. 2011. Lgr5 intestinal stem cells have high telomerase activity and randomly segregate their chromosomes. *EMBO J* **30:** 1104–1109.

Shih IM, Wang TL, Traverso G, Romans K, Hamilton SR, Ben-Sasson S, Kinzler KW, Vogelstein B. 2001. Top-down morphogenesis of colorectal tumors. *Proc Natl Acad Sci* **98:** 2640–2645.

Snippert HJ, van Es JH, van den Born M, Begthel H, Stange DE, Barker N, Clevers H. 2009. Prominin-1/CD133 marks stem cells and early progenitors in mouse small intestine. *Gastroenterology* **136:** 2187.e1–2194.e1.

Snippert HJ, van der Flier LG, Sato T, van Es JH, van den Born M, Kroon-Veenboer C, Barker N, Klein AM, van Rheenen J, Simons BD, et al. 2010. Intestinal crypt homeostasis results from neutral competition between symmetrically dividing Lgr5 stem cells. *Cell* **143:** 134–144.

Soreide K, Janssen EA, Soiland H, Korner H, Baak JP. 2006. Microsatellite instability in colorectal cancer. *Br J Surg* **93:** 395–406.

Su LK, Kinzler KW, Vogelstein B, Preisinger AC, Moser AR, Luongo C, Gould KA, Dove WF. 1992. Multiple intestinal neoplasia caused by a mutation in the murine homolog of the APC gene. *Science* **256:** 668–670.

Takashima S, Kadowaki M, Aoyama K, Koyama M, Oshima T, Tomizuka K, Akashi K, Teshima T. 2011. The Wnt agonist R-spondin1 regulates systemic graft-versus-host disease by protecting intestinal stem cells. *J Exp Med* **208:** 285–294.

Takeda N, Jain R, LeBoeuf MR, Wang Q, Lu MM, Epstein JA. 2011. Interconversion between intestinal stem cell populations in distinct niches. *Science* **334:** 1420–1424.

Taylor RW, Barron MJ, Borthwick GM, Gospel A, Chinnery PF, Samuels DC, Taylor GA, Plusa SM, Needham SJ, Greaves LC, et al. 2003. Mitochondrial DNA mutations in human colonic crypt stem cells. *J Clin Invest* **112:** 1351–1360.

Tetsu O, McCormick F. 1999. β-catenin regulates expression of cyclin D1 in colon carcinoma cells. *Nature* **398:** 422–426.

Thompson M, Fleming KA, Evans DJ, Fundele R, Surani MA, Wright NA. 1990. Gastric endocrine cells share a clonal origin with other gut cell lineages. *Development* **110:** 477–481.

Tian H, Biehs B, Warming S, Leong KG, Rangell L, Klein OD, de Sauvage FJ. 2011. A reserve stem cell population in small intestine renders Lgr5-positive cells dispensable. *Nature* **478:** 255–259.

Tomita H, Yamada Y, Oyama T, Hata K, Hirose Y, Hara A, Kunisada T, Sugiyama Y, Adachi Y, Linhart H, et al. 2007. Development of gastric tumors in $Apc^{Min/+}$ mice by the activation of the β-catenin/Tcf signaling pathway. *Cancer Res* **67:** 4079–4087.

Ushijima T, Sasako M. 2004. Focus on gastric cancer. *Cancer Cell* **5:** 121–125.

van der Flier LG, Sabates-Bellver J, Oving I, Haegebarth A, De Palo M, Anti M, Van Gijn ME, Suijkerbuijk S, van de Wetering M, Marra G, et al. 2007. The intestinal Wnt/TCF signature. *Gastroenterology* **132:** 628–632.

van der Flier LG, Haegebarth A, Stange DE, van de Wetering M, Clevers H. 2009a. OLFM4 is a robust marker for stem cells in human intestine and marks a subset of colorectal cancer cells. *Gastroenterology* **137:** 15–17.

van der Flier LG, van Gijn ME, Hatzis P, Kujala P, Haegebarth A, Stange DE, Begthel H, van den Born M, Guryev V, Oving I, et al. 2009b. Transcription factor achaete scute-like 2 controls intestinal stem cell fate. *Cell* **136:** 903–912.

van de Wetering M, Sancho E, Verweij C, de Lau W, Oving I, Hurlstone A, van der Horn K, Batlle E, Coudreuse D, Haramis AP, et al. 2002. The β-catenin/TCF-4 complex imposes a crypt progenitor phenotype on colorectal cancer cells. *Cell* **111:** 241–250.

van Es JH, Jay P, Gregorieff A, van Gijn ME, Jonkheer S, Hatzis P, Thiele A, van den Born M, Begthel H, Brabletz T, et al. 2005. Wnt signalling induces maturation of Paneth cells in intestinal crypts. *Nat Cell Biol* **7:** 381–386.

Vermeulen L, De Sousa EMF, van der Heijden M, Cameron K, de Jong JH, Borovski T, Tuynman JB, Todaro M, Merz C, Rodermond H, et al. 2010. Wnt activity defines colon cancer stem cells and is regulated by the microenvironment. *Nat Cell Biol* **12:** 468–476.

Wang Y, Giel-Moloney M, Rindi G, Leiter AB. 2007. Enteroendocrine precursors differentiate independently of Wnt and form serotonin expressing adenomas in response to active β-catenin. *Proc Natl Acad Sci* **104:** 11328–11333.

Wicha MS, Liu S, Dontu G. 2006. Cancer stem cells: An old idea—A paradigm shift. *Cancer Res* **66:** 1886–1890.

Winton DJ, Ponder BA. 1990. Stem-cell organization in mouse small intestine. *Proc Biol Sci* **241:** 13–18.

Zeilstra J, Joosten SP, Dokter M, Verwiel E, Spaargaren M, Pals ST. 2008. Deletion of the WNT target and cancer stem cell marker CD44 in $Apc^{Min/+}$ mice attenuates intestinal tumorigenesis. *Cancer Res* **68:** 3655–3661.

Zhu L, Gibson P, Currle DS, Tong Y, Richardson RJ, Bayazitov IT, Poppleton H, Zakharenko S, Ellison DW, Gilbertson RJ. 2009. Prominin 1 marks intestinal stem cells that are susceptible to neoplastic transformation. *Nature* **457:** 603–607.

Wnt Signaling in Bone Development and Disease: Making Stronger Bone with Wnts

Jean B. Regard[1], Zhendong Zhong[2], Bart O. Williams[2], and Yingzi Yang[1]

[1]National Human Genome Research Institute, National Institutes of Health, Bethesda, Maryland 20892

[2]Center for Skeletal Disease Research, Van Andel Research Institute, Grand Rapids, Michigan 49503

Correspondence: yingzi@mail.nih.gov

The skeleton as an organ is widely distributed throughout the entire vertebrate body. Wnt signaling has emerged to play major roles in almost all aspects of skeletal development and homeostasis. Because abnormal Wnt signaling causes various human skeletal diseases, Wnt signaling has become a focal point of intensive studies in skeletal development and disease. As a result, promising effective therapeutic agents for bone diseases are being developed by targeting the Wnt signaling pathway. Understanding the functional mechanisms of Wnt signaling in skeletal biology and diseases highlights how basic and clinical studies can stimulate each other to push a quick and productive advancement of the entire field. Here we review the current understanding of Wnt signaling in critical aspects of skeletal biology such as bone development, remodeling, mechanotransduction, and fracture healing. We took special efforts to place fundamentally important discoveries in the context of human skeletal diseases.

The skeleton has many important functions related to human health. Aside from the classical functions of the skeleton in structural support and movement, the bone matrix forms a major reservoir of calcium and other inorganic ions, and bone cells are active regulators of calcium homeostasis. Recent data suggest that bone cells can secrete hormones (e.g., FGF23 and osteocalcin) and likely play a physiologically significant role in regulating phosphate and energy homeostasis. It has emerged that Wnt signaling plays a major role controlling multiple aspects of skeletal development and maintenance. Thus, understanding how the Wnt pathway controls skeletal growth and homeostasis has broad implications for human health and disease.

Cartilage and bone define the skeleton and are produced by chondrocytes and osteoblasts, respectively. During embryonic development, bones are formed by two distinct processes: intramembranous and endochondral ossification (Fig. 1A). A number of cranial bones and the lateral portion of the clavicles are formed by intramembranous ossification. In this process, mesenchymal progenitor cells condense and differentiate directly into bone-forming osteoblasts. The majority of bones in our body are formed by endochondral ossification, during which mesenchymal progenitor cells condense and differentiate first into cartilage-forming chondrocytes to generate an avascular template of the future bone. Chondrocytes in these

Cite this article as *Cold Spring Harb Perspect Biol* doi: 10.1101/cshperspect.a007997

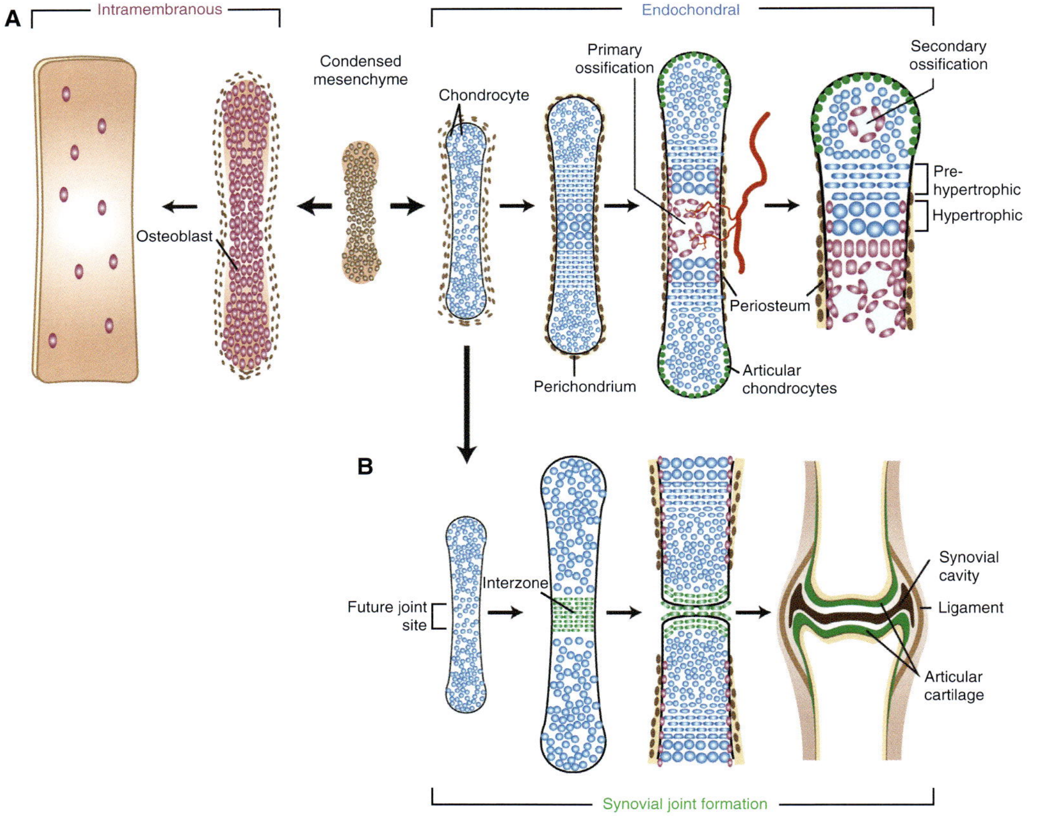

Figure 1. Mechanisms of skeleton formation. (*A*) Bones can form by either intramembranous or endochondral ossification. Both processes are initiated by the condensation of mesenchymal cells. During intramembranous ossification, mesenchymal cells differentiate directly into osteoblasts and deposit bone. During endochondral ossification, mesenchymal cells differentiate into chondrocytes and first make a cartilage intermediate. Chondrocytes in the center of the bone initiate a growth plate, stop proliferating, and undergo hypertrophy. Hypertrophic chondrocytes mineralize their matrix and undergo apoptosis, attracting blood vessels and osteoblasts that remodel the intermediate into bone. (*B*) The first histologic sign of synovial joint formation is the gathering and flattening of cells, forming the interzone. Cavitation occurs within the presumptive joint separating the two cartilaginous structures. Remodeling and maturation proceed to give rise to the mature synovial joint. Wnt signaling plays a significant role in controlling almost all aspects of skeleton formation. Osteoblasts (purple); chondrocytes (blue); osteochondroprogenitor cells (brown).

templates undergo a program of proliferation and progressive cellular maturation. Eventually, they exit the cell cycle and become pre-hypertrophic, then terminally differentiating into hypertrophic chondrocytes, which are eliminated ultimately by apoptosis. Hypertrophic chondrocytes produce a matrix that is calcified and functions as a scaffold for new bone formation. Concomitant with chondrocyte hypertrophy, osteoclasts, osteoblasts, and blood vessels migrate in from perichondral regions and remodel this template into bone.

The developing skeletal elements are often segmented to form joints, which are required to support mobility. Synovial joints, which allow movement via smooth articulation between bones, form when chondrogenic cells in a newly formed cartilage undergo a program of dedifferentiation and flattening to form an interzone (Fig. 1B). Cavitation occurs within the flattened

Cite this article as *Cold Spring Harb Perspect Biol* doi: 10.1101/cshperspect.a007997

cells, allowing physical separation of the skeletal elements, and the formation of the synovial cavity. Cells and tissues in and around the interzone are remodeled at the same time to form the articular cartilage and other joint structures. Failure to form or maintain joints leads to joint fusion or osteoarthritis, a major skeletal disease.

Following its formation, bone remains a regenerative tissue and is maintained during postnatal life by continuous remodeling. This highly active, homeostatic process is required for its functions and is controlled by three cell types: osteoblasts on the bone surface that deposit new bone matrix; osteocytes embedded in bone that are terminally differentiated from osteoblasts and function as mechanical and metabolic sensors; and the matrix-resorbing osteoclasts (Fig. 2). Osteoblasts are derived from mesenchymal stem cells (MSCs), whereas osteoclasts differentiate from hematopoietic progenitors. Decreased bone mass may be due to reduced osteoblast function or elevated osteoclast activity, and, conversely, increased bone mass may result from increased osteoblast function or decreased osteoclast activity. The precise balance of formation and resorption is critical for maintaining normal bone mass, and alterations in this balance lead to common bone diseases such as osteoporosis and osteopetrosis.

There are two major bone types, cortical and trabecular, which show different anatomical properties (Fig. 2). Cortical (or compact) bone is the solid, densely packed bone that forms the outer layer of most bones and gives strength and rigidity. Trabecular (or cancellous) bone is present mostly in the marrow cavities of long bones and is the dominant bone type in vertebral bodies. Trabecular bone forms a porous, cobweb-like network of trabeculae whose large surface area is thought to facilitate the metabolic activity of bones mediated by osteoblasts and osteoclasts. Trabeculae are sites of active remodeling and will often orient in the direction of mechanical loading, dissipating the energy of loading and adding to bone strength. It is trabecular bone, rather than cortical bone, which is most severely affected in osteoporosis.

The Wnt/β-catenin pathway plays a major role in controlling skeletal development and homeostasis, which are the focus of this work. We focus not only on differentiation of skeletal cells and formation of skeletal tissues, but also on the

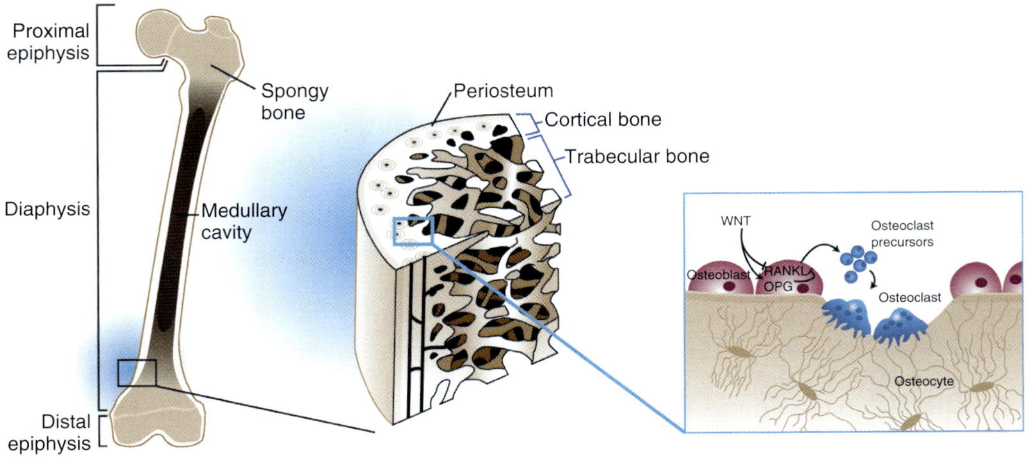

Figure 2. Anatomy of bone. Cortical and trabecular bone represent the two major forms of bone. Osteoblasts (dark purple) are present on the surface and form new bone. Osteocytes (brown) are terminally differentiated osteoblasts that have become embedded in bone and communicate information to one another and to cells on the surface to regulate bone homeostasis. Osteoclasts (blue) are of hematopoietic origin and catabolize bone. A major function of Wnt/β-catenin signaling in osteoblasts is to suppress RANKL and to promote OPG production, thereby inhibiting osteoclast formation.

role of the Wnt/β-catenin signaling pathway on bone homeostasis, mechanotransduction, and wound healing, paying particular attention to human and mouse studies.

WNT/β-CATENIN SIGNALING IN HUMAN SKELETAL DISEASES

The importance of the Wnt/β-catenin signaling pathway in patterning, development, and maintenance of the human skeleton is underscored by the number of human bone diseases associated with mutations in this pathway (Table 1). Early in skeletal development, Wnt signaling controls pattern formation before skeletal elements are laid down, and abnormal Wnt signaling is associated with defects of the human skeleton. At the more extreme end of the spectrum, loss of WNT3 leads to tetra-amelia, a severe congenital defect in which all four limbs fail to form, as well as craniofacial defects (Niemann et al. 2004). At the milder end of the spectrum, mutations in the Wnt/β-catenin pathway inhibitor AXIN2 leads to a dominant, familial form of tooth agenesis, in which a large number of permanent teeth fail to form (Table 1) (Lammi et al. 2004).

The first human genetic evidence revealing the importance of the Wnt/β-catenin pathway in bone formation came from the work of Gong et al. (2001) that identified several loss-of-function (LOF) mutations in the low-density lipoprotein receptor-related protein 5 (LRP5) associated with autosomal recessive osteoporosis pseudoglioma (OPPG) syndrome. Patients suffer from a dramatic reduction in trabecular bone leading to multiple fractures and bone deformities. LRP5 loss-of-function mutations have subsequently been shown to be associated with juvenile osteoporosis (Saarinen et al. 2007). Conversely, gain-of-function mutations of LRP5 lead to the opposite phenotype: autosomal-dominant high bone mass and enhanced bone strength (Boyden et al. 2002; Little et al. 2002; Van Wesenbeeck et al. 2003). Loss-of-function mutations in the closely related LRP6 lead to an autosomal-dominant familiar form of coronary artery disease associated with osteoporosis (Mani et al. 2007).

Mutations in the secreted Wnt/β-catenin signaling antagonist sclerostin (SOST) cause two overlapping diseases: sclerosteosis and van Buchem disease (Balemans et al. 2001, 2002; Brunkow et al. 2001). Sclerosteosis is a rare autosomal-recessive skeletal dysplasia mostly affecting the Afrikaner population of South Africa. It is caused by loss-of-function mutations in the *SOST* gene and is marked by hand malformations, unrestrained bone growth leading to thickened and sclerotic bones, and high bone mass. Bone overgrowth is generalized but most severe in the cranium, leading to increased intracranial pressure and pinching of cranial nerves causing facial palsy and loss of hearing and smell. Interestingly, heterozygous carriers of these *SOST* mutations have increased bone mineral density, suggesting that one affected allele is sufficient to confer a skeletal phenotype and the effects are dominant (Gardner et al. 2005). Most patients with van Buchem disease come from a fishing village in the Netherlands and present with a similar, although milder, sclerosing bone dysplasia affecting mainly cranial bones. Unlike sclerosteosis, van Buchem disease is not caused by mutations in the *SOST* gene coding sequence, but rather by a deletion of an enhancer region downstream causing reduced expression of SOST.

Although it has long been recognized that bone mass and predisposition to osteoporosis are highly heritable, the mechanism and genes involved have been poorly defined. Several recent genomic screens (both genome-wide association studies as well as more targeted screens) have identified polymorphisms in or close to Wnt/β-catenin signaling components that correlate with bone mineral density and further strengthen the importance of this pathway in human bone physiology. Polymorphisms in LRP5, LRP6, GPR177 (Wntless), SFRP1, SOST, and RSPO3 have all been reported to be associated with bone parameter variations in humans (Uitterlinden et al. 2004; van Meurs et al. 2006; Sims et al. 2008; Duncan et al. 2011). Studies using mouse genetics have revealed causative and fundamentally important roles of Wnt/β-catenin signaling in skeletal biology and further established in vivo experimental models to

 Cite this article as *Cold Spring Harb Perspect Biol* doi: 10.1101/cshperspect.a007997

Table 1. Human and mouse genetic analysis of Wnt signaling in skeletal biology

Human gene	Human disease	Skeletal phenotype	References
LRP5	Osteoporosis-pseudoglioma syndrome	Low bone mass	Gong et al. 2001
LRP5	Juvenile osteoporosis	Low bone mass	Gong et al. 2001; Saarinen et al. 2007
LRP5	Osteopetrosis	High bone mass	Boyden et al. 2002; Little et al. 2002; Van Wesenbeeck et al. 2003
LRP6	Osteoporosis	Low bone mass	Mani et al. 2007
SOST	Sclerosteosis, Van Buchem disease	High bone mass, bone overgrowth	Balemans et al. 2001, 2002; Brunkow et al. 2001
AXIN2	Tooth agenesis	Fewer permanent teeth	Lammi et al. 2004
WNT3	Tetra-amelia	Absence of all four limbs	Niemann et al. 2004

Mouse allele	Skeletal phenotype	References
β-Catenin conditional LOF		
Dermo1-cre	Reduced osteoblastogenesis, ectopic chondrogenesis, survival to birth	Day et al. 2005; Hu et al. 2005
Prx1-cre	Reduced osteoblastogenesis, ectopic chondrogenesis, survival to birth	Hill et al. 2005
Col1a1(2.3)-cre	Normal osteoblastogenesis, increased osteoclastogenesis and bone resorption, reduced bone mass, normal life span	Glass et al. 2005
Osteocalcin-cre	Reduced osteoblastogenesis, increased osteoclastogenesis, reduced bone mass, survival to 5 wk	Holmen et al. 2005
DMP1-cre	Normal osteoblastogenesis, increased osteoclastogenesis, reduced bone mass, survival to 3 mo	Kramer et al. 2010
β-Catenin conditional GOF		
Prx1-cre	Reduced differentiation of osteoblasts and chondrocytes	Hill et al. 2005
Col2-cre	Reduced chondrocyte differentiation, impaired joint formation	Guo et al. 2004
Col1a1(2.3)-cre	High bone mass, reduced osteoclastogenesis	Glass et al. 2005
APC conditional LOF		
Osteocalcin-cre	High bone mass, increased osteoblastogenesis, reduced osteoclastogenesis, survival to 2 wk	Holmen et al. 2005
LRP5 KO	Postnatal low bone mass, reduced osteoblastogenesis	Kato et al. 2002; Fujino et al. 2003; Holmen et al. 2004
LRP5 conditional LOF		
Prx1-cre	Reduced bone mass in limbs, normal bone mass in spine	Cui et al. 2011
Villin-cre	Reduced bone mass	Yadav et al. 2008
Villin-cre	Normal bone mass	Cui et al. 2011
Col1a1(2.3)-cre	Normal bone mass	Yadav et al. 2008
DMP1-cre	Reduced bone mass	Cui et al. 2011
LRP5 GOF	Increased bone mass, bone formation rate, bone strength	Babij et al. 2003; Yadav et al. 2008; Cui et al. 2011
Villin-cre	Increased bone mass	Yadav et al. 2008
Villin-cre	Normal bone mass	Cui et al. 2011

Continued

Table 1. *Continued*

Mouse allele	Skeletal phenotype	Reference
DMP1-cre	Increased bone mass	Cui et al. 2011
LRP6 KO	Interacts genetically with Lrp5 to reduce bone mass	Holmen et al. 2004
Ringelschwanz (Rs) (LRP6 LOF hypomorph)	Delayed ossification, reduced bone mass	Kokubu et al. 2004
Crooked tail (Cd) (LRP6 GOF hypomorph)	Delayed ossification	Carter et al. 2005
Axin2 KO	Increased osteoblastogenesis and maturation, craniosynostosis	Yu et al. 2005
Dkk1 Het	Increased bone formation	Morvan et al. 2006
Doubleridge (Dkk1 LOF hypomorph)	Increased bone mass	MacDonald et al. 2007
Sfrp1 KO	Increased trabecular bone mass, increased osteoblastogenesis	Bodine et al. 2004
Sfrp3 KO	Stiffer bone with increased cortical bone thickness and density	Lories et al. 2007
Sost KO	Increased bone formation	Li et al. 2008
Gsk3β Het	Increased bone formation	Kugimiya et al. 2007
Wnt10b KO	Reduced bone mass	Bennett et al. 2005
Fzd9 KO	Reduced bone formation	Albers et al. 2011
Naked cuticle 1 and 2 KO	Craniofacial abnormalities	Zhang et al. 2007
Lef1 Het	Reduced bone mass in females, decreased osteoblast function	Noh et al. 2009
Tcf1 KO	Reduced bone mass, increased osteoclastogenesis	Glass et al. 2005

investigate the pathological mechanisms of human mutations in Wnt pathway components during human bone formation and homeostasis (see below).

WNT/β-CATENIN SIGNALING IN OSTEOBLAST DIFFERENTIATION AND MATURATION

The mouse has proven to be a productive animal model for the genetic dissection of Wnt/β-catenin signaling in bone biology (Fig. 3), and work from a number of groups has shown that Wnt/β-catenin signaling regulates multiple aspects of skeletal development, controlling the differentiation and function of MSCs, chondrocytes, osteoblasts, and osteoclasts. During skeletogenesis, conditional removal of β-catenin in mesenchymal progenitor cells leads to decreased osteoblast differentiation (Day et al. 2005; Hill et al. 2005; Hu et al. 2005; Rodda and McMahon 2006). Cells surrounding the developing bones are capable of expressing early markers of the osteoblast lineage (i.e., Runx2), but fail to express Osterix, a marker of osteoblast commitment. Indeed, removal of β-catenin in osteoblast progenitor cells promoted chondrogenesis in vitro and ectopic chondrocyte differentiation at the expense of osteoblasts in vivo (Day et al. 2005; Hill et al. 2005). This suggests that, in vivo in an osteochondral progenitor cell, high Wnt/β-catenin signaling promotes osteoblast differentiation and inhibits chondrogenesis. Indeed, ectopic activation of Wnt/β-catenin signaling suppresses chondrogenic differentiation from mesenchymal progenitors in culture (Rudnicki and Brown 1997; Day et al. 2005; Reinhold et al. 2006). However, the effect of Wnt/β-catenin signaling on osteoblast formation appears to

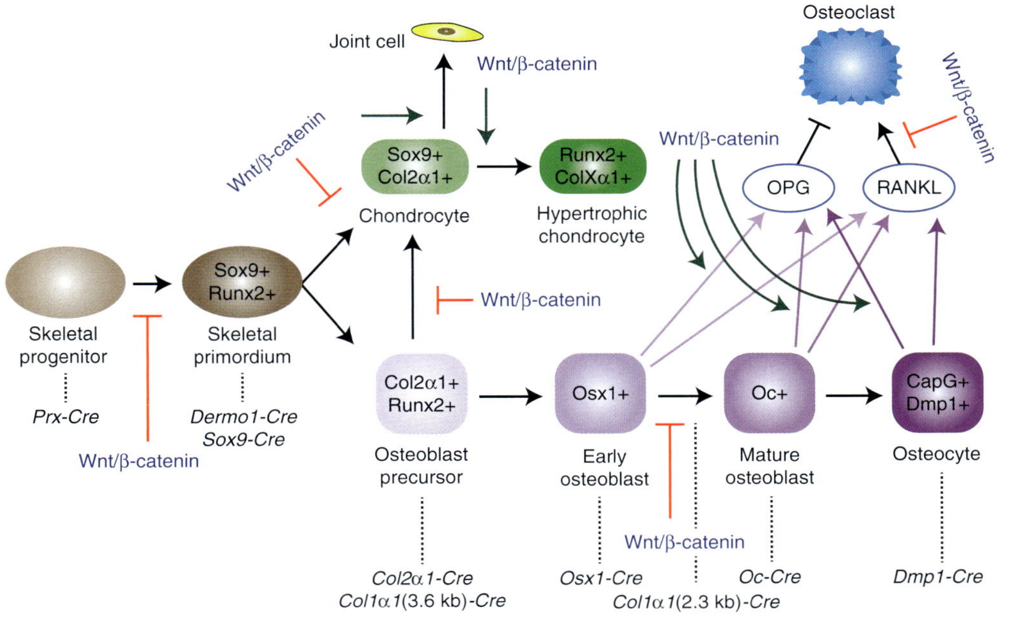

Figure 3. Genetic dissection of skeletal cells using mouse genetics. Schematic diagram of the differentiation of chondrocytes and osteoblasts from a common osteochondral progenitor cell. Overlaid are the corresponding mouse Cre-recombinase lines that allow for the in vivo dissection of signaling pathways. Regulation of cell differentiation by Wnt/β-catenin signaling is indicated (green for activation and red for inhibition). (Figure created from data from Day et al. 2005, Hill et al. 2005, Hu et al. 2005, Rodda and McMahon 2006, Zylstra et al. 2008, and Bonewald 2011.)

depend on the developmental stage of the cells. In vitro activation of the Wnt/β-catenin signaling pathway in MSCs promotes their proliferation and inhibits osteoblastic differentiation (Boland et al. 2004; de Boer et al. 2004; Cho et al. 2006; Regard et al. 2011). Once MSCs have been committed to the osteoblast lineage, Wnt/β-catenin signaling promotes their growth and differentiation, but can also block their terminal differentiation into mature osteoblasts (Kahler and Westendorf 2003; Kahler et al. 2006, 2008; Rodda and McMahon 2006; Eijken et al. 2008). This stage-dependent function of Wnt/β-catenin signaling in bone development is also manifested in human diseases. For instance, fibrous dysplasia (FD) caused by activated Gαs protein is due to inhibition of osteoblast differentiation, and maturation resulted from up-regulated Wnt/β-catenin signaling in MSCs and/or early osteoblasts (Regard et al. 2011).

WNT/β-CATENIN SIGNALING IN CARTILAGE AND SYNOVIAL JOINT FORMATION

Removal of β-catenin from chondrocytes disrupts normal endochondral ossification and leads to neonatal lethality (Akiyama et al. 2004; Day et al. 2005). Mutant knockout mice displayed delayed formation of hypertrophic chondrocytes, reduced chondrocyte proliferation and survival, delays in blood vessel invasion, ossification, and cleft palates. Thus, β-catenin inhibits chondrogenesis, but, once cartilage has formed, it promotes maturation and survival of chondrocytes in the growth plate. Removal of the Wnt antagonist secreted Frizzled-related protein 1 (sFRP1) in mice led to a diminished growth plate and increased calcification in the hypertrophic zone, and accelerated differentiation of hypertrophic chondrocytes (Gaur et al. 2006). A similar phenotype is seen in Axin2-null mice that were runted and showed evidence of accelerated chondrocyte maturation (Dao et al. 2010). Constitutive activation of β-catenin in chondrocytes and skeletal mesenchyme also caused neonatal lethality with severe chondrodysplasia (Akiyama et al. 2004; Guo et al. 2004; Hill et al. 2005). Activation led to small, hypoplastic skeletal elements with disorganized growth plates. It should be noted that although Wnt/β-catenin signaling plays a major role in promoting chondrocyte differentiation, maturation, and survival (Mak et al. 2006), chondrocytes still mature eventually (despite a significant delay), and most chondrocytes do survive, suggesting there are other pathways in cartilage that can support chondrocyte hypertrophy and survival independently of Wnt/β-catenin signaling.

Wnt signaling has also been shown to play an inductive role in synovial joint formation. Topgal, a transgenic LacZ readout of endogenous Wnt/β-catenin signaling, marks sites of joint formation (Guo et al. 2004). Wnt4, Wnt9a (formerly known as Wnt14), and Wnt16 are expressed early in the development of interzones and can function as markers of synovial joint formation (Guo et al. 2004). However, inactivation of both Wnt4 and Wnt9a leads to only limited joint fusion (Spater et al. 2006), suggesting that removing more Wnt ligands (such as Wnt16) may be required to address the importance of Wnts in this process. β-Catenin is required for the development of synovial joints, because its removal before interzone formation led to a reduction in markers of synovial joint formation and resulted in joint fusions (Guo et al. 2004; Spater et al. 2006). Importantly, removal of β-catenin from the joint region following interzone formation led to a very mild joint phenotype, indicating that Wnt/β-catenin signaling is particularly important early in joint induction (Fig. 3) (Koyama et al. 2008).

Different from the growth plate cartilage, articular cartilage coats the ends of long bones. Although growth plate cartilage matures and turns over quickly, articular cartilage undergoes minimal turnover and is essentially maintained as a permanent tissue. Pathways that control the maintenance of joints and articular cartilage are particularly important for the treatment of osteoarthritis (OA), which is characterized by inappropriate hypertrophy of articular chondrocytes and degradation of extracellular matrix leading to loss of articular cartilage. Much of the available data suggest that activation of Wnt/β-catenin signaling leads to loss of a stable chondrocyte phenotype and destruction of cartilaginous tissues. Genome-wide analysis in human has shown an association between a functional single nucleotide polymorphism (SNP) in secreted Frizzled-related protein 3 (sFRP3, also called FRZB), a Wnt antagonist (Loughlin et al. 2004). Inactivation of sfrp3 in mice resulted in grossly normal cartilage development; however, mutant mice showed increased cartilage damage in models of OA (Lories et al. 2007). In addition, conditional activation of β-catenin postnatally in articular chondrocytes leads to an OA-like phenotype, suggesting that Wnt/β-catenin signaling continues to be important in maintaining synovial joints (Zhu et al. 2009).

WNT/β-CATENIN SIGNALING IN OSTEOBLAST FUNCTION

Wnt/β-catenin signaling remains important throughout life, playing a central role in controlling bone mass. In mature osteoblasts in vivo, the Wnt/β-catenin signaling pathway continues to be important, and a major function appears to be in the control of osteoclast formation. Deletion of β-catenin in osteoblasts using the Cre recombinase driven by either the human Osteocalcin or Collagen 1α1(2.3 kb) promoters caused severe low bone mass with strikingly increased osteoclastogenesis, whereas constitutive activation of β-catenin under the same conditions resulted in dramatically increased bone deposition with reduced osteoclast formation (Glass et al. 2005; Holmen et al. 2005). Consistent with these, Wnt7b$^{-/-}$ and Wnt10b$^{-/-}$ mice have produced low bone mass phenotypes (Bennett et al. 2005; Tu et al. 2007). Because it is known that multiple Wnts and Fzds are expressed in the developing bone, it is likely that functional redundancy is masking the relative in vivo importance of these factors. It is likely that work with mice carrying conditional deletions

 Cite this article as *Cold Spring Harb Perspect Biol* doi: 10.1101/cshperspect.a007997

in either the Porcupine or Wntless/GPR177/Evi genes, which are both specifically required for secretion of Wnt ligands from Wnt-producing cells (Hausmann et al. 2007; Lorenowicz and Korswagen 2009), may help to address these questions (Fu et al. 2011; Stefater et al. 2011).

Bone formation has to be coupled to bone resorption in a delicate balance in order to maintain a healthy bone. This is achieved through an interaction between osteoblasts and osteoclasts. Osteoclast differentiation is regulated by at least three genes expressed in osteoblasts: macrophage colony-stimulating factor (M-CSF), receptor activator of nuclear factor κ-B ligand (RANKL) and osteoprotegrin (Opg). RANKL is expressed on the surface of osteoblasts and stimulates osteoclast maturation through interactions with its receptor, RANK, which is expressed at the surface of osteoclast precursors. Opg is a secreted decoy receptor for RANKL that functions to inhibit osteoclastogenesis by disrupting RANKL–RANK interaction (Fig. 2). Both RANKL and Opg expression can be controlled by Wnt/β-catenin signaling. Opg has been shown to be a direct target of β-catenin transcriptional activation, and loss of β-catenin leads to decreased Opg expression and increased osteoclast activity (Glass et al. 2005; Holmen et al. 2005). β-Catenin-deficient osteoblasts have also been documented to express higher RANKL expression (Holmen et al. 2005). Conversely, activation of Wnt/β-catenin signaling by stabilizing β-catenin increased Opg expression and bone formation (Glass et al. 2005). In addition, activating the pathway by removing adenomatous polyposis coli (Apc) in osteoblasts showed reduced RANKL expression and elevated Opg levels (Holmen et al. 2005). Lef1 and Tcf1 are major downstream mediators of Wnt/β-catenin signaling in osteoblasts, because their loss is associated with reduced bone mass and, in the case of Tcf1, increased osteoclastogenesis (Glass et al. 2005; Noh et al. 2009).

Wnt/β-catenin signaling activity is tightly regulated by secreted antagonists, one of which is Dkk1, which also plays a critical role in bone formation. $Dkk1^{-/-}$ mice die in utero because of lack of head formation and also show limb defects with altered cell proliferation and survival (Mukhopadhyay et al. 2001). $Dkk1^{+/-}$ mice, however, survive and show higher numbers of osteoblasts and increased bone mass, demonstrating that partial loss of Dkk1 is sufficient to generate skeletal phenotypes (Morvan et al. 2006). Indeed, a Dkk1 allelic series including a hypomorphic Dkk1 allele (doubleridge; $Dkk1^{d}$) showed that bone mass correlates inversely with Dkk1 levels (MacDonald et al. 2007). Consistent with Dkk1 suppressing bone formation in vivo, transgenic mice expressing Dkk1 under control of osteoblast promoters are osteopenic (Li et al. 2006; Guo et al. 2010). In addition, a Dkk1-neutralizing antibody promotes bone formation when injected into mice (Yaccoby et al. 2007). Kremens, the Dkk1 co-receptors, also play physiologically significant roles in bone formation. Ablation of both kremens (Krm1 and Krm2) increased trabecular bone volume twofold and was associated with increased osteoblast number and function (Ellwanger et al. 2008). Krm2 is likely the more relevant gene in bone, because $Krm2^{-/-}$ mice developed a high bone mass phenotype and overexpression of Krm2 in bone led to severe osteoporosis with decreased osteoblast and elevated osteoclast differentiation and function (Schulze et al. 2010).

Sost is another secreted Wnt antagonist that is specifically expressed in osteocytes, and its overexpression in bone results in decreased osteoblast numbers and reduced bone formation rate (Winkler et al. 2003). $Sost^{-/-}$ mice show high bone mass in both cortical and trabeular bone compartments due to increased osteoblast numbers and normal osteoclast functions (Li et al. 2008). Sost-neutralizing antibodies have also been shown to promote bone formation and increase bone mass in a rat model of osteoporosis (Li et al. 2009).

Secreted Frizzled-related proteins (Sfrps) bind Wnt ligands and are Wnt pathway antagonists proving to be important in bone formation. Although widely expressed during development, phenotypes of the $Sfrp1^{-/-}$ mice appear to be largely restricted to skeletal tissues (Bodine et al. 2004; Morvan et al. 2006; Li et al. 2008). $Sfrp1^{-/-}$ mice have normal cortical bone but increased trabecular bone associated with decreased osteoblast apoptosis and enhanced osteoblast function. Although modest, $Sfrp3^{-/-}$ mice show increased cortical bone thickness and density and demonstrated increased stiffness relative to controls (Lories et al. 2007).

Axin2, an intracellular negative regulator of Wnt/β-catenin signaling, also regulates bone formation. Loss of Axin2 leads to enhanced proliferation and differentiation of osteoblasts, and $Axin2^{-/-}$ mice develop craniosynostosis, a premature closure of cranial sutures (Yu et al. 2005). This phenotype can be rescued by reducing levels of β-catenin (Liu et al. 2007). Compound $Axin2^{-/-}$; $Axin1^{+/-}$ mice have recently been shown to develop severe craniofacial and axial skeleton defects (Dao et al. 2010), again highlighting the important pro-osteogenic functions of the Wnt/β-catenin signaling pathway in skeletal development.

Gsk3β is another important intracellular negative regulator of the pathway. Although $Gsk3\beta^{-/-}$ mice die around embryonic days 13.5–14.5 (E13.5–E14.5), $Gsk3\beta^{+/-}$ mice survive and show increased bone formation (Kugimiya et al. 2007). Similarly, lithium chloride (LiCl), a Gsk3β inhibitor, increased bone mass and osteoblast function in mice and rescued the low bone mass phenotype of the $Lrp5^{-/-}$ mice (Clement-Lacroix et al. 2005). Consistent with this, LiCl treatment in humans is associated with a decreased incident of fractures (Vestergaard et al. 2005; Wilting et al. 2007). All of these results indicate that Wnt/β-catenin signaling in osteoblasts regulates both bone formation and resorption.

Despite the important roles of Wnt/β-catenin signaling in osteoblasts, one should not forget that it also acts directly in osteoclasts (Modarresi et al. 2009; Wei et al. 2011). Wnt/β-catenin signaling blocks osteoclast differentiation by inhibiting RANKL signaling. Osteoclast differentiation is enhanced in β-catenin heterozygous mice, whereas stabilizing β-catenin in osteoclast precursors inhibits osteoclast differentiation.

Although the functional importance of Lrp5 in regulating bone mass is clear, the precise mechanism has become controversial. Studies in mouse models confirmed the human findings: Mice globally deficient for Lrp5 have decreased bone mass, whereas mice carrying the gain-of-function (GOF) Lrp5-G171V point mutations under the control of an osteoblast promoter displayed increased bone mass (Kato et al. 2002; Babij et al. 2003; Cui et al. 2011). The controversy originates from the model in which Lrp5 functions within the enterochromaffin cells of the small intestine to regulate the production of serotonin (Yadav et al. 2008). Increased serotonin secretion is associated with loss of Lrp5 and reduction in osteoblast activity, leading to lower bone mass. In support of this model, Yadav et al. (2008) found that conditional deletion of Lrp5 in the intestine led to low bone mass, whereas conditional deletion within the osteoblast had no discernible phenotype. Furthermore, intestinal-specific expression of the gain-of-function Lrp5-G171V allele led to increased bone mass, but osteoblast-specific expression of the same allele had no effect (Yadav et al. 2008). However, this model has recently been challenged when the characterization of independently derived mouse models carrying similar alterations in the *Lrp5* locus suggested that Lrp5 did, as has been expected, function within the osteoblast and not in the intestine to control bone mass (Cui et al. 2011). In addition, recent work has shown that Lrp5 acts redundantly with Lrp6 within the osteochondral progenitor cells to mediate commitment to the osteoblast lineage (Joeng et al. 2011). The reasons for these different observations regarding Lrp5 function are currently unclear and will undoubtedly be the subject of additional investigation.

WNT/β-CATENIN SIGNALING IN OSTEOCYTES

Osteocytes represent >90%–95% of all bone cells and communicate through extensive dendritic processes present in canaliculi with one another, with cells on the bone surface, and in the bone marrow (Fig. 2). Osteocytes can recruit osteoclast precursors to stimulate bone resorption and regulate osteoblastic differentiation from a mesenchymal stem cell (Kamioka et al. 2001; Zhao et al. 2002; Heino et al. 2004).

Similar to mice with deletions of β-catenin in osteoblasts, mice with deletion of β-catenin induced by the DMP1-cre (which is highly enriched in osteocytes) showed a severe low bone mass phenotype, which was associated with increased osteoclast activity caused by decreased Opg expression in osteocytes (Kramer et al. 2010).

Mechanical stress leads to the movement of bone fluid through the lacuno-canalicular system, generating shear stress that the osteocyte senses. Osteocytes were first proposed to be mechanosensory cells 20 years ago, when an increased ^{3}H-uridine level was detected in osteocytes following mechanical loading in vivo (Pead et al. 1988). More recently, it has been shown that, in mice, ablation of osteocytes caused resistance to unloading-induced bone loss, providing the first direct evidence for the role of osteocytes in mechanotransduction (Tatsumi et al. 2007). Several reports have shown that β-catenin signaling is activated as a normal physiological response to mechanical loading in bone (Robinson et al. 2006). In the Lrp5$^{-/-}$ mice, the osteogenic response to mechanical loading of the ulna was dramatically reduced, with normal recruitment and activation of osteoblasts at the mechanically strained surface (Sawakami et al. 2006). This suggests that Lrp5$^{-/-}$ osteocytes may have a normal primary response to mechanic loading, whereas the intrinsic osteogenic abilities of osteoblasts and/or osteocytes are reduced in the absence of Lrp5. Thus, the osteoporotic phenotype of the Lrp5$^{-/-}$ mouse could be due to inadequate processing of signals derived from mechanical stimulation. Consistent with this model, mice expressing the Lrp5 G171V mutation specifically in bone showed a higher sensitivity to mechanical loading with stronger β-catenin activation, and experienced significantly less bone loss induced by unloading (Akhter et al. 2004; Robinson et al. 2006; Saxon et al. 2011).

Osteocytes may be involved in a feedback signaling system to maintain bone-lining cells in a quiescent state under conditions of low mechanical load (Marotti et al. 1992; Martin 2000; Metz et al. 2003). A candidate to mediate such a feedback mechanism is Sost, which is specifically expressed in mature osteocytes and plays a negative regulatory role in bone formation (van Bezooijen et al. 2004; Poole et al. 2005). Sost has been reported to function as both an inhibitor of BMP signaling and as an inhibitor of Wnt/β-catenin signaling by binding directly to LRP5 and LRP6 (Li et al. 2005; Semenov et al. 2005). Ulnar loading decreased *Sost* expression and increased bone formation rate in a strain dose-dependent manner (Robling et al. 2008). Sost$^{-/-}$ mice were resistant to mechanical unloading-induced bone loss and failed to show decreased Wnt/β-catenin signaling during unloading (Lin et al. 2009). Thus, Sost may represent an important primary response to mechanical loading. Mechanical loading may regulate bone mass by modulating the expression levels of this Wnt inhibitor.

In vitro models to mimic physiologic shear stress include steady fluid flow, pulsatile fluid flow, or oscillatory flow to generate shear stress on osteoblasts or osteoblast-like cells (Reich et al. 1990; Williams et al. 1994; Hillsley and Frangos 1997; Jacobs et al. 1998). There are limitations for these kinds of in vitro fluid flow shear stress on osteoblast cultures (Bonewald 2011), but they provide valuable information toward understanding the in vivo response to such forces. Consistent with the in vivo findings, in vitro studies also show that β-catenin signaling is required for the response to fluid shear stress (Lau et al. 2006; Robinson et al. 2006; Sawakami et al. 2006). In addition, some in vitro studies suggest that osteocytes support osteoclast formation through a response to fluid shear stress (Kim et al. 2006; You et al. 2008).

WNT/β-CATENIN SIGNALING IN BONE FRACTURE HEALING

We briefly cover the function of Wnt/β-catenin signaling in bone fracture healing here because this topic is covered in more detail in Whyte et al. (2012). Adult bone has the ability to regenerate after injury or fracture. In most aspects, fracture healing recapitulates key steps of embryonic bone development (Bruder et al. 1994); however, there are also differences between fracture healing and embryonic bone development.

For example, embryonic bone development does not go through an inflammatory response. Fracture healing involves the participation of hematopoietic cells, immune cells, and mesenchymal stem cells that are recruited from the surrounding tissues and the circulation. In classical histological terms, fracture healing can be divided into direct and indirect healing (Dimitriou et al. 2005). The majority of fracture healing is through an indirect healing process, which consists of three main phases: (1) the inflammatory stage, (2) the repair stage, and (3) the remodeling stage (Chen and Alman 2009). A microarray study specifies the process further into five specific physiological events: inflammation, intramembranous ossification, chondrogenesis, endochondral ossification, and remodeling (Hadjiargyrou et al. 2002).

Wnt signaling may be involved throughout the healing process by regulating a wide variety of cell-fate decisions associated with osteogenesis. Notably, the β-catenin signaling pathway is significantly activated in the fracture callus (Hadjiargyrou et al. 2002; Chen et al. 2007). Consistent with this, studies on the Topgal mouse, an in vivo LacZ reporter for Wnt/β-catenin signaling, showed that LacZ activity was activated specifically at the site of injury (Kim et al. 2007).

Because mouse models with gain of function in Wnt signaling show high bone mass phenotypes, it is not surprising to find that the mice with an activated form of β-catenin in osteoblasts and the $Axin2^{-/-}$ mouse show dramatically enhanced bone healing (Chen et al. 2007; Minear et al. 2010). However, in a fracture-healing model involving both endochondral and intramembranous ossification, Wnt/β-catenin signaling plays varying roles in different phases of fracture repair. For example, either increases or decreases in Wnt/β-catenin signaling during early stages of healing inhibit the differentiation of mesenchymal stem cells into osteoblasts. At later stages, Wnt/β-catenin signaling positively regulates osteoblasts after the undifferentiated cells have committed to the osteoblast lineage (Chen et al. 2007). Both recombinant Wnt3a and LiCl can enhance healing when the treatment starts after the fracture (Chen et al. 2007; Minear et al. 2010). Similarly, an anti-Sost antibody can enhance bone healing in rats (Agholme et al. 2010). Conversely, Wnt inhibition in the injury site by adenoviral expression of DKK1 can prevent the differentiation of osteoblastic cells (Kim et al. 2007).

Although it is clear that Wnt/β-catenin signaling plays a crucial role during fracture healing, this does not rule out the possibility that β-catenin-independent pathways are also involved. It has been reported that Wnts that often do not signal via the β-catenin pathway are upregulated during fracture healing (for review, see Chen and Alman 2009). In addition, periodontal ligament stem cells, a recently identified population of mesenchymal stem cells with bone-regeneration ability, were reported to have a lower osteogenic differentiation in inflammatory microenvironments, which might be due to inhibition of non-canonical Wnt pathways (Liu et al. 2011).

WNT IN BONE CANCER AND METASTASIS

The function of Wnt/β-catenin signaling within the skeletal microenvironment has drawn the attention of not only academic, but also industry scientists and clinicians targeting this pathway to treat skeletal tumors. DKK1 has been proposed to be an important regulator in prostate cancer bone metastasis because alterations in DKK1 expression are linked to several tumor types with a propensity to grow in bone including prostate carcinoma, breast carcinoma, and multiple myeloma (Forget et al. 2007). Prostate cancer cells overexpressing DKK1 have a higher potential to metastasize into bone and generate osteolytic lesions (Thudi et al. 2010). DKK1 upregulation is a frequent event in mammary cancer, and increased serum levels of DKK1 are associated with the presence of bone metastasis in patients (Voorzanger-Rousselot et al. 2007; Bu et al. 2008). In multiple myeloma, the serum level of DKK1 is positively correlated with the presence of bone lesions, and anti-DKK1 antibody may be an effective adjunct in the treatment of multiple myeloma–associated bone disease (Tian et al. 2003; Yaccoby et al. 2007; Fulciniti et al. 2009; Heath et al. 2009). Based

 Cite this article as *Cold Spring Harb Perspect Biol* doi: 10.1101/cshperspect.a007997

on these observations, several pharmaceutical companies have agents at various stages of clinical trials that target this pathway for treatment of bone metastases. For example, antibodies or small molecules that target DKK1 or SOST are in various stages of development for treating multiple myeloma and skeletal metastases (Rey and Ellies 2010).

CONCLUDING REMARKS

In this article, we attempt to summarize the current understanding of Wnt/β-catenin signaling in skeletal tissues. Work from many groups in the past decade has clearly shown the importance of the Wnt/β-catenin pathway in both embryonic and postnatal bone formation and homeostasis. Modulating this pathway is widely recognized as an important area of therapeutic development for osteoporosis, fracture healing, and tissue engineering. Moving forward, important questions still remain regarding the precise mechanism(s) by which Wnt/β-catenin signaling regulates skeletal tissues and how this pathway interacts with other major pathways (e.g., BMP, Hedgehog, FGF, etc.) in the skeleton.

ACKNOWLEDGMENTS

We thank Darryl Leja and Julia Fekecs for preparing the figures. This work is supported by the Division of Intramural Research of the National Institutes of Health, DHHS, to J.B.R. and Y.Y. and NIH R01AR053293 to B.O.W. and Z.Z.

REFERENCES

*Reference is also in this collection.

Agholme F, Li X, Isaksson H, Ke HZ, Aspenberg P. 2010. Sclerostin antibody treatment enhances metaphyseal bone healing in rats. *J Bone Miner Res* **25:** 2412–2418.

Akhter MP, Wells DJ, Short SJ, Cullen DM, Johnson ML, Haynatzki GR, Babij P, Allen KM, Yaworsky PJ, Bex F, et al. 2004. Bone biomechanical properties in LRP5 mutant mice. *Bone* **35:** 162–169.

Akiyama H, Lyons JP, Mori-Akiyama Y, Yang X, Zhang R, Zhang Z, Deng JM, Taketo MM, Nakamura T, Behringer RR, et al. 2004. Interactions between Sox9 and β-catenin control chondrocyte differentiation. *Genes Dev* **18:** 1072–1087.

Albers J, Schulze J, Beil FT, Gebauer M, Baranowsky A, Keller J, Marshall RP, Wintges K, Friedrich FW, Priemel M, et al. 2011. Control of bone formation by the serpentine receptor Frizzled-9. *J Cell Biol* **192:** 1057–1072.

Babij P, Zhao W, Small C, Kharode Y, Yaworsky PJ, Bouxsein ML, Reddy PS, Bodine PV, Robinson JA, Bhat B, et al. 2003. High bone mass in mice expressing a mutant LRP5 gene. *J Bone Miner Res* **18:** 960–974.

Balemans W, Ebeling M, Patel N, Van Hul E, Olson P, Dioszegi M, Lacza C, Wuyts W, Van Den Ende J, Willems P, et al. 2001. Increased bone density in sclerosteosis is due to the deficiency of a novel secreted protein (SOST). *Hum Mol Genet* **10:** 537–543.

Balemans W, Patel N, Ebeling M, Van Hul E, Wuyts W, Lacza C, Dioszegi M, Dikkers FG, Hildering P, Willems PJ, et al. 2002. Identification of a 52 kb deletion downstream of the *SOST* gene in patients with van Buchem disease. *J Med Genet* **39:** 91–97.

Bennett CN, Longo KA, Wright WS, Suva LJ, Lane TF, Hankenson KD, MacDougald OA. 2005. Regulation of osteoblastogenesis and bone mass by Wnt10b. *Proc Natl Acad Sci* **102:** 3324–3329.

Bodine PV, Zhao W, Kharode YP, Bex FJ, Lambert AJ, Goad MB, Gaur T, Stein GS, Lian JB, Komm BS. 2004. The Wnt antagonist secreted Frizzled-related protein-1 is a negative regulator of trabecular bone formation in adult mice. *Mol Endocrinol* **18:** 1222–1237.

Boland GM, Perkins G, Hall DJ, Tuan RS. 2004. Wnt 3a promotes proliferation and suppresses osteogenic differentiation of adult human mesenchymal stem cells. *J Cell Biochem* **93:** 1210–1230.

Bonewald LF. 2011. The amazing osteocyte. *J Bone Miner Res* **26:** 229–238.

Boyden LM, Mao J, Belsky J, Mitzner L, Farhi A, Mitnick MA, Wu D, Insogna K, Lifton RP. 2002. High bone density due to a mutation in LDL-receptor-related protein 5. *N Engl J Med* **346:** 1513–1521.

Bruder SP, Fink DJ, Caplan AI. 1994. Mesenchymal stem cells in bone development, bone repair, and skeletal regeneration therapy. *J Cell Biochem* **56:** 283–294.

Brunkow ME, Gardner JC, Van Ness J, Paeper BW, Kovacevich BR, Proll S, Skonier JE, Zhao L, Sabo PJ, Fu Y, et al. 2001. Bone dysplasia sclerosteosis results from loss of the SOST gene product, a novel cystine knot-containing protein. *Am J Hum Genet* **68:** 577–589.

Bu G, Lu W, Liu CC, Selander K, Yoneda T, Hall C, Keller ET, Li Y. 2008. Breast cancer-derived Dickkopf1 inhibits osteoblast differentiation and osteoprotegerin expression: Implication for breast cancer osteolytic bone metastases. *Int J Cancer* **123:** 1034–1042.

Carter M, Chen X, Slowinska B, Minnerath S, Glickstein S, Shi L, Campagne F, Weinstein H, Ross ME. 2005. Crooked tail (Cd) model of human folate-responsive neural tube defects is mutated in Wnt coreceptor lipoprotein receptor-related protein 6. *Proc Natl Acad Sci* **102:** 12843–12848.

Chen Y, Alman BA. 2009. Wnt pathway, an essential role in bone regeneration. *J Cell Biochem* **106:** 353–362.

Chen Y, Whetstone HC, Lin AC, Nadesan P, Wei Q, Poon R, Alman BA. 2007. β-Catenin signaling plays a disparate role in different phases of fracture repair: Implications for therapy to improve bone healing. *PLoS Med* **4:** e249.

Cho HH, Kim YJ, Kim SJ, Kim JH, Bae YC, Ba B, Jung JS. 2006. Endogenous Wnt signaling promotes proliferation and suppresses osteogenic differentiation in human adipose derived stromal cells. *Tissue Eng* **12:** 111–121.

Clement-Lacroix P, Ai M, Morvan F, Roman-Roman S, Vayssiere B, Belleville C, Estrera K, Warman ML, Baron R, Rawadi G. 2005. Lrp5-independent activation of Wnt signaling by lithium chloride increases bone formation and bone mass in mice. *Proc Natl Acad Sci* **102:** 17406–17411.

Cui Y, Niziolek PJ, MacDonald BT, Zylstra CR, Alenina N, Robinson DR, Zhong Z, Matthes S, Jacobsen CM, Conlon RA, et al. 2011. Lrp5 functions in bone to regulate bone mass. *Nat Med* **17:** 684–691.

Dao DY, Yang X, Flick LM, Chen D, Hilton MJ, O'Keefe RJ. 2010. Axin2 regulates chondrocyte maturation and axial skeletal development. *J Orthop Res* **28:** 89–95.

Day TF, Guo X, Garrett-Beal L, Yang Y. 2005. Wnt/β-catenin signaling in mesenchymal progenitors controls osteoblast and chondrocyte differentiation during vertebrate skeletogenesis. *Dev Cell* **8:** 739–750.

de Boer J, Siddappa R, Gaspar C, van Apeldoorn A, Fodde R, van Blitterswijk C. 2004. Wnt signaling inhibits osteogenic differentiation of human mesenchymal stem cells. *Bone* **34:** 818–826.

Dimitriou R, Tsiridis E, Giannoudis PV. 2005. Current concepts of molecular aspects of bone healing. *Injury* **36:** 1392–1404.

Duncan EL, Danoy P, Kemp JP, Leo PJ, McCloskey E, Nicholson GC, Eastell R, Prince RL, Eisman JA, Jones G, et al. 2011. Genome-wide association study using extreme truncate selection identifies novel genes affecting bone mineral density and fracture risk. *PLoS Genet* **7:** e1001372.

Eijken M, Meijer IM, Westbroek I, Koedam M, Chiba H, Uitterlinden AG, Pols HA, van Leeuwen JP. 2008. Wnt signaling acts and is regulated in a human osteoblast differentiation dependent manner. *J Cell Biochem* **104:** 568–579.

Ellwanger K, Saito H, Clement-Lacroix P, Maltry N, Niedermeyer J, Lee WK, Baron R, Rawadi G, Westphal H, Niehrs C. 2008. Targeted disruption of the Wnt regulator Kremen induces limb defects and high bone density. *Mol Cell Biol* **28:** 4875–4882.

Forget MA, Turcotte S, Beauseigle D, Godin-Ethier J, Pelletier S, Martin J, Tanguay S, Lapointe R. 2007. The Wnt pathway regulator DKK1 is preferentially expressed in hormone-resistant breast tumours and in some common cancer types. *Br J Cancer* **96:** 646–653.

Fu J, Ivy Yu HM, Maruyama T, Mirando AJ, Hsu W. 2011. Gpr177/mouse Wntless is essential for Wnt-mediated craniofacial and brain development. *Dev Dyn* **240:** 365–371.

Fujino T, Asaba H, Kang MJ, Ikeda Y, Sone H, Takada S, Kim DH, Ioka RX, Ono M, Tomoyori H, et al. 2003. Low-density lipoprotein receptor-related protein 5 (LRP5) is essential for normal cholesterol metabolism and glucose-induced insulin secretion. *Proc Natl Acad Sci* **100:** 229–234.

Fulciniti M, Tassone P, Hideshima T, Vallet S, Nanjappa P, Ettenberg SA, Shen Z, Patel N, Tai YT, Chauhan D, et al. 2009. Anti-DKK1 mAb (BHQ880) as a potential therapeutic agent for multiple myeloma. *Blood* **114:** 371–379.

Gardner JC, van Bezooijen RL, Mervis B, Hamdy NA, Lowik CW, Hamersma H, Beighton P, Papapoulos SE. 2005. Bone mineral density in sclerosteosis; affected individuals and gene carriers. *J Clin Endocrinol Metab* **90:** 6392–6395.

Gaur T, Rich L, Lengner CJ, Hussain S, Trevant B, Ayers D, Stein JL, Bodine PV, Komm BS, Stein GS, et al. 2006. Secreted Frizzled related protein 1 regulates Wnt signaling for BMP2 induced chondrocyte differentiation. *J Cell Physiol* **208:** 87–96.

Glass DA II, Bialek P, Ahn JD, Starbuck M, Patel MS, Clevers H, Taketo MM, Long F, McMahon AP, Lang RA, et al. 2005. Canonical Wnt signaling in differentiated osteoblasts controls osteoclast differentiation. *Dev Cell* **8:** 751–764.

Gong Y, Slee RB, Fukai N, Rawadi G, Roman-Roman S, Reginato AM, Wang H, Cundy T, Glorieux FH, Lev D, et al. 2001. LDL receptor-related protein 5 (LRP5) affects bone accrual and eye development. *Cell* **107:** 513–523.

Guo X, Day TF, Jiang X, Garrett-Beal L, Topol L, Yang Y. 2004. Wnt/β-catenin signaling is sufficient and necessary for synovial joint formation. *Genes Dev* **18:** 2404–2417.

Guo J, Liu M, Yang D, Bouxsein ML, Saito H, Galvin RJ, Kuhstoss SA, Thomas CC, Schipani E, Baron R, et al. 2010. Suppression of Wnt signaling by Dkk1 attenuates PTH-mediated stromal cell response and new bone formation. *Cell Metab* **11:** 161–171.

Hadjiargyrou M, Lombardo F, Zhao S, Ahrens W, Joo J, Ahn H, Jurman M, White DW, Rubin CT. 2002. Transcriptional profiling of bone regeneration. Insight into the molecular complexity of wound repair. *J Biol Chem* **277:** 30177–30182.

Hausmann G, Banziger C, Basler K. 2007. Helping Wingless take flight: How WNT proteins are secreted. *Nat Rev Mol Cell Biol* **8:** 331–336.

Heath DJ, Chantry AD, Buckle CH, Coulton L, Shaughnessy JD Jr, Evans HR, Snowden JA, Stover DR, Vanderkerken K, Croucher PI. 2009. Inhibiting Dickkopf-1 (Dkk1) removes suppression of bone formation and prevents the development of osteolytic bone disease in multiple myeloma. *J Bone Miner Res* **24:** 425–436.

Heino TJ, Hentunen TA, Vaananen HK. 2004. Conditioned medium from osteocytes stimulates the proliferation of bone marrow mesenchymal stem cells and their differentiation into osteoblasts. *Exp Cell Res* **294:** 458–468.

Hill TP, Spater D, Taketo MM, Birchmeier W, Hartmann C. 2005. Canonical Wnt/β-catenin signaling prevents osteoblasts from differentiating into chondrocytes. *Dev Cell* **8:** 727–738.

Hillsley MV, Frangos JA. 1997. Alkaline phosphatase in osteoblasts is down-regulated by pulsatile fluid flow. *Calcif Tissue Int* **60:** 48–53.

Holmen SL, Giambernardi TA, Zylstra CR, Buckner-Berghuis BD, Resau JH, Hess JF, Glatt V, Bouxsein ML, Ai M, Warman ML, et al. 2004. Decreased BMD and limb deformities in mice carrying mutations in both Lrp5 and Lrp6. *J Bone Miner Res* **19:** 2033–2040.

Holmen SL, Zylstra CR, Mukherjee A, Sigler RE, Faugere MC, Bouxsein ML, Deng L, Clemens TL, Williams BO.

2005. Essential role of β-catenin in postnatal bone acquisition. *J Biol Chem* **280:** 21162–21168.

Hu H, Hilton MJ, Tu X, Yu K, Ornitz DM, Long F. 2005. Sequential roles of Hedgehog and Wnt signaling in osteoblast development. *Development* **132:** 49–60.

Jacobs CR, Yellowley CE, Davis BR, Zhou Z, Cimbala JM, Donahue HJ. 1998. Differential effect of steady versus oscillating flow on bone cells. *J Biomech* **31:** 969–976.

Joeng KS, Schumacher CA, Zylstra-Diegel CR, Long F, Williams BO. 2011. Lrp5 and Lrp6 redundantly control skeletal development in the mouse embryo. *Dev Biol* **359:** 222–229.

Kahler RA, Westendorf JJ. 2003. Lymphoid enhancer factor-1 and β-catenin inhibit Runx2-dependent transcriptional activation of the osteocalcin promoter. *J Biol Chem* **278:** 11937–11944.

Kahler RA, Galindo M, Lian J, Stein GS, van Wijnen AJ, Westendorf JJ. 2006. Lymphocyte enhancer-binding factor 1 (Lef1) inhibits terminal differentiation of osteoblasts. *J Cell Biochem* **97:** 969–983.

Kahler RA, Yingst SM, Hoeppner LH, Jensen ED, Krawczak D, Oxford JT, Westendorf JJ. 2008. Collagen 11a1 is indirectly activated by lymphocyte enhancer-binding factor 1 (Lef1) and negatively regulates osteoblast maturation. *Matrix Biol* **27:** 330–338.

Kamioka H, Honjo T, Takano-Yamamoto T. 2001. A three-dimensional distribution of osteocyte processes revealed by the combination of confocal laser scanning microscopy and differential interference contrast microscopy. *Bone* **28:** 145–149.

Kato M, Patel MS, Levasseur R, Lobov I, Chang BH, Glass DA II, Hartmann C, Li L, Hwang TH, Brayton CF, et al. 2002. Cbfa1-independent decrease in osteoblast proliferation, osteopenia, and persistent embryonic eye vascularization in mice deficient in Lrp5, a Wnt coreceptor. *J Cell Biol* **157:** 303–314.

Kim CH, You L, Yellowley CE, Jacobs CR. 2006. Oscillatory fluid flow-induced shear stress decreases osteoclastogenesis through RANKL and OPG signaling. *Bone* **39:** 1043–1047.

Kim JB, Leucht P, Lam K, Luppen C, Ten Berge D, Nusse R, Helms JA. 2007. Bone regeneration is regulated by wnt signaling. *J Bone Miner Res* **22:** 1913–1923.

Kokubu C, Heinzmann U, Kokubu T, Sakai N, Kubota T, Kawai M, Wahl MB, Galceran J, Grosschedl R, Ozono K, et al. 2004. Skeletal defects in *ringelschwanz*mutant mice reveal that Lrp6 is required for proper somitogenesis and osteogenesis. *Development* **131:** 5469–5480.

Koyama E, Shibukawa Y, Nagayama M, Sugito H, Young B, Yuasa T, Okabe T, Ochiai T, Kamiya N, Rountree RB, et al. 2008. A distinct cohort of progenitor cells participates in synovial joint and articular cartilage formation during mouse limb skeletogenesis. *Dev Biol* **316:** 62–73.

Kramer I, Halleux C, Keller H, Pegurri M, Gooi JH, Weber PB, Feng JQ, Bonewald LF, Kneissel M. 2010. Osteocyte Wnt/β-catenin signaling is required for normal bone homeostasis. *Mol Cell Biol* **30:** 3071–3085.

Kugimiya F, Kawaguchi H, Ohba S, Kawamura N, Hirata M, Chikuda H, Azuma Y, Woodgett JR, Nakamura K, Chung UI. 2007. GSK-3β controls osteogenesis through regulating Runx2 activity. *PLoS ONE* **2:** e837.

Lammi L, Arte S, Somer M, Jarvinen H, Lahermo P, Thesleff I, Pirinen S, Nieminen P. 2004. Mutations in *AXIN2* cause familial tooth agenesis and predispose to colorectal cancer. *Am J Hum Genet* **74:** 1043–1050.

Lau KH, Kapur S, Kesavan C, Baylink DJ. 2006. Up-regulation of the Wnt, estrogen receptor, insulin-like growth factor-I, and bone morphogenetic protein pathways in C57BL/6J osteoblasts as opposed to C3H/HeJ osteoblasts in part contributes to the differential anabolic response to fluid shear. *J Biol Chem* **281:** 9576–9588.

Li X, Zhang Y, Kang H, Liu W, Liu P, Zhang J, Harris SE, Wu D. 2005. Sclerostin binds to LRP5/6 and antagonizes canonical Wnt signaling. *J Biol Chem* **280:** 19883–19887.

Li J, Sarosi I, Cattley RC, Pretorius J, Asuncion F, Grisanti M, Morony S, Adamu S, Geng Z, Qiu W, et al. 2006. Dkk1-mediated inhibition of Wnt signaling in bone results in osteopenia. *Bone* **39:** 754–766.

Li X, Ominsky MS, Niu QT, Sun N, Daugherty B, D'Agostin D, Kurahara C, Gao Y, Cao J, Gong J, et al. 2008. Targeted deletion of the *sclerostin*gene in mice results in increased bone formation and bone strength. *J Bone Miner Res* **23:** 860–869.

Li X, Ominsky MS, Warmington KS, Morony S, Gong J, Cao J, Gao Y, Shalhoub V, Tipton B, Haldankar R, et al. 2009. Sclerostin antibody treatment increases bone formation, bone mass, and bone strength in a rat model of postmenopausal osteoporosis. *J Bone Miner Res* **24:** 578–588.

Lin C, Jiang X, Dai Z, Guo X, Weng T, Wang J, Li Y, Feng G, Gao X, He L. 2009. Sclerostin mediates bone response to mechanical unloading through antagonizing Wnt/β-catenin signaling. *J Bone Miner Res* **24:** 1651–1661.

Little RD, Carulli JP, Del Mastro RG, Dupuis J, Osborne M, Folz C, Manning SP, Swain PM, Zhao SC, Eustace B, et al. 2002. A mutation in the LDL receptor-related protein 5 gene results in the autosomal dominant high-bone-mass trait. *Am J Hum Genet* **70:** 11–19.

Liu B, Yu HM, Hsu W. 2007. Craniosynostosis caused by Axin2 deficiency is mediated through distinct functions of β-catenin in proliferation and differentiation. *Dev Biol* **301:** 298–308.

Liu N, Shi S, Deng M, Tang L, Zhang G, Liu N, Ding B, Liu W, Liu Y, Shi H, et al. 2011. High levels of β-catenin signaling reduce osteogenic differentiation of stem cells in inflammatory microenvironments through inhibition of the noncanonical Wnt pathway. *J Bone Miner Res* **26:** 2082–2095.

Lorenowicz MJ, Korswagen HC. 2009. Sailing with the Wnt: Charting the Wnt processing and secretion route. *Exp Cell Res* **315:** 2683–2689.

Lories RJ, Peeters J, Bakker A, Tylzanowski P, Derese I, Schrooten J, Thomas JT, Luyten FP. 2007. Articular cartilage and biomechanical properties of the long bones in Frzb-knockout mice. *Arthritis Rheum* **56:** 4095–4103.

Loughlin J, Dowling B, Chapman K, Marcelline L, Mustafa Z, Southam L, Ferreira A, Ciesielski C, Carson DA, Corr M. 2004. Functional variants within the secreted Frizzled-related protein 3 gene are associated with hip osteoarthritis in females. *Proc Natl Acad Sci* **101:** 9757–9762.

MacDonald BT, Joiner DM, Oyserman SM, Sharma P, Goldstein SA, He X, Hauschka PV. 2007. Bone mass is inversely proportional to Dkk1 levels in mice. *Bone* **41:** 331–339.

Mak KK, Chen MH, Day TF, Chuang PT, Yang Y. 2006. Wnt/β-catenin signaling interacts differentially with Ihh signaling in controlling endochondral bone and synovial joint formation. *Development* **133:** 3695–3707.

Mani A, Radhakrishnan J, Wang H, Mani MA, Nelson-Williams C, Carew KS, Mane S, Najmabadi H, Wu D, Lifton RP. 2007. LRP6 mutation in a family with early coronary disease and metabolic risk factors. *Science* **315:** 1278–1282.

Marotti G, Ferretti M, Muglia MA, Palumbo C, Palazzini S. 1992. A quantitative evaluation of osteoblast–osteocyte relationships on growing endosteal surface of rabbit tibiae. *Bone* **13:** 363–368.

Martin RB. 2000. Does osteocyte formation cause the nonlinear refilling of osteons? *Bone* **26:** 71–78.

Metz LN, Martin RB, Turner AS. 2003. Histomorphometric analysis of the effects of osteocyte density on osteonal morphology and remodeling. *Bone* **33:** 753–759.

Minear S, Leucht P, Jiang J, Liu B, Zeng A, Fuerer C, Nusse R, Helms Ja. 2010. Wnt proteins promote bone regeneration. *Sci Transl Med* **2:** 29ra30.

Modarresi R, Xiang Z, Yin M, Laurence J. 2009. WNT/β-catenin signaling is involved in regulation of osteoclast differentiation by human immunodeficiency virus protease inhibitor ritonavir: Relationship to human immunodeficiency virus-linked bone mineral loss. *Am J Pathol* **174:**123–135.

Morvan F, Boulukos K, Clement-Lacroix P, Roman Roman S, Suc-Royer I, Vayssiere B, Ammann P, Martin P, Pinho S, Pognonec P, et al. 2006. Deletion of a single allele of the Dkk1 gene leads to an increase in bone formation and bone mass. *J Bone Miner Res* **21:** 934–945.

Mukhopadhyay M, Shtrom S, Rodriguez-Esteban C, Chen L, Tsukui T, Gomer L, Dorward DW, Glinka A, Grinberg A, Huang SP, et al. 2001. Dickkopf1 is required for embryonic head induction and limb morphogenesis in the mouse. *Dev Cell* **1:** 423–434.

Niemann S, Zhao C, Pascu F, Stahl U, Aulepp U, Niswander L, Weber JL, Muller U. 2004. Homozygous WNT3 mutation causes tetra-amelia in a large consanguineous family. *Am J Hum Genet* **74:** 558–563.

Noh T, Gabet Y, Cogan J, Shi Y, Tank A, Sasaki T, Criswell B, Dixon A, Lee C, Tam J, et al. 2009. *Lef1* haploinsufficient mice display a low turnover and low bone mass phenotype in a gender- and age-specific manner. *PLoS ONE* **4:** e5438.

Pead MJ, Suswillo R, Skerry TM, Vedi S, Lanyon LE. 1988. Increased ^{3}H-uridine levels in osteocytes following a single short period of dynamic bone loading in vivo. *Calcif Tissue Int* **43:** 92–96.

Poole KE, van Bezooijen RL, Loveridge N, Hamersma H, Papapoulos SE, Lowik CW, Reeve J. 2005. Sclerostin is a delayed secreted product of osteocytes that inhibits bone formation. *FASEB J* **19:** 1842–1844.

Regard JB, Cherman N, Palmer D, Kuznetsov SA, Celi FS, Guettier JM, Chen M, Bhattacharyya N, Wess J, Coughlin SR, et al. 2011. Wnt/β-catenin signaling is differentially regulated by Gα proteins and contributes to fibrous dysplasia. *Proc Natl Acad Sci* **108:** 20101–20106.

Reich KM, Gay CV, Frangos JA. 1990. Fluid shear stress as a mediator of osteoblast cyclic adenosine monophosphate production. *J Cell Physiol* **143:** 100–104.

Reinhold MI, Kapadia RM, Liao Z, Naski MC. 2006. The Wnt-inducible transcription factor Twist1 inhibits chondrogenesis. *J Biol Chem* **281:** 1381–1388.

Rey JP, Ellies DL. 2010. Wnt modulators in the biotech pipeline. *Dev Dyn* **239:** 102–114.

Robinson JA, Chatterjee-Kishore M, Yaworsky PJ, Cullen DM, Zhao W, Li C, Kharode Y, Sauter L, Babij P, Brown EL, et al. 2006. Wnt/β-catenin signaling is a normal physiological response to mechanical loading in bone. *J Biol Chem* **281:** 31720–31728.

Robling AG, Niziolek PJ, Baldridge La, Condon KW, Allen MR, Alam I, Mantila SM, Gluhak-Heinrich J, Bellido TM, Harris SE, et al. 2008. Mechanical stimulation of bone in vivo reduces osteocyte expression of Sost/sclerostin. *J Biol Chem* **283:** 5866–5875.

Rodda SJ, McMahon AP. 2006. Distinct roles for Hedgehog and canonical Wnt signaling in specification, differentiation and maintenance of osteoblast progenitors. *Development* **133:** 3231–3244.

Rudnicki JA, Brown AM. 1997. Inhibition of chondrogenesis by Wnt gene expression in vivo and in vitro. *Dev Biol* **185:** 104–118.

Saarinen A, Valimaki VV, Valimaki MJ, Loyttyniemi E, Auro K, Uusen P, Kuris M, Lehesjoki AE, Makitie O. 2007. The A1330V polymorphism of the low-density lipoprotein receptor-related protein 5 gene (LRP5) associates with low peak bone mass in young healthy men. *Bone* **40:** 1006–1012.

Sawakami K, Robling AG, Ai M, Pitner ND, Liu D, Warden SJ, Li J, Maye P, Rowe DW, Duncan RL, et al. 2006. The Wnt co-receptor LRP5 is essential for skeletal mechanotransduction but not for the anabolic bone response to parathyroid hormone treatment. *J Biol Chem* **281:** 23698–23711.

Saxon LK, Jackson BF, Sugiyama T, Lanyon LE, Price JS. 2011. Analysis of multiple bone responses to graded strains above functional levels, and to disuse, in mice in vivo show that the human Lrp5 G171V High Bone Mass mutation increases the osteogenic response to loading but that lack of Lrp5 activity reduces it. *Bone* **49:**184–193.

Schulze J, Seitz S, Saito H, Schneebauer M, Marshall RP, Baranowsky A, Busse B, Schilling AF, Friedrich FW, Albers J, et al. 2010. Negative regulation of bone formation by the transmembrane Wnt antagonist Kremen-2. *PLoS ONE* **5:** e10309.

Semenov M, Tamai K, He X. 2005. SOST is a ligand for LRP5/LRP6 and a Wnt signaling inhibitor. *J Biol Chem* **280:** 26770–26775.

Sims AM, Shephard N, Carter K, Doan T, Dowling A, Duncan EL, Eisman J, Jones G, Nicholson G, Prince R, et al. 2008. Genetic analyses in a sample of individuals with high or low BMD shows association with multiple Wnt pathway genes. *J Bone Miner Res* **23:** 499–506.

Spater D, Hill TP, Gruber M, Hartmann C. 2006. Role of canonical Wnt-signalling in joint formation. *Eur Cell Mater* **12:** 71–80.

Stefater JA III, Lewkowich I, Rao S, Mariggi G, Carpenter AC, Burr AR, Fan J, Ajima R, Molkentin JD, Williams BO, et al. 2011. Regulation of angiogenesis by a non-canonical Wnt–Flt1 pathway in myeloid cells. *Nature* **474:** 511–515.

Tatsumi S, Ishii K, Amizuka N, Li M, Kobayashi T, Kohno K, Ito M, Takeshita S, Ikeda K. 2007. Targeted ablation of osteocytes induces osteoporosis with defective mechanotransduction. *Cell Metab* **5:** 464–475.

Thudi NK, Martin CK, Murahari S, Shu ST, Lanigan LG, Werbeck JL, Keller ET, McCauley LK, Pinzone JJ, Rosol TJ. 2010. Dickkopf-1 (DKK-1) stimulated prostate cancer growth and metastasis and inhibited bone formation in osteoblastic bone metastases. *Prostate* **71:** 615–625.

Tian E, Zhan F, Walker R, Rasmussen E, Ma Y, Barlogie B, Shaughnessy JD Jr. 2003. The role of the Wnt-signaling antagonist DKK1 in the development of osteolytic lesions in multiple myeloma. *N Engl J Med* **349:** 2483–2494.

Tu X, Joeng KS, Nakayama KI, Nakayama K, Rajagopal J, Carroll TJ, McMahon AP, Long F. 2007. Noncanonical Wnt signaling through G protein-linked PKCδ activation promotes bone formation. *Dev Cell* **12:** 113–127.

Uitterlinden AG, Arp PP, Paeper BW, Charmley P, Proll S, Rivadeneira F, Fang Y, van Meurs JB, Britschgi TB, Latham JA, et al. 2004. Polymorphisms in the sclerosteosis/van Buchem disease gene (SOST) region are associated with bone-mineral density in elderly whites. *Am J Hum Genet* **75:** 1032–1045.

van Bezooijen RL, Roelen BA, Visser A, van der Wee-Pals L, de Wilt E, Karperien M, Hamersma H, Papapoulos SE, ten Dijke P, Lowik CW. 2004. Sclerostin is an osteocyte-expressed negative regulator of bone formation, but not a classical BMP antagonist. *J Exp Med* **199:** 805–814.

van Meurs JB, Rivadeneira F, Jhamai M, Hugens W, Hofman A, van Leeuwen JP, Pols HA, Uitterlinden AG. 2006. Common genetic variation of the low-density lipoprotein receptor-related protein 5 and 6 genes determines fracture risk in elderly white men. *J Bone Miner Res* **21:** 141–150.

Van Wesenbeeck L, Cleiren E, Gram J, Beals RK, Benichou O, Scopelliti D, Key L, Renton T, Bartels C, Gong Y, et al. 2003. Six novel missense mutations in the LDL receptor-related protein 5 (LRP5) gene in different conditions with an increased bone density. *Am J Hum Genet* **72:** 763–771.

Vestergaard P, Rejnmark L, Mosekilde L. 2005. Reduced relative risk of fractures among users of lithium. *Calcif Tissue Int* **77:** 1–8.

Voorzanger-Rousselot N, Goehrig D, Journe F, Doriath V, Body JJ, Clezardin P, Garnero P. 2007. Increased Dickkopf-1 expression in breast cancer bone metastases. *Br J Cancer* **97:** 964–970.

Wei W, Zeve D, Suh JM, Wang X, Du Y, Zerwekh JE, Dechow PC, Graff JM, Wan Y. 2011. Biphasic and dosage-dependent regulation of osteoclastogenesis by β-catenin. *Mol Cell Biol* **31:** 4706–4719.

* Whyte JL, Smith AA, Helms JA. 2012. Wnt signaling and injury repair. *Cold Spring Harb Perspect Biol* doi: 10.1101/cshperspect.a008078.

Williams JL, Iannotti JP, Ham A, Bleuit J, Chen JH. 1994. Effects of fluid shear stress on bone cells. *Biorheology* **31:** 163–170.

Wilting I, de Vries F, Thio BM, Cooper C, Heerdink ER, Leufkens HG, Nolen WA, Egberts AC, van Staa TP. 2007. Lithium use and the risk of fractures. *Bone* **40:** 1252–1258.

Winkler DG, Sutherland MK, Geoghegan JC, Yu C, Hayes T, Skonier JE, Shpektor D, Jonas M, Kovacevich BR, Staehling-Hampton K, et al. 2003. Osteocyte control of bone formation via sclerostin, a novel BMP antagonist. *EMBO J* **22:** 6267–6276.

Yaccoby S, Ling W, Zhan F, Walker R, Barlogie B, Shaughnessy JD Jr. 2007. Antibody-based inhibition of DKK1 suppresses tumor-induced bone resorption and multiple myeloma growth in vivo. *Blood* **109:** 2106–2111.

Yadav VK, Ryu JH, Suda N, Tanaka KF, Gingrich JA, Schutz G, Glorieux FH, Chiang CY, Zajac JD, Insogna KL, et al. 2008. Lrp5 controls bone formation by inhibiting serotonin synthesis in the duodenum. *Cell* **135:** 825–837.

You L, Temiyasathit S, Lee P, Kim CH, Tummala P, Yao W, Kingery W, Malone AM, Kwon RY, Jacobs CR. 2008. Osteocytes as mechanosensors in the inhibition of bone resorption due to mechanical loading. *Bone* **42:** 172–179.

Yu HM, Jerchow B, Sheu TJ, Liu B, Costantini F, Puzas JE, Birchmeier W, Hsu W. 2005. The role of Axin2 in calvarial morphogenesis and craniosynostosis. *Development* **132:** 1995–2005.

Zhang S, Cagatay T, Amanai M, Zhang M, Kline J, Castrillon DH, Ashfaq R, Oz OK, Wharton KA Jr. 2007. Viable mice with compound mutations in the Wnt/Dvl pathway antagonists nkd1 and nkd2. *Mol Cell Biol* **27:** 4454–4464.

Zhao S, Zhang YK, Harris S, Ahuja SS, Bonewald LF. 2002. MLO-Y4 osteocyte-like cells support osteoclast formation and activation. *J Bone Miner Res* **17:** 2068–2079.

Zhu M, Tang D, Wu Q, Hao S, Chen M, Xie C, Rosier RN, O'Keefe RJ, Zuscik M, Chen D. 2009. Activation of β-catenin signaling in articular chondrocytes leads to osteoarthritis-like phenotype in adult β-catenin conditional activation mice. *J Bone Miner Res* **24:** 12–21.

Zylstra CR, Wan C, VanKoevering KK, Sanders AK, Lindvall C, Clemens TL, Williams BO. 2008. Gene targeting approaches in mice: Assessing the roles of LRP5 and LRP6 in osteoblasts. *J Musculoskelet Neuronal Interact* **8:** 291–293.

Wnt Signaling in the Vertebrate Central Nervous System: From Axon Guidance to Synaptic Function

Patricia C. Salinas

Department of Cell and Developmental Biology, University College London, London WC1E 6BT, United Kingdom

Correspondence: p.salinas@ucl.ac.uk

Regulation of cell signaling by Wnt proteins is critical for the formation of neuronal circuits. Wnts modulate axon pathfinding, dendritic development, and synaptic assembly. Through different receptors, Wnts activate diverse signaling pathways that lead to local changes on the cytoskeleton or global cellular changes involving nuclear function. Recently, a link between neuronal activity, essential for the formation and refinement of neuronal connections, and Wnt signaling has been uncovered. Indeed, neuronal activity regulates the release of Wnt and the localization of their receptors. Wnts mediate synaptic structural changes induced by neuronal activity or experience. New emerging evidence suggests that dysfunction in Wnt signaling contributes to neurological disorders. In this article, the attention is focused on the function of Wnt signaling in the formation of neuronal circuits in the vertebrate central nervous system.

The formation of neuronal connections requires the navigation of axons to their appropriate synaptic targets, the formation of terminal branches, and the assembly of functional synapses. These processes greatly depend on the proper dialogue between axons and their environment as they navigate to their target, and between axons and their postsynaptic dendrites during synapse assembly. A combination of secreted molecules and transmembrane proteins modulates these processes. Studies over the last 10 years have revealed an essential role for Wnt signaling in axon pathfinding, dendritic development, and synapse assembly in both central and peripheral nervous systems. Wnts also modulate basal synaptic transmission and the structural and functional plasticity of synapses in the central nervous system. Studies of Wnts in the nervous system have significantly contributed to our current understanding of the molecular mechanisms that control neuronal circuit assembly. These studies have also shed light into fundamental aspects of cell signaling such as novel mechanisms of protein secretion (Korkut et al. 2009) and receptor dynamics (Sahores et al. 2010). Here I review the mechanisms by which Wnts modulate axon guidance and synapse formation in the vertebrate central nervous system. I also discuss the increasing evidence in support for a role of Wnts in basal

Cite this article as *Cold Spring Harb Perspect Biol* doi: 10.1101/cshperspect.a008003

synaptic transmission, synaptic plasticity, and neurological disorders.

AXON NAVIGATION AND TERMINAL REMODELING

Wnt signaling controls neuronal polarity, promotes axon outgrowth, and regulates the navigation of axons to their final synaptic targets. The first clear evidence that Wnts regulate axon guidance came from studies using loss-of-function approaches in different development systems including vertebrates (Salinas and Zou 2008; Sanchez-Camacho and Bovolenta 2009; Fradkin et al. 2010). Wnts induce a range of responses including repulsion, attraction, and promoting axon branching.

Axon Guidance

In the vertebrate nervous system, Wnts also regulate the direction of axon growth and guidance. In the spinal cord, commissural axons turn anteriorly after crossing the midline following a posterior-to-anterior gradient of attractive cues. Indeed, Wnt4 and Wnt7b form a gradient (anterior high–posterior low) in the floor plate. Gain of function of Wnt4 using cultured open-book explants from the spinal cord reveals that Wnt4 is an attractive cue for commissural axons after midline crossing (Fig. 1) (Lyuksyutova et al. 2003). This response seems to be mediated by the Frizzled-3 (Fz3) receptor, because *Fz3* mutant mice show anterior–posterior guidance defects after commissural axons have crossed the midline (Lyuksyutova et al. 2003).

In contrast to the attractive activity of Wnt4, other Wnts function as repulsive signals. Wnt1 and Wnt5a form a high-anterior–to–low-posterior gradient along the dorsal spinal cord and repel corticospinal axons that descend along the spinal cord (Liu et al. 2005). During the initial part of their trajectory, these axons are insensitive to the repulsive activity of Wnts owing to the lack of expression of the Ryk receptor. After crossing the midline and entering the rostral part of the spinal cord, these axons begin to express Ryk. Importantly, injection of antibodies against Ryk affects the posterior growth of corticospinal axons (Liu et al. 2005). Together, these results strongly suggest that two opposing gradients guide axons in the spinal cord. Early in development, a ventral gradient of Wnt4 attracts commissural axons in a posterior-to-anterior direction. Postnatally, corticospinal axons, which descend along the dorsal funiculus, experience a repulsing gradient of Wnt1 and Wnt5a as they grow down the spinal cord (Fig. 1C).

How does Wnt5a regulate the behavior of commissural axons? Wnt5a activates a non-canonical Wnt signaling pathway that requires Fz3 and the four-pass transmembrane protein Van Gogh/strabismus (Vang) and Jun-N-terminal kinase (JNK) to promote axon outgrowth (Shafer et al. 2011). Analyses in cell lines led to the surprising finding that the core scaffold protein Dishevelled-1 (Dvl1) antagonizes Vang function. Indeed, expression of Dvl1 decreases the ability of Wnt5a to activate JNK by increasing Fz3 phosphorylation and decreasing its internalization. Because receptor internalization is crucial for signaling, Dvl1 blocks Wnt5a signaling. In contrast, expression of Vang2 decreases Fz3 phosphorylation and promotes its internalization (Shafer et al. 2011). Thus, Vang2 and Dvl1 induce opposing effects on Fz3 localization. Although the effect of Dvl1 on the behavior of commissural axons has not been examined, the localization of Vang2 and Fz3 suggests that Vang2 induces the internalization of Fz3 and activation of the PCP pathway by antagonizing Dvl1 function. It would be of interest to know whether Dvl1 also plays a negative role in Wnt signaling in other cellular processes.

Activation of a non-canonical pathway through the Ryk receptor regulates axon guidance. In the corpus callosum, Ryk mutant cortical axons are able to cross the midline, but they show aberrant trajectories on the contralateral side. Wnt5a, which is expressed around the area of the corpus callosum, repels Ryk-expressing axons in cortical explants (Fig. 1A,B) (Keeble et al. 2006). Knockdown of Ryk decreases calcium transients and increases misrouting of post-crossing axons (Hutchins et al. 2011). Interestingly, blockade of IP3 receptors and TRP channels reduces the rate of outgrowth, but only inhibition of TRP induces guidance

 Cite this article as *Cold Spring Harb Perspect Biol* doi: 10.1101/cshperspect.a008003

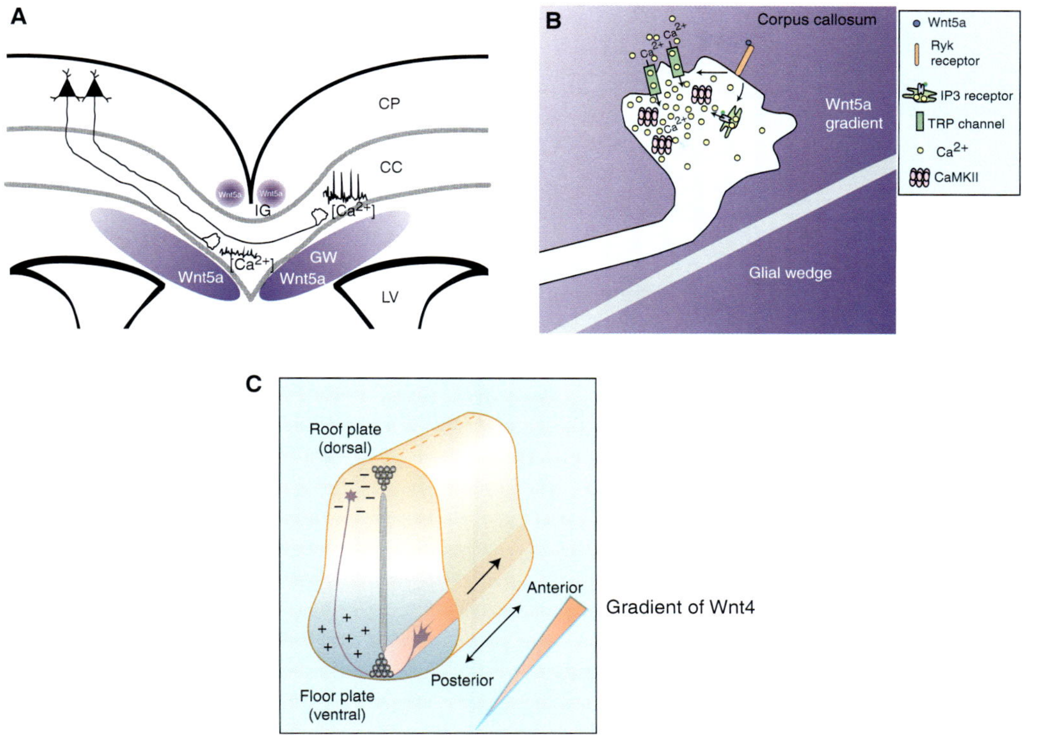

Figure 1. Gradients of Wnts regulate axon guidance in the spinal cord. (*A*,*B*) In the cortex, a repulsive Wnt5 expressed in indusium griseum (IG) and in the glial wedge (GW) guides cortical axons through the corpus callosum (CC). Pyramidal cortical axons project to the contralateral side of the cortex by crossing the corpus callosum. After crossing to the contralateral side, these axons are repelled by Wnt5a, possibly by the expression of Ryk receptor in post-crossing axons. Wnt5a through Ryk induces the opening of IP3 and TRP channels to transiently activate CaMKII, resulting in increased axon outgrowth and repulsion. (*C*) A gradient of Wnt4 in the floor plate controls the direction of commissural axons. After midline crossing, commissural axons turn in an anterior direction following the attracting gradient of Wnt4. (Panels *A* and *B* are from Hutchins and Kalil 2011; reprinted, with permission, from John Wiley & Sons © 2011. Panel *C* is from Imondi and Thomas 2003; reprinted, with permission, from the author.)

defects. Moreover, expression of a specific CaMKII inhibitory protein induces outgrowth and pathfinding defects (Hutchins et al. 2011). These results are consistent with previous studies showing that Wnt5a/Ryk signaling through CaMKII regulates axon outgrowth and turning of dissociated cortical neurons (Li et al. 2009). Together, these results show that activation of Ryk receptors by different Wnts can trigger different axon guidance responses.

Although a role for Wnts in axon guidance has been shown in few brain areas, the broad but specific expression of different Wnts and their receptors in the nervous system suggests that these molecules play a general role in axon guidance. Studies addressing the contribution of Wnts together with other axon guidance molecules will be invaluable in understanding how axon behavior is controlled in normal and regenerative conditions.

Formation of Topographical Maps

Wnts also contribute to the formation of topographical maps in the brain. In the visual system, retinal ganglion cells (RGCs), organized in a two-dimensional (2D) layer, project their axons through the optic nerve into the brain in an orderly manner. In the chick, axons from the dorsal retina project to the lateral part of the tectum, whereas axons from the ventral retina project to the more medial tectum. This retinotopic map permits an object projected into the retina to be properly represented in the tectum. Molecules such as Eph and Ephrin guide RGC axons to their appropriate terminal zone in the tectum (McLaughlin and O'Leary 2005). EphrinA molecules have been implicated in the formation of an anterior–posterior topographic map through their repulsive activity. In addition, a medial-to-lateral gradient of EphB in the retina with a medial–lateral gradient of EphrinB1 in the tectum controls patterning through attraction (Hindges et al. 2002). Theoretical models propose that molecular gradients with opposing activities are required for the formation of this topographical map. Indeed, Wnts contribute to the formation of this retinotopic map. In the tectum, Wnt3 is expressed in a gradient from medial high to lateral low as observed for EphrinB1 (Fig. 2). In the retina, the Wnt3 receptor Ryk is expressed in a gradient from ventral high to dorsal low following the same pattern of EphB (Schmitt et al. 2006). In contrast to the attraction elicited by EphB and EphrinB1, Wnt3–Ryk induces repulsion (Schmitt et al.

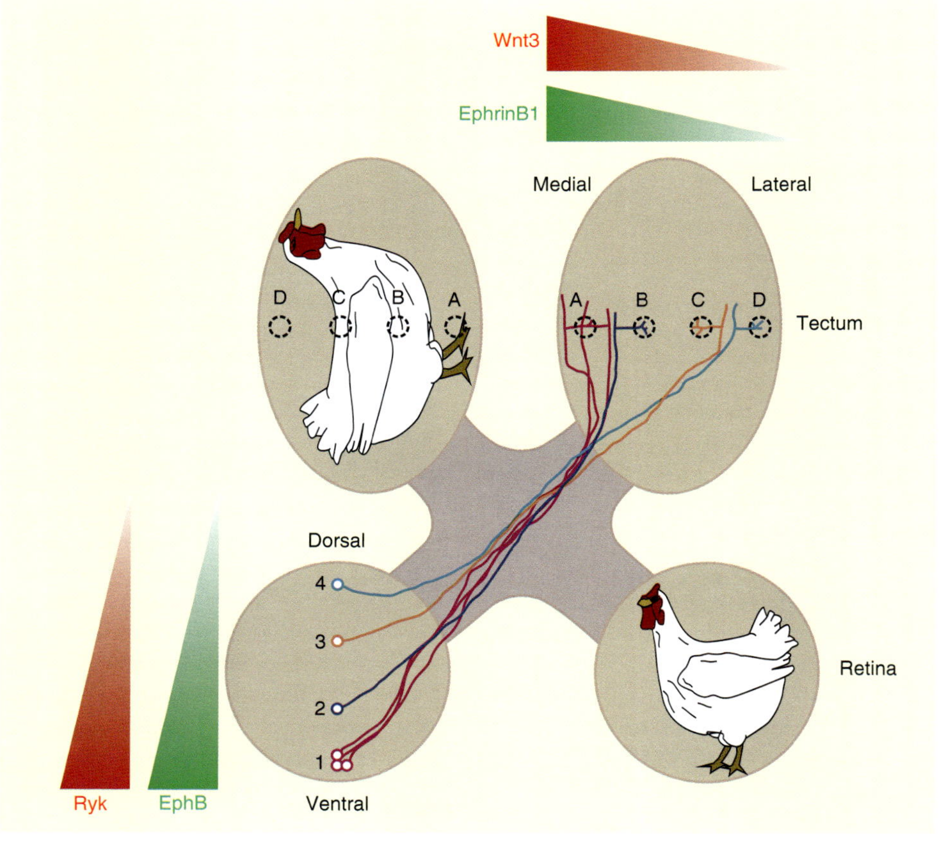

Figure 2. Wnt3–Ryk signaling controls the formation of topographical maps. In the retina, retinal ganglia cell axons express the Ryk receptor in a gradient from ventral high to dorsal low in the same manner as EphB. These axons project to the contralateral side and into the tectum whether they follow two opposing forces. Both Wnt3 and EphrinB1 are expressed in a gradient from medial high to lateral low. EphrinB–EphB1 forms an attractive medial-directing activity. In contrast, Wnt3–Ryk forms an opposing lateral-directing activity. (Figure is from Luo 2006; reprinted, with permission, from Nature © 2006.)

Cite this article as *Cold Spring Harb Perspect Biol* doi: 10.1101/cshperspect.a008003

2006). Thus, an attracting gradient of EphB–EphrinB1 and an opposing repulsing gradient of Ryk–Wnt3 regulate the formation of the retinotopic map. The graded distribution of Wnts and their receptors in different brain regions strongly suggests that Wnts will also contribute to the formation of topographical maps in other sensory systems such as somatosensory and auditory systems.

Axon Terminal Remodeling

Wnt signaling promotes the terminal remodeling of axons before synapses begin to assemble. In cerebellar mossy fiber axons and DRG neurons, activation of the canonical Wnt pathway induces growth cone pausing, axon branching, and extensive enlargement of growth cones (Hall et al. 2000; Krylova et al. 2002). This behavior is distinct from the typical attractive or repulsive responses induced by axon guidance molecules. In contrast, this terminal remodeling is consistent with the response of axons as they arrive at their target fields and begin to form synapses. Studies in cultured neurons and at the *Drosophila* NMJ have shown that Wnt signaling through Dvl inhibits Gsk3 by a divergent canonical Wnt pathway that does not require transcription (Packard et al. 2002; Miech et al. 2008; Purro et al. 2008). In fact, terminal remodeling is achieved through profound changes on the axonal cytoskeleton. Wnt increases microtubule stability and induces the formation of microtubule loops within the growth cone, which contribute to growth cone pausing and enlargement (Ciani et al. 2004). Time-lapse analyses of the behavior of microtubule plus ends labeled with EB3-GFP reveal that Wnt3a induces the loss of microtubule growth direction, resulting in the formation of looped microtubules within the growth cone (Purro et al. 2008). This profound change in microtubule organization leads to growth cone pausing and enlargement. Two findings support a role for APC, a Wnt canonical component and a microtubule binding protein, in mediating growth cone remodeling. First, Wnt3a decreases the level of endogenous APC at the plus end of microtubules. Second, ShRNA knockdown of APC induces microtubule looping and axon remodeling, thus mimicking the effect of Wnt3a (Purro et al. 2008). Thus, Wnts regulate the organization of microtubules resulting in the terminal remodeling of growth cones, a crucial process in the conversion of actively growing growth cones into synaptic boutons.

SYNAPSE FORMATION

Synapse formation requires the proper apposition between the presynaptic terminal, containing the machinery necessary for neurotransmitter release, and the membrane of the postsynaptic cells containing neurotransmitter receptors and signaling molecules. Time-lapse studies have shown that presynaptic release sites form first, and they are followed by the accumulation of postsynaptic markers (Okabe et al. 2001). Although presynaptic release sites can assemble without clear evidence of postsynaptic assembly, these "orphan" sites are not stable. Time-lapse studies of dendritic spines, mushroom-like protrusions that receive excitatory inputs, revealed that these postsynaptic structures are very dynamic (Trachtenberg et al. 2002; Dunaevsky and Mason 2003), but their stability depends on their apposition to presynaptic boutons (Yoshihara et al. 2009). Thus, synaptic assembly requires the proper dialogue between the presynaptic terminal or bouton and its postsynaptic target cells. Several studies now show that bidirectional Wnt signaling contributes to this *trans*-synaptic dialogue.

Presynaptic Assembly

Studies using cultured neurons and analyses of *Wnt*-null mutant mice have led to the discovery that Wnts promote synapse formation. Wnt7a is expressed in cerebellar granule cells as they begin to contact their presynaptic mossy fiber axons (Hall et al. 2000). The mossy fiber–granule cell synapse is characterized by the interdigitation of several granule cell dendrites into a single mossy fiber axon, resulting in the formation of a complex synaptic structure, the glomerular rosette (Hamori and Somogyi 1983). A combination of gain-of-function studies using mossy fiber explants and analyses of *Wnt7a* mutant mice showed that Wnt7a, released by granule cells, acts on mossy fiber axons to promote the formation of complex glomerular rosettes and the accumulation of synaptic components to future presynaptic sites (Hall et al. 2000). Consistent with its role in presynaptic assembly, Wnt7a promotes synaptic vesicle recycling (Ahmad-Annuar et al. 2006; Cerpa et al. 2008). Together these findings show that Wnt7a promotes the assembly of synaptic boutons.

How does Wnt signaling regulate synaptic assembly? Different receptors mediate the synaptogenic activity of Wnts in hippocampal neurons (Davis et al. 2008; Varela-Nallar et al. 2009; Sahores et al. 2010). Gain of function of Fz1 or Fz5 induces the clustering of synaptic vesicle proteins and active-zone proteins. In contrast, exposure to the soluble CRD domain of Fz1 or Fz5 or shRNA knockdown of Fz5 shows that these receptors are required for presynaptic assembly (Varela-Nallar et al. 2009; Sahores et al. 2010). Moreover, silencing of Fz5 blocks the ability of Wnt7a to promote presynaptic assembly, indicating that this receptor is required for Wnt7a function (Sahores et al. 2010). Other Wnt receptors also regulate the formation of synapses. For example, Wnt5a binds to and requires Ror tyrosine receptors to promote presynaptic assembly in cultured hippocampal neurons (Paganoni et al. 2010). Although the in vivo role of Frizzled and Ror receptors remains to be established, these studies show that in hippocampal cells three different receptors—Fz1, Fz5, and Ror—are involved in presynaptic assembly. These findings pose the question as to whether these molecules act as receptors for different Wnt ligands or whether they function in different subcellular compartments in hippocampal neurons.

Wnt signaling triggers a transcription-independent pathway to promote bouton assembly. In cultured hippocampal neurons, Wnt7a activates a signaling cascade that requires Dishevelled-1 (Dvl1) and inhibition of Gsk3β for presynaptic assembly (Fig. 3). Dvl1 is present at

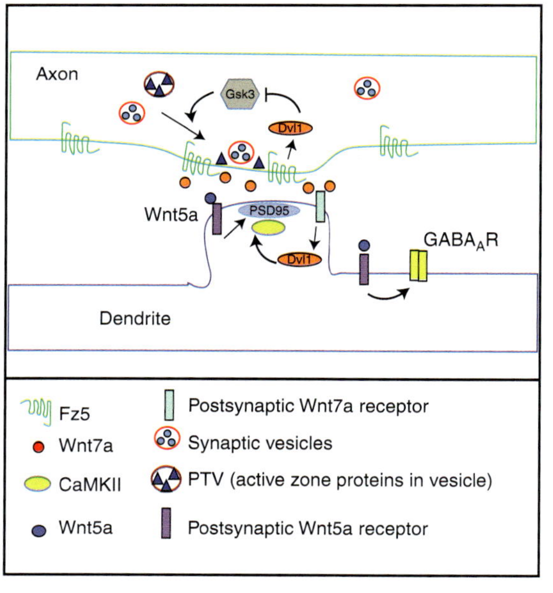

Figure 3. Wnts promote synapse formation. On axons, binding of Wnt7a to presynaptic Fz5 leads to Dvl activation, resulting in the inhibition of GSK3β. This, in turn, promotes the recruitment of synaptic vesicles and active-zone proteins to future synaptic sites. On the postsynaptic dendrite, Wnt5a triggers the clustering of PSD95 and promotes the formation of excitatory synapses. Wnt5a also increases the clustering of GABA$_A$R at inhibitory synapses.

 Cite this article as *Cold Spring Harb Perspect Biol* doi: 10.1101/cshperspect.a008003

both sides of the synapse, suggesting that Wnt signaling might act presynaptically and postsynaptically (Ahmad-Annuar et al. 2006; Ciani et al. 2011). In axons, gain of function of Dvl1 increases the number of neurotransmitter release sites and the clustering of presynaptic proteins. Conversely, axons from *Dvl1* mutant neurons, which show fewer presynaptic sites, partially respond to exogenous Wnt7a, showing that Dvl1 is required by Wnt7a to promote presynaptic assembly (Ahmad-Annuar et al. 2006). The use of pharmacological inhibitors (Hall et al. 2000) and specific antagonists (Davis et al. 2008) of the canonical Wnt signaling pathway shows that axon remodeling and presynaptic assembly are achieved through inhibition of Gsk3. Similarly to the fly (Miech et al. 2008), blockade of transcription does not suppress the ability of Wnt7a to promote presynaptic assembly in cultured hippocampal neurons (EM Dickins and PC Salinas, unpubl.). How, then, do Wnts promote presynaptic assembly? A possible hypothesis is that Wnt signaling might promote the formation of synaptic boutons by modulating the microtubule cytoskeleton, resulting in changes in the transport of organelles or synaptic components to future synaptic sites. Future analyses using time-lapse recordings of fluorescently labeled synaptic molecules will provide important insights into the mechanism of Wnt-mediated synaptic assembly.

Postsynaptic Assembly

Studies at the *Drosophila* neuromuscular junction (NMJ) have clearly shown that Wg promotes the recruitment of both presynaptic and postsynaptic components (Packard et al. 2002). At vertebrate central synapses, Wnts also promote the recruitment of postsynaptic components. For example, Wnt5a increases the clustering of the postsynaptic scaffold protein PSD95 (Farias et al. 2009) and postsynaptic γ-aminobutyric acid type a ($GABA_A$) receptors (Cuitino et al. 2010). Wnt7a or postsynaptic expression of Dvl1 increases the clustering of PSD95 and the colocalization of presynaptic and postsynaptic markers in hippocampal neurons (Ciani et al. 2011). Given that Wnt7a signals to the presynaptic axon through the Fz5 receptor, these studies show that Wnt7a signals bidirectionally to the axon and dendrite to promote synapse formation.

What is the mechanism responsible for postsynaptic assembly? Pharmacological studies have shown that Wnt5a promotes PSD95 clustering through JNK (Farias et al. 2009). In contrast, recruitment of $GABA_A$ receptors requires CaMKII activity (Cuitino et al. 2010). Although it is unclear whether Wnt5a directly regulates JNK and CaMKII on the postsynaptic side, Wnt5a rapidly increases the levels of Ca^{2+} in dendrites, suggesting that Wnt5a directly signals to the postsynaptic side (Varela-Nallar et al. 2010). Studies of Wnt7a and Dvl1 have shown that Wnt7a directly signals postsynaptically, because *Dvl1* mutant dendrites fail to respond to Wnt7a. This postsynaptic effect of Wnt7a is mediated through the local activation CaMKII (Fig. 3) (Ciani et al. 2011). Thus, activation of CaMKII promotes both the assembly of postsynaptic structures of inhibitory and excitatory synapses. This poses the question of how activation of CaMKII can result in the assembly of two very different types of synapses. How is this achieved? Further studies will provide the answer.

Excitatory versus Inhibitory Synapses

In cultured hippocampal neurons, Wnt5a modulates inhibitory synapses by increasing the surface levels and the retention of $GABA_A$ receptors (Fig. 3) (Cuitino et al. 2010). Consistently, Wnt5a increases the amplitude of miniature inhibitory postsynaptic currents (mIPSCs). These results suggest that Wnt5a promotes the maturity of inhibitory synapses. Intriguingly, short exposure to Wnt5a does not affect the frequency of mIPSCs or the number of presynaptic sites (Cuitino et al. 2010). Time-course studies showed that Wnt5a first signals to the postsynaptic side to then induce presynaptic changes. Whether this effect is direct on the presynaptic axon remains to be determined.

Wnt5a also promotes the assembly of excitatory synapses as indicated by the increased clustering of PSD95, a marker of excitatory postsynaptic sites (Farias et al. 2009). Although initial studies suggested that Wnt5a increases the formation of dendritic spines (Varela-Nallar et al. 2010), a recent publication suggests that Wnt5a does not affect spine formation (Cerpa et al. 2011). Recordings of fEPSCs showed that Wnt5a promotes glutamatergic transmission by increasing NMDA currents in hippocampal neurons. However, AMPA currents are unaffected (Cerpa et al. 2011). A Wnt/Ca^{2+} signaling pathway regulates these postsynaptic changes as evident by the rapid increase in Ca^{2+} levels and the requirement of CaMKII activity (Varela-Nallar et al. 2010; Cerpa et al. 2011). Although loss-of-function approaches are needed to establish the in vivo role of Wnt5a at excitatory and inhibitory synapses, together these studies show that Wnt5a is a pansynaptogenic factor that promotes the assembly of both excitatory and inhibitory synapses.

In contrast to Wnt5a, Wnt7a specifically promotes the formation of excitatory synapses without affecting inhibitory ones (Ciani et al. 2011). Within 1 h, Wnt7a significantly increases the number of excitatory synapses and the frequency and amplitude of mEPSCs. However, the number of inhibitory sites and mIPSCs remains unchanged. In addition, Wnt7a increases the number and size of dendritic spines, postsynaptic structures that received excitatory inputs. In vivo analyses of the *Wnt7a* and *Dvl1* double-mutant mice showed that Wnt7a–Dvl1 signaling is required for the formation and growth of spines at the CA3 and CA1 regions of the hippocampus. Cell-autonomous activation of the Wnt pathway in dendrites by expression of Dvl1 revealed that Wnt7a directly signals to dendrites to promote spine growth. Consistently, *Dvl1* mutant dendrites do not respond to Wnt7a (Ciani et al. 2011). Interestingly, postsynaptic activation of Wnt signaling does not affect spine number, suggesting that this effect of Wnt7a is mediated through another mechanism. Analyses of the localization of Dvl1 and the use of a local reporter of CaMKII activity suggest that Wnt7a through postsynaptic Dvl1 locally activates CaMKII in dendritic spines (Ciani et al. 2011) to promote spine growth (Fig. 3). Thus, Wnt7a specifically promotes the formation of excitatory synapses by regulating dendritic spines.

Together, the studies on Wnt5a and Wnt7a suggest that Wnts differentially modulate distinct types of synapses. The specific effect of Wnt7a on excitatory synapses suggests that increased Wnt7a signaling could contribute to neurological disorders such as epilepsy. Future studies on Wnts will shed light into the mechanisms that control the balance between excitability and inhibition of neuronal circuits.

NEURONAL ACTIVITY ON WNT RELEASE AND RECEPTOR LOCALIZATION

Neuronal activity modulates neuronal circuit formation by regulating cell migration, axon guidance, dendritic development, and the formation and pruning of synapses (Budnik and Salinas 2011; West and Greenberg 2011). Several molecules including BDNF, neurotransmitter receptors, and adhesion molecules are regulated by activity (Meyer-Franke et al. 1998; Bozdagi et al. 2000; Du et al. 2000; Hayashi et al. 2000; Tanaka et al. 2000). However, little is known about the mechanisms by which neuronal activity sculptures neuronal networks. New studies show that Wnts and their receptors are under the control of neuronal activity and that they contribute to activity-dependent circuit formation and refinement.

The first indication that neuronal activity regulates Wnts came from studies in cultured hippocampal neurons. Depolarization with KCl, which promotes dendritic development, increases Wnt activity in the media of cultured neurons, indicating that neural activity modulates the release, synthesis, or transcription of Wnts (Yu and Malenka 2003). Subsequent studies revealed that activation of NMDA receptors increases the transcription of Wnt2, a protein expressed in the hippocampus. Importantly, blockade of Wnt2 compromised dendritic development (Wayman et al. 2006). Further evidence comes from studies showing that the levels of Wnt7a/b protein increase in the hippocampus of animals exposed to an enriched environment (Gogolla et al. 2009).

Changes in neuronal activity also influence the localization and trafficking of Frizzled receptors. High-frequency stimulation, which induces synaptic potentiation, increases the levels of surface Fz5 (sFz5) without affecting the total level of the receptor in hippocampal neurons. Conversely, low-frequency stimulation (LFS), which elicits long-term depression, decreases the levels of sFz5. Importantly, HFS increases, whereas LFS decreases the amount of sFz5 receptor at synapses and the percentage of synapses containing Fz5 (Sahores et al. 2010). Given that Fz localization is regulated by Wnt ligands (Mathew et al. 2005; Klassen and Shen 2007; Mosca and Schwarz 2010) and that activity promotes the expression of Wnts (Wayman et al. 2006; Korkut et al. 2009), the contribution of Wnts in Fz5 localization was investigated. Wnt blockade with the secreted Wnt antagonist Sfrps or with the extracellular domain of Fz5 showed that endogenous Wnts expressed by hippocampal neurons mediate the effect of neuronal activity on Fz5 localization. Thus, patterned activity induced by HFS regulates the release or expression of Wnts, which then promote the trafficking of Fz5 to the cell surface and to the synapse (Fig. 4). Future studies focused on the mechanisms that control the trafficking of Fz receptors to synapses will provide important insights into the effect of ligands on their receptor localization and also how neuronal activity modulates synapse formation and function.

STRUCTURAL AND FUNCTIONAL PLASTICITY OF SYNAPSES

The expression of Wnts, their receptors, and several components of the Wnt pathway in the mature brain suggests that Wnts could also regulate synaptic plasticity in the adult. The first hint that synaptic function is modulated by Wnt signaling came from analyses of *Wnt7a; Dvl1* double-mutant mice. In these mice, the complexity of the glomerular rosettes is affected,

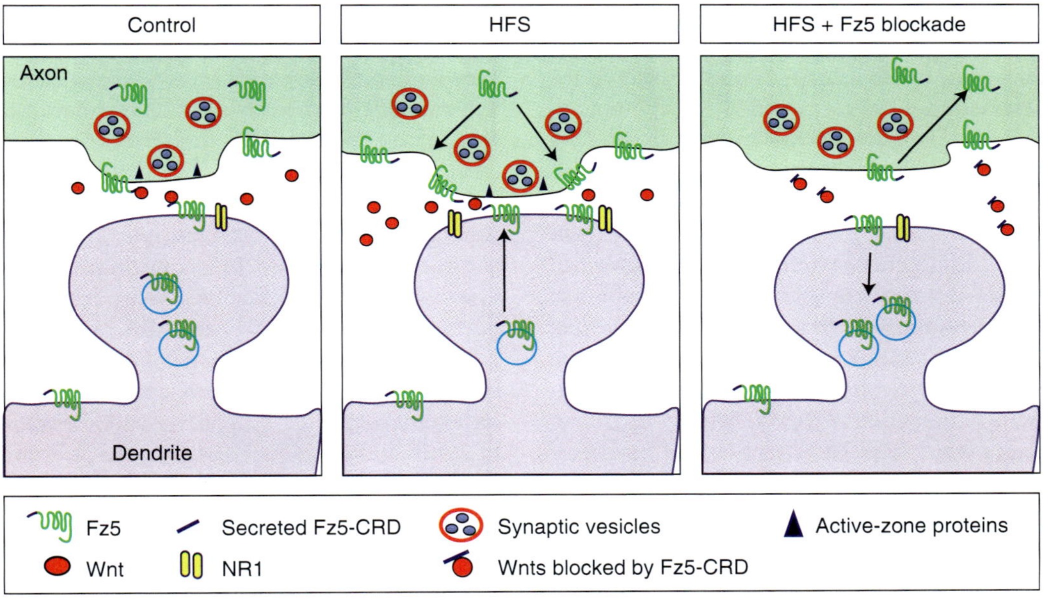

Figure 4. Neuronal activity regulates the membrane insertion of Fz5 receptors and their localization to synapses. In hippocampal neurons, surface Fz5 is localized to synaptic and extrasynaptic sites under basal conditions. Increased neuronal firing induced by high-frequency stimulation (HFS) increases the mobilization of Fz5 to the cell surface and in particular to the surface of synapses. Blockade of endogenous Wnt factors with the soluble extracellular domain of Fz5 (Fz5-CRD) suppresses the effect of HFS on Fz5 localization. Importantly, Fz5-CRD completely blocks activity-induced synapse formation as indicated by the decrease in the colocalization of presynaptic vesicle markers and NMDARs.

but the presynaptic active zones appear normal. Interestingly, electrophysiological recordings showed defects in the frequency of mEPSCs, suggesting a potential defect in neurotransmitter release (Ahmad-Annuar et al. 2006). A more recent study has reinforced this hypothesis. Electrophysiological recording of brain slices showed that Wnt7a decreases pair pulse facilitation at the hippocampal CA3–CA1 synapse, suggesting a role for Wnt7a in the regulation of neurotransmitter release (Cerpa et al. 2008). This process could be regulated at different levels of the synaptic vesicle cycle, from changes in the reserved pool of vesicles to the rate of docking, fusion, or endocytosis. Dynamics studies of synaptic vesicles labeled with the lipophilic dye FM-1-43 suggest that Wnt7a increases synaptic vesicle exocytosis (Cerpa et al. 2008). However, further detailed characterization is required to fully establish a role for Wnt signaling in neurotransmitter release.

Wnt signaling regulates synaptic strength through postsynaptic changes. As discussed before, analyses of Wnt signaling–deficient mutant mice revealed that Wnt7a signaling is essential for spine formation and growth in the hippocampus. Increased spine growth has been correlated with an increase in synaptic strength (Yuste and Bonhoeffer 2001; Matsuzaki et al. 2004; Bourne and Harris 2008). Indeed, *Wnt7a/Dvl1* double-mutant mice show deficits in synaptic strength in the hippocampus. Conversely, gain-of-function studies with exogenous Wnt7a or expression of Dvl1 in dendrites show that Wnt7a-Dvl1 signaling increases synaptic strength at excitatory synapses in a CaMKII-dependent manner (Ciani et al. 2011). Thus, Wnt7a signaling modulates the structural and synaptic plasticity of neuronal circuits in the hippocampus.

Several studies suggest that Wnt signaling also modulates long-term potentiation. In acute brain slices, blockade of Wnt signaling with specific antibodies against Wnt3a or the secreted extracellular domain of Frizzled-8 weakly decreases long-term potentiation (Chen et al. 2006). Conversely, activation of the canonical Wnt signaling with lithium or with Wnt3a mildly increases potentiation. A more recent study suggests that Wnt5a, through a non-canonical Wnt pathway, modulates the induction of synaptic potentiation at the CA3–CA1 synapse (Cerpa et al. 2011). Thus, Wnts, through the activation of canonical or non-canonical pathways, regulate long-term potentiation. Further studies are essential to elucidate the contribution of Wnt signaling to maintenance of long-term synaptic plasticity.

Wnt signaling also contributes to the structural plasticity of synaptic structures in the central nervous system. In the hippocampus, mossy fiber axons form complex terminal protrusions with dendritic spines of the CA3 region (Rollenhagen et al. 2007). When mice are exposed to an enriched environment (EE), these structures become extensively remodeled as mossy fiber terminals interdigitate extensively with CA3 dendrites (Gogolla et al. 2009). Because this modeling is similar to that observed by Wnt7a in the cerebellar mossy fibers (Hall et al. 2000), the role of Wnts was investigated. Three pieces of data support the role of Wnt7a in EE-mediated remodeling. First, Wnt7a protein levels are elevated in the hippocampus of EE animals. Second, blockade of Wnt7a by injection of the Wnt antagonist Sfrp1 suppresses EE-mediated axon remodeling. Third, injection of exogenous Wnt7a into the hippocampus mimics the effect of EE (Gogolla et al. 2009). These studies raise the interesting possibility that Wnts might also modulate structural changes following sensory experience. Indeed, Wnt signaling mediates visual experience-dependent plasticity. In the optic tectum, the receptive field (RF), which depends on the number of retinal ganglia cell axons, decreases in size during development in an activity-dependent manner. Acute blockade of Wnts or infusion of Wnts in the *Xenopus* tectum interferes with this developmental reduction in RF (Lim et al. 2010), suggesting that Wnt signaling regulates the refinement of neuronal connections during sensory experience-dependent plasticity.

Activity-mediated synapse formation is also modulated by Wnt proteins through the Fz5 receptor. As mentioned above, HFS increases the level of Fz5 receptor at the surface and at synapses (Sahores et al. 2010). This increased level of Fz5 at synapses is correlated with the formation and growth of synapses. Importantly, blockade of endogenous Wnts with Sfrps or with the soluble extracellular domain of the receptor Fz5 completely suppresses the effect of HFS on synapse formation. Thus, Wnt signaling through Fz5 significantly contributes to activity-mediated synapse formation in hippocampal neurons.

NEUROLOGICAL DISORDERS AND NEURODEGENERATION

Mounting evidence suggests a link between dysfunction of Wnt signaling and neurological disorders as well as neurodegenerative diseases. Mutations in the *disrupted in schizophrenia* (*DISC1*) gene were found in a Scottish family with a high incidence of schizophrenia, major depression, and bipolar disorder (Porteous and Millar 2006). Two recent studies showed that DISC1 modulates canonical Wnt signaling. DISC1 interacts with and inhibits Gsk3β activity, resulting in increased β-catenin signaling (Mao et al. 2009). Moreover, DISC1 also interacts with the scaffold protein DIX domain-containing protein (DIXDC1), which together with DSC1 modulates canonical Wnt signaling (Singh et al. 2010). In addition, Gsk3 inhibitors restore the behavioral defects manifested in DISC1-deficient mutant mice (Mao et al. 2009). Because different signaling pathways regulate Gsk3 activity, further studies are needed to establish whether deficiency in Wnt signaling contributes to schizophrenia.

Deregulation of Wnt signaling might contribute to the changes in mood and behavior characteristic of psychiatric disorders such as manic depression. The first evidence that Wnt signaling contributes to social behavior came from analyses of *Dvl1* mutant mice, which show defects in social interactions and deficits in the acoustic startled reflex (Lijam et al. 1997). Further support came from the finding that lithium, used for the treatment of bipolar affective disorders (O'Brien and Klein 2007), is a direct inhibitor of Gsk3 (Klein and Melton 1996; Stambolic et al. 1996). Although many molecular targets of lithium have been identified, the mechanism by which lithium modulates mood is currently unclear. Nevertheless, Gsk3 remains a strong candidate because heterozygous Gsk3 mutant mice show a similar phenotype to mice treated with lithium (Beaulieu et al. 2004; O'Brien et al. 2004). These findings open the doors for developing new therapeutic strategies for the treatment of mood disorders.

In the aging brain, deficiency in canonical Wnt signaling has been associated with Alzheimer's disease (AD). The link between increased Gsk3 activity and AD has been known (Hernandez et al. 2010). Yet only recently, a genetic link between decreased Wnt signaling and AD has been presented. Genome-wide studies have identified a broad susceptibility region for later-onset Alzheimer's disease in chromosome 12, which contains the LRP6 locus (D'Introno et al. 2006). Importantly, a common variant of LRP6, which shows decreased levels of canonical Wnt signaling in heterologous cells, has been associated with late-onset AD (De Ferrari et al. 2007). Further evidence comes from the finding that Dickkopf-1 (Dkk1), the secreted Wnt antagonist that specifically blocks canonical Wnt signaling through binding to LRP6, is elevated in AD brain biopsies and in the brain of transgenic mouse models for AD (Caricasole et al. 2004; Rosi et al. 2010). Together, these findings suggest that decreased canonical Wnt signaling might contribute to AD pathogenesis. Future studies are essential to determine a role of Wnt signaling in this devastating neurodegenerative disease.

CONCLUSIONS

In the last few years, great progress has been made in understanding the role of Wnt signaling in the formation of neuronal circuits during development. It is now evident that Wnts also regulate different aspects of synaptic plasticity, through presynaptic and postsynaptic mechanisms. Moreover, Wnts mediate activity-dependent processes such as synapse formation, synaptic remodeling, and circuit refinement. Finally, there is increased evidence suggesting a potential link between dysfunction of Wnt signaling and developmental disorders and bipolar affective disorders. Future studies on Wnts in the developing, mature, and aging brain will provide crucial information for developing therapeutic approaches for the treatment and/or prevention of neurological disorders and neurodegeneration.

ACKNOWLEDGMENTS

I thank members of my laboratory for their useful and insightful comments on the manuscript. I also thank the MRC, Wellcome Trust, BBSRC, ARUK, and EU for supporting our work.

REFERENCES

Ahmad-Annuar A, Ciani L, Simeonidis I, Herreros J, Fredj NB, Rosso SB, Hall A, Brickley S, Salinas PC. 2006. Signaling across the synapse: A role for Wnt and Dishevelled in presynaptic assembly and neurotransmitter release. *J Cell Biol* **174:** 127–139.

Beaulieu JM, Sotnikova TD, Yao WD, Kockeritz L, Woodgett JR, Gainetdinov RR, Caron MG. 2004. Lithium antagonizes dopamine-dependent behaviors mediated by an AKT/glycogen synthase kinase 3 signaling cascade. *Proc Natl Acad Sci* **101:** 5099–5104.

Bourne JN, Harris KM. 2008. Balancing structure and function at hippocampal dendritic spines. *Ann Rev Neurosci* **31:** 47–67.

Bozdagi O, Shan W, Tanaka H, Benson DL, Huntley GW. 2000. Increasing numbers of synaptic puncta during late-phase LTP: N-Cadherin is synthesized, recruited to synaptic sites, and required for potentiation. *Neuron* **28:** 245–259.

Budnik V, Salinas PC. 2011. Wnt signaling during synaptic development and plasticity. *Curr Opin Neurobiol* **21:** 151–159.

Caricasole A, Copani A, Caraci F, Aronica E, Rozemuller AJ, Caruso A, Storto M, Gaviraghi G, Terstappen GC, Nicoletti F. 2004. Induction of Dickkopf-1, a negative modulator of the Wnt pathway, is associated with neuronal degeneration in Alzheimer's brain. *J Neurosci* **24:** 6021–6027.

Cerpa W, Godoy JA, Alfaro I, Farias GG, Metcalfe MJ, Fuentealba R, Bonansco C, Inestrosa NC. 2008. Wnt-7a modulates the synaptic vesicle cycle and synaptic transmission in hippocampal neurons. *J Biol Chem* **283:** 5918–5927.

Cerpa M, Gambrill A, Inestrosa N, Barria A. 2011. Regulation of NMDA-receptor synaptic transmission by Wnt signaling. *J Neurosci* **31:** 9466–9471.

Chen J, Park CS, Tang SJ. 2006. Activity-dependent synaptic Wnt release regulates hippocampal long term potentiation. *J Biol Chem* **281:** 11910–11916.

Ciani L, Krylova O, Smalley MJ, Dale TC, Salinas PC. 2004. A divergent canonical WNT-signaling pathway regulates microtubule dynamics: Dishevelled signals locally to stabilize microtubules. *J Cell Biol* **164:** 243–253.

Ciani L, Boyle K, Dickins E, Sahores M, Anane D, Lopes DM, Gibb A, Salinas PC. 2011. Wnt7a signaling promotes dendritic spine growth and synaptic strength through Ca^{2+}/calmodulin-dependent protein kinase II. *Proc Natl Acad Sci* **108:** 10732–10737.

Cuitino L, Godoy JA, Farias GG, Couve A, Bonansco C, Fuenzalida M, Inestrosa NC. 2010. Wnt-5a modulates recycling of functional $GABA_A$ receptors on hippocampal neurons. *J Neurosci* **30:** 8411–8420.

Davis EK, Zou Y, Ghosh A. 2008. Wnts acting through canonical and noncanonical signaling pathways exert opposite effects on hippocampal synapse formation. *Neural Dev* **3:** 32.

De Ferrari GV, Papassotiropoulos A, Biechele T, Wavrant De-Vrieze F, Avila ME, Major MB, Myers A, Saez K, Henriquez JP, Zhao A, et al. 2007. Common genetic variation within the low-density lipoprotein receptor-related protein 6 and late-onset Alzheimer's disease. *Proc Natl Acad Sci* **104:** 9434–9439.

D'Introno A, Solfrizzi V, Colacicco AM, Capurso C, Amodio M, Todarello O, Capurso A, Kehoe PG, Panza F. 2006. Current knowledge of chromosome 12 susceptibility genes for late-onset Alzheimer's disease. *Neurobiol Aging* **27:** 1537–1553.

Du J, Feng L, Yang F, Lu B. 2000. Activity- and Ca^{2+}-dependent modulation of surface expression of brain-derived neurotrophic factor receptors in hippocampal neurons. *J Cell Biol* **150:** 1423–1434.

Dunaevsky A, Mason CA. 2003. Spine motility: A means towards an end? *Trends Neurosci* **26:** 155–160.

Farias GG, Alfaro IE, Cerpa W, Grabowski CP, Godoy JA, Bonansco C, Inestrosa NC. 2009. Wnt-5a/JNK signaling promotes the clustering of PSD-95 in hippocampal neurons. *J Biol Chem* **284:** 15857–15866.

Fradkin LG, Noordermeer JN, Nusse R. 1995. The *Drosophila* Wnt protein DWnt-3 is a secreted glycoprotein localized on the axon tracts of the embryonic CNS. *Dev Biol* **168:** 202–213.

Fradkin LG, Dura JM, Noordermeer JN. 2010. Ryks: New partners for Wnts in the developing and regenerating nervous system. *Trends Neurosci* **33:** 84–92.

Gogolla N, Galimberti I, Deguchi Y, Caroni P. 2009. Wnt signaling mediates experience-related regulation of synapse numbers and mossy fiber connectivities in the adult hippocampus. *Neuron* **62:** 510–525.

Hall AC, Lucas FR, Salinas PC. 2000. Axonal remodeling and synaptic differentiation in the cerebellum is regulated by WNT-7a signaling. *Cell* **100:** 525–535.

Hamori J, Somogyi J. 1983. Differentiation of cerebellar mossy fiber synapses in the rat: A quantitative electron microscope study. *J Comp Neurol* **220:** 365–377.

Hayashi Y, Shi SH, Esteban JA, Piccini A, Poncer JC, Malinow R. 2000. Driving AMPA receptors into synapses by LTP and CaMKII: Requirement for GluR1 and PDZ domain interaction. *Science* **287:** 2262–2267.

Hernandez F, Gomez de Barreda E, Fuster-Matanzo A, Lucas JJ, Avila J. 2010. GSK3: A possible link between β amyloid peptide and tau protein. *Exp Neurol* **223:** 322–325.

Hindges R, McLaughlin T, Genoud N, Henkemeyer M, O'Leary DD. 2002. EphB forward signaling controls directional branch extension and arborization required for dorsal–ventral retinotopic mapping. *Neuron* **35:** 475–487.

Hutchins BI, Li L, Kalil K. 2011. Wnt/calcium signaling mediates axon growth and guidance in the developing corpus callosum. *Dev Neurobiol* **71:** 269–283.

Imondi R, Thomas JB. 2003. Neuroscience. The ups and downs of Wnt signaling. *Science* **302:** 1903–1904.

Keeble TR, Halford MM, Seaman C, Kee N, Macheda M, Anderson RB, Stacker SA, Cooper HM. 2006. The Wnt receptor Ryk is required for Wnt5a-mediated axon guidance on the contralateral side of the corpus callosum. *J Neurosci* **26:** 5840–5848.

Klassen MP, Shen K. 2007. Wnt signaling positions neuromuscular connectivity by inhibiting synapse formation in *C. elegans*. *Cell* **130:** 704–716.

Klein PS, Melton DA. 1996. A molecular mechanism for the effect of lithium on development. *Proc Natl Acad Sci* **93:** 8455–8459.

Korkut C, Ataman B, Ramachandran P, Ashley J, Barria R, Gherbesi N, Budnik V. 2009. *Trans*-synaptic transmission of vesicular Wnt signals through Evi/Wntless. *Cell* **139:** 393–404.

Krylova O, Herreros J, Cleverley KE, Ehler E, Henriquez JP, Hughes SM, Salinas PC. 2002. WNT-3, expressed by motoneurons, regulates terminal arborization of neurotrophin-3-responsive spinal sensory neurons. *Neuron* **35:** 1043–1056.

Li L, Hutchins BI, Kalil K. 2009. Wnt5a induces simultaneous cortical axon outgrowth and repulsive axon guidance through distinct signaling mechanisms. *J Neurosci* **29:** 5873–5883.

Lijam N, Paylor R, McDonald MP, Crawley JN, Deng CX, Herrup K, Stevens KE, Maccaferri G, McBain CJ, Sussman DJ, et al. 1997. Social interaction and sensorimotor gating abnormalities in mice lacking Dvl1. *Cell* **90:** 895–905.

Lim BK, Cho SJ, Sumbre G, Poo MM. 2010. Region-specific contribution of ephrin-B and Wnt signaling to receptive field plasticity in developing optic tectum. *Neuron* **65:** 899–911.

Liu Y, Shi J, Lu CC, Wang ZB, Lyuksyutova AI, Song XJ, Zou Y. 2005. Ryk-mediated Wnt repulsion regulates posterior-directed growth of corticospinal tract. *Nat Neurosci* **8:** 1151–1159.

Luo L. 2006. Developmental neuroscience: Two gradients are better than one. *Nature* **439:** 23–24.

Lyuksyutova AI, Lu CC, Milanesio N, King LA, Guo N, Wang Y, Nathans J, Tessier-Lavigne M, Zou Y. 2003. Anterior–posterior guidance of commissural axons by Wnt-frizzled signaling. *Science* **302:** 1984–1988.

Mao Y, Ge X, Frank CL, Madison JM, Koehler AN, Doud MK, Tassa C, Berry EM, Soda T, Singh KK, et al. 2009. Disrupted in schizophrenia 1 regulates neuronal progenitor proliferation via modulation of GSK3β/β-catenin signaling. *Cell* **136:** 1017–1031.

Mathew D, Ataman B, Chen J, Zhang Y, Cumberledge S, Budnik V. 2005. Wingless signaling at synapses is through cleavage and nuclear import of receptor DFrizzled2. *Science* **310:** 1344–1347.

Matsuzaki M, Honkura N, Ellis-Davies GC, Kasai H. 2004. Structural basis of long-term potentiation in single dendritic spines. *Nature* **429:** 761–766.

McLaughlin T, O'Leary DD. 2005. Molecular gradients and development of retinotopic maps. *Annu Rev Neurosci* **28:** 327–355.

Meyer-Franke A, Wilkinson GA, Kruttgen A, Hu M, Munro E, Hanson MG Jr, Reichardt LF, Barres BA. 1998. Depolarization and cAMP elevation rapidly recruit TrkB to the plasma membrane of CNS neurons. *Neuron* **21:** 681–693.

Miech C, Pauer HU, He X, Schwarz TL. 2008. Presynaptic local signaling by a canonical wingless pathway regulates development of the *Drosophila* neuromuscular junction. *J Neurosci* **28:** 10875–10884.

Mosca TJ, Schwarz TL. 2010. The nuclear import of Frizzled2-C by Importins-β11 and α2 promotes postsynaptic development. *Nat Neurosci* **13**, 935–943.

O'Brien WT, Klein PS. 2007. Regulation of glycogen synthase kinase-3 in patients with affective disorders. *Biol Psychiatry* **61:** 139–141.

O'Brien WT, Harper AD, Jove F, Woodgett JR, Maretto S, Piccolo S, Klein PS. 2004. Glycogen synthase kinase-3β haploinsufficiency mimics the behavioral and molecular effects of lithium. *J Neurosci* **24:** 6791–6798.

Okabe S, Miwa A, Okado H. 2001. Spine formation and correlated assembly of presynaptic and postsynaptic molecules. *J Neurosci* **21:** 6105–6114.

Packard M, Koo ES, Gorczyca M, Sharpe J, Cumberledge S, Budnik V. 2002. The *Drosophila* Wnt, wingless, provides an essential signal for pre- and postsynaptic differentiation. *Cell* **111:** 319–330.

Paganoni S, Bernstein J, Ferreira A. 2010. Ror1–Ror2 complexes modulate synapse formation in hippocampal neurons. *Neuroscience* **165:** 1261–1274.

Porteous DJ, Millar JK. 2006. Disrupted in schizophrenia 1: Building brains and memories. *Trends Mol Med* **12:** 255–261.

Purro SA, Ciani L, Hoyos-Flight M, Stamatakou E, Siomou E, Salinas PC. 2008. Wnt regulates axon behavior through changes in microtubule growth directionality: A new role for adenomatous polyposis coli. *J Neurosci* **28:** 8644–8654.

Rollenhagen A, Satzler K, Rodriguez EP, Jonas P, Frotscher M, Lubke JH. 2007. Structural determinants of transmission at large hippocampal mossy fiber synapses. *J Neurosci* **27:** 10434–10444.

Rosi MC, Luccarini I, Grossi C, Fiorentini A, Spillantini MG, Prisco A, Scali C, Gianfriddo M, Caricasole A, Terstappen GC, et al. 2010. Increased Dickkopf-1 expression in transgenic mouse models of neurodegenerative disease. *J Neurochem* **112:** 1539–1551.

Sahores M, Gibb A, Salinas PC. 2010. Frizzled-5, a receptor for the synaptic organizer Wnt7a, regulates activity-mediated synaptogenesis. *Development* **137:** 2215–2225.

Salinas PC, Zou Y. 2008. Wnt signaling in neural circuit assembly. *Annu Rev Neurosci* **31:** 339–358.

Sanchez-Camacho C, Bovolenta P. 2009. Emerging mechanisms in morphogen-mediated axon guidance. *Bioessays* **31:** 1013–1025.

Schmitt AM, Shi J, Wolf AM, Lu CC, King LA, Zou Y. 2006. Wnt–Ryk signalling mediates medial–lateral retinotectal topographic mapping. *Nature* **439:** 31–37.

Shafer B, Onishi K, Lo C, Colakoglu G, Zou Y. 2011. Vangl2 promotes Wnt/planar cell polarity-like signaling by antagonizing Dvl1-mediated feedback inhibition in growth cone guidance. *Dev Cell* **20:** 177–191.

Singh KK, Ge X, Mao Y, Drane L, Meletis K, Samuels BA, Tsai LH. 2010. Dixdc1 is a critical regulator of DISC1 and embryonic cortical development. *Neuron* **67:** 33–48.

Stambolic V, Ruel L, Woodgett JR. 1996. Lithium inhibits glycogen synthase kinase-3 activity and mimics wingless signalling in intact cells. *Curr Biol* **6:** 1664–1668.

Tanaka H, Shan W, Phillips GR, Arndt K, Bozdagi O, Shapiro L, Huntley GW, Benson DL, Colman DR. 2000. Molecular modification of *N*-cadherin in response to synaptic activity. *Neuron* **25:** 93–107.

Trachtenberg JT, Chen BE, Knott GW, Feng G, Sanes JR, Welker E, Svoboda K. 2002. Long-term in vivo imaging of experience-dependent synaptic plasticity in adult cortex. *Nature* **420:** 788–794.

Varela-Nallar L, Grabowski CP, Alfaro IE, Alvarez AR, Inestrosa NC. 2009. Role of the Wnt receptor Frizzled-1 in presynaptic differentiation and function. *Neural Dev* **4:** 41.

Varela-Nallar L, Alfaro IE, Serrano FG, Parodi J, Inestrosa NC. 2010. Wingless-type family member 5A (Wnt-5a) stimulates synaptic differentiation and function of glutamatergic synapses. *Proc Natl Acad Sci* **107:** 21164–21169.

Wayman GA, Impey S, Marks D, Saneyoshi T, Grant WF, Derkach V, Soderling TR. 2006. Activity-dependent dendritic arborization mediated by CaM-kinase I activation and enhanced CREB-dependent transcription of Wnt-2. *Neuron* **50:** 897–909.

West AE, Greenberg ME. 2011. Neuronal activity-regulated gene transcription in synapse development and cognitive function. *Cold Spring Harb Perspect Biol* doi: 10.1101/cshperspect.a005744.

Yoshihara Y, De Roo M, Muller D. 2009. Dendritic spine formation and stabilization. *Curr Opin Neurobiol* **19:** 146–153.

Yu X, Malenka RC. 2003. β-Catenin is critical for dendritic morphogenesis. *Nat Neurosci* **6:** 1169–1177.

Yuste R, Bonhoeffer T. 2001. Morphological changes in dendritic spines associated with long-term synaptic plasticity. *Annu Rev Neurosci* **24:** 1071–1089.

Wnt Signaling and Forebrain Development

Susan J. Harrison-Uy[1] and Samuel J. Pleasure[1,2]

[1]Department of Neurology, University of California, San Francisco, California 94143

[2]Programs in Neuroscience and Developmental Biology, Eli and Edythe Broad Center of Regeneration Medicine and Stem Cell Research, University of California, San Francisco, California 94143

Correspondence: sam.pleasure@ucsf.edu

Components of the Wnt signaling pathway are expressed in a tightly regulated and spatially specific manner during development of the forebrain, and Wnts are key regulators of regional forebrain identity. Wnt signaling from the cortical hem regulates the expansion and cell-type specification of the adjacent neuroepithelium and, in conjunction with Bmp, Fgf, and Shh signaling, controls dorsal–ventral forebrain patterning. Subsequently, Wnt signaling dynamically regulates the behavior of cortical progenitor cells, initially promoting the expansion of radial glia progenitor cells and later inducing neurogenesis by promoting terminal differentiation of intermediate progenitor cells. A role for Wnt signaling in cell-type specification has also been proposed.

During the initial specification of the telencephalon, discrete signaling centers are established through the actions of a series of interconnected signaling pathways. These signaling centers impart regional identity to the surrounding neuroepithelium, resulting in functional subdivision along the dorsal/ventral and rostral/caudal axes. These regional subdivisions allow diverse neural cell types to emerge from adjacent progenitor populations. Wnts are secreted glycoproteins that have been identified as key regulators of regional identity in the early developing forebrain. During subsequent cortical development, Wnt signaling in cortical progenitor cells (PCs) has also been implicated in cell cycle exit, neuronal differentiation, and establishing the laminar identity of excitatory neurons.

Although the various Wnt signaling pathways are described in great detail elsewhere in this collection, there are a few details that are relevant for this article that bear reiterating. In the canonical Wnt signaling pathway, secreted Wnt ligands bind to a Frizzled (Fzd)/low-density lipoprotein receptor-related protein (LRP) complex at the cell membrane. This complex recruits Dishevelled and inhibits the action of a pathway leading to the degradation of β-catenin. When degradation is blocked, β-catenin accumulates in the cytoplasm and subsequently translocates to the nucleus. In the nucleus, β-catenin interacts with the transcription factors T-cell factor (TCF)/lymphoid-enhancer factor (LEF) to modulate transcription of target genes (for review, see Logan and Nusse 2004; MacDonald et al. 2009; van Amerongen and Nusse 2009). There are also multiple non-canonical Wnt signaling pathways essentially independent of β-catenin, including the Wnt/calcium

Cite this article as *Cold Spring Harb Perspect Biol* doi: 10.1101/cshperspect.a008094

pathway (for review, see Kuhl et al. 2000) and the planer cell polarity pathway (for review, see Seifert and Mlodzik 2007).

WNT LIGAND AND SIGNALING COMPONENT EXPRESSION IN THE DEVELOPING FOREBRAIN

Wnt ligands are expressed in discrete and overlapping patterns in the developing forebrain. By embryonic day 9.5 (E9.5) in the mouse, the neural tube has closed, and the telencephalon is located at the anterior segment of the neural tube. The expression of several Wnts can be detected in the forebrain at this age: *Wnt1*, *Wnt3*, *Wnt3a*, *Wnt4*, and *Wnt7b* in dorsal regions and *Wnt5a* and *Wnt7a* in ventral regions (Parr et al. 1993). Some of these genes have broad expression patterns, such as *Wnt7b*, whereas others, such as *Wnt3a*, show restricted expression patterns (Parr et al. 1993). Although *LRP5/6* and several of the Frizzled receptors, including *Fzd1*, *Fzd2*, and *Fzd3*, are generally ubiquitously expressed in the early telencephalon (Pinson et al. 2000; Fischer et al. 2007), other Fzd receptors have more restricted patterns: *Fzd4* and *Fzd7* (dorsal telencephalic/-diencephalic border), *Fzd5* and *Fzd8* (ventral telencephalon), *Fzd9* and *Fzd10* (dorsal telencephalon), and *Fzd6* (dorsomedial telencephalon) (Fischer et al. 2007).

By E12.5, differentiated neurons have begun to appear in the cortex, and the PCs adjacent to the ventricles are regionally specified. Those in the dorsal cortex will produce excitatory neurons, followed by oligodendrocytes and astrocytes (Gorski et al. 2002). PCs in the ventral forebrain produce inhibitory interneurons, oligodendrocytes, and astrocytes (Wonders and Anderson 2005; Kessaris et al. 2006; Ono et al. 2008). The hippocampal anlage is located in the medial wall of the forebrain, adjacent to an important signaling center, the cortical hem (Grove et al. 1998). At this stage, expression of *Wnt2b*, *Wnt3a*, and *Wnt5a* is restricted to the cortical hem, whereas *Wnt7a* is expressed throughout the cerebral cortical ventricular zone but excluded from the cortical hem (Grove et al. 1998). *Wnt7b* and *Wnt8b* are expressed by the cortical hem and hippocampal anlage, although *Wnt7b* expression extends throughout the cortex in postmitotic neurons (Kim et al. 2001a; Theil 2005). *Wnt7a* and *Wnt7b* are expressed by PCs in the ventral telencephalon (Grove et al. 1998; Theil 2005; Fotaki et al. 2011). *Fzd1* and *Fzd2* are expressed in PCs in both the dorsal and ventral telencephalon, whereas *Fzd3* is ubiquitously expressed in the telencephalon (Fischer et al. 2007). *Fzd10* expression is restricted to the cortical hem (Kim et al. 2001b; Fischer et al. 2007). *Fzd5&8* and *Fzd9* have complementary expression patterns: *Fzd9* is strongly expressed in the hippocampal anlage with weak expression dorsal and lateral regions, whereas *Fzd5&8* is expressed in a strong-lateral to low-medial gradient (Kim et al. 2001b; Fischer et al. 2007). Figure 1 summarizes Wnt ligand (Fig. 1A) and Fzd receptor (Fig. 1B) expression in the mouse forebrain at E12.5.

Canonical Wnt signaling has been mapped in the developing forebrain using BAT-gal reporter mice, in which expression of *β-galactosidase* is under the control of seven TCF/LEF binding sites, thereby reflecting canonical Wnt signaling (Maretto et al. 2003). These mice showed that Wnt signaling is active in cortical PCs during early forebrain development in a high-medial to low-lateral pattern, and that as development progresses, cortical Wnt signaling decreases (Backman et al. 2005). Electroporation studies of Wnt-responsive elements driving fluorescent markers at midneurogenesis indicate that Wnt signaling is active in radial glia (RG) PCs and a subpopulation of intermediate PCs (IPCs); is down-regulated in differentiating cells as they migrate away from the ventricular zone; but is then reactivated in cortical neurons 24 h after they enter the cortical plate (Fig. 1C) (Woodhead et al. 2006; Munji et al. 2011). Immunohistochemistry studies identified that β-catenin is localized to the apical (ventricular) surface of cortical PCs, and colocalizes with E-cadherin and N-cadherin, markers of apical adherens junctions (Junghans et al. 2005; Stocker and Chenn 2006).

Wnt ligand expression changes postnatally, resulting in de novo expression or decreased expression of Wnt ligands. *Wnt2b* and *Wnt5a*,

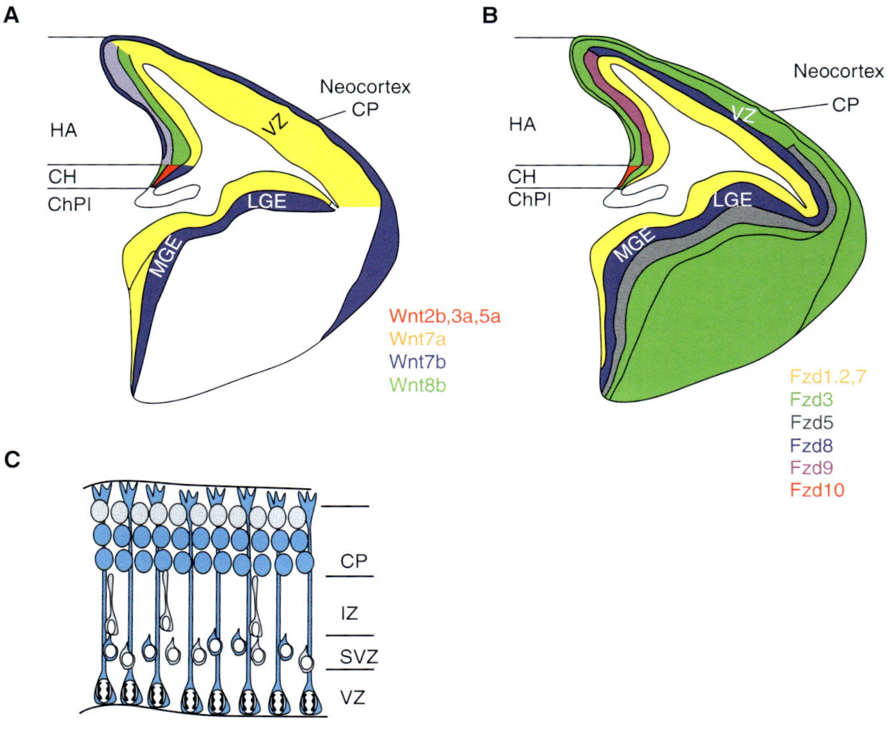

Figure 1. Expression of Wnt signaling components at E12.5 and E14.5. (*A*) Expression of Wnt ligands in the E12.5 telencephalon. (*B*) Expression of Fzd receptors in the E12.5 telencephalon. (*C*) At E14.5, Wnt signaling is active (blue) in radial glia in the VZ, a subpopulation of IPCs in the SVZ, down-regulated in migrating neurons in the IZ. Twenty-four hours after cells reach the CP, they become Wnt responsive. Abbreviations: ChPl, choroid plexus; CH, cortical hem; HA, hippocampal anlage; VZ, ventricular zone; CP, cortical plate; LGE, lateral ganglionic eminence; MGE, medial ganglionic eminence; VZ, ventricular zone; SVZ, subventricular zone; IZ, intermediate zone.

which are confined to the cortical hem during embryonic development, are expressed in complementary layers in the postnatal somatosensory cortex (*Wnt2b*: layers IV and VI; *Wnt 5a*: layers II/III and V) (Shimogori et al. 2004b). *Wnt2b* expression continues in layer VI in the visual and auditory cortex, whereas *Wnt5a* expression in layers II/III and V is observed in the parietal and temporal cortex (Shimogori et al. 2004b). *Wnt7b*, which is expressed in the developing cortical plate (Kim et al. 2001a; Theil 2005), is expressed in a "complex lamina pattern" in the adult cortex (Shimogori et al. 2004b). Expression of *Wnt7a* is extinguished in the adult neocortex (Shimogori et al. 2004b). These studies indicate a dynamic regulation of members of the Wnt signaling pathway in the developing and adult forebrain and suggest potential roles in several stage-specific developmental events in the forebrain.

WNT SIGNALING AND FOREBRAIN PATTERNING

The developing forebrain is initially composed of a relatively homogeneous population of PCs. As development progresses, key signaling centers pattern the neuroepithelium, influencing the fate of the cells generated from different regions. As mentioned above, the cortical hem is a transient embryonic neuroepithelial structure located in the medial wall, between the hippocampal anlage and the chorid plexus, characterized by expression of Wnt and bone morphogenic protein (Bmp) family members (Furuta et al. 1997; Grove et al. 1998). It is now apparent that signaling from the cortical hem influences development of the adjacent hippocampal and cortical neuroepithelium and probably has long-range effects on the dorso–ventral patterning of the forebrain. These events and the role of the hem have been studied by analyzing Wnt signaling mutants and mice with altered cortical hem development.

Gli3, a downstream effector of Sonic hedgehog signaling, is disrupted in the mouse mutant *extra-toes*, resulting in severe alteration of the developing telencephalon (Hui and Joyner 1993; Franz 1994). In these mutant mice, the cortical hem fails to develop, and cortical hem-specific *Wnt* and *Bmp* expression is absent or severely reduced (Grove et al. 1998; Tole et al. 2000). Hippocampal markers *Ephb1* and *KA1* are absent from the medial wall of the developing telencephalon (Tole et al. 2000), and the hippocampus is not histologically evident in *extra-toes* mutants (Theil et al. 1999). Interestingly, although the forebrain choroid plexus does not express *Gli3*, it also fails to develop when *Gli3* is disrupted, suggesting a role for signaling from the cortical hem in regulating the induction of the choroid plexus as well (Franz 1994; Grove et al. 1998). These studies suggest that the Wnt-rich cortical hem controls the development of adjacent cortical structures—the developing hippocampus and choroid plexus. It is likely that Bmp signaling collaborates with Wnts, at least in choroid plexus development, because mutants lacking almost all Bmp signaling in the developing cortex lack any evidence of lateral ventricle choroid plexus development (Hebert et al. 2002).

Analysis of embryos deficient for Wnt3a and Lef1 confirm a role for Wnt signaling in the development of the hippocampus. In the absence of Wnt3a, the cortical hem and choroid plexus are specified and Bmp signaling is maintained; however, the medial wall of the telencephalon is shorter (from E10.75 onward), and the choroid plexus has an abnormal morphology (at E12.5) (Lee et al. 2000). The hippocampus and dentate gyrus are not present at E18.5 in *Wnt3a* mutants, and markers of these structures, *KA1* and *Steel*, are largely absent (remnants of CA1, CA3, and the subiculum can be detected at caudal levels) (Lee et al. 2000). Deletion of the transcription factor *Lef1* (*Lef1* knockout) results in essentially complete loss of the dentate gyrus and abnormal development of the CA fields of the hippocampus (Galceran et al. 2000). Disruption of *Lef1* via insertion of a *LacZ* gene ($Lef1^{lacZ/lacZ}$ mutants), which in addition to disrupting the normal function of Lef1 generates a Lef1-lacZ fusion dominant interfering protein, results in a complete loss of both the dentate gyrus and hippocampus (Galceran et al. 2000). Proliferation studies using 5-bromo-2′-deoxyuridine (BrdU) to label proliferating cells identified a decrease in the number of $BrdU^+$ cells in the medial wall of the developing telencephalon at E10.5 (*Wnt3a* mutants) and E11.5 (*Lef1* mutants), suggesting that Wnt signaling regulates the expansion of the hippocampal PC population by promoting proliferation of PCs (Galceran et al. 2000; Lee et al. 2000). This was confirmed by in vitro studies that showed that Wnt3a treatment of hippocampal PC cultures decreased the cell cycle length and increased the number of proliferating cells (Yoshinaga et al. 2010). These changes in the dynamics of PCs lead to an increase in the number of neurons after Wnt3a stimulation (Yoshinaga et al. 2010). Consistent with these studies, inactivation of β-catenin using *D6-Cre*, results in a shorter medial wall of the telencephalon and severe disruption of hippocampal fields CA1 and CA2 (Machon et al. 2003). Because *D6-Cre* results in the inactivation of β-catenin in the subiculum and CA1 fields but not in the cortical hem, CA3, and dentate gyrus (Machon et al. 2003), it supports a role for Wnt signaling in the target areas receiving Wnts expressed by the cortical hem. Interestingly, cell proliferation at E12.5 and E13.5 was not altered, which suggests that, whereas Wnt signaling may regulate the initial expansion of the neural epithelium giving rise to the hippocampus, ongoing proliferation of PCs may be independent of β-catenin-mediated signaling (Machon et al. 2003). In *Wnt3a*-mutants, *Lef1* knockout, and mice with *D6-Cre* conditional inactivation of β-catenin, molecular

 Cite this article as *Cold Spring Harb Perspect Biol* doi: 10.1101/cshperspect.a008094

markers of hippocampal cell types remain, suggesting that additional signaling pathways may influence cell fate specification. The complete loss of hippocampal and dentate gyrus markers in the *Lef1*$^{lacZ/lacZ}$ mutants raises the possibility that Lef1-mediated interactions may play a role in these alternative pathways.

Lhx2 is a homeobox gene expressed by cortical PCs throughout the cortex but excluded from the cortical hem (Nakagawa et al. 1999). Disruption of the *Lhx2* gene results in a smaller telencephalon and expansion of the cortical hem and choroid plexus (Bulchand et al. 2001; Monuki et al. 2001). Although previous studies suggested that the cortical hem is required for expansion of the hippocampal neuroepithelium, the hippocampus fails to form in *Lhx2* mutants (Bulchand et al. 2001). This apparent contradiction in function of the cortical hem in regulating hippocampal specification was examined using *Lhx2* mosaic models (Mangale et al. 2008). This study suggests that Lhx2 functions as a cortical selector gene. Medially Lhx2-negative tissue adopts a Wnt-expressing cortical hem fate and influences the surrounding Lhx2-positive tissue to adopt a hippocampal (medial regions) or cortical (lateral regions) fate (Mangale et al. 2008). This is consistent with studies showing that sustained Wnt signaling in cortical PCs (via expression of a β-catenin/Lef1 fusion protein) results in ectopic expression of hippocampal markers in the lateral cortex (Machon et al. 2007). Thus, Wnt signals from the cortical hem acting on adjacent neuroepithelium are required for the initial expansion of the hippocampal anlage and influence the subsequent specification of hippocampal cell types.

In addition to a role for the Wnt-rich cortical hem in specifying the adjacent hippocampus, this signaling center can have long-range effects establishing the molecular identity of PCs in the dorsal/ventral axis. In *extra-toes* mutant mice, wherein Wnt and Bmp signaling in the cortical hem is disrupted, the cortical neuroepithelium expresses markers of the ventral telencephalon (Tole et al. 2000; Kuschel et al. 2003). These changes in *extra-toes* mutants are associated with ectopic expression of *Fgf8* at the dorsal midline (Kuschel et al. 2003). Fgf signaling from the anterior telencephalon influences patterning along the anterior–posterior axis (Fukuchi-Shimogori and Grove 2001), and ectopic Fgf expression in the midline via in utero electroporation leads to decreased *Wnt* expression in the cortical hem (Shimogori et al. 2004a). The anterior *Fgf8* expression is normally restricted by cortical hem Bmp signaling (Shimogori et al. 2004a). Similarly, in transgenic mice, ectopically expressing Sonic hedgehog in cortical PCs, *Wnt* expression in the cortical hem is absent, *Fgf8* expression is expanded, and markers of the ventral telencephalon are expressed in the cortical neuroepithelium (Huang et al. 2007). Therefore, it seems that cross-regulation between Wnt, Bmp, FgF, and Shh signaling, produced by discrete signaling centers, is critical in controlling dorsal–ventral forebrain patterning.

Further evidence for the role of Wnt signaling in dorsal–ventral patterning was obtained from transgenic mice with altered expression of β-catenin. Loss of β-catenin in neural PCs from E8 resulted in expression of ventral restricted markers (*Gsh2*, *Mash1*, *Dlx2/5*) in cortical PCs (Backman et al. 2005). Loss of β-catenin in neural PCs from E11 did not alter the fate of PCs in the dorsal or ventral telencephalon, suggesting that the role of Wnt signaling in patterning the forebrain is relatively early (Backman et al. 2005). Expression of constitutively active β-catenin from E8 or E11 results in partial expression of dorsal-restricted markers in the ventral forebrain (most notably *Ngn2*) and a decrease in expression of ventral-restricted markers, suggesting that Wnt signaling may repress the molecular identity of ventral PC populations even at subsequent stages (Backman et al. 2005). The mechanism of Wnt regulation of dorsal/ventral patterning is not clear, but one likely possibility is the direct regulation of homeobox genes that are effectors of patterning. *Emx2* is a direct target of Wnt signaling, and in *extra-toes* mutants, *Emx1* expression is absent and *Emx2* expression is decreased (Theil et al. 1999, 2002; Tole et al. 2000). In β-catenin loss-of-function mutants (E8), expression of *Emx1* and *Emx2* is decreased, and in β-catenin loss-of-function mutants (E11), *Emx2* expression is decreased

(Backman et al. 2005). Because *Emx2* and *Pax6* are expressed in opposing gradients and function in an opposing manner to regulate early positional identity in the developing forebrain (Bishop et al. 2002), this is probably one of the ways that Wnt signaling regulates cortical PC regional identity. Alterations in Wnt signaling components are also found in *Pax6* mutants (Cho and Dressler 1998; Kim et al. 2001a; Muzio et al. 2002), and alterations in *Pax6* expression are observed in β-catenin gain-of-function mutants (Backman et al. 2005), suggesting that Wnt signaling may be both upstream and downstream developmentally of the key regulators of dorsal/ventral patterning.

WNT REGULATION OF RADIAL GLIA AND INTERMEDIATE PROGENITOR CELL FATE

The properties of cortical PCs have been characterized using lineage tracing, live cell imaging, and genetic manipulation studies (Kriegstein and Noctor 2004; Guillemot 2005; Hevner 2006). Radially oriented cortical PCs (also called radial glia, RG) are located in a proliferative zone adjacent to the lateral ventricles. Early in cortical development, RGs undergo symmetric proliferative divisions, resulting in the production of two daughter RGs, thus expanding the PC population (Takahashi et al. 1996). As neurogenesis progresses RGs can undergo asymmetric divisions generating an RG and either a neuron or an intermediate progenitor cell (IPC) (Noctor et al. 2001, 2004; Haubensak et al. 2004; Miyata et al. 2004). IPCs migrate away from the ventricular surface and typically undergo symmetric divisions, resulting in the production of two neurons, thus expanding neuronal output (Noctor et al. 2004). By mid-corticogenesis, terminal differentiation of IPCs represent the major means of neuron production (Noctor et al. 2004).

The continued expression of components of the Wnt signaling pathway after patterning events are complete suggests a role in subsequent developmental processes. In the presence of constitutively active Wnt signaling, via expression of a stabilized form of β-catenin, the forebrain is enlarged owing to a lateral expansion of the PC population and failure of asymmetric neuronal output (Chenn and Walsh 2002; Wrobel et al. 2007). Activation of Wnt signaling from E10.5, via expression of a β-catenin/Lef1 fusion protein, also inhibits cortical neural differentiation, whereas inactivation of β-catenin from E11 promoted terminal differentiation of PCs and produced early neurogenesis (Machon et al. 2007). Inactivation of β-catenin at E13.5 in cortical PCs by in utero electroporation resulted in an increase in the number of electroporated cells exiting the cell cycle, prematurely differentiating, and populating the cortical plate, compared with controls (Woodhead et al. 2006). These studies suggest a role for Wnt signaling in promoting symmetric proliferative divisions during early cortical development, expanding the PC population. In the event of failed Wnt signaling in the RG population, these cells instead apparently terminally differentiate and become neurons.

A role for Wnt signaling in regulating the maturation of RGs into IPCs has also been proposed. Some studies indicate that these effects on IPC dynamics may be predominantly secondary. Activation of Wnt signaling from E10.5 results in decreased IPC number, and IPC cell number is increased when Wnt signaling is lost from E10.5 (Machon et al. 2007; Mutch et al. 2010). Expression of constitutively active β-catenin from E10.5 expands the RG population by promoting symmetric proliferative divisions, and initially IPC number is decreased; however, as neurogenesis progresses, IPC number increases (Wrobel et al. 2007). Therefore, IPC production may just be delayed in these models and the observed alterations in the IPC population are secondary to the role for Wnt signaling in promoting the expansion of RG. However, because there is a precursor–product relationship between RGs and IPCs, it can be difficult to isolate the effects of Wnts on IPC cells using genetic methodology lacking appropriately specific tools.

In marked contrast, other studies using nongenetic methodologies suggested that Wnt signaling may, under some circumstances, promote neuronal differentiation rather than preserving PC identity. In primary neuronal cultures,

 Cite this article as *Cold Spring Harb Perspect Biol* doi: 10.1101/cshperspect.a008094

Wnt7a or constitutively active β-catenin promoted neuronal differentiation (Hirabayashi et al. 2004). In utero electroporation of constitutively active β-catenin at E13.5 resulted in an increase in the number of cells that had migrated to the cortical plate after 48 h, whereas inhibition of Wnt signaling led to an increase in the number of electroporated cells in the VZ after 48 h (Hirabayashi et al. 2004). Consistent with this role of Wnt signaling in promoting neurogenesis, expression of constitutively active β-catenin from E12.5 increased IPC number at the expense of RG number (Kuwahara et al. 2010). These findings may be due to a role of Wnt signaling in activating the pro-neural genes *Ngn1&2* and *N-Myc* (Hirabayashi et al. 2004; Israsena et al. 2004; Kuwahara et al. 2010).

In reconciliation of these two opposing roles of Wnt signaling in regulating cortical PCs, recent studies suggest that this pathway may have distinct effects on RG and IPCs. In utero electroporation of Wnt3a at E13.5 promotes expansion of the RG population, consistent with earlier studies; however, these studies identified an additional role for Wnt signaling in promoting the terminal differentiation of IPCs (Munji et al. 2011). Wnt3a promoted IPCs to differentiate into neurons, although these prematurely differentiated neurons failed to migrate properly to the cortical plate (Munji et al. 2011). Thus, the most consistent model is a temporal and lineage-dependent model for Wnt action, whereby the capacity of PCs to respond to Wnt signals changes as the PCs mature. Early in neurogenesis PCs (RGs) respond to Wnt stimulation by proliferating symmetrically, whereas later in development, although Wnt stimulation can still promote symmetric proliferation of RGs, it also induces neuronal differentiation by promoting terminal neurogenesis of IPCs (Fig. 2) (Hirabayashi et al. 2004; Hirabayashi and Gotoh 2005; Munji et al. 2011). These studies suggest a complex temporal and context-dependent control of Wnt signaling in distinct PC types.

WNT SIGNALING AND CELL ADHESION

In cortical PCs, β-catenin is localized to adherens junctions, associated with proteins that regulate cell adhesion, including N-cadherin and E-cadherin (Junghans et al. 2005; Stocker and Chenn 2006). Loss of β-catenin from E9.0 or expression of a constitutively active form of β-catenin from E10.5 results in disruption of the integrity of the neuroepithelium architecture and loss of adherens junctions at the apical

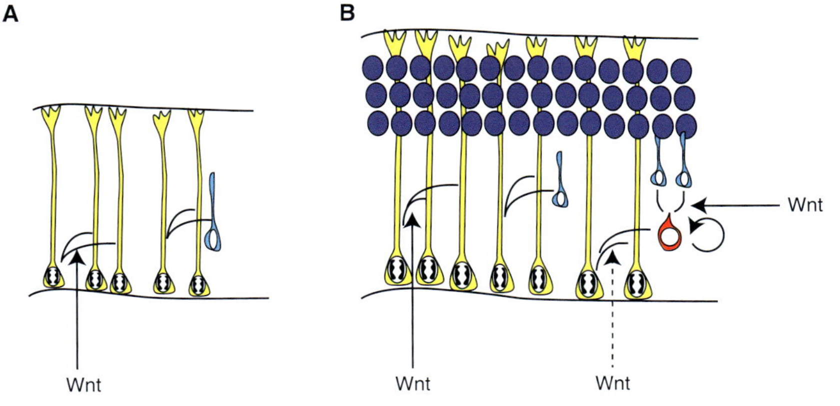

Figure 2. Wnt signaling promotes RG self-renewal and the differentiation of IPCs. (*A*) Early in cortical development (before E11.5), Wnt signaling promotes RG symmetric proliferative divisions, expanding the PC population. (*B*) During later cortical development (from E13.5), although Wnt signaling can still promote symmetric proliferative divisions, it also induces neurogenesis by promoting terminal neurogenesis of intermediate PCs. Wnt signaling may also promote the production of IPCs from RGs (*B*). (Yellow) RG; (red) IPC; (light blue) migrating neuron; (dark blue) mature neuron.

surface (Junghans et al. 2005; Wrobel et al. 2007). Focal loss of E-cadherin or N-cadherin results in increased neuronal differentiation and is accompanied by a loss of Wnt signaling (Stocker and Chenn 2006; Zhang et al. 2010). This disruption in neurogenesis can be rescued by expression of a constitutively active form of β-catenin (Stocker and Chenn 2006; Zhang et al. 2010). Thus, in addition to a potential role for β-catenin in mediating cell–cell adhesion interactions in the developing forebrain, the interaction of β-catenin with cadherin molecules at adherens junctions may function to influence the dynamics of cell division type.

WNT AND LAMINATION

The cortex is a laminar structure containing six layers of cells that show distinct molecular characteristics, which is populated in an inside-out manner with deep layers born before more-superficial layers (Angevine and Sidman 1961; Molyneaux et al. 2007). Wnt signaling has been implicated in laminar cell fate specification of cortical excitatory neurons. Activation of Wnt signaling in cortical PCs from E10.5 results in a loss of superficial cortical layers, whereas deep cortical layers are still generated (Wrobel et al. 2007). In utero electroporation of constitutively active β-catenin at E13.5 promotes a deep layer fate and inhibits the production of a superficial layer fate, whereas inhibition of Wnt signaling with a dominant-negative TCF construct had an opposite effect (Wrobel et al. 2007; Mutch et al. 2010). At E15.5, activation of Wnt signaling promoted a deep cortical fate at the expense of a superficial cortical fate (Mutch et al. 2010). In contrast, another study using the Wnt inhibitor Dkk1 identified that in utero electroporation at E13.5 inhibited the production of deep cortical layers, whereas in utero electroporation at E15.5 inhibited the production of superficial cortical layers, suggesting that Wnt signaling generally promotes neurogenesis, rather than exerting a specific role in laminar fate specification (Munji et al. 2011). All of these methods alter the dynamics of cortical PCs, and the phenotypes observed may reflect secondary effects, rather than a direct role for Wnt signaling in neural cell-type specification. Disruption of Wnt signaling specifically in postmitotic neurons should address the role of Wnt signaling in cortical laminar fate.

WNT AND VENTRAL FOREBRAIN NEUROGENESIS

Similar to its role in cortical PCs, Wnt signaling in the ventral forebrain regulates the expansion of PCs in the medial ganglionic eminence (Gulacsi and Anderson 2008). In the developing spinal cord, Wnt signaling regulates oligodendrocyte maturation (Fancy et al. 2009; Feigenson et al. 2009; Ye et al. 2009). The first wave of ventrally derived oligodendrocytes is produced in the medial ganglionic eminence/anterior entopeduncular area from E11.5 (Kessaris et al. 2006). Loss of the non-canonical Wnt receptor Ryk results in an increase in the production of early-born oligodendrocytes at the expense of GABAergic interneurons (Zhong et al. 2011). These studies also identified that Wnt3a stimulation promotes a neuronal fate over an oligodendrocyte fate, and Ryk is required for this response (Zhong et al. 2011). These findings suggest that in addition to regulating the expansion of PCs in the ventral forebrain, Wnt signaling controls cell-type specification as well.

CONCLUSIONS

Wnts play numerous roles at distinct developmental stages depending on the context of the receiving cells and the source and identities of the ligands. Following forebrain specification, Wnts contribute to anterior–posterior and dorsal–ventral patterning, similar to their role in other CNS regions and organ systems. The biggest contribution of Wnts in this process comes from the cortical hem, a signaling center that is a transient developmental source of several Wnt ligands. Once these domains are established, Wnts regulate the dynamic behavior of neural progenitor cells, first supporting symmetric, expansive stem cell divisions and later driving the differentiation of limited potential progenitors. Wnts are also involved in the cell fate decisions that these progenitors make as they differentiate. Thus, the future prospects for understanding

 Cite this article as *Cold Spring Harb Perspect Biol* doi: 10.1101/cshperspect.a008094

how Wnts regulate all of these events in the nervous system are really dependent on understanding with which other signaling events Wnts intersect at each of these nodal steps. The next few years will certainly bring many of these interactions to light.

ACKNOWLEDGMENTS

This work is supported by grants from NIDA: R01 DA017627, NIMH: R01 MH077694, CIRM, and the National Multiple Sclerosis Society. We thank Julie Siegenthaler for her help with the figures.

REFERENCES

Angevine JB Jr, Sidman RL. 1961. Autoradiographic study of cell migration during histogenesis of cerebral cortex in the mouse. *Nature* **192:** 766–768.

Backman M, Machon O, Mygland L, van den Bout CJ, Zhong W, Taketo MM, Krauss S. 2005. Effects of canonical Wnt signaling on dorso–ventral specification of the mouse telencephalon. *Dev Biol* **279:** 155–168.

Bishop KM, Rubenstein JL, O'Leary DD. 2002. Distinct actions of Emx1, Emx2, and Pax6 in regulating the specification of areas in the developing neocortex. *J Neurosci* **22:** 7627–7638.

Bulchand S, Grove EA, Porter FD, Tole S. 2001. LIM-homeodomain gene *Lhx2*regulates the formation of the cortical hem. *Mech Dev* **100:** 165–175.

Chenn A, Walsh CA. 2002. Regulation of cerebral cortical size by control of cell cycle exit in neural precursors. *Science* **297:** 365–369.

Cho EA, Dressler GR. 1998. TCF-4 binds β-catenin and is expressed in distinct regions of the embryonic brain and limbs. *Mech Dev* **77:** 9–18.

Fancy SP, Baranzini SE, Zhao C, Yuk DI, Irvine KA, Kaing S, Sanai N, Franklin RJ, Rowitch DH. 2009. Dysregulation of the Wnt pathway inhibits timely myelination and remyelination in the mammalian CNS. *Genes Dev* **23:** 1571–1585.

Feigenson K, Reid M, See J, Crenshaw EB III, Grinspan JB. 2009. Wnt signaling is sufficient to perturb oligodendrocyte maturation. *Mol Cell Neurosci* **42:** 255–265.

Fischer T, Guimera J, Wurst W, Prakash N. 2007. Distinct but redundant expression of the Frizzled Wnt receptor genes at signaling centers of the developing mouse brain. *Neuroscience* **147:** 693–711.

Fotaki V, Price DJ, Mason JO. 2011. Wnt/β-catenin signaling is disrupted in the *extra-toes* ($Gli3_{Xt/Xt}$) mutant from early stages of forebrain development, concomitant with anterior neural plate patterning defects. *J Comp Neurol* **519:** 1640–1657.

Franz T. 1994. *Extra-toes* (*Xt*) homozygous mutant mice demonstrate a role for the *Gli-3* gene in the development of the forebrain. *Acta Anat (Basel)* **150:** 38–44.

Fukuchi-Shimogori T, Grove EA. 2001. Neocortex patterning by the secreted signaling molecule FGF8. *Science* **294:** 1071–1074.

Furuta Y, Piston DW, Hogan BL. 1997. Bone morphogenetic proteins (BMPs) as regulators of dorsal forebrain development. *Development* **124:** 2203–2212.

Galceran J, Miyashita-Lin EM, Devaney E, Rubenstein JL, Grosschedl R. 2000. Hippocampus development and generation of dentate gyrus granule cells is regulated by LEF1. *Development* **127:** 469–482.

Gorski JA, Talley T, Qiu M, Puelles L, Rubenstein JL, Jones KR. 2002. Cortical excitatory neurons and glia, but not GABAergic neurons, are produced in the Emx1-expressing lineage. *J Neurosci* **22:** 6309–6314.

Grove EA, Tole S, Limon J, Yip L, Ragsdale CW. 1998. The hem of the embryonic cerebral cortex is defined by the expression of multiple Wnt genes and is compromised in Gli3-deficient mice. *Development* **125:** 2315–2325.

Guillemot F. 2005. Cellular and molecular control of neurogenesis in the mammalian telencephalon. *Curr Opin Cell Biol* **17:** 639–647.

Gulacsi AA, Anderson SA. 2008. β-Catenin-mediated Wnt signaling regulates neurogenesis in the ventral telencephalon. *Nat Neurosci* **11:** 1383–1391.

Haubensak W, Attardo A, Denk W, Huttner WB. 2004. Neurons arise in the basal neuroepithelium of the early mammalian telencephalon: A major site of neurogenesis. *Proc Natl Acad Sci* **101:** 3196–3201.

Hebert JM, Mishina Y, McConnell SK. 2002. BMP signaling is required locally to pattern the dorsal telencephalic midline. *Neuron* **35:** 1029–1041.

Hevner RF. 2006. From radial glia to pyramidal-projection neuron: Transcription factor cascades in cerebral cortex development. *Mol Neurobiol* **33:** 33–50.

Hirabayashi Y, Gotoh Y. 2005. Stage-dependent fate determination of neural precursor cells in mouse forebrain. *Neurosci Res* **51:** 331–336.

Hirabayashi Y, Itoh Y, Tabata H, Nakajima K, Akiyama T, Masuyama N, Gotoh Y. 2004. The Wnt/β-catenin pathway directs neuronal differentiation of cortical neural precursor cells. *Development* **131:** 2791–2801.

Huang X, Litingtung Y, Chiang C. 2007. Ectopic Sonic hedgehog signaling impairs telencephalic dorsal midline development: Implication for human holoprosencephaly. *Hum Mol Genet* **16:** 1454–1468.

Hui CC, Joyner AL. 1993. A mouse model of Greig cephalopolysyndactyly syndrome: The *extra-toes^J* mutation contains an intragenic deletion of the *Gli3*gene. *Nat Genet* **3:** 241–246.

Israsena N, Hu M, Fu W, Kan L, Kessler JA. 2004. The presence of FGF2 signaling determines whether β-catenin exerts effects on proliferation or neuronal differentiation of neural stem cells. *Dev Biol* **268:** 220–231.

Junghans D, Hack I, Frotscher M, Taylor V, Kemler R. 2005. β-Catenin-mediated cell-adhesion is vital for embryonic forebrain development. *Dev Dyn* **233:** 528–539.

Kessaris N, Fogarty M, Iannarelli P, Grist M, Wegner M, Richardson WD. 2006. Competing waves of oligodendrocytes in the forebrain and postnatal elimination of an embryonic lineage. *Nat Neurosci* **9:** 173–179.

Kim AS, Anderson SA, Rubenstein JL, Lowenstein DH, Pleasure SJ. 2001a. *Pax-6* regulates expression of *SFRP-2* and *Wnt-7b* in the developing CNS. *J Neurosci* **21:** RC132–RC136.

Kim AS, Lowenstein DH, Pleasure SJ. 2001b. Wnt receptors and Wnt inhibitors are expressed in gradients in the developing telencephalon. *Mech Dev* **103:** 167–172.

Kriegstein AR, Noctor SC. 2004. Patterns of neuronal migration in the embryonic cortex. *Trends Neurosci* **27:** 392–399.

Kuhl M, Sheldahl LC, Park M, Miller JR, Moon RT. 2000. The Wnt/Ca^{2+} pathway: A new vertebrate Wnt signaling pathway takes shape. *Trends Genet* **16:** 279–283.

Kuschel S, Ruther U, Theil T. 2003. A disrupted balance between Bmp/Wnt and Fgf signaling underlies the ventralization of the Gli3 mutant telencephalon. *Dev Biol* **260:** 484–495.

Kuwahara A, Hirabayashi Y, Knoepfler PS, Taketo MM, Sakai J, Kodama T, Gotoh Y. 2010. Wnt signaling and its downstream target N-myc regulate basal progenitors in the developing neocortex. *Development* **137:** 1035–1044.

Lee SM, Tole S, Grove E, McMahon AP. 2000. A local Wnt-3a signal is required for development of the mammalian hippocampus. *Development* **127:** 457–467.

Logan CY, Nusse R. 2004. The Wnt signaling pathway in development and disease. *Annu Rev Cell Dev Biol* **20:** 781–810.

MacDonald BT, Tamai K, He X. 2009. Wnt/β-catenin signaling: Components, mechanisms, and diseases. *Dev Cell* **17:** 9–26.

Machon O, van den Bout CJ, Backman M, Kemler R, Krauss S. 2003. Role of β-catenin in the developing cortical and hippocampal neuroepithelium. *Neuroscience* **122:** 129–143.

Machon O, Backman M, Machonova O, Kozmik Z, Vacik T, Andersen L, Krauss S. 2007. A dynamic gradient of Wnt signaling controls initiation of neurogenesis in the mammalian cortex and cellular specification in the hippocampus. *Dev Biol* **311:** 223–237.

Mangale VS, Hirokawa KE, Satyaki PR, Gokulchandran N, Chikbire S, Subramanian L, Shetty AS, Martynoga B, Paul J, Mai MV, et al. 2008. Lhx2 selector activity specifies cortical identity and suppresses hippocampal organizer fate. *Science* **319:** 304–309.

Maretto S, Cordenonsi M, Dupont S, Braghetta P, Broccoli V, Hassan AB, Volpin D, Bressan GM, Piccolo S. 2003. Mapping Wnt/β-catenin signaling during mouse development and in colorectal tumors. *Proc Natl Acad Sci* **100:** 3299–3304.

Miyata T, Kawaguchi A, Saito K, Kawano M, Muto T, Ogawa M. 2004. Asymmetric production of surface-dividing and non-surface-dividing cortical progenitor cells. *Development* **131:** 3133–3145.

Molyneaux BJ, Arlotta P, Menezes JR, Macklis JD. 2007. Neuronal subtype specification in the cerebral cortex. *Nat Rev Neurosci* **8:** 427–437.

Monuki ES, Porter FD, Walsh CA. 2001. Patterning of the dorsal telencephalon and cerebral cortex by a roof plate–Lhx2 pathway. *Neuron* **32:** 591–604.

Munji RN, Choe Y, Li G, Siegenthaler JA, Pleasure SJ. 2011. Wnt signaling regulates neuronal differentiation of cortical intermediate progenitors. *J Neurosci* **31:** 1676–1687.

Mutch CA, Schulte JD, Olson E, Chenn A. 2010. β-Catenin signaling negatively regulates intermediate progenitor population numbers in the developing cortex. *PLoS ONE* **5:** e12376.

Muzio L, DiBenedetto B, Stoykova A, Boncinelli E, Gruss P, Mallamaci A. 2002. Emx2 and Pax6 control regionalization of the pre-neuronogenic cortical primordium. *Cereb Cortex* **12:** 129–139.

Nakagawa Y, Johnson JE, O'Leary DD. 1999. Graded and areal expression patterns of regulatory genes and cadherins in embryonic neocortex independent of thalamocortical input. *J Neurosci* **19:** 10877–10885.

Noctor SC, Flint AC, Weissman TA, Dammerman RS, Kriegstein AR. 2001. Neurons derived from radial glial cells establish radial units in neocortex. *Nature* **409:** 714–720.

Noctor SC, Martinez-Cerdeno V, Ivic L, Kriegstein AR. 2004. Cortical neurons arise in symmetric and asymmetric division zones and migrate through specific phases. *Nat Neurosci* **7:** 136–144.

Ono K, Takebayashi H, Ikeda K, Furusho M, Nishizawa T, Watanabe K, Ikenaka K. 2008. Regional- and temporal-dependent changes in the differentiation of Olig2 progenitors in the forebrain, and the impact on astrocyte development in the dorsal pallium. *Dev Biol* **320:** 456–468.

Parr BA, Shea MJ, Vassileva G, McMahon AP. 1993. Mouse Wnt genes exhibit discrete domains of expression in the early embryonic CNS and limb buds. *Development* **119:** 247–261.

Pinson KI, Brennan J, Monkley S, Avery BJ, Skarnes WC. 2000. An LDL-receptor-related protein mediates Wnt signalling in mice. *Nature* **407:** 535–538.

Seifert JR, Mlodzik M. 2007. Frizzled/PCP signalling: A conserved mechanism regulating cell polarity and directed motility. *Nat Rev Genet* **8:** 126–138.

Shimogori T, Banuchi V, Ng HY, Strauss JB, Grove EA. 2004a. Embryonic signaling centers expressing BMP, WNT and FGF proteins interact to pattern the cerebral cortex. *Development* **131:** 5639–5647.

Shimogori T, VanSant J, Paik E, Grove EA. 2004b. Members of the Wnt, Fz, and Frp gene families expressed in postnatal mouse cerebral cortex. *J Comp Neurol* **473:** 496–510.

Stocker AM, Chenn A. 2006. Differential expression of α-E-catenin and α-N-catenin in the developing cerebral cortex. *Brain Res* **1073–1074:** 151–158.

Takahashi T, Nowakowski RS, Caviness VS Jr. 1996. The leaving or Q fraction of the murine cerebral proliferative epithelium: A general model of neocortical neuronogenesis. *J Neurosci* **16:** 6183–6196.

Theil T. 2005. Gli3 is required for the specification and differentiation of preplate neurons. *Dev Biol* **286:** 559–571.

Theil T, Alvarez-Bolado G, Walter A, Ruther U. 1999. Gli3 is required for Emx gene expression during dorsal telencephalon development. *Development* **126:** 3561–3571.

Theil T, Aydin S, Koch S, Grotewold L, Ruther U. 2002. Wnt and Bmp signalling cooperatively regulate graded Emx2 expression in the dorsal telencephalon. *Development* **129:** 3045–3054.

Tole S, Ragsdale CW, Grove EA. 2000. Dorsoventral patterning of the telencephalon is disrupted in the mouse mutant *extra-toes*^{J}. *Dev Biol* **217:** 254–265.

van Amerongen R, Nusse R. 2009. Towards an integrated view of Wnt signaling in development. *Development* **136:** 3205–3214.

Wonders C, Anderson SA. 2005. Cortical interneurons and their origins. *Neuroscientist* **11:** 199–205.

Woodhead GJ, Mutch CA, Olson EC, Chenn A. 2006. Cell-autonomous β-catenin signaling regulates cortical precursor proliferation. *J Neurosci* **26:** 12620–12630.

Wrobel CN, Mutch CA, Swaminathan S, Taketo MM, Chenn A. 2007. Persistent expression of stabilized β-catenin delays maturation of radial glial cells into intermediate progenitors. *Dev Biol* **309:** 285–297.

Ye F, Chen Y, Hoang T, Montgomery RL, Zhao XH, Bu H, Hu T, Taketo MM, van Es JH, Clevers H, et al. 2009. HDAC1 and HDAC2 regulate oligodendrocyte differentiation by disrupting the β-catenin–TCF interaction. *Nat Neurosci* **12:** 829–838.

Yoshinaga Y, Kagawa T, Shimizu T, Inoue T, Takada S, Kuratsu J, Taga T. 2010. Wnt3a promotes hippocampal neurogenesis by shortening cell cycle duration of neural progenitor cells. *Cell Mol Neurobiol* **30:** 1049–1058.

Zhang J, Woodhead GJ, Swaminathan SK, Noles SR, McQuinn ER, Pisarek AJ, Stocker AM, Mutch CA, Funatsu N, Chenn A. 2010. Cortical neural precursors inhibit their own differentiation via N-cadherin maintenance of β-catenin signaling. *Dev Cell* **18:** 472–479.

Zhong J, Kim HT, Lyu J, Yoshikawa K, Nakafuku M, Lu W. 2011. The Wnt receptor Ryk controls specification of GABAergic neurons versus oligodendrocytes duringtelencephalon development. *Development* **138:** 409–419.

Wnt Signaling in Normal and Malignant Hematopoiesis

William Lento[1], Kendra Congdon[2], Carlijn Voermans[3], Marcie Kritzik[4], and Tannishtha Reya[4]

[1]Division of Cellular Therapy, Department of Medicine, Duke University Medical Center, Durham, North Carolina 27710

[2]Duke Brain Tumor Immunotherapy Program, Duke University Medical Center, Durham, North Carolina 27710

[3]Department of Hematopoiesis, Sanquin Research and Landsteiner Laboratory, Academic Medical Center, University of Amsterdam, Amsterdam, The Netherlands

[4]Department of Pharmacology, University of California, San Diego School of Medicine, La Jolla, California 92093

Correspondence: treya@ucsd.edu

One of the most remarkable characteristics of stem cells is their ability to perpetuate themselves through self-renewal while concomitantly generating differentiated cells. In the hematopoietic system, stem cells balance these mechanisms to maintain steady-state hematopoiesis for the lifetime of the organism, and to effectively regenerate the system following injury. Defects in the proper control of self-renewal and differentiation can be potentially devastating and contribute to the development of malignancies. In this review, we trace the emerging role of Wnt signaling as a critical regulator of distinct aspects of self-renewal and differentiation, its contribution to the maintenance of homeostasis and regeneration, and how the pathway can be hijacked to promote leukemia development. A better understanding of these processes could pave the way to enhancing recovery after injury and to developing better therapeutic approaches for hematologic malignancies.

Hematopoietic stem cells (HSCs) give rise to all the cells of the blood. The ability of HSCs to successfully balance self-renewal to permit regeneration with differentiation to produce mature blood cells is critical to survival. Regulation of this balance recapitulates events that occur during embryonic development, and thus often uses many of the key signaling pathways found in ontogeny. In this context the Wnt pathway, which is a key player in pattern formation and other aspects of early development, has emerged as an important new regulator of many aspects of hematopoiesis, including stem cell establishment, homeostatic maintenance, and regeneration. Furthermore, Wnt signaling can also influence many differentiated cells of the blood. Given the close connections between the self-renewal that occurs during normal development and the uncontrolled self-renewal that can occur in cancers, Wnt signaling has also been studied in the context of cancers of the blood and has been shown to contribute to

Cite this article as *Cold Spring Harb Perspect Biol* doi: 10.1101/cshperspect.a008011

the aberrant renewal that drives leukemia formation. As with all pleiotropic factors, the Wnt pathway has a multifaceted influence on hematopoiesis. The diversity of effects comes in part from the complex nature of the signal itself: the family represents 19 different ligands and up to ten different receptors, and the nature of the ligand and receptor combination can activate either canonical or noncanonical cascades. Further, distinct cellular contexts provided by the multitude of developing and mature blood lineage cells can lead to differential outcomes following activation or inhibition of the pathway. Despite this complexity, experiments from many different angles are leading to consensus views of the nature of Wnt signaling in the hematopoietic system and how it can contribute to hematopoietic stem cell and progenitor cell function in specific contexts, raising the possibility that if correctly modulated it could be used to enhance hematopoietic regeneration. Perhaps even more exciting is the emerging role of Wnt signaling in driving the growth of numerous leukemias and lymphomas, suggesting that its effective blockade could have a profound impact on the treatment of hematologic malignancies in the clinic.

HSCs

Activation of Wnt Signaling during Hematopoiesis

During development, hematopoiesis occurs at distinct anatomical sites. Murine hematopoietic development begins in the extraembryonic yolk sac at E7.5, continues in the chorioallantoic placenta, para-aortic splanchnopleura, and aorta–gonad–mesonephros (AGM) at E9 and the fetal liver at E10, and culminates in the colonization of the bone marrow at E15 (Fig. 1) (Medvinsky et al. 2011). In adult life all hematopoiesis stably occurs in the bone marrow.

At each of these sites in fetal and adult life the maturation of blood cells from the hematopoietic stem cell follows a series of commitment steps (Fig. 2). Long-term HSCs lie at the apex of the hierarchy and display an extensive capacity for self-renewal, allowing them to reconstitute the hematopoietic system in a recipient for the long term. These cells give rise to the "short-term" HSCs, which have more limited self-renewal capacity and are only able to support reconstitution for a few weeks following transplantation. These short-term HSCs give rise to multipotent progenitor cells, which have no self-

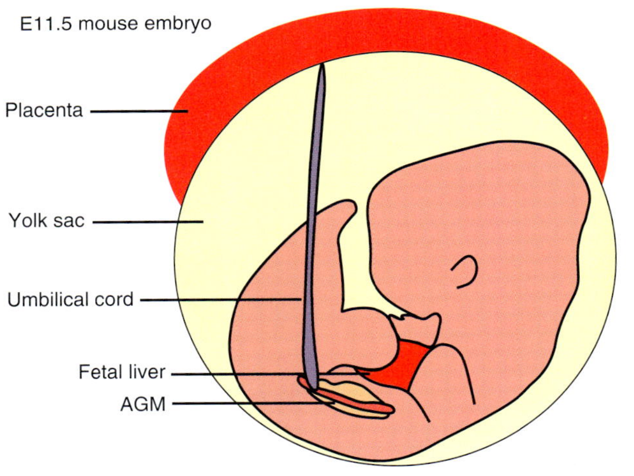

Figure 1. Murine embryonic hematopoiesis occurs simultaneously in several organs during development. The mouse embryo generates de novo hematopoietic stem cells (HSCs) in the placenta, yolk sac, and aorta–gonad–mesonephros (AGM) beginning at E7.5. These HSCs then migrate into and seed various organs at the start of definitive hematopoiesis around E11. Normal adult hematopoiesis only occurs in the bone marrow. (Figure created from data in Medvinsky et al. 2011.)

 Cite this article as *Cold Spring Harb Perspect Biol* doi: 10.1101/cshperspect.a008011

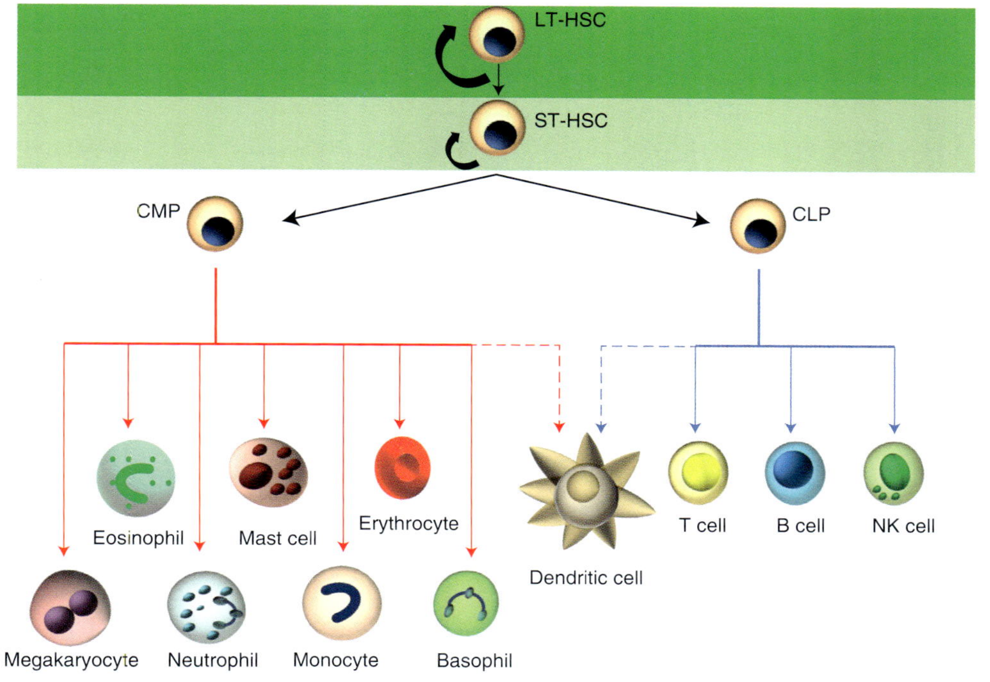

Figure 2. The hematopoietic stem cell hierarchy in adult bone marrow. All differentiated blood cells are generated from a small pool of self-renewing long-term hematopoietic stem cells (LT-HSCs) (dark green area). LT-HSCs are capable of both self-renewal and differentiation. When LT-HSCs differentiate, they create short-term HSCs (ST-HSCs) with limited self-renewal potential (light green area). The ST-HSCs produce progenitors that give rise to multipotent non-self-renewing common myeloid progenitors (CMP) and common lymphoid progenitors (CLP). These progenitor populations generate mature hematopoietic cells such as macrophages, B and T cells, and red blood cells (erythrocytes). The dashed lines indicate partial lineage connections.

renewal capacity but which can differentiate to the full complement of hematopoietic cells, in part by stepping through semicommitted states such as the common myeloid progenitor, a precursor of cells like macrophages and granulocytes, and the common lymphoid progenitor, a precursor of B, T, and NK cells.

Early evidence that Wnt signaling may be involved in orchestrating hematopoietic ontogeny came from studies showing that Wnt ligands, including Wnt-5a and Wnt-10b, are expressed by the murine fetal liver (Austin et al. 1997). Subsequent studies, including observations in both mouse and *Xenopus* embryos that Wnt signaling is found in sites of primitive erythropoiesis (Cheng et al. 2008; Tran et al. 2010), as well as the demonstration that Wnt ligands and frizzled receptors are expressed in the murine yolk sac, aorta–gonad–mesonephros, and fetal liver, support a role for Wnt signaling in developmental hematopoiesis (Corrigan et al. 2009). More recently, studies using zebrafish have shown that early embryonic expression of the noncanonical ligand Wnt-16 in the somites is required to initiate definitive hematopoiesis in the prehematopoietic mesoderm and dorsal aorta (Clements et al. 2011).

In the adult bone marrow, Wnts are expressed in the microenvironment surrounding the HSCs (Table 1). For example, studies performed on human bone marrow indicate that Wnt-2b, Wnt-5a, and Wnt-10b are expressed in unfractionated bone marrow cells, and that Wnt-5a is expressed in populations enriched for HSCs and progenitors (Van den Berg et al. 1998). In the mouse bone marrow, Wnt-5a is

Table 1. Expression of Wnt pathway elements in the bone marrow microenvironment

Cell source	Wnt ligands	FZD/LRP	sFRP/Dkk/ WIF	Transcription factors	References
Osteoblasts	Wnt2, Wnt2B2, Wnt3A, Wnt4, Wnt5AQ, Wnt10A, Wnt11	FZD2, FZD3, FZD4, FZD5, FZD6	sFRP1, sFRP2, sFRP3, sFRP4	TCF1, TCF2, TCF3, TCF4, TLE1, CTBP1, β-catenin	Spencer et al. 2006
Mesenchymal stem cell	Wnt2B, Wnt5A, Wnt2B, Wnt11	FZD1, FZD2, FZD3, FZD4, FZD5, FZD6	sFRP2, sFRP2, Dkk2	β-catenin, TCF1	Dufourcq et al. 2008; Li et al. 2010; Qui et al. 2011
Endothelial cells	Wnt2, Wnt2b, Wnt3, Wnt5A, Wnt7A, Wnt 11, Wnt14	FZD3, FZD4, FZD6, LRP5, LRP6	sFRP1, sFRP3, Dkk1, Dkk2, Dkk3	β-catenin, TCF1, Lef1	Li et al. 2010; Planutiene et al. 2011

expressed by mature B cells (Liang et al. 2003), whereas Wnt-10b is produced by myeloid cells, erythrocytes, and immature B cells (Congdon et al. 2008). Moreover, the precise expression of Wnt ligand seems to be regulated dynamically during hematopoietic regeneration; whereas Wnt-3a, Wnt-5a, and Wnt-10b are all expressed in the homeostatic murine bone marrow microenvironment (Reya et al. 2000), Wnt-10b is specifically up-regulated in the injured murine bone marrow microenvironment (Congdon et al. 2008), suggesting that specific Wnts may be particularly adept at or used during regeneration. The receptors for these ligands are also extensively expressed (Table 1), with long-term human HSCs expressing Frizzled 6 (Wagner et al. 2004). These studies collectively indicate that Wnt signaling components are found in both fetal and adult sites of hematopoiesis. Further, the possibility that ligand and receptor expression is indicative of physiologic utilization is supported by reporter expression in HSCs (Nemeth et al. 2007).

Gain-of-Function Studies

In an effort to understand whether the observed expression of Wnt/Fz is functionally important, numerous gain-of-function and loss-of-function experiments have been performed in model organisms as well as in human cells. Gain-of-function experiments to assess Wnt signaling in hematopoiesis have been approached in several ways, including exposure to soluble Wnt proteins, inhibition of GSK-3β, and activation of β-catenin.

Independent studies have shown that the addition of Wnt-3a can increase hematopoietic output in murine embryonic stem cell cultures and that canonical Wnt signaling can promote proliferation of committed hematopoietic progenitors in human embryonic stem cell cultures (Corrigan et al. 2009; Goessling et al. 2009). These studies are consistent with reports of Wnt-10b-induced proliferation of cultured murine fetal liver hematopoietic progenitor cells (Austin et al. 1997). Signaling via Wnt-11 can also increase the number of hematopoietic progenitors, possibly implicating noncanonical Wnts in hematopoietic lineage specification (Vijayaragavan et al. 2009).

Wnt signaling has also been triggered in murine and human HSCs using inhibitors of GSK-3β, a strategy that leads to activation of β-catenin. In vivo administration of GSK-3β inhibitors has been shown to promote the engraftment of HSCs in bone marrow transplantation models (Trowbridge et al. 2006), and short-term pretreatment of human HSC with a GSK-3β inhibitor can enhance the engraftment of human hematopoietic progenitor cells in xenograft models (Ko et al. 2011). In wild-type bone marrow, GSK-3β inhibition resulted in transient expansion of phenotypic hematopoietic stem cells, but concurrently resulted in progressive depletion of the long-term repopulation ability of these HSCs (Huang et al. 2009).

Studies based on the more direct activation of β-catenin have also provided important insight into the role of Wnt signaling in hematopoiesis. Transduction of the constitutively active form of β-catenin into HSCs of transgenic mice expressing the prosurvival protein Bcl-2 increased the number of hematopoietic stem cells during culture in vitro and enhanced reconstitution in vivo (Reya et al. 2003). Activated β-catenin can also reprogram CLPs to display more characteristics of undifferentiated multipotent cells (Baba et al. 2005) and its expression in murine bone marrow cells dramatically enhanced the ability of these cells to expand in culture. The cells maintained many primitive characteristics and responsiveness to normal signals, although they appeared arrested in an undifferentiated state (Baba et al. 2006). Interestingly, using a genetic approach, mice expressing stabilized β-catenin in the hematopoietic system were generated; these mice showed expansion of HSCs, arrested differentiation, and subsequent defects in hematopoietic reconstitution (Kirstetter et al. 2006; Scheller et al. 2006). By combining β-catenin activation with inhibition of apoptosis through PTEN deletion, recent studies have also shown that induction of prosurvival pathways in combination with Wnt activation drives the self-renewal and long-term reconstitution ability of HSCs (Perry et al. 2011). This study together with previous work (Reya et al. 2003) suggests that the cellular context of β-catenin activation and whether it is coexpressed with a survival factor may dictate its ultimate effect on expansion or exhaustion. Additionally, the levels at which β-catenin is overexpressed may also dictate the outcome. This was most effectively shown in recent studies in which a genetic series of adenomatous polyposis coli (APC) alleles was used to activate β-catenin at different levels, resulting in either expansion of HSCs at low doses or in their exhaustion at high doses (Luis et al. 2011).

Work from model organisms such as zebrafish is consistent with a role for Wnt signaling in the induction of long-term HSCs (Goessling et al. 2009). Prostaglandin E_2 regulates hematopoietic stem cell specification in zebrafish as well (North et al. 2007) and promotes increased proliferation, survival, homing, and engraftment of murine HSC (Hoggatt et al. 2009; Durand and Zon 2010). Importantly, PGE_2 can activate β-catenin-responsive Wnt reporter activity in the zebrafish AGM, the site of long-term hematopoietic stem cell potential. Studies in mice further showed that PGE_2-dependent Wnt signaling enhanced hematopoietic progenitor cell generation from differentiating embryonic stem cells. Cumulatively, the work over the last decade indicates that the Wnt pathway is active in HSCs and that triggering it either at low levels or at high levels in the context of survival cues can lead to expansion of the stem cell pool.

Loss-of-Function Studies

To define whether Wnt signaling may be required for HSC specification, growth, and function, many studies targeting deletion of β-catenin and other components of the Wnt signaling pathway have been undertaken. Germline deletion of Wnt-3a leads to a dramatic reduction in the numbers of hematopoietic stem and progenitor cells in the fetal liver and results in embryonic lethality at E12.5 (Luis et al. 2009). Interestingly, the impaired self-renewal capacity of Wnt-3a null cells is not reversed by a normal microenvironment after transplantation, suggesting that the developmental absence of Wnt-3a may affect multiple aspects of stem cell function. Subsequent studies using a Vav-Cre transgene to conditionally delete β-catenin in the hematopoietic system of developing mice supported these findings. Although HSCs were established in these mice, they were deficient in their long-term maintenance and reconstitution capacity in transplantation studies (Zhao et al. 2007). Interestingly, mice in which an IFN-inducible Mx-Cre transgene was used to delete β-catenin in bone marrow progenitor cells did not display defects in the ability of HSC to self-renew or reconstitute hematopoietic lineages. The different consequences of β-catenin deletion in vivo are very likely attributable to the different contexts in which the deletions were initiated. The impact of deleting β-catenin in embryonic versus adult life could lead to the differences observed; alternatively, interferon (IFN) exposure, which can activate HSCs (Baldridge et al. 2010) in the Mx-Cre mice, could change the transcriptional context within the HSCs in which β-catenin is deleted and mask the effects of β-catenin loss.

Wnt Signaling in Lymphocyte Development

As stem cells differentiate to generate functional cells of the blood, they give rise to committed cells of both myeloid and lymphoid lineages. Components of the Wnt pathway, such as the downstream transcription factors Lef and Tcf, are expressed in developing lymphocytes (Oosterwegel et al. 1991; Travis et al. 1991; van Genderen et al. 1994). Gain- and loss-of-function approaches indicate that Wnt/β-catenin signaling is a key regulator of T-cell development during thymocyte differentiation (Molenaar et al. 1996; Schilham et al. 1998; Nusse 1999; Ioannidis et al. 2001; Staal et al. 2001; Mulroy et al. 2002, 2003; Weerkamp et al. 2006). Overexpression of Dkk1, a Wnt antagonist that prevents interactions between Wnt and Frizzled, blocks T-cell development at the double-negative ($CD4^-/CD8^-$) stage in fetal thymic organ cultures (Weerkamp et al. 2006). Furthermore, in vivo genetic deletion of β-catenin in T cells using Lck-Cre causes a maturation defect in double-positive T cells at the β-selection checkpoint and significantly reduces the peripheral splenic T-cell population (Xu et al. 2003). The Wnt/β-catenin signaling pathways also continue to influence T cells post differentiation (Gattinoni and Restifo 2010). In cytotoxic T cells, Wnt signaling promotes the formation of $CD8^+$ memory T cells by inhibiting effector cell differentiation. In the helper T-cell lineage, Wnt signaling enhances the survival of regulatory $CD4^+$ T cells and influences $CD4^+$ Th-cell polarization, favoring the formation of Th2 over Th1 cells (Yu et al. 2009; Notani et al. 2010). In the B-cell lineage, genetic studies using Lef-1-deficient mice show that loss of Lef-1 leads to defects in pro-B-cell proliferation and survival in vitro and in vivo and that purified Wnt3a can induce proliferation of immature B cells (Reya et al. 2000). Consistent with this, B-cell defects were reported in the absence of Wnt receptors in Frizzled 9-deficient mice (Ranheim et al. 2005). The fact that activated β-catenin can lead to B-cell leukemia in transplant assays in the long term is consistent with the possibility that activation of this pathway promotes the growth and survival of B cells.

A few studies have also examined the role of the noncanonical Wnt signaling pathway in lymphocyte development. Wnt-5a-deficient mice display abnormal B-cell development; in this setting, Wnt-5a was found to signal at least in part through the noncanonical Wnt/Ca^{2+} pathway to interfere with pro-B-cell responses to IL-7 (Liang et al. 2003). The possibility that Wnt-5a may oppose canonical signaling was also suggested by the increased canonical signaling observed in thymocytes and distal limb buds from Wnt-5a$^{-/-}$ mice (Topol et al. 2003; Liang et al. 2007). In addition, whereas canonical Wnt signaling mediated by Wnt-3a overexpression in stromal cells from osteopetrotic mice inhibited B-cell and plasmacytoid dendritic cell development, Wnt-5a increased B-cell lymphopoiesis (Malhotra et al. 2008). Interestingly, in this setting the addition of Wnt-3a triggered increased expression of stem cell markers, suggesting the possibility that these culture conditions can induce an undifferentiated state.

These studies cumulatively indicate that Wnt signaling is important not only in HSCs but also in more committed lymphoid progenitor cells, including pro-B cells and pre-B cells in the bone marrow, as well as in the most immature subsets of thymocytes and in mature T cells (Staal and Clevers 2005). Emerging studies also indicate that Wnt signaling can influence nonlymphoid lineages. For example, GSK-3β negatively regulates megakaryocyte differentiation and platelet production from primary human bone marrow cells in vitro. In addition, numerous Wnt canonical effectors have been detected in human platelets, and Wnt-3a negatively regulates platelet function, including adhesion, activation, granule secretion, and aggregation (Steele et al. 2009).

LEUKEMIA

Wnt Signaling in Myeloid Leukemia

In addition to its role in normal hematopoietic function, activation of Wnt signaling is intricately involved in cancer formation in blood cells. Several studies in a variety of leukemic cell lines and patient samples support this possibility. For example, pre-B-cell leukemia lines carrying the E2A-PbX translocation overexpress Wnt proteins (McWhirter et al. 1999), and survival of these cells in vitro can be inhibited by blocking Wnt signaling (Mazieres et al. 2005). Similarly, studies of acute myelogenous leukemia show that associated fusion proteins enhance replating efficiency of HSCs and that this is abrogated on inhibition of the Wnt pathway (Muller-Tidow et al. 2004; Zheng et al. 2004). Most importantly, primary cells from patients with chronic myelogenous leukemia (CML) display activated Wnt signaling and a dependence on this pathway for their growth in vitro (Jamieson et al. 2004). An investigation into the molecular basis of Wnt activation in CML patient samples identified a novel missplicing of GSK-3β RNA that deletes exons 8 and 9 (Abrahamsson et al. 2009). The resulting truncated GSK-3β protein lacks a critical axin binding domain and is therefore unable to phosphorylate β-catenin to initiate its degradation. These experiments were complemented with genetic experiments showing that the Wnt pathway was required for CML initiation and maintenance in vivo. Specifically, whether β-catenin activity was essential for initiation and maintenance of leukemia stem cells was tested by transducing β-catenin null stem cells with BCR-ABL translocation and transferring them into recipient mice (Zhao et al. 2007). Overall, the loss of β-catenin led to a significant reduction in leukemia incidence and increase in leukemia latency in vivo (Zhao et al. 2007). Interestingly, BCR-ABL-induced B-ALL still occurred, suggesting that the dependence of leukemia on β-catenin may differ with cellular context and the cell of origin for the leukemia. The possibility that Wnt is an independent regulator of CML is particularly important because current evidence indicates that although Gleevec blocks BCR-ABL in CML cancer stem cells, these cells are no longer dependent on BCR-ABL and continue to propagate the tumor. Thus, Wnt signaling could represent an effective ABL-independent target for pharmacological intervention in CML.

Following the work performed in CML, the dependence of acute myelogenous leukemia on β-catenin in vivo was also tested. Importantly, the genetic deletion of β-catenin led to reduced leukemia incidence and a blockade of leukemia stem cell self-renewal (Wang et al. 2010) in mouse models of acute myelogenous leukemia driven either with a combination of the oncogenes Hoxa9 and Meis1 or by viral delivery of the translocation MLL-AF9. Interestingly, activation of β-catenin directly imparted self-renewal capacity to otherwise nonrenewing granulocyte-macrophage progenitors when expressed in conjunction with HoxA9/Meis1a (Wang et al. 2010).

The loss of β-catenin can also affect AML driven by other oncogenes. β-catenin deletion in preleukemia stem cells showed that β-catenin is required for initiation of mixed lineage leukemia-eleven nineteen leukemia (MLL-ENL)-driven AML (Yeung et al. 2010). Viral knockdown of β-catenin in patient samples decreases colony formation and reduces the frequency of long-term initiating cells in vitro. Additionally, β-catenin activation is required for drug resistance to GSK3 inhibitors in MLL-ENL leukemia stem cells (Yeung et al. 2010). These data collectively indicate that propagation of acute myeloid leukemia requires intact activity of β-catenin for leukemia stem cell self-renewal and resistance to GSK3 inhibitors and suggests that directly targeting β-catenin could impact multiple aspects of leukemia pathogenesis. It is important to note that in leukemia driven by MLL-AF4 and MLL-AF5 translocations, GSK3 inhibition paradoxically inhibits leukemia clonogenic potential through regulation of the cell-cycle inhibitor $p27^{kip1}$ (Wang et al. 2008). These data suggest that specific MLL translocations may differentially use Wnt signaling, and that the context set up by the different oncogenic alleles governs leukemia stem cell behavior in distinct ways.

Wnt Signaling in T-Cell Leukemia and Lymphoma

To study whether activation of β-catenin can lead to T-cell leukemia, transgenic mice in which activated β-catenin is driven by Lck (LckCre-$Ctnnb^{Dex3}$) or CD4 (CD4Cre-$Ctnnb^{Dex3}$) have been generated (Guo et al. 2007). These mice display $CD4^{+}CD8^{+}$ thymic lymphoma within 60 days. In other studies in which β-catenin was driven with the Lck promoter, tumors only arose in the absence of p53 (Xu et al. 2008). The fact that PTEN mutant fetal liver stem cells drive a combination of acute myeloid leukemia and $CD4^{+}CD8^{+}$ T-acute lymphoblastic leukemia (T-ALL) also suggests an oncogenic role for β-catenin in c-$Kit^{mid}CD3^{+}lineage^{-}$ T-ALL (Guo et al. 2008). Consistent with a role for β-catenin in mediating self-renewal of the leukemia stem cell population, β-catenin heterozygous T-ALLs show premature exhaustion and lose tumor-initiating ability. Although most studies have focused on aspects of β-catenin and its role in leukemia, the noncanonical β-catenin-independent Wnt pathway may also modulate leukemia propagation and survival. A synthetic lethal screen for short hairpin RNAs (shRNAs) that cooperate with imatinib mesylate to induce apoptosis identified the Wnt/Ca^{+}/NFAT (nuclear factor associated with T cells) signaling pathway as required for CML survival in vitro (Gregory et al. 2010) Interestingly, the requirement for NFAT signaling was also shown in B-cell acute lymphoblastic leukemia survival in vivo (Gregory et al. 2010).

SUMMARY

The activation of Wnt signaling from development to oncogenesis in the hematopoietic system is indicative of a crucial role for the pathway. In many ways, this is an exciting time, with opportunities to pursue a deeper understanding of the mechanisms Wnt signaling controls and its interactions with other key developmental signals. Importantly, this understanding should provide a strong foundation for initiating concerted efforts to identify therapeutics in the Wnt cascade that could prove broadly effective in enhancing stem-cell based therapies and targeting leukemia.

REFERENCES

Abrahamsson AE, Geron I, Gotlib J, Dao K-HT, Barroga CF, Newton IG, Giles FJ, Durocher J, Creusot RS, Karimi M, et al. 2009. Glycogen synthase kinase 3β missplicing contributes to leukemia stem cell generation. *Proc Natl Acad Sci* **106:** 3925–3959.

Austin TW, Solar GP, Ziegler FC, Liem L, Matthews W. 1997. A role for the Wnt gene family in hematopoiesis: Expansion of multilineage progenitor cells. *Blood* **89:** 3624–3635.

Baba Y, Garrett KP, Kincade PW. 2005. Constitutively active β-catenin confers multilineage differentiation potential on lymphoid and myeloid progenitors. *Immunity* **23:** 599–609.

Baba Y, Yokota T, Spits H, Garrett KP, Hayashi S, PW K. 2006. Constitutively active β-catenin promotes expansiion for multipotent hematopoietic progenitors in culture. *J Immunol* **4:** 2294–2303.

Baldridge MT, King KY, Boles NC, Weksberg DC, Goodell MA. 2010. Quiescent haematopoietic stem cells are activated by IFN-γ in response to chronic infection. *Nature* **465:** 793–797.

Cheng X, Huber TL, Chen VC, Gadue P, Keller GM. 2008. Numb mediates the interaction between Wnt and Notch to modulate primitive erythropoietic specification from the hemangioblast. *Development* **135:** 3447–3458.

Clements WK, Kim AD, Ong KG, Moore JC, Lawson ND, Traver D. 2011. A somitic Wnt16/Notch pathway specifies haematopoietic stem cells. *Nature* **474:** 220–225.

Congdon KL, Voermans C, Ferguson EC, DiMascio LN, Uqoezwa M, Zhao C, Reya T. 2008. Activation of Wnt signaling in hematopoietic regeneration. *Stem Cells* **26:** 1202–1210.

Corrigan PM, Dobbin E, Freeburn RW, Wheadon H. 2009. Patterns of Wnt/Fzd/LRP gene expression during embryonic hematopoiesis. *Stem Cells Dev* **18:** 759–772.

Dufourcq P, Descamps B, Tojais NF, Leroux L, Oses P, Daret D, Moreau C, Lamazière JM, Couffinhal T, Duplàa C. 2008. Secreted frizzled-related protein-1 enhances mesenchymal stem cell function in angiogenesis and contributes to neovessel maturation. *Stem Cells* **26:** 2991–3001.

Durand EM, Zon LI. 2010. Newly emerging roles for prostaglandin E2 regulation of hematopoiesis and hematopoietic stem cell engraftment. *Curr Opin Hematol* **17:** 308–312.

Gattinoni L, Ji Y, Restifo NP. 2010. Wnt/β-catenin signaling in T-cell immunity and cancer immunotherapy. *Clin Cancer Res* **16:** 4695–4701.

 Cite this article as *Cold Spring Harb Perspect Biol* doi: 10.1101/cshperspect.a008011

Goessling W, North TE, Loewer S, Lord AM, Lee S, Stoick-Cooper CL, Weidinger G, Puder M, Daley GQ, Moot RT, et al. 2009. Genetic interaction of PGE2 and Wnt signaling regulates developmental specification of stem cells and regeneration. *Cell* **136:** 1136–1147.

Gregory M, Phang T, Neviani P, Alvarez-Calderon F, Eide C, O'Hare T, Zaberezhnyv V, Williams RT, Druker BJ, Perrotti D, et al. 2010. Wnt/Ca^{2+}/NFAT signaling maintains survival of Ph^{+} leukemia cells upon inhibition of Bcr-Abl. *Cancer Cell* **18:** 74–87.

Guo Z, Dose M, Kovalovsky D, Chang R, O'Neil J, Look AT, von Boehmer H, Khazaie K, Gounari F. 2007. β-Catenin stabilization stalls the transition from double-positive to single-positive stage and predisposes thymocytes to malignant transformation. *Blood* **109:** 5463–5472.

Guo W, Lasky JL, Chang C-J, Mosessian S, Lewis X, Xiao Y, Yeh JE, Chen JY, Iruela-Arispe ML, Varella-Garcia M, et al. 2008. Multi-genetic events collaboratively contribute to Pten-null leukaemia stem-cell formation. *Nature* **453:** 529–533.

Hoggatt J, Singh P, Sampath J, Pelus LM. 2009. Prostaglandin E2 enhances hematopoietic stem cell homing, survival, and proliferation. *Blood* **113:** 5444–5455.

Huang J, Zhang Y, Bersenev A, O'Brien WT, Tong W, Emerson SG, Klein PS. 2009. Pivotal role for glycogen synthase kinase-3 in hematopoietic stem cell homeostasis in mice. *J Clin Invest* **12:** 3519–3529.

Ioannidis V, Beermann F, Clevers H, Held W. 2001. The β-catenin-TCF-1 pathway ensures $CD4^{+}CD8^{+}$ thymocyte survival. *Nat Immunol* **2:** 691–697.

Jamieson CH, Ailles LE, Dylla SJ, Muijtjens MS, Jones C, Zehnder JL, Gotlib J, Li K, Manz MG, Keating A, et al. 2004. Granulocyte–macrophage progenitors in chronic myelogenous leukemia are candidate leukemia stem cells in blast-crisis CML. *New Engl J Med* **351:** 657–667.

Kirstetter P, Anderson K, Porse BT, Jacobsen SEW, Nerlov C. 2006. Activation of the canonical Wnt pathway leads to loss of hematopoietic stem cell repopulation and multilineage differentiation block. *Nat Immunol* **7:** 1048–1056.

Ko KH, Holmes T, Palladinetti P, Song E, Nordon R, O'Brien TA, Dolnikov A. 2011. GSK-3β inhibition promotes engraftment of ex vivo-expanded hematopoietic stem cells and modulates gene expression. *Stem Cells* **1:** 101–118.

Li H, Daculsi R, Grellier M, Bareille R, Bourget C, Amedee J. 2010. Role of neural-cadherin in early osteoblastic differentiation of human bone marrow stromal cells cocultured with human umbilical vein endothelial cells. *Am J Physiol Cell Physiol* **299:** C422–C430.

Liang H, Chen Q, Coles AH, Anderson SJ, Pihan G, Bradley A, Gerstein R, Jurecic R, Jones SN. 2003. Wnt5a inhibits B cell proliferation and functions as a tumor suppressor in hematopoietic tissue. *Cancer Cell* **4:** 349–360.

Liang H, Coles AH, Zhu Z, Zayas J, Jurecic R, Kang J, Jones SN. 2007. Noncanonical Wnt signaling promotes apoptosis in thymocyte development. *J Exp Med* **204:** 3077–3084.

Luis TC, Weerkamp F, Naber BAE, Baert MRM, de Haas EFE, Nikolic T, Heuvelmans S, De Krijger RR, van Dongen JJ, Staal FJ. 2009. Wnt3a deficiency irreversibly impairs hematopoietic stem cell self-renewal and leads to defects in progenitor cell differentiation. *Blood* **113:** 546–554.

Luis TC, Naber BA, Fibbe WE, van Dongen JJ, Staal FJ. 2011. Canonical Wnt signaling regulates hematopoiesis in a dosage-dependent fashion. *Cell Stem Cell* **9:** 345–356.

Malhotra S, Baba Y, Garrett KP, Staal FJ, Gerstein R. 2008. Contrasting responses of lymphoid progenitors to canonical and noncanonical Wnt signals. *J Immunol* **181:** 3955–3964.

Mazieres J, You L, He B, Xu Z, Lee AY, Makami I, McCormick F, Jablons DM. 2005. Inhibition of Wnt16 in human acute lymphoblastoid leukemia cells containing the t(1;19) translocation induces apoptosis. *Oncogene* **24:** 5396–5400.

McWhirter JR, Neuteboom ST, Wancewicz EV, Monia BP, Downing JR, Murre C. 1999. Oncogenic homeodomain transcription factor E2A-Pbx1 activates a novel WNT gene in pre-B acute lymphoblastoid leukemia. *Proc Natl Acad Sci* **96:** 11464–11469.

Medvinsky A, Rybtsov S, Taoudi S. 2011. Embryonic origin of the adult hematopoietic system: Advances and questions. *Development* **138:** 1017–1031.

Molenaar M, van de Wetering M, Oosterwegel M, Peterson-Maduro J, Godsave S, Korinek V, Roose J, Destrée O, Clevers H. 1996. XTcf-3 transcription factor mediates β-catenin-induced axis formation in *Xenopus* embryos. *Cell* **86:** 391–399.

Muller-Tidow C, Steffen B, Cauvet T, Tickenbrock L, Ji P, Diederichs S, Sargin B, Köhler G, Stelljes M, Puccetti E, et al. 2004. Translocation products in acute myeloid leukemia activate the Wnt signaling pathway in hematopoietic cells. *Mol Cell Biol* **24:** 2890–2904.

Mulroy T, McMahon JA, Burakoff SJ, McMahon AP, Sen J. 2002. Wnt-1 and Wnt-4 regulate thymic cellularity. *Eur J Immunol* **32:** 967–971.

Mulroy T, Xu Y, Sen JM. 2003. β-catenin expression enhances generation of mature thymocytes. *Int Immunol* **15:** 1485–1494.

Nemeth MJ, Topol L, Anderson SM, Yang Y, Bodine DM. 2007. Wnt5a inhibits canonical Wnt signaling in hematopoietic stem cells and enhances repopulation. *Proc Natl Acad Sci* **104:** 15436–15441.

North TE, Goessling W, Walkley CR, Lengerke C, Kopani KR, Lord AM, Weber GJ, Bowman TV, Jang IH, Grosser T, et al. 2007. Prostaglandin E2 regulates vertebrate haematopoietic stem cell homeostasis. *Nature* **447:** 1007–1011.

Notani D, Gottimukkala KP, Jayani RS, Limaye AS, Damle MV, Mehta S, Purbey PK, Joseph J, Galande S. 2010. Global regulator SATB1 recruits β-catenin and regulates T(H)2 differentiation in Wnt-dependent manner. *PLoS Biol* **8:** 1.

Nusse R. 1999. WNT targets: Repression and activation. *Trends Genet* **15:** 1–3.

Oosterwegel M, van de Wetering M, Dooijes D, Klomp L, Winoto A, Georgopoulos K, Meijlink F, Clevers H. 1991. Cloning of murine TCF-1, a T cell-specific transcription factor interacting with functional motifs in the CD3-e and TCR-a enhancers. *J Exp Med* **173:** 1133–1142.

Perry JM, He XC, Sugimura R, Grindley JC, Haug JS, Ding S, Li L. 2011. Cooperation between both Wnt/β-catenin and PTEN/PI3K/Akt signaling promotes primitive hematopoietic stem cell self-renewal and expansion. *Genes Dev* **25:** 1928–1942.

Planutiene M, Planutis K, Holcombe RF. 2011. Lymphoid enhancer-binding factor 1, a representative of vertebrate-specific Lef1/Tcf1 sub-family, is a Wnt-β-catenin pathway target gene in human endothelial cells which regulates matrix metalloproteinase-2 expression and promotes endothelial cell invasion. *Vasc Cell* **3:** 28.

Qiu W, Chen L, Kassem M. 2011. Activation of non-canonical Wnt/JNK pathway by Wnt3a is associated with differentiation fate determination of human bone marrow stromal (mesenchymal) stem sells. *Biochem Biophys Res Commun* **413:** 98–104.

Ranheim EA, Kwan HC, Reya T, Wang YK, Weissman IL, Francke U. 2005. Frizzled 9 knock-out mice have abnormal B-cell development. *Blood* **105:** 2487–2494.

Reya T, O'Riordan M, Okamura R, Devaney E, Willert K, Nusse R, Grosschedl R. 2000. Wnt signaling regulates B lymphocyte proliferation through a LEF-1 dependent mechanism. *Immunity* **13:** 15–24.

Reya T, Duncan AW, Ailles L, Domen J, Scherer DC, Willert K, Hintz L, Nusse R, Weissman IL. 2003. A role for Wnt signalling in self-renewal of haematopoietic stem cells. *Nature* **423:** 409–414.

Scheller M, Huelsken J, Rosenbauer F, Taketo MM, Birchmeier W, Tenen DG, Leutz A. 2006. Hematopoietic stem cell and multilineage defects generated by constitutive β-catenin activation. *Nat Immunol* **7:** 1037–1047.

Schilham MW, Wilson A, Moerer P, Benaissa-Trouw BJ, Cumano A, Clevers HC. 1998. Critical involvement of Tcf-1 in expansion of thymocytes. *J Immunol* **161:** 3984–3991.

Spencer GJ, Utting JC, Etheridge SL, Arnett TR, Genever PG. 2006. Wnt signaling in osteoblasts regulates expression of the receptor activator of NFκB ligand and inhibits osteoclastogenesis in vitro. *J Cell Sci* **119:** 1283–1296.

Staal FJ, Clevers H. 2005. WNT signalling and haematopoiesis: A WNT-WNT situation. *Nat Rev Immunol* **5:** 21–30.

Staal FJ, Meeldijk J, Moerer P, Jay P, van de Weerdt BC, Vainio S, Nolan GP, Clevers H. 2001. Wnt signaling is required for thymocyte development and activates Tcf-1 mediated transcription. *Eur J Immunol* **31:** 285–293.

Steele BM, Harper MT, Macaulay IC, Morrell CN, Perez-Tamayo A, Foy M, Habas R, Poole AW, Fitzgerald DJ, Maguire PB. 2009. Canonical Wnt signaling negatively regulates platelet function. *Proc Natl Acad Sci* **106:** 1936–1941.

Topol L, Jiang X, Choi H, Garrett-Beal L, Carolan PJ. 2003. Wnt-5a inhibits the canonical Wnt pathway by promoting GSK-3-independent β-catenin degradation. *J Cell Biol* **162:** 899–908.

Tran HT, Sekkali B, Van Imschoot G, Janssens S, Vleminckx K. 2010. Wnt/β-catenin signaling is involved in the induction and maintenance of primitive hematopoiesis in the vertebrate embryo. *Proc Natl Acad Sci* **107:** 16160–16165.

Travis A, Amsterdam A, Belanger C, Grosschedl R. 1991. LEF-1, a gene encoding a lymphoid-specific protein with an HMG domain regulates T-cell receptor a enhancer function. *Genes Dev* **5:** 880–894.

Trowbridge JJ, Xenocostas A, Moon RT, Bhatia M. 2006. Glycogen synthase kinase-3 is an in vivo regulator of hematopoietic stem cell repopulation. *Nat Med* **1:** 89–98.

Van Den Berg DJ, Sharma AK, Bruno E, Hoffman R. 1998. Role of members of the Wnt gene family in human hematopoiesis. *Blood* **92:** 3189–3202.

van Genderen C, Okamura RM, Fariñas I, Quo RG, Parslow TG, Bruhn L, Grosschedl R. 1994. Development of several organs that require inductive epithelial-mesenchymal interactions is impaired in LEF-1 deficient mice. *Genes Dev* **8:** 2691–2703.

Vijayaragavan K, Szabo E, Bosse M, Menendez P, Weisel K, Bhatia M. 2009. Noncanonical Wnt signaling orchestrates early developmental events toward hematopoietic cell fate from human embryonic stem cells. *Cell Stem Cell* **4:** 248–262.

Wagner W, Ansorge A, Wirkner U, Eckstein V, Schwager C, Blake J, Miesala K, Selig J, Saffrich R, Ansorge W, et al. 2004. Molecular evidence for stem cell function of the slow dividing fraction among human hematopoietic progenitor cells by genome wide analysis. *Blood* **104:** 675–686.

Wang Z, Smith KS, Murphy M, Piloto O, Somervaille TCP, Cleary ML. 2008. Glycogen synthase kinase 3 in MLL leukaemia maintenance and targeted therapy. *Nature* **455:** 1205–1209.

Wang Y, Krivtsov AV, Sinha AU, North TE, Goessling W, Feng Z, Zon LI, Armstrong SA. 2010. The Wnt/β-catenin pathway is required for the development of leukemia stem cells in AML. *Science* **327:** 1650–1653.

Weerkamp F, Baert MR, Naber BA, Koster EE, de Haas EF, Atkuri KR, van Dongen JJ, Herzenberg LA, Staal FJ. 2006. Wnt signaling in the thymus is regulated by differential expression of intracellular signaling molecules. *Proc Natl Acad Sci* **103:** 3322–3326.

Xu Y, Banerjee D, Huelsken J, Birchmeier W, Sen J. 2003. Deletion of β-catenin impairs T cell development. *Nat Immunol* **4:** 1177–1182.

Xu M, Yu Q, Subrahmanyam R, Difilippantonio MJ, Ried T, Sen JM. 2008. β-Catenin expression results in p53-independent DNA damage and oncogene-induced senescence in prelymphomagenic thymocytes in vivo. *Mol Cell Biol* **28:** 1713–1723.

Yeung J, Esposito MT, Gandillet A, Zeisig BB, Griessinger E, Bonnet D, So CW. 2010. β-Catenin mediates the establishment and drug resistance of MLL leukemic stem cells. *Cancer Cell* **18:** 606–618.

Yu Q, Sharma A, Oh SY, Moon HG, Hossain MZ, Salay TM, Leeds KE, Du H, Wu B, Waterman ML, et al. 2009. T cell factor 1 initiates the T helper type 2 fate by inducing the transcription factor GATA-3 and repressing interferon-γ. *Nat Immunol* **9:** 992–999.

Zhao C, Blum J, Chen A, Kwon HY, Jung SH, Cook JM, Lagoo A, Reya T. 2007. Loss of β-catenin impairs the renewal of normal and CML stem cells in vivo. *Cancer Cell* **12:** 528–541.

Zheng X, Beissert T, Kukoc-Zivojnov N, Puccetti E, Altschmied J, Strolz C, Boehrer S, Gul H, Schneider O, Ottmann OG, et al. 2004. γ-Catenin contributes to leukemogenesis induced by AML-associated translocation products by increasing the self-renewal of very primitive progenitor cells. *Blood* **103:** 3535–3543.

Wnt Signaling in Skin Development, Homeostasis, and Disease

Xinhong Lim and Roel Nusse

Department of Developmental Biology, Stanford University School of Medicine, Stanford, California 94305

Correspondence: xinhonglim@gmail.com

The skin and its appendages constitute the largest organ of the body. Its stratified epithelia offer protection from environmental stresses such as dehydration, irradiation, mechanical trauma, and pathogenic infection, whereas its appendages, like hair and sebaceous glands, help regulate body temperature as well as influence animal interaction and social behavior through camouflage and sexual signaling. To respond to and function effectively in a dynamic external environment, the skin and its appendages possess a remarkable ability to regenerate in a carefully controlled fashion. When this finely tuned homeostatic process is disrupted, skin diseases such as cancers may result. At present, the molecular signals that orchestrate cell proliferation, differentiation, and patterning in the skin remain incompletely understood. It is increasingly apparent that many morphogenetic pathways with key roles in development are also important in regulating skin biology. Of these, Wnt signaling has emerged as the dominant pathway controlling the patterning of skin and influencing the decisions of embryonic and adult stem cells to adopt the various cell lineages of the skin and its appendages, as well as subsequently controlling the function of differentiated skin cells. Here we will review established concepts and present recent advances in our understanding of the diverse roles that Wnt signaling plays in skin development, homeostasis, and disease.

WNT IN THE DEVELOPMENT OF SKIN AND ITS APPENDAGES (FIG. 1)

Wnt signaling is involved at the very earliest stages of skin development (Fuchs 2007). Following gastrulation, embryonic cells of the ectoderm and the mesoderm will differentiate to form the epidermis and dermis, respectively.

The ectoderm layer that emerges at the surface of the embryo is able to specify either the nervous system or the skin epithelium. The decision of which fate to adopt is strongly influenced by Wnt signaling, which inhibits the ectoderm's responsiveness to fibroblast growth factors (FGFs). Without FGF signaling, the ectodermal cells are able to express bone morphogenetic proteins (BMPs), which block neural induction and direct the cells to differentiate into keratin-expressing cells (keratinocytes) that form the epidermis (Wilson and Hemmati-Brivanlou 1995; Wilson et al. 2001).

The mesoderm, on the other hand, segments into somites, which eventually give rise to the dermis. Wnt signals are important for somitogenesis, and also instruct the lateral plate

Cite this article as *Cold Spring Harb Perspect Biol* doi: 10.1101/cshperspect.a008029

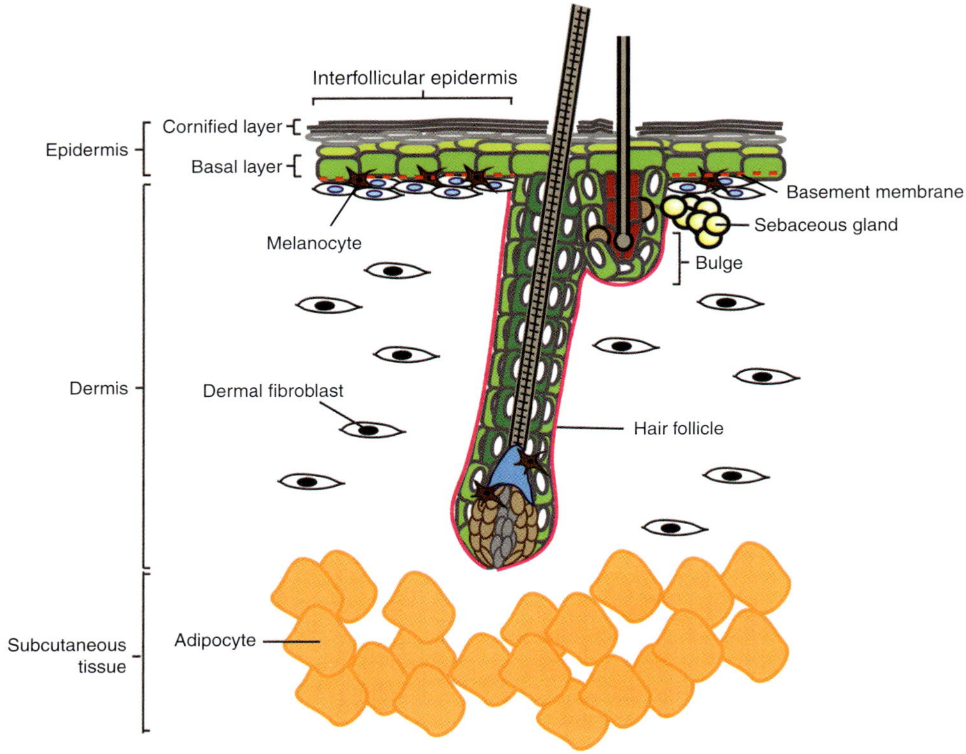

Figure 1. Anatomy of mammalian skin. The skin is a complex composite of multiple miniorgans. The topmost layer is a stratified epidermis consisting of keratinocytes that differentiate as they progress from the basal layer to the suprabasal and eventually form dead enucleated squames that are sloughed off from the cornified layer. Multiple different appendages branch from the epidermal layer, including hair follicles (HFs), sebaceous glands, and sweat glands. The dermis, which lies below the epidermis and surrounds the HFs, is composed of fibroblasts of mesenchymal origin that reside in a fibrous matrix that they produce. Melanocytes are interspersed among the keratinocytes of the epidermis and HF (bulge and matrix), where they pigment the skin and hair. Below all these is a subcutaneous layer of adipocytes that serve to insulate the animal.

mesoderm and somite-derived cells to form the mesenchymal cells populating the dermis (Atit et al. 2006; Ohtola et al. 2008).

When the embryonic epidermis and dermis are both in place, complex signaling begins between them that leads to fate changes in both tissue layers, ultimately resulting in epidermal patterning and the development of skin appendages like hair follicles (HFs) or feathers (Hardy 1992). Hair morphogenesis begins with an initial signal from the dermis ("first dermal message") that initiates the aggregation of the epidermis at regularly spaced intervals, leading to the formation of epidermal placodes that eventually develop into hair. These placodes then emit an epidermal signal that instructs underlying mesenchymal cells to cluster and form "dermal condensates." A "second dermal message" from the dermal condensates induces epidermal placode cells to rapidly divide downward and invade the dermis, enwrapping the dermal condensate, which becomes the HF dermal papilla. Further cross talk between epithelial cells of the developing hair leads to their proliferation and differentiation to form the inner root sheath and hair shaft of the mature HF.

Although the precise nature of the epidermal placode-inducing "first dermal message" remains poorly understood, several studies suggest that Wnts may constitute both the "first

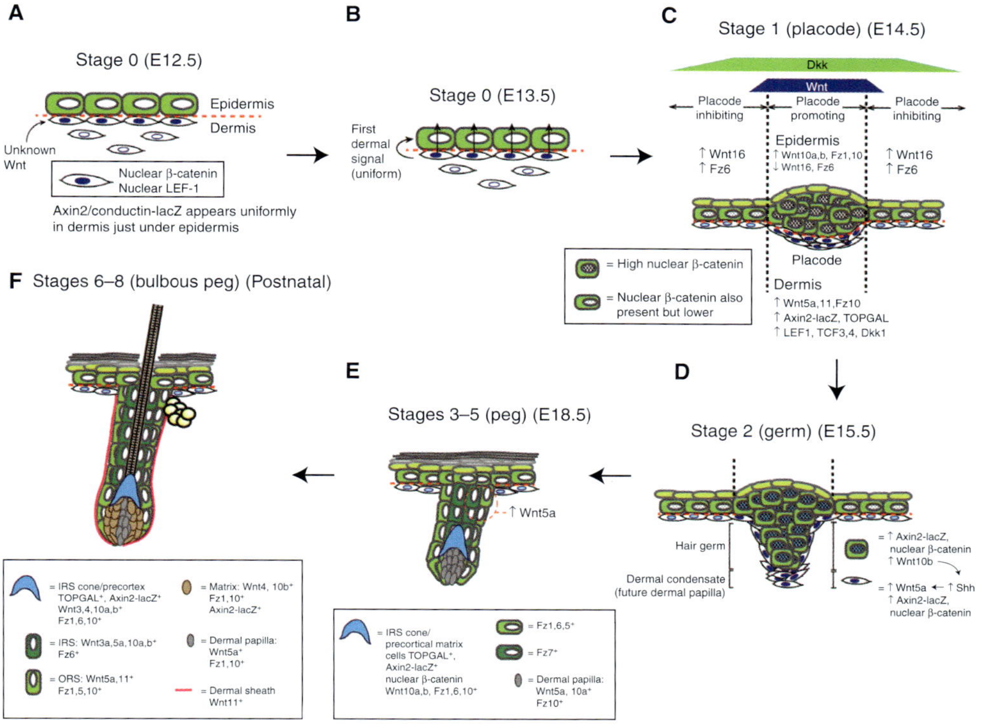

Figure 2. Wnt signaling during skin development and hair follicle morphogenesis. Hair follicle (HF) morphogenesis proceeds in three phases: induction (*A–C*), organogenesis (*D,E*), and cytodifferentiation (*F*), which can be further separated into eight stages (for a detailed review of these stages see Paus et al. 1999 and Schneider et al. 2009). The morphology and molecular events occurring at each stage are depicted. Data for developmental timings are from Blanpain and Fuchs (2006) and Zhang et al. (2009). Expression data are from DasGupta and Fuchs (1999), Reddy et al. (2001, 2004), and Zhang et al. (2009). Induction: (*A*) Stage 0 (E12.5): Before any visible hair placode formation, dermal fibroblasts appear to uniformly receive an unidentified Wnt signal from an unknown source, resulting in nuclear localization of β-catenin. (*B*) Stage 0 (E13.5): Dermal fibroblasts then produce the first dermal signal, which is likely also a Wnt. (*C*) Stage 1 (E14.5): On receiving the first dermal signal, Wnt/β-catenin signaling becomes active in the epidermis and is especially elevated at placode-forming regions. Reaction–diffusion reactions between slower-diffusing activating Wnts and faster-diffusing inhibitory Dkks produce a regular array of these high-Wnt placode-forming regions that eventually result in patterned hair growth (see also Box 1). The nascent epidermal placode cells proliferate and accumulate and also start producing Wnt ligands like Wnt10b. Organogenesis: (*D*) Stage 2 (E15.5): As the placode develops into a visible hair germ, the Wnt ligands it produces induce underlying fibroblasts to form a dermal condensate that will eventually develop into a follicular dermal papilla. This process is also dependent on Shh pathway activity in the dermal condensates. (*E*) Stages 3–5 (E18.5): As the hair germ grows into a bulbous hair peg, differentiating epithelial layers become apparent. Wnt/β-catenin is evident in these layers. Cytodifferentiation: (*F*) Stages 6–8 (bulbous peg) (early postnatal): The morphology and biochemistry of the growing HF resembles that of a postnatal anagen follicle. Wnt/β-catenin signaling also seems to be involved in determining the specific follicular lineages that progenitor keratinocytes adopt, with strong and complex expression of Wnt ligands, receptors, and pathway reporters in the various follicular layers.

dermal message" and the upstream signal stimulating its expression (Fig. 2A,B). This is evidenced by the fact that Wnt/β-catenin signaling is first activated in the dermis and subsequently in the epidermis. In developing chick feathers, nuclear localization of β-catenin is observed transiently in the dense dermis below the feather tract 2 days before placode development

(Noramly et al. 1999). Formation of this feather-inducing dermis is dependent on a signal emanating from the dorsal neural tube, likely to be Wnt1 (Olivera-Martinez et al. 2001). In mice, the LEF/TCF transcription factor family member LEF-1 is detected in the mesenchyme of the mouse whisker (vibrissa) pad before discernible whisker development and is required for subsequent vibrissa but not dorsal back skin (pelage) follicle formation (van Genderen et al. 1994; Kratochwil et al. 1996). Corroborating this observation of Wnt/β-catenin pathway activation first occurring in the dermis, in Axin2-lacZ transgenic Wnt-reporter mice (where a bacterial β-galactosidase [lacZ] reporter gene was knocked into the conductin/Axin2 locus), lacZ reporter activity is present in the dermis at E12.5, before placode development (Zhang et al. 2009). In both mouse and chick, Wnt/β-catenin signaling activity appears uniformly in the dermis (Noramly et al. 1999; Zhang et al. 2009), suggesting that the "first dermal message" is uniform rather than being localized in a periodic fashion (Millar 2002).

Various genetic studies further underscore the importance of Wnt/β-catenin signaling in the early dermis for subsequent placode initiation (Table 1). Ectopic epidermal expression of Dkk1, the diffusible inhibitor of Wnt/β-catenin signaling, results in complete loss of β-catenin and LEF-1 expression in the dermis and abrogation of subsequent placode formation (Andl et al. 2002). Similarly, Porcupine (Porcn) heterozygous-null embryonic mouse skins display a general loss of LEF-1-positive placodes, suggesting that fully functional Wnt ligand production and secretion is essential for placode formation (Barrott et al. 2011). Interestingly, a subset of these Porcn$^{\Delta/+}$ mice also displays large areas of dermal atrophy, such that the internal organs are visible through an epidermal monolayer.

However, even though all this evidence strongly suggests that the first dermal message and its upstream initiating signal are Wnt ligands, the precise source and identity of these Wnt ligands remains unknown.

In response to the first dermal signal (Fig. 2B), Wnt/β-catenin signaling is activated in the epidermis to promote placode fate (Fig. 2C). Both Wnt ligands and receptors are expressed in the E14.5 mouse epidermis, including Wnts 3, 4, 6, 7a, 7b, 10a, 10b, and 16 (Reddy et al. 2001) and Frizzleds 1, 6, and 10 (Reddy et al. 2004). By one day following the appearance of nuclear β-catenin in the dense chick dermis, nuclear β-catenin becomes apparent in the entire epidermis and is specifically up-regulated in regions that will form the placodes (Noramly et al. 1999), whereas in mice, nuclear β-catenin is observed in dermal condensates (Zhang et al. 2009). Further corroborating these observations, Wnt-reporter mouse lines like Axin2-lacZ and TOPGAL (containing the lacZ reporter gene under the control of lymphoid enhancer factor/T-cell factor (LEF/TCF) consensus binding motifs and a minimal *c-fos* promoter) (DasGupta and Fuchs 1999) also display up-regulated and restricted reporter activity within developing placodes (Zhang et al. 2009).

When stabilized β-catenin (Gat et al. 1998; Noramly et al. 1999) or a Wnt ligand (Wnt7a) (Widelitz et al. 1999) is overexpressed in mouse and chick epidermis, ectopic HFs and feather buds form, demonstrating that activation of the Wnt/β-catenin signaling pathway is sufficient to direct epidermal cells toward an appendage fate. In contrast, mouse epidermis carrying a loss-of-function mutation in β-catenin failed to develop placodes (Huelsken et al. 2001), and a LEF-1-null mutation in mouse epidermis resulted in arrested development and loss of vibrissae as well as two-thirds of the pelage follicles (van Genderen et al. 1994; Kratochwil et al. 1996), suggesting that Wnt/β-catenin is necessary for placode formation.

After the placodes have formed, epithelial placode cells signal to underlying mesenchymal cells to form the dermal condensate that ultimately develops into the dermal papilla. Again, Wnt/β-catenin signaling is important for these processes (Fig. 2D). Dermal condensates do not develop if epithelial β-catenin is not present (Huelsken et al. 2001), and the presence of Axin2-lacZ and TOPGAL reporter activity in the dermal condensates suggests that Wnt/β-catenin signaling is taking place there (DasGupta and Fuchs 1999; Zhang et al. 2009).

 Cite this article as *Cold Spring Harb Perspect Biol* doi: 10.1101/cshperspect.a008029

Table 1. Summary of mouse models for the functional study of Wnt/β-catenin signaling in the skin

					Hair phenotype			
Mutation	References	Description	Induction	Examination	Morphogenesis	Postnatal hair cycle	IFE phenotype	Notes
K14-ΔN87β-catenin	Gat et al. 1998	Constitutively stable β-catenin (human K14 promoter; transgene)	E9.5 onward	P24, P90–120	No defects described during embryogenesis	Misangled follicles	Thickening; cell morphology changes	K14 activity by E9.5; strong K14 transgene activity by E14.5
ΔNB-catenin	Zhu and Watt 1999	Constitutively stable β-catenin (in vitro)	Constitutive	NA	NA	NA	Increased proportion of stem cells	Stem cell population shows higher endogenous cytoplasmic β-catenin and TOPFLASH activity
K14-ΔNΔCβ-catenin	DasGupta et al. 2002	Dominant-negative constitutively stable β-catenin without transcriptional effector domain	E9.5 onward	P21–28	No defects described during embryogenesis	Formation of epidermal and sebaceous cysts; loss of TOPGAL reporter expression in bulge	Thickening; also, $TOPGAL^{+}$ IFE resembling de novo hair germs (similar to Gat et al. 1998; Van Mater et al. 2003)	
K5-S33Y-β-catenin-ER	Van Mater et al. 2003	Tamoxifen-inducible constitutively active β-catenin (bovine K5 promoter; transgene)	Starting from P50, either single 4OHT dose or daily administration for 1, 3, 7, 14 d	P51–64	NA	Early induction of anagen; hyperplastic anagen HFs, hyperkeratosis	Little to no effect	4OHT has no significant effect on progression to anagen in wild-type mice (Chanda et al. 2000)
K14-ΔNβ-catenin-ER	Lo Celso et al. 2004	Tamoxifen-inducible constitutively active β-catenin (transgene)	Starting from P42, tamoxifen daily up to 14 d	P51–64	NA	Early induction of anagen; abnormal HFs that progress into trichofolliculomas	Hyperproliferative epidermis (pelage and dorsal paw); ectopic hair follicle rudiments in otherwise hairless ventral paw skin	

Continued

Table 1. *Continued*

					Hair phenotype			
Mutation	References	Description	Induction	Examination	Morphogenesis	Postnatal hair cycle	IFE phenotype	Notes
K14-ΔNβ-catenin-ER	Silva-Vargas et al. 2005	Tamoxifen-inducible constitutively active β-catenin (transgene)	Starting from P42–56, tamoxifen daily up to 21 d	P63–77	NA	With increasing induction of ΔNβ-catenin, altered timing and location of ectopic HF formation	None reported	
K14-ΔN87β-catenin (hemizygous)	Lowry et al. 2005	Constitutively stable β-catenin (human K14 promoter; transgene)	E9.5 onward	P28, P49, P55	No defects described during embryogenesis	Increased proliferation during second telogen, and subsequent early anagen entry (by P55)	None reported	
K14-Cre(neo)/K14-Cre(ΔNeo);β-cateninΔex3-6	Huelsken et al. 2001	Inducible β-catenin loss-of-function; Cre-recombinase knocked into endogenous mouse K14 locus	K14-Cre(neo): strong expression by E15; K14-Cre(neo and Δneo): strong expression after birth	P8, P16, P30	No defects described during embryogenesis	Shaggy fur and fewer zigzag HFs until P16, hair loss to baldness from P17 onward; bulge cells lose stem cell markers	Epidermal cysts expressing some bulge stem cell markers; increased proliferation in basal epidermal cells	
K14-Cre(neo);β-cateninΔex3-6	Teulière et al. 2004	Inducible β-catenin loss-of-function; Cre-recombinase knocked into endogenous mouse K14 locus	E15 onward	P30	No defects described during embryogenesis	Presumably similar to that described above in Huelsken et al. (2001)	Thickening; proliferating cells in basal and suprabasal layers; differentiation defects	
K14-CreER;Β-cateninΔex2-6	Lowry et al. 2005	Inducible β-catenin loss of function (human K14 promoter; transgene)	BrdU at P28 for 72 h; tamoxifen injection at P44	P57	NA	Loss of bulge label-retaining cells and CD34 expression, possibly converted to IFE fate	None reported	

$LEF1^{-/-}$	van Genderen et al. 1994	LEF-1 knockout	Constitutive	P16	Follicle forms normally but in reduced numbers	At P3–9: follicles are short and rudimentary (poorly differentiated); at P16, lack of whiskers and bodily HFs	None reported; lacks dermal fat	Melanocytes present, although pigmentation lost
K14-LEF1	Zhou et al. 1995	LEF-1 overexpression (human K14 promoter fused to human LEF-1 cDNA; transgene)	E9.5 onward	P28	Refer to postnatal hair section	Irregularly clustered, misoriented, and rudimentary follicles	None reported	Ectopic hair formation in oral epithelium
K14-ΔNLef1	Merrill et al. 2001	LEF-1 unable to interact with β-catenin (transgene)	E9.5 onward	9 mo	No defects described during embryogenesis	Progressively sparse hair coat; hair germ cells adopt sebocyte fate and form sebaceous cysts	None reported	
K14-ΔNLef1	Niemann et al. 2002	Lef1 unable to interact with β-catenin (different deletion from Merrill et al. 2001; transgene)	E9.5 onward	P42–10 mo	No defects described during embryogenesis	Delayed entry into first postnatal hair cycle; progressive hair loss; hair loss complete only at 10 mo	Formation of epithelial cysts but no sebaceous differentiation (different from Merrill et al. 2001); light IFE thickening not confined to cyst regions; delayed but normal IFE wound healing; sebaceous adenoma/sebeomas and squamous papillomas/carcinomas observed at ~3 mo	Transgene expression at 100× of endogenous LEF1; effects on hair only after first postnatal cycle, possibly because transgene not strongly expressed during development; LEF1 is not expressed in postnatal IFE

Continued

Table 1. *Continued*

					Hair phenotype			
Mutation	References	Description	Induction	Examination	Morphogenesis	Postnatal hair cycle	IFE phenotype	Notes
K14-Tcf3	Merrill et al. 2001	Overexpression of wild-type mouse Tcf3 (transgene)	E9.5 onward	P0	Epidermal differentiation defect and impaired barrier function	No hair phenotype, although increased nuclear β-catenin in ORS and bulge	Late-stage epidermal differentiation defect; impaired barrier function	
K14-Cre;Tcf3-fl/fl	Nguyen et al. 2009	Conditional Tcf3 knockout (K14-Cre transgene)	E9.5 onward	P0, adult	No defects	No defects	No defects	
K14-myc-ΔNTcf4	Nguyen et al. 2009	Overexpression of Tcf4 lacking β-catenin-interacting domain (transgene)	E9.5 onward	P0	Epidermal differentiation defect and impaired barrier function	None reported	Epidermal differentiation defects similar to those stemming from K14-Tcf3 overexpression (Merrill et al. 2001)	
K14-Cre;Tcf3-fl/fl;Tcf4$^{-/-}$	Nguyen et al. 2009	Conditional Tcf3 and Tcf4 knockout (K14-Cre transgene)	E9.5 onward	P0 and grafted skin	Refer to postnatal hair and IFE sections	Lack whiskers at birth; HFs initiated, but defective downgrowth; Tcf3/4-null skin grafts are hairless	Thinner skin and flattened basal cells; grafted Tcf3/4-null skin shrank in overall area, although epidermis is slightly thickened; loss of clonogenic skin stem cells	
K14-Dkk1	Andl et al. 2002	Dkk1 overexpression (transgene)	E9.5 onward	E14.5, E15.5, E18.5, P1, P7, P13, P17, P28, P33, P52	No or few primary, secondary, and vibrissae follicles develop	Loss of ventral HFs and thin ventral dermis; sparse dorsal guard HFs; reduced dermal fat	No differentiation or proliferation defects	

K5rtTA;tet-Wnt1	Castilho et al. 2009	Inducible mouse Wnt1 overexpression (transgene)	Fed Dox pellets since birth	P7, P14, P17, P21, and >1 yr	No defects	Premature entry into anagen and sustained anagen with increased follicular density, subsequently progressing into hair loss; increased proliferation in $K5^{+}$ basal ORS cells	Thickening at P90	Wnt1 overexpression may induce HF stem cell to exit from quiescence, causing proliferation and subsequent exhaustion of HF stem cell population
$Wnt5a^{-/-}$	Hu et al. 2010	Wnt5a knockout originally described in Yamaguchi et al. (1999)	Constitutive	Grafted skin	No defects described during embryogenesis	Follicular abnormalities in skin grafts; dermal papilla cells induce extensive cyst formation instead of hair	None reported	$Wnt5a^{-/-}$ mice die at birth (Yamaguchi et al. 1999)
$ROR2^{-/-}$	Hu et al. 2010	ROR2-knockout originally described in Takeuchi et al. (2000)	Constitutive	Grafted skin	No defects described during embryogenesis	Altered biochemistry in grafted follicles	None reported	$ROR2^{-/-}$ mice die at birth (Takeuchi et al. 2000)
Sox2-Cre; $Porcn^{-/+}$ (females)	Barrott et al. 2011	Porcupine knockout, gene deleted during epiblast stage	Epiblast	E15.5–E18.5	Refer to postnatal hair and IFE sections	Loss of HFs and Lef1 expression in nascent HFs	Epidermal monolayer, dermal atrophy but otherwise apparently normal keratinocyte differentiation; loss of Lef1 expression in basal epidermis	

All K14- and K5-promoter-driven transgenes are overexpressed in the basal epidermal and follicular bulge and ORS cells.
IFE, interfollicular epidermis; ORS, outer root sheath; ER, estrogen receptor; HF, hair follicle.

Additionally, knocking out Sonic hedgehog (Shh) in embryos results in the loss of Wnt5a expression in dermal condensates and the inhibition of dermal papilla maturation (Reddy et al. 2001). From this, it may seem that Wnt5a plays a role in dermal papilla development. However, the deletion of Wnt5a in the skin appears to have no effects on HF morphogenesis (Hu et al. 2010).

Epidermal keratinocytes surrounding the placodes form stretches of hairless interfollicular epidermis (IFE) that evenly space hair follicles (HFs) apart into regular arrays (Fig. 2C, Box 1). This array structure suggests the presence of global spatial patterning mechanisms that either promote or inhibit HF fate. The periodicity of placode formation and subsequent formation of a regular, patterned array of HFs is believed to arise via a Turing model of reaction–diffusion involving competition between diffusible placode-promoting and placode-inhibiting morphogens (Jiang et al. 2004; Schmidt-Ullrich and Paus 2005). Such a reaction–diffusion system might result from differences in the rates of diffusion between placode promoters and inhibitors, setting up a "short-range activation, long-range inhibition" phenomenon that produces the diffusion-driven instability needed to produce Turing patterns (Maini et al. 2006). Both Wnt ligands such as Wnt10a and Wnt10b (Reddy et al. 2001) and inhibitors such as Dkk4 are expressed in the placodes and are secreted into the extracellular space, and can hence potentially participate in a reaction–diffusion system (Sick et al. 2006). Because it is smaller in molecular size, Dkk4 could potentially traverse a larger distance than the Wnt ligands and diffuse beyond the placode, reducing Wnt/β-catenin signaling in the IFE adjacent to the placodes and refining the placode boundaries (Zhang et al. 2009). Conversely, the larger and hydrophobic Wnt ligands do not move freely in an aqueous extracellular environment and hence remain within the placode and exert short-range effects to focally up-regulate Wnt/β-catenin signaling in the placode. These differences in the levels of Wnt/β-catenin signaling could thus result in the emergence of a periodic array of placodes (Box 1) (Sick et al. 2006). In further support of the notion that optimal levels of Wnt/β-catenin signaling exist that are required for proper placode patterning, constitutive expression of LEF1 or stabilized β-catenin in the basal epidermis causes irregular spacing and tight clustering of hair placodes and follicles (Zhou et al. 1995; Närhi et al. 2008).

Although the Turing model is elegant and produces remarkably accurate predictions of HF array patterns, it is based on the premise of freely diffusible and interacting chemicals. Because Wnt and Dkk are not freely diffusible in the extracellular milieu, and they do not directly interact but instead bind to cells, it is unclear if using reaction–diffusion models to describe their interaction is really representative of actual in vivo events. Additionally, it seems likely that mechanical forces may interact with these diffusible morphogens to influence spatial patterning, although this possibility remains relatively unexplored in the mammalian skin (Morelli et al. 2012).

WNT IN HF

In the postnatal skin, mature HFs undergo repeated cycles of growth (anagen), death/regression (catagen), and rest (telogen) in distinct morphological states (reviewed in Schneider et al. 2009). Similar to HF morphogenesis, the initiation of a new growth phase and the resulting proliferation, differentiation, and progressive downgrowth of the follicle is likely to involve

BOX 1. FORMATION OF PLACODE ARRAY

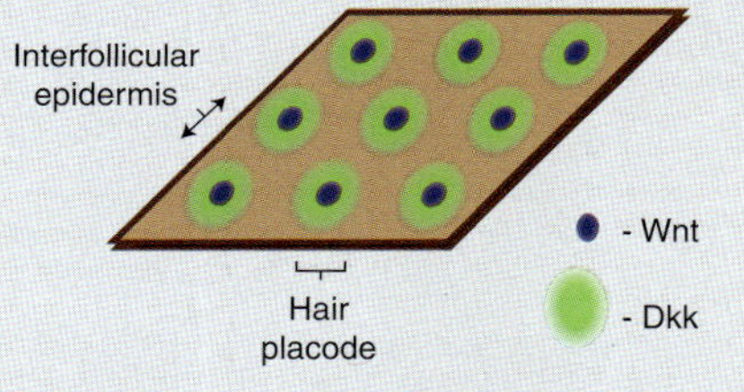

Reaction–diffusion reactions between a slowly diffusing Wnt and a faster diffusing Wnt inhibitor Dkk1 set up a regular array of nodes of elevated Wnt signaling that are the sites of placode formation.

extensive reciprocal interactions between dermal and epidermal cells (Oliver and Jahoda 1988; Schmidt-Ullrich and Paus 2005). Numerous molecular studies suggest that the same signaling pathways that are active during HF morphogenesis are reused in cycling postnatal hair. In particular, Wnt/β-catenin signaling is known to play a critical role at multiple stages of the hair cycle, from the earliest stages at the transition from rest to growth, to influencing lineage decisions during HF differentiation (Fig. 3).

During the telogen phase of hair cycle, the resting HF is ensheathed by layers of epithelial keratinocytes in a structure called the bulge (Fig. 3A). Numerous lineage tracing and functional genetic studies have elegantly shown that follicular stem cells are keratinocytes that reside in the outermost bulge layer, expressing cytokeratin-14 (K14) (Vasioukhin et al. 1999) and -15 (K15) (Morris et al. 2004), Shh (Levy et al. 2005), and the transcription factors Sox9 (Vidal et al. 2005; Sick et al. 2006) and Lhx2 (Rhee et al. 2006). DNA and histone label-retaining assays further suggest that many bulge cells cycle slowly (Cotsarelis et al. 1990; Tumbar et al. 2004), and remain largely quiescent during telogen, not taking up any BrdU label or expressing proliferation markers (Greco et al. 2009).

It is conventionally believed that little or no Wnt/β-catenin signaling takes place during most of telogen (Fig. 3A), and that Wnt/β-catenin signaling is up-regulated only at the end of telogen, promoting entry into anagen (Fig. 3B). Wnt ligand expression is not detected in telogen follicles, and Wnt10a and 10b only become expressed at anagen onset in the dermal papilla and secondary hair germ, respectively (Reddy et al. 2001). This is supported by the observation that β-catenin is generally confined to the membranes of bulge cells through most of telogen, and nuclear β-catenin only becomes apparent in hair germ cells just before the follicle enters anagen (Fig. 3B) (Lowry et al. 2005; Greco et al. 2009). The TOPGAL Wnt/β-catenin signaling reporter is generally inactive in telogen follicles (DasGupta and Fuchs 1999; Plikus et al. 2008), although TOPGAL-positive bulge cells are sometimes detected (DasGupta and Fuchs 1999; Merrill et al. 2001). In further support of the notion that the bulge is a Wnt-inhibited environment, microarray gene expression analyses show that prospectively isolated telogen follicular bulge cells and slow-cycling cells express secreted repressors of Wnt/β-catenin signaling such as Sfrp1, Wif1, and Dkk3 (Blanpain et al. 2004; Morris et al. 2004; Tumbar et al. 2004), as well as Tcf3, which is believed to repress Wnt/β-catenin signaling in the absence of nuclear β-catenin (Nguyen et al. 2006, 2009). Interestingly, isolated human bulge cells also express slightly elevated levels of Wif1 and Dkk3, supporting the findings in mice (Ohyama et al. 2006).

Loss- and gain-of-function mutations in β-catenin provide further genetic support for Wnt/β-catenin signaling as an inducer of anagen onset (Table 1). Expression of constitutively stable β-catenin in basal epidermal keratinocytes leads to hyperactive Wnt/β-catenin signaling and is sufficient to induce anagen onset (Van Mater et al. 2003) and ectopic HF formation (Gat et al. 1998; Lo Celso et al. 2004), whereas loss-of-functional β-catenin blocked entry into anagen when induced postnatally (Huelsken et al. 2001).

The decision to exit telogen and enter anagen appears to result from the antagonistic interplay between BMP, TGF-β, and Wnt/β-catenin signaling in follicular stem cells. Although Wnt/β-catenin signaling and TGF-β (Oshimori and Fuchs 2012) promote anagen onset, BMP signaling is thought to inhibit it. This growth inhibitory role of BMP is supported by studies in mice where BMP receptor 1a was conditionally ablated in basal keratinocytes, leading to continuous activation and aberrant proliferation of follicular stem cells and the eventual loss of slow-cycling cells from the bulge (Kobielak et al. 2003, 2007; Andl et al. 2004; Yuhki et al. 2004). BMPR1a loss in bulge cells also results in increased levels of Lef1 and stabilized β-catenin, providing evidence that BMP signaling may inhibit anagen onset by preventing the up-regulation of Wnt/β-catenin signaling in follicular stem cells. This mutual antagonism divides telogen into two phases: refractory, which is BMP-high and Wnt-low, and competent telogen, which is BMP-low and Wnt-high (Plikus

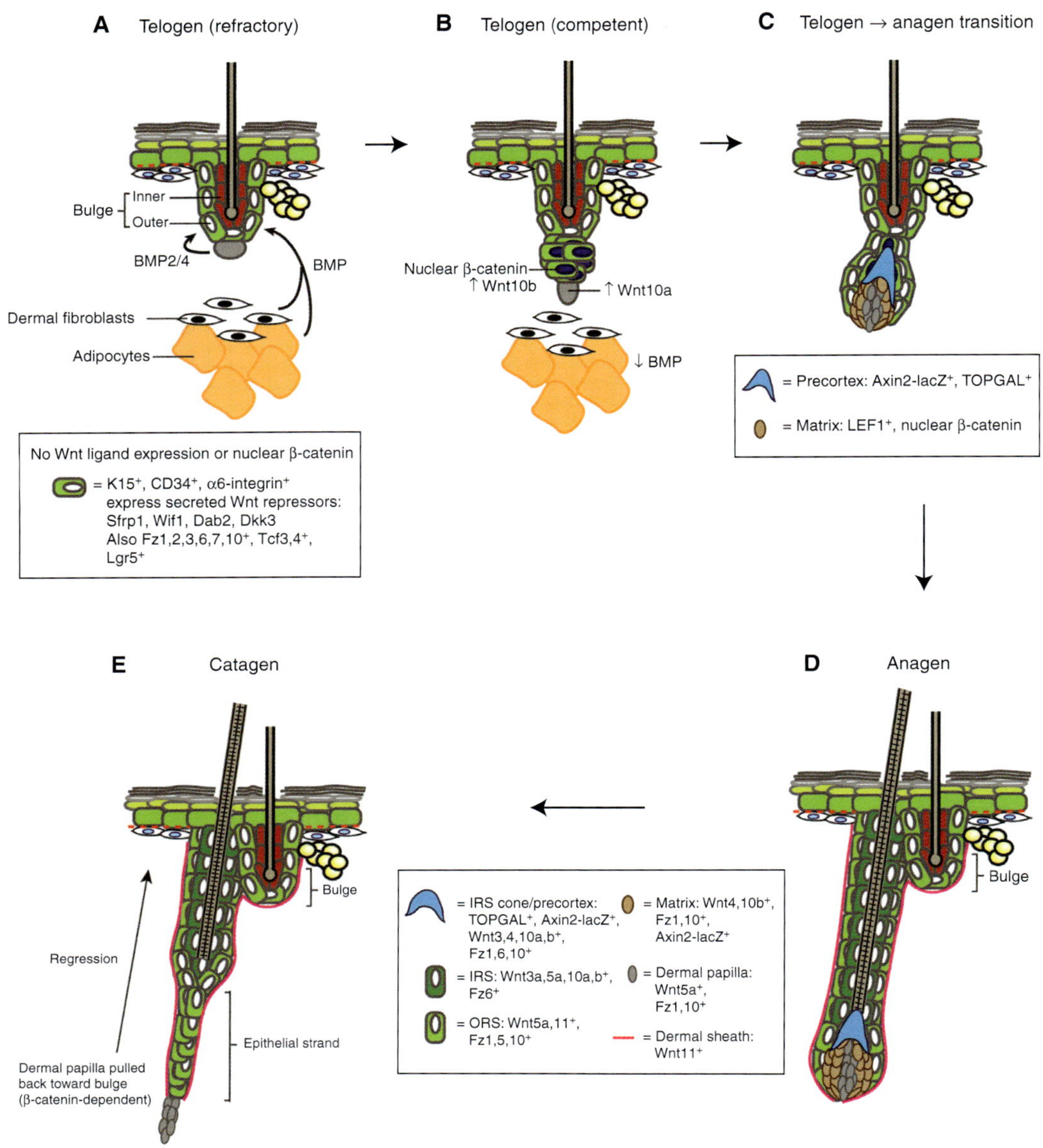

Figure 3. Wnt signaling during hair follicle cycling. Postnatal hair follicles (HFs) cycle repeatedly through rest (telogen), growth (anagen), and death/regression (catagen). The morphology and molecular events taking place at each stage of the hair cycle are depicted (for a comprehensive histological review of hair cycle stages see Müller-Röver et al. 2001). Expression data are from DasGupta and Fuchs (1999), Reddy et al. (2001, 2004), Blanpain et al. (2004), Morris et al. (2004), Tumbar et al. (2004), and Zhang et al. (2009). (*A*) During refractory telogen (BMP-high, Wnt-low) (Plikus et al. 2008), BMP is secreted from adipocytes and dermal fibroblasts, as well as the dermal papilla, to maintain the HF bulge in a low-growth, quiescent state. The bulge is thought to also be in a Wnt-inhibited state, by expressing various Wnt pathway inhibitors. (*B*) In the competent telogen phase (BMP-low, Wnt-high), BMP expression in adipocytes and dermal cells is reduced, allowing bulge cells to respond to anagen-inducing cues. One such cue is likely to involve an as-yet-unidentified Wnt. At this time, Wnt/β-catenin signaling becomes up-regulated in bulge and secondary hair germ cells, and they produce Wnt ligands such as Wnt10b. (*See facing page for legend.*)

et al. 2008). During the refractory phase, dermal cells such as fibroblasts, adipocytes, and dermal papillae express high levels of BMPs 2 and 4, inhibiting the ability of follicular epithelial stem cells to respond to growth stimuli (Plikus et al. 2008). In the competent phase, BMP expression in the dermal macroenvironment is reduced (Plikus et al. 2008) and the dermal papilla secretes noggin (Botchkarev et al. 2001; Zhang et al. 2006), turning BMP signaling off in bulge cells and allowing them to respond to anagen-inducing cues and up-regulate Wnt signaling.

Although most studies place the role of Wnt/β-catenin signaling in the activation of stem cells to transit from telogen to anagen, some evidence exists for the presence of active Wnt/β-catenin signaling throughout telogen. Both human and mouse bulge cells may be Wnt-responsive, expressing Frizzled receptors (Reddy et al. 2004; Tumbar et al. 2004; Ohyama et al. 2006) as well as Wnt-signaling-related transcription factors (Tumbar et al. 2004; Nguyen et al. 2006, 2009). They also express Wnt target genes such as the leucine-rich G-protein-coupled receptor 5 (Lgr5) (Jaks et al. 2008), initially identified as a β-catenin/TCF4 target gene in colon cancer (van de Wetering et al. 2002) and one of the most highly expressed genes in follicular bulge cells (Morris et al. 2004). If Wnt/β-catenin signaling is indeed present during the telogen phase, then the initiation of anagen could be the result of an increase in relative levels of Wnt/β-catenin signaling, rather than activation from a state of no signaling. The manner in which quantitative differences in Wnt signaling levels are interpreted to produce distinct cell fates remains an unanswered question of fundamental importance.

Despite the well-established role for Wnt/β-catenin signaling as an anagen inducer, the cellular source and precise Wnt ligand(s) involved remain unknown. As detected by in situ hybridization, some of the earliest Wnt ligands expressed during the telogen–anagen transition include Wnt5a and Wnt10a in the dermal papilla, and Wnt10b in the secondary hair germ (Reddy et al. 2001). Examination of Shh-knockout skin showed that Wnt10a and 10b expression were unaffected, whereas Wnt5a expression was lost, placing Wnt10a and 10b as upstream of Shh, which in turn regulates Wnt5a expression (Reddy et al. 2001). Although in vitro and skin reconstitution and grafting experiments suggest that Wnt10b is sufficient to promote hair development and growth (Ouji et al. 2006, 2008), Wnt10b-null mice are viable and show no reported skin phenotypes (Stevens et al. 2010). There are no known Wnt10a-null mice, although Wnt10a mutations are associated with human follicular hyperkeratosis and abnormal hair patterning (Adaimy et al. 2007; Bohring et al. 2009; Nawaz et al. 2009). Although Wnt5a is downstream from Wnts10a and 10b, its expression is necessary for normal progression of anagen, as ablation of Wnt5a in the dermal papilla results in impaired HF induction and the formation of epithelial cysts (Hu et al. 2010). Interestingly, continued Wnt/β-catenin signaling within the dermal papilla is also required for normal progression through anagen, possibly by regulating the secretion of anagen-inducing signals, or by influencing the response to those

Figure 3. (*Continued*) (*C*) As the follicle transitions from telogen to anagen and starts growing, bulge/secondary hair germ keratinocytes proliferate and migrate along the developing outer root sheath (ORS) to enwrap the dermal papilla and push it downward, away from the bulge. Wnt/β-catenin signaling continues to be active in these keratinocytes that form the developing matrix and the precortex. (*D*) The HF continues to grow downward in anagen. Distinct, differentiated layers (sheaths) are observed, each expressing different Wnt ligands and receptors. The significance of these complex expression patterns remains unknown. $Lgr5^+$ follicle stem cells are found throughout the ORS and likely migrate to the matrix, where they differentiate into keratinocytes populating the various lineages of the anagen HF. Wnt/β-catenin signaling is also active in the matrix and is thought to be important for specifying these lineages. It is unclear if particular Wnts are responsible for specifying particular anagen lineages. (*E*) As the follicle exits anagen and enters catagen, it stops growing and starts dying. The follicle regresses and forms an epithelial strand that slowly pulls the dermal papilla upward to the level of the bulge, a process that is dependent on β-catenin (van Genderen et al. 1994).

inductive signals or follicular maturation following anagen onset (Enshell-Seijffers et al. 2010).

As the follicle enters the growth (anagen) phase, BMP signaling is inhibited and Wnt/β-catenin signaling is ramped up (Fig. 3B). Nuclear β-catenin accumulates in secondary hair germ cells, concomitant with large-scale alterations in histone modifications and the derepression and subsequent expression of cell cycling, extracellular matrix-remodeling, and HF fate specification genes (Lowry et al. 2005; Greco et al. 2009; Lien et al. 2011). Bulge cells also begin to proliferate, resulting in growth of the follicle and migration of bulge/hair germ cells along the growing ORS (Fig. 3C) (Greco et al. 2009; Hsu et al. 2011). As the follicle continues to grow downward, the bulge becomes more distant from the dermal papilla (DP), which becomes surrounded by the concentric layers of epithelial keratinocytes that form the outer root sheath (ORS), inner root sheath (IRS), and matrix (Mx) and the medulla, cortex, and cuticle of the hair shaft.

Many Wnt ligands (Reddy et al. 2001), Frizzled receptors (Reddy et al. 2004), and the effector Dishevelled (Millar et al. 1999) are expressed in overlapping patterns throughout the anagen follicle, suggesting that Wnt/β-catenin signaling continues to be active throughout the progression of anagen (Fig. 3D). However, in most anagen follicular compartments, the precise role of Wnt signaling in orchestrating the follicular differentiation program remains unclear. The most well-examined role for Wnt/β-catenin signaling is in the anagen matrix, where it likely plays a critical role in influencing the differentiation decisions of matrix cells along the various hair lineages. Matrix cells express LEF1 (Zhou et al. 1995; Merrill et al. 2001) and accumulate nuclear β-catenin (Rabbani et al. 2011), suggesting that active Wnt/β-catenin signaling is taking place. Additionally, Wnt-reporter genes like TOPGAL (DasGupta and Fuchs 1999) turn on in the apical region of the matrix known as the precortex, consisting of cells about to differentiate into the cortex of the hair shaft. As follicular keratinocytes differentiate to form the hair shaft, they express multiple keratin genes, which have TCF/LEF binding sites in their promoters and are thought to be Wnt target genes (Zhou et al. 1995; Merrill et al. 2001; Fuchs 2007). Additional Wnt target genes such as Sonic Hedgehog (Gat et al. 1998; Lo Celso et al. 2004; Silva-Vargas et al. 2005) and the Notch ligand Jagged1 (Estrach et al. 2006) are also expressed in the anagen matrix and function to promote keratinocyte proliferation and differentiation. Within the anagen DP, Wnt/β-catenin signaling continues to be required to maintain matrix cell proliferation and differentiation, possibly by influencing the secretion of various secreted growth factors like FGFs 7 and 10, and insulinlike growth factor (IGF) (Enshell-Seijffers et al. 2010).

Quantitative differences in the levels of Wnt/β-catenin signaling influence the decisions of follicular cells to adopt particular lineages. This is evident from transgenic mouse experiments involving epidermal loss and gain of function of β-catenin signaling. For instance, expression of constitutively stable β-catenin promotes differentiation toward the hair lineage (Gat et al. 1998; Lo Celso et al. 2004), whereas inhibition of Wnt/β-catenin signaling in the basal epidermis directs epidermal keratinocytes toward the interfollicular and sebocyte lineages, resulting in progressive hair loss (Merrill et al. 2001; Niemann et al. 2002). Interestingly, K14-mediated ectopic expression of Wnt3 in the ORS results in the formation of fragile hair shafts and up-regulated expression of hair shaft intermediate filament proteins, suggesting that modulating the levels of Wnt signaling can result in more subtle differences in cellular phenotypes (Millar et al. 1999).

Much less is known about the involvement of Wnt/β-catenin signaling during catagen (Fig. 3E). There are no known published studies on the expression of various Wnt pathway components, and most studies on Wnt pathway mutations do not describe catagen defects (for example, see Millar et al. 1999), although there is some evidence that β-catenin is required for the normal regression of the HF and expression of catagen-stage genes (Huelsken et al. 2001).

Apart from controlling the hair cycle telogen–anagen transition and follicular lineage decisions, Wnt/β-catenin signaling may also influence the direction in which hairs grow.

 Cite this article as *Cold Spring Harb Perspect Biol* doi: 10.1101/cshperspect.a008029

HFs emerge at an angle to the skin surface, and in mice, they point in an anterior to posterior direction (reviewed in Millar 2002). The polarity of follicular orientation may be regulated by the Wnt target gene Shh, which is asymmetrically expressed at only one side of the hair matrix (Gat et al. 1998), and has been shown to have a functional role in regulating feather bud orientation in embryonic chicken skin (Ting-Berreth and Chuong 1996a,b). Up-regulated Wnt/β-catenin signaling can disrupt the normal polar orientation of follicles, as observed in embryonic chick (Noramly et al. 1999; Widelitz et al. 1999) and mouse skin (Zhou et al. 1995; Gat et al. 1998). This may be the result of hyperactive β-catenin signaling inducing symmetrical expression of Shh in both sides of the follicle matrix (Gat et al. 1998). Additionally, the Wnt receptor and planar cell polarity gene *Frizzled 6* is also asymmetrically polarized in epidermal keratinocytes to orient developing HFs, a process that is independent of placode formation or epidermal stratification (Devenport and Fuchs 2008; Devenport et al. 2011). It remains unclear if or how Wnt ligands are actually involved in the process of setting up epidermal tissue polarity.

Sebocytes are specialized skin cells that produce and accumulate lipids (sebum), eventually releasing these lipids to the skin surface. Sebocytes exist either in sebaceous glands associated with a HF (forming the pilosebaceous unit), or independently in specialized glands (Schneider and Paus 2010).

Numerous studies suggest that inhibition of Wnt/β-catenin signaling may be necessary for specification of the sebocyte lineage (reviewed in Fuchs and Horsley 2008). In mice, K14-driven expression of dominant–negative LEF-1 (Niemann et al. 2002) results in progressive hair loss and concomitant induction of hair-follicle-derived sebaceous tumors. Corroborating this, one-third of human sebaceous tumors were found to contain mutations in LEF-1 that result in similar impairment of β-catenin binding (Takeda et al. 2006). Additionally, the balance between Wnt and Hedgehog signaling is critical for proper sebocyte fate determination (Lo Celso et al. 2008).

Although most studies largely support a role for repression of Wnt/β-catenin signaling in adoption of the sebocyte lineage, some deviations from this paradigm may exist. K14-driven overexpression of TCF-3 induced a Wnt-inhibited state but also resulted in repression of pro-adipogenic genes like peroxisome proliferator-activated receptor-γ (PPAR-γ) and loss of sebaceous glands (Nguyen et al. 2006). These findings underscore the complexity of the cross talk among signaling pathways in the regulation of adipogenesis and sebaceous gland development.

WNT IN INTERFOLLICULAR EPIDERMIS

Conflicting evidence exists as to whether Wnt/β-catenin signaling is involved in IFE development and adult IFE maintenance. As discussed above, the constitutive attenuation of Wnt/β-catenin signaling in the epidermis throughout development as well as postnatally, either by loss-of-function mutations in β-catenin (Huelsken et al. 2001), LEF1 (Merrill et al. 2001; Niemann et al. 2002), or Porcupine (Porcn) (Barrott et al. 2011), generally results in the loss of hair morphogenesis or cycling. In contrast, the IFE in these mice appears mostly normal.

However, many Wnt/β-catenin signaling mutations do result in IFE phenotypes (Table 1). For instance, mice overexpressing stabilized β-catenin or Wnt1 in the epidermis displayed hyperthickening together with altered HF structures (Lo Celso et al. 2004; Castilho et al. 2009). Mice with β-catenin null mutations in skin display IFE thickening in young adult mice (Huelsken et al. 2001; Teulière et al. 2004), and when β-catenin null skins were grafted onto nude mouse recipients, the grafts also showed epidermal hyperthickening (Nguyen et al. 2009). Constitutive overexpression of Tcf-3 in the skin resulted in embryonic lethality with defective late-stage epidermal differentiation and impaired skin barrier function (Merrill et al. 2001). These observations suggest that an optimal level of β-catenin signaling is somehow required for IFE maintenance, by either affecting progenitor proliferation or differentiation into more mature keratinocytes.

Consistent with this, experiments with human keratinocytes suggest that Wnt/β-catenin signaling may indeed play a role in maintaining the IFE stem cell pool and promoting adoption of the IFE lineage. In cultured human keratinocytes, putative IFE stem cells showed increased clone-forming ability and expressed higher levels of noncadherin-associated, cytoplasmic β-catenin than putative transit-amplifying cells (Zhu and Watt 1999). Retroviral transduction of constitutively stable β-catenin increased the proportion of putative stem cells to ~90% of the proliferative subpopulation with no effect on terminal differentiation or alterations to cell-cycle kinetics. Conversely, induced expression of a dominant–negative form of β-catenin stimulated cells to exit from the stem cell compartment (Zhu and Watt 1999). These experiments provide evidence for a role of Wnt/β-catenin signaling in maintaining the IFE stem cell pool, while also working coordinately with other signals to influence subsequent differentiation decisions into terminal IFE, hair, or sebocytic lineages.

If Wnt/β-catenin signaling is indeed involved in IFE homeostasis, then what is the precise level of Wnt signaling needed? This remains to be addressed in detail, although closer examination of some of the loss-of-function studies above reveals that Wnt/β-catenin signaling may not have been completely suppressed, potentially allowing a low level of Wnt/β-catenin signaling to continue that might suffice to maintain the IFE. For example, Andl and colleagues examined only mice moderately overexpressing Dkk1 or that were mosaic for transgene expression (which showed sparse hair or bald patches with apparently normal IFE when analyzed up to 60 days postnatally) because mice strongly expressing the K14-Dkk1 transgene were not viable, precluding analysis of the effects of strong Dkk1 overexpression on postnatal IFE maintenance (Andl et al. 2002). The expression levels of Dkk1 in these surviving transgenic animals may not have completely abrogated Wnt/β-catenin signaling in all IFE cells, leaving a low level of Wnt signaling or a population of unaffected IFE progenitors that is sufficient for IFE maintenance. Similarly, Huelsken and colleagues managed to mostly but not completely delete the wild-type β-catenin allele (Fig. 2) in the skin (Huelsken et al. 2001). Any small amount of wild-type β-catenin remaining might provide enough signaling activity to maintain the IFE, accounting for a relatively normal IFE in older animals (Teulière et al. 2004). Consistent with this idea that some β-catenin signaling might remain, hair loss only occurred in patches, and the mice generally developed hair normally up till postnatal day 16. Other plausible explanations for a lack of IFE phenotype in Wnt/β-catenin signaling mutants might include the presence of other unidentified proteins redundant with the mutated signaling components that could substitute for their function in the IFE but not hair keratinocytes.

Many other signaling pathways intersect with Wnt/β-catenin signaling to regulate the adult IFE, sometimes with seemingly contradictory effects. One example is the protooncogene c-Myc (Watt et al. 2008), a known Wnt/β-catenin signaling targeting gene and the key effector of β-catenin/TCF-induced intestinal proliferation (Klaus and Birchmeier 2008). In the skin, Myc overexpression results in epidermal hyperplasia (Waikel et al. 2001) similar to that observed in epidermis overexpressing Wnt1 (Castilho et al. 2009). At the same time, however, Myc and β-catenin appear to exert rather opposite effects. Constitutively high levels of β-catenin signaling induce ectopic HF formation, whereas constitutive expression of Myc promotes differentiation into IFE and sebaceous gland (reviewed in Watt et al. 2008). Clearly, whether or not Myc is indeed a β-catenin/TCF target gene in the skin remains to be addressed.

Taken together, these experiments provide evidence for a role of Wnt/β-catenin signaling in maintaining the IFE stem cell pool. Based on the studies above, it is tempting to speculate that Wnt/β-catenin signaling is ubiquitous in the epidermis, and it is the precise amount of Wnt/β-catenin signaling, coupled with other signaling inputs, that may influence the specific epidermal lineage decisions. In this scenario, relatively lower (but present) levels of Wnt/β-catenin signaling may maintain IFE stem cells and promote IFE fate, intermediate levels

 Cite this article as *Cold Spring Harb Perspect Biol* doi: 10.1101/cshperspect.a008029

promote sebocyte differentiation, whereas high levels direct IFE cells to form hair (as first proposed in Niemann and Watt 2002; Niemann et al. 2002). In essence, any epidermal stem cell can adopt any of the epidermal lineages, given the right signals at the appropriate levels and at the right time (Silva-Vargas et al. 2005). Consistent with this, following a full-thickness skin injury in postnatal skin, a Wnt-dependent wound signal can induce the de novo formation of HFs in the healed skin through a process that closely resembles that of embryonic follicular morphogenesis (Ito et al. 2007), underscoring the multipotency of interfollicular keratinocytes and the importance of high-Wnt signaling in promoting adoption of the HF lineage.

WNT IN MELANOCYTES

Skin pigmentation plays important physiological and evolutionary roles, by protecting animals against mutagenic UV radiation and allowing camouflage, mimicry, and social communication. Pigmentation results from the deposition of melanin pigment from melanin-producing melanocytes into recipient skin cells. Melanocytes arise from neural crest-derived melanoblasts, and are maintained in adulthood by melanocyte stem cells. The embryonic development and tissue regeneration of melanocytes share many common cellular and genetic events. Among them, Wnt/β-catenin signaling is known to govern many aspects of melanocyte biology, from developmental specification of the melanocyte lineage, to subsequent maintenance of the adult melanocyte regenerative unit (comprising adult stem and progenitor cells) and even melanoma pathogenesis.

During embryonic development, multipotent trunk neural crest cells migrate from the neural plate to the epidermis and dermis, undergoing gradual lineage restriction to form committed precursor melanoblasts, which subsequently produce the adult melanocyte stem and progenitor cells and their differentiated, pigmented melanocyte progeny (Fig. 4) (Rawles 1947), reviewed in White and Zon (2008). Dopachrome tautomerase (DCT, also known as Tyrosinase-related protein 2, TRP-2) -expressing melanoblasts are not present in mice mutant for both Wnt-1 and Wnt-3a, providing one of the earliest indications that Wnt/β-catenin signaling might be important for melanoblast specification (Fig. 4A) (Ikeya et al. 1997). Subsequent studies in zebrafish showed that activated β-catenin signaling directed neural crest cells toward pigment–cell fates, at the expense of neuronal and glial cell fates (Dorsky et al. 1998). Transgenic overexpression of Wnt1 in neural crest explant cultures was sufficient to expand the number of melanoblasts and increase differentiation to pigmented melanocytes, showing in vitro that Wnt signaling can direct melanocyte lineage specification and differentiation decisions (Dunn et al. 2000).

Much mechanistic insight into the way Wnt/β-catenin signaling drives the melanocyte lineage has emerged from studies of melanocyte-specific microphthalmia-associated transcription factor (MITF), a gene that regulates melanocyte differentiation and causes auditory–pigmentary diseases like Waardenburg syndrome type 2 (WS2) when mutated in humans (Saito et al. 2003). A promoter/enhancer region of *nacre,* the zebrafish homolog of MITF, is bound by LEF1 in vitro and contains TCF/LEF binding sites that are required for MITF promoter–reporter gene expression (Dorsky et al. 2000). By adding Wnt-3a to cultured melanocytes, Takeda and colleagues (2000) found that exogenous Wnt recruited β-catenin and LEF-1 to the MITF promoter. Together, these studies provide strong evidence that Wnt/β-catenin signaling directly regulates the transcription of melanocyte-specific genes like MITF. The MITF protein itself directly interacts with LEF-1, and MITF and LEF-1 function synergistically to up-regulate DCT expression. Thus, by functioning as both a Wnt target gene as well as a nuclear effector of the Wnt pathway, MITF participates in a feed-forward loop to efficiently propagate Wnt signals in melanocytes (Yasumoto et al. 2002).

How is the sensitive balance between self-renewal and differentiation of lineage-restricted melanocyte stem cells controlled? Studies in mouse HFs and cell culture suggest that Wnt signaling interacts with Pax3 and MITF to

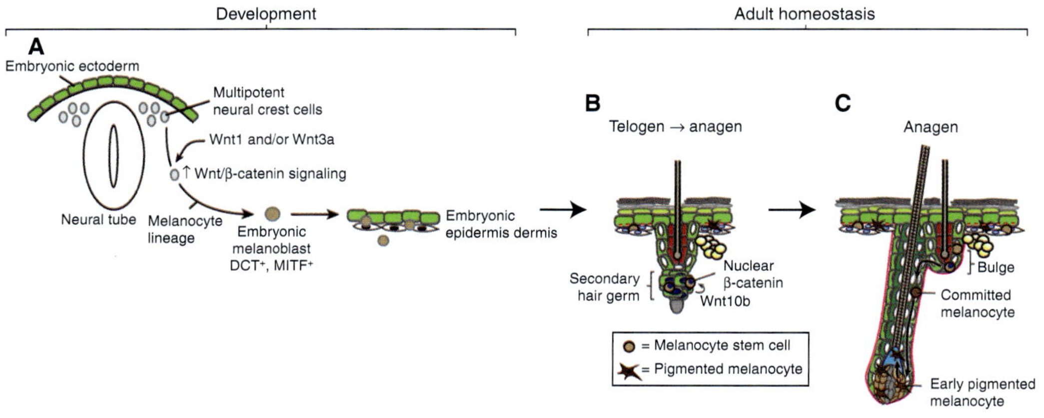

Figure 4. Wnt signaling during melanocyte development and homeostasis. (*A*) As the embryo develops, multipotent neural crest cells receive Wnt signals that instruct them to adopt the melanocyte lineage and form melanoblasts. These melanoblasts migrate and populate the embryonic epidermis and developing hair follicles (HFs). (*B*) Embryonic melanoblasts eventually give rise to adult melanocyte stem cells, which reside in the telogen bulge in close proximity to follicular stem cells. Follicular and melanocyte stem cells regulate each other and coordinately activate Wnt/β-catenin signaling at the onset of anagen (Rabbani et al. 2011). (*C*) The activation of Wnt signaling in melanocyte stem cells causes them to proliferate and differentiate into melanocytes, which then migrate along the outer root sheath (ORS) to enter the matrix. In the matrix, they potentially receive additional Wnt signals that cause them to transfer pigment to differentiating cortical keratinocytes, producing a pigmented hair shaft.

mediate stem cell self-renewal versus differentiation decisions (Lang et al. 2005). In HF bulge melanocyte stem cells, Pax3 cooperates with Sox10 to promote MITF expression, hence directing these cells toward a melanocyte fate. Interestingly, however, Pax3 simultaneously prevents further differentiation of these cells, by associating with Groucho corepressors and competing with MITF for binding to an enhancer element in the DCT promoter, causing strong repression of DCT transcription. This results in a buildup of Pax3-expressing melanocyte precursors that accumulate MITF and so are primed to express further differentiation genes when the right signals are received. Activated nuclear β-catenin displaces Pax3 and the Groucho corepressors from the enhancer, allowing MITF binding and activating DCT expression. In this way, low levels of Wnt/β-catenin signaling maintain melanocyte stem cells in an undifferentiated state, whereas increased Wnt/β-catenin signaling promotes terminal differentiation to mature melanocytes.

If Wnt/β-catenin signaling is required for proper differentiation of melanocytes, where does the Wnt signal come from? Unpigmented DCT-expressing HF melanocyte stem cells are present in the secondary hair germ near the follicular bulge (Nishimura et al. 2002), which incidentally is also where K15$^+$ and Lgr5$^+$ follicular stem cells reside (Fig. 4B). Because of the proximity of these two stem cell populations and the fact that they act coordinately to regenerate pigmented hair with each hair cycle, it has been proposed that follicular and melanocyte stem cells may regulate each other. Indeed, several recent studies show that follicular stem cells provide adhesion (Tanimura et al. 2011) and extrinsic signals, including Wnt, necessary for melanocyte stem cell maintenance (Nishimura et al. 2010; Rabbani et al. 2011). At the onset of anagen, secondary hair germ keratinocytes stabilize β-catenin and subsequently express Wnt10b, activating Wnt/β-catenin signaling in adjacent melanocyte stem cells and promoting their proliferation and differentiation to melanocytes (Fig. 4B,C). Strikingly, hyperactive Wnt/β-catenin signaling in melanocyte stem cells by DCT-mediated expression of a constitutively stabilized β-catenin resulted in exhaustion of the stem cell compartment and premature hair graying (Rabbani et al. 2011).

Wnt/β-catenin signaling may even be involved in melanocyte function. LEF-1 knockout mouse HFs lack melanin even though melanocytes are still present, suggesting that LEF-1 may be involved in regulating the ability of melanocytes to synthesize or transfer pigment to adjacent hair keratinocytes (van Genderen et al. 1994). Wnt5a mutant mice also show reduced pigment deposition in the hair cortex, possibly by decreasing FoxN1 expression (Hu et al. 2010). As such, it is tempting to speculate that Wnt signals in the matrix and precortex instruct the melanocytes there to deposit pigment in differentiating cortical keratinocytes, resulting in pigmented hair shafts.

Wnt/β-catenin signaling may also have a role in controlling different relative pigmentation of the same skin. In humans, palmoplantar epidermis contains much fewer melanocytes and is subsequently much less pigmented than nonpalmoplantar epidermis. Compared to nonpalmoplantar dermal fibroblasts, palmoplantar dermal fibroblasts secrete higher levels of the Wnt-antagonist Dkk1, which acts on cultured melanocytes to decrease β-catenin expression and suppress transcription of MITF (Yamaguchi et al. 2004). As such, reduced Wnt/β-catenin signaling in palmoplantar skin may account for the reduced numbers of melanocytes there.

What is the role of Wnt/β-catenin signaling in the pathogenesis of melanocyte diseases? Hyperactive β-catenin signaling seems to initiate melanoma genesis, as evidenced by the fact that coexpression of stabilized nuclear β-catenin (bcat^{sta}) with constitutively active N-ras in mouse melanocytes leads to enhanced melanoma formation in vivo and increased immortalization of melanocytes in vitro (Delmas et al. 2007). Consistent with this, as observed in multiple immunohistochemistry studies of human melanomas, benign melanocytic neoplasms (nevi) and many primary melanoma tumors have up-regulated levels of nuclear β-catenin. Interestingly, however, the loss of Wnt/β-catenin signaling may aid the progression of melanoma. As the melanomas progress to become more malignant and metastatic, nuclear β-catenin levels appear to decrease (Kageshita et al. 2001; Maelandsmo et al. 2003; Bachmann et al. 2005; Chien et al. 2009). In fact, loss of nuclear β-catenin in malignant tumors was associated with decreased patient survival (Kageshita et al. 2001; Maelandsmo et al. 2003; Bachmann et al. 2005; Chien et al. 2009). The suppression of Wnt/β-catenin signaling may be achieved through the *BRAF* oncogene, which is constitutively activated by the $BRAF^{V600E}$ mutation in the majority of melanomas (Davies et al. 2002; Biechele et al. 2012). Knocking down *Axin1*, a negative regulator of Wnt/β-catenin signaling, synergized with inhibition of $BRAF^{V600E}$ to induce apoptosis in previously resistant melanoma cell lines, suggesting that a combination of enhanced Wnt/β-catenin signaling with other therapies may be a valuable strategy in the treatment of melanoma (Biechele et al. 2012).

WNT IN EPIDERMAL AND HAIR DISEASES

Given the importance of Wnt/β-catenin signaling in skin development and homeostasis, it is not surprising that perturbations in signaling pathway activity can lead to developmental defects and diseases like cancer (reviewed in Watt and Collins 2008; Ambler and Määttä 2009; Blanpain and Fuchs 2009).

Prolonged activation of Wnt/β-catenin signaling leads to a variety of tumors in transgenic mice (Owens and Watt 2003; Klaus and Birchmeier 2008). In the hair, constitutive expression of stabilized β-catenin leads to the formation of pilomatricomas, which are densely packed benign tumors with a center of hair shaft cells surrounded by matrix cells, or trichofolliculomas displaying multilineage differentiation (Gat et al. 1998) that regress when the pathway activation is stopped (Lo Celso et al. 2004). Strikingly, many human pilomatricomas also harbor β-catenin-stabilizing mutations, confirming the relevance of the mouse findings (Chan et al. 1999). Conversely, when Wnt/β-catenin signaling is reduced via expression of the K14-ΔNLef1 transgene in mice, sebaceous gland tumors develop with high frequency, consistent with the sebocyte-promoting property of the transgene (Niemann et al. 2002). This occurs in human tumors as well, as examination of human sebaceous tumors revealed that one-third of them contained mutations in the amino terminus of LEF-1 that similarly blocked β-catenin binding (Takeda et al. 2006). Taken together, these studies suggest that an optimal level of Wnt/β-catenin signaling exists that is required for skin homeostasis, and that either hyperactivation or attenuation of signaling can lead to tumor formation.

As stem cells are long-lived residents in tissue, they are able to accumulate many mutations that might confer tumor-initiating properties, and are hence attractive candidates for tumor-initiating cells or cancer stem cells. It is plausible that some of these mutations may affect Wnt signaling. In a two-step chemical carcinogenesis model of squamous cell carcinoma, tumor-propagating cells expressing the follicular bulge cell marker CD34 also showed nuclear β-catenin, suggesting the presence of active Wnt signaling. The tumors were functionally dependent on β-catenin signaling for their maintenance, as loss of β-catenin resulted in complete regression of the tumors (Malanchi et al. 2008). However, the source of the Wnt ligands for these tumors remains unknown. It is tempting to speculate that a candidate source of Wnt signals might be fibroblasts residing in the tumor stroma, which are known to produce Wnt ligands to support tumors of the mammary epithelia (Bhowmick et al. 2004; Kalluri and Zeisberg 2006). In a genetic mouse model of basal cell carcinoma (BCC), constitutive Hedgehog signaling in a small subset of BCC tumor cells induces Wnt/β-catenin signaling in the BCC tumor cells and surrounding stromal cells, as evidenced by their expression of various Wnt ligands and target genes (Yang et al. 2008). Blocking Wnt/β-catenin signaling via conditional and concomitant expression of Dkk1 resulted in suppression of tumor formation (Yang et al. 2008). Consistent with these animal model observations, human BCC samples also show nuclear accumulation of β-catenin (Salto-Tellez et al. 2006, and references therein), supporting a role for Wnt/β-catenin signaling in BCC etiology. Together, these studies show the potentially crucial function of Wnt/β-catenin signaling in development and progression of various skin cancers.

In concert with other signaling pathways, altered Wnt/β-catenin signaling may determine the specific types of tumors formed (Watt and Collins 2008). In wild-type skin, dimethyl benzanthracene/tetradecanoyl-phorbol acetate (DMBA/TPA) chemical carcinogenesis induces papillomas possessing Ha-Ras mutations that progress into squamous cell carcinomas. In K14-ΔNLef1 mice, the same DMBA/TPA treatment protocol results in sebaceous gland tumors (Niemann et al. 2007). Expression of constitutively active β-catenin in the absence of vitamin D receptor (VDR) leads to the formation of BCC-like lesions with up-regulated Ptch1 expression rather than pilomatricomas (Pálmer et al. 2008). Human trichofolliculoma cells have high levels of β-catenin and VDR, whereas infiltrative human BCCs have high β-catenin levels and low VDR levels (Pálmer et al. 2008). Clearly, altered Wnt/β-catenin signaling with concomitant mutations in other signaling pathways can yield very different tumor types.

Some human skin diseases are known to result from mutations in Wnt genes themselves. Wnt10a missense mutations have been associated with a rare form of ectodermal dysplasia called odonto-onycho-dermal (OOD) dysplasia, which results in dry hair and skin that may result from reduced numbers of sebaceous glands, lesions, and excessive sweating of palms and soles, as well as abnormal thickening of the cornified layer of the epidermis (Adaimy et al. 2007; Bohring et al. 2009; Nawaz et al. 2009). Patients with heterozygotic mutations also presented with some of the OOD phenotype, suggesting that the precise levels of nonmutated Wnt10a are crucial for normal skin homeostasis (Bohring et al. 2009).

CONCLUSIONS AND PERSPECTIVES

From the large body of evidence reviewed here, it is clear that Wnt signaling is important right from the earliest stages of skin development by initiating, patterning, and propagating hair morphogenesis, and continues to be involved postnatally by controlling hair cycling and possibly IFE maintenance, and even managing hair and skin pigmentation.

The fate of skin cells appears to be exquisitely sensitive to the precise levels of Wnt/β-catenin signaling: High-Wnt signaling instructs keratinocytes to form hair, whereas intermediate to low levels of Wnt signaling specify the sebocyte and interfollicular epidermis lineages. When this delicate balance of Wnt/β-catenin signaling is upset, skin cells are driven excessively to form particular lineages, leading to developmental defects or cancer.

Although much has been discovered, several outstanding questions remain.

1. What is the Wnt ligand that initiates hair morphogenesis, and where does it come from?
2. If Wnts and their inhibitors are not freely diffusible in the milieu of the embryonic skin, how exactly is an array of foci of upregulated Wnt signaling established to form hair placodes? What are the identities of the patterning Wnts and their inhibitors? Are they distinct from the morphogenesis-initiating Wnts, and where do they come from?
3. How exactly are the levels of Wnt/β-catenin signaling modulated to specify the various skin lineages?
4. Do either or both follicular stem cells and their niche cells require Wnt signaling for their function?
5. Which Wnt ligands and receptors expressed in the anagen follicle specify which different anagen follicular keratinocyte fates?
6. Does Wnt/β-catenin signaling have a broader role in the progression of catagen?
7. How precisely does Wnt/β-catenin signaling influence the polarity of HFs, and how is this polarity set up in the first place?
8. What exactly is the role that Wnt/β-catenin signaling plays in IFE homeostasis?

As work continues at the intersection of skin and Wnt biology, it seems likely that these and other questions will be addressed, yielding exciting, interesting, and perhaps unexpected new insights into the roles that Wnt/β-catenin signaling plays in the development, homeostasis, and diseases of the skin.

ACKNOWLEDGMENTS

We thank the Agency for Science, Technology and Research (A*STAR), Singapore, and the Howard Hughes Medical Institute (HHMI) for financial support. X.L. is a recipient of a National Science Scholarship from A*STAR, and R.N. is an Investigator of the HHMI.

REFERENCES

Adaimy L, Chouery E, Megarbane H, Mroueh S, Delague V, Nicolas E, Belguith H, de Mazancourt P, Megarbane A. 2007. Mutation in WNT10A is associated with an autosomal recessive ectodermal dysplasia: The odonto-onycho-dermal dysplasia. *Am J Hum Genet* **81:** 821–828.

Ambler CA, Määttä A. 2009. Epidermal stem cells: Location, potential and contribution to cancer. *J Pathol* **217:** 206–216.

Andl T, Reddy ST, Gaddapara T, Millar SE. 2002. WNT signals are required for the initiation of hair follicle development. *Dev Cell* **2:** 643–653.

Andl T, Ahn K, Kairo A, Chu EY, Wine-Lee L, Reddy ST, Croft NJ, Cebra-Thomas JA, Metzger D, Chambon P, et al. 2004. Epithelial Bmpr1a regulates differentiation and proliferation in postnatal hair follicles and is essential for tooth development. *Development* **131:** 2257–2268.

Atit R, Sgaier SK, Mohamed OA, Taketo MM, Dufort D, Joyner AL, Niswander L, Conlon RA. 2006. β-Catenin activation is necessary and sufficient to specify the dorsal dermal fate in the mouse. *Dev Biol* **296:** 164–176.

Bachmann IM, Straume O, Puntervoll HE, Kalvenes MB, Akslen LA. 2005. Importance of P-cadherin, β-catenin, and Wnt5a/frizzled for progression of melanocytic tumors and prognosis in cutaneous melanoma. *Clin Cancer Res* **11:** 8606–8614.

Barrott JJ, Cash GM, Smith AP, Barrow JR, Murtaugh LC. 2011. Deletion of mouse Porcn blocks Wnt ligand secretion and reveals an ectodermal etiology of human focal dermal hypoplasia/Goltz syndrome. *Proc Natl Acad Sci* **108:** 12752–12757.

Bhowmick NA, Neilson EG, Moses HL. 2004. Stromal fibroblasts in cancer initiation and progression. *Nature* **432:** 332–337.

Biechele TL, Kulikauskas RM, Toroni RA, Lucero OM, Swift RD, James RG, Robin NC, Dawson DW, Moon RT, Chien AJ. 2012. Wnt/β-catenin signaling and AXIN1 regulate apoptosis triggered by inhibition of the mutant Kinase BRAFV600E in human melanoma. *Sci Signal* **5:** ra3.

Blanpain C, Fuchs E. 2006. Epidermal stem cells of the skin. *Annu Rev Cell Dev Biol* **22:** 339–373.

Blanpain C, Fuchs E. 2009. Epidermal homeostasis: A balancing act of stem cells in the skin. *Nat Rev Mol Cell Biol* **10:** 207–217.

Blanpain C, Lowry WE, Geoghegan A, Polak L, Fuchs E. 2004. Self-renewal, multipotency, and the existence of two cell populations within an epithelial stem cell niche. *Cell* **118:** 635–648.

Bohring A, Stamm T, Spaich C, Haase C, Spree K, Hehr U, Hoffmann M, Ledig S, Sel S, Wieacker P, et al. 2009. WNT10A mutations are a frequent cause of a broad spectrum of ectodermal dysplasias with sex-biased manifestation pattern in heterozygotes. *Am J Hum Genet* **85:** 97–105.

Botchkarev VA, Botchkareva NV, Nakamura M, Huber O, Funa K, Lauster R, Paus R, Gilchrest BA. 2001. Noggin is required for induction of the hair follicle growth phase in postnatal skin. *FASEB J* **15:** 2205–2214.

Castilho RM, Squarize CH, Chodosh LA, Williams BO, Gutkind JS. 2009. mTOR mediates Wnt-induced epidermal stem cell exhaustion and aging. *Cell Stem Cell* **5:** 279–289.

Chan EF, Gat U, McNiff JM, Fuchs E. 1999. A common human skin tumour is caused by activating mutations in β-catenin. *Nat Genet* **21:** 410–413.

Chanda S, Robinette CL, Couse JF, Smart RC. 2000. 17β-Estradiol and ICI-182780 regulate the hair follicle cycle in mice through an estrogen receptor-α pathway. *Am J Physiol Endocrinol Metab* **278:** E202–E210.

Chien AJ, Moore EC, Lonsdorf AS, Kulikauskas RM, Rothberg BG, Berger AJ, Major MB, Hwang ST, Rimm DL, Moon RT. 2009. Activated Wnt/β-catenin signaling in melanoma is associated with decreased proliferation in patient tumors and a murine melanoma model. *Proc Natl Acad Sci* **106:** 1193–1198.

Cotsarelis G, Sun T-T, Lavker RM. 1990. Label-retaining cells reside in the bulge area of pilosebaceous unit: Implications for follicular stem cells, hair cycle, and skin carcinogenesis. *Cell* **61:** 1329–1337.

DasGupta R, Fuchs E. 1999. Multiple roles for activated LEF/TCF transcription complexes during hair follicle development and differentiation. *Development* **126:** 4557–4568.

DasGupta R, Rhee H, Fuchs E. 2002. A developmental conundrum: A stabilized form of β-catenin lacking the transcriptional activation domain triggers features of hair cell fate in epidermal cells and epidermal cell fate in hair follicle cells. *J Cell Biol* **158:** 331–344.

Davies H, Bignell GR, Cox C, Stephens P, Edkins S, Clegg S, Teague J, Woffendin H, Garnett MJ, Bottomley W, et al. 2002. Mutations of the BRAF gene in human cancer. *Nature* **417:** 949–954.

Delmas V, Beermann F, Martinozzi S, Carreira S, Ackermann J, Kumasaka M, Denat L, Goodall J, Luciani F, Viros A, et al. 2007. β-catenin induces immortalization of melanocytes by suppressing p16INK4a expression and cooperates with N-Ras in melanoma development. *Genes Dev* **21:** 2923–2935.

Devenport D, Fuchs E. 2008. Planar polarization in embryonic epidermis orchestrates global asymmetric morphogenesis of hair follicles. *Nat Cell Biol* **10:** 1257–1268.

Devenport D, Oristian D, Heller E, Fuchs E. 2011. Mitotic internalization of planar cell polarity proteins preserves tissue polarity. *Nat Cell Biol* **13:** 893–902.

Dorsky RI, Moon RT, Raible DW. 1998. Control of neural crest cell fate by the Wnt signalling pathway. *Nature* **396:** 370–373.

Dorsky RI, Raible DW, Moon RT. 2000. Direct regulation of nacre, a zebrafish MITF homolog required for pigment cell formation, by the Wnt pathway. *Genes Dev* **14:** 158–162.

Dunn KJ, Williams BO, Li Y, Pavan WJ. 2000. Neural crest-directed gene transfer demonstrates Wnt1 role in melanocyte expansion and differentiation during mouse development. *Proc Natl Acad Sci* **97:** 10050–10055.

Enshell-Seijffers D, Lindon C, Kashiwagi M, Morgan BA. 2010. β-catenin activity in the dermal papilla regulates morphogenesis and regeneration of hair. *Dev Cell* **18:** 633–642.

Estrach S, Ambler CA, Lo Celso C, Hozumi K, Watt FM. 2006. Jagged 1 is a β-catenin target gene required for ectopic hair follicle formation in adult epidermis. *Development* **133:** 4427–4438.

Fuchs E. 2007. Scratching the surface of skin development. *Nature* **445:** 834.

Fuchs E, Horsley V. 2008. More than one way to skin. *Genes Dev* **22:** 976–985.

Gat U, DasGupta R, Degenstein L, Fuchs E. 1998. De Novo hair follicle morphogenesis and hair tumors in mice expressing a truncated β-catenin in skin. *Cell* **95:** 605–614.

Greco V, Chen T, Rendl M, Schober M, Pasolli HA, Stokes N, Cruz-Racelis Dela J, Fuchs E. 2009. A two-step mechanism for stem cell activation during hair regeneration. *Cell Stem Cell* **4:** 155–169.

Hardy MH. 1992. The secret life of the hair follicle. *Trends Genet* **8:** 55–61.

Hsu Y-C, Pasolli HA, Fuchs E. 2011. Dynamics between stem cells, niche, and progeny in the hair follicle. *Cell* **144:** 92–105.

Hu B, Lefort K, Qiu W, Nguyen BC, Rajaram RD, Castillo E, He F, Chen Y, Angel P, Brisken C, et al. 2010. Control of hair follicle cell fate by underlying mesenchyme through a CSL-Wnt5a-FoxN1 regulatory axis. *Genes Dev* **24:** 1519–1532.

Huelsken J, Vogel R, Erdmann B, Cotsarelis G, Birchmeier W. 2001. β-Catenin controls hair follicle morphogenesis and stem cell differentiation in the skin. *Cell* **105:** 533–545.

Ikeya M, Lee SM, Johnson JE, McMahon AP, Takada S. 1997. Wnt signalling required for expansion of neural crest and CNS progenitors. *Nature* **389:** 966–970.

Ito M, Yang Z, Andl T, Cui C, Kim N, Millar SE, Cotsarelis G. 2007. Wnt-dependent de novo hair follicle regeneration in adult mouse skin after wounding. *Nature* **447:** 316–320.

Jaks V, Barker N, Kasper M, van Es JH, Snippert HJ, Clevers H, Toftgård R. 2008. Lgr5 marks cycling, yet long-lived, hair follicle stem cells. *Nat Genet* **40:** 1291–1299.

Jiang T-X, Widelitz RB, Shen W-M, Will P, Wu D-Y, Lin C-M, Jung H-S, Chuong C-M. 2004. Integument pattern formation involves genetic and epigenetic controls: Feather arrays simulated by digital hormone models. *Int J Dev Biol* **48:** 117–135.

Kageshita T, Hamby CV, Ishihara T, Matsumoto K, Saida T, Ono T. 2001. Loss of β-catenin expression associated with disease progression in malignant melanoma. *Br J Dermatol* **145:** 210–216.

Kalluri R, Zeisberg M. 2006. Fibroblasts in cancer. *Nat Rev Cancer* **6:** 392–401.

Klaus A, Birchmeier W. 2008. Wnt signalling and its impact on development and cancer. *Nat Rev Cancer* **8:** 387–398.

Kobielak K, Pasolli HA, Alonso L, Polak L, Fuchs E. 2003. Defining BMP functions in the hair follicle by conditional ablation of BMP receptor IA. *J Cell Biol* **163:** 609–623.

Kobielak K, Stokes N, la Cruz de J, Polak L, Fuchs E. 2007. Loss of a quiescent niche but not follicle stem cells in the absence of bone morphogenetic protein signaling. *Proc Natl Acad Sci* **104:** 10063–10068.

Kratochwil K, Dull M, Farinas I, Galceran J, Grosschedl R. 1996. Lef1 expression is activated by BMP-4 and regulates inductive tissue interactions in tooth and hair development. *Genes Dev* **10:** 1382–1394.

Lang D, Lu M-M, Huang L, Engleka KA, Zhang M, Chu EY, Lipner S, Skoultchi A, Millar SE, Epstein JA. 2005. Pax3 functions at a nodal point in melanocyte stem cell differentiation. *Nature* **433:** 884–887.

Levy V, Lindon C, Harfe BD, Morgan BA. 2005. Distinct stem cell populations regenerate the follicle and interfollicular epidermis. *Dev Cell* **9:** 855–861.

Lien W-H, Guo X, Polak L, Lawton LN, Young RA, Zheng D, Fuchs E. 2011. Genome-wide maps of histone modifications unwind in vivo chromatin states of the hair follicle lineage. *Cell Stem Cell* **9:** 219–232.

Lo Celso C, Prowse DM, Watt FM. 2004. Transient activation of β-catenin signalling in adult mouse epidermis is sufficient to induce new hair follicles but continuous activation is required to maintain hair follicle tumours. *Development* **131:** 1787–1799.

Lo Celso C, Berta MA, Braun KM, Frye M, Lyle S, Zouboulis CC, Watt FM. 2008. Characterization of bipotential epidermal progenitors derived from human sebaceous gland: Contrasting roles of c-Myc and β-catenin. *Stem Cells* **26:** 1241–1252.

Lowry WE, Blanpain C, Nowak JA, Guasch G, Lewis L, Fuchs E. 2005. Defining the impact of β-catenin/Tcf transactivation on epithelial stem cells. *Genes Dev* **19:** 1596–1611.

Maelandsmo GM, Holm R, Nesland JM, Fodstad Ø, Flørenes VA. 2003. Reduced β-catenin expression in the cytoplasm of advanced-stage superficial spreading malignant melanoma. *Clin Cancer Res* **9:** 3383–3388.

Maini PK, Baker RE, Chuong C-M. 2006. Developmental biology. The Turing model comes of molecular age. *Science* **314:** 1397–1398.

Malanchi I, Peinado H, Kassen D, Hussenet T, Metzger D, Chambon P, Huber M, Hohl D, Cano A, Birchmeier W, et al. 2008. Cutaneous cancer stem cell maintenance is dependent on β-catenin signalling. *Nature* **452:** 650–653.

Merrill BJ, Gat U, DasGupta R, Fuchs E. 2001. Tcf3 and Lef1 regulate lineage differentiation of multipotent stem cells in skin. *Genes Dev* **15:** 1688–1705.

Millar SE. 2002. Molecular mechanisms regulating hair follicle development. *J Invest Dermatol* **118:** 216.

Millar SE, Willert K, Salinas PC, Roelink H, Nusse R, Sussman DJ, Barsh GS. 1999. WNT signaling in the control of hair growth and structure. *Dev Biol* **207:** 133–149.

Morelli LG, Uriu K, Ares S, Oates AC. 2012. Computational approaches to developmental patterning. *Science* **336:** 187–191.

Morris RJ, Liu Y, Marles L, Yang Z, Trempus C, Li S, Lin JS, Sawicki JA, Cotsarelis G. 2004. Capturing and profiling adult hair follicle stem cells. *Nat Biotechnol* **22:** 411–417.

Müller-Röver S, Handjiski B, van der Veen C, Eichmüller S, Foitzik K, McKay IA, Stenn KS, Paus R. 2001. A comprehensive guide for the accurate classification of murine hair follicles in distinct hair cycle stages. *J Invest Dermatol* **117:** 3–15.

Närhi K, Järvinen E, Birchmeier W, Taketo MM, Mikkola ML, Thesleff I. 2008. Sustained epithelial β-catenin activity induces precocious hair development but disrupts hair follicle down-growth and hair shaft formation. *Development* **135:** 1019–1028.

Nawaz S, Klar J, Wajid M, Aslam M, Tariq M, Schuster J, Baig SM, Dahl N. 2009. *WNT10A* missense mutation associated with a complete Odonto-Onycho-Dermal Dysplasia syndrome. *Eur J Hum Genet* **17:** 1600–1605.

Nguyen H, Rendl M, Fuchs E. 2006. Tcf3 governs stem cell features and represses cell fate determination in skin. *Cell* **127:** 171–183.

Nguyen H, Merrill BJ, Polak L, Nikolova M, Rendl M, Shaver TM, Pasolli HA, Fuchs E. 2009. Tcf3 and Tcf4 are essential for long-term homeostasis of skin epithelia. *Nat Genet* **41:** 1068–1075.

Niemann C, Watt FM. 2002. Designer skin: Lineage commitment in postnatal epidermis. *Trends Cell Biol* **12:** 185–192.

Niemann C, Owens DM, Hülsken J, Birchmeier W, Watt FM. 2002. Expression of ΔNLef1 in mouse epidermis results in differentiation of hair follicles into squamous epidermal cysts and formation of skin tumours. *Development* **129:** 95–109.

Niemann C, Owens DM, Schettina P, Watt FM. 2007. Dual role of inactivating Lef1 mutations in epidermis: Tumor promotion and specification of tumor type. *Cancer Res* **67:** 2916–2921.

Nishimura EK, Jordan SA, Oshima H, Yoshida H, Osawa M, Moriyama M, Jackson IJ, Barrandon Y, Miyachi Y, Nishikawa S-I. 2002. Dominant role of the niche in melanocyte stem-cell fate determination. *Nature* **416:** 854–860.

Nishimura EK, Suzuki M, Igras V, Du J, Lonning S, Miyachi Y, Roes J, Beermann F, Fisher DE. 2010. Key roles for transforming growth factor β in melanocyte stem cell maintenance. *Cell Stem Cell* **6:** 130–140.

Noramly S, Freeman A, Morgan BA. 1999. β-catenin signaling can initiate feather bud development. *Development* **126:** 3509–3521.

Ohtola J, Myers J, Akhtar-Zaidi B, Zuzindlak D, Sandesara P, Yeh K, Mackem S, Atit R. 2008. β-catenin has sequential roles in the survival and specification of ventral dermis. *Development* **135:** 2321–2329.

Ohyama M, Terunuma A, Tock CL, Radonovich MF, Pise-Masison CA, Hopping SB, Brady JN, Udey MC, Vogel JC. 2006. Characterization and isolation of stem cell-enriched human hair follicle bulge cells. *J Clin Invest* **116:** 249–260.

Oliver RF, Jahoda CA. 1988. Dermal-epidermal interactions. *Clin Dermatol* **6:** 74–82.

Olivera-Martinez I, Thélu J, Teillet MA, Dhouailly D. 2001. Dorsal dermis development depends on a signal from the dorsal neural tube, which can be substituted by Wnt-1. *Mech Dev* **100:** 233–244.

Oshimori N, Fuchs E. 2012. Paracrine TGF-β signaling counterbalances BMP-mediated repression in hair follicle stem cell activation. *Cell Stem Cell* **10:** 63–75.

Ouji Y, Yoshikawa M, Shiroi A, Ishizaka S. 2006. Promotion of hair follicle development and trichogenesis by Wnt-10b in cultured embryonic skin and in reconstituted skin. *Biochem Biophys Res Commun* **345:** 581–587.

Ouji Y, Yoshikawa M, Moriya K, Nishiofuku M, Matsuda R, Ishizaka S. 2008. Wnt-10b, uniquely among Wnts, promotes epithelial differentiation and shaft growth. *Biochem Biophys Res Commun* **367:** 299–304.

Owens DM, Watt FM. 2003. Contribution of stem cells and differentiated cells to epidermal tumours. *Nat Rev Cancer* **3:** 444–451.

Pálmer HG, Anjos-Afonso F, Carmeliet G, Takeda H, Watt FM. 2008. The vitamin D receptor is a Wnt effector that controls hair follicle differentiation and specifies tumor type in adult epidermis. *PLoS ONE* **3:** e1483.

Paus R, Müller-Röver S, van der Veen C, Maurer M, Eichmüller S, Ling G, Hofmann U, Foitzik K, Mecklenburg L, Handjiski B. 1999. A comprehensive guide for the recognition and classification of distinct stages of hair follicle morphogenesis. *J Invest Dermatol* **113:** 523–532.

Plikus MV, Mayer JA, la Cruz de D, Baker RE, Maini PK, Maxson R, Chuong C-M. 2008. Cyclic dermal BMP signalling regulates stem cell activation during hair regeneration. *Nature* **451:** 340–344.

Rabbani P, Takeo M, Chou W, Myung P, Bosenberg M, Chin L, Taketo MM, Ito M. 2011. Coordinated activation of Wnt in epithelial and melanocyte stem cells initiates pigmented hair regeneration. *Cell* **145:** 941–955.

Rawles ME. 1947. Origin of pigment cells from the neural crest in the mouse embryo. *Physiol Zool* **20:** 248–266.

Reddy S, Andl T, Bagasra A, Lu MM, Epstein DJ, Morrisey EE, Millar SE. 2001. Characterization of Wnt gene expression in developing and postnatal hair follicles and identification of Wnt5a as a target of Sonic hedgehog in hair follicle morphogenesis. *Mech Dev* **107:** 69–82.

Reddy ST, Andl T, Lu M-M, Morrisey EE, Millar SE. 2004. Expression of Frizzled genes in developing and postnatal hair follicles. *J Invest Dermatol* **123:** 275–282.

Rhee H, Polak L, Fuchs E. 2006. Lhx2 maintains stem cell character in hair follicles. *Science* **312:** 1946.

Saito H, Yasumoto K-I, Takeda K, Takahashi K, Yamamoto H, Shibahara S. 2003. Microphthalmia-associated transcription factor in the Wnt signaling pathway. *Pigment Cell Res* **16:** 261–265.

Salto-Tellez M, Peh BK, Ito K, Tan SH, Chong PY, Han HC, Tada K, Ong WY, Soong R, Voon DC, et al. 2006. RUNX3 protein is overexpressed in human basal cell carcinomas. *Oncogene* **25:** 7646–7649.

Schmidt-Ullrich R, Paus R. 2005. Molecular principles of hair follicle induction and morphogenesis. *Bioessays* **27:** 247–261.

Schneider MR, Paus R. 2010. Sebocytes, multifaceted epithelial cells: Lipid production and holocrine secretion. *Int J Biochem Cell Biol* **42:** 181–185.

Schneider MR, Schmidt-Ullrich R, Paus R. 2009. The hair follicle as a dynamic miniorgan. *Curr Biol* **19:** R132–R142.

Sick S, Reinker S, Timmer J, Schlake T. 2006. WNT and DKK determine hair follicle spacing through a reaction-diffusion mechanism. *Science* **314:** 1447.

Silva-Vargas V, Lo Celso C, Giangreco A, Ofstad T, Prowse DM, Braun KM, Watt FM. 2005. β-catenin and Hedgehog signal strength can specify number and location of hair follicles in adult epidermis without recruitment of bulge stem cells. *Dev Cell* **9:** 121–131.

Stevens JR, Miranda-Carboni GA, Singer MA, Brugger SM, Lyons KM, Lane TF. 2010. Wnt10b deficiency results in age-dependent loss of bone mass and progressive reduction of mesenchymal progenitor cells. *J Bone Miner Res* **25:** 2138–2147.

Takeda K, Yasumoto K, Takada R, Takada S, Watanabe K, Udono T, Saito H, Takahashi K, Shibahara S. 2000. Induction of melanocyte-specific microphthalmia-associated transcription factor by Wnt-3a. *J Biol Chem* **275:** 14013–14016.

Takeda H, Lyle S, Lazar AJF, Zouboulis CC, Smyth I, Watt FM. 2006. Human sebaceous tumors harbor inactivating mutations in LEF1. *Nat Med* **12:** 395–397.

Takeuchi S, Takeda K, Oishi I, Nomi M, Ikeya M, Itoh K, Tamura S, Ueda T, Hatta T, Otani H, et al. 2000. Mouse Ror2 receptor tyrosine kinase is required for the heart development and limb formation. *Genes Cells* **5:** 71–78.

Tanimura S, Tadokoro Y, Inomata K, Binh NT, Nishie W, Yamazaki S, Nakauchi H, Tanaka Y, McMillan JR, Sawamura D, et al. 2011. Hair follicle stem cells provide a functional niche for melanocyte stem cells. *Cell Stem Cell* **8:** 177–187.

Teulière J, Faraldo M, Shtutman M, Birchmeier W, Huelsken J, Thiery J, Glukhova M. 2004. β-Catenin-dependent and -independent effects of ΔN-plakoglobin on epidermal growth and differentiation. *Mol Cell Biol* **24:** 8649–8661.

Ting-Berreth SA, Chuong CM. 1996a. Local delivery of TGF β2 can substitute for placode epithelium to induce mesenchymal condensation during skin appendage morphogenesis. *Dev Biol* **179:** 347–359.

Ting-Berreth SA, Chuong CM. 1996b. Sonic Hedgehog in feather morphogenesis: Induction of mesenchymal condensation and association with cell death. *Dev Dyn* **207:** 157–170.

Tumbar T, Guasch G, Greco V, Blanpain C, Lowry WE, Rendl M, Fuchs E. 2004. Defining the epithelial stem cell niche in skin. *Science* **303:** 359–363.

van de Wetering M, Sancho E, Verweij C, de Lau W, Oving I, Hurlstone A, van der Horn K, Batlle E, Coudreuse D, Haramis AP, et al. 2002. The β-catenin/TCF-4 complex imposes a crypt progenitor phenotype on colorectal cancer cells. *Cell* **111:** 241–250.

van Genderen C, Okamura RM, Farinas I, Quo RG, Parslow TG, Bruhn L, Grosschedl R. 1994. Development of several organs that require inductive epithelial-mesenchymal interactions is impaired in LEF-1-deficient mice. *Genes Dev* **8:** 2691–2703.

Van Mater D, Kolligs FT, Andrzej A Dlugosz, Fearon ER. 2003. Transient activation of β-catenin signaling in cutaneous keratinocytes is sufficient to trigger the active growth phase of the hair cycle in mice. *Genes Dev* **17:** 1219–1224.

Vasioukhin V, Degenstein L, Wise B, Fuchs E. 1999. The magical touch: Genome targeting in epidermal stem cells induced by tamoxifen application to mouse skin. *Proc Natl Acad Sci* **96:** 8551–8556.

Vidal VPI, Chaboissier M-C, Lützkendorf S, Cotsarelis G, Mill P, Hui C-C, Ortonne N, Ortonne J-P, Schedl A. 2005. Sox9 is essential for outer root sheath differentiation and the formation of the hair stem cell compartment. *Curr Biol* **15:** 1340–1351.

Waikel RL, Kawachi Y, Waikel PA, Wang XJ, Roop DR. 2001. Deregulated expression of c-Myc depletes epidermal stem cells. *Nat Genet* **28:** 165–168.

Watt FM, Collins CA. 2008. Role of β-catenin in epidermal stem cell expansion, lineage selection, and cancer. *Cold Spring Harbor Symp Quant Biol* **73:** 503–512.

Watt FM, Frye M, Benitah SA. 2008. MYC in mammalian epidermis: How can an oncogene stimulate differentiation? *Nat Rev Cancer* **8:** 234–242.

White RM, Zon LI. 2008. Melanocytes in development, regeneration, and cancer. *Cell Stem Cell* **3:** 242–252.

Widelitz RB, Jiang TX, Chen CW, Stott NS, Jung HS, Chuong CM. 1999. Wnt-7a in feather morphogenesis: Involvement of anterior-posterior asymmetry and proximal-distal elongation demonstrated with an in vitro reconstitution model. *Development* **126:** 2577–2587.

Wilson PA, Hemmati-Brivanlou A. 1995. Induction of epidermis and inhibition of neural fate by Bmp-4. *Nature* **376:** 331–333.

Wilson SI, Rydström A, Trimborn T, Willert K, Nusse R, Jessell TM, Edlund T. 2001. The status of Wnt signalling regulates neural and epidermal fates in the chick embryo. *Nature* **411:** 325–330.

Yamaguchi TP, Bradley A, McMahon AP, Jones S. 1999. A Wnt5a pathway underlies outgrowth of multiple structures in the vertebrate embryo. *Development* **126:** 1211–1223.

Yamaguchi Y, Itami S, Watabe H, Yasumoto K-I, Abdel-Malek ZA, Kubo T, Rouzaud F, Tanemura A, Yoshikawa K, Hearing VJ. 2004. Mesenchymal-epithelial interactions in the skin: Increased expression of dickkopf1 by palmoplantar fibroblasts inhibits melanocyte growth and differentiation. *J Cell Biol* **165:** 275–285.

Yang SH, Andl T, Grachtchouk V, Wang A, Liu J, Syu L-J, Ferris J, Wang TS, Glick AB, Millar SE, et al. 2008. Pathological responses to oncogenic Hedgehog signaling in skin are dependent on canonical Wnt/β3-catenin signaling. *Nat Genet* **40:** 1130–1135.

Yasumoto K-I, Takeda K, Saito H, Watanabe K-I, Takahashi K, Shibahara S. 2002. Microphthalmia-associated transcription factor interacts with LEF-1, a mediator of Wnt signaling. *EMBO J* **21:** 2703–2714.

Yuhki M, Yamada M, Kawano M, Iwasato T, Itohara S, Yoshida H, Ogawa M, Mishina Y. 2004. BMPR1A signaling is necessary for hair follicle cycling and hair shaft differentiation in mice. *Development* **131:** 1825–1833.

Zhang J, He XC, Tong W-G, Johnson T, Wiedemann LM, Mishina Y, Feng JQ, Li L. 2006. Bone morphogenetic protein signaling inhibits hair follicle anagen induction by restricting epithelial stem/progenitor cell activation and expansion. *Stem Cells* **24:** 2826–2839.

Zhang Y, Tomann P, Andl T, Gallant NM, Huelsken J, Jerchow B, Birchmeier W, Paus R, Piccolo S, Mikkola ML, et al. 2009. Reciprocal requirements for EDA/EDAR/NF-κB and Wnt/β-catenin signaling pathways in hair follicle induction. *Dev Cell* **17:** 49–61.

Zhou P, Byrne C, Jacobs J, Fuchs E. 1995. Lymphoid enhancer factor 1 directs hair follicle patterning and epithelial cell fate. *Genes Dev* **9:** 700–713.

Zhu AJ, Watt FM. 1999. β-catenin signalling modulates proliferative potential of human epidermal keratinocytes independently of intercellular adhesion. *Development* **126:** 2285–2298.

Wnt Signaling in Mammary Glands: Plastic Cell Fates and Combinatorial Signaling

Caroline M. Alexander, Shruti Goel, Saja A. Fakhraldeen, and Soyoung Kim

McArdle Laboratory for Cancer Research, University of Wisconsin–Madison, Madison, Wisconsin 53706-1599

Correspondence: alexander@oncology.wisc.edu

The mouse mammary gland is an outstanding developmental model that exemplifies the activities of many of the effector pathways known to organize mammalian morphogenesis; furthermore, there are well-characterized methods for the specific genetic manipulation of various mammary epithelial cell components. Among these signaling pathways, Wnt signaling has been shown to generate plasticity of fate determination, expanding the genetic programs available to cells in the mammary lineage. It is responsible first for the appearance of the mammary fate in embryonic ectoderm and then for maintaining bi-potential basal stem cells in adult mammary ductal trees. Recent technical developments have led to the separate analysis of various mammary epithelial cell subpopulations, spurring the investigation of Wnt-dependent interactions. Although Wnt signaling was shown to be oncogenic for mouse mammary epithelium even before being identified as the principle oncogenic driver for gut epithelium, conclusive data implicating this pathway as a tumor driver for breast cancer lag behind, and we examine potential reasons.

There are several excellent recent reviews that cover aspects of Wnt signaling and mammary gland development and transformation (Boras-Granic and Wysolmerski 2008; van Amerongen and Nusse 2009; Incassati et al. 2010; Roarty and Rosen 2010; Wend et al. 2010; Jarde and Dale 2011). The aim of this article is to focus attention on the open questions in this area. There are a remarkable number of tools available to assist with this, given the focus on breast cancer research in the past 10 years. They include large collections of human breast cancer cell lines (Neve et al. 2006; Hoeflich et al. 2009; Hollestelle et al. 2010) and dozens of strains of mice that are useful for analysis of different aspects of Wnt signaling and biology (van Amerongen and Berns 2006). This battery of genetic tools includes transgenic mice that express (or induce conditional ablation of) genes in either of the two main mammary epithelial cell types that comprise the mammary gland (the organization of the mammary gland is shown in Fig. 1). These are the luminal cells—typically targeted by one of three drivers, namely, MMTV LTR (mouse mammary tumor virus long terminal repeat), WAP (whey acidic protein promoter, expressed during milk production/terminal differentiation), or BLG (β-lactoglobulin, another milk whey protein) promoters)—and the basal cells (typically targeted using keratin-5 or -14 promoters, also expressed in other stratified epithelia, notably skin). The definitions of luminal

Cite this article as *Cold Spring Harb Perspect Biol* doi: 10.1101/cshperspect.a008037

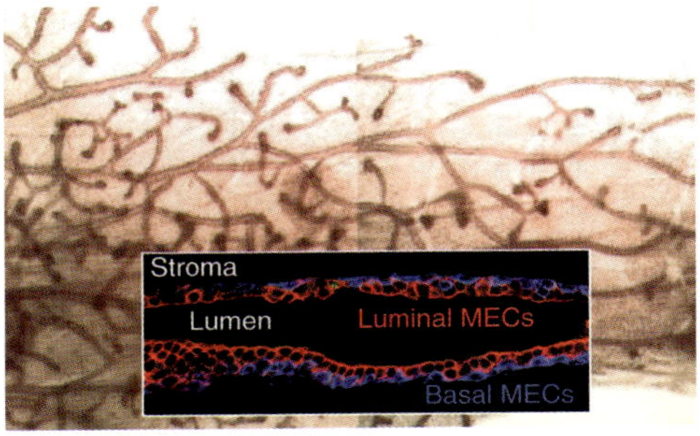

Figure 1. The mouse mammary gland, organization, and cell types. The (10) mouse mammary gland(s) comprise fat pads attached to the ventral mouse skin, colonized by a branched tree of hollow, epithelial mammary ducts (stained with carmine red) that are connected to the nipple. During pregnancy, there is massive proliferation of lobuloalveolar side branches to colonize the interstitial spaces between ducts, becoming filled with milk, expressed by oxytocin-induced contraction of myoepithelial cells. (*Inset*) An immunofluorescent stain of a longitudinal cross section of a non-pregnant duct, stained with a basal cell anti-cytokeratin (CK5) together with a luminal cell anti-cytokeratin (CK8), to show the bilayered structure of the epithelium.

and basal cells vary between studies because they are based on the analytical method used. These may be (1) location in tissue sections (facing the lumen or adherent to the basement membrane); (2) expression of molecular markers usually associated with basal cells (e.g., expression of basal-specific cytokeratin 5 [CK 5] or p63, or luminal-specific CK8 or Muc1); or (3) copurification with subgroups of cells isolated by flow cytometry (e.g., using Lin/CD29/CD24 or Lin/EpCAM/CD49f). These latter cell populations are often subsequently typed using their expression of histological markers or mRNA profile to generate a "luminal" or "basal" descriptor. The cell groups described by these means are not always the same, and this becomes important to experimental interpretation.

Not all cells in a given lineage are identical. For luminal cells, perhaps one of their most obvious differences is their expression of ERα (estrogen receptor-α). Thus, for both mouse and human, ~15% of luminal cells express ERα (at any one time), irrespective of their stage of development (Clarke 2003; Mastroianni et al. 2009). ERα-positive and -negative cells can be purified by flow cytometry and analyzed separately (Kendrick et al. 2008; Lim et al. 2010). The ERα-negative luminal cell subpopulation contains a proliferative activity, leading it to be labeled "luminal progenitor cells" (a heterogeneous group), whereas the ERα-positive group is non-clonogenic in vitro and therefore labeled "mature" (Table 1).

Using lacZ reporter strains, the heterogeneity of expression of MMTV- and WAP-driven transgenes is obvious, but the basis for this is unknown (e.g., Wagner et al. 2001). Indeed, the MMTV LTR is a remarkable 1200-bp gene expression motif that condenses all of the key elements of mammary regulation, to include the specification of mammary fate (expression starting early in the ectoderm of embryogenesis), together with the hormone inducibility that appears during puberty in females (Rouault et al. 2007), with further up-regulation during pregnancy and lactation (Mink et al. 1990). Cre expression has effects on mammary morphogenesis and lactation; therefore, experiments that use this tool need to be interpreted with caution (Chan et al. 2007; Robinson and Hennighausen 2011).

There are several Wnt reporter lines that show highly regulated (but slightly different) expression patterns in mammary gland (including a number of strains based on Wnt response elements [WRE] driving lacZ [for review, see Barolo 2006], and two newer strains based on WRE-GFP, perhaps more useful for purifying Wnt-responsive cell types [Currier et al. 2010; Ferrer-Vaquer et al. 2010]). Because of the redundant expression of many Fzd (8/10), both Ror receptors (2/2) and Wnt ligand types (11/19) in mammary gland (Table 1), initial experiments have focused on producing gain or loss of function for canonical Wnt signaling. Thus, overexpression of the forerunner Wnt ligand species, Wnt1- (using an MMTV-Wnt1 construct) (Table 1) (Tsukamoto et al. 1988) induced ductal hyperplasia (not correctly described as mid-pregnant equivalent) and basaloid tumors (Li et al. 2003a; Liu et al. 2004; Vaillant et al. 2008), whereas the cell-autonomous Wnt signaling effector

ΔNβ-catenin (using MMTV-LTR) induced later onset hyperplasia (luminal in origin) and luminal-type tumors (Imbert et al. 2001) or ductal hyperplasia and basaloid tumors when expressed by a CK5 construct (Teuliere et al. 2005). Loss of function for β-catenin, induced by overexpression of Axin or a dominant-negative β-catenin effector in luminal cells, has implicated β-catenin in lobulo-alveolar development during pregnancy (Hsu et al. 2001; Tepera et al. 2002). This is despite a lack of clear evidence for activation of canonical Wnt reporter activity during pregnancy (Chu et al. 2004; Boras-Granic et al. 2006a). If a canonical βcat/TCF signaling is indeed activated in luminal cells, this presents a paradox. Cell surface presentation of Lrp is considered to be the limiting factor for canonical Wnt signaling (Brennan and Brown 2004), and Wnt1 responses are confined to Lrp5-positive cells (Badders et al. 2009; Teissedre et al. 2009; Baker et al. 2010; Kim et al. 2011). Only basal cells appear to be competent to generate a canonical response in response to Wnt ligands (they express the obligate canonical Lrp5 and Lrp6 receptors). Therefore, the identity of the cell surface receptor able to mediate β-catenin/TCF-induced luminal cell division is not yet known.

Lrp6 knockout (KO) mice die at birth, and the functionality of Lrp6 KO mammary rudiments has not yet been evaluated, although Lrp6 hypomorphs ($Lrp6^{+/-}$) are underbranched (Lindvall et al. 2009). Although dominant-negative TCF molecules have been used effectively in other cell types (van de Wetering et al. 2002), their use has not yet been reported for mammary gland. Table 1 summarizes the expression data and functional information that are available to describe Wnt signaling effectors from studies of normal and transgenic mice and human tissues (including some observations from cell lines).

The vast majority of data that describe the molecular basis of signaling via Lrp5/6 derives from ectopic expression of Wnt components in HEK293T cells (Brennan et al. 2004; Cong et al. 2004; Zeng et al. 2005, 2008). Thus, the specific molecular complexes that form in mammary epithelial cells have yet to be described (Fig. 2). For example, there are convincing data for the activation of LRP (at the PPSP motif) by a range of different kinases, including GSK, PKA, Pftk1, Grk5/6, and CK1 (Niehrs and Shen 2010), and for the control of kinase recruitment by local phospho-lipid effectors such as P(4,5)IP2 (Tanneberger et al. 2011); however, the relative significance of each of these mechanisms for mammary epithelial cells is not yet known.

PLASTICITY OF MAMMARY EPITHELIAL CELL FATE REQUIRES A WNT SIGNAL

When mammary precursor cells are set aside from the embryonic ectoderm, there is a pulse of Wnt signaling that can be visualized by the expression of Wnt reporter genes (the lacZ Wnt reporter TOPgal) (Chu et al. 2004). If dkk1 is overexpressed in the basal keratinocyte layer (using a CK14 promoter) or when Lef1 function is missing (Boras-Granic et al. 2006b; Boras-Granic and Wysolmerski 2008), mammary specification is inhibited. Wnt10b is expressed at the earliest time during differentiation of the mammary placode (Veltmaat et al. 2004; Boras-Granic and Wysolmerski 2008), suggesting that Wnt10b could be the placode specifier. Thus Wnt signaling is essential for increasing the plasticity of embryonic skin cells so that they initiate mammary-specific programs. The induction of plasticity is a typical function associated with pulses of Wnt signaling during mammalian developmental processes (Wend et al. 2010; Domyan and Sun 2011).

Somatic stem cells are different from the majority of cells in mammalian organs, owing to their plasticity and pluripotency. For mammary gland, ductal mammary stem cells are defined by their capacity to regenerate a new mammary gland comprising a bilayered epithelial ductal network following transplantation to cleared fat pads. These stem cells have been shown to comprise a specific subpopulation of basal epithelial cells (with high expression of integrins α6 and β1, and other cell surface-associated antigens such as epithelial cell adhesion molecule [EpCAM] and CD24, a GPI-linked sialoglycoprotein). When ectopic Wnt signals are present, mammary stem cells comprise a larger fraction of the total population (Li et al. 2003b; Liu et al.

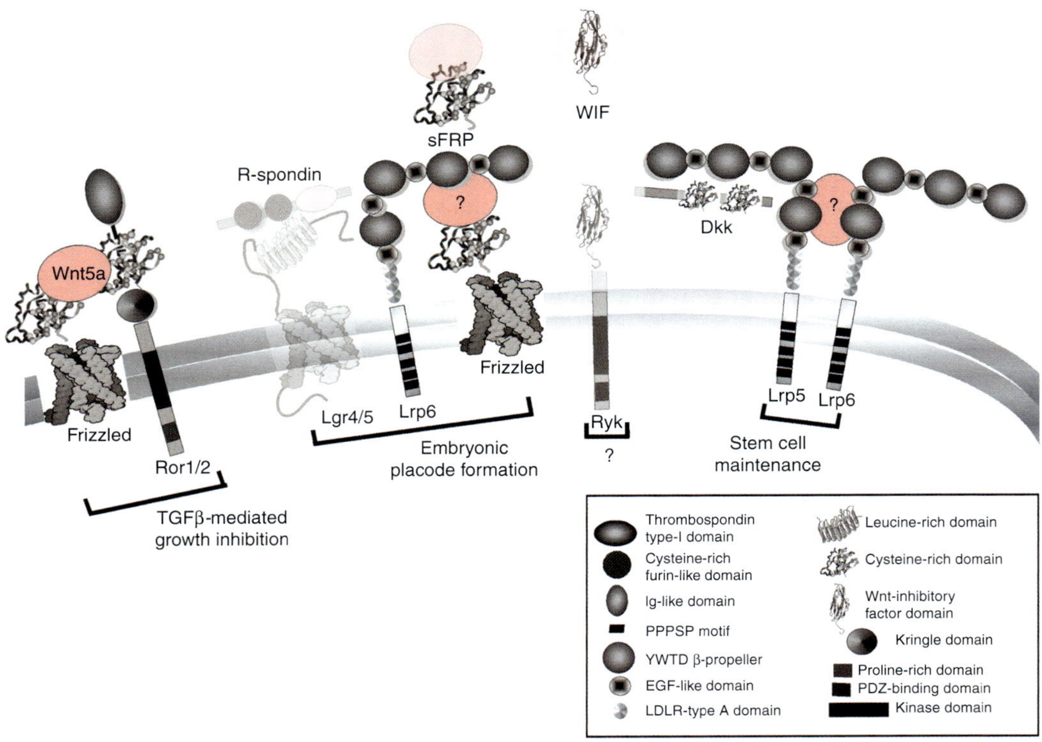

Figure 2. Cell surface Wnt signaling components shown to function during various phases of mammary gland growth and development. Structural and function information is drawn from data from He et al. (2004), Liepinsh et al. (2006), Kikuchi et al. (2007), Green et al. (2008), Bourhis et al. (2010), and Weis (2011). The information presented here can be cross-referenced to Table 1 for expression data. Physiological activities are detailed in the text. The Lgr4/5/RSpo component of embryonic placode formation is likely but has not yet been shown (hence it is grayed out). Although both Lrp5 and Lrp6 are required to maintain mammary stem cells, the Lrp5-6 heterodimer shown here is implied from our data (Goel et al. 2012a). It is not known whether the inhibitors shown (dkks, WIF, and sFRPs) are functional, but several are expressed. When components are illustrated with ribbon diagrams, their structure is known in some detail. (Wnt ligands are shown in pink.)

2004; Shackleton et al. 2006), and undifferentiated cells (lacking expression of marker proteins) are evident in mammary glands (Liu et al. 2004). There are data to support a role for Wnt proteins as inhibitors of differentiation. Analysis of a mouse mammary cell line (HC11 cells) showed that Wnt3a inhibited differentiation (assayed as β-casein expression) in response to endocrine effectors (including prolactin/dexamethasone/insulin), and that was reversible by sFRP4 expression (Constantinou et al. 2008). Furthermore, cells from Wnt1-induced glands show higher expression of the ΔN-isoform of the p53 family member, p63 (Badders et al. 2009), and less TA-p63; a profile typically associated with dividing progenitors of basal epithelial cell types and functionally implicated in tissue senescence and aging (Wu et al. 2003).

Ductal mammary stem cells require an Lrp5-dependent, Wnt-associated signal for their induction or maintenance, and their activity is much reduced when the Lrp5 receptor species is missing (Lindvall et al. 2006; Badders et al. 2009). After the embryonic phase of development, there is little evidence for the expression of TCF-consensus concatamer-based Wnt reporters, including BAT-gal (Chu et al. 2004; Boras-Granic and Wysolmerski 2008; Lindvall et al. 2009). This is somewhat of a conundrum, given the importance of canonical Wnt signaling to

 Cite this article as *Cold Spring Harb Perspect Biol* doi: 10.1101/cshperspect.a008037

adult stem cells. However, more recently, Zeng and Nusse (2010) and R van Amerongen and R Nusse (pers. comm.) identified rare $lacZ^+$ cells in postnatal glands of $Axin2^{LacZ}$ mice. These cells were quantified using flow cytometric analysis, separated using a substrate loading approach, and found to be enriched in basal stem cells (Zeng and Nusse 2010). Furthermore, the addition of high concentrations of Wnt3a ligand sustained a long-lived mammary stem cell activity (measured by clonogenicity in vitro and in vivo) when added to cultures of primary mammary epithelial cells.

Note that the absence of Lrp5 does not significantly affect embryonic mammary placode development (even though that too is Wnt dependent), nor does it affect lobulo-alveolar development during pregnancy. The Wnt factors that regulate these processes are therefore likely to be different and Lrp5 independent. The role of Lgr4/5 and RSpo proteins in the maintenance of other stem cell types is described later. This together with the expression pattern of RSpo proteins during embryogenesis (Table 1) suggests that Lgr/RSpo may be the embryonic counterpart of the Lrp5-Lrp6 complex that is essential to the adult gland (Fig. 2).

Furthermore, in the absence of stem cells (due either to the absence of Lrp5 or β1 integrin), ductal outgrowth is almost normal, as indeed are all other functions associated with mammary gland growth and differentiation (Lindvall et al. 2006; Taddei et al. 2008; Badders et al. 2009). This may appear to be counterintuitive. However, recently, it has emerged that bi-potential stem cell activity is not activated during normal mammary development. Thus, Van Keymeulen et al. (2011) genetically tagged CK14-expressing or CK5-expressing mammary epithelial basal cells and showed that this cell population contained bipotent stem cell activity when transferred to cleared fat pads. When tagged during puberty, the bi-potential activity (stem cell activity) of these basal cells was not expressed throughout ductal extension or pregnancy, deduced from the lack of tagged luminal cell daughters (Van Keymeulen et al. 2011). Vice versa, CK8-expressing or CK18-expressing tagged luminal cells did not appear in the basal cell compartment. In fact, the bipotency of basal cell stem cells could be suppressed during fat pad transfer if luminal cells were included at a cell ratio of one per five basal cells. These investigators propose that there are long-lived monopotent basal and luminal stem/progenitor cell types that organize the morphogenesis of mammary gland after birth. It is implied that the activity of each stem/progenitor activity cell activity would primarily affect the directly related lineage. Indeed, the loss of basal stem cell activity (by targeted gene mutations) typically results in a decrease in the proportion of basal cells (compared with luminal cells), without affecting overall mammary gland proliferation (Taddei et al. 2008; Badders et al. 2009). Using other constructs and reporters, some of these results have been reproduced by other investigators (R Van Amerongen, R Nusse, and J Jonkers, pers. comm.). This represents a profound change in our view of stem cell function during adult organogenesis.

What, then, is "stem cell" activity scored by fat pad transfer, and is it a relevant assay? Stem cell activity is scored by assaying the outgrowth of mammary ductal trees after inoculating limiting doses of dissociated, singlized cells from mammary glands. The regenerative activity must therefore be able to survive cell isolation, to subsequently differentiate in vivo, and to create both basal and luminal mammary epithelial cells. Without ectopic endocrine support (e.g., pituitary extract), outgrowths of purified luminal cells are limited in extent (Kamiya et al. 1998; Visvader and Smith 2011). Given the evidence for luminal stem/progenitor cells provided by these studies, it is important to be able to assay this activity somehow, preferably in vitro. Perhaps it is this activity that is measured by the mammosphere assay (spheres contain no dividing basal cells) (S Kim and C Alexander, unpubl.).

IS THERE A WNT SIGNAL THAT CREATES THE MAMMARY STEM CELL NICHE?

Given that mammary stem cells are dependent on Wnt signaling, the question then turns to the identity of the ligand that interacts with Lrp5/

Fzd to maintain these cells. Lrp5 is expressed and presented, alongside Lrp6, on all basal mammary epithelial cells (within the limits of detection) (Badders et al. 2009), meaning that the stem cells are not alone in being able to respond to Wnt ligands. Indeed, it is interesting that Lrp5 is specifically required as a maintenance factor for stem cells, given that Lrp6 is often described as the "better," more potent receptor (MacDonald et al. 2008). However, we have shown that a subgroup of Wnt ligands (including Wnt1 and Wnt9b) requires both Lrp5 and Lrp6 to signal in some cell contexts (Goel et al. 2012b). One of this subgroup is presumably responsible for adult MaSC function (Fig. 2).

It is likely that there is a localized source of a Wnt ligand that is responsible for the stem cell niche. The phenotype of $Lrp5^{-/-}$ mammary glands maps to the mammary epithelial cells, and not the $Lrp5^{-/-}$ fat pads (NM Badders and CM Alexander, unpubl.). Most of the Wnt ligands expressed in mammary gland (Table 1) are known to have a short range of effect in vivo (due to binding to heparan sulfates or their lipidation). In fact, the diffusion range of a Wnt ligand has been visualized in mouse limb bud, where Wnt is produced by the surface ectoderm. In this case, the range appears to be approximately five cells, through the relatively loosely packed cells of the limb bud (and assuming no influence of cross gradients of other molecules) (ten Berge et al. 2008). The ligand responsible for stem cell maintenance may be expressed by other epithelial cells, or by a stromal host cell (for candidates, see Table 1).

WNT RESPONSES: A SUM TOTAL

Given the variety of Wnt signaling receptors (illustrated in Fig. 2) together with the fact that these receptors are commonly coexpressed, any one Wnt ligand is likely to generate several different signals in any one cell. The signaling outcome will depend on the absolute concentration of each ligand and receptor species, along with their relative affinity. Whether these pathways synergize with one another, or antagonize each other, depends on the context and molecules involved (Semenov et al. 2007; van Amerongen and Nusse 2009). Thus, for melanoma cells, non-canonical Wnt signaling, mediated by Wnt5a/Ror2, promotes malignant changes in the presence of high levels of canonical signaling, and these pathways are considered to be synergistic during tumor development (O'Connell and Weeraratna 2009).

For mammary gland, the "old-fashioned" distinction of canonical and non-canonical Wnt ligands appears to hold up, presumably because of the specific repertoire of cell surface receptors expressed (van Amerongen et al. 2008). Thus, Wnt5a, although it can acquire canonical signaling activity when Lrp5 and Fzd4 are present (Mikels and Nusse 2006), has no βcat/TCF trans-activation activity in mammary gland (not surprising because Fzd4 is not expressed) (Table 1). Indeed, as shown for other cell types, Wnt5a not only does not induce Wnt reporter expression, it inhibits Wnt3A/Wnt1 canonical responses. In fact, TGF-β (a well-known mammary gland growth inhibitor) induces Wnt5a expression (directly via Smad binding sites on the promoter), and Wnt5a may be necessary to limit and control mammary gland growth. Thus, when Wnt5a is absent (Wnt5a KO), mammary gland development is hyperproliferative (Roarty and Serra 2007), whereas in the presence of ectopic Wnt5a (produced by slow release beads), ductal extension is inhibited. These phenotypes are the inverse of gain of function of canonical Wnt activity. It is puzzling then that this inhibition is not reproduced in transgenic MMTV-Wnt5a strains (Baxley et al. 2011).

How does a non-canonical ligand inhibit the canonical response? Various ideas have support; the first relies on competition; thus, if Wnt5a competes effectively for Fzd binding (in association with Ror1/2) (Mikels et al. 2009; Grumolato et al. 2011), the amount of the rate-limiting LRP receptor available to other ligands could be reduced. The second is that Wnt5a may generate an intracellular signal that is inhibitory for canonical responses; thus, when Ror1/2 binds Wnt ligands, Rac is activated (leading to JNK activation) and Fzd2 is internalized and inactivated by a clathrin-dependent mechanism (Sato et al. 2010). Witte et al. (2010) suggest that Ror2 generates an inhibitory signal via CK1-mediated dvl phosphorylation and that CK1 constitutes a switch that directs the signaling output (studies of HEK293 and Cos cells) (Foldynova-Trantirkova et al. 2010; Witte et al. 2010). Indeed, the effects of the morphogenetic planar cell polarity pathway, typically mediated by ligands such as Wnt11, are largely unexplored for mammary gland. During development, it is often true that multiple Wnt signals are perceived by cells simultaneously (see the examples drawn from genetic analysis of *Caenorhabditis elegans*, summarized by van Amerongen and Nusse 2009). These signals are key to tube building and the convergent extension processes associated with polarization of epithelial layers (Kouros-Mehr and Werb 2006; Miller and McCrea 2010). In addition, it is likely that there are polarized cell divisions that occur either in the plane of the basal and luminal layers, or that cross between them. Evidence suggests that Notch signaling and p53 are important to determining these division planes, and both pathways are highly cross-regulated by, or interact with, Wnt signaling (Callahan and Egan 2004; Collu and Brennan 2007; Bouras et al. 2008; Cicalese et al. 2009; Yan et al. 2010).

Wnt signaling in other tissue contexts also serves to organize the growth and development of more differentiated cells that emerge from the stem cell compartment. Thus, for intestinal cells, a battery of genes are induced by βcat/TCF signaling (van de Wetering et al. 2002), and these can be divided functionally into at least three programs (maintenance of progenitor phenotype, promotion of differentiation of Paneth cells in the crypt, and compartmentalization of differentiated intestinal cells from progenitors). The latter is regulated by ephrins (EphB2 and EphB3). Thus, in Apc^{min} mice, when ephrins are missing and cell sorting is defective, tumor development proceeds even more rapidly (Batlle et al. 2002). For the retinotectum, opposing gradients of ephrin and Wnt signals (via Ryk or Fzd) organize and map the cell function (Schmitt et al. 2006). For mammary gland, the ephrin Robo1 is a suppressor of mammary branching and basal cell division (Macias et al. 2011) and promotes the accumulation of cytoplasmic E-cadherin and β-catenin. Whether ephrin expression is downstream and/or upstream of Wnt signals in this cell type is not fully understood. Other Wnt target genes identified in intestine by van de Wetering et al. (2002) include the orphan GPCR receptors Lgr4 and Lgr5 (indeed, for intestine, an Lgr5-dependent tag serves to identify intestinal stem cells in studies of growth and pathogenesis) (Barker et al. 2007). Lgr4/5 functions to feed forward to amplify the stem cell function in this tissue, by binding RSpo proteins, forming a complex with Lrp/Fzd molecules that potentiates a canonical Wnt (Wnt3a) signal (Carmon et al. 2011; de Lau et al. 2011). Because all of the same components are also expressed in mammary gland, the same principle may apply (Fig. 2).

It is possible that Wnt signaling could act indirectly on mammary morphogenesis and function via its effects on non-epithelial components of the mammary gland, including blood vessels, macrophages, fibroblasts, and adipocytes. For example, endothelial cell maturation is highly Wnt dependent (Parmalee and Kitajewski 2008), including a relatively little characterized pathway that depends on the Norrin ligand and Fzd4/Lrp5 (Ye et al. 2010). Macrophages and other immunomodulatory cells are well known to affect mammary development, involution, and tumor progression (Schwertfeger et al. 2006; Schaale et al. 2011; O'Brien et al. 2012). Fibroblasts are activated by local Wnt signals to initiate a desmoplastic response (Xu et al. 2000) that elicits a mobilization and proliferation response from circulating cells and bone marrow precursors (Kim et al. 2008). These cells are recruited to the tumor, where they participate in the angiogenic and inflammatory response, creating leaky vessels and macrophage activation, neither of which is associated with the tumors that develop in response to the intracellular Wnt effector β-catenin (Kim et al. 2008). Adipocyte differentiation is controlled by Wnt proteins (specifically Wnt10b), and adipocyte differentiation/fat content varies substantially throughout pregnancy and lactation (Prestwich and Macdougald 2007). Indeed, it is likely that the controlled spacing and branching of mammary ducts in the fat pad represents an interaction between the host tissue and the expanding ductal epithelial cell population (Sakakura 1987).

Furthermore, responses to Wnt signals can be modulated by cross talk from other effectors and pathways, including metabolic checkpoints such as AMPK, mitochondrial function, autophagy, and lipid effectors (Inoki et al. 2006; Gao et al. 2010; Yoon et al. 2010; Kang et al. 2011). Prostaglandins (Goessling et al. 2009) have been shown to cross-talk to Wnt signaling to maintain hematopoietic stem cells, and there are other pathways that regulate the turnover and metabolism of β-catenin (Incassati et al. 2010). Although the significance of these is largely unknown for normal mammary development and tumor initiation, the interaction of Wnt signaling with EGFR or with nuclear hormone receptors suggests that these could be particularly important for dissecting mammary behaviors (Schlange et al. 2007; Mastroianni et al. 2009; Roarty and Rosen 2010).

WNT4: A WNT PROBLEM CHILD?

Wnt4 knockout glands were reported to show a delay of lobulo-alveolar development (Brisken et al. 2000) during pregnancy, implying that this ligand could be a mammary epithelial cell growth factor with similar properties to Wnt1 (and Wnt10b). Indeed, when mammary epithelial cells were transduced with a Wnt4 expression vector, the outgrowths were hyperplastic (Bradbury et al. 1995). However, unlike Wnt1, Wnt4 does not induce mammary gland hyperplasia when expressed as an MMTV-based transgene. This is despite effective expression levels (equivalent to mid-pregnancy) and evidence of Wnt4-induced gene expression (e.g., Wnt16) (Kim et al. 2009), both in vitro (in HC11 cells) and in vivo in virgin glands. In contrast to Wnt1 (and the other canonical ligands), Wnt4 does not induce axin2 expression or TOP-FLASH activation in mammary epithelial cells or fibroblasts, and does not induce the typical developmental phenotypes associated with canonical Wnt ligand expression in zebrafish embryos.

It may be that Wnt4 is just one arm of a progesterone-induced program, making Wnt4 necessary but not sufficient. Other progesterone-induced components, such as RANKL (Beleut et al. 2010) and/or the differentiation/specification factor bHLH protein Id4 (Fernandez-Valdivia et al. 2005), could also be required for an effective response. Wnt4 may be a prime candidate to explain the regulation of mammary stem cell function by endocrine factors. Thus, recent reports describe a dynamic twofold expansion and retraction of the stem cell activity during the 4-d estrus cycle (Joshi et al. 2010), and an 11-fold difference in activity in estrogen/progesterone positive/negative mammary environments (Asselin-Labat et al. 2010). These hormonal dynamics closely resemble the expression of Wnt4 mRNA and protein expression in luminal mammary epithelial cells (Table 1) (Silberstein et al. 2006).

Information derived from other Wnt4-dependent processes, such as kidney development, could provide a useful precedent for mammary gland. For kidney, several Wnt proteins are expressed by the epithelial ureteric bud, including Wnt6, 7b, 9b, and 11, and two Wnt proteins by the morphogenic mesenchyme, Wnt 2b and Wnt4. Of these, Wnt4, 9b, and 11 are essential, and none show redundant functions, suggesting that the nature of the signal generated by each is unique (Carroll et al. 2005; Pulkkinen et al. 2008). In MDCK and other kidney cell types, Wnt4 can induce a canonical signal, but the receptor mediating this signaling event is still mysterious (although Wnt4 binds Fzd6, this is not the canonical signaling reaction) (Lyons et al. 2004). Interestingly, Bernard et al. (2008) suggest that Wnt4 elicits a distinct response from that induced by Wnt5a, namely, accumulation of β-catenin at the cell membranes (of HEK293T cells). These, or other factors, may be important to understanding the molecular mechanism of Wnt4 activity.

IS WNT SIGNALING A DRIVER FOR BREAST TUMOR GROWTH?

Wnt signaling has been known to be highly oncogenic for mouse mammary glands since 1982, when Nusse and Varmus showed that mouse mammary tumors induced by MMTV

 Cite this article as *Cold Spring Harb Perspect Biol* doi: 10.1101/cshperspect.a008037

were caused by activation of Wnt ligand gene expression. In fact, not only were Wnt1, Wnt3, Wnt3A, and Wnt10b activated to induce mammary tumors, so indeed were Wnt-pathway coactivators RSpo2 and RSpo3 (Nusse and Varmus 1982; Nusse 1988; Theodorou et al. 2007; Callahan and Smith 2008). Of these, only Wnt10b and the RSpo ligands are usually expressed in significant amounts in mammary gland (Table 1). It may be significant that other Wnt components are not on that list—including Lrp6, even though overexpressed Lrp6 can induce hyperplasia in mouse models (Lindvall et al. 2009; Liu et al. 2010; Zhang et al. 2010a).

However, although it is true that the vast majority of colorectal tumors have a Wnt signaling driver (initiated by mutations in one of three genes in the canonical Wnt signaling pathway, namely, inactivating mutations of the tumor suppressors Apc or axin or activating mutations of the oncogene β-catenin), there is little evidence to suggest that breast tumors share this etiology. Mice with an Apc mutation (Apc^{min}), although they develop intestinal polyps, do not spontaneously develop breast tumors. (They are, however, sensitized to tumor development, revealed by the enhanced rate of tumorigenesis after the administration of carcinogens [Moser et al. 1993].) Indeed, unlike mammary glands expressing Wnt ligands, Apc^{min} mammary glands are not hyperplastic, so this loss of function is insufficient to generate a growth signal. The mechanisms governing inactivation of Wnt signaling components may be different for mammary cells; for example, data derived from combination microarray–CGH suggest that the expression of genes in 5q22.2 is reduced in basaloid breast tumors (Geyer et al. 2011). This interval includes Apc. Perhaps total inactivation of the typical Wnt tumor suppressors by mutation is lethal for mammary epithelial cells.

Preneoplastic lesions with ectopic Wnt signaling often require cooperating mutation(s) to progress. Perhaps the lack of Wnt mutations in breast tumors could be explained if these cooperating mutations did not arise readily during breast cancer development, or were not tolerated. For example, gut tissues in Apc^{min} mice (after loss of heterozygosity) accumulate β-catenin, and this is sufficient to prevent differentiation and induce morphological changes in tissue architecture. Note that this tissue is remarkable for its high rate of continuous growth, which has no parallel in adult mammary gland. However, nuclear β-catenin does not arise until later in tumor development, after Ras is mutated (Phelps et al. 2009). Ras is rarely mutated in any type of breast tumor, although Ras mutations are relatively common in breast tumor cell lines (Hollestelle et al. 2007). Ectopic Ras activity is often detected by surveillance checkpoints (Zhu et al. 2005), and cells are eliminated in response (typically a p53-dependent process). Thus, ras activation has to be carefully titrated to contribute effectively to tumor development. Perhaps it is no coincidence that the Fgf signaling pathway is frequently activated during Wnt-induced tumor development (where Ras is activated as a downstream effector in a regulated manner) (Callahan and Smith 2008). In developmental models, when Wnt and Fgf are coordinately ex-pressed, the signal generated is distinct from either alone (ten Berge et al. 2008). This may also explain the observation that human basaloid breast tumor aggression is associated with (and dependent on) hyperactivation of Fgf signaling (Turner et al. 2010a,b; Sharpe et al. 2011). Similarly, activation of the tyrosine kinase Ron has been associated with β-catenin phosphorylation (Y^{654} and Y^{670}), together with enhanced nuclear translocation and transactivation activity (Wagh et al. 2011). Furthermore, when a Wnt signal is present, two reports suggest that Akt activation is sufficient to activate the nuclear translocation of β-catenin, in human and mouse breast tumor epithelial cells (Korkaya et al. 2009; Zhang et al. 2010b). Akt has been shown to phosphorylate β-catenin (S-552) to enhance nuclear translocation (He et al. 2007), and an inhibitor of Akt signaling (perifosine) inhibits the accumulation and translocation of β-catenin, Wnt signaling reporters, and stem cell activity, in normal cells and tumor cells deficient in PTEN and/or p53. Up-regulation of Akt signaling in the absence of Wnt signals is unlikely to cross-talk to Wnt end points, despite the fact that they share a signaling component (GSK3β) (Korkaya et al. 2009; Ng et al. 2009); indeed, the data to date suggest that the pools of GSK3β allocated to these pathways are different and separate (Ding et al. 2000).

It is useful to outline the limitations of assays designed to test whether Wnt signaling is a tumor driver, in the following sections.

Wnt-Dependent Subpopulations Are Obscured by General Wnt Signatures

The tumorigenic effects of ectopic Wnt signaling may be associated with only a subpopulation of (stem/progenitor) cells (especially early in tumor development), whereas the tumor majority shows a different Wnt signaling response. This has been illustrated in a Wnt1–Lrp5-dependent transgenic tumor model system (Gunther et al. 2003; Lindvall et al. 2006). These tumors contain a minority "responder" cell population that depends on a paracrine canonical Wnt ligand (Lrp5-positive cells) from the cell majority (Kim et al. 2011). This cell majority is therefore unable to generate a canonical Wnt signal (Lrp5-negative; visualized by axin2 expression), but is nonetheless recruited to grow in a Wnt1-dependent manner. Thus, the Wnt signature is not activated in the cell majority, and the tumor cell population grows in response to a field effect, with so far unknown mediators.

A "field effect" for growth is common for other tumor drivers, such as the estrogen-ERα axis. Thus, for ERα-positive tumors, the tumorigenic ERα receptor is typically expressed by a minority population (ERα-positive cells comprise 1%–100% of tumor cells, depending on the exact pathological criteria), but the Ki67/mitotic index is raised throughout (Clarke 2003; Schlange et al. 2007; Mastroianni et al. 2009). Extrapolating from this idea, it is clear that transcriptional signatures of whole populations of mixed cell types could be difficult to interpret (Huang et al. 2005). The markers c-myc and cyclinD1 that characterize Wnt-dependent transcription for intestinal cells are not prominent (or required for tumorigenesis) for mammary gland (Huang et al. 2005), and indeed are not specific to Wnt signaling. Data also suggest that activation of Akt signaling may be a component of Wnt ligand-mediated oncogenesis, and this outcome may make ectopic Wnt signaling difficult to recognize by a genetic signature (Korkaya et al. 2009; Zhang et al. 2010b).

Nuclear β-Catenin

As discussed above, a combination of signaling activities may be required to generate nuclear β-catenin. Nuclear β-catenin has been convincingly observed in some breast tumors (Geyer et al. 2011). However, even when Wnt signaling is known to be the tumor driver, it may not be associated with significant immunostaining of nuclear β-catenin, thus the breast tumors identified with ectopic Wnt signaling by this assay may underrepresent the prevalence of this tumor driver. For example, although the growth of Wnt1-induced breast tumors is known to require the continuous presence of Wnt1/Lrp5, there is little support for the presence of nuclear β-catenin (data not shown), despite high expression of Wnt reporters in responder cells (Kim et al. 2011). Furthermore, basaloid breast tumors (in particular) are demonstrably heterogeneous with respect to basal and luminal cell subpopulations (Rakha et al. 2009; Kim et al. 2012), and the Wnt-dependent subpopulation may be rare.

Loss of E-Cadherin

Loss of function for the tumor suppressor E-cadherin (characteristic of many breast tumors) can lead to redistribution or elimination of β-catenin, an E-cadherin binding partner (van de Wetering et al. 2001; Incassati et al. 2010). However, this is not necessarily accompanied by the functional recruitment of β-catenin to the Wnt signaling pathway (Gumbiner 1998; Geyer et al. 2011).

These issues do not eliminate Wnt signaling as a significant tumor driver. Indeed, there is a large body of literature that describes a disproportionate expression of mRNA for positive effectors of Wnt signaling in tumors, either with respect to normal tissue, sometimes with respect to a larger group of breast tumors (for review, see

Incassati et al. 2010). All of these studies have the caveat that the cell composition of tumors is different from subtype to subtype (e.g., the luminal vs. basaloid subtypes, where the latter are mixtures of both basal and luminal cells), and the samples are seldom microdissected, perhaps including contributions from cells that may express high amounts of Wnt signaling components (such as macrophages). However, the accumulated data suggest that there are significant expression of Wnt ligands (summarized by Benhaj et al. 2006), loss of non-canonical signaling (Jonsson et al. 2002; Serra et al. 2011), and shutdown of expression of extracellular inhibitors. Numerous extracellular inhibitors of Wnt signaling are expressed by mammary gland (including 2/4 dkk species, WIF1, and 3/4 sFRP species) (Fig. 2; Table 1). Thus, Ugolini et al. (2001) originally described the epigenetic inhibition of sFRP1 expression in ERα-positive tumors, and Schlange et al. (2007) showed that autocrine Wnt signaling (measured by dvl phosphorylation) was inhibited by pan siRNA to Dvl species, and by the ectopic administration of sFRP1, for a group of breast tumor cells lines that represent all the major subtypes of breast tumor. This relied partly on transactivation of EGFR (and ERK phosphorylation), mediated by Wnt-induced MMP activation and the release of erbB2 ligands from extracellular matrix. Note that this study (and others) may not implicate canonical Wnt signaling, but Wnt ligand/Fzd-dependent growth and invasion. Alterations in mRNA species for Wnt signaling components require cautious interpretation, because the effects of so-called inhibitors (in particular) are often difficult to predict a priori. Thus, loss of dkk3 (reported in breast tumors) (Suzuki et al. 2008; Veeck et al. 2008) induces apoptosis of lung cancer cells (Jung et al. 2010), and ectopic dkk3 can potentiate Wnt signaling (probably by binding and disabling Kremen) (Nakamura and Hackam 2010).

However, despite all these caveats, Geyer et al. (2011) have recently published clear pictures of the localization and amount of β-catenin (using two different monoclonal antibodies) visualized for a cohort of 245 breast tumors, and showed that nuclear β-catenin is prevalent in a subpopulation of tumor epithelial cells in approximately one-third of basaloid breast tumors. Basaloid breast tumors comprise about one-sixth of total breast tumors and show high levels of expression of basal cell markers (both protein and mRNA) alongside luminal cell markers. The Wnt receptor, Lrp6, is overexpressed in some of these tumors (by approximately twofold to fivefold) (Lindvall et al. 2009; Liu et al. 2010), and inhibition of Lrp6 in MDA-MB231 basaloid cells in culture inhibits TOP-FLASH ($10\times - 100\times$), growth ($2\times$), Axin2 expression ($2\times$), colony formation ($5\times$) in vitro, and tumor growth in vivo (Matsuda et al. 2009).

The focus is now on the inhibitors of Wnt signaling for their possible use in tumor treatment, and given the data described above, the most likely target cohort may be patients with basaloid tumors. Drug development of inhibitors of Wnt signaling tend to assay HEK293 cells, and they have been highly successful in producing molecules that inhibit signaling outside and inside of the cell (Barker and Clevers 2006; Dasgupta 2009). It may be possible to exploit unique aspects of Wnt signaling in breast tumor cells to enhance these more generic approaches.

CONCLUDING REMARKS

This is not a comprehensive review, in part because it focuses on the cell surface components of Wnt signaling, and does not detail how that signal is modified by (1) cytoplasmic modifiers; (2) nuclear uptake mechanisms for β-catenin; (3) the nature of the TCF species expressed, together with the associated transcription factors (Archbold et al. 2012); or (4) the epigenetic chromatin status (Hatzis et al. 2008; Gu et al. 2009; Wend et al. 2010). This article mentions these mechanisms only in passing because there is little specific information published to describe the importance of these aspects with respect to mammary gland or mammary cell lines. We have focused instead on some of the outstanding questions likely to be our preoccupation in the near future, such as the evident plasticity of fate induced by Wnt signaling (indicated by transitions between basal and luminal

Table 1. Characterized expression and function of cell surface Wnt signaling components in mammary cells and tissues

Gene	Array	mRNA	Protein	Cell/tissue type	Function	Citation
Wnt1	X					All sources
		X (N)		Normal mouse MG	Tumorigenic in transgenic mice (MMTV-Wnt1)	Tsukamoto et al. 1988
					Induces branching and lobuloalveolar hyperplasia	
		✓ (R)		T-47D cells		Suzuki et al. 2008
					Highly transforming in C57MG cells	Wong et al. 1994
					Transfection into HC11 cells and transplantation into cleared fat pads leads to fibrotic outgrowths	Humphreys and Rosen 1997
Wnt2 (Wnt2a)	✓	✓ (I)		TEB stroma		Kouros-Mehr and Werb 2006
		✓ (R, I, N)		Basal cells and TEBs	mRNA decreases 10× –40× during pregnancy and lactation	Buhler et al. 1993
					Transforms C57MG cells	Wong et al. 1994
		✓ (RP, I)		Normal human breast fibroblasts and human breast tumor epithelium	Low expression levels in normal tissue and overexpression (5× –8×) in some carcinomas	Dale et al. 1996
					Transfection into HC11 cells and transplantation into cleared fat pads leads to fibrotic outgrowths	Humphreys and Rosen 1997
	✓			Mouse breast cancer cell line bone metastasis model	Significantly down-regulated at tumor–bone interface	Sadanandam et al. 2011
Wnt2b (Wnt13)	X					All sources
		✓ (R)		Normal human mammoplasty tissue and infiltrating ductal carcinoma biopsies	Low expression levels in normal tissue and overexpression (5× –8×) in some carcinomas	Bergstein et al. 1995
Wnt3	X					All sources
		✓ (N)		Normal mouse mammary gland	MMTV CIS	Roelink et al. 1990; Theodorou et al. 2007; Callahan and Smith 2008
		✓ (RP)		Normal human breast tissue, MTSV1-7, BT20, MCF7adr (ER^-, $EGFR^+$)		Huguet et al. 1994

Cite this article as *Cold Spring Harb Perspect Biol* doi: 10.1101/cshperspect.a008037

Wnt3a	X					All sources
					Highly transforming in C57MG cells	Wong et al. 1994
					MMTV CIS	Roelink et al. 1990; Theodorou et al. 2007; Callahan and Smith 2008
Wnt4	✓			Mature luminal		Lim et al. 2010
	✓			Differentiated luminal		Grigoriadis et al. 2006
	✓			Luminal ER^+ cells		Kendrick et al. 2008
		✓		NDE		Kendrick et al. 2008
	✓	✓ (I)		TEBs/mature ducts		Kouros-Mehr and Werb 2006
		✓ (I)		Luminal		Brisken et al. 2000
			X (W)	C3H 10T1/2 cells	Plays a role in branching during pregnancy	Bradbury et al. 1995; Brisken et al. 2000
		✓ (R)		Luminal cells from mammary glands treated with estrogen	Mediates progesterone-induced stem cell expansion	Joshi et al. 2010
		✓ (N)		Highest expression in virgin also in cell lines C57MG, C127I, and NMuMG cells		Gavin and McMahon 1992
		✓ (RP)		MDA415		Huguet et al. 1994
Wnt5a	✓	✓		Luminal ER^+ cells		Kendrick et al. 2008
	✓			Mature luminal cells		Lim et al. 2010
	✓			Luminal cells		Grigoriadis et al. 2006
	✓	✓ (I)		TEB epithelium		Kouros-Mehr and Werb 2006
		✓ (N)		Weakly expressed in 12-wk-old virgin and induced during pregnancy		Weber-Hall et al. 1994
		✓ (N)		Whole mammary gland	Expression detectable during pregnancy (peak at 10 days), no expression during lactation	Buhler et al. 1993
					Required for MG development and TGFβ-mediated inhibition of ductal growth.	Roarty and Serra 2007
					Wnt5a KO mice show accelerated ductal morphogenesis	

Continued

Table 1. *Continued*

Gene	Array	mRNA	Protein	Cell/tissue type	Function	Citation
	✓ (N)			Virgin, increase during pregnancy; C57MG and C127I cells		Gavin and McMahon 1992
	✓ (N)			Human breast carcinoma	Does not transform C57MG cells	Iozzo et al. 1995
	✓ (RP, I)			Little to no expression in normal breast tissue Increased expression in benign proliferations and invasive breast cancer ISH signal localized to epithelial compartment		Lejeune et al. 1995
					Overexpression results in lactation defect in MMTV-Wnt5a mice	Baxley et al. 2011
Wnt5b	✓			Bipotent CFCs		Raouf et al. 2008
	✓			Luminal		Grigoriadis et al. 2006
	✓	✓ (I)		TEBs/mature ducts		Kouros-Mehr and Werb 2006
		✓ (N)		Whole mammary gland	Expression detectable during pregnancy (d10-18), no expression during lactation	Buhler et al. 1993
					Transforms C57MG cells	Wong et al. 1994
		✓ (N)		Virgin mammary gland, increases during pregnancy		Gavin and McMahon 1992
Wnt6		✓		Basal cells		Kendrick et al. 2008
	✓	✓ (I)		TEBs/mature ducts		Kouros-Mehr and Werb 2006
		✓ (N)		Expressed in virgin, increase during pregnancy		Gavin and McMahon 1992
Wnt7a	X					All sources
					Highly transforming in C57MG cells	Wong et al. 1994
		✓ (N)		NMuMG cell line		Gavin and McMahon 1992
Wnt7b	✓	✓		Luminal ER+ cells		Kendrick et al. 2008
		✓		Luminal ER- cells		Kendrick et al. 2008
	✓			Mature luminal cells		Lim et al. 2010
	✓	✓ (I)		TEB epithelium		Kouros-Mehr and Werb 2006
					Does not alter MG development Transforms C57MG cells	Wong et al. 1994; Naylor et al. 2000

					Transfection into HC11 cells and transplantation into cleared fat pads leads to fibrotic outgrowths and palpable adenocarcinomas	Humphreys and Rosen 1997
	✓ (N)			Highest expression in virgin glands		Gavin and McMahon 1992
	✓ (RP)			Normal human breast tissue, MTSV1-7, MCF7, ZR75, T47D, MCF7adr, MDA231, MDA361, BT20, MDA415, MDA457, MDAMB157, ZR9B11		Huguet et al. 1994
Wnt9a (Wnt14)	✓			Luminal cells		Grigoriadis et al. 2006
		✓		Primary human breast cancer	Overexpressed in one of nine cases of breast cancer	Kirikoshi et al. 2001
Wnt9b (Wnt14b Wnt15)		✓		Whole mammary gland	Transforms C57MG cells (weaker than Wnt1)	Qian et al. 2003
		✓		Primary human breast cancer		Kirikoshi et al. 2001
Wnt10a	✓	✓		Basal/myoepithelial		Kendrick et al. 2008
Wnt10b (Wnt12)					Tumorigenic in transgenic mice (MMTV-Wnt10b) MG hyperplasia, increased proliferation and branching	Lane and Leder 1997
		✓ (RP,R)		MCF-7 Adr^r, MDA-MB-435, and MDA-MB-157 cells Human breast samples, and some primary human breast carcinomas		Bui et al. 1997
		✓ (N)		Virgin mouse mammary glands (not pregnant)		Lee et al. 1995
		✓		Mammary ridge (E11.5) Dual abdominal MG anlagen (12.5 dpc)		Christiansen et al. 1995
Wnt11	X					All sources
					Transforms C57MG cells	Christiansen et al. 1996
			✓ (W)	MDA-MB-231	ERRα- and β-catenin-regulated gene Involved in cancer cell migration	Dwyer et al. 2010
Wnt16	✓				Induced by Wnt4	Kim et al. 2009

Continued

Table 1. *Continued*

Gene	Array	mRNA	Protein	Cell/tissue type	Function	Citation
RSpo1	✓			Luminal ER-		Kendrick et al. 2008
		✓ (I)		Mammary mesenchyme (E15.5)		Nam et al. 2006, 2007
					KD might result in absence of ductal side-branching and alveolar formation	Chadi et al. 2009
RSpo2	X					Not in adult tissues
		✓ (R)		Detected at very low levels early in MG development but not detectable in MG of day 15 pregnant mice		Lowther et al. 2005
		X (R)	X (W, ICC)	HC11, C57MG mouse cell lines		Klauzinska et al. 2011
		X (R)		HC11 cells		SA Fakhraldeen and CM Alexander, unpubl.
RSpo3	X					Not in adult tissues
					CIS for MMTV	Theodorou et al. 2007; Callahan and Smith 2008
Fzd1	✓	✓		Basal/myoepithelial		Kendrick et al. 2008
	✓			Down-regulated in mature luminal cells		Lim et al. 2010
	✓			Mature myoepithelial cells		Raouf et al. 2008
	✓	✓ (I)		TEB stroma		Kouros-Mehr and Werb 2006
		✓ (R)		HMEC, MDA-MB-468, MDA-MB-453, MCF-7, T-47D, BT-20, and BT-474 cells		Benhaj et al. 2006
Fzd2	✓	✓		Basal/myoepithelium		Kendrick et al. 2008
	✓	✓ (I)		TEB		Kouros-Mehr and Werb 2006
		✓ (R)		MDA-MB-468, MDA-MB-453, MCF-7, T47D, BT-20, and BT-474 cells		Benhaj et al. 2006
Fzd3	✓			Basal/myo		Kendrick et al. 2008
	✓			Myo		Grigoriadis et al. 2006
		✓ (R)		MDA-MB-231 and SUM-159 cells	Functions in MDA-MB-231 cell motility	Valastyan et al. 2009
			✓ (W)	MDA-MB-231 cells		
		✓ (R)		HMEC, MDA-MB-468, MCF-7, T-47D, BT-20, and BT-474 cells		Benhaj et al. 2006

Cite this article as *Cold Spring Harb Perspect Biol* doi: 10.1101/cshperspect.a008037

Fzd4	X					All sources
Fzd5	✓			Luminal		Grigoriadis et al. 2006
		✓ (R)		MDA-MB-468 cells		Benhaj et al. 2006
Fzd6	✓			Basal/myoepithelium		Jones et al. 2004
	✓			Basal/myoepithelium		Grigoriadis et al. 2006
		✓ (R)		HMEC, MDA-MB-468, MDA-MB-453, MCF-7, T-47D, BT-20, and BT-474 cells		Benhaj et al. 2006
Fzd7	✓	✓		Basal/myoepithelium		Kendrick et al. 2008
	✓			Luminal-restricted CFCs		Raouf et al. 2008
		✓ (R)		HMEC, MDA-MB-468, MDA-MB-453, T-47D, and BT-20 cells		Benhaj et al. 2006
					Plays a role in proliferation and invasiveness of TNBC cell lines	Yang et al. 2011
Fzd8	✓	✓		Basal/myoepithelium		Kendrick et al. 2008
	✓			MaSC-enriched fraction		Lim et al. 2010
		✓ (R)		HMEC and MDA-MB-453 cells		Benhaj et al. 2006
		X (R)		MDA-MB-468		
Fzd9	✓			Differentiated luminal cells		Raouf et al. 2008
Fzd10	X					
Lrp5	✓				Bipotent CFCs	
			✓ (F, IHC)	Basal cells		Badders et al. 2009
		✓			Translation induced in Wnt1-induced luminal cells	Kim et al. 2011
		✓ (R)		HMEC, MDA-MB-468, MDA-MB-453, MCF-7, T-47D, BT-20, and BT-474 cells		Benhaj et al. 2006
		✓ (R)		Breast cancer specimens and normal breast tissue specimens		Bjorklund et al. 2009
					Mammary stem cell maintenance	Lindvall et al. 2006
					Maintenance of basal cell population	Badders et al. 2009

Continued

Table 1. *Continued*

Gene	Array	mRNA	Protein	Cell/tissue type	Function	Citation
Lrp6	✓			Luminal		Grigoriadis et al. 2006
			✓ (W)	SUM1315 cells		DiMeo et al. 2009
		✓ (R)		HMEC, MDA-MB-468, MDA-MB-453, MCF-7, T-47D, BT-20, and BT-474 cells		Benhaj et al. 2006
		✓ (R)		Breast tumor, MDA-MB-231 and MDA-MB-468 cells		Li et al. 2004
					OE in mice (MMTV-Lrp6) induces hyper-lobular development and increased TEB number	Zhang et al. 2010a
			✓ (F, IHC)	Basal in adult Basal and luminal in newborns	Loss interferes with mammary placode, fat pad, and branching development during embryogenesis Heterozygosity for inactivating mutation leads to reduced TEBs and branches	Badders et al. 2009; Lindvall et al. 2009
Ror1	✓			Luminal CFCs		Raouf et al. 2008
	✓			Myoepithelium/basal cells		Kendrick et al. 2008
Ror2		✓ (R)	✓ (IHC)	Expressed in basal/myoepithelial cells and luminal cells of mouse mammary gland		K Roarty and JM Rosen, pers. comm.
		✓ (R)		Expressed in breast cancer cell lines (MCF7 and MDA-MB231) and brain metastases of brCA in vivo		Klemm et al. 2011
Lgr4	✓			Basal cells		Kendrick et al. 2008
	X			Normal luminal epithelium		Grigoriadis et al. 2006
Lgr5	✓			Basal cells		Kendrick et al. 2008
Dkk1		✓		Basal cells		Kendrick et al. 2008
	✓			NDE		Kendrick et al. 2008
	✓			Luminal cells		Grigoriadis et al. 2006
			✓ (W)	SUM1315 cells		DiMeo et al. 2009
					OE prevents mammary placode development	Chu et al. 2004
		X (R)		HC11 mouse cells		SA Fakhraldeen and CM Alexander, unpubl.

Cite this article as *Cold Spring Harb Perspect Biol* doi: 10.1101/cshperspect.a008037

Dkk2	X					All sources
		X (R)		HC11 mouse cells		SA Fakhraldeen and CM Alexander, unpubl.
Dkk3	✓	✓		Bipotent CFCs		Raouf et al. 2008
		✓		Differentiated basal cells		Raouf et al. 2008
	✓			MaSC-enriched		Lim et al. 2010
	✓			Myoepithelium/basal cells		Jones et al. 2004
	✓			Myoepithelium/basal cells		Grigoriadis et al. 2006
	✓			Myoepithelium/basal cells		Kendrick et al. 2008
		X (R)		HC11 cells		SA Fakhraldeen and CM Alexander, unpubl.
Dkk4	X					All sources
		X (R)				SA Fakhraldeen and CM Alexander, unpubl.
WIF1	✓	✓		Myoepithelium/basal cells		Kendrick et al. 2008
	✓			MaSC-enriched		Lim et al. 2010
		✓ (R)		T47D, normal human breast tissue	Targeted for epigenetic silencing in human breast cancer	Ai et al. 2006
			✓ (IHC)	Normal breast tissue	Down-regulated in invasive ductal breast carcinoma	Wissmann et al. 2003
	✓			Mouse breast cancer cell line bone metastasis model	Increased expression at tumor-bone interface	Sadanandam et al. 2011
SFRP1	✓	✓		Myoepithelium/basal cells		Kendrick et al. 2008
	✓			Down-regulated in mature luminal		Lim et al. 2010
					Maintains/reduces canonical signaling levels	Cowling et al. 2007
					mRNA expression lost in >80% of invasive breast carcinomas (medullary)	Ugolini et al. 2001
			✓ (IHC)	Mammary epithelial cells	Tumor suppressive function Lost in tumors Loss associated with poor prognosis	Klopocki et al. 2004

Continued

Table 1. *Continued*

Gene	Array	mRNA	Protein	Cell/tissue type	Function	Citation
			X (W)	SUM1315 cells		DiMeo et al. 2009
	✓			Mature luminal		Grigoriadis et al. 2006
		✓ (R)		Normal breast tissue		Suzuki et al. 2008
				MDA-MB-435S, MDA-MB-436, MDA-MB-468, MDA-MB-157, MDA-MB-361, and ZR-75-1 cells		
		X (R)		MCF-7, MDA-MB-231, T47D, and SK-BR-3 cells		
SFRP2	✓			Myoepithelium/basal cells		Grigoriadis et al. 2006
		✓ (R)		Normal breast tissue and MDA-MB-157 cells		Suzuki et al. 2008
		X (R)		MCF-7, MDA-MB-231, MDA-MB-435S, MDA-MB-468, T-47D, SK-BR-3, MDA-MB-453, and ZR-75-1 cells		
SFRP4		✓		Luminal		Grigoriadis et al. 2006
	✓			Mouse breast cancer cell line bone metastasis model	Increased expression at tumor-bone interface	Sadanandam et al. 2011
SFRP5	X					All sources
WISP1	✓			Myoepithelium/basal cells		Grigoriadis et al. 2006
	✓			Myoepithelium/basal cells		Kendrick et al. 2008
WISP2	✓			Luminal		Grigoriadis et al. 2006
		✓ (R)	✓ (W)	MCF7, T47D, ZR-75.1, SKBr3	Knockdown promotes proliferation (including E-independent) of MCF7 cells	Fritah et al. 2008
		X (R)	X (W)	HMEC, MDA-MB-231	Inducing expression inhibits proliferation of MCF7 and MDA-MB-231 cells	

Cite this article as *Cold Spring Harb Perspect Biol* doi: 10.1101/cshperspect.a008037

WISP3	✓	Bipotent CFCs	Grigoriadis et al. 2006
Naked1	✓	Myoepithelium/basal cells	Kendrick et al. 2008
Naked2	✓	Bipotent CFCs	Raouf et al. 2008
	✓	Myoepithelium/basal cells	Kendrick et al. 2008

The expression patterns of Wnt-receptor signaling components are summarized, deduced from significant expression by mRNA species by microarray analysis (Array), or more specific analysis by qPCR/RT-PCR (R), in situ hybridization (I), RNase protection (RP), or Northern blotting (N), and of cognate proteins, shown by flow cytometry (F), immunohistochemical/cytochemical localization (IHC/ICC), or Western blotting (W). Functional information for each component is summarized.

Specificity of expression in mammary gland cell types: The array results reported are from several separate studies, and the cell fractions and gene chips are as follows:

1. (Raouf et al. 2008) Normal human mammary epithelial cells were sorted into four fractions. Both two stem/progenitor-enriched fractions (a lbipotent basal stem cell and a luminal-restricted luminal progenitor; both scored as colony-forming cells in vitro, CFCs) showed high EpCAM expression compared with their mature myoepithelial and luminal counterparts. Two mature cell types extracted from the low-EpCAM fraction were luminal cells (high Muc1/CD133, low Thy1/CD10) and myoepithelial cells (high Thy1 and low Muc1). mRNA was analyzed using Affymetrix human X3P GeneChip arrays or PCR-LongSAGE libraries.

2. (Lim et al. 2010) Normal mouse mammary glands were sorted into three epithelial cell fractions, and one stromal. A basal stem cell–enriched fraction (MaSC) was separated as β1 integrin (CD29)-high, CD24-low, and CD61-positive. Luminal cells were isolated as CD29-low (CD24-positive), and the progenitors were CD61-negative, whereas the mature cells were CD61-positive. For normal human mammary gland cells, the sorting parameters were different, but the cell populations were broadly similar. Three fractions described were a basal, mammary stem cell–enriched (MaSC) fraction—α6 integrin (CD49f)-high, EpCAM-negative—and two luminal fractions—EpCAM-positive, the progenitor fraction CD49f-positive, and the mature fraction CD49f-negative. mRNA was analyzed using Illumina MouseWG-6 v 2.0 BeadChips.

3. (Kendrick et al. 2008) Normal mouse mammary glands were sorted into three epithelial cell fractions, broadly similar to those described by Lim et al. (2010), although the cell surface markers used are different, as is the preparation and exclusion of other non-epithelial cell types. A basal cell fraction (stem/myoepithelial) was collected as CD24-low and Sca1-negative. Two luminal cell fractions both expressed CD24; the ERα-negative, luminal progenitor cell fraction was isolated as Sca1-low; and the mature, ERα-positive cells were Sc11-positive. mRNA analysis was performed used Mouse Affymetrix Mouse Expression MOE430 2.0 arrays.

4. (Grigoriadis et al. 2006) This is the only study to use cultured primary cells. Cells from normal human breast and from primary tumors were separated by double immunomagnetic sorting, either as EMA^+/integrin β4-negative luminal cells, or $CD10^+$/BerEP4-negative myoepithelial cells. The mRNA was analyzed using MPSS and Affymetrix Human Genome U133 Plus 2.0 GeneChip, CodeLink Human Whole Genome Bioarray, Agilent Whole Human Genome Oligo Microarray 44K cDNA array, and 20K cDNA microarray (constructed at the Breakthrough Breast Cancer Research Center, UK).

Wnt ligand and Fzd receptor expression in embryonic mammary tissues has been previously described (Chu et al. 2004).

CFC, Colony-forming cell; NDE, not differentially expressed; TNBC, triple negative breast cancer; OE, overexpression; KO, knockout; ISH, in situ hybridization; IHC, immunohistochemistry; CIS, common integration site (applied to MMTV retroviral tagging studies); ✓, detectable expression; X, no detectable expression.

mammary epithelial cell fates), the molecules that regulate the mammary stem cell niche, the combinatorial outputs that are offered by the multiplicity of Wnt signals perceived by mammary epithelial cells, the difficulties of conclusively implicating Wnt signaling in tumor growth or tumor cell survival, and the potential Wnt responses of the Lrp-deficient luminal cell population. Key Wnt signaling components that discriminate luminal and basal cell function, such as Lrp5, are regulated at the posttranscriptional level (Joshi et al. 2010; Kim et al. 2011), and this requires more investigation. Much of the information available to describe the regulation of Wnt signaling involves ectopic expression of various components. At many levels, this does not reflect the obvious competition for binding partners that regulates the transcriptional response and the response to Wnt ligands at the cell surface. The next generation of data will focus more heavily on studying the function of Wnt signaling components when expressed at endogenous levels. Given the new generation of Wnt signaling inhibitors, there are new horizons for investigations of mechanism and therapeutic potential.

ACKNOWLEDGMENTS

This work is supported by the BCRP/DOD Era of Hope Scholars Award W81XWH (to C.M.A., S.G., S.K., and S.A.F.).

REFERENCES

Ai L, Tao Q, Zhong S, Fields CR, Kim WJ, Lee MW, Cui Y, Brown KD, Robertson KD. 2006. Inactivation of Wnt inhibitory factor-1 (WIF1) expression by epigenetic silencing is a common event in breast cancer. *Carcinogenesis* **27:** 1341–1348.

Archbold HC, Yang YX, Chen L, Cadigan KM. 2012. How do they do Wnt they do?: Regulation of transcription by the Wnt/β-catenin pathway. *Acta Physiol (Oxf)* **204:** 74–109.

Asselin-Labat ML, Vaillant F, Sheridan JM, Pal B, Wu D, Simpson ER, Yasuda H, Smyth GK, Martin TJ, Lindeman GJ, et al. 2010. Control of mammary stem cell function by steroid hormone signalling. *Nature* **465:** 798–802.

Badders NM, Goel S, Clark RJ, Klos KS, Kim S, Bafico A, Lindvall C, Williams BO, Alexander CM. 2009. The Wnt receptor, Lrp5, is expressed by mouse mammary stem cells and is required to maintain the basal lineage. *PLoS ONE* **4:** e6594.

Baker R, Kent CV, Silbermann RA, Hassell JA, Young LJ, Howe LR. 2010. Pea3 transcription factors and Wnt1-induced mouse mammary neoplasia. *PloS ONE* **5:** e8854.

Barker N, Clevers H. 2006. Mining the Wnt pathway for cancer therapeutics. *Nat Rev Drug Discov* **5:** 997–1014.

Barker N, van Es JH, Kuipers J, Kujala P, van den Born M, Cozijnsen M, Haegebarth A, Korving J, Begthel H, Peters PJ, et al. 2007. Identification of stem cells in small intestine and colon by marker gene Lgr5. *Nature* **449:** 1003–1007.

Barolo S. 2006. Transgenic Wnt/TCF pathway reporters: All you need is Lef? *Oncogene* **25:** 7505–7511.

Batlle E, Henderson JT, Beghtel H, van den Born MM, Sancho E, Huls G, Meeldijk J, Robertson J, van de Wetering M, Pawson T, et al. 2002. β-Catenin and TCF mediate cell positioning in the intestinal epithelium by controlling the expression of EphB/ephrinB. *Cell* **111:** 251–263.

Baxley SE, Jiang W, Serra R. 2011. Misexpression of Wingless-related MMTV integration site 5A in mouse mammary gland inhibits the milk ejection response and regulates connexin43 phosphorylation. *Biol Reprod* **85:** 907–915.

Beleut M, Rajaram RD, Caikovski M, Ayyanan A, Germano D, Choi Y, Schneider P, Brisken C. 2010. Two distinct mechanisms underlie progesterone-induced proliferation in the mammary gland. *Proc Natl Acad Sci* **107:** 2989–2994.

Benhaj K, Akcali KC, Ozturk M. 2006. Redundant expression of canonical Wnt ligands in human breast cancer cell lines. *Oncol Rep* **15:** 701–707.

Bergstein I, Schultz R, Osborne MP, Welcsh PL, Bowcock AM, Brown AM. 1995. Investigation of the possible role of WNT genes in human breast cancer. *Ann NY Acad Sci* **768:** 257.

Bernard P, Fleming A, Lacombe A, Harley VR, Vilain E. 2008. Wnt4 inhibits β-catenin/TCF signalling by redirecting β-catenin to the cell membrane. *Biol Cell* **100:** 167–177.

Bjorklund P, Svedlund J, Olsson AK, Akerstrom G, Westin G. 2009. The internally truncated LRP5 receptor presents a therapeutic target in breast cancer. *PLoS ONE* **4:** e4243.

Boras-Granic K, Wysolmerski JJ. 2008. Wnt signaling in breast organogenesis. *Organogenesis* **4:** 116–122.

Boras-Granic K, Chang H, Grosschedl R, Hamel PA. 2006a. Lef1 is required for the transition of Wnt signaling from mesenchymal to epithelial cells in the mouse embryonic mammary gland. *Dev Biol* **295:** 219–231.

Boras-Granic K, Grosschedl R, Hamel PA. 2006b. Genetic interaction between Lef1 and Alx4 is required for early embryonic development. *Int J Dev Biol* **50:** 601–610.

Bouras T, Pal B, Vaillant F, Harburg G, Asselin-Labat ML, Oakes SR, Lindeman GJ, Visvader JE. 2008. Notch signaling regulates mammary stem cell function and luminal cell-fate commitment. *Cell Stem Cell* **3:** 429–441.

Bourhis E, Tam C, Franke Y, Bazan JF, Ernst J, Hwang J, Costa M, Cochran AG, Hannoush RN. 2010. Reconstitution of a frizzled8·Wnt3a·LRP6 signaling complex reveals multiple Wnt and Dkk1 binding sites on LRP6. *J Biol Chem* **285:** 9172–9179.

Bradbury JM, Edwards PA, Niemeyer CC, Dale TC. 1995. Wnt-4 expression induces a pregnancy-like growth pattern in reconstituted mammary glands in virgin mice. *Dev Biol* **170:** 553–563.

Brennan KR, Brown AM. 2004. Wnt proteins in mammary development and cancer. *J Mammary Gland Biol Neoplasia* **9:** 119–131.

Brennan K, Gonzalez-Sancho JM, Castelo-Soccio LA, Howe LR, Brown AM. 2004. Truncated mutants of the putative Wnt receptor LRP6/Arrow can stabilize β-catenin independently of Frizzled proteins. *Oncogene* **23:** 4873–4884.

Brisken C, Heineman A, Chavarra T, Elenbaas B, Tan J, Dey SK, McMahon AP, Weinberg R. 2000. Essential function of Wnt-4 in mammary gland development downstream of progesterone signaling. *Genes Dev* **14:** 650–654.

Buhler TA, Dale TC, Kieback C, Humphreys RC, Rosen JM. 1993. Localization and quantification of Wnt-2 gene expression in mouse mammary development. *Dev Biol* **155:** 87–96.

Bui TD, Rankin J, Smith K, Huguet EL, Ruben S, Strachan T, Harris AL, Lindsay S. 1997. A novel human Wnt gene, *WNT10B*, maps to 12q13 and is expressed in human breast carcinomas. *Oncogene* **14:** 1249–1253.

Callahan R, Egan SE. 2004. Notch signaling in mammary development and oncogenesis. *J Mammary Gland Biol Neoplasia* **9:** 145–163.

Callahan R, Smith GH. 2008. Common integration sites for MMTV in viral induced mouse mammary tumors. *J Mammary Gland Biol Neoplasia* **13:** 309–321.

Carmon KS, Gong X, Lin Q, Thomas A, Liu Q. 2011. R-spondins function as ligands of the orphan receptors LGR4 and LGR5 to regulate Wnt/β-catenin signaling. *Proc Natl Acad Sci* **108:** 11452–11457.

Carroll TJ, Park JS, Hayashi S, Majumdar A, McMahon AP. 2005. Wnt9b plays a central role in the regulation of mesenchymal to epithelial transitions underlying organogenesis of the mammalian urogenital system. *Dev Cell* **9:** 283–292.

Chadi S, Buscara L, Pechoux C, Costa J, Laubier J, Chaboissier MC, Pailhoux E, Vilotte JL, Chanat E, Le Provost F. 2009. R-spondin1 is required for normal epithelial morphogenesis during mammary gland development. *Biochem Biophys Res Commun* **390:** 1040–1043.

Chan EL, Peace BE, Toney K, Kader SA, Pathrose P, Collins MH, Waltz SE. 2007. Homozygous K5Cre transgenic mice have wavy hair and accelerated malignant progression in a murine model of skin carcinogenesis. *Mol Carcinog* **46:** 49–59.

Christiansen JH, Dennis CL, Wicking CA, Monkley SJ, Wilkinson DG, Wainwright BJ. 1995. Murine Wnt-11 and Wnt-12 have temporally and spatially restricted expression patterns during embryonic development. *Mech Dev* **51:** 341–350.

Christiansen JH, Monkley SJ, Wainwright BJ. 1996. Murine WNT11 is a secreted glycoprotein that morphologically transforms mammary epithelial cells. *Oncogene* **12:** 2705–2711.

Chu E, Hens J, Andl T, Kairo A, Yamaguchi TP, Brisken C, Glick A, Wysolmerski JJ, Millar SE. 2004. Canonical WNT signaling promotes mammary placode development and is essential for initiation of mammary gland morphogenesis. *Development* **131:** 4819–4829.

Cicalese A, Bonizzi G, Pasi CE, Faretta M, Ronzoni S, Giulini B, Brisken C, Minucci S, Di Fiore PP, Pelicci PG. 2009. The tumor suppressor p53 regulates polarity of self-renewing divisions in mammary stem cells. *Cell* **138:** 1083–1095.

Clarke RB. 2003. Steroid receptors and proliferation in the human breast. *Steroids* **68:** 789–794.

Collu GM, Brennan K. 2007. Cooperation between Wnt and Notch signalling in human breast cancer. *Breast Cancer Res* **9:** 105.

Cong F, Schweizer L, Varmus H. 2004. Wnt signals across the plasma membrane to activate the β-catenin pathway by forming oligomers containing its receptors, Frizzled and LRP. *Development* **131:** 5103–5115.

Constantinou T, Baumann F, Lacher MD, Saurer S, Friis R, Dharmarajan A. 2008. SFRP-4 abrogates Wnt-3a-induced β-catenin and Akt/PKB signalling and reverses a Wnt-3a-imposed inhibition of in vitro mammary differentiation. *J Mol Signal* **3:** 10.

Cowling VH, D'Cruz CM, Chodosh LA, Cole MD. 2007. c-Myc transforms human mammary epithelial cells through repression of the Wnt inhibitors DKK1 and SFRP1. *Mol Cell Biol* **27:** 5135–5146.

Currier N, Chea K, Hlavacova M, Sussman DJ, Seldin DC, Dominguez I. 2010. Dynamic expression of a LEF-EGFP Wnt reporter in mouse development and cancer. *Genesis* **48:** 183–194.

Dale TC, Weber-Hall SJ, Smith K, Huguet EL, Jayatilake H, Gusterson BA, Shuttleworth G, O'Hare M, Harris AL. 1996. Compartment switching of WNT-2 expression in human breast tumors. *Cancer Res* **56:** 4320–4323.

Dasgupta R. 2009. Functional genomic approaches targeting the wnt signaling network. *Curr Drug Targets* **10:** 620–631.

de Lau W, Barker N, Low TY, Koo BK, Li VS, Teunissen H, Kujala P, Haegebarth A, Peters PJ, van de Wetering M, et al. 2011. Lgr5 homologues associate with Wnt receptors and mediate R-spondin signalling. *Nature* **476:** 293–297.

DiMeo TA, Anderson K, Phadke P, Fan C, Perou CM, Naber S, Kuperwasser C. 2009. A novel lung metastasis signature links Wnt signaling with cancer cell self-renewal and epithelial–mesenchymal transition in basal-like breast cancer. *Cancer Res* **69:** 5364–5373.

Ding VW, Chen RH, McCormick F. 2000. Differential regulation of glycogen synthase kinase 3β by insulin and Wnt signaling. *J Biol Chem* **275:** 32475–32481.

Domyan ET, Sun X. 2011. Patterning and plasticity in development of the respiratory lineage. *Dev Dyn* **240:** 477–485.

Dwyer MA, Joseph JD, Wade HE, Eaton ML, Kunder RS, Kazmin D, Chang CY, McDonnell DP. 2010. WNT11 expression is induced by estrogen-related receptor α and β-catenin and acts in an autocrine manner to increase cancer cell migration. *Cancer Res* **70:** 9298–9308.

Fernandez-Valdivia R, Mukherjee A, Mulac-Jericevic B, Conneely OM, DeMayo FJ, Amato P, Lydon JP. 2005. Revealing progesterone's role in uterine and mammary gland biology: Insights from the mouse. *Semin Reprod Med* **23:** 22–37.

Ferrer-Vaquer A, Piliszek A, Tian G, Aho RJ, Dufort D, Hadjantonakis AK. 2010. A sensitive and bright single-cell resolution live imaging reporter of Wnt/ss-catenin signaling in the mouse. *BMC Dev Biol* **10:** 121.

Foldynova-Trantirkova S, Sekyrova P, Tmejova K, Brumovska E, Bernatik O, Blankenfeldt W, Krejci P, Kozubik A, Dolezal T, Trantirek L, et al. 2010. Breast cancer-specific mutations in CK1ε inhibit Wnt/β-catenin and activate the Wnt/Rac1/JNK and NFAT pathways to decrease cell adhesion and promote cell migration. *Breast Cancer Res* **12:** R30.

Fritah A, Saucier C, De Wever O, Bracke M, Bieche I, Lidereau R, Gespach C, Drouot S, Redeuilh G, Sabbah M. 2008. Role of WISP-2/CCN5 in the maintenance of a differentiated and noninvasive phenotype in human breast cancer cells. *Mol Cell Biol* **28:** 1114–1123.

Gao C, Cao W, Bao L, Zuo W, Xie G, Cai T, Fu W, Zhang J, Wu W, Zhang X, et al. 2010. Autophagy negatively regulates Wnt signalling by promoting Dishevelled degradation. *Nat Cell Biol* **12:** 781–790.

Gavin BJ, McMahon AP. 1992. Differential regulation of the Wnt gene family during pregnancy and lactation suggests a role in postnatal development of the mammary gland. *Mol Cell Biol* **12:** 2418–2423.

Geyer FC, Lacroix-Triki M, Savage K, Arnedos M, Lambros MB, Mackay A, Natrajan R, Reis-Filho JS. 2011. β-Catenin pathway activation in breast cancer is associated with triple-negative phenotype but not with CTNNB1 mutation. *Mod Pathol* **24:** 209–231.

Goel S, Fakhraldeen SA, Chin EN, Alexander CM. 2012a. Some Wnt ligands require both Lrp receptors (Lrp5 and Lrp6) for normal canonical signaling. *J Biol Chem* (in press).

Goel S, Chin EN, Fakhraldeen SA, Berry SM, Beebe DJ, Alexander CM. 2012b. Both Lrp5 and Lrp6 receptors are required to respond to physiological Wnt ligands in mammary epithelial cells (and fibroblasts). *J Biol Chem* doi: 10.1074/jbc.M112.362137.

Goessling W, North TE, Loewer S, Lord AM, Lee S, Stoick-Cooper CL, Weidinger G, Puder M, Daley GQ, Moon RT, et al. 2009. Genetic interaction of PGE2 and Wnt signaling regulates developmental specification of stem cells and regeneration. *Cell* **136:** 1136–1147.

Green JL, Kuntz SG, Sternberg PW. 2008. Ror receptor tyrosine kinases: Orphans no more. *Trends Cell Biol* **18:** 536–544.

Grigoriadis A, Mackay A, Reis-Filho JS, Steele D, Iseli C, Stevenson BJ, Jongeneel CV, Valgeirsson H, Fenwick K, Iravani M, et al. 2006. Establishment of the epithelial-specific transcriptome of normal and malignant human breast cells based on MPSS and array expression data. *Breast Cancer Res* **8:** R56.

Grumolato L, Liu G, Mong P, Mudbary R, Biswas R, Arroyave R, Vijayakumar S, Economides KD, Aaronson SA. 2011. Canonical and noncanonical Wnts use a common mechanism to activate completely unrelated coreceptors. *Genes Dev* **24:** 2517–2530.

Gu B, Sun P, Yuan Y, Moraes RC, Li A, Teng A, Agrawal A, Rheaume C, Bilanchone V, Veltmaat JM, et al. 2009. Pygo2 expands mammary progenitor cells by facilitating histone H3 K4 methylation. *J Cell Biol* **185:** 811–826.

Gumbiner BM. 1998. Propagation and localization of Wnt signaling. *Curr Opin Genet Dev* **8:** 430–435.

Gunther EJ, Moody SE, Belka GK, Hahn KT, Innocent N, Dugan KD, Cardiff RD, Chodosh LA. 2003. Impact of p53 loss on reversal and recurrence of conditional Wnt-induced tumorigenesis. *Genes Dev* **17:** 488–501.

Hatzis P, van der Flier LG, van Driel MA, Guryev V, Nielsen F, Denissov S, Nijman IJ, Koster J, Santo EE, Welboren W, et al. 2008. Genome-wide pattern of TCF7L2/TCF4 chromatin occupancy in colorectal cancer cells. *Mol Cell Biol* **28:** 2732–2744.

He X, Semenov M, Tamai K, Zeng X. 2004. LDL receptor-related proteins 5 and 6 in Wnt/β-catenin signaling: Arrows point the way. *Development* **131:** 1663–1677.

He XC, Yin T, Grindley JC, Tian Q, Sato T, Tao WA, Dirisina R, Porter-Westpfahl KS, Hembree M, Johnson T, et al. 2007. PTEN-deficient intestinal stem cells initiate intestinal polyposis. *Nat Genet* **39:** 189–198.

Hoeflich KP, O'Brien C, Boyd Z, Cavet G, Guerrero S, Jung K, Januario T, Savage H, Punnoose E, Truong T, et al. 2009. In vivo antitumor activity of MEK and phosphatidylinositol 3-kinase inhibitors in basal-like breast cancer models. *Clin Cancer Res* **15:** 4649–4664.

Hollestelle A, Elstrodt F, Nagel JH, Kallemeijn WW, Schutte M. 2007. Phosphatidylinositol-3-OH kinase or RAS pathway mutations in human breast cancer cell lines. *Mol Cancer Res* **5:** 195–201.

Hollestelle A, Nagel JH, Smid M, Lam S, Elstrodt F, Wasielewski M, Ng SS, French PJ, Peeters JK, Rozendaal MJ, et al. 2010. Distinct gene mutation profiles among luminal-type and basal-type breast cancer cell lines. *Breast Cancer Res Treat* **121:** 53–64.

Hsu W, Shakya R, Costantini F. 2001. Impaired mammary gland lymphoid development caused by inducible expression of Axin in transgenic mice. *J Cell Biol* **155:** 1055–1064.

Huang S, Li Y, Chen Y, Podsypanina K, Chamorro M, Olshen AB, Desai KV, Tann A, Petersen D, Green JE, et al. 2005. Changes in gene expression during the development of mammary tumors in MMTV–*Wnt-1* transgenic mice. *Genome Biol* **6:** R84.

Huguet EL, McMahon JA, McMahon AP, Bicknell R, Harris AL. 1994. Differential expression of human Wnt genes 2, 3, 4, and 7B in human breast cell lines and normal and disease states of human breast tissue. *Cancer Res* **54:** 2615–2621.

Humphreys RC, Rosen JM. 1997. Stably transfected HC11 cells provide an in vitro and in vivo model system for studying Wnt gene function. *Cell Growth Differ* **8:** 839–849.

Imbert A, Eelkema R, Jordan S, Feiner H, Cowin P. 2001. DN89b-catenin induces precocious development, differentiation, and neoplasia in mammary gland. *J Cell Biol* **153:** 555–568.

Incassati A, Chandramouli A, Eelkema R, Cowin P. 2010. Key signaling nodes in mammary gland development and cancer: β-catenin. *Breast Cancer Res* **12:** 213.

Inoki K, Ouyang H, Zhu T, Lindvall C, Wang Y, Zhang X, Yang Q, Bennett C, Harada Y, Stankunas K, et al. 2006. TSC2 integrates Wnt and energy signals via a coordinated phosphorylation by AMPK and GSK3 to regulate cell growth. *Cell* **126:** 955–968.

Iozzo RV, Eichstetter I, Danielson KG. 1995. Aberrant expression of the growth factor Wnt-5A in human malignancy. *Cancer Res* **55:** 3495–3499.

Jarde T, Dale T. 2011. Wnt signalling in murine postnatal mammary gland development. *Acta Physiol (Oxf)* **204:** 118–127.

Jones C, Mackay A, Grigoriadis A, Cossu A, Reis-Filho JS, Fulford L, Dexter T, Davies S, Bulmer K, Ford E, et al. 2004. Expression profiling of purified normal human luminal and myoepithelial breast cells: Identification of novel prognostic markers for breast cancer. *Cancer Res* **64:** 3037–3045.

Jonsson M, Dejmek J, Bendahl PO, Andersson T. 2002. Loss of Wnt-5a protein is associated with early relapse in invasive ductal breast carcinomas. *Cancer Res* **62:** 409–416.

Joshi PA, Jackson HW, Beristain AG, Di Grappa MA, Mote PA, Clarke CL, Stingl J, Waterhouse PD, Khokha R. 2010. Progesterone induces adult mammary stem cell expansion. *Nature* **465:** 803–807.

Jung IL, Kang HJ, Kim KC, Kim IG. 2010. Knockdown of the Dickkopf 3 gene induces apoptosis in a lung adenocarcinoma. *Int J Mol Med* **26:** 33–38.

Kamiya K, Gould MN, Clifton KH. 1998. Quantitative studies of ductal versus alveolar differentiation from rat mammary clonogens. *Proc Soc Exp Biol Med* **219:** 217–225.

Kang DW, Choi K-Y, Min DS. 2011. Phospholipase D meets Wnt signaling: A new target for cancer therapy. *Cancer Res* **71:** 293–297.

Kendrick H, Regan JL, Magnay FA, Grigoriadis A, Mitsopoulos C, Zvelebil M, Smalley MJ. 2008. Transcriptome analysis of mammary epithelial subpopulations identifies novel determinants of lineage commitment and cell fate. *BMC Genomics* **9:** 591.

Kikuchi A, Yamamoto H, Kishida S. 2007. Multiplicity of the interactions of Wnt proteins and their receptors. *Cell Signal* **19:** 659–671.

Kim YC, Clark RJ, Ranheim EA, Alexander CM. 2008. Wnt1 expression induces short-range and long-range cell recruitments that modify mammary tumor development and are not induced by a cell-autonomous β-catenin effector. *Cancer Res* **68:** 10145–10153.

Kim YC, Clark RJ, Pelegri F, Alexander CM. 2009. Wnt4 is not sufficient to induce lobuloalveolar mammary development. *BMC Dev Biol* **9:** 55.

Kim S, Goel S, Alexander CM. 2011. Differentiation generates paracrine cell pairs that maintain basaloid mouse mammary tumors: Proof of concept. *PLoS ONE* **6:** e19310.

Kim S, Roopra AS, Alexander CM. 2012. A phenotypic mouse model of basaloid breast tumors. *PLoS ONE* **7:** e30979.

Kirikoshi H, Sekihara H, Katoh M. 2001. Expression of WNT14 and WNT14B mRNAs in human cancer, up-regulation of WNT14 by IFNγ and up-regulation of WNT14B by β-estradiol. *Int J Oncol* **19:** 1221–1225.

Klauzinska M, Baljinnyam B, Raafat A, Rodriguez-Canales J, Strizzi L, Greer YE, Rubin JS, Callahan R. 2011. Rspo2/Int7 regulates invasiveness and tumorigenic properties of mammary epithelial cells. *J Cell Physiol* **227:** 1960–1971.

Klemm F, Bleckmann A, Siam L, Chuang HN, Rietkotter E, Behme D, Schulz M, Schaffrinski M, Schindler S, Trumper L, et al. 2011. β-Catenin-independent WNT signaling in basal-like breast cancer and brain metastasis. *Carcinogenesis* **32:** 434–442.

Klopocki E, Kristiansen G, Wild PJ, Klaman I, Castanos-Velez E, Singer G, Stohr R, Simon R, Sauter G, Leibiger H, et al. 2004. Loss of SFRP1 is associated with breast cancer progression and poor prognosis in early stage tumors. *Int J Oncol* **25:** 641–649.

Korkaya H, Paulson A, Charafe-Jauffret E, Ginestier C, Brown M, Dutcher J, Clouthier SG, Wicha MS. 2009. Regulation of mammary stem/progenitor cells by PTEN/Akt/β-catenin signaling. *PLoS Biol* **7:** e1000121.

Kouros-Mehr H, Werb Z. 2006. Candidate regulators of mammary branching morphogenesis identified by genome-wide transcript analysis. *Dev Dyn* **235:** 3404–3412.

Lane TF, Leder P. 1997. Wnt-10b directs hypermorphic development and transformation in mammary glands of male and female mice. *Oncogene* **15:** 2133–2144.

Lee FS, Lane TF, Kuo A, Shackleford GM, Leder P. 1995. Insertional mutagenesis identifies a member of the Wnt gene family as a candidate oncogene in the mammary epithelium of int-2/Fgf-3 transgenic mice. *Proc Natl Acad Sci* **92:** 2268–2272.

Lejeune S, Huguet EL, Hamby A, Poulsom R, Harris AL. 1995. Wnt5a cloning, expression, and up-regulation in human primary breast cancers. *Clin Cancer Res* **1:** 215–222.

Li Y, Welm B, Podsypanina K, Huang S, Chamorro M, Zhang X, Rowlands T, Egeblad M, Cowin P, Werb Z, et al. 2003a. Evidence that transgenes encoding components of the Wnt signaling pathway preferentially induce mammary cancers from progenitor cells. *Proc Natl Acad Sci* **100:** 15853–15858.

Li Y, Welm B, Podsypanina K, Huang S, Chamorro M, Zhang X, Rowlands T, Egeblad M, Cowin P, Werb Z, et al. 2003b. Evidence that transgenes encoding components of the Wnt signaling pathway preferentially induced mammary cancers from progenitor cells. *Proc Natl Acad Sci* **100:** 15853–15858.

Li Y, Lu W, He X, Schwartz AL, Bu G. 2004. LRP6 expression promotes cancer cell proliferation and tumorigenesis by altering β-catenin subcellular distribution. *Oncogene* **23:** 9129–9135.

Liepinsh E, Banyai L, Patthy L, Otting G. 2006. NMR structure of the WIF domain of the human Wnt-inhibitory factor-1. *J Mol Biol* **357:** 942–950.

Lim E, Wu D, Pal B, Bouras T, Asselin-Labat ML, Vaillant F, Yagita H, Lindeman GJ, Smyth GK, Visvader JE. 2010. Transcriptome analyses of mouse and human mammary cell subpopulations reveal multiple conserved genes and pathways. *Breast Cancer Res* **12:** R21.

Lindvall C, Evans NC, Zylstra CR, Li Y, Alexander CM, Williams BO. 2006. The Wnt signaling receptor Lrp5 is required for mammary ductal stem cell activity and Wnt1-induced tumorigenesis. *J Biol Chem* **281:** 35081–35087.

Lindvall C, Zylstra CR, Evans N, West RA, Dykema K, Furge KA, Williams BO. 2009. The Wnt co-receptor Lrp6 is required for normal mouse mammary gland development. *PloS ONE* **4:** e5813.

Liu Y, McDermott SP, Khwaja SS, Alexander CM. 2004. The transforming ability of Wnt effectors correlates with their ability to induce the accumulation of mammary progenitor cells. *Proc Natl Acad Sci* **101:** 4158–4163.

Liu CC, Prior J, Piwnica-Worms D, Bu G. 2010. LRP6 overexpression defines a class of breast cancer subtype and is a target for therapy. *Proc Natl Acad Sci* **107:** 5136–5141.

Lowther W, Wiley K, Smith GH, Callahan R. 2005. A new common integration site, Int7, for the mouse mammary tumor virus in mouse mammary tumors identifies a gene whose product has furin-like and thrombospondin-like sequences. *J Virol* **79:** 10093–10096.

Lyons JP, Mueller UW, Ji H, Everett C, Fang X, Hsieh JC, Barth AM, McCrea PD. 2004. Wnt-4 activates the canonical β-catenin-mediated Wnt pathway and binds Frizzled-6 CRD: Functional implications of Wnt/β-catenin activity in kidney epithelial cells. *Exp Cell Res* **298:** 369–387.

MacDonald BT, Yokota C, Tamai K, Zeng X, He X. 2008. Wnt signal amplification via activity, cooperativity, and regulation of multiple intracellular PPPSP motifs in the Wnt co-receptor LRP6. *J Biol Chem* **283:** 16115–16123.

Macias H, Moran A, Samara Y, Moreno M, Compton JE, Harburg G, Strickland P, Hinck L. 2011. SLIT/ROBO1 signaling suppresses mammary branching morphogenesis by limiting basal cell number. *Dev Cell* **20:** 827–840.

Mastroianni M, Kim S, Kim YC, Esch A, Wagner C, Alexander CM. 2009. Wnt signaling can substitute for estrogen to induce division of ERα-positive cells in a mouse mammary tumor model. *Cancer Lett* **289:** 23–31.

Matsuda Y, Schlange T, Oakeley EJ, Boulay A, Hynes NE. 2009. WNT signaling enhances breast cancer cell motility and blockade of the WNT pathway by sFRP1 suppresses MDA-MB-231 xenograft growth. *Breast Cancer Res* **11:** R32.

Mikels AJ, Nusse R. 2006. Purified Wnt5a protein activates or inhibits β-catenin–TCF signaling depending on receptor context. *PLoS Biol* **4:** e115.

Mikels A, Minami Y, Nusse R. 2009. Ror2 receptor requires tyrosine kinase activity to mediate Wnt5A signaling. *J Biol Chem* **284:** 30167–30176.

Miller RK, McCrea PD. 2010. Wnt to build a tube: Contributions of Wnt signaling to epithelial tubulogenesis. *Dev Dyn* **239:** 77–93.

Mink S, Ponta H, Cato AC. 1990. The long terminal repeat region of the mouse mammary tumour virus contains multiple regulatory elements. *Nucleic Acids Res* **18:** 2017–2024.

Moser AR, Mattes EM, Dove WF, Lindstrom MJ, Haag JD, Gould MN. 1993. ApcMin, a mutation in the murine Apc gene, predisposes to mammary carcinomas and focal alveolar hyperplasias. *Proc Natl Acad Sci* **90:** 8977–8981.

Nakamura RE, Hackam AS. 2010. Analysis of Dickkopf3 interactions with Wnt signaling receptors. *Growth Factors* **28:** 232–242.

Nam JS, Turcotte TJ, Smith PF, Choi S, Yoon JK. 2006. Mouse cristin/R-spondin family proteins are novel ligands for the Frizzled 8 and LRP6 receptors and activate β-catenin-dependent gene expression. *J Biol Chem* **281:** 13247–13257.

Nam JS, Turcotte TJ, Yoon JK. 2007. Dynamic expression of R-spondin family genes in mouse development. *Gene Expr Patterns* **7:** 306–312.

Naylor S, Smalley MJ, Robertson D, Gusterson BA, Edwards PA, Dale TC. 2000. Retroviral expression of Wnt-1 and Wnt-7b produces different effects in mouse mammary epithelium. *J Cell Sci* **113:** 2129–2138.

Neve RM, Chin K, Fridlyand J, Yeh J, Baehner FL, Fevr T, Clark L, Bayani N, Coppe JP, Tong F, et al. 2006. A collection of breast cancer cell lines for the study of functionally distinct cancer subtypes. *Cancer Cell* **10:** 515–527.

Ng SS, Mahmoudi T, Danenberg E, Bejaoui I, de Lau W, Korswagen HC, Schutte M, Clevers H. 2009. Phosphatidylinositol 3-kinase signaling does not activate the wnt cascade. *J Biol Chem* **284:** 35308–35313.

Niehrs C, Shen J. 2010. Regulation of Lrp6 phosphorylation. *Cell Mol Life Sci* **67:** 2551–2562.

Nusse R. 1988. The *int* genes in mammary tumorigenesis and in normal development. *Trends Genet* **4:** 291–295.

Nusse R, Varmus HE. 1982. Many tumors induced by the mouse mammary tumor virus contain a provirus integrated in the same region of the host genome. *Cell* **31:** 99–109.

O'Brien J, Martinson H, Durand-Rougely C, Schedin P. 2012. Macrophages are crucial for epithelial cell death and adipocyte repopulation during mammary gland involution. *Development* **139:** 269–275.

O'Connell MP, Weeraratna AT. 2009. Hear the Wnt Ror: How melanoma cells adjust to changes in Wnt. *Pigment Cell Melanoma Res* **22:** 724–739.

Parmalee NL, Kitajewski J. 2008. Wnt signaling in angiogenesis. *Curr Drug Targets* **9:** 558–564.

Phelps RA, Chidester S, Dehghanizadeh S, Phelps J, Sandoval IT, Rai K, Broadbent T, Sarkar S, Burt RW, Jones DA. 2009. A two-step model for colon adenoma initiation and progression caused by APC loss. *Cell* **137:** 623–634.

Prestwich TC, Macdougald OA. 2007. Wnt/β-catenin signaling in adipogenesis and metabolism. *Curr Opin Cell Biol* **19:** 612–617.

Pulkkinen K, Murugan S, Vainio S. 2008. Wnt signaling in kidney development and disease. *Organogenesis* **4:** 55–59.

Qian J, Jiang Z, Li M, Heaphy P, Liu YH, Shackleford GM. 2003. Mouse Wnt9b transforming activity, tissue-specific expression, and evolution. *Genomics* **81:** 34–46.

Rakha EA, El-Sayed ME, Reis-Filho J, Ellis IO. 2009. Pathobiological aspects of basal-like breast cancer. *Breast Cancer Res Treat* **113:** 411–422.

Raouf A, Zhao Y, To K, Stingl J, Delaney A, Barbara M, Iscove N, Jones S, McKinney S, Emerman J, et al. 2008. Transcriptome analysis of the normal human mammary cell commitment and differentiation process. *Cell Stem Cell* **3:** 109–118.

Roarty K, Rosen JM. 2010. Wnt and mammary stem cells: Hormones cannot fly wingless. *Curr Opin Pharmacol* **10:** 643–649.

Roarty K, Serra R. 2007. Wnt5a is required for proper mammary gland development and TGF-β-mediated inhibition of ductal growth. *Development* **134:** 3929–3939.

Robinson GW, Hennighausen L. 2011. MMTV-Cre transgenes can adversely affect lactation: Considerations for conditional gene deletion in mammary tissue. *Anal Biochem* **412:** 92–95.

Roelink H, Wagenaar E, Lopes da Silva S, Nusse R. 1990. Wnt-3, a gene activated by proviral insertion in mouse mammary tumors, is homologous to int-1/Wnt-1 and is normally expressed in mouse embryos and adult brain. *Proc Natl Acad Sci* **87:** 4519–4523.

Rouault F, Nejad Asl SB, Rungaldier S, Fuchs E, Salmons B, Gunzburg WH. 2007. Promoter complex in the central part of the mouse mammary tumor virus long terminal repeat. *J Virol* **81:** 12572–12581.

Sadanandam A, Futakuchi M, Lyssiotis CA, Gibb WJ, Singh RK. 2011. A cross-species analysis of a mouse model of breast cancer-specific osteolysis and human bone metastases using gene expression profiling. *BMC Cancer* **11:** 304.

Sakakura T. 1987. Mammary embryogenesis. In *The mammary gland* (ed. Neville MC, Daniel CW), pp. 37–66. Plenum, New York.

Sato A, Yamamoto H, Sakane H, Koyama H, Kikuchi A. 2010. Wnt5a regulates distinct signalling pathways by binding to Frizzled2. *EMBO J* **29:** 41–54.

Schaale K, Neumann J, Schneider D, Ehlers S, Reiling N. 2011. Wnt signaling in macrophages: Augmenting and inhibiting mycobacteria-induced inflammatory responses. *Eur J Cell Biol* **90:** 553–559.

Schlange T, Matsuda Y, Lienhard S, Huber A, Hynes NE. 2007. Autocrine WNT signaling contributes to breast cancer cell proliferation via the canonical WNT pathway and EGFR transactivation. *Breast Cancer Res* **9:** R63.

Schmitt AM, Shi J, Wolf AM, Lu CC, King LA, Zou Y. 2006. Wnt–Ryk signalling mediates medial–lateral retinotectal topographic mapping. *Nature* **439:** 31–37.

Schwertfeger KL, Rosen JM, Cohen DA. 2006. Mammary gland macrophages: Pleiotropic functions in mammary development. *J Mammary Gland Biol Neoplasia* **11:** 229–238.

Semenov MV, Habas R, Macdonald BT, He X. 2007. SnapShot: Noncanonical Wnt signaling pathways. *Cell* **131:** 1378.

Serra R, Easter SL, Jiang W, Baxley SE. 2011. Wnt5a as an effector of TGFβ in mammary development and cancer. *J Mammary Gland Biol Neoplasia* **16:** 157–167.

Shackleton M, Vaillant F, Simpson KJ, Stingl J, Smyth GK, Asselin-Labat ML, Wu L, Lindeman GJ, Visvader JE. 2006. Generation of a functional mammary gland from a single stem cell. *Nature* **439:** 84–88.

Sharpe R, Pearson A, Herrera-Abreu MT, Johnson DA, Mackay A, Welti JC, Natrajan R, Reynolds AR, Reis-Filho JS, Ashworth A, et al. 2011. FGFR signalling promotes the growth of triple-negative and basal-like breast cancer cell lines both in vitro and in vivo. *Clin Cancer Res* **17:** 5275–5286.

Silberstein GB, Van Horn K, Hrabeta-Robinson E, Compton J. 2006. Estrogen-triggered delays in mammary gland gene expression during the estrous cycle: Evidence for a novel timing system. *J Endocrinol* **190:** 225–239.

Suzuki H, Toyota M, Carraway H, Gabrielson E, Ohmura T, Fujikane T, Nishikawa N, Sogabe Y, Nojima M, Sonoda T, et al. 2008. Frequent epigenetic inactivation of Wnt antagonist genes in breast cancer. *Br J Cancer* **98:** 1147–1156.

Taddei I, Deugnier MA, Faraldo MM, Petit V, Bouvard D, Medina D, Fassler R, Thiery JP, Glukhova MA. 2008. β1 integrin deletion from the basal compartment of the mammary epithelium affects stem cells. *Nat Cell Biol* **10:** 716–722.

Tanneberger K, Pfister AS, Brauburger K, Schneikert J, Hadjihannas MV, Kriz V, Schulte G, Bryja V, Behrens J. 2011. Amer1/WTX couples Wnt-induced formation of PtdIns(4,5)P2 to LRP6 phosphorylation. *EMBO J* **30:** 1433–1443.

Teissedre B, Pinderhughes A, Incassati A, Hatsell SJ, Hiremath M, Cowin P. 2009. MMTV-Wnt1 and -ΔN89β-catenin induce canonical signaling in distinct progenitors and differentially activate Hedgehog signaling within mammary tumors. *PloS ONE* **4:** e4537.

ten Berge D, Brugmann SA, Helms JA, Nusse R. 2008. Wnt and FGF signals interact to coordinate growth with cell fate specification during limb development. *Development* **135:** 3247–3257.

Tepera SB, McCrea PD, Rosen JM. 2002. A β-catenin survival signal is required for normal lobular development in the mammary gland. *J Cell Sci* **116:** 1137–1149.

Teuliere J, Faraldo MM, Deugnier MA, Shtutman M, Ben-Ze'ev A, Thiery JP, Glukhova MA. 2005. Targeted activation of β-catenin signaling in basal mammary epithelial cells affects mammary development and leads to hyperplasia. *Development* **132:** 267–277.

Theodorou V, Kimm MA, Boer M, Wessels L, Theelen W, Jonkers J, Hilkens J. 2007. MMTV insertional mutagenesis identifies genes, gene families and pathways involved in mammary cancer. *Nat Genet* **39:** 759–769.

Tsukamoto AS, Grosschedl R, Guzman RC, Parslow T, Varmus HE. 1988. Expression of the *int*-1 gene in transgenic mice is associated with mammary gland hyperplasia and adenocarcinomas in male and female mice. *Cell* **55:** 619–625.

Turner N, Lambros MB, Horlings HM, Pearson A, Sharpe R, Natrajan R, Geyer FC, van Kouwenhove M, Kreike B, Mackay A, et al. 2010a. Integrative molecular profiling of triple negative breast cancers identifies amplicon drivers and potential therapeutic targets. *Oncogene* **29:** 2013–2023.

Turner N, Pearson A, Sharpe R, Lambros M, Geyer F, Lopez-Garcia MA, Natrajan R, Marchio C, Iorns E, Mackay A, et al. 2010b. FGFR1 amplification drives endocrine therapy resistance and is a therapeutic target in breast cancer. *Cancer Res* **70:** 2085–2094.

Ugolini F, Charafe-Jauffret E, Bardou VJ, Geneix J, Adelaide J, Labat-Moleur F, Penault-Llorca F, Longy M, Jacquemier J, Birnbaum D, et al. 2001. WNT pathway and mammary carcinogenesis: Loss of expression of candidate tumor suppressor gene SFRP1 in most invasive carcinomas except of the medullary type. *Oncogene* **20:** 5810–5817.

Vaillant F, Asselin-Labat ML, Shackleton M, Forrest NC, Lindeman GJ, Visvader JE. 2008. The mammary progenitor marker CD61/β3 integrin identifies cancer stem cells in mouse models of mammary tumorigenesis. *Cancer Res* **68:** 7711–7717.

Valastyan S, Reinhardt F, Benaich N, Calogrias D, Szasz AM, Wang ZC, Brock JE, Richardson AL, Weinberg RA. 2009. A pleiotropically acting microRNA, miR-31, inhibits breast cancer metastasis. *Cell* **137:** 1032–1046.

van Amerongen R, Berns A. 2006. Knockout mouse models to study Wnt signal transduction. *Trends Genet* **22:** 678–689.

van Amerongen R, Nusse R. 2009. Towards an integrated view of Wnt signaling in development. *Development* **136:** 3205–3214.

van Amerongen R, Mikels A, Nusse R. 2008. Alternative wnt signaling is initiated by distinct receptors. *Sci Signal* **1:** pre9.

van de Wetering M, Barker N, Harkes IC, van der Heyden M, Dijk NJ, Hollestelle A, Klijn JG, Clevers H, Schutte M. 2001. Mutant E-cadherin breast cancer cells do not display constitutive Wnt signaling. *Cancer Res* **61:** 278–284.

van de Wetering M, Sancho E, Verweij C, de Lau W, Oving I, Hurlstone A, van der Horn K, Batlle E, Coudreuse D, Haramis AP, et al. 2002. The β-catenin/TCF4 complex imposes a crypt progenitor phenotype on colorectal cancer cells. *Cell* **111:** 241–250.

Van Keymeulen A, Rocha AS, Ousset M, Beck B, Bouvencourt G, Rock J, Sharma N, Dekoninck S, Blanpain C. 2011. Distinct stem cells contribute to mammary gland development and maintenance. *Nature* **479:** 189–193.

Veeck J, Bektas N, Hartmann A, Kristiansen G, Heindrichs U, Knuchel R, Dahl E. 2008. Wnt signalling in human breast cancer: Expression of the putative Wnt inhibitor Dickkopf-3 (*DKK3*) is frequently suppressed by promoter hypermethylation in mammary tumours. *Breast Cancer Res* **10:** R82.

Veltmaat JM, Van Veelen W, Thiery JP, Bellusci S. 2004. Identification of the mammary line in mouse by Wnt10b expression. *Dev Dyn* **229:** 349–356.

Visvader JE, Smith GH. 2011. Murine mammary epithelial stem cells: Discovery, function, and current status. *Cold Spring Harb Perspect Biol* **3:** a004879.

Wagh PK, Gray JK, Zinser GM, Vasiliauskas J, James L, Monga SP, Waltz SE. 2011. β-Catenin is required for Ron receptor-induced mammary tumorigenesis. *Oncogene* **30:** 3694–3704.

Wagner KU, McAllister K, Ward T, Cavis B, Wiseman R, Hennighausen L. 2001. Spatial and temporal expression of the Cre gene under the control of the MMTV-LTR in different lines of transgenic mice. *Transgenic Res* **10:** 545–553.

Weber-Hall SJ, Phippard DJ, Niemeyer CC, Dale TC. 1994. Developmental and hormonal regulation of Wnt gene expression in the mouse mammary gland. *Differentiation* **57:** 205–214.

Weis W. 2011. *Proceedings of WNT2011*, June 29-July 3, Los Angeles, CA.

Wend P, Holland JD, Ziebold U, Birchmeier W. 2010. Wnt signaling in stem and cancer stem cells. *Semin Cell Dev Biol* **21:** 855–863.

Wissmann C, Wild PJ, Kaiser S, Roepcke S, Stoehr R, Woenckhaus M, Kristiansen G, Hsieh JC, Hofstaedter F, Hartmann A, et al. 2003. WIF1, a component of the Wnt pathway, is down-regulated in prostate, breast, lung, and bladder cancer. *J Pathol* **201:** 204–212.

Witte F, Bernatik O, Kirchner K, Masek J, Mahl A, Krejci P, Mundlos S, Schambony A, Bryja V, Stricker S. 2010. Negative regulation of Wnt signaling mediated by CK1-phosphorylated Dishevelled via Ror2. *FASEB J* **24:** 2417–2426.

Wong GT, Gavin BJ, McMahon AP. 1994. Differential transformation of mammary epithelial cells by Wnt genes. *Mol Cell Biol* **14:** 6278–6286.

Wu G, Nomoto S, Hoque MO, Dracheva T, Osada M, Lee CC, Dong SM, Guo Z, Benoit N, Cohen Y, et al. 2003. ΔNp63α and TAp63α regulate transcription of genes with distinct biological functions in cancer and development. *Cancer Res* **63:** 2351–2357.

Xu L, Corcoran RB, Welsh JW, Pennica D, Levine AJ. 2000. WISP-1 is a Wnt-1- and β-catenin-responsive oncogene. *Genes Dev* **14:** 585–595.

Yan H, Blackburn AC, McLary SC, Tao L, Roberts AL, Xavier EA, Dickinson ES, Seo JH, Arenas RB, Otis CN, et al. 2010. Pathways contributing to development of spontaneous mammary tumors in BALB/c-*Trp53*$^{+/-}$ mice. *Am J Pathol* **176:** 1421–1432.

Yang L, Wu X, Wang Y, Zhang K, Wu J, Yuan YC, Deng X, Chen L, Kim CC, Lau S, et al. 2011. FZD7 has a critical role in cell proliferation in triple negative breast cancer. *Oncogene* **30:** 4437–4446.

Ye X, Wang Y, Nathans J. 2010. The Norrin/Frizzled4 signaling pathway in retinal vascular development and disease. *Trends Mol Med* **16:** 417–425.

Yoon JC, Ng A, Kim BH, Bianco A, Xavier RJ, Elledge SJ. 2010. Wnt signaling regulates mitochondrial physiology and insulin sensitivity. *Genes Dev* **24:** 1507–1518.

Zeng YA, Nusse R. 2010. Wnt proteins are self-renewal factors for mammary stem cells and promote their long-term expansion in culture. *Cell Stem Cell* **6:** 568–577.

Zeng X, Tamai K, Doble B, Li S, Huang H, Habas R, Okamura H, Woodgett J, He X. 2005. A dual-kinase mechanism for Wnt co-receptor phosphorylation and activation. *Nature* **438:** 873–877.

Zeng X, Huang H, Tamai K, Zhang X, Harada Y, Yokota C, Almeida K, Wang J, Doble B, Woodgett J, et al. 2008. Initiation of Wnt signaling: Control of Wnt coreceptor Lrp6 phosphorylation/activation via Frizzled, Dishevelled and Axin functions. *Development* **135:** 367–375.

Zhang J, Li Y, Liu Q, Lu W, Bu G. 2010a. Wnt signaling activation and mammary gland hyperplasia in MMTV-LRP6 transgenic mice: Implication for breast cancer tumorigenesis. *Oncogene* **29:** 539–549.

Zhang M, Atkinson RL, Rosen JM. 2010b. Selective targeting of radiation-resistant tumor-initiating cells. *Proc Natl Acad Sci* **107:** 3522–3527.

Zhu Y, Guignard F, Zhao D, Liu L, Burns DK, Mason RP, Messing A, Parada LF. 2005. Early inactivation of p53 tumor suppressor gene cooperating with NF1 loss induces malignant astrocytoma. *Cancer Cell* **8:** 119–130.

Wnt Signaling in Neuromuscular Junction Development

Kate Koles and Vivian Budnik

Department of Neurobiology, University of Massachusetts Medical School, Worcester, Massachusetts 01605
Correspondence: vivian.budnik@umassmed.edu

Wnt proteins are best known for their profound roles in cell patterning, because they are required for the embryonic development of all animal species studied to date. Besides regulating cell fate, Wnt proteins are gaining increasing recognition for their roles in nervous system development and function. New studies indicate that multiple positive and negative Wnt signaling pathways take place simultaneously during the formation of vertebrate and invertebrate neuromuscular junctions. Although some Wnts are essential for the formation of NMJs, others appear to play a more modulatory role as part of multiple signaling pathways. Here we review the most recent findings regarding the function of Wnts at the NMJ from both vertebrate and invertebrate model systems.

Wnt proteins are evolutionarily conserved, secreted lipo-glycoproteins involved in a wide range of developmental processes in all metazoan organisms examined to date. In addition to governing many embryonic developmental processes, Wnt signaling is also involved in nervous system maintenance and function, and deregulation of Wnt signaling pathways occurs in many neurodegenerative and psychiatric diseases (De Ferrari and Inestrosa 2000; Caricasole et al. 2005; Okerlund and Cheyette 2011). The first link between Wnt signaling and synapse development was established by Salinas and colleagues in the vertebrate nervous system (Lucas and Salinas 1997; Hall et al. 2000) and by Budnik and colleagues at the invertebrate neuromuscular junction (NMJ) (Packard et al. 2002). Since then, Wnt signaling has emerged as an essential regulator of synaptic development and function in both central and peripheral synapses. Although important roles for Wnt signaling have become known from studies in both the central and peripheral nervous system, this article is concerned with the role of Wnts at the NMJ.

WNT LIGANDS, RECEPTORS, AND WNT SIGNALING PATHWAYS

Wnts and their receptors comprise a large family of proteins. (For a description of these families, see He [2012], Willert and Nusse [2012], and www.stanford.edu/group/nusselab/cgi-bin/wnt.) Frizzled receptors form a functional heterocomplex with members of the low-density lipoprotein receptor (LDLR)-related protein (LRP) family, Lrp5 and Lrp6 (Pinson et al. 2000; Tamai et al. 2000) and its *Drosophila* homolog, Arrow (Wehrli et al. 2000). Furthermore, there are families of proteins belonging to the unconventional Wnt receptors, such as

Cite this article as *Cold Spring Harb Perspect Biol* doi: 10.1101/cshperspect.a008045

members of the ROR (receptor tyrosine kinase-like orphan receptor) (Green et al. 2008) and Derailed (Drl)/Ryk (related to receptor tyrosine kinase) families (Fradkin et al. 2010). Adding to this diversity of ligands and receptors is the number of distinct signaling pathways activated by Wnt ligand/receptor interactions, namely, the canonical Wnt pathway (Logan and Nusse 2004), the divergent canonical Wnt pathway (Ciani et al. 2004), the planar cell polarity (PCP) pathway (Mlodzik 2002), the calcium Wnt signaling pathway (Kuhl et al. 2000), and the Frizzled nuclear import (FNI) Wnt pathway (Speese and Budnik 2007). Given that most of the transduction pathways are described in Cadigan and Peifer (2012), He (2012), Stamos and Weis (2012), and van Amerongen (2012), only a brief description of the pathways is provided in this review, except for the FNI pathway, which so far has only been characterized in the nervous system. In the canonical Wnt pathway, Wnt binding to Frizzled receptors activates Dishevelled (Dvl), which inhibits the phosphorylation of β-catenin by glycogen synthase kinase 3β (Gsk3β), through the disassembly of a destruction complex, which normally triggers the degradation of cytoplasmic β-catenin. Wnt binding to Frizzled receptors disrupts this complex, leading to the stabilization of β-catenin, its import to the nucleus, and the activation of Wnt target genes (Fig. 1A). In the divergent canonical Wnt pathway, Dvl regulates the Gsk3β-dependent phosphorylation of microtubule-associated proteins (MAPs), which control the stability of microtubule networks (Fig. 1B) (Lucas et al. 1998; Ciani et al. 2004). The PCP Wnt pathway regulates tissue and cell polarity through a Dvl-dependent activation of small GTPases, such as Rho, Rac, and Jun amino-terminal kinase (Jnk) (Fig. 1C). In the Wnt calcium pathway, cell fate decisions and cell movement choices are made via a Dvl-dependent increase in intracellular calcium levels, which leads to the activation of protein kinase C (PKC) and calcium/calmodulin-dependent protein kinase II (CaMKII), the nuclear import of the nuclear factor of activated T-cells (NFAT) and concomitant changes in gene expression (Fig. 1D). Finally, in the FNI Wnt pathway, Frizzled-2 receptors are endocytosed and trafficked toward the nucleus, where a carboxy-terminal fragment of Frizzed-2 is cleaved and imported into the nucleus (Fig. 1E) (Mathew et al. 2005). Within the nucleus, DFz2C fragments associate with large ribonucleoprotein (RNP) granules, which exit the nucleus through a novel mechanism of nuclear envelope budding, akin to the nuclear egress of Herpes-type viruses (Roller 2008; Speese et al. 2012). The large RNPs contain transcripts encoding postsynaptic proteins, and because these transcripts are also localized at postsynaptic sites, it is presumed that they are involved in local translation (Speese et al. 2012). Indeed, genetic manipulations interfering with the FNI pathway lead to abnormal synaptic growth (Mathew et al. 2005; Speese et al. 2012) (discussed below). A similar pathway of receptor endocytosis, cleavage, and fragment import to the nucleus is observed for the atypical Wnt receptor Ryk (Lyu et al. 2008). However, in this case, the pathway is required for normal cortical neurogenesis, as opposed to synapse development, and the exact role of Ryk receptor fragments within the nucleus has not been investigated. Most importantly, through alternative ligands, receptors, and signaling pathways, Wnts have considerable potential to regulate development differentially.

Figure 1. Overview of Wnt signaling pathways. Wnts signal through Frizzled receptors that can form a complex with Lrp-5/6 receptors. With the exception of the FNI pathway, these signaling pathways activate the scaffolding protein Dishevelled (Dvl). The different pathways diverge downstream from Dvl. (*A*) In the canonical pathway, Dvl disrupts the "destruction complex" composed of Axin, Adenomatous Polyposis Coli (APC), and Gsk3β. This prevents the phosphorylation of β-catenin by Gsk3β, which leads to its stabilization and subsequent nuclear import. In the nucleus, β-catenin associates with the transcription factor T-cell factor/lymphoid enhancer factor 1 (TCF/Lef1) and regulates gene expression. The secreted proteins Dickkopf1 (Dkk1) and secreted Frizzled-related protein (Sfrp) inhibit this pathway. (*Legend continues on facing page.*)

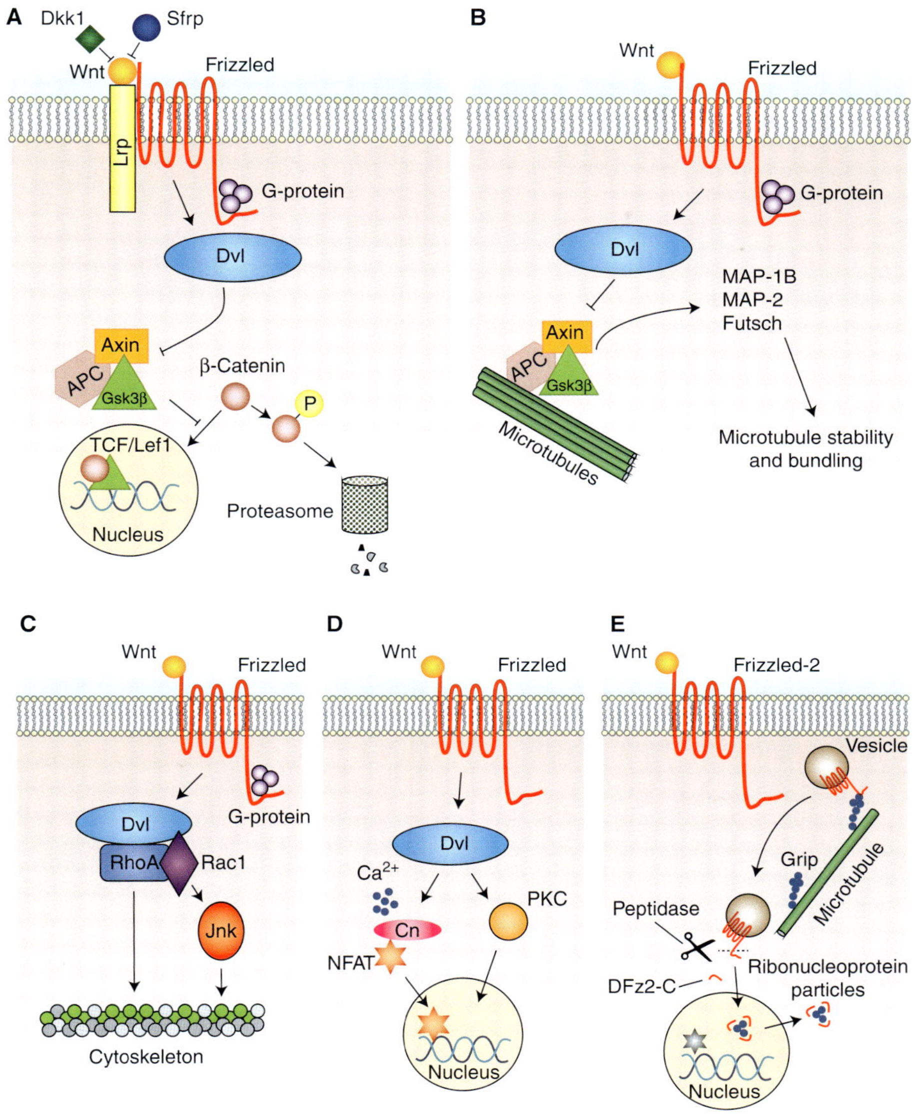

Figure 1. (*Continued*) (*B*) Following Wnt activation of the divergent canonical pathway, Dvl inhibits Gsk3β, which leads to the decreased phosphorylation of microtubule-associated proteins (MAPs), such as Tau, MAP-1B, MAP-2, or the related *Drosophila* protein Futsch. (*C*) In the planar cell polarity pathway, Dvl activates the small Rho GTPases RhoA and Rac1, and c-Jun–amino-terminal kinase (Jnk), which leads to the regulation of actin and microtubule cytoskeletons. (*D*) In the Wnt calcium pathway, Dvl activation leads to a rapid increase in intracellular calcium levels and subsequent activation of calcium-sensitive enzymes such as protein kinase C (PKC), calcium/calmodulin-dependent protein kinase II (CaMKII) and the Ca^{2+}-activated phosphatase calcineurin (Cn). Calcineurin directly binds to and dephosphorylates nuclear factor of activated T-cells (NFAT), which, in turn, allows their nuclear translocation where they regulate gene expression. (*E*) In the Frizzled nuclear import pathway (FNI), Wnt binding to Frizzled-2 (DFz2) leads to its endocytosis and Grip-dependent trafficking to the periphery of the nucleus. Here the carboxyl terminus of DFz2 is cleaved and imported. In *Drosophila*, DFz2-C organizes ribonucleoprotein (RNP) particles containing specific transcripts, which then egress from the nucleus and are likely translated in the cytoplasm. A similar nuclear import pathway has also been described for the atypical Ryk receptor during cortical neurogenesis (Lyu et al. 2008).

BRIEF OVERVIEW OF VERTEBRATE NMJ DEVELOPMENT

Vertebrate NMJs are among the longest and best-studied peripheral synaptic units because of their simplicity and accessibility. Motor neuron axons innervate the muscle at the end plate, where axon terminals branch into fine synaptic varicosities, apposed to foldings of the muscle membrane, the junctional folds. The crests of these folds appose presynaptic active zones, the sites of acetylcholine (ACh)–containing vesicle exocytosis at presynaptic terminals, and contain a high density of neurotransmitter receptors. In contrast, the valleys of the junctional folds and their basement membrane are enriched in acetylcholinesterase (AChE), which rapidly degrades acetylcholine, thus terminating ACh-induced responses.

Even before nerve innervation, AChR clusters form in a broad band at the muscle midline, a process called AChR prepatterning (Yang et al. 2001). These "preformed" clusters are also called aneural AChR clusters to underscore their formation independently of innervation. Once the motor axon interacts with the muscle, AChR clusters form at postsynaptic regions exactly apposed to the sites of nerve contact. These mature clusters (also called neural AChR clusters) are formed by the recruitment of aneural AChR clusters as well as by the targeted accumulation of newly synthesized AChRs (Sanes and Lichtman 2001).

Despite its relative simplicity, the vertebrate NMJ has provided us with remarkable insight into the mechanisms of synapse development. Adding to the multiple players found to participate in NMJ development is the recent discovery that, not totally unexpectedly, these also include members of the Wnt family.

An important initial discovery in the quest to understand NMJ development was the identification of Agrin, originally isolated from the electric organ of the pacific electric ray (Smith et al. 1987) as a protein that could trigger AChR clustering in cultured myotubes (Wallace 1986). These prolific initial studies also led to the model that Agrin, a heparan sulfate proteoglycan, was secreted by neurons, accumulated in the basal lamina at the synaptic cleft, where it triggered a molecular cascade in the muscle. This signaling cascade led to the phosphorylation of the β-subunit of AChR and the redistribution of AChRs to the NMJs in association with a 43-kDa AChR-associated protein. This 43-kDa AChR-associated protein was subsequently identified as Rapsyn (receptor-associated protein at synapses) (Frail et al. 1988). Muscles and Schwann cells were also known to express Agrin, but only the neuronal isoform was sufficiently potent to induce AChR clustering (Gesemann et al. 1995). Interestingly, *agrin*$^{-/-}$ mice were able to form aneural AChR clusters before innervation, indicating that Agrin was not required for prepatterning (Lin et al. 2001). During NMJ maturation, the preformed AChR clusters outside of the innervated area are dispersed upon acetylcholine release, and from this perspective, Agrin, which prevents the dispersal of neural AChR clusters, can also be considered as an "anti-declustering factor," in addition to its AChR clustering role (Lin et al. 2005; Kummer et al. 2006).

One of the key milestones after the identification of Agrin was cloning its muscle-specific receptor, aptly named muscle-specific kinase (MuSK) (Jennings et al. 1993; Valenzuela et al. 1995; DeChiara et al. 1996; Glass et al. 1996). Unlike Agrin, MuSK was required for both neural and aneural AChR clustering in mice and zebrafish (Yang et al. 2001; Zhang et al. 2004). Surprisingly, however, during neural AChR clustering, Agrin did not interact directly with MuSK, suggesting the existence of an MASC (myotube-associated specificity component) (Glass et al. 1996).

WNT SIGNALING AT THE VERTEBRATE NMJ

It was around this time that a link between NMJ development and Wnt signaling was suspected, because the newly cloned MuSK contained a conserved cysteine-rich domain (CRD) found in Frizzled receptors, and this domain was known to be required for Wnt ligand binding (Masiakowski and Yancopoulos 1998). However, experimental evidence supporting the ability of Wnts to bind the MuSK CRD domain was

 Cite this article as *Cold Spring Harb Perspect Biol* doi: 10.1101/cshperspect.a008045

only obtained a decade later (Jing et al. 2009). During the intervening decade, a more pressing challenge had been the identification of the mysterious MASC.

First, Weatherbee et al. (2006) reported that NMJ abnormalities in Lrp4$^{-/-}$ mice had striking similarities to those observed in Musk$^{-/-}$ mutants. Lrp4 is an Lrp family member with an extracellular domain resembling that of Lrp5 and 6/Arrow, but that appears to antagonize Wnt signaling during digit formation in mice (Johnson et al. 2005). Lrp4 has also been shown to interact with postsynaptic density protein 95 (PSD-95) in the brain (Tian et al. 2006). Subsequently, two independent groups reported that Lrp4 was the long-sought Agrin binding receptor (Kim et al. 2008; Zhang et al. 2008). These studies showed that Lrp4 was expressed at NMJs in myotubes, where it bound selectively to the neuronal isoform of Agrin and was necessary for Agrin-induced activation of MuSK and AChR clustering. Indeed, recent detailed structural studies by Zhang and colleagues have now revealed that several of the extracellular domains on Lrp4 serve as ligands for MuSK and that Agrin dramatically stimulates this interaction (Gomez and Burden 2011; Zhang et al. 2011). Following Agrin-stimulated Lrp4 binding to MuSK, the receptor is tyrosine-phosphorylated and recruits Dok-7 (Inoue et al. 2009), a muscle-specific adaptor protein. Like MuSK, Dok-7 is essential for NMJ formation, because in its absence NMJs fail to form, and Dok-7 overexpression dramatically stimulates AChR clustering (Okada et al. 2006). Dok-7 acts through two different pathways determined by its amino-terminal and carboxy-terminal protein interactions. First, Dok-7 dimerizes (Bergamin et al. 2010), and its amino terminus stimulates MuSK activity. However, this activity of Dok-7 only minimally affects postsynaptic differentiation (Hallock et al. 2010). Second, Dok-7, via its carboxyl terminus, also recruits two adaptor proteins, CT10 regulators of kinase, Crk and Crk-L, downstream from MuSK activation (Hallock et al. 2010; Burden 2011), which are required in the muscle for normal presynaptic and postsynaptic NMJ differentiation. In Crk and Crk-L double mutants, motor axon terminals occupied a larger muscle area than controls and had significantly reduced synapse numbers, with several muscle fibers lacking innervation. These mutants die at birth because of the inability to inflate their lungs. Interestingly, many patients with congenital myasthenia have mutations in the carboxyl terminus of Dok-7, and these results provide insight into the possible molecular pathways affected in this disease. Together with Agrin, Rapsyn, MuSK, and Lrp4, Dok-7 represents an indispensable component of the molecular pathway essential for NMJ formation in vertebrates. Numerous additional molecules and pathways involved in the assembly of NMJ have been revealed over the past few years. These are discussed in excellent reviews (Kummer et al. 2006; Wu et al. 2010).

WNTs AS POSITIVE REGULATORS OF NMJ DEVELOPMENT

The first experimental evidence implicating Wnt signaling in NMJ development was provided when a yeast two-hybrid screen geared to identify intracellular binding partners of MuSK led to the identification of Dishevelled1 (Dvl1), a scaffolding protein activated by several Wnt signaling pathways (Luo et al. 2002). Gain- and loss-of-function experiments in vivo and in vitro showed that Dvl1 was required for the Agrin-mediated clustering of AChRs. In addition, the Dvl1-interacting protein p21-activated kinase 1 (PAK1) was identified, and it was shown that its activation by Agrin required Dvl1. MuSK, Dvl, and PAK1 formed a ternary complex required for AChR clustering in muscle. Furthermore, expression of a Dvl1 mutant transgene lacking the DEP (Dishevelled-Egl-10-Pleckstrin) domain in muscle cells of *Xenopus* nerve–muscle cocultures led to decreased spontaneous synaptic current frequency and amplitude, indicating a potential muscle-derived retrograde signal that could influence presynaptic neurotransmitter release (Luo et al. 2002).

Subsequently, it was shown that the β-subunit but not the α1-, α3-, α7-, or ε-subunits of AChR interacted with APC, and its function was required in Agrin-induced AChR clustering (Wang et al. 2003). APC is a well-known regulator of β-catenin, a key effector of canonical

Wnt signaling (Aoki and Taketo 2007). APC is also known to interact with microtubules, actin filaments, and EB1 (Su et al. 1995), a microtubule plus-end tracking protein that negatively regulates APC–actin interaction (Reilein and Nelson 2005; Moseley et al. 2007). It was suggested that APC supported a cross talk between Agrin- and Wnt-mediated signaling pathways, by coordinating the functions of actin and the microtubule cytoskeleton. Notably, the APC–AChR interaction was not unique to muscle-specific AChRs, because APC also regulated the synaptic localization of neuronal α3-nicotinic AChRs by binding to PSD-93, β-catenin, and EB1 (Temburni et al. 2004; Rosenberg et al. 2008). Furthermore, APC functions presynaptically as well, where it increases the clustering and membrane localization of the α7-nicotinic AChR in response to Wnt-7a signaling (Farias et al. 2007). All of these data implicate APC as a Wnt downstream effector that is crucially involved in the localization of select AChRs both at neuronal and NMJ synapses.

More recently, it was found that Rapsyn binds to β-catenin (Zhang et al. 2007). Rapsyn has been known to associate with AChRs in an Agrin-dependent manner (Moransard et al. 2003), and, indeed, β-catenin associated with AChR following Agrin treatment. This Agrin-triggered association did not occur in *rapsyn*$^{-/-}$ mice, indicating that Rapsyn is required for β-catenin association with AChR. Reducing β-catenin levels in C2C12 myotubes by shRNA resulted in decreased AChR clustering following Agrin treatment (Zhang et al. 2007; but see Li et al. 2008). To understand the role of β-catenin in vivo, β-catenin was conditionally knocked out only in the muscles of mice. Surprisingly, in addition to muscle-specific phenotypes, such as the increased size and wider distribution of AChR clusters, primary nerve branches were mislocalized, evoked and spontaneous neurotransmitter release was decreased, and short-term plasticity was abnormal in these mice (Li et al. 2008). In contrast, Agrin-induced AChR clustering was unaffected in cultured muscles from β-catenin$^{-/-}$ mice. Most importantly, β-catenin conditional knockout in neurons did not have any effect on NMJ morphology or function, implying that muscle β-catenin is involved in presynaptic development and function in an as-yet-unidentified retrograde pathway (Fig. 3) (Li et al. 2008).

In addition to Wnt signaling pathway components, some of the Wnt ligands have also been identified. Whereas Wnt1 (Luo et al. 2002), Wnt4, Wnt6, and Wnt7b (Zhang et al. 2007) were found to have no effect on AChR clustering, Wnt3, which is expressed in mouse motor neurons at the time of innervation (Krylova et al. 2002), enhanced AChR clustering in developing chicken wing NMJs and the clustering ability of Agrin in mouse cultured myotubes (Henriquez et al. 2008). Dickkopf-1 (Dkk1), which specifically blocks canonical Wnt signaling (Fig. 1A) (Niehrs 2006), had no effect, whereas secreted Frizzled-related protein 1 (Sfrp1), which blocks both canonical and non-canonical Wnt signaling (Rattner et al. 1997), reduced the number of AChRs induced by Wnt3, indicating that Wnt3 is likely to mediate this effect via a non-canonical Wnt signaling pathway in myotubes, even though Wnt3 can activate canonical signaling (Barrow et al. 2003). It was also found that Wnt3 induced a quick, Agrin-independent transient up-regulation of Rac1 activity, which led to the rapid formation of AChR microclusters (Fig. 2A). These microclusters were subsequently stabilized into full-sized clusters in an Agrin-dependent manner, in a process involving Rho signaling (Fig. 2B) (Weston et al. 2003; Henriquez et al. 2008). These observations led to a model in which Wnt3 released from motor neurons induces the first wave of Rac1 activation and subsequent AChR microclustering, which is then consolidated into mature AChR clusters via Agrin-mediated signaling. Nevertheless, the muscle receptor(s) through which Wnt3 signals this activation remain to be identified.

A role for Wnts in both AChR prepatterning and axon guidance is also provided by studies in zebrafish. As discussed above, before the arrival of motor nerve terminals to the muscle field, AChRs spontaneously cluster in the mid-region of the muscle, where prospective NMJs eventually form, and this prepatterning depends on MuSK (Lin et al. 2001; Yang et al. 2001).

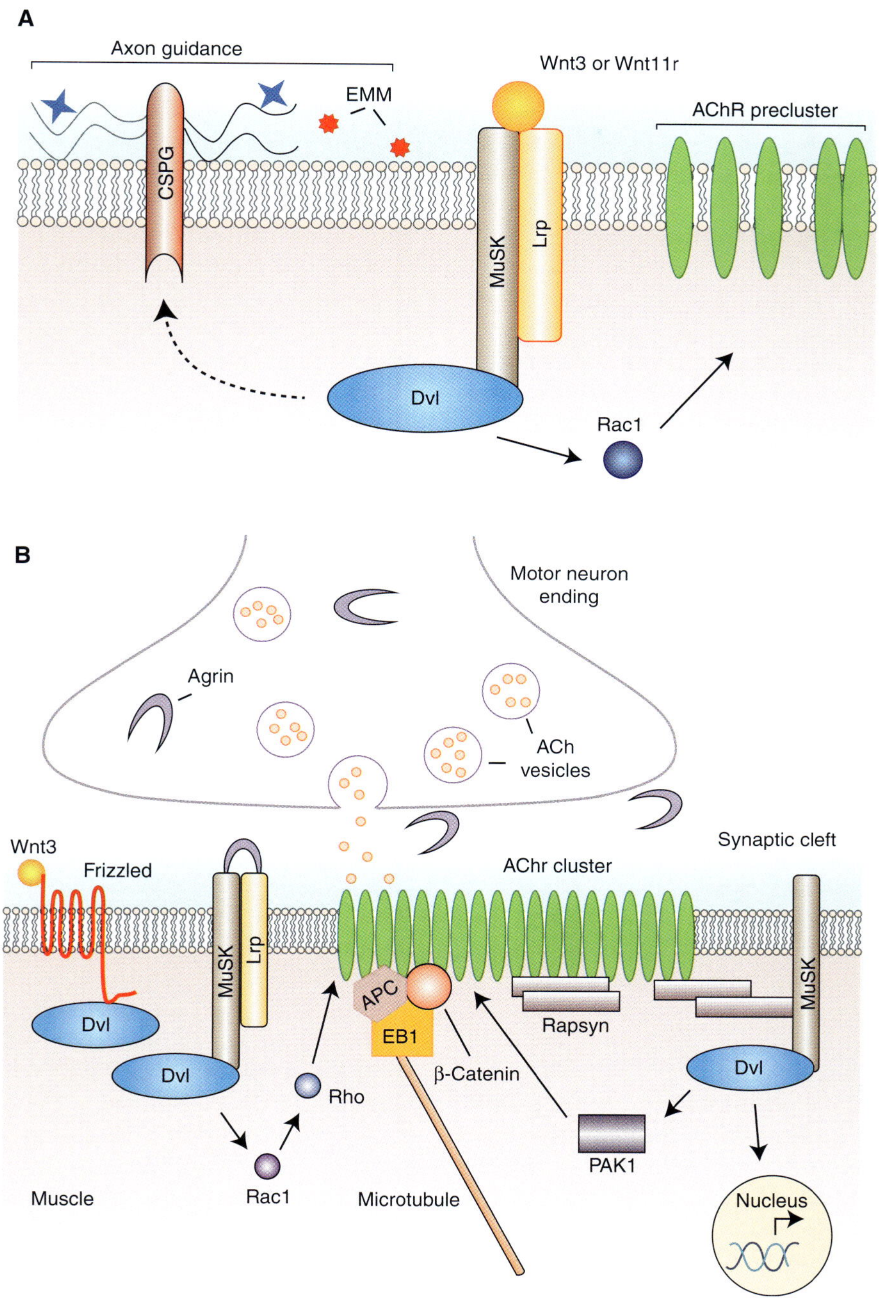

Figure 2. Wnt signaling at vertebrate neuromuscular junctions. (*A*) Mouse Wnt3 and zebrafish Wnt11r function during prepatterning. Wnt11r signaling through Unplugged/MuSK before innervation leads to aneural AChR clustering and also via a parallel signaling pathway to changes in extracellular matrix modifications (EMMs), possibly via chondroitin sulfate proteoglycans (CSPGs) that eventually influence incoming motor neuron axonal growth cone guidance. Wnt3 signaling leads to the rapid and transient activation of Rac1 and AChR microcluster formation, which (*B*) in the presence of neural agrin at the time of innervation are stabilized in exact opposition to the presynaptic terminal.

Many of these aneural clusters are incorporated into neural AChRs clusters upon nerve arrival (Flanagan-Steet et al. 2005). However, the function of prepatterning has been a matter of debate, partly owing to the observations that some nerves can ignore preformed AChR clusters and initiate NMJ development at new areas (Anderson and Cohen 1977), raising the possibility that the nerve, or nerve-derived factors, are instructive for NMJ formation. Elegant studies by Jing et al. (2009) address this long-standing dilemma. Live imaging of wild-type and MusK$^{-/-}$ mutants (which in zebrafish is called "unplugged") revealed that neuronal growth cones explored a wider muscle area and invaded muscle territories that they usually avoid. Because *unplugged*/MuSK is a muscle-specific gene, these data indicate that muscle plays a role in restricting the area that growth cones explore during innervation. Based on expression time and pattern (spinal cord and adaxial somites) in addition to its mutant phenotype, the investigators identified the nonconventional Wnt11r, as required for AChR prepatterning on the adaxial muscles. Live imaging of motor growth cones revealed that *wnt11r* morpholino-induced abnormalities in prepatterning were virtually identical to those observed in *unplugged*/MuSK$^{-/-}$ mutants, suggesting that Wnt11r might act through *unplugged*/MuSK. In contrast, neurally induced AChR clusters were formed and developed normally in *wnt11r* morpholino-treated animals, suggesting that Wnt11r function is restricted to prepatterning. That Wnt11r and Unplugged/MuSK were mutually required for prepatterning was confirmed by showing genetic interactions between the genes. In addition, by using rescue experiments with the short (containing the CRD and Kringle domains) or the long (containing three amino-terminal Ig domains in addition to CRD and Kringle domains) Unplugged, it was shown that the CRD domain of Unplugged/MuSK was required for physical interaction between Wnt11r and Unplugged. Expressing a Dvl isoform that interferes with non-canonical Wnt signaling also led to prepatterning defects, implicating it in the signaling pathway downstream from Wnt11r.

Interestingly, in the absence of prepatterning, neurally induced AChR clusters and functional NMJs were formed, albeit being slightly smaller. However, in the absence of prepatterning, axonal growth cones were not directed to the narrower mid-region of the prospective NMJs, and they explored much broader muscle areas. The investigators conclude that one of the functions of prepatterning is to guide incoming axons (perhaps via the Wnt11r or Unplugged signaling-induced modifications to the extracellular matrix) and to restrict muscle AChRs cluster formation to a common muscle area for efficient NMJ formation (Fig. 2A) (Jing et al. 2009).

NEGATIVE REGULATION OF NMJs BY WNT

Negative regulation of AChR clustering by Wnts has also been documented. Wnt3a expression in the muscle gradually increases during the phase of embryogenesis that coincides with NMJ development (Wang et al. 2008). However, pretreatment of cultured myotubes with Wnt3a-containing conditioned medium before Agrin application significantly reduced AChR aggregation. Moreover, transfection of C2C12 myotubes with Wnt3a, but not Wnt5, Wnt7a, or Wnt11, reduced Agrin-mediated AChR clustering. Transfection with Dkk1, which antagonizes the canonical Wnt pathway, increased AChR clustering, indicating that canonical Wnt signaling negatively regulates NMJ development. The stability of AChR clusters was also reduced when Wnt3a was applied after Agrin-induced clustering. The effects of Wnt3a were mediated via β-catenin and resulted in reduced Rapsyn expression. Reduction of β-catenin by shRNA resulted in decreased Agrin-dependent AChR clustering, likely because of the disrupted β-catenin–Rapsyn interaction (Zhang et al. 2008) or the increased Rapsyn expression (Yoshihara and Hall 1993; Wang et al. 2008). Downstream from Wnt3a, β-catenin is known to function as a cotranscription factor with TCF (Korinek et al. 1997). However, dominant negative TCF had no effect on AChR clustering, indicating that β-catenin function is independent of TCF-mediated

 Cite this article as *Cold Spring Harb Perspect Biol* doi: 10.1101/cshperspect.a008045

transcription (Zhang et al. 2007; Li et al. 2008). These results indicate that muscle Wnt3a disperses AChR clusters by down-regulating Rapsyn expression in a β-catenin-dependent manner (Fig. 3). In a larger context, these observations highlight the importance of both positive and negative Wnt signals in sculpting the mature NMJ.

In summary, Wnt proteins have been found to play multiple roles in all stages of vertebrate NMJ formation, both in the form of positive and negative signaling pathways. At least two Wnts, muscle Wnt3 and non-muscle Wnt11r, acting through non-canonical pathways, appear to function during prepatterning predominantly by modulating AChR precluster formation, which are subsequently stabilized into mature AChRs in an agrin-dependent manner. In contrast, Wnt3a induces the dispersal of AChR clusters in the presence of Agrin, suggesting that the balance of positive and negative signaling pathways is at work during NMJ development.

WNT SIGNALING AT INVERTEBRATE NMJs

Two invertebrate systems, *Drosophila melanogaster* (fruit fly) and *Caenorhabditis elegans* (roundworm), have been influential in uncovering the function of Wnt family molecules at the neuromuscular junction (NMJ). The ease of genetic manipulation in these relatively simple and highly accessible model organisms has provided a wealth of groundbreaking insight into the development and plasticity of NMJs.

THE ROLE OF WNTs AT THE LARVAL NMJ OF *DROSOPHILA*

Drosophila larvae offer an exceptionally accessible model system to study the development and plasticity of NMJs. The larval body wall muscles are innervated by excitatory glutamatergic motor neurons that terminate in stereotypical NMJs with easily discernible synaptic boutons. In response to the rapid growth of the larva and the consequent growth of its body wall muscle

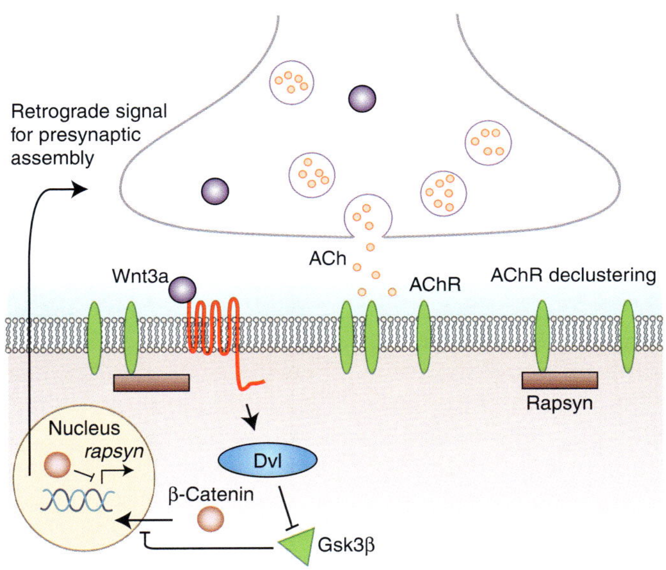

Figure 3. Negative regulation of AChR clustering by Wnt3a. Muscle Wnt3a initiates a Wnt signaling pathway in the muscle that leads to decreased Rapsyn levels. This is achieved through the β-catenin-mediated negative regulation of *rapsyn* gene expression, and decreased Rapsyn levels result in the destabilization and dispersal of AChR clusters. In addition, Wnt3a, via β-catenin, also triggers an unknown retrograde signal that affects presynaptic function and/or assembly.

cells after hatching from the egg, NMJs show a 10-fold increase in synaptic bouton number and size during larval development (Gorczyca et al. 1993; Schuster et al. 1996). In addition, larval NMJs undergo structural and functional modifications in response to experience and electrical activity (Budnik et al. 1990; Sigrist et al. 2003; Steinert et al. 2006; Ataman et al. 2008; Koon et al. 2011), offering a powerful experimental system to study the molecular processes underlying synaptic growth and plasticity.

A role for *Drosophila* Wnt1, Wingless (Wg), in NMJ development was first discovered by Packard et al. (2002), when they found that Wg was present at synaptic boutons of the NMJ and that transient reduction in Wg secretion prevented NMJ expansion during muscle growth (Packard et al. 2002). NMJs from these temperature-sensitive *wg* mutants (wg^{ts}) shifted to restrictive temperature for 12–18 h during late larval development, also had NMJs containing unusually large and irregularly shaped boutons. These abnormal boutons also displayed abnormal distribution of glutamate receptor subunit, DGluRIIA, and the PDZ protein of the PSD95-family Discs-Large (Dlg). Most strikingly, several synaptic boutons, later called "ghost boutons" (Ataman et al. 2006), were filled with synaptic vesicles yet lacked mitochondria, presynaptic active zones, and the postsynaptic subsynaptic reticulum (SSR) (Packard et al. 2002). These ghost boutons were later found to be devoid of any postsynaptic protein and lacked most labeling for the active-zone protein Bruchpilot (Brp) (Ataman et al. 2006, 2008; Korkut et al. 2009). Live imaging of NMJs during acute elevation of synaptic activity, which resulted in rapid formation of ghost boutons, together with the observation that ghost boutons are also observed at wild-type NMJs, albeit at very low frequencies (Ataman et al. 2008), led to the conclusion that ghost boutons are a normal undifferentiated stage during the development of synaptic boutons. In Wnt mutants, the differentiation of ghost boutons is blocked, leading to a reduced number of boutons and the accumulation of ghost boutons.

The presynaptic microtubule cytoskeleton at the NMJ was also altered, because labeling NMJs with the microtubule-associated protein 1B (MAP-1B)–related protein Futsch, which is phosphorylated by Gsk3β/Shaggy (Sgg) (Gogel et al. 2006), revealed abnormal microtubule bundling (Packard et al. 2002). Wild-type NMJs display a characteristic proportion of microtubule loops (thought to be stable microtubule bundles), splayed microtubule bundles, and fragmented microtubules (both likely representing microtubules undergoing remodeling) throughout NMJ branches (Roos et al. 2000; Packard et al. 2002; Ruiz-Canada et al. 2004). However, the proportion of looped, splayed, and fragmented microtubules was altered in the *wg* mutant (Packard et al. 2002), again suggesting an abnormal development of the NMJ cytoskeleton in the absence of Wg signaling.

These defects were shown to be mediated by the Wg receptor *Drosophila* Frizzled-2 (DFz2) (Packard et al. 2002; Ataman et al. 2006) and its coreceptor Arrow/Lrp5/6 (low-density lipoprotein receptor-related protein) (Miech et al. 2008). These receptors are likely to function both presynaptically in the motor neurons and postsynaptically in the muscle. Efforts to delineate the exact signaling pathway triggered by Wg release led to the finding that Wg, likely released by synaptic boutons (Packard et al. 2002), activated alternative signaling pathways in the presynaptic motor neurons and the postsynaptic muscles (Ataman et al. 2008; Miech et al. 2008). Activation of the Wg pathway in the presynaptic compartment appeared to activate a divergent canonical Wnt pathway involving Gsk3β, which regulated the presynaptic microtubule cytoskeleton (Ataman et al. 2008; Miech et al. 2008). On the other hand, activation of postsynaptic muscle DFz2 receptors turned on a novel Wg-signaling pathway, the so-called Frizzled nuclear import pathway (FNI) (Fig. 1E), which was required for building the postsynaptic apparatus (Mathew et al. 2005; Ataman et al. 2006, 2008; Mosca and Schwarz 2010; Speese et al. 2012).

Using antibodies specifically recognizing the carboxyl terminus and amino terminus of DFz2, it was shown that in postsynaptic muscles, the DFz2 receptor was endocytosed and trafficked to the periphery of the muscle nucleus, likely in an

 Cite this article as *Cold Spring Harb Perspect Biol* doi: 10.1101/cshperspect.a008045

endocytic vesicle, where its cytoplasmic carboxy-terminal fragment was cleaved. Whereas the 8-kDa carboxy-terminal fragment of DFz2 (DFz2C) was imported into the nucleus, the amino-terminal fragment remained at the nuclear periphery. Within the nucleus, DFz2C formed submicrometer-sized foci (Mathew et al. 2005). Overexpressing Wg in motor neurons, which increased the levels of secreted Wg around synaptic boutons, led to an increase in the number of nuclear DFz2C foci in the muscles, and decreasing Wg secretion in *wg* mutants or interfering with DFz2 signaling in muscles led to the opposite effect (Mathew et al. 2005).

The cleavage of DFz2 was found to be required for Wg signaling in the muscle. Expression of a DFz2 transgene in the muscle, carrying a mutation that eliminated the cleavage site, in a *dfz2* mutant background did not rescue the reduced bouton number observed in *dfz2* mutants (Mathew et al. 2005). Importantly, attempts to circumvent the nuclear import pathway by expressing a tagged DFz2C fragment failed, because this did not rescue the reduced bouton number in *dfz2* mutants (Mathew et al. 2005; but see Mosca and Schwarz 2010). These findings led the investigators to suggest that an interaction between the full-length DFz2 and Wg was likely required to make DFz2C competent for nuclear import and signaling.

The journey of DFz2 from the plasma membrane to the nucleus involved a direct interaction between the carboxyl terminus of DFz2C, which contains a PDZ binding sequence, and the 7 PDZ domain protein dGRIP (glutamate receptor interacting protein) (Ataman et al. 2006). In the vertebrate nervous system, GRIP mediates the synaptic trafficking and scaffolding of several proteins, including AMPA (Dong et al. 1997) and EphB (Hoogenraad et al. 2005) receptors. In *Drosophila*, mutations in *dgrip* or selective postsynaptic reduction in dGRIP levels led to phenotypes that highly resembled mutations in Wnt signaling (Ataman et al. 2006). These included a drastically reduced number of nuclear DFz2C foci, a reduced number of synaptic boutons, and an increase in the number of ghost boutons. Analysis of the trafficking of DFz2 receptors by using an antibody against the extracellular amino terminus of DFz2, which binds to the receptor in vivo, allowed the researchers to follow receptor trafficking through pulse-chase experiments. In wild-type larvae, DFz2 receptors localized at the postsynaptic region traffic toward the nucleus over time. However, when dGRIP levels were reduced, DFz2 accumulated at the postsynaptic region and failed to be transported toward the nucleus (Ataman et al. 2006). These observations suggested that dGRIP was required to transport DFz2C from the muscle plasma membrane to the nucleus.

Further insight into the FNI pathway was provided by Mosca and Schwarz (2010), who showed that the nuclear import of DFz2C requires the nucleoporin-associated proteins Importin-β11 (Imp-β11) and Importin-α2 (Imp-α2). In the absence of either of these two importins, synapses displayed phenotypes that resembled the alterations found at NMJs of *wg*, *dfz2*, and *dgrip* mutants, that is, a reduced number of synaptic boutons, an excessive number of ghost boutons, and a reduced thickness of the subsynaptic reticulum (SSR) (Mosca and Schwarz 2010). These results indicate that importins and therefore nuclear entry of DFz2C fragments through the canonical nuclear import pathway are key players in the FNI Wnt pathway.

More recently, Speese et al. (2012) have begun to unravel the function of DFz2C foci within the nucleus. The foci are composed of large ribonucleoprotein (RNP) particles bearing synaptic protein transcripts. These granules localize to the space between the inner and outer nuclear membrane and require Lamin C, a nuclear protein linked to muscular dystrophies (Worman and Bonne 2007). *lamC* mutations mimicked NMJ phenotypes of mutations interfering with the FNI pathway. Notably, it was shown that the granules exit the nucleus via a budding mechanism akin to the nuclear egress of Herpes-type viruses (Mettenleiter et al. 2006), a process previously thought to be exclusive for these viruses. Like HSV nuclear egress, this process required PKC, which is known to disrupt the lamin through phosphorylation. It was suggested that nuclear envelope budding is an endogenous nuclear export mechanism for large

RNPs (Speese et al. 2012). Interestingly, the nuclear DFz2C foci contained transcripts encoding synaptic proteins, such as Par6 and CaMKII, and some of the transcripts were also found at the NMJ. These observations suggested that the large RNPs were required for local translation of the postsynaptic apparatus during new synaptic bouton formation (Speese et al. 2012). Although the exact function of DFz2C fragments within the large RNP granules is not known, the authors speculate that the PDZ binding domain of DFz2C may serve as a zipcode to properly target large RNPs to synaptic sites. Several PDZ-containing scaffolding proteins are localized at postsynaptic sites and are required for synaptic development at the NMJ (Lahey et al. 1994; Thomas et al. 1997; Zito et al. 1997; Mathew et al. 2002). Furthermore, proteins containing PDZ-binding motifs are known to be targeted to postsynaptic sites (Zito et al. 1999). Although the generality of the FNI signaling pathway for other systems and tissues is presently unknown, nuclear DFz2C foci are also observed in Schneider-2 cells and at *Drosophila* larval salivary glands (Speese et al. 2012), where the nuclear envelope granules were first observed along with epithelial cells of the midgut (Gay 1956; Hochstrasser and Sedat 1987). In addition, these foci are observed in peripheral glial cells in the larva (SD Speese and V Budnik, unpubl.). Notably, granules accumulating between the outer and inner nuclear membrane resembling the DFz2-decorated large RNP granules have been observed in diverse contexts, from plants (Dickinson 1971) to mammals, where they are particularly prevalent at early embryonic stages (Hadek and Swift 1962; Szollosi 1965). Indeed, that such granules might reflect an alternate mode of nucleocytoplasmic export has been previously proposed (Gay 1956; Szollosi and Szollosi 1988), but it had not been experimentally validated until now. However, whether these granules are also associated with Frizzled receptor fragments is currently unknown.

As discussed above, Wg also turns on a presynaptic signaling pathway with resemblance to the canonical Wnt signaling pathway (Ataman et al. 2008; Miech et al. 2008). Several components of the canonical Wnt signaling pathway can be detected in presynaptic boutons (Packard et al. 2002; Miech et al. 2008), including Dishevelled (Dvl), Gsk3β/Shaggy, DFz2 and its coreceptor Arrow/Lrp5/6, but not Armadillo/β-catenin (Miech et al. 2008). Mutations in *arrow/Lrp5/6*, a receptor thought to be involved in canonical Wnt signaling (He et al. 2004), showed phenotypes similar to those observed in *wg* mutants such as reduced bouton number, increase in large (>5 μm), irregularly shaped boutons, and decrease in the number of boutons with presynaptic Futsch loops (Miech et al. 2008). However, the presence of ghost boutons was not examined in those studies. All of these phenotypes were rescued by the presynaptic expression of Arrow. Notably, expressing an *arrow* transgene in muscles of *arr* mutants also rescued the reduced bouton number while not rescuing the presynaptic microtubule phenotypes. In addition, Arrow overexpression in neurons resulted in an increase in the number of bouton and satellite boutons, similarly to what was shown for presynaptic Wg overexpression (Packard et al. 2002; Miech et al. 2008). Subsequent experiments indicated that Wg binding to presynaptic Arrow leads to the activation of a divergent canonical pathway. Indeed, overexpression of a dominant negative Dvl in neurons, but not in muscle, led to the prototypical *wg* mutant phenotypes, whereas overexpression of dominant negative Gsk3β/Sgg, as expected, resulted in the opposite phenotype, mimicking Wg overexpression (Franco et al. 2004; Miech et al. 2008). No phenotypic changes were observed upon dominant negative TCF/pangolin or Armadillo/β-catenin overexpression in neurons (Miech et al. 2008), suggesting that the dominant negative Dvl and Gsk3β overexpression phenotypes did not result from interfering with canonical Wnt signaling. Thus, by activating alternative signaling pathways, a single Wnt simultaneously coordinates presynaptic and postsynaptic development.

An important avenue of inquiry examined the involvement of Wnt signaling during synaptic plasticity. Ataman et al. (2008) showed that larval NMJs respond to repeated or patterned stimulation administered via high-potassium depolarization, motor nerve stimulation,

 Cite this article as *Cold Spring Harb Perspect Biol* doi: 10.1101/cshperspect.a008045

or blue light stimulation of channelrhodopsin-2 (Nagel et al. 2002) expressing motor neurons, by extending new "synaptopods," dynamic filopodia-like structures that are observed, albeit at lower frequency, in unstimulated animals (Ataman et al. 2008). In addition, NMJs respond to these stimuli by forming new ghost boutons. Analysis by time-lapse microscopy revealed that a subset of these ghost boutons could subsequently mature by acquiring active-zone and postsynaptic proteins, suggesting that ghost boutons are an undifferentiated state of mature boutons. In addition to the above morphological changes, the frequency of spontaneous miniature synaptic potentials (mEJPs) was potentiated by these stimulation paradigms. The above changes required at least five stimulation bouts separated by a resting period, and intact transcriptional and translational machinery. Most importantly, Wg signaling was necessary for these changes downstream from motor neuron stimulation, because removing just one copy of the *wg* gene was sufficient to eliminate the structural and functional changes elicited by patterned stimulation. Furthermore, neuronal overexpression of Wg, which increased Wg secretion by synaptic boutons, reduced the number of cycles required to initiate the structural and functional changes (Ataman et al. 2008). Thus, these studies show that Wg not only plays a developmental role during NMJ expansion, but is also involved in acute, activity-dependent remodeling of the NMJ.

Another interesting aspect of the Wg signaling pathway uncovered recently is the observation that Wg secretion from motor neuron axonal terminals requires the evolutionarily conserved segment polarity protein Evenness interrupted (Evi)/Wntless (Wls)/Sprinter (Spr) (Banziger et al. 2006; Bartscherer et al. 2006; Goodman et al. 2006). Similarly to observations in embryos, cell culture, and developing wing discs (Banziger et al. 2006; Bartscherer et al. 2006; Goodman et al. 2006), Wg release from motor neuron terminals at the NMJ requires Evi (Fig. 4A) (Korkut et al. 2009). Mutations in Evi or specific Evi down-regulation in motor neurons led to drastic reduction in Wg release from synaptic boutons, and phenocopied the wg^{ts} phenotypes. Because Evi, a multipass transmembrane protein, was shown to cross the synapse with Wg (Korkut et al. 2009), it was suggested that Evi-containing microvesicles, possibly exosomes, play a role in Wg transfer at the synapse (Korkut et al. 2009). This mechanism addressed a long-lasting problem in the transmission of Wnt signals, this being in the formation of gradients during pattern formation in the embryo or in the traffic of Wnts from presynaptic sites to relatively distant sites at the subsynaptic reticulum, a complex system of muscle-derived postsynaptic membranes, where DFz2 receptors are localized. Although the primary amino acid sequence of Wnts predicts a hydrophilic molecule, Wnts are actually hydrophobic in nature, because of posttranslational lipid modifications (Willert et al. 2003; Takada et al. 2006). Thus, Wnts are not readily diffusible in the extracellular space. Several mechanisms have been documented to be involved in the diffusion of Wnt signals, including the association of Wnt proteins with lipoprotein particles (Panakova et al. 2005), the transmission of argosomes (Greco et al. 2001), cytonemes (Ramirez-Weber and Kornberg 1999), and, very recently, Secreted Wg-interacting molecule (Swim), a member of the Lipocalin family that is suggested to maintain Wg in a soluble form for long-range signaling (Mulligan et al. 2012). The studies at the NMJ suggest yet another mechanism, the shuttling of Wnts associated with Evi, which is released in vesicles (Fig. 4A). Most recently, analysis of the Evi vesicles showed that they indeed correspond to exosomes (K Koles and V Budnik, unpubl.). Exosomes are released by the fusion of multivesicular bodies with the plasma membrane, thus releasing their vesicular content. Our initial studies indicate that Evi-exosome release from presynaptic terminals requires Rab11, Myosin 5, and Syntaxin-1A, suggesting that exosome release is an integral component of the synaptic bouton release machinery.

Most recently, Wg signaling has been implicated in negatively regulating homeostatic compensation (Marie et al. 2010). Neurotransmitter release and postsynaptic muscle depolarization are tightly coupled processes, and their

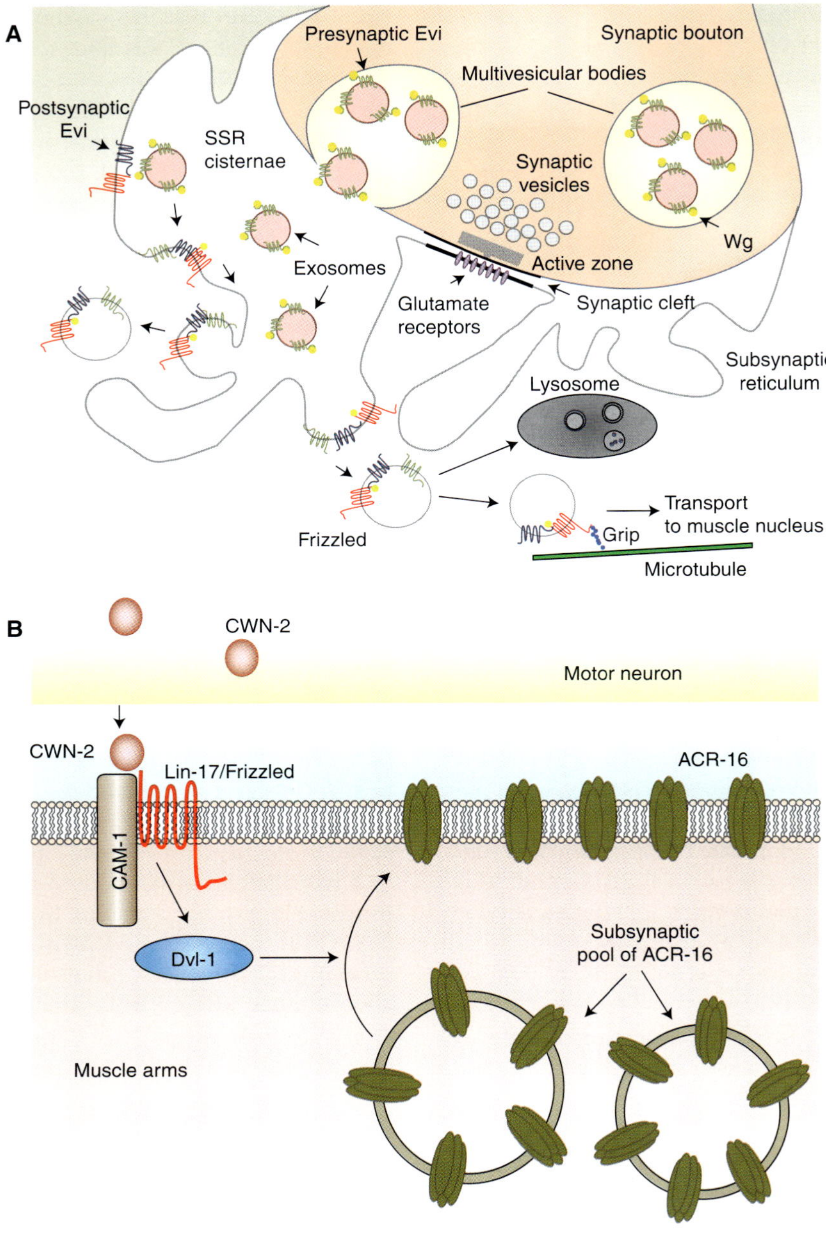

Figure 4. Evi-mediated synaptic release of *Drosophila* Wg and regulation of ACR-16 receptor abundance in postsynaptic muscle membranes by CWN-2 in *C. elegans*. (*A*) Schematic representation of Evi-mediated Wg release at *Drosophila* larval NMJs. Wg carried by Evi-exosomes is sorted into multivesicular bodies in synaptic terminals that fuse with the synaptic membrane, releasing their content to the cisternae of underlying subsynaptic reticulum. Potential fates of postsynaptic exosomes are shown, such as fusion with the postsynaptic membrane and subsequent endocytosis. The endocytosed DFz2/Wg/Evi-containing vesicles can be either sorted and targeted to lysosomes or, via DFz2 binding to Grip, transported near the muscle nucleus (also see Fig. 1E). (*B*) In *C. elegans*, motor neurons release CWN-2, which binds to the heteromeric CAM-1/Lin-17 receptor and activates Dishevelled-1. Activated Dvl-1, in turn, leads to the mobilization and membrane delivery of subsynaptic ACR-16 receptor pools.

coordination is required to maintain synaptic efficacy. The process of synaptic homeostasis involves the presynaptic (increase in neurotransmitter release) or postsynaptic (abundance of glutamate receptors) compensatory adjustment to changes in postsynaptic or presynaptic parameters, resulting in the restoration of wild-type synaptic efficacy. Synaptic homeostasis is thought to occur in two phases, an induction phase that is transcription and protein synthesis independent, and a maintenance phase that requires transcription and the presynaptic BMP-signaling pathway (Goold and Davis 2007). Marie et al. (2010) have recently shown that *Goosberry* (*Gsb*) is required for the maintenance phase of this homeostatic compensation process. Using the $GluRIIA^{sp16}$-null mutant background (Petersen et al. 1997), which leads to a compensatory increase in presynaptic neurotransmitter release due to a decrease in the amplitude of spontaneous mEPSPs, the investigators discovered that this compensation was significantly reduced in heterozygous gsb^{01155} mutants (Marie et al. 2010). Interestingly, *Gsb* also functions early during embryonic development to inhibit Wg signaling in the Wg-secreting cell (Duman-Scheel et al. 1997). Indeed, when the investigators examined the possibility of whether *Gsb* may function in homeostatic compensation in motor neurons by inhibiting presynaptic Wg signaling, they found that replacing one copy of the wild-type *wg* gene with a temperature-sensitive wg^{ts} allele in the $gsb^{01155}/+$ $GluRIIA^{sp16}$ background restored homeostatic compensation (Marie et al. 2010). Therefore, in addition to its other structural and functional roles at the NMJ, these results raise the interesting possibility that Wg, along with BMP signaling, regulates synaptic homeostasis (Marie et al. 2010).

In addition to Wingless, two additional Wnt family members—Wnt5, its postsynaptic receptor Derailed (Drl) (Liebl et al. 2008), and Wnt2 (Liebl et al. 2010)—were also shown to regulate NMJ development. Mutations in *Wnt5* led to structural as well as functional defects at the NMJ. Bouton numbers and NMJ areas were significantly decreased in *wnt5* mutants, and the number and density of active zones, as assessed by Bruchpilot (Brp) immunolabeling, was also significantly lower. Interestingly, only neuronal Wnt5 expression was able to rescue bouton numbers to wild type, but either neuronal or muscle-specific rescue restored the number and density of active zones, suggesting a potential retrograde function for Wnt5 in regulating active-zone formation. Electrophysiological examination of *wnt5* mutant synapses revealed a 40% decrease in evoked excitatory junctional currents (EJCs) and a similar decrease in the frequency of mEJCs (Liebl et al. 2008). Because the amplitude of mEJCs (which reflects postsynaptic glutamate receptor function) was not affected in *wnt5* mutants, these data point toward the presynaptic function of Wnt5. Earlier studies have shown that Wnt5 can bind Derailed (Drl) (Yoshikawa et al. 2003). Indeed, *drl* mutant NMJs manifest similar structural defects to *wnt5* mutants, such as a significant decrease in bouton number (Liebl et al. 2008). This phenotype could be rescued by postsynaptic expression of Drl. However, there were also differences in the phenotypes of *drl* and *wnt5* mutant larvae. *drl* mutations, unlike *wnt5* mutations, did not result in a decrease in active-zone density, nor in functional changes in synaptic transmission, suggesting that some aspects of Wnt5 function are Drl independent. It is interesting to note, however, that presynaptic expression of Wnt5 in the wild-type background led to a significant increase in bouton number, and that this increase was dependent on Drl, because in the *drl* mutant background no change in bouton number was observed upon neuronal Wnt5 overexpression. In summary, Wnt5 is posed to play a complex structural and functional role at the larval NMJ, in part by signaling via the Drl receptor, while also showing Drl-independent effects.

Subsequently, the same researchers have shown that *Drosophila Wnt2*, the ortholog of mammalian *Wnt7b*, is required for proper NMJ development and for the maintenance of proper Brp density and intensity at synapses (Liebl et al. 2010). In the absence of Wnt2, presynaptic arbors branch more frequently, although the number of boutons is unchanged, indicating that unlike other Wnts, Wnt2 negatively regulates

NMJ branching. The total number of Brp puncta did not change in *Wnt2* mutants. However, the density of Brp puncta decreased significantly, as expected from the increased NMJ arbor size in these mutants. A combination of rescue experiments with Wnt2 overexpression and tissue-specific knockdown of *Wnt2* suggests that it functions in the muscle, most likely in a retrograde manner to negatively regulate presynaptic growth. In addition to the structural phenotypes, *Wnt2* mutants showed a decrease in the mean intensity of Synaptobrevin (Syb) and Brp immunoreactivity. In agreement with these observations, the amplitude of evoked excitatory junctional potentials was also reduced, whereas the frequency and amplitude of spontaneous vesicle release were unaffected. It is as yet unclear what receptor Wnt2 signals through at the NMJ. Previous studies implicated Fz, DFz2, and Fz3 as potential receptors for Wnt2 (Wu and Nusse 2002), but genetic studies did not affirm these possibilities at the larval NMJ (Liebl et al. 2010).

THE ROLE OF WNTs AT *C. elegans* NMJs

The muscles of *C. elegans* are innervated by excitatory cholinergic or inhibitory GABAergic neurons. Unlike *Drosophila* or vertebrate NMJs, muscles extend "muscle arms" toward the nerve cord to form en passant synapses (Dixon and Roy 2005). Two types of acetylcholine receptors (AChRs) have been identified in *C. elegans*, the N-type, likely a homomer of ACR-16 subunits (Francis et al. 2005) and activated by nicotine, and the L-type channels, which are heteropentamers and sensitive to levamisole. Two distinct mechanisms have evolved to localize these two types of receptors. N-type AChRs require CAM-1, a MuSK-related orphan receptor tyrosine kinase (Francis et al. 2005), and N-type AChRs require an extracellular scaffold in the synaptic cleft, composed of secreted Lev-9 (Gendrel et al. 2009), the extracellular domain of Lev-10 (Gally et al. 2004), and a small, secreted single immunoglobulin containing OIG-4 (Rapti et al. 2011).

Although in vertebrates and *Drosophila*, initial discoveries revealed pro-synaptogenic functions for Wnts (even though subsequent experiments identified negatively acting Wnt signals as well), the first *C. elegans* Wnt signaling pathway involved in regulating synapse formation served to restrict the position of synapses on axons. For a while these findings led to the notion that Wnts were inhibitory in *C. elegans*; however, some of the latest discoveries seem to dispel this notion (see below).

The first Wnt discovered to function in *C. elegans* neuromuscular junction formation was Lin-44 (Klassen and Shen 2007). Lin-44/Wnt is secreted from four hypodermal cells during embryonic and larval development. In *lin-44/wnt* mutants, ectopic synapses are formed, indicating that *lin-44/wnt* inhibits synapse formation. Lin-44/Wnt functions in DA9 motor neurons, where it recruits the Wnt receptor Lin-17/Frizzled to a subdomain of the DA9 axon via binding to the cysteine-rich domain (CRD) of Lin-17/Frizzled. Lin-44/Wnt cooperates with another *C. elegans* Wnt, EGL-20/Wnt, and by binding to and signaling through Lin-17/Frizzled, it prevents synapse formation at this region of the DA9 axon (Klassen and Shen 2007). Interestingly, except for Dvl, none of the known canonical, planar cell polarity, or calcium-dependent Wnt signaling pathway effectors were found to be involved in this process, suggesting that novel effectors mediate the local inhibition of synaptogenesis on DA9 motor neuron axons.

An additional aspect of Wnt signaling regulation at the *C. elegans* NMJ was revealed by the findings of Babu et al. (2011). In an RNAi screen for cell surface Ig-domain-containing (IgSF) proteins that are involved in modulating sensitivity to aldicarb (a cholinesterase inhibitor), Babu and colleagues identified RIG-3, a GPI-anchored IgSF-domain protein. Decreased levels of this protein, which is normally expressed in cholinergic neurons, led to heightened sensitivity to aldicarb. When the investigators examined the underlying mechanisms for this increased sensitivity, they found that in the absence of RIG-3, the delivery of ACR-16 (the homomer-forming subunit of nicotinic AChRs) from muscle mobile stores to the muscle membrane was significantly increased following aldicarb treatment. Furthermore, this increase was the result of increased levels of

CAM-1, which is an unconventional Wnt receptor of the Ror-type receptor tyrosine kinase family (Green et al. 2008) involved in ACR-16 delivery to synaptic sites (Francis et al. 2005). Interestingly, CAM-1 binds Wnt ligands, and mutations that reduce secreted Wnt levels, such as *mig-14*, led to the elimination of aldicarb-induced hypersensitivity in *rig-3*, *mig-14* double mutants, indicating that CAM-1 was mediating this phenotype in a Wnt-dependent manner. These data show that RIG-3, a presynaptic molecule, negatively regulates CAM-1-mediated Wnt signaling at ACR-16-containing synapses following aldicarb treatment and thus may play a role in stabilizing synapses and synaptic circuits.

As mentioned above, the first discovery of *C. elegans* Wnt signaling in NMJ formation revealed the negative regulation of synapse formation, suggesting that Wnts may be inhibitory in *C. elegans*. However, a new study from Jensen et al. (2012) seems to dispel this notion. In a series of elegant and groundbreaking studies, the researchers showed that neuronally released CWN-2 (but not CWN-1 or EGL-20), by binding to the CAM-1/Lin-17 heteromeric receptor (but not those containing MIG-1 or CFZ-2), resulted in the increased postsynaptic muscle membrane delivery of ARC-16 (Fig. 4B). This signaling required ongoing CWN-2 release, because transient CWN-2 pulses led to only transient increases in postsynaptic ACR-16. For the first time in *C. elegans*, the investigators also showed that this Wnt signaling pathway was activity dependent. Channelrhodopsin2 stimulation of neurons led to an increased translocation of ACR-16 to the muscle membrane and a corresponding increase in ACR-16-mediated current, and these effects required the neuronal expression of CWN-2. In summary, an ongoing, activity-dependent CWN-2 release and signaling through CAM-1/Lin-17 receptors is required for activity-dependent synaptic plasticity and the maintenance of proper synaptic transmission in adult *C. elegans* neuromuscular junctions. These observations also point out the similarities of Wnt function at the NMJ across species. The activity-dependent release and anterograde function of CWN-2 are strikingly similar to what has been described for Wg at *Drosophila* larval NMJs (Ataman et al. 2008). In a broader context, these new results highlight that Wnts do not only function during development, but are also required throughout the life span of organisms for proper synaptic functioning.

CONCLUDING REMARKS

In the last decade, we have witnessed a large number of new discoveries regarding the function of Wnt signaling at neuromuscular synapse formation and function. From these findings, it is apparent that Wnts are intricately involved in almost every aspect of NMJ development, maintenance, and function, including activity-dependent plasticity. They can play both essential and modulatory roles and promote or inhibit synapse formation. As surmised by Speese and Budnik (2007) and van Amerongen and Nusse (2009), the likely scenario of Wnt action is that of a parallel, combinatorial network of signals emanating from the neuron, muscle, and surrounding non-neuronal cells that work in anterograde, retrograde, and autocrine manner to effect and coordinate the developmental and synaptic plasticity-induced processes.

One of the key challenges of the Wnt signaling field will be to reconcile data from in vivo and in vitro experiments. For example, Dvl1 was identified via its MuSK interaction and was shown to be required for Agrin-induced AChR clustering, yet the *Dvl1* mutant mouse phenotypes are surprisingly mild compared with the Dvl1 knockdown in C2C12 cell cultures (Luo et al. 2002). Furthermore, in muscle-specific β-catenin$^{-/-}$ knockout muscle cultures, Agrin-mediated AChR clustering was not affected (Li et al. 2008), unlike in the C2C12 cultured cell line, where β-catenin levels were reduced via shRNA treatment (Zhang et al. 2007; Wang et al. 2008). These data highlight the difficulties involved in translating the cell culture findings to in vivo situations in a living/developing organism. These data illustrate that multiple pathways may converge at the level of β-catenin and that the preceding developmental history of the cell could lead to diverse compensatory mechanisms or adaptations that may lead to conflicting results if these factors are not controlled for.

Among the exciting future challenges will be the identification of retrograde signaling mechanisms that were revealed through examining the muscle specific β-catenin$^{-/-}$ NMJs (Li et al. 2008). Likewise, because many aspects of Wnt signaling are conserved across species, it will be exciting to test whether the continuous, ongoing requirement for *C. elegans* Wnt, CWN-2, in synaptic plasticity (Jensen et al. 2012) is also at play in other model organisms, such as the fly, zebrafish, or other vertebrates. Furthermore, the contribution to NMJ development and function of terminal Schwann cells in vertebrates (Feng and Ko 2008; Sugiura and Lin 2011) is poised to lead to novel insights.

In addition, future research is likely to lead to an integrated model of multiple Wnt signaling pathways (van Amerongen and Nusse 2009) that are acting simultaneously and synergistically from multiple tissue sources (nerve, muscle, somites, glia) to sculpt the synapse. Current efforts naturally focus on delineating single, rather than multiple, linear signaling pathway effects (e.g., as described above, Wnt1, Wnt5, and Wnt2 are all involved in *Drosophila* larval NMJ development); however, their integration into a dynamic, spatio-temporal map of signaling events will transform our understanding of NMJ development and function. Given how far the field has come in the last few decades and the continuous technical advances allowing more nuanced experiments to be performed, future developments in the field of Wnt signaling at the neuromuscular synapse are bound to be exciting.

ACKNOWLEDGMENTS

We thank Dr. Cassandra Brewer for critical review of the manuscript. V.B. is supported by NIH grants RO1 MH070000 and R01 NS063228.

REFERENCES

Anderson MJ, Cohen MW. 1977. Nerve-induced and spontaneous redistribution of acetylcholine receptors on cultured muscle cells. *J Physiol* **268:** 757–773.

Aoki K, Taketo MM. 2007. Adenomatous polyposis coli (APC): A multi-functional tumor suppressor gene. *J Cell Sci* **120:** 3327–3335.

Ataman B, Ashley J, Gorczyca D, Gorczyca M, Mathew D, Wichmann C, Sigrist SJ, Budnik V. 2006. Nuclear trafficking of *Drosophila* Frizzled-2 during synapse development requires the PDZ protein dGRIP. *Proc Natl Acad Sci* **103:** 7841–7846.

Ataman B, Ashley J, Gorczyca M, Ramachandran P, Fouquet W, Sigrist SJ, Budnik V. 2008. Rapid activity-dependent modifications in synaptic structure and function require bidirectional Wnt signaling. *Neuron* **57:** 705–718.

Babu K, Hu Z, Chien SC, Garriga G, Kaplan JM. 2011. The immunoglobulin super family protein RIG-3 prevents synaptic potentiation and regulates Wnt signaling. *Neuron* **71:** 103–116.

Banziger C, Soldini D, Schutt C, Zipperlen P, Hausmann G, Basler K. 2006. Wntless, a conserved membrane protein dedicated to the secretion of Wnt proteins from signaling cells. *Cell* **125:** 509–522.

Barrow JR, Thomas KR, Boussadia-Zahui O, Moore R, Kemler R, Capecchi MR, McMahon AP. 2003. Ectodermal Wnt3/β-catenin signaling is required for the establishment and maintenance of the apical ectodermal ridge. *Genes Dev* **17:** 394–409.

Bartscherer K, Pelte N, Ingelfinger D, Boutros M. 2006. Secretion of Wnt ligands requires Evi, a conserved transmembrane protein. *Cell* **125:** 523–533.

Bergamin E, Hallock PT, Burden SJ, Hubbard SR. 2010. The cytoplasmic adaptor protein Dok7 activates the receptor tyrosine kinase MuSK via dimerization. *Mol Cell* **39:** 100–109.

Budnik V, Zhong Y, Wu CF. 1990. Morphological plasticity of motor axons in *Drosophila* mutants with altered excitability. *J Neurosci* **10:** 3754–3768.

Burden SJ. 2011. SnapShot: Neuromuscular junction. *Cell* **144:** 826–826.e1.

Cadigan KM, Peifer M. 2012. Wnt signaling from development to disease: Insights from model systems. *Cold Spring Harb Perspect Biol* doi: 10.1101.cshperspect.a002881.

Caricasole A, Bakker A, Copani A, Nicoletti F, Gaviraghi G, Terstappen GC. 2005. Two sides of the same coin: Wnt signaling in neurodegeneration and neuro-oncology. *Biosci Rep* **25:** 309–327.

Ciani L, Krylova O, Smalley MJ, Dale TC, Salinas PC. 2004. A divergent canonical WNT-signaling pathway regulates microtubule dynamics: Dishevelled signals locally to stabilize microtubules. *J Cell Biol* **164:** 243–253.

DeChiara TM, Bowen DC, Valenzuela DM, Simmons MV, Poueymirou WT, Thomas S, Kinetz E, Compton DL, Rojas E, Park JS, et al. 1996. The receptor tyrosine kinase MuSK is required for neuromuscular junction formation in vivo. *Cell* **85:** 501–512.

De Ferrari GV, Inestrosa NC. 2000. Wnt signaling function in Alzheimer's disease. *Brain Res Brain Res Rev* **33:** 1–12.

Dickinson HG. 1971. Nucleo-cytoplasmic interaction following meiosis in the young microspores of *Lilium longiflorum*; events at the nuclear envelope. *Grana* **11:** 117–127.

Dixon SJ, Roy PJ. 2005. Muscle arm development in *Caenorhabditis elegans*. *Development* **132:** 3079–3092.

Dong H, O'Brien RJ, Fung ET, Lanahan AA, Worley PF, Huganir RL. 1997. GRIP: A synaptic PDZ domain-containing protein that interacts with AMPA receptors. *Nature* **386:** 279–284.

Duman-Scheel M, Li X, Orlov I, Noll M, Patel NH. 1997. Genetic separation of the neural and cuticular patterning functions of gooseberry. *Development* **124:** 2855–2865.

Farias GG, Valles AS, Colombres M, Godoy JA, Toledo EM, Lukas RJ, Barrantes FJ, Inestrosa NC. 2007. Wnt-7a induces presynaptic colocalization of α7-nicotinic acetylcholine receptors and adenomatous polyposis coli in hippocampal neurons. *J Neurosci* **27:** 5313–5325.

Feng Z, Ko CP. 2008. The role of glial cells in the formation and maintenance of the neuromuscular junction. *Ann NY Acad Sci* **1132:** 19–28.

Flanagan-Steet H, Fox MA, Meyer D, Sanes JR. 2005. Neuromuscular synapses can form in vivo by incorporation of initially aneural postsynaptic specializations. *Development* **132:** 4471–4481.

Fradkin LG, Dura JM, Noordermeer JN. 2010. Ryks: New partners for Wnts in the developing and regenerating nervous system. *Trends Neurosci* **33:** 84–92.

Frail DE, McLaughlin LL, Mudd J, Merlie JP. 1988. Identification of the mouse muscle 43,000-dalton acetylcholine receptor-associated protein (RAPsyn) by cDNA cloning. *J Biol Chem* **263:** 15602–15607.

Francis MM, Evans SP, Jensen M, Madsen DM, Mancuso J, Norman KR, Maricq AV. 2005. The Ror receptor tyrosine kinase CAM-1 is required for ACR-16-mediated synaptic transmission at the *C. elegans* neuromuscular junction. *Neuron* **46:** 581–594.

Franco B, Bogdanik L, Bobinnec Y, Debec A, Bockaert J, Parmentier ML, Grau Y. 2004. Shaggy, the homolog of glycogen synthase kinase 3, controls neuromuscular junction growth in *Drosophila*. *J Neurosci* **24:** 6573–6577.

Gally C, Eimer S, Richmond JE, Bessereau JL. 2004. A transmembrane protein required for acetylcholine receptor clustering in *Caenorhabditis elegans*. *Nature* **431:** 578–582.

Gay H. 1956. Nucleocytoplasmic relations in *Drosophila*. *Cold Spring Harb Symp Quant Biol* **21:** 257–269.

Gendrel M, Rapti G, Richmond JE, Bessereau JL. 2009. A secreted complement-control-related protein ensures acetylcholine receptor clustering. *Nature* **461:** 992–996.

Gesemann M, Denzer AJ, Ruegg MA. 1995. Acetylcholine receptor-aggregating activity of agrin isoforms and mapping of the active site. *J Cell Biol* **128:** 625–636.

Glass DJ, Bowen DC, Stitt TN, Radziejewski C, Bruno J, Ryan TE, Gies DR, Shah S, Mattsson K, Burden SJ, et al. 1996. Agrin acts via a MuSK receptor complex. *Cell* **85:** 513–523.

Gogel S, Wakefield S, Tear G, Klambt C, Gordon-Weeks PR. 2006. The *Drosophila* microtubule associated protein Futsch is phosphorylated by Shaggy/Zeste-white 3 at an homologous GSK3β phosphorylation site in MAP1B. *Mol Cell Neurosci* **33:** 188–199.

Gomez AM, Burden SJ. 2011. The extracellular region of Lrp4 is sufficient to mediate neuromuscular synapse formation. *Dev Dyn* **240:** 2626–2633.

Goodman RM, Thombre S, Firtina Z, Gray D, Betts D, Roebuck J, Spana EP, Selva EM. 2006. Sprinter: A novel transmembrane protein required for Wg secretion and signaling. *Development* **133:** 4901–4911.

Goold CP, Davis GW. 2007. The BMP ligand Gbb gates the expression of synaptic homeostasis independent of synaptic growth control. *Neuron* **56:** 109–123.

Gorczyca M, Augart C, Budnik V. 1993. Insulin-like receptor and insulin-like peptide are localized at neuromuscular junctions in *Drosophila*. *J Neurosci* **13:** 3692–3704.

Greco V, Hannus M, Eaton S. 2001. Argosomes: A potential vehicle for the spread of morphogens through epithelia. *Cell* **106:** 633–645.

Green JL, Kuntz SG, Sternberg PW. 2008. Ror receptor tyrosine kinases: Orphans no more. *Trends Cell Biol* **18:** 536–544.

Hadek R, Swift H. 1962. Nuclear extrusion and intracisternal inclusions in the rabbit blastocyst. *J Cell Biol* **13:** 445–451.

Hall AC, Lucas FR, Salinas PC. 2000. Axonal remodeling and synaptic differentiation in the cerebellum is regulated by WNT-7a signaling. *Cell* **100:** 525–535.

Hallock PT, Xu CF, Park TJ, Neubert TA, Curran T, Burden SJ. 2010. Dok-7 regulates neuromuscular synapse formation by recruiting Crk and Crk-L. *Genes Dev* **24:** 2451–2461.

He X. 2012. Wnt receptors and signaling mechanisms. *Cold Spring Harb Perspect Biol* doi: 10.1101/cshperspect.a007880.

He X, Semenov M, Tamai K, Zeng X. 2004. LDL receptor-related proteins 5 and 6 in Wnt/β-catenin signaling: Arrows point the way. *Development* **131:** 1663–1677.

Henriquez JP, Webb A, Bence M, Bildsoe H, Sahores M, Hughes SM, Salinas PC. 2008. Wnt signaling promotes AChR aggregation at the neuromuscular synapse in collaboration with agrin. *Proc Natl Acad Sci* **105:** 18812–18817.

Hochstrasser M, Sedat JW. 1987. Three-dimensional organization of *Drosophila melanogaster* interphase nuclei. II. Chromosome spatial organization and gene regulation. *J Cell Biol* **104:** 1471–1483.

Hoogenraad CC, Milstein AD, Ethell IM, Henkemeyer M, Sheng M. 2005. GRIP1 controls dendrite morphogenesis by regulating EphB receptor trafficking. *Nat Neurosci* **8:** 906–915.

Inoue A, Setoguchi K, Matsubara Y, Okada K, Sato N, Iwakura Y, Higuchi O, Yamanashi Y. 2009. Dok-7 activates the muscle receptor kinase MuSK and shapes synapse formation. *Sci Signal* **2:** ra7. doi: 10.1126/scisignal.2000113.

Jennings CG, Dyer SM, Burden SJ. 1993. Muscle-specific trk-related receptor with a kringle domain defines a distinct class of receptor tyrosine kinases. *Proc Natl Acad Sci* **90:** 2895–2899.

Jensen M, Hoerndli FJ, Brockie PJ, Wang R, Johnson E, Maxfield D, Francis MM, Madsen DM, Maricq AV. 2012. Wnt signaling regulates acetylcholine receptor translocation and synaptic plasticity in the adult nervous system. *Cell* (in press).

Jing L, Lefebvre JL, Gordon LR, Granato M. 2009. Wnt signals organize synaptic prepattern and axon guidance through the zebrafish unplugged/MuSK receptor. *Neuron* **61:** 721–733.

Johnson EB, Hammer RE, Herz J. 2005. Abnormal development of the apical ectodermal ridge and polysyndactyly in Megf7-deficient mice. *Hum Mol Genet* **14:** 3523–3538.

Kim N, Stiegler AL, Cameron TO, Hallock PT, Gomez AM, Huang JH, Hubbard SR, Dustin ML, Burden SJ. 2008. Lrp4 is a receptor for Agrin and forms a complex with MuSK. *Cell* **135:** 334–342.

Klassen MP, Shen K. 2007. Wnt signaling positions neuromuscular connectivity by inhibiting synapse formation in *C. elegans*. *Cell* **130:** 704–716.

Koon AC, Ashley J, Barria R, Dasgupta S, Brain R, Waddell S, Alkema MJ, Budnik V. 2011. Autoregulatory and paracrine control of synaptic and behavioral plasticity by octopaminergic signaling. *Nat Neurosci* **14:** 190–199.

Korinek V, Barker N, Morin PJ, van Wichen D, de Weger R, Kinzler KW, Vogelstein B, Clevers H. 1997. Constitutive transcriptional activation by a β-catenin-Tcf complex in $APC^{-/-}$ colon carcinoma. *Science* **275:** 1784–1787.

Korkut C, Ataman B, Ramachandran P, Ashley J, Barria R, Gherbesi N, Budnik V. 2009. Trans-synaptic transmission of vesicular Wnt signals through Evi/Wntless. *Cell* **139:** 393–404.

Krylova O, Herreros J, Cleverley KE, Ehler E, Henriquez JP, Hughes SM, Salinas PC. 2002. WNT-3, expressed by motoneurons, regulates terminal arborization of neurotrophin-3-responsive spinal sensory neurons. *Neuron* **35:** 1043–1056.

Kuhl M, Sheldahl LC, Park M, Miller JR, Moon RT. 2000. The Wnt/Ca^{2+} pathway: A new vertebrate Wnt signaling pathway takes shape. *Trends Genet* **16:** 279–283.

Kummer TT, Misgeld T, Sanes JR. 2006. Assembly of the postsynaptic membrane at the neuromuscular junction: Paradigm lost. *Curr Opin Neurobiol* **16:** 74–82.

Lahey T, Gorczyca M, Jia XX, Budnik V. 1994. The *Drosophila* tumor suppressor gene *dlg*is required for normal synaptic bouton structure. *Neuron* **13:** 823–835.

Li XM, Dong XP, Luo SW, Zhang B, Lee DH, Ting AK, Neiswender H, Kim CH, Carpenter-Hyland E, Gao TM, et al. 2008. Retrograde regulation of motoneuron differentiation by muscle β-catenin. *Nat Neurosci* **11:** 262–268.

Liebl FL, Wu Y, Featherstone DE, Noordermeer JN, Fradkin L, Hing H. 2008. Derailed regulates development of the *Drosophila* neuromuscular junction. *Dev Neurobiol* **68:** 152–165.

Liebl FL, McKeown C, Yao Y, Hing HK. 2010. Mutations in Wnt2 alter presynaptic motor neuron morphology and presynaptic protein localization at the *Drosophila* neuromuscular junction. *PLoS ONE* **5:** e12778.

Lin W, Burgess RW, Dominguez B, Pfaff SL, Sanes JR, Lee KF. 2001. Distinct roles of nerve and muscle in postsynaptic differentiation of the neuromuscular synapse. *Nature* **410:** 1057–1064.

Lin W, Dominguez B, Yang J, Aryal P, Brandon EP, Gage FH, Lee KF. 2005. Neurotransmitter acetylcholine negatively regulates neuromuscular synapse formation by a Cdk5-dependent mechanism. *Neuron* **46:** 569–579.

Logan CY, Nusse R. 2004. The Wnt signaling pathway in development and disease. *Annu Rev Cell Dev Biol* **20:** 781–810.

Lucas FR, Salinas PC. 1997. WNT-7a induces axonal remodeling and increases synapsin I levels in cerebellar neurons. *Dev Biol* **192:** 31–44.

Lucas FR, Goold RG, Gordon-Weeks PR, Salinas PC. 1998. Inhibition of GSK-3β leading to the loss of phosphorylated MAP-1B is an early event in axonal remodelling induced by WNT-7a or lithium. *J Cell Sci* **111:** 1351–1361.

Luo ZG, Wang Q, Zhou JZ, Wang J, Luo Z, Liu M, He X, Wynshaw-Boris A, Xiong WC, Lu B, et al. 2002. Regulation of AChR clustering by Dishevelled interacting with MuSK and PAK1. *Neuron* **35:** 489–505.

Lyu J, Yamamoto V, Lu W. 2008. Cleavage of the Wnt receptor Ryk regulates neuronal differentiation during cortical neurogenesis. *Dev Cell* **15:** 773–780.

Marie B, Pym E, Bergquist S, Davis GW. 2010. Synaptic homeostasis is consolidated by the cell fate gene gooseberry, a *Drosophila pax3/7* homolog. *J Neurosci* **30:** 8071–8082.

Masiakowski P, Yancopoulos GD. 1998. The Wnt receptor CRD domain is also found in MuSK and related orphan receptor tyrosine kinases. *Curr Biol* **8:** R407.

Mathew D, Gramates LS, Packard M, Thomas U, Bilder D, Perrimon N, Gorczyca M, Budnik V. 2002. Recruitment of scribble to the synaptic scaffolding complex requires GUK-holder, a novel DLG binding protein. *Curr Biol* **12:** 531–539.

Mathew D, Ataman B, Chen J, Zhang Y, Cumberledge S, Budnik V. 2005. Wingless signaling at synapses is through cleavage and nuclear import of receptor DFrizzled2. *Science* **310:** 1344–1347.

Mettenleiter TC, Klupp BG, Granzow H. 2006. Herpesvirus assembly: A tale of two membranes. *Curr Opin Microbiol* **9:** 423–429.

Miech C, Pauer HU, He X, Schwarz TL. 2008. Presynaptic local signaling by a canonical wingless pathway regulates development of the *Drosophila* neuromuscular junction. *J Neurosci* **28:** 10875–10884.

Mlodzik M. 2002. Planar cell polarization: Do the same mechanisms regulate *Drosophila* tissue polarity and vertebrate gastrulation? *Trends Genet* **18:** 564–571.

Moransard M, Borges LS, Willmann R, Marangi PA, Brenner HR, Ferns MJ, Fuhrer C. 2003. Agrin regulates rapsyn interaction with surface acetylcholine receptors, and this underlies cytoskeletal anchoring and clustering. *J Biol Chem* **278:** 7350–7359.

Mosca TJ, Schwarz TL. 2010. The nuclear import of Frizzled2-C by Importins-β11 and α2 promotes postsynaptic development. *Nat Neurosci* **13:** 935–943.

Moseley JB, Bartolini F, Okada K, Wen Y, Gundersen GG, Goode BL. 2007. Regulated binding of adenomatous polyposis coli protein to actin. *J Biol Chem* **282:** 12661–12668.

Mulligan KA, Fuerer C, Ching W, Fish M, Willert K, Nusse R. 2012. Secreted Wingless-interacting molecule (Swim) promotes long-range signaling by maintaining Wingless solubility. *Proc Natl Acad Sci* **109:** 370–377.

Nagel G, Ollig D, Fuhrmann M, Kateriya S, Musti AM, Bamberg E, Hegemann P. 2002. Channelrhodopsin-1: A light-gated proton channel in green algae. *Science* **296:** 2395–2398.

Niehrs C. 2006. Function and biological roles of the Dickkopf family of Wnt modulators. *Oncogene* **25:** 7469–7481.

Okada K, Inoue A, Okada M, Murata Y, Kakuta S, Jigami T, Kubo S, Shiraishi H, Eguchi K, Motomura M, et al. 2006. The muscle protein Dok-7 is essential for neuromuscular synaptogenesis. *Science* **312:** 1802–1805.

Okerlund ND, Cheyette BN. 2011. Synaptic Wnt signaling—a contributor to major psychiatric disorders? *J Neurodev Disord* **3:** 162–174.

Packard M, Koo ES, Gorczyca M, Sharpe J, Cumberledge S, Budnik V. 2002. The Drosophila Wnt, Wingless, provides an essential signal for pre- and postsynaptic differentiation. *Cell* **111:** 319–330.

Panakova D, Sprong H, Marois E, Thiele C, Eaton S. 2005. Lipoprotein particles are required for Hedgehog and Wingless signalling. *Nature* **435:** 58–65.

Petersen SA, Fetter RD, Noordermeer JN, Goodman CS, DiAntonio A. 1997. Genetic analysis of glutamate receptors in *Drosophila* reveals a retrograde signal regulating presynaptic transmitter release. *Neuron* **19:** 1237–1248.

Pinson KI, Brennan J, Monkley S, Avery BJ, Skarnes WC. 2000. An LDL-receptor-related protein mediates Wnt signalling in mice. *Nature* **407:** 535–538.

Ramirez-Weber FA, Kornberg TB. 1999. Cytonemes: Cellular processes that project to the principal signaling center in *Drosophila* imaginal discs. *Cell* **97:** 599–607.

Rapti G, Richmond J, Bessereau JL. 2011. A single immunoglobulin-domain protein required for clustering acetylcholine receptors in *C. elegans*. *EMBO J* **30:** 706–718.

Rattner A, Hsieh JC, Smallwood PM, Gilbert DJ, Copeland NG, Jenkins NA, Nathans J. 1997. A family of secreted proteins contains homology to the cysteine-rich ligand-binding domain of frizzled receptors. *Proc Natl Acad Sci* **94:** 2859–2863.

Reilein A, Nelson WJ. 2005. APC is a component of an organizing template for cortical microtubule networks. *Nat Cell Biol* **7:** 463–473.

Roller RJ. 2008. Nuclear egress of herpes viruses. *Virol Sin* **23:** 406–415.

Roos J, Hummel T, Ng N, Klambt C, Davis GW. 2000. *Drosophila* Futsch regulates synaptic microtubule organization and is necessary for synaptic growth. *Neuron* **26:** 371–382.

Rosenberg MM, Yang F, Giovanni M, Mohn JL, Temburni MK, Jacob MH. 2008. Adenomatous polyposis coli plays a key role, in vivo, in coordinating assembly of the neuronal nicotinic postsynaptic complex. *Mol Cell Neurosci* **38:** 138–152.

Ruiz-Canada C, Ashley J, Moeckel-Cole S, Drier E, Yin J, Budnik V. 2004. New synaptic bouton formation is disrupted by misregulation of microtubule stability in aPKC mutants. *Neuron* **42:** 567–580.

Sanes JR, Lichtman JW. 2001. Induction, assembly, maturation and maintenance of a postsynaptic apparatus. *Nat Rev Neurosci* **2:** 791–805.

Schuster CM, Davis GW, Fetter RD, Goodman CS. 1996. Genetic dissection of structural and functional components of synaptic plasticity. I. Fasciclin II controls synaptic stabilization and growth. *Neuron* **17:** 641–654.

Sigrist SJ, Reiff DF, Thiel PR, Steinert JR, Schuster CM. 2003. Experience-dependent strengthening of *Drosophila* neuromuscular junctions. *J Neurosci* **23:** 6546–6556.

Smith MA, Yao YM, Reist NE, Magill C, Wallace BG, McMahan UJ. 1987. Identification of agrin in electric organ extracts and localization of agrin-like molecules in muscle and central nervous system. *J Exp Biol* **132:** 223–230.

Speese SD, Budnik V. 2007. Wnts: Up-and-coming at the synapse. *Trends Neurosci* **30:** 268–275.

Speese S, Ashley J, Jokhi V, Nunnari J, Barria R, Li Y, Ataman B, Koon AC, Chang HY, Li Q, et al. 2012. Nuclear envelope budding enables large ribonucleoprotein particle export during synaptic Wnt signaling. *Cell* (in press).

Stamos LJ, Weis WL. 2012. The beta-catenin destruction complex. *Cold Spring Harb Perspect Biol* doi: 10.1101.cshperspect.a007898.

Steinert JR, Kuromi H, Hellwig A, Knirr M, Wyatt AW, Kidokoro Y, Schuster CM. 2006. Experience-dependent formation and recruitment of large vesicles from reserve pool. *Neuron* **50:** 723–733.

Su LK, Burrell M, Hill DE, Gyuris J, Brent R, Wiltshire R, Trent J, Vogelstein B, Kinzler KW. 1995. APC binds to the novel protein EB1. *Cancer Res* **55:** 2972–2977.

Sugiura Y, Lin W. 2011. Neuron–glia interactions: The roles of Schwann cells in neuromuscular synapse formation and function. *Biosci Rep* **31:** 295–302.

Szollosi D. 1965. Extrusion of nucleoli from pronuclei of the rat. *J Cell Biol* **25:** 545–562.

Szollosi MS, Szollosi D. 1988. "Blebbing" of the nuclear envelope of mouse zygotes, early embryos and hybrid cells. *J Cell Sci* **91:** 257–267.

Takada R, Satomi Y, Kurata T, Ueno N, Norioka S, Kondoh H, Takao T, Takada S. 2006. Monounsaturated fatty acid modification of Wnt protein: Its role in Wnt secretion. *Dev Cell* **11:** 791–801.

Tamai K, Semenov M, Kato Y, Spokony R, Liu C, Katsuyama Y, Hess F, Saint-Jeannet JP, He X. 2000. LDL-receptor-related proteins in Wnt signal transduction. *Nature* **407:** 530–535.

Temburni MK, Rosenberg MM, Pathak N, McConnell R, Jacob MH. 2004. Neuronal nicotinic synapse assembly requires the adenomatous polyposis coli tumor suppressor protein. *J Neurosci* **24:** 6776–6784.

Thomas U, Kim E, Kuhlendahl S, Koh YH, Gundelfinger ED, Sheng M, Garner CC, Budnik V. 1997. Synaptic clustering of the cell adhesion molecule fasciclin II by discs-large and its role in the regulation of presynaptic structure. *Neuron* **19:** 787–799.

Tian QB, Suzuki T, Yamauchi T, Sakagami H, Yoshimura Y, Miyazawa S, Nakayama K, Saitoh F, Zhang JP, Lu Y, et al. 2006. Interaction of LDL receptor-related protein 4 (LRP4) with postsynaptic scaffold proteins via its C-terminal PDZ domain-binding motif, and its regulation by Ca/calmodulin-dependent protein kinase II. *Eur J Neurosci* **23:** 2864–2876.

Valenzuela DM, Stitt TN, DiStefano PS, Rojas E, Mattsson K, Compton DL, Nunez L, Park JS, Stark JL, Gies DR, et al. 1995. Receptor tyrosine kinase specific for the skeletal muscle lineage: expression in embryonic muscle, at the neuromuscular junction, and after injury. *Neuron* **15:** 573–584.

Van Amerongen R. 2012. Alternative Wnt pathways and receptors. *Cold Spring Harb Perspect Biol* doi: 10.1101. cshperspect.a007914.

van Amerongen R, Nusse R. 2009. Towards an integrated view of Wnt signaling in development. *Development* **136:** 3205–3214.

Wallace BG. 1986. Aggregating factor from Torpedo electric organ induces patches containing acetylcholine receptors, acetylcholinesterase, and butyrylcholinesterase on cultured myotubes. *J Cell Biol* **102:** 783–794.

Wang J, Jing Z, Zhang L, Zhou G, Braun J, Yao Y, Wang ZZ. 2003. Regulation of acetylcholine receptor clustering by the tumor suppressor APC. *Nat Neurosci* **6:** 1017–1018.

Wang J, Ruan NJ, Qian L, Lei WL, Chen F, Luo ZG. 2008. Wnt/β-catenin signaling suppresses Rapsyn expression and inhibits acetylcholine receptor clustering at the neuromuscular junction. *J Biol Chem* **283:** 21668–21675.

Weatherbee SD, Anderson KV, Niswander LA. 2006. LDL-receptor-related protein 4 is crucial for formation of the neuromuscular junction. *Development* **133:** 4993–5000.

Wehrli M, Dougan ST, Caldwell K, O'Keefe L, Schwartz S, Vaizel-Ohayon D, Schejter E, Tomlinson A, DiNardo S. 2000. *arrow* encodes an LDL-receptor-related protein essential for Wingless signalling. *Nature* **407:** 527–530.

Weston C, Gordon C, Teressa G, Hod E, Ren XD, Prives J. 2003. Cooperative regulation by Rac and Rho of agrin-induced acetylcholine receptor clustering in muscle cells. *J Biol Chem* **278:** 6450–6455.

Willert K, Nusse R. 2012. Wnt Proteins. *Cold Spring Harb Perspect Biol* doi: 10.1101/cshperspect.a007864.

Willert K, Brown JD, Danenberg E, Duncan AW, Weissman IL, Reya T, Yates JR III, Nusse R. 2003. Wnt proteins are lipid-modified and can act as stem cell growth factors. *Nature* **423:** 448–452.

Worman HJ, Bonne G. 2007. "Laminopathies": A wide spectrum of human diseases. *Exp Cell Res* **313:** 2121–2133.

Wu CH, Nusse R. 2002. Ligand receptor interactions in the Wnt signaling pathway in *Drosophila*. *J Biol Chem* **277:** 41762–41769.

Wu H, Xiong WC, Mei L. 2010. To build a synapse: Signaling pathways in neuromuscular junction assembly. *Development* **137:** 1017–1033.

Yang X, Arber S, William C, Li L, Tanabe Y, Jessell TM, Birchmeier C, Burden SJ. 2001. Patterning of muscle acetylcholine receptor gene expression in of motor innervation. *Neuron* **30:** 399–410.

Yoshihara CM, Hall ZW. 1993. Increased expression of the 43-kD protein disrupts acetylcholine receptor clustering in myotubes. *J Cell Biol* **122:** 169–179.

Yoshikawa S, McKinnon RD, Kokel M, Thomas JB. 2003. Wnt-mediated axon guidance via the *Drosophila* Derailed receptor. *Nature* **422:** 583–588.

Zhang J, Lefebvre JL, Zhao S, Granato M. 2004. Zebrafish unplugged reveals a role for muscle-specific kinase homologs in axonal pathway choice. *Nat Neurosci* **7:** 1303–1309.

Zhang B, Luo S, Dong XP, Zhang X, Liu C, Luo Z, Xiong WC, Mei L. 2007. β-Catenin regulates acetylcholine receptor clustering in muscle cells through interaction with Rapsyn. *J Neurosci* **27:** 3968–3973.

Zhang B, Luo S, Wang Q, Suzuki T, Xiong WC, Mei L. 2008. LRP4 serves as a coreceptor of agrin. *Neuron* **60:** 285–297.

Zhang W, Coldefy AS, Hubbard SR, Burden SJ. 2011. Agrin binds to the N-terminal region of Lrp4 protein and stimulates association between Lrp4 and the first immunoglobulin-like domain in muscle-specific kinase (MuSK). *J Biol Chem* **286:** 40624–40630.

Zito K, Fetter RD, Goodman CS, Isacoff EY. 1997. Synaptic clustering of Fascilin II and Shaker: Essential targeting sequences and role of Dlg. *Neuron* **19:** 1007–1016.

Zito K, Parnas D, Fetter RD, Isacoff EY, Goodman CS. 1999. Watching a synapse grow: Noninvasive confocal imaging of synaptic growth in *Drosophila*. *Neuron* **22:** 719–729.

Wnt Signaling in Cancer

Paul Polakis

Genentech, Inc., South San Francisco, California 94608

Correspondence: polakis.paul@gene.com

Aberrant regulation of the Wnt signaling pathway is a prevalent theme in cancer biology. From the earliest observation that Wnt overexpression could lead to malignant transformation of mouse mammary tissue to the most recent genetic discoveries gleaned from tumor genome sequencing, the Wnt pathway continues to evolve as a central mechanism in cancer biology. This article summarizes the evidence supporting a role for Wnt signaling in human cancer. This includes a review of the genetic mutations affecting Wnt pathway components, as well as some of epigenetic mechanisms that alter expression of genes relevant to Wnt. I also highlight some research on the cooperativity of Wnt with other signaling pathways in cancer. Finally, some emphasis is placed on laboratory research that provides a proof of concept for the therapeutic inhibition of Wnt signaling in cancer.

The starting point of the field was the discovery that mammary tumors arising in mice infected with the murine mammary tumor virus was often caused by the activation of the murine int-1 gene, later called Wnt1. It was later shown that the Wnt gene resembled the fly wingless gene, a secreted factor controlling a signaling cascade that included GSK3 and armadillo, the fly version of mammalian β-catenin. The importance of this pathway in human cancer became very clear when the human tumor suppressor adenomatous polyposis coli (APC) protein was found in association with β-catenin. The finding that APC could downregulate β-catenin and Wnt-1 could upregulate it, provided further support for the Wnt cancer connection. Ultimately, the TCF transcription factors that associated with β-catenin completed the understanding of a basic signaling pathway that could account for the potent tumorigenic effects of Wnt (reviewed by Klaus and Birchmeier 2008).

ONCOGENES AND TUMOR SUPPRESSORS

As in many other oncogenic signaling pathways, constituents of Wnt signaling can roughly be subdivided into positive and negatively acting components. By and large, the negatively acting, suppressing components are found mutated to a loss of function status in cancer, while the positive components are activated (Fig. 1). Among the suppressing components of Wnt signaling, APC stands as the most frequently mutated gene in human cancers. Genetic defects in APC are the cause of familial adenomatous polyosis, a heritable syndrome in which affected individuals develop hundreds of polyps in the large intestine at an early age and ultimately succumb to colorectal cancer (Clements et al.

Cite this article as *Cold Spring Harb Perspect Biol* doi: 10.1101/cshperspect.a008052

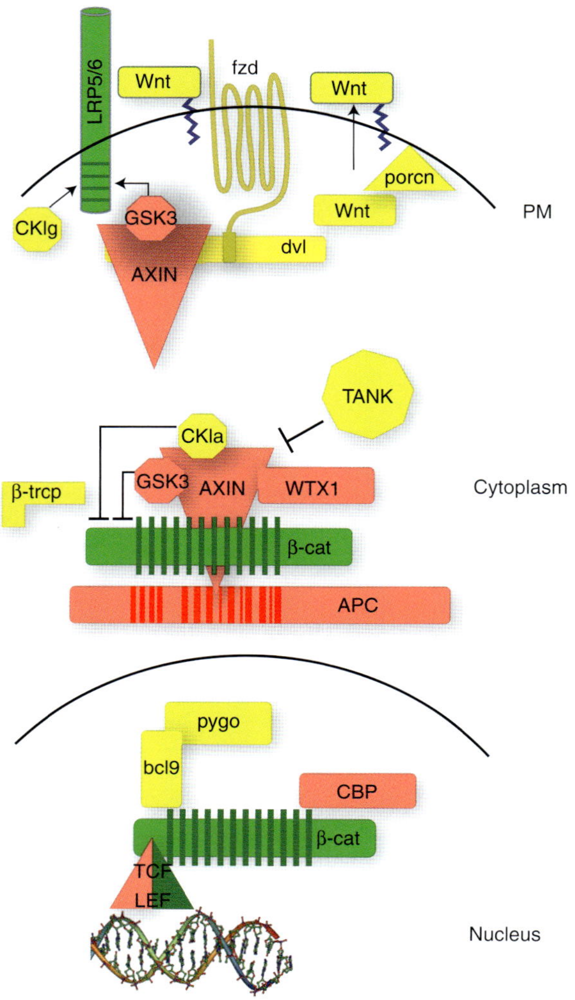

Figure 1. Tumor suppressors and oncogenes in the Wnt pathway. Diagram of a basic Wnt signaling pathway in which oncogenes are depicted in green and tumor suppressors in red.

2003). APC is also mutated in the vast majority of all sporadic colorectal cancers. Loss of function in both alleles is required for tumorigenesis and that loss is structurally linked to the protein's ability to regulate β-catenin protein stability (Polakis 2007).

Specifically, the truncating mutations in APC remove all binding sites for Axin, a scaffold that also binds β-catenin and recruits the protein kinases GSK3 and CKI, both essential for marking β-catenin for destruction facilitated by the E3 ubiquitin ligase β-TRCP (Fig 1). Axins I and II are also tumor suppressors found mutated in both sporadic cancers, particularly hepatocellular and some colorectal, as well as in some familial cancer syndromes (Lammi et al. 2004; Salahshor and Woodgett 2005; Marvin et al. 2011). Regulation of β-catenin also fails when β-catenin itself contains mutations that prevent it from being marked for destruction by the kinases (Polakis 2007). These mutations are found with significant frequency in hepatocellular cancers and medulloblastoma.

More recently, WTX has joined APC, Axin, and β-TrCP as a part of the so-called β-catenin destruction complex (Major et al. 2007). This is particularly intriguing as WTX is a tumor suppressor associated with the pediatric renal cancer Wilm's tumor, which is also commonly associated with β-catenin mutations (Huff 2011). That both WTX and β-catenin mutations coexist in some Wilm's tumors suggests the two genes are not strictly functionally redundant (Ruteshouser et al. 2008). In a recent developmental study in mice, germline inactivation of WTX resulted in the accumulation of multipotent mesenchymal precursor cells resulting from aberrant β-catenin activation (Moisan et al. 2011). Although these animals did not develop renal tumors, it was proposed that the expansion of these mesenchymal progenitors could increase the target population of cells susceptible to transformation by additional genetic insults.

One might expect the occurrence of inactivating mutations in the GSK3 genes as they are critical to β-catenin regulation. Although mutations in the alleles coding for GSK3α and β have not been associated with cancer, the Jamieson lab found an in-frame splice deletion affecting the kinase domain of GSK3β in chronic mylogenous leukemia (Abrahamsson et al. 2009). Aberrant splicing has also been described for the Wnt coreceptor LRP5 in parathyroid and breast cancers. Here missplicing deletes the region of LRP5 that interacts with the secreted Wnt signaling repressor DDK1 (Bjorklund et al. 2009).

Frameshift mutations in the 3′ region of the gene encoding the β-catenin-binding transcription factor TCF4 are extremely common in colorectal cancers with microsatellite instability (Cuilliere-Dartigues et al. 2006). These mutations were proposed to be activating as they ablate binding of the transcriptional repressor CtBP to TCF4. Paradoxically, mutations in TCF4, identified through genomic sequencing efforts of colorectal cancers, appear to be inactivating by compromising its ability to repress cell growth (Sjoblom et al. 2006; Tang et al. 2008). This is consistent with recent findings from the Cappechi laboratory in which TCF4 haploinsuffciency promotes cell proliferation in the mouse intestine and enhances tumorigenesis in an APC mutant mouse model (Angus-Hill et al. 2011). In a colorectal cancer genome sequencing effort, the Meyerson laboratory recently reported a rare, but recurring rearrangement in which the TCF4 gene TCFL72 was fused to VTI1A (Bass et al. 2011). The specific function of the VTI1A-TCF7L2 fusion was not elucidated, but knockdown of the transcript inhibited anchorage independent cell growth, indicative of a positive acting tumorigenic function.

Mutations in the gene coding for the CREB binding protein (CREBBP/CBP) have been identified in acute lypmhoblastic leukemia and B-cell lymphoma (Mullighan et al. 2011; Pasqualucci et al. 2011). This is of interest to Wnt signaling as CBP associates directly with the carboxy-terminal sequence in β-catenin and CBP/β-catenin mediated transcription has been proposed to be critical for stem cell/progenitor cell maintenance and proliferation (Teo and Kahn 2011). Accordingly, CBP is currently a clinical target for small molecule inhibition in Wnt-driven cancers. CBP is a transcriptional coactivator with histone acetyltransferase activity and associates with a wide variety of transcription factors in controlling facets of development and cellular homeostasis. Given this pleiotropy, the relevance of the CBP mutations to Wnt signaling, per se, is difficult to gauge. Moreover, the mutations identified in the hematopoietic cancers impair the histone acetyltransferase activity, which seem inconsistent with a role in facilitating active Wnt signaling. Table 1 summarizes some of the salient DNA and mRNA aberrations that potentially contribute to Wnt-driven tumorigenesis.

EPIGENETIC MODULATION OF Wnt SIGNALING

Recurrent mutations in a gene that alter the products' function in a relevant manner typically offer the clearest evidence for its involvement in oncogenic transformation. Nevertheless, it is apparent that epigenetic modifications in DNA and chromatin also play a role in tumor progression by modulating the suppressor or activator components of a signaling pathway (Rodriguez-Paredes and Esteller 2011). The impact of DNA methylation on Wnt-driven tumorigenesis was

Table 1. Genetic alterations in Wnt signaling in cancer

Affected gene	DNA/mRNA alteration	Functional outcome	Cancer type	Reference
CTNNBl (b-catenin)	Missense/in-frame deletion	Enhanced protein stability	Hepatocellular/ Medulloblastoma	Polakis 2007
APC (APC)	Truncation	Reduced regulatory activity	Colorectal/gastric	Clements et al. 2003
Axins (Axin I, Axin II)	Truncation/ missense	Reduced regulatory activity	Hepatocellular/ colorectal	Salahshor and Woodgett 2005
CREBP(CBP)	Truncation/ missense	Inactive acetyltransferase	Lymphoma/ leukemia	Teo and Kahn 2011
GSK3b	Missplicing, in-frame deletion	Inactive kinase	Leukemia	Abrahamsson et al. 2009
LRP5	Missplicing, in-frame deletion	Loss of repression by DKK1	Breast/parathyroid	Bjorklund et al. 2009
TCF7L2 (TCF4)	Missense/deletion/ truncation	Loss of repression	Colorectal	Cuilliere-Dartigues et al. 2006
TCF7L2 (TCF4)	Fusion with VT11A gene	Unclear	Colorectal	Bass 2011
FAB123B (WTX)	Truncation/ deletion	Loss of function	Wilm's tumor	Ruteshouser et al. 2008

reported by Laird et al., who found that genetic depletion and pharmacological suppression of DNA methyl transferase effectively reduced the tumor burden in the APC min mouse (Laird et al. 1995). Subsequent studies with this model revealed that deficiency of the methyl–CpG binding repressor Mbd2 also severely repressed tumor formation by attenuating Wnt signaling (Sansom et al. 2003; Phesse et al. 2008). These studies point to the silencing of Wnt antagonists by DNA hypermethylation. Indeed, genome-wide analysis of promotor DNA methylation in human medulloblastoma, colon, and pancreatic cancers showed that Wnt pathway inhibitors are common targets of this mechanism of gene repression (Kongkham et al. 2010; Vincent et al. 2011; Suzuki et al. 2002). In particular, the promotor region of the gene coding for the secreted frizzled related protein 1 (sFRP1), that binds to and inhibits Wnt ligands, was commonly identified in these studies. Members of the sFRP family have also been reported to be silenced by DNA methylation in numerous other cancers including hepatocellular cancer, breast cancer, and melanoma (Suzuki et al. 2008; Takagi et al. 2008; Ekstrom et al. 2011).

In addition to the direct methylation of DNA, many Wnt signaling genes and targets undergo chromatin modifications that affect gene transcription. Considering that DNA hypermethylation of promotor regions is frequently accompanied by chromatin modifications, it is not surprising that the sSFRP1 gene is also a target of chromatin modifying factors (McGarvey et al. 2006). In the case of AML, it was proposed that chromatin modification resulted from the targeting of the sFRP1 gene by the RUNX-ETO fusion protein (Cheng et al. 2011b). A RUNX binding site was identified in the 5′ regulatory region of the sFRP1 gene and this region was also detected in chromatin immunoprecipitated with antibodies specific to ETO. The ETO portion of this fusion recruits multiple chromatin modifying factors that can repress gene activation and, in the case of sFRP1, drive Wnt signaling and cell proliferation.

An additional epigenetic mechanism underlying the suppression of numerous Wnt antagonists in hepatocellular cancer was reported by Cheng et al (2011a). In this study, chromatin immunoprecipitation/microarray analysis was used to show that many genes coding for Wnt inhibitors, including sFRP5, Axin2, and NKD2 were occupied by EZH2, a subunit of the polycomb repressor complex 2 (PRC2). EZH2 is a methyl transferase that deposits the repressive H3K27me3 marks on chromatin (Rodriguez-Paredes and Esteller 2011). Knockdown of EZH2 in HCC cells resulted in increased expression of NKD2 and a concomitant reduction in Wnt reporter signaling. This is particularly intriguing because somatic mutations in EZH2, as well as another H3k27 methyltransferase, UTX, have been detected in a variety of human cancers (van Haaften et al. 2009; Morin et al. 2011).

DACT3 is an additional inhibitor of Wnt signaling that is repressed in colon cancers by histone modification (Jiang et al. 2008). DACT3 is a member of the FDO/DPR gene family that inhibits Wnt signaling by interacting with DVL proteins. Derepression of DACT3 inhibited Wnt signaling and promoted apoptosis of colorectal cancer cells, an effect that could be rescued by knockdown of DACT3 mRNA. Unlike sFRP1 in colon cancer cells, the DACT3 DNA promotor region does not appear to be hypermethylated. Thus, DACT3 derepression might not require reversal of DNA methylation, but perhaps may respond singularly to treatment with an inhibitor of histone modification such as that represented by the Trichostatin A compound.

In contrast to gene silencing, modification of chromatin can also enable gene activation. The carboxyl terminus of β-catenin is known to recruit numerous chromatin-remodeling subunits, such as set1 and paf1, that mediate the H3K4me3 histone marks associated with highly active genes (Willert and Jones 2006). Although many of these sorts of modifiers are commonly associated with gene activation by transcription factors, the Clevers laboratory reported that the MLLT10/AF10-DOT1L complex was principally dedicated to Wnt signaling (Mahmoudi et al. 2011). They discovered that this leukemia-associated chromatin-modifying complex was recruited to Wnt target genes in a β-catenin/TCF4-dependent manner. DOT1L is a methyl transferase responsible for H3K79me3 modification associated with gene activation. The study showed a requirement for MLLT10/AF10-DOT1L in Wnt target gene expression in colon cancer cells and in normal intestinal homeostasis. MLLT10/AF10 expression was localized to the proliferative compartment of the intestine and its knockdown rescued defects in APC mutant zebrafish. On the basis of this specificity, the investigators concluded that the DOT1L enzyme might represent a colon cancer target.

COOPERATIVITY WITH OTHER SIGNALING PATHWAYS

Studies involving the use genetically engineered mouse models of cancer, as well as expansive human studies on nonsteroidal anti-inflammatory drugs (NSAIDS) and cancer risk, have firmly established a role for prostaglandin signaling in colorectal tumorigenesis (Oshima and Taketo 2002; Chan et al. 2007). The cyclooxygenase 2 (COX2) enzyme, which catalyzes the production of PGE2 is targeted by NSAIDS and its deletion in the APCmin mouse negatively impacts tumor burden. Mechanistic proposals for the interaction between prostaglandin and Wnt signaling include COX2 as a direct target gene of Wnt signaling, PPARDelta as a convergent focal point for the two pathways, and stimulation of the Wnt pathway by crosstalk from the EP2 prostaglandin receptor (Araki et al. 2003; Wang et al. 2004; Castellone et al. 2005). The levels of PGE2 are downregulated by 15-Hydroxyprostaglandin dehydrogenase (15-PGDH), which catalyzes the rate-limiting step in prostaglandin degradation. Therefore, it acts opposite to COX2, and knockout of 15-PGDH in the APCmin mouse dramatically increases tumor burden, further fortifying the case for prostaglandins as intestinal tumor promotors (Myung et al. 2006). In human cancers, 15-PGDH is consistently downregulated, although it is highly expressed in normal human intestinal epithelium. In a recent study, the Paraskeva lab offered an additional mechanism for the interaction of the prostaglandin and Wnt signals by showing that the 15-PGDH gene is a target of β-catenin-dependent repression (Smartt et al. 2011). This study documented 15-PGDH gene promotor occupancy by β-catenin/TCF4, in vivo activation of 15-PDGH expression by conditional knockdown of β-catenin, and inhibition of 15-PGDH expression in cultured cells treated with Wnt ligand.

Cooperativity between the androgen receptor (AR) and Wnt signaling was proposed as a mechanism for the potentiation of EGF receptor tyrosine kinase family signaling in ER-/Her2+ breast cancers (Ni et al. 2011). In this study, the Wnt7b gene was occupied and activated by AR on stimulation of breast cancer cells with DHT. This led to activation of canonical Wnt signaling in which the stabilized β-catenin cooperated with AR to turn on the Her3 gene. Her3 is the ligand binding receptor that partners with the catalytically active Her2 tyrosine kinase to activate the Pi3K pathway. The paper showed that the upregulation of HER3, activation of HER2/HER3 signaling and enhancement of cell growth by AR activation was dependent on β-catenin.

The EGF receptor tyrosine kinase has had a long history of cooperating with Wnt signaling, resulting in a litany of mechanisms to account for their interaction. These include direct binding of β-catenin to EGFR, modulation of the cadherin-β-catenin interface by EGFR, activation of β-catenin-dependent signaling by EGFR through PI3K signaling as well as the EGFR gene itself as a Wnt target (review by Hu and Li 2011). However, in a very surprising finding by the Lu lab, the M2 splice form of pyruvate kinase (PKM2) was found to translocate to the nucleus in response to EGFR signaling, where it binds to tyrosine phosphorylated of β-catenin and activates β-catenin-dependent target genes (Fig. 2A) (Yang et al. 2011b). PKM2 has received considerable attention lately, although mainly because of a renewed interest in the role of metabolism and aerobic glycolysis in cancer (Christofk et al. 2008). However, in the Lu study, knockdown of PKM2 reduced tumor cell proliferation by EGFR, which correlated with a failure to activate the myc and cyclin D1 genes, both targets of Wnt signaling. Moreover, chromatin immunoprecipitation was used to show binding of PKM2 to the cyclin D1 promotor in a β-catenin-dependent manner. Nevertheless, PKM2 deficiency did interfere with the activation of cyclin D1 by Wnt ligands. Thus, EGF and Wnt receptors both employ β-catenin to activate some common target genes, yet engage independent effectors in route to the nucleus.

Another surprise from the Lu study was that the tyrosine phosphorylation of β-catenin resulting from EGFR activation was mediated by src and occurred on β-catenin residue Y-333. Previously, EGFR-mediated phoshorylation of β-catenin on Y-654 had been reported to affect its interaction with cadherins and also correlated with an increase in β-catenin-mediated transcriptional activity (Lilien and Balsamo 2005). The relevance of Y-654 phosphorylation was finally tested in vivo by the Smits laboratory, who corroborated its importance in binding of cadherin to β-catenin, but also found a positive impact on Wnt signaling (van Veelen et al. 2011). When the constitutive phosphorylation was mimicked by knocking in the Y654E mutation, the homozygotes were embryonic lethal, whereas 38% of the heterozygotes presented with intestinal tumors by 16–18 months of age. On the background of a germline APC mutation, the compound heterozygotes were embryonic lethal, suggesting that half of a dose of wildtype APC was not sufficient to regulate the excess signaling produced by a single Y654E β-catenin allele. An intestine-specific allele of the Y654E β-catenin mutant was then generated and when crossed to an APC mutant mouse, increased intestinal tumor multiplicity was observed in male animals.

Given the interplay between morphogenic signals in developing embryos, the interaction of these pathways might be expected in cancer. In a particularly intriguing example of Wnt and notch pathway cooperativity, Fre et al. (2009) showed that the expansion of intestinal progenitor cells by constitutive Notch signaling was dependent on Wnt. Notch and Wnt signaling also synergized to induce intestinal adenomas in compound mice, with a dramatic increase in colonic tumors, which are typically rare in mice deficient only in APC. The impact of Notch in the intestine was consistent with previous work showing that its activation or inhibition increased or decreased the population of proliferating/progenitor cells, respectively, at the expense of the differentiated population (Fre et al. 2005; van Es et al. 2005). The Srivastava

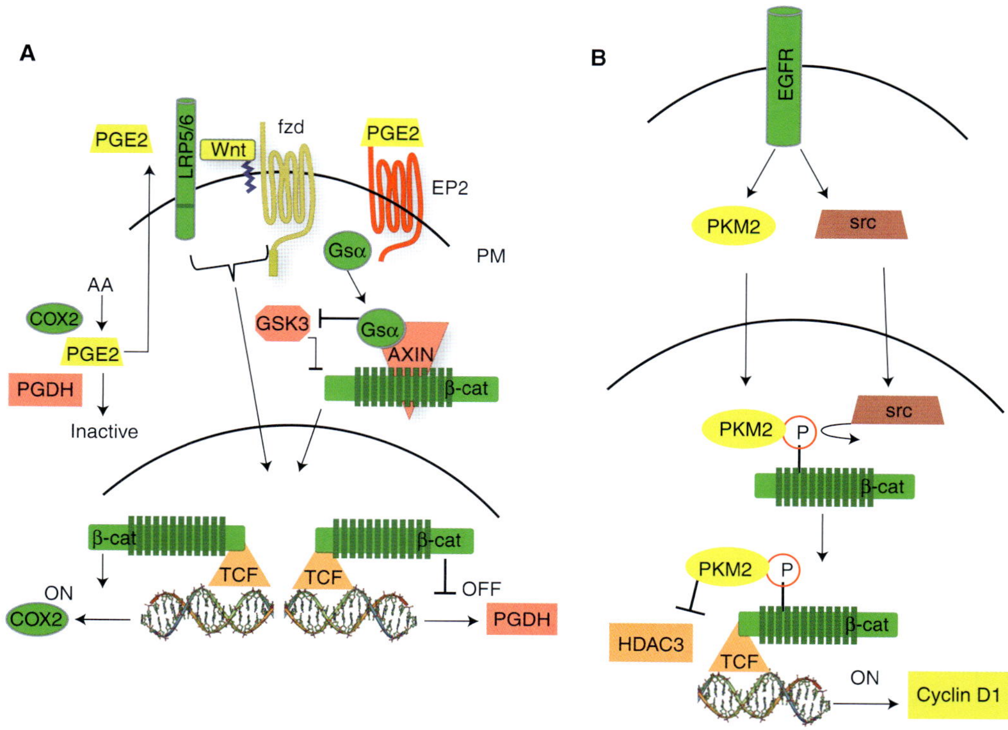

Figure 2. Signaling pathways that cooperate with Wnt signaling. (*A*) Wnt and prostaglandin signaling. Activation of canonical Wnt signaling drives expression of COX2, which catalyzes production of PGE2, and repression of 15-PGDH, which catalyzes the inactivation of PGE2. PGE2 activates the prostaglandin GPCR receptor EP2 releasing the Gsα subunit that displaces GSK3 from Axin, resulting in the stabilization of β-catenin. (*B*) Activation of β-catenin by EGFR. EGFR activity promotes the activation of src and the translocation of PKM2 into the nucleus, in which src phosphorylates residue Y333 on β-catenin. The phosphorylated β-catenin binds PKM2 and the resulting complex associates with TCF in which kinase-active PKM2 displaces HDAC3 to turn on the cyclin D1 target gene.

laboratory recently offered an unexpected mechanism for cooperativity between Wnt and Notch signaling (Kwon et al. 2011). Here the ICD domain of Notch was shown to directly associate with and thereby titrate the unphosphorylated pool of active β-catenin, rendering it unavailable for transcriptional activity. Accordingly, the knockdown of Notch1 exacerbated the activation of Wnt target genes, whereas chemical inhibition of γ-secretase or expression of a noncleavable Notch1 mutant, was inhibitory to Wnt signaling. It was concluded that the active β-catenin associated with Notch was ultimately directed to the lysosome for destruction in a Numb-dependent manner.

EXPERIMENTAL EVIDENCE FOR A CRITICAL ROLE FOR WNT IN HUMAN CANCER

A variety of preclinical experiments suggest that inhibition of Wnt signaling can thwart tumor cell growth and survival. One of the first such experiments involved the induced expression of wild type APC in a colorectal cancer cell line harboring only mutant alleles. On induction, the growth of cultured HT29 cells was inhibited by the wild-type APC, which was attributed to an increase in apoptosis (Morin et al. 1996). Similarly, ectopic expression of Axin1, an additional tumor suppressor in the Wnt pathway, promoted apoptosis in colorectal and hepatocellular cancer cell lines containing mutations in either β-catenin, APC or Axin1 (Satoh et al. 2000). Although encouraging, the reanimation of wild-type tumor suppressors is inferior to a proof of principle in which a positive acting element is interfered with. Early examples of this included the ectopic expression of a dominant negative TCF-4. This amino-terminally deleted TCF-4 binds Wnt target genes nonproductively as a result of its inability to associate with β-catenin. Consequently, inhibition of target gene activation in colorectal cancer cells expressing the dominant negative TCF4 resulted in their growth arrest (Tetsu and McCormick 1999; van de Wetering et al. 2002).

Based on the genetics of Wnt signaling in cancer, β-catenin would be considered a high value drug target. Two tumor suppressors, Axin and APC, negatively regulate β-catenin, and β-catenin itself is oncogenically activated by gain-of-function mutations. Accordingly, interfering with β-catenin, by various methods, produces antitumorigenic affects. Gottardi et al. (2001), exploited the mutually exclusive binding of β-catenin to E-cadherin and the TCF transcription factors by expressing cadherin chimeras that block the latter interaction. This was accompanied by growth inhibition of SW480 colorectal cancer cells. Suppression of β-catenin by antisense oligonucleotides was also found to inhibit tumorgenicity of colon cancer cells in vitro and in vivo (Green et al. 2001; Roh et al. 2001). Kim et al. (2002) employed a somatic cell gene targeting strategy to show that β-catenin expression was essential for clonal growth of the APC mutant and β-catenin mutant colorectal cancer lines, DLD-1 and SW48, respectively. Curiously, a second β-catenin mutant cell line, HCT116, was unaffected by the gene targeting.

The advent of interfering RNA technologies ushered in further validation of β-catenin as a potential cancer target. In one of the first examples of this, siRNA knockdown of β-catenin reduced β-catenin-dependent transcription and diminished the anchorage independent growth of colon cancer cells and their ability to generate tumors in mice (Verma et al. 2003). Unlike the gene targeting approach, the tumorigenecity of the HCT116 cell line was negatively impacted by the siRNA. In a separate study, knockdown of β-catenin by inducible shRNA strongly inhibited proliferation of HCT116 and Ls174t colorectal cancer cells, both of which harbor mutant β-catenin alleles (Mologni et al. 2010). However, induction of apoptosis was insignificant with knockdown of either β-catenin or K-ras individually, whereas the combination produced a strong effect. The effect of this combination on in vivo tumor growth was again far more effective than knockdown of either transcript alone.

In addition to its role in gene activation, β-catenin is essential to cadherin function in forming cell–cell contacts. To some extent this complicates the interpretation of β-catenin mRNA knockdown experiments. To overcome this, Cong et al. (2003), devised a fusion protein that artificially recruited the E3 ubiquitin ligase β-TrCP to β-catenin. This resulted in the selective destruction of the Wnt signaling pool, but not the cadherin-bound fraction. Colon cells expressing this construct lost their tumorigenic potential, including their ability to form tumors in mice.

Inhibitors of Wnt signaling might also be effective in hepatocellular cancers, where mutations in Wnt pathway components are relatively common. In the HepG2 human hepatoma cell line, RNAi mediated knockdown of β-catenin decreased cell viability, proliferation and growth in soft agar (Zeng et al. 2007). Although the HepG2 cell line carries a mutant β-catenin allele, a second cell line, Hep3b, is wild type, yet similar inhibitory effects were observed with it. Ashihara et al. (2009) provided an additional example in which interference with Wnt signaling impaired the tumorigencity of cancer cells lacking Wnt pathway mutations. Here, siRNA knockdown of β-catenin significantly reduced the tumor burden resulting from the inoculation of RPMI8226 myeloma cells in irradiated nude mice. Thus, interference with Wnt signaling in cancers lacking pathway mutations can be effective and suggests that aberrant activation of Wnt receptors should also be entertained as an oncogenic mechanism.

Mammary cancer is a prominent example in which Wnt pathway mutations are very rare, yet hyperactive signaling is apparent, particularly in cancers classified as basal-like or triple negative type (Lin et al. 2000; Chung et al. 2004; Howe and Brown 2004; Khramtsov et al. 2010; Geyer et al. 2011). Expression of the Wnt receptor FZD7 is characteristic of these types of breast cancer (Sorlie et al. 2003). Accordingly, Yang, et al. recently reported a proof of principle experiment in which knockdown of FZD7, in cell line models of triple negative breast cancer, reduced expression of Wnt target genes, inhibited tumorigenesis in vitro and greatly retarded the capacity of the MDA-MD-231 cell line to form tumors in mice (Yang et al. 2011a). This suggests that Wnt ligands might drive certain breast cancers and is consistent with previous work from the Hynes laboratory (Matsuda et al. 2009). Here the secreted frizzled related protein (sFRP1), which binds to and effectively competes with FZD receptors for Wnt ligands, was ectopically expressed in the MDA-MB-231 cell line. The sFRP1 expressing cells struggled to form tumors upon inoculation into the mammary fat pads of mice and their propensity to metastasize to lung was greatly impaired. The source of Wnt ligand in these studies could arise from either host cells and/or the cancer cells, themselves. The latter possibility was first made apparent by Bafico et al. (2004), when they reported autocrine Wnt signaling in a panel of breast and ovarian cancer cell lines, MDA-MB-231 among them. These cells were identified by the presence of a signaling pool of uncomplexed β-catenin, a hallmark of Wnt signaling. This pool of β-catenin, as well as the signal from a Wnt reporter gene, was reduced on expression or addition of the soluble Wnt inhibitors sFRP1 or DKK1. These investigators subsequently reported autocrine signaling in non-small cell lung cancers, as well (Akiri et al. 2009).

Work from the Massague laboratory encourages us to target Wnt signaling in lung cancer, which again typically lacks pathway mutations (Nguyen et al. 2009). In this study, a Wnt target gene signature was associated with metastatic lesions derived from human lung cancer models grown in mice. Expression of dominant negative TCF4 or TCF1 in the metastasis-prone derivatives of the PC9 and H2030 nonsmall lung cancer lines provided a proof-of-principle. The expression of these dominant negative transcription factors had little impact on the growth rate of the pulmonary tumors, but retarded their metastatic spread to bone and brain.

Although the knockdown of FZD7 or the ectopic expression of sFRP1 implicate Wnt ligands and in cancer, they still involve genetic manipulation, and thus remain one step removed from a pharmacological proof of concept. A more pharmacologically relevant approach was provided by De Almeida et al. (2007), whom injected tumor-bearing mice with a fusion protein consisting of the Fc region IgG fused to the extra cellular domain of the FZD8 (FZD8CRD) Wnt receptor. That the FZD8CRD molecule was active in vivo was initially shown by its dramatic inhibitory effect in the MMTV-Wnt-1 model of mammary cancer. Secondly, the FZD8CRD showed significant tumor inhibition of two nonengineered cancer cell lines, the N-TERA2 human testicular cancer and the PA1 human ovarian cancer cell line. The PA1 was among those cells identified by Bafico et al. (2004) as having autocrine Wnt signaling. In an additional pharmacological test, You et al. (2004), inhibited the growth of a human melanoma xenograft model by administering a monoclonal antibody to Wnt2.

The impact of β-catenin knockdown in cancers without identifiable genetic lesions in Wnt signaling could relate to the purported role of Wnt in the establishment and maintenance of so-called cancer stem cells (CSCs) (Reya and Clevers 2005; Malanchi and Huelsken 2009). There is considerable evidence for this in hematopoetic cancers, in which evidence for a cellular hierarchy is well established. Granulocytic-macrophage progenitors isolated from CML patients show attributes of excessive Wnt signaling relative to normal progenitors and the introduction of the Wnt pathway inhibitor Axin decreased their capacity for proliferation and self-renewal (Jamieson et al. 2004). In a mouse model of CML, introduction of BCR-ABL into marrow cells from conditional β-catenin$^{-/-}$ knockout animals significantly reduced the onset of CML-like disease relative to that observed with wild-type marrow cells (Zhao et al. 2007). Nevertheless, recipients of the transformed β-catenin$^{-/-}$ cells went on to develop disease resembling acute lymphoblastic leukemia. In another mouse model of leukemia, initiated by expression of a mixed lineage leukemia (MLL) fusion protein, the conditional ablation of the β-catenin gene prevented preleukemic stem cells from inducing leukemia in grafted animals (Wang et al. 2010). In this same study, shRNA knock down of β-catenin in leukemia stem cells severely delayed the onset of leukemia in recipient hosts.

Recent evidence also points to a role for Wnt signaling in cutaneous cancer stem cells. The Huelsken laboratory found many similarities between normal skin stem cells and cancer stem cells isolated from murine epidermal tumors induced by carcinogens (Malanchi et al. 2008). The cancer stem cells appeared enriched for Wnt signaling and conditional knockdown of β-catenin effectively eliminated this population of cells resulting in complete regression of the epidermal tumors. Although this shows that a mouse epidermal tumor is indeed impacted by inhibition of Wnt signaling, the investigators also provided some evidence that Wnt signaling is exacerbated in human squamous cell carcinomas.

SUMMARY

Wnt signaling clearly contributes to human tumor progression. This is best defined by germline and somatic mutations in Wnt pathway components that deregulate signaling. The rapidly evolving field of epigenetics also continues to implicate Wnt signaling in cancer by identifying suppressors and activators whose expression is modulated by DNA and chromatin modifications. New mechanisms of cooperativity between Wnt signaling and other pathways underscore the flexibility available to cancer cells for maintaining growth and survival. Finally, a large body of experimental evidence strongly encourages us to develop therapies that specifically interfere with Wnt signaling in the cancer clinic.

REFERENCES

Abrahamsson AE, Geron I, Gotlib J, Dao KH, Barroga CF, Newton IG, Giles FJ, Durocher J, Creusot RS, Karimi M, et al. 2009. Glycogen synthase kinase 3β missplicing contributes to leukemia stem cell generation. *Proc Natl Acad Sci* **106:** 3925–3929.

Akiri G, Cherian MM, Vijayakumar S, Liu G, Bafico A, Aaronson SA. 2009. Wnt pathway aberrations including autocrine Wnt activation occur at high frequency in human non-small-cell lung carcinoma. *Oncogene* **28:** 2163–2172.

Angus-Hill ML, Elbert KM, Hidalgo J, Capecchi MR. 2011. T-cell factor 4 functions as a tumor suppressor whose disruption modulates colon cell proliferation and tumorigenesis. *Proc Natl Acad Sci* **108:** 4914–4919.

Araki Y, Okamura S, Hussain SP, Nagashima M, He P, Shiseki M, Miura K, Harris CC. 2003. Regulation of cyclooxygenase-2 expression by the Wnt and ras pathways. *Cancer Res* **63:** 728–734.

Ashihara E, Kawata E, Nakagawa Y, Shimazaski C, Kuroda J, Taniguchi K, Uchiyama H, Tanaka R, Yokota A, Takeuchi M, et al. 2009. β-catenin small interfering RNA successfully suppressed progression of multiple myeloma in a mouse model. *Clin Cancer Res* **15:** 2731–2738.

Bafico A, Liu G, Goldin L, Harris V, Aaronson SA. 2004. An autocrine mechanism for constitutive Wnt pathway activation in human cancer cells. *Cancer Cell* **6:** 497–506.

Bass AJ, Lawrence MS, Brace LE, Ramos AH, Drier Y, Cibulskis K, Sougnez C, Voet D, Saksena G, Sivachenko A, et al. 2011. Genomic sequencing of colorectal adenocarcinomas identifies a recurrent VTI1A-TCF7L2 fusion. *Nat Genet* **43:** 964–968.

Bjorklund P, Svedlund J, Olsson AK, Akerstrom G, Westin G. 2009. The internally truncated LRP5 receptor presents a therapeutic target in breast cancer. *PLoS ONE* **4:** e4243.

Castellone MD, Teramoto H, Williams BO, Druey KM, Gutkind JS. 2005. Prostaglandin E2 promotes colon cancer cell growth through a Gs-axin-β-catenin signaling axis. *Science* **310:** 1504–1510.

Chan AT, Ogino S, Fuchs CS. 2007. Aspirin and the risk of colorectal cancer in relation to the expression of COX-2. *N Engl J Med* **356:** 2131–2142.

Cheng AS, Lau SS, Chen Y, Kondo Y, Li MS, Feng H, Ching AK, Cheung KF, Wong HK, Tong JH, et al. 2011a. EZH2-mediated concordant repression of Wnt antagonists promotes β-catenin-dependent hepatocarcinogenesis. *Cancer Res* **71:** 4028–4039.

Cheng CK, Li L, Cheng SH, Ng K, Chan NP, Ip RK, Wong RS, Shing MM, Li CK, Ng MH. 2011b. Secreted-frizzled related protein 1 is a transcriptional repression target of the t(8;21) fusion protein in acute myeloid leukemia. *Blood* **118:** 6638–6648.

Christofk HR, Vander Heiden MG, Harris MH, Ramanathan A, Gerszten RE, Wei R, Fleming MD, Schreiber SL, Cantley LC. 2008. The M2 splice isoform of pyruvate kinase is important for cancer metabolism and tumour growth. *Nature* **452:** 230–233.

Chung GG, Zerkowski MP, Ocal IT, Dolled-Filhart M, Kang JY, Psyrri A, Camp RL, Rimm DL. 2004. β-Catenin and p53 analyses of a breast carcinoma tissue microarray. *Cancer* **100:** 2084–2092.

Clements WM, Lowy AM, Groden J. 2003. Adenomatous polyposis coli/β-catenin interaction and downstream targets: Altered gene expression in gastrointestinal tumors. *Clin Colorectal Cancer* **3:** 113–120.

Cong F, Zhang J, Pao W, Zhou P, Varmus H. 2003. A protein knockdown strategy to study the function of β-catenin in tumorigenesis. *BMC Mol Biol* **4:** 10.

Cuilliere-Dartigues P, El-Bchiri J, Krimi A, Buhard O, Fontanges P, Flejou JF, Hamelin R, Duval A. 2006. TCF-4 isoforms absent in TCF-4 mutated MSI-H colorectal cancer cells colocalize with nuclear CtBP and repress TCF-4-mediated transcription. *Oncogene* **25:** 4441–4448.

DeAlmeida VI, Miao L, Ernst JA, Koeppen H, Polakis P, Rubinfeld B. 2007. The soluble Wnt receptor Frizzled8CRD-hFc inhibits the growth of teratocarcinomas in vivo. *Cancer Res* **67:** 5371–5379.

Ekstrom EJ, Sherwood V, Andersson T. 2011. Methylation and loss of Secreted Frizzled-Related Protein 3 enhances melanoma cell migration and invasion. *PLoS ONE* **6:** e18674.

Fre S, Huyghe M, Mourikis P, Robine S, Louvard D, Artavanis-Tsakonas S. 2005. Notch signals control the fate of immature progenitor cells in the intestine. *Nature* **435:** 964–968.

Fre S, Pallavi SK, Huyghe M, Lae M, Janssen KP, Robine S, Artavanis-Tsakonas S, Louvard D. 2009. Notch and Wnt signals cooperatively control cell proliferation and tumorigenesis in the intestine. *Proc Natl Acad Sci* **106:** 6309–6314.

Geyer FC, Lacroix-Triki M, Savage K, Arnedos M, Lambros MB, MacKay A, Natrajan R, Reis-Filho JS. 2011. β-Catenin pathway activation in breast cancer is associated with triple-negative phenotype but not with CTNNB1 mutation. *Mod Pathol* **24:** 209–231.

Gottardi CJ, Wong E, Gumbiner BM. 2001. E-cadherin suppresses cellular transformation by inhibiting β-catenin signaling in an adhesion-independent manner. *J Cell Biol* **153:** 1049–1060.

Green DW, Roh H, Pippin JA, Drebin JA. 2001. β-catenin antisense treatment decreases β-catenin expression and tumor growth rate in colon carcinoma xenografts. *J Surg Res* **101:** 16–20.

Howe LR, Brown AM. 2004. Wnt signaling and breast cancer. *Cancer Biol Ther* **3:** 36–41.

Hu T, Li C. 2011. Convergence between Wnt-β-catenin and EGFR signaling in cancer. *Mol Cancer* **9:** 236.

Huff V. 2011. Wilms' tumours: About tumour suppressor genes, an oncogene and a chameleon gene. *Nat Rev Cancer* **11:** 111–121.

Jamieson CH, Ailles LE, Dylla SJ, Muijtjens M, Jones C, Zehnder JL, Gotlib J, Li K, Manz MG, Keating A, et al. 2004. Granulocyte-macrophage progenitors as candidate leukemic stem cells in blast-crisis CML. *N Engl J Med* **351:** 657–667.

Jiang X, Tan J, Li J, Kivimae S, Yang X, Zhuang L, Lee PL, Chan MT, Stanton LW, Liu ET, et al. 2008. DACT3 is an epigenetic regulator of Wnt/β-catenin signaling in colorectal cancer and is a therapeutic target of histone modifications. *Cancer Cell* **13:** 529–541.

Khramtsov AI, Khramtsova GF, Tretiakova M, Huo D, Olopade OI, Goss KH. 2010. Wnt/β-catenin pathway activation is enriched in basal-like breast cancers and predicts poor outcome. *Am J Pathol* **176:** 2911–2920.

Kim JS, Crooks H, Foxworth A, Waldman T. 2002. Proof-of-principle: oncogenic β-catenin is a valid molecular target for the development of pharmacological inhibitors. *Mol Cancer Ther* **1:** 1355–1359.

Klaus A, Birchmeier W. 2008. Wnt signalling and its impact on development and cancer. *Nat Rev Cancer* **8:** 387–398.

Kongkham PN, Northcott PA, Croul SE, Smith CA, Taylor MD, Rutka JT. 2010. The SFRP family of WNT inhibitors function as novel tumor suppressor genes epigenetically silenced in medulloblastoma. *Oncogene* **29:** 3017–3024.

Kwon C, Cheng P, King IN, Andersen P, Shenje L, Nigam V, Srivastava D. 2011. Notch post-translationally regulates β-catenin protein in stem and progenitor cells. *Nat Cell Biol* **13:** 1244–1251.

Laird PW, Jackson-Grusby L, Fazeli A, Dickinson SL, Jung WE, Li E, Weinberg RA, Jaenisch R. 1995. Suppression of intestinal neoplasia by DNA hypomethylation. *Cell* **81:** 197–205.

Lammi L, Arte S, Somer M, Jarvinen H, Lahermo P, Thesleff I, Pirinen S, Nieminen P. 2004. Mutations in AXIN2 cause familial tooth agenesis and predispose to colorectal cancer. *Am J Hum Genet* **74:** 1043–1050.

Lilien J, Balsamo J. 2005. The regulation of cadherin-mediated adhesion by tyrosine phosphorylation/dephosphorylation of β-catenin. *Curr Opin Cell Biol* **17:** 459–465.

Lin SY, Xia W, Wang JC, Kwong KY, Spohn B, Wen Y, Pestell RG, Hung MC. 2000. β-catenin, a novel prognostic marker for breast cancer: Its roles in cyclin D1 expression and cancer progression. *Proc Natl Acad Sci* **97:** 4262–4266.

Mahmoudi T, Boj SF, Hatzis P, Li VS, Taouatas N, Vries RG, Teunissen H, Begthel H, Korving J, Mohammed S, et al. 2011. The leukemia-associated Mllt10/Af10-Dot1l are Tcf4/β-catenin coactivators essential for intestinal homeostasis. *PLoS Biol* **8:** e1000539.

Major MB, Camp ND, Berndt JD, Yi X, Goldenberg SJ, Hubbert C, Biechele TL, Gingras AC, Zheng N, Maccoss MJ, et al. 2007. Wilms tumor suppressor WTX negatively regulates WNT/β-catenin signaling. *Science* **316:** 1043–1046.

Malanchi I, Huelsken J. 2009. Cancer stem cells: Never Wnt away from the niche. *Curr Opin Oncol* **21:** 41–46.

Malanchi I, Peinado H, Kassen D, Hussenet T, Metzger D, Chambon P, Huber M, Hohl D, Cano A, Birchmeier W, et al. 2008. Cutaneous cancer stem cell maintenance is dependent on β-catenin signalling. *Nature* **452:** 650–653.

Marvin ML, Mazzoni SM, Herron CM, Edwards S, Gruber SB, Petty EM. 2011. AXIN2-associated autosomal dominant ectodermal dysplasia and neoplastic syndrome. *Am J Med Genet A* **155A:** 898–902.

Matsuda Y, Schlange T, Oakeley EJ, Boulay A, Hynes NE. 2009. WNT signaling enhances breast cancer cell motility and blockade of the WNT pathway by sFRP1 suppresses MDA-MB-231 xenograft growth. *Breast Cancer Res* **11:** R32.

McGarvey KM, Fahrner JA, Greene E, Martens J, Jenuwein T, Baylin SB. 2006. Silenced tumor suppressor genes reactivated by DNA demethylation do not return to a fully euchromatic chromatin state. *Cancer Res* **66:** 3541–3549.

Moisan A, Rivera MN, Lotinun S, Akhavanfard S, Coffman EJ, Cook EB, Stoykova S, Mukherjee S, Schoonmaker JA, Burger A, et al. 2011. The WTX tumor suppressor regulates mesenchymal progenitor cell fate specification. *Dev Cell* **20:** 583–596.

Mologni L, Dekhil H, Ceccon M, Purgante S, Lan C, Cleris L, Magistroni V, Formelli F, Gambacorti-Passerini CB. 2010. Colorectal tumors are effectively eradicated by combined inhibition of {β}-catenin, KRAS, and the oncogenic transcription factor ITF2. *Cancer Res* **70:** 7253–7263.

Morin PJ, Vogelstein B, Kinzler KW. 1996. Apoptosis and APC in colorectal tumorigenesis. *Proc Natl Acad Sci* **93:** 7950–7954.

Morin RD, Johnson NA, Severson TM, Mungall AJ, An J, Goya R, Paul JE, Boyle M, Woolcock BW, Kuchenbauer F, et al. 2011. Somatic mutations altering EZH2 (Tyr641) in follicular and diffuse large B-cell lymphomas of germinal-center origin. *Nat Genet* **42:** 181–185.

Mullighan CG, Zhang J, Kasper LH, Lerach S, Payne-Turner D, Phillips LA, Heatley SL, Holmfeldt L, Collins-Underwood JR, Ma J, et al. 2011. CREBBP mutations in relapsed acute lymphoblastic leukaemia. *Nature* **471:** 235–239.

Myung SJ, Rerko RM, Yan M, Platzer P, Guda K, Dotson A, Lawrence E, Dannenberg AJ, Lovgren AK, Luo G, et al. 2006. 15-Hydroxyprostaglandin dehydrogenase is an in vivo suppressor of colon tumorigenesis. *Proc Natl Acad Sci* **103:** 12098–12102.

Nguyen DX, Chiang AC, Zhang XH, Kim JY, Kris MG, Ladanyi M, Gerald WL, Massague J. 2009. WNT/TCF signaling through LEF1 and HOXB9 mediates lung adenocarcinoma metastasis. *Cell* **138:** 51–62.

Ni M, Chen Y, Lim E, Wimberly H, Bailey ST, Imai Y, Rimm DL, Liu XS, Brown M. 2011. Targeting androgen receptor in estrogen receptor-negative breast cancer. *Cancer Cell* **20:** 119–131.

Oshima M, Taketo MM. 2002. COX selectivity and animal models for colon cancer. *Curr Pharm Des* **8:** 1021–1034.

Pasqualucci L, Dominguez-Sola D, Chiarenza A, Fabbri G, Grunn A, Trifonov V, Kasper LH, Lerach S, Tang H, Ma J, et al. 2011. Inactivating mutations of acetyltransferase genes in B-cell lymphoma. *Nature* **471:** 189–195.

Phesse TJ, Parry L, Reed KR, Ewan KB, Dale TC, Sansom OJ, Clarke AR. 2008. Deficiency of Mbd2 attenuates Wnt signaling. *Mol Cell Biol* **28:** 6094–6103.

Polakis P. 2007. The many ways of Wnt in cancer. *Curr Opin Genet Dev* **17:** 45–51.

Reya T, Clevers H. 2005. Wnt signalling in stem cells and cancer. *Nature* **434:** 843–850.

Rodriguez-Paredes M, Esteller M. 2011. Cancer epigenetics reaches mainstream oncology. *Nat Med* **17:** 330–339.

Roh H, Green DW, Boswell CB, Pippin JA, Drebin JA. 2001. Suppression of β-catenin inhibits the neoplastic growth of APC-mutant colon cancer cells. *Cancer Res* **61:** 6563–6568.

Ruteshouser EC, Robinson SM, Huff V. 2008. Wilms tumor genetics: Mutations in WT1, WTX, and CTNNB1 account for only about one-third of tumors. *Genes Chromosomes Cancer* **47:** 461–470.

Salahshor S, Woodgett JR. 2005. The links between axin and carcinogenesis. *J Clin Pathol* **58:** 225–236.

Sansom OJ, Berger J, Bishop SM, Hendrich B, Bird A, Clarke AR. 2003. Deficiency of Mbd2 suppresses intestinal tumorigenesis. *Nat Genet* **34:** 145–147.

Satoh S, Daigo Y, Furukawa Y, Kato T, Miwa N, Nishiwaki T, Kawasoe T, Ishiguro H, Fujita M, Tokino T, et al. 2000. AXIN1 mutations in hepatocellular carcinomas, and growth suppression in cancer cells by virus-mediated transfer of AXIN1. *Nat Genet* **24:** 245–250.

Sjoblom T, Jones S, Wood LD, Parsons DW, Lin J, Barber TD, Mandelker D, Leary RJ, Ptak J, Silliman N, et al. 2006. The consensus coding sequences of human breast and colorectal cancers. *Science* **314:** 268–274.

Smartt HJ, Greenhough A, Ordonez-Moran P, Talero E, Cherry CA, Wallam CA, Parry L, Al Kharusi M, Roberts HR, Mariadason JM, et al. 2011. β-catenin represses expression of the tumour suppressor 15-prostaglandin dehydrogenase in the normal intestinal epithelium and colorectal tumour cells. *Gut* doi: 10.1136/gutjnl-2011-300817.

Sorlie T, Tibshirani R, Parker J, Hastie T, Marron JS, Nobel A, Deng S, Johnsen H, Pesich R, Geisler S, et al. 2003. Repeated observation of breast tumor subtypes in independent gene expression data sets. *Proc Natl Acad Sci* **100:** 8418–8423.

Suzuki H, Gabrielson E, Chen W, Anbazhagan R, van Engeland M, Weijenberg MP, Herman JG, Baylin SB. 2002. A genomic screen for genes upregulated by demethylation and histone deacetylase inhibition in human colorectal cancer. *Nat Genet* **31:** 141–149.

Suzuki H, Toyota M, Carraway H, Gabrielson E, Ohmura T, Fujikane T, Nishikawa N, Sogabe Y, Nojima M, Sonoda T, et al. 2008. Frequent epigenetic inactivation of Wnt antagonist genes in breast cancer. *Br J Cancer* **98:** 1147–1156.

Takagi H, Sasaki S, Suzuki H, Toyota M, Maruyama R, Nojima M, Yamamoto H, Omata M, Tokino T, Imai K, et al. 2008. Frequent epigenetic inactivation of SFRP genes in hepatocellular carcinoma. *J Gastroenterol* **43:** 378–389.

Tang W, Dodge M, Gundapaneni D, Michnoff C, Roth M, Lum L. 2008. A genome-wide RNAi screen for Wnt/β-catenin pathway components identifies unexpected roles for TCF transcription factors in cancer. *Proc Natl Acad Sci* **105:** 9697–9702.

Teo JL, Kahn M. 2011. The Wnt signaling pathway in cellular proliferation and differentiation: A tale of two coactivators. *Adv Drug Deliv Rev* **62:** 1149–1155.

Tetsu O, McCormick F. 1999. β-catenin regulates expression of cyclin D1 in colon carcinoma cells. *Nature* **398:** 422–426.

van de Wetering M, Sancho E, Verweij C, de Lau W, Oving I, Hurlstone A, van der Horn K, Batlle E, Coudreuse D, Haramis AP, et al. 2002. The β-catenin/TCF-4 complex imposes a crypt progenitor phenotype on colorectal cancer cells. *Cell* **111:** 241–250.

van Es JH, van Gijn ME, Riccio O, van den Born M, Vooijs M, Begthel H, Cozijnsen M, Robine S, Winton DJ, Radtke F, et al. 2005. Notch/γ-secretase inhibition turns proliferative cells in intestinal crypts and adenomas into goblet cells. *Nature* **435:** 959–963.

van Haaften G, Dalgliesh GL, Davies H, Chen L, Bignell G, Greenman C, Edkins S, Hardy C, O'Meara S, Teague J, et al. 2009. Somatic mutations of the histone H3K27 demethylase gene UTX in human cancer. *Nat Genet* **41:** 521–523.

van Veelen W, Le NH, Helvensteijn W, Blonden L, Theeuwes M, Bakker ER, Franken PF, van Gurp L, Meijlink F, van der Valk MA, et al. 2011. β-catenin tyrosine 654 phosphorylation increases Wnt signalling and intestinal tumorigenesis. *Gut* **60:** 1204–1212.

Verma UN, Surabhi RM, Schmaltieg A, Becerra C, Gaynor RB. 2003. Small interfering RNAs directed against β-catenin inhibit the in vitro and in vivo growth of colon cancer cells. *Clin Cancer Res* **9:** 1291–1300.

Vincent A, Omura N, Hong SM, Jaffe A, Eshleman J, Goggins M. 2011. Genome-wide analysis of promoter methylation associated with gene expression profile in pancreatic adenocarcinoma. *Clin Cancer Res* **17:** 4341–4354.

Wang D, Mann JR, DuBois RN. 2004. WNT and cyclooxygenase-2 cross-talk accelerates adenoma growth. *Cell Cycle* **3:** 1512–1515.

Wang Y, Krivtsov AV, Sinha AU, North TE, Goessling W, Feng Z, Zon LI, Armstrong SA. 2010. The Wnt/β-catenin pathway is required for the development of leukemia stem cells in AML. *Science* **327:** 1650–1653.

Willert K, Jones KA. 2006. Wnt signaling: Is the party in the nucleus? *Genes Dev* **20:** 1394–1404.

Yang L, Wu X, Wang Y, Zhang K, Wu J, Yuan YC, Deng X, Chen L, Kim CC, Lau S, et al. 2011a. FZD7 has a critical role in cell proliferation in triple negative breast cancer. *Oncogene* **43:** 4437–4446.

Yang W, Xia Y, Ji H, Zheng Y, Liang J, Huang W, Gao X, Aldape K, Lu Z. 2011b. Nuclear PKM2 regulates β-catenin transactivation upon EGFR activation. *Nature* **480:** 118–122.

You L, He B, Xu Z, Uematsu K, Mazieres J, Mikami I, Reguart N, Moody TW, Kitajewski J, McCormick F, et al. 2004. Inhibition of Wnt-2-mediated signaling induces programmed cell death in non-small-cell lung cancer cells. *Oncogene* **23:** 6170–6174.

Zeng G, Apte U, Cieply B, Singh S, Monga SP. 2007. siRNA-mediated β-catenin knockdown in human hepatoma cells results in decreased growth and survival. *Neoplasia* **9:** 951–959.

Zhao C, Blum J, Chen A, Kwon HY, Jung SH, Cook JM, Lagoo A, Reya T. 2007. Loss of β-catenin impairs the renewal of normal and CML stem cells in vivo. *Cancer Cell* **12:** 528–541.

Wnt Signaling and Injury Repair

Jemima L. Whyte, Andrew A. Smith, and Jill A. Helms

Division of Plastic and Reconstructive Surgery, Department of Surgery, Stanford School of Medicine, Stanford, California 94305

Correspondence: jhelms@stanford.edu

Wnt signaling is activated by wounding and participates in every subsequent stage of the healing process from the control of inflammation and programmed cell death, to the mobilization of stem cell reservoirs within the wound site. In this review we summarize recent data elucidating the roles that the Wnt pathway plays in the injury repair process. These data provide a foundation for potential Wnt-based therapeutic strategies aimed at stimulating tissue regeneration.

The Oxford English Dictionary's definition of repair is to "restore to good or proper condition by replacing or fixing parts." In contrast, regeneration is the process of complete renewal, characterized by a full restoration of form and function. Why do some organisms regenerate tissues after injury, while others do not? And is it possible to induce regeneration in cases where repair is the norm? In this article we discuss some of the recent discoveries that have been made by investigators seeking to understand these differences, and how they have used this information to develop therapeutic strategies aimed at stimulating a regenerative response to acute injury.

The "reproduction of lost parts" has long been a subject of intense curiosity. In the 18th century the Swiss naturalist Abraham Trembley demonstrated that when the Cnidarian polyp *Hydra* was divided into small fragments, each piece "grew again into perfect *hydrae*" (Trembley 1744). The Scottish biologist and mathematician D'Arcy Thompson was fascinated by these early observations, and wrote that "the ability of an animal to regenerate lost parts [had] excited the interest of crowned heads. . .ambassadors, and state-couriers, who carried it through Europe" (Thompson 1884). Some years later the Vice President of the British Medical Association voiced a similar interest and wrote, "In these days of strife and stress, when fire and water play havoc with men's lives and limbs, limb regeneration. . .would be of inestimable value." Despite the fact that 238 years have elapsed since those initial observations, our curiosity about regeneration—and the possibility that we could harness this regenerative potential to aid in the healing of our own tissues—has not waned.

Our own regenerative capacities are remarkably limited. Early observers (Dinsmore 1992) who were interested in distinguishing "repair" from "regeneration" recognized that animals did not respond uniformly to an act of injury. In some cases, adult animals mount an injury response that results in the complete regeneration of the damaged tissue or organ; the appendage

Cite this article as *Cold Spring Harb Perspect Biol* doi: 10.1101/cshperspect.a008078

of the aquatic axolotl or the fins of a zebrafish are excellent examples (see Fig. 1) (Poss et al. 2000; Straube et al. 2004; Kawakami et al. 2006; Stewart et al. 2009; Satoh et al. 2010). In other cases, the same tissue or organ in a different species had a very limited regenerative response; for example, reptiles seem to lack this regenerative response, at least in their limbs (Galis et al. 2003), but most species retain the ability to regenerate their tails (McLean and Vickaryous 2011). In contrast, most mammalian tissues respond to injury by generating scar tissue (Harty et al. 2003), which is composed of granulation and fibrous tissues that are poorly organized and lack functionality.

One might interpret these observations to mean that an animal's regenerative capacity is species specific, but this is an oversimplification; some animals begin life with a robust regenerative response that wanes over time. Tadpoles, for instance, exhibit a regenerative capacity only up to metamorphosis; after this developmental event frogs cannot regenerate an amputated limb (Kawakami et al. 2006; Yokoyama et al. 2011). Avian embryos have some ability to regenerate a damaged retina but this potential is lost by the time they are hatchlings (Fischer and Reh 2000; Fischer and Reh 2001). Mammalian embryos also appear to possess an early regenerative ability that erodes by birth (Rinkevich et al. 2011). Why is there such variation in regenerative ability of animals, and between different stages of postnatal life? Such a question has obvious clinical importance, because if we understood the key features that distinguish human healing from that of our amphibian ancestors, then we could potentially "jump-start" a similar regenerative program (Kragl et al. 2009).

The observation that "simple animals" such as sponges and flatworms regenerate, whereas more complex organisms do not (Gurtner et al. 2008), is the basis for some theories that propose there is an association between regenerative capacity and tissue complexity (Sanchez Alvarado 2000). Many examples challenge this theory: the teleost, avian, and mammalian retinas exhibit very similar cellular architecture (Reh and Levine 1998). Yet fish can completely regenerate the neural retina (Cameron et al. 1999; Cameron 2000; Raymond et al. 2006), as can birds up to the hatchling stage (Coulombre and Coulombre 1965, 1970), but rodents and primates retain no such inherent regenerative capacity (Tropepe et al. 2000). Another theory proposes that regenerative abilities wane as the immune system matures (Mescher and Neff 2005, 2006). This theory is largely based on the observation that fetal wounds heal without scar formation, and this developmental period is associated with an immature immune system. A third theory posits that the loss of regenerative ability has evolved to curtail inappropriate cell division and malignant transformation (Gardiner 2005; Levesque et al. 2010). Perhaps most relevant to humans is the observation that across multiple species, regenerative capacity diminishes with age (reviewed in Silva and Conboy 2008; but see Eguchi et al. 2011 for conflicting data). The inference from these observations is that aging depletes a stem cell population

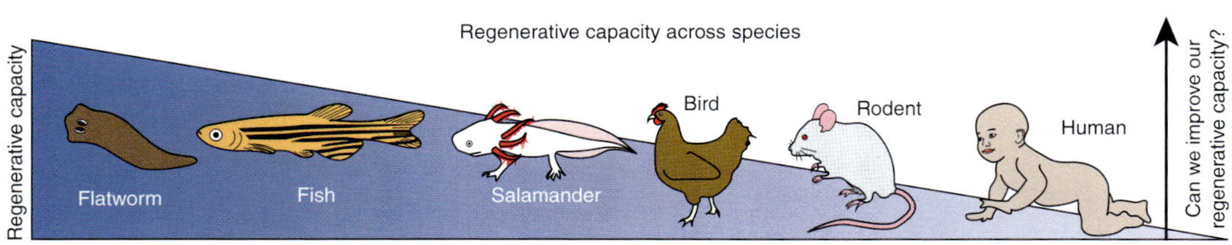

Figure 1. Regenerative capacity across species. The ability to regenerate tissues varies across species. Planarians can regenerate their entire body from a single fragment, fish and salamanders can regenerate their fins and limbs, while mammals have a relatively poor repair mechanism resulting in inadequate tissue replacement and scarring. A common goal of doctors and scientists is to develop biologic therapies to increase mammalian regenerative potential.

from which new body parts arise. But this interpretation has been repeatedly challenged, which leaves open an obvious question: if they are not from a reservoir, where do the cells come from that reform the missing or damaged tissues?

New data demonstrate that adult cells at the edges of the wound dedifferentiate to generate the new tissues (Straube et al. 2004; Kragl et al. 2009; Azevedo et al. 2011). This feature is not unique to mammals: The wound blastema of an axolotl is composed of tissue-specific progenitor cells that arise from the partial dedifferentiation of neurons, cartilage, and muscle (Kragl et al. 2009). What factors are responsible for stimulating this dedifferentiation of adult cells to a stemlike state or the recruitment of stem/progenitor cells into the regenerating tissue? Growing evidence implicates the Wnt pathway in this critical event.

In the following sections we present a summation of recent data supporting a model whereby the act of injury triggers the endogenous Wnt pathway, and that this pathway activity is essential for subsequent healing. Most data suggest that within a damaged tissue, the endogenous Wnt signal activates tissue-resident stem cells, and these cells contribute to the repair and/or regeneration of the damaged tissue. Augmenting this endogenous Wnt signal appears to enhance the healing response; consequently, a number of approaches are being tested that aim to activate Wnt signaling in a local, transient manner to stimulate tissue regeneration in humans.

WNT SIGNALING IS NECESSARY FOR TISSUE REGENERATION

Generally speaking, when Wnt signaling is reduced in an animal with robust regenerative capacities, their inherent regenerative abilities are impaired. For example, blocking Wnt signaling in planarians disrupts the polarity of the regenerate, resulting in inappropriate anatomical generation (i.e., the posterior half of the flatworm regenerates another posterior [tail] segment, and the anterior half generates another anterior [head] segment) (Gurley et al. 2008; Petersen and Reddien 2008, 2009; Yazawa et al. 2009). Likewise, blocking Wnt signaling after amputation of the dorsal fin in zebrafish impairs their normal fin regeneration (Kawakami et al. 2006). Inhibiting Wnt signaling in the eyes of animals with continually regenerating retinas results in an abrupt cessation in this regenerative ability (Kubo et al. 2003; Cho and Cepko 2006; Stephens et al. 2010; Ramachandran et al. 2011).

Which aspects of these regenerative responses are Wnt dependent? The early consensus seems to be that Wnt signaling blockade disrupts the recruitment of stem/progenitor cells to the injury site (Liu et al. 2007b; Das et al. 2008; Denayer et al. 2008; Stephens et al. 2010; Ramachandran et al. 2011) and adversely affects the proliferative phase of the healing process (Qyang et al. 2007; Stoick-Cooper et al. 2007). The formation of a blastema, the mass of multipotent cells at the tip of the wounded limb (Bryant et al. 2002), depends upon recruitment and proliferation (Agata et al. 2007). Consequently, perturbations in Wnt signaling in these regenerating animals manifest as disruptions in the aggregation of morphologically undifferentiated cells that comprise the blastema.

In mammalian organs and tissues with limited regenerative capacity, Wnt signaling is still necessary for the repair process. For example, when Wnt signaling is blocked after skeletal fracture, the result is a nonunion of the injured bone (Chen et al. 2007; Kim et al. 2007; Leucht et al. 2008a). Inhibition of Wnt signaling during skin wounding prevents formation of epithelial appendages including hair and sweat glands, which results in prominent scarring of the epidermis (Ito et al. 2007). Myocardial infarctions in mammals produce a region of scar tissue surrounding the occluded vessels and if Wnt signaling is repressed, the scarring worsens and the result is myocardial rupture (Chen et al. 2004).

If Wnt signaling is required for tissue regeneration, can elevating Wnt signaling in animals with limited regenerative capacity improve the healing response? Current data support this hypothesis. For example, when the limb of a postmetamorphic frog is amputated, the result is a "spike," an outgrowth that lacks any discernible digits. This spike can be converted into a fully functional limb, complete with all appendages, provided that Wnt signaling is elevated at the time of amputation (Kawakami et al. 2006) (although, see Yokoyama et al. 2011 for a differing opinion). Mammalian skin wounds typically heal by scarring; however, elevating Wnt signaling within the wound site leads to the growth of new hair follicles, a hallmark of a fully functional epidermis (Ito et al. 2007).

Even in diseases that are characterized by tissue destruction, elevating Wnt signaling can induce a repair of sorts. For example, multiple myeloma is a cancer of the plasma cells in the bone marrow that is distinguished by severe osteolytic lesions (Sezer 2005). This osteolysis has been linked to elevated levels of Dkk1 in the bone marrow cavity (Tian et al. 2003; Kaiser et al. 2008). Phase I/II clinical trials are now under way to test the efficacy of enhancing Wnt signaling via antibody-mediated repression of Dkk1 as a means to stimulate new bone formation in multiple myeloma patients (Yaccoby et al. 2007; Fulciniti et al. 2009). Likewise, some inflammatory bowel diseases are characterized by reduced Wnt pathway activity (Wehkamp et al. 2007) and reduced turnover of stem cells in the intestinal crypt (Sato et al. 2009). In these types of debilitating chronic injury states, transiently elevating Wnt signaling may be beneficial (reviewed in Anderson and Wong 2010). Indeed, such a Wnt-based therapeutic approach has been attempted. Oral mucositis is a frequent occurrence following chemotherapy, and is characterized by inflammation and ulceration of the mucosal lining of the digestive tract. Investigators demonstrated that they could stimulate regeneration of the mucous membranes in mice using systemic administration of the Wnt agonist, R-spondin (Zhao et al. 2009a).

Collectively, these data demonstrate that endogenous Wnt signaling is a prerequisite for tissue repair, but there are obvious caveats. Most experimental methods used to study Wnt signaling in tissue healing rely on techniques that, in general, produce unrestrained Wnt pathway activation. A well-documented effect of constitutively active Wnt signaling, however, is cancer (Chan et al. 1999; de Lau et al. 2007; Vermeulen et al. 2010), so experimental methods to test the role of Wnt signaling in healing must take this into account. Likewise, increasing Wnt signaling, via genetic deletion of negative regulators such as Ror2 (Brunetti-Pierri et al. 2008) and Axin2 (Liu et al. 2007a; Minear et al. 2010), or mutations in Wnt coreceptors such as Lrp6 (Boyden et al. 2002) or Wnt ligands themselves (Liang et al. 2003), have their own adverse sequelae. Clearly, what is required is a means to activate the pathway in a local, transient manner with the goal of restoring the endogenous Wnt response that is necessary for normal healing. But when should the Wnt pathway be activated? And what, precisely, does Wnt signaling contribute to the injury repair response? The answers to these questions will undoubtedly inform strategies that aim to take advantage of the Wnt pathway for regenerative medicine therapies, and thus we review recent data addressing these questions in the following paragraphs.

WNT SIGNALING AT THE INITIAL STAGE OF THE INJURY RESPONSE

When the body is injured, a cascade of events is set into motion that have as their most fundamental objective the cessation of bleeding, followed by an attempt to wall off the injured tissues from the rest of the body to maintain homeostasis. This vascular response occurs within minutes of tissue damage, and is largely due to the hypoxic condition that develops in tissues deprived of a vascular supply.

The activation of β-catenin-dependent Wnt signaling appears to be one of the initial molecular responses to injury. This activation is typically rapid and spatially restricted, in most cases to the site of damage. For example, severing a planarian in half leads to the rapid up-regulation in Wnt gene expression specifically at the cut site (Petersen and Reddien 2009). Wnt signaling is also initiated by bone fracture (Chen et al. 2007; Kim et al. 2007; Leucht et al. 2008a), by exposure to ionizing radiation (Gurung et al. 2009), by inhalation injury to the lung (Villar et al. 2011; also reviewed in Beers and Morrisey 2011), skin wounding, laser damage to the retina, myocardial infarction, and stroke (see Fig. 2, showing unpublished results from multiple laboratories).

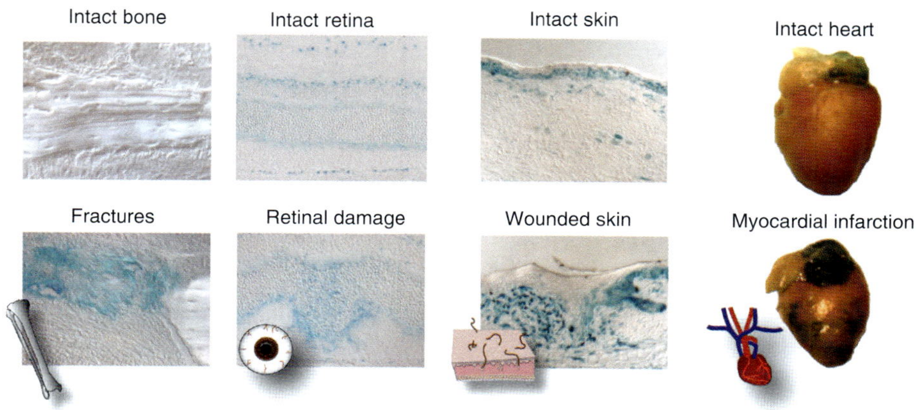

Figure 2. Wnt signaling is activated in response to injury. Damage in most tissues up-regulates Wnt signaling transiently at the injury site. While too much Wnt signaling has been shown to be deleterious, a number of studies have shown that the exogenous addition of Wnt signaling stimulates healing in a number of different injuries, including bone fractures, retinal damage, skin wounding, and myocardial infarction. (Wnt response seen here in blue, by X-gal staining of murine $Axin^{LacZ/+}$ tissue.) (Heart images are thanks to Professor Joseph Wu.)

What features of injury lead to activation of the endogenous Wnt pathway? One intriguing possibility is oxygen tension: Embryonic stem cells—but not differentiated cells—respond to hypoxic conditions by up-regulating the expression of hypoxia-inducible factor (HIF)-1α and in turn, HIF1α regulates expression of the Wnt target genes Lef1 and Tcf1 (Mazumdar et al. 2010). If this in vitro observation is validated in vivo, then changes in oxygen availability and, by extension, HIF-1α activity, may be the direct trigger between injury and activation of the Wnt pathway.

INFLAMMATION AND WNT SIGNALING

Within minutes of tissue injury, platelets aggregate at a wound site to form a clot that both protects surrounding tissues from the invasion of microorganisms, and acts as a reservoir for growth factors that will eventually serve to recruit cells to repair the damage (reviewed in Midwood et al. 2004; Nurden et al. 2008). Although the role of Wnt signaling in platelet regulation during wound healing in vivo is unclear, one study suggests that β-catenin-dependent Wnt signaling inhibits platelet aggregation, at least in vitro (Steele et al. 2009).

Within hours of injury, the inflammatory response ensues: Neutrophils, monocytes, and lymphocytes invade the wound bed in an attempt to phagocytose cell debris and release growth factors and cytokines (Newton et al. 2004; Martin and Leibovich 2005). The inflammatory response is crucial for fighting infection but a state of chronic inflammation is largely considered to be detrimental, both to healing and to the program of regeneration. Some speculate that the mammalian immune system has evolved in such a way as to optimize tissue repair through fibrosis but that these modifications have been detrimental to the program of tissue regeneration (Mescher and Neff 2005). However, direct demonstration of a causal relationship between the inflammatory response and tissue regeneration is lacking. By selectively depleting cells of the neutrophil lineage from a wound site, investigators demonstrated that inflammation can retard wound closure (Dovi et al. 2003; Martin and Leibovich 2005), but how this affects tissue regeneration remains unknown.

The role(s) of Wnt signaling in the inflammatory response is poorly understood. β-catenin-independent Wnt signaling may be a proinflammatory stimulus, based on the observation that Wnt5a expression is increased in the sera of patients suffering from severe sepsis (Pereira et al. 2009). β-Catenin-independent Wnt signaling has also been implicated in a number of chronic inflammatory diseases such as rheumatoid arthritis (Sen et al. 2000) and atherosclerosis (Polzer et al. 2008), but in these cases a causal link between Wnt function and etiology of the inflammatory disease has not been established.

β-Catenin-dependent Wnt signaling, on the other hand, may inhibit or restrict the inflammatory response, based on the observation that inhibitors of the pathway increase the expression of some genes associated with inflammation (Kim et al. 2010). A more direct link has come with the demonstration that by elevating Wnt signaling through inhibition of Dkk1, the bone-resorbing phenotype observed in a mouse model of rheumatoid arthritis could be replaced with the bone-forming pattern of osteoarthritis (Diarra et al. 2007; reviewed in Baker-LePain et al. 2011). The theory that a reduction in Wnt signaling causes inflammatory bone loss is further supported by the observation that osteolytic lesions in multiple myeloma are associated with unrestricted Dkk1 activity (Tian et al. 2003). But it must be emphasized that these effects of Wnt signaling are largely attributed to a pro-osteogenic influence and not because Wnt signaling directly impacts inflammatory mediators. For example, in the condition of rheumatoid arthritis, tumor necrosis factor (TNF) contributes to the disease pathology because the cytokine increases the activity of bone-resorbing osteoclasts; Wnt signals counter this effect by stimulating osteoblast activity.

WNT SIGNALING AND THE PROLIFERATIVE PHASE OF WOUND REPAIR

Within days or months of an injury, the proliferative phase develops, and it is here that the difference between repair and regeneration becomes most obvious. In the repair program, granulation tissue accumulates at the wound site, which is composed of new blood vessels and highly proliferative fibroblasts that produce copious amounts of fibronectin, collagen type III, glycosaminoglycans, elastin, glycoprotein, proteoglycans and hyaluronan (Buchanan et al. 2009). This matrix appears to support cell migration but it is transient; gradually, a collagen type III-rich matrix that provides tensile strength to the wound site replaces the granulation tissue. In the regeneration program, the hyaluronan content of the extracellular matrix is much higher, and investigators speculate that this may curtail or reduce that amount of collagen deposited at the wound site by overproliferative fibroblasts (Carre et al. 2010). In support of this theory, investigators showed that hyaluronan removal results in fibrotic scarring (West et al. 1997), and that Wnt3a induces hyaluronan synthesis (Larson et al. 2010).

In the intestine, Wnt signaling maintains crypt architecture (reviewed in Clevers 2006; Sansom et al. 2007; van der Flier and Clevers 2008) and does so by being a positive stimulus for proliferation. Investigators demonstrated that the Wnt inhibitor Dkk1 is induced by inflammatory cytokines during colitis. Dkk1 causes further tissue damage by stimulating epithelial cell apoptosis. Blocking Dkk1 function results in elevated Wnt signaling and the induction of cell proliferation, which promotes wound repair after colitis (Koch et al. 2011). The experimental approach in this particular study, however, resulted in abolition of Dkk1 function and therefore unrestricted Wnt signaling, which had its own set of side effects, namely, hyperproliferation of the epithelial cells lining the crypt (Koch et al. 2011). These and other studies in skin wound healing (Ito et al. 2007) and bone formation (French et al. 2004; Friedman et al. 2009) and repair (Kim et al. 2007) emphasize that the proliferative effects of β-catenin-dependent Wnt signaling must be transient and localized to be beneficial in the wound repair/regeneration process.

In the final stages of wound repair, extensive extracellular matrix remodeling occurs. In repair, the granulation tissue is converted to mature scar tissue through collagen synthesis and catabolism (Chang et al. 2004). Type III collagen is gradually degraded and is replaced by the stronger type I collagen. Proapoptotic signals at the injury site can regulate collagen degradation both by decreasing fibroblast numbers and by initiation of collagenase activity (Rai et al.

2005). These collagen fibers are rearranged, cross-linked, and aligned providing increased strength to the wound. Over time, this densely packed collagen develops into an inelastic white collagen scar. Regeneration, on the other hand, has no such fibrous tissue. A schematic of the stages of wound repair is shown in Figure 3.

CHRONIC INJURY, CANCER, AND WNT SIGNALING

Multiple lines of evidence demonstrate that overactivation of the Wnt pathway causes hyperproliferation and cancer (reviewed in Barker and Clevers 2006). The majority of these cancers are due to loss of function mutations in Wnt pathway inhibitors. There are also clear examples of an intimate relationship between chronic injury and cancer initiation (see Fig. 4). For example, smoking inflicts chronic injury of the trachea and lung tissue, resulting in lung cancer (Kersting et al. 2000), extensive sun exposure leads to prolonged inflammation of the dermis and can result in melanomas and other skin cancers (El Ghissassi et al. 2009), and chronic alcohol consumption causes inflammation and death of hepatocytes, which can result in liver cancer (McKillop and Schrum 2005).

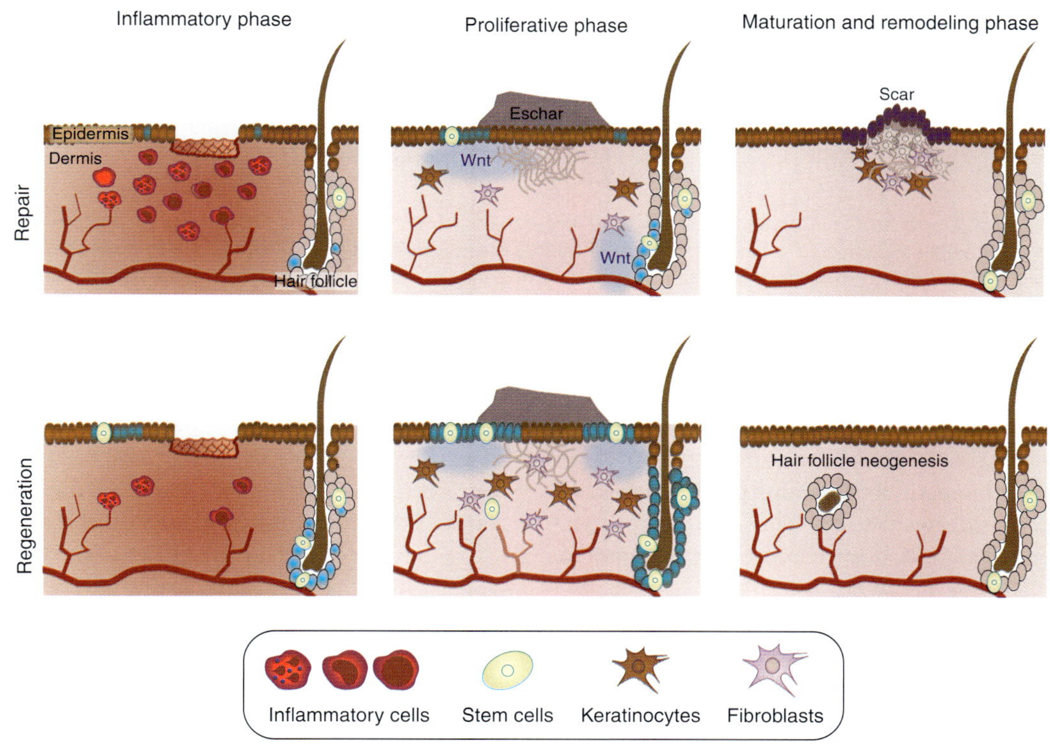

Figure 3. Wnt signaling drives tissue repair. In both repair and regeneration, wound healing is characterized by three main phases. During the inflammatory phase there is an influx of inflammatory cells and local Wnt signaling begins to increase. During the proliferative phase a scab (eschar) is formed and the wound is reepithelialized. This phase includes an increased local Wnt response, extracellular matrix deposition, angiogenesis, and the recruitment and proliferation of multiple cell types including stem cells, keratinocytes, and fibroblasts. The third phase, the maturation and remodeling phase, is characterized by extensive extracellular matrix remodeling. Whereas repair leads to scarring, or functionally inadequate tissue, increased Wnt signaling during regeneration leads to a reduced inflammatory response, increased angiogenesis, increased cellular proliferation and recruitment, as well as differences in matrix composition, leading to the complete reconstruction of the original tissue architecture. (Increase in blue color represents an increase in Wnt signaling.)

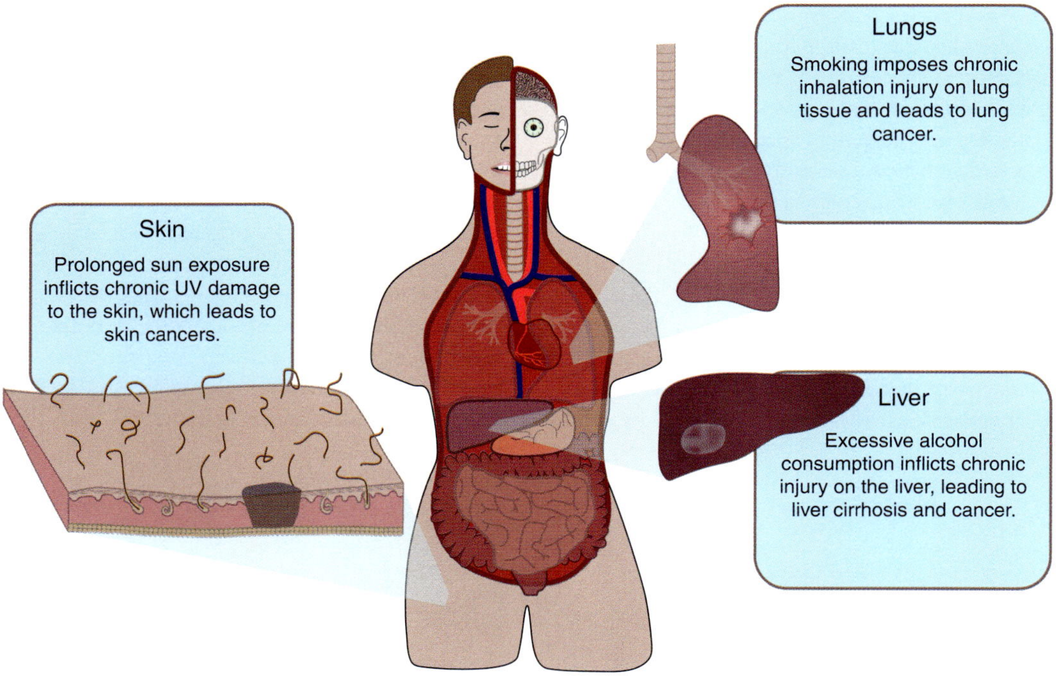

Figure 4. Does chronic injury lead to prolonged Wnt signaling and cancer? Injury results in a local increase in Wnt signaling and chronic injury leads to cancer. Many cancers also have up-regulated Wnt signaling. The Wnt pathway is best known to promote proliferation; an intriguing possibility is that chronic injury may induce a prolonged Wnt response leading to uncontrolled cell proliferation and cancer. In the lungs, cigarette smokers inflict chronic injury to the trachea and lung tissue, which may result in lung cancer. Relating to the skin, prolonged UV exposure leads to chronic inflammation of the dermis, which may result in skin cancer. In the liver, chronic alcohol consumption subjects the tissue to constant damage, which may lead to liver cancer.

Abundant data demonstrate that injury locally activates Wnt signaling (see previous paragraphs), but does chronic injury cause inappropriately prolonged Wnt signaling, thus heightening the risk of cell transformation at the site of injury? Growing data support this hypothesis: Wnt1 is overexpressed in nonsmall cell lung carcinoma (He et al. 2004) and some melanoma cell lines exhibit abnormally high levels of β-catenin (Rubinfeld et al. 1997). More recently, investigators demonstrated that melanoma cells have evolved a mechanism to activate β-catenin-dependent signaling that does not require high levels of Wnt ligands (Sinnberg et al. 2010; reviewed in Vaid et al. 2011). Thus, the connection between protracted, repeated injury, persistent Wnt pathway activation, and tumorogenesis is compelling, and represents a novel hypothesis to explain the molecular etiologies of injury-related cancers.

HARNESSING THE WNT PATHWAY FOR THERAPEUTIC APPLICATIONS IN INJURY REPAIR

There is significant interest in finding ways in which the Wnt pathway can be modulated for therapeutic purposes (see Fig. 5). A number of small molecules have been identified that can activate or inhibit the pathway but thus far, they all depend upon Wnt ligands to exert their effects (reviewed in Leucht et al. 2008b). Recombinant proteins such as the Wnt agonist R-spondin (Kazanskaya et al. 2004; Kim et al. 2008) have been used to treat oral mucositis (Zhao et al. 2009a), and recombinant Wnt3a

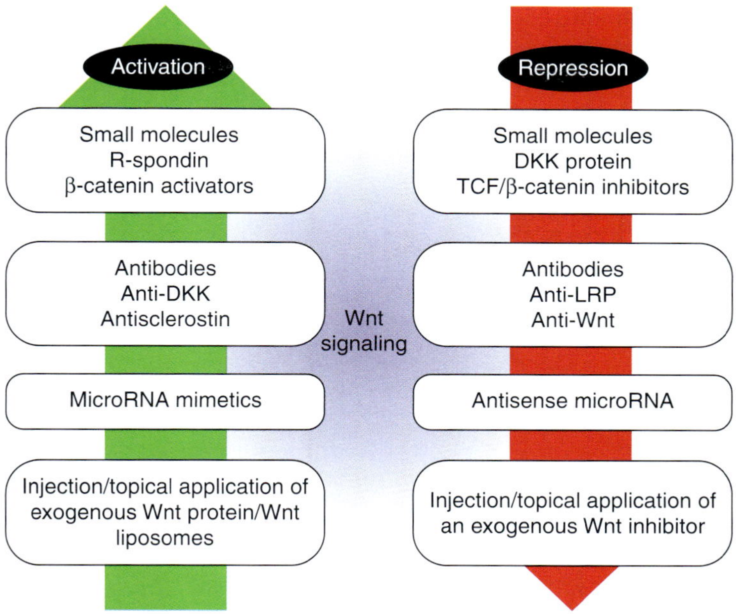

Figure 5. Harnessing Wnt signaling for therapeutic applications in injury repair. There are many biological ways to activate or repress the Wnt signaling pathway; these may be therapeutic options for increasing a healing response or reducing uncontrolled proliferation. The use of small molecules, microRNAs, genetically modified cells, as well as ligands and inhibitors, are all potential future therapeutics for both activating and repressing the Wnt signaling pathway.

protein has been used to transiently elevate Wnt signaling in bone injury sites and in cases of implant osseointegration (Zhao et al. 2009b; Minear et al. 2010; Popelut et al. 2010). In these cases, the hydrophobic Wnt protein has been packaged into a lipid vesicle that maintains its activity in vivo (Morrell et al. 2008).

Anti-Dkk1 antibodies have been employed to treat the osteolytic lesions associated with multiple myeloma (Fulciniti et al. 2009), and antisclerostin antibodies are in phase I trials for the treatment of rheumatoid arthritis (Choi et al. 2009). Although microRNAs and small interfering RNAs have been identified that effectively block activity of Wnt pathway components (Korpal et al. 2008; Thatcher et al. 2008; Hashimi et al. 2009), it is unclear whether they have the ability to regulate Wnt signaling at sites of injury. Such an approach is likely to be challenging since a single miRNA can have multiple mRNA targets, each target can be regulated by several miRNAs, and the effects of a specific miRNA are variable depending on the cell context.

CONCLUDING REMARKS

Converting tissue repair to tissue regeneration remains a lofty goal, but growing evidence suggests it is a realistic objective. We now know that almost every adult tissue harbors stem cells with potential to regenerate damaged or diseased tissues. But there are problems: adult stem cells are usually found in low abundance, and their response to stress and aging typically diminishes their ability to self-renew and proliferate. Endogenous Wnt signaling regulates stem cell recruitment and differentiation during the wound repair process, which suggests that modulation of Wnt signaling can positively influence the injury repair process. Strict attention will have to be paid to controlling Wnt signaling in a spatial and temporal manner in order to avoid the well-documented untoward effects of unrestrained Wnt pathway activity.

ACKNOWLEDGMENTS

We thank Wilfred Manzano and Edward Wang for their help in compiling the manuscript. This work is supported by grants from the California Institute for Regenerative Medicine (CIRM) to J.A.H. and to A.A.S.

REFERENCES

Agata K, Saito Y, Nakajima E. 2007. Unifying principles of regeneration I: Epimorphosis versus morphallaxis. *Dev Growth Differ* **49:** 73–78.

Anderson EC, Wong MH. 2010. Caught in the Akt: Regulation of Wnt signaling in the intestine. *Gastroenterology* **139:** 718–722.

Azevedo AS, Grotek B, Jacinto A, Weidinger G, Saude L. 2011. The regenerative capacity of the zebrafish caudal fin is not affected by repeated amputations. *PLoS ONE* **6:** e22820.

Baker-LePain JC, Nakamura MC, Lane NE. 2011. Effects of inflammation on bone: An update. *Curr Opin Rheumatol* **23:** 389–395.

Barker N, Clevers H. 2006. Mining the Wnt pathway for cancer therapeutics. *Nat Rev Drug Discov* **5:** 997–1014.

Beers MF, Morrisey EE. 2011. The three R's of lung health and disease: Repair, remodeling, and regeneration. *J Clin Invest* **121:** 2065–2073.

Boyden LM, Mao J, Belsky J, Mitzner L, Farhi A, Mitnick MA, Wu D, Insogna K, Lifton RP. 2002. High bone density due to a mutation in LDL-receptor-related protein 5. *N Engl J Med* **346:** 1513–1521.

Brunetti-Pierri N, Del Gaudio D, Peters H, Justino H, Ott CE, Mundlos S, Bacino CA. 2008. Robinow syndrome: Phenotypic variability in a family with a novel intragenic ROR2 mutation. *Am J Med Genet A* **146A:** 2804–2809.

Buchanan EP, Longaker MT, Lorenz HP. 2009. Fetal skin wound healing. *Adv Clin Chem* **48:** 137–161.

Cameron DA. 2000. Cellular proliferation and neurogenesis in the injured retina of adult zebrafish. *Vis Neurosci* **17:** 789–797.

Cameron DA, Vafai H, White JA. 1999. Analysis of dendritic arbors of native and regenerated ganglion cells in the goldfish retina. *Vis Neurosci* **16:** 253–261.

Carre AL, James AW, MacLeod L, Kong W, Kawai K, Longaker MT, Lorenz HP. 2010. Interaction of wingless protein (Wnt), transforming growth factor-β1, and hyaluronan production in fetal and postnatal fibroblasts. *Plast Reconstr Surg* **125:** 74–88.

Chan EF, Gat U, McNiff JM, Fuchs E. 1999. A common human skin tumour is caused by activating mutations in β-catenin. *Nat Genet* **21:** 410–413.

Chang HY, Sneddon JB, Alizadeh AA, Sood R, West RB, Montgomery K, Chi JT, van de Rijn M, Botstein D, Brown PO. 2004. Gene expression signature of fibroblast serum response predicts human cancer progression: Similarities between tumors and wounds. *PLoS Biol* **2:** E7.

Chen L, Wu Q, Guo F, Xia B, Zuo J. 2004. Expression of Dishevelled-1 in wound healing after acute myocardial infarction: Possible involvement in myofibroblast proliferation and migration. *J Cell Mol Med* **8:** 257–264.

Chen Y, Whetstone HC, Lin AC, Nadesan P, Wei Q, Poon R, Alman BA. 2007. β-Catenin signaling plays a disparate role in different phases of fracture repair: Implications for therapy to improve bone healing. *PLoS Med* **4:** e249.

Cho SH, Cepko CL. 2006. Wnt2b/β-catenin-mediated canonical Wnt signaling determines the peripheral fates of the chick eye. *Development* **133:** 3167–3177.

Choi Y, Arron JR, Townsend MJ. 2009. Promising bone-related therapeutic targets for rheumatoid arthritis. *Nat Rev Rheumatol* **5:** 543–548.

Clevers H. 2006. Wnt/β-catenin signaling in development and disease. *Cell* **127:** 469–480.

Coulombre JL, Coulombre AJ. 1965. Regeneration of neural retina from the pigmented epithelium in the chick embryo. *Dev Biol* **12:** 79–92.

Coulombre JL, Coulombre AJ. 1970. Influence of mouse neural retina on regeneration of chick neural retina from chick embryonic pigmented epithelium. *Nature* **228:** 559–560.

Das AV, Bhattacharya S, Zhao X, Hegde G, Mallya K, Eudy JD, Ahmad I. 2008. The canonical Wnt pathway regulates retinal stem cells/progenitors in concert with Notch signaling. *Dev Neurosci* **30:** 389–409.

de Lau W, Barker N, Clevers H. 2007. WNT signaling in the normal intestine and colorectal cancer. *Front Biosci* **12:** 471–491.

Denayer T, Locker M, Borday C, Deroo T, Janssens S, Hecht A, van Roy F, Perron M, Vleminckx K. 2008. Canonical Wnt signaling controls proliferation of retinal stem/progenitor cells in postembryonic Xenopus eyes. *Stem Cells* **26:** 2063–2074.

Diarra D, Stolina M, Polzer K, Zwerina J, Ominsky MS, Dwyer D, Korb A, Smolen J, Hoffmann M, Scheinecker C, et al. 2007. Dickkopf-1 is a master regulator of joint remodeling. *Nat Med* **13:** 156–163.

Dinsmore CE. 1992. The foundations of contemporary regeneration research: Historical perspectives. *Monogr Dev Biol* **23:** 1–27.

Dovi JV, He LK, DiPietro LA. 2003. Accelerated wound closure in neutrophil-depleted mice. *J Leukoc Biol* **73:** 448–455.

Eguchi G, Eguchi Y, Nakamura K, Yadav MC, Millan JL, Tsonis PA. 2011. Regenerative capacity in newts is not altered by repeated regeneration and ageing. *Nat Commun* **2:** 384.

El Ghissassi F, Baan R, Straif K, Grosse Y, Secretan B, Bouvard V, Benbrahim-Tallaa L, Guha N, Freeman C, Galichet L, et al. 2009. A review of human carcinogens—part D: Radiation. *Lancet Oncol* **10:** 751–752.

Fischer AJ, Reh TA. 2000. Identification of a proliferating marginal zone of retinal progenitors in postnatal chickens. *Dev Biol* **220:** 197–210.

Fischer AJ, Reh TA. 2001. Muller glia are a potential source of neural regeneration in the postnatal chicken retina. *Nat Neurosci* **4:** 247–252.

French DM, Kaul RJ, D'Souza AL, Crowley CW, Bao M, Frantz GD, Filvaroff EH, Desnoyers L. 2004. WISP-1 is an osteoblastic regulator expressed during skeletal development and fracture repair. *Am J Pathol* **165:** 855–867.

Friedman MS, Oyserman SM, Hankenson KD. 2009. Wnt11 promotes osteoblast maturation and mineralization through R-spondin 2. *J Biol Chem* **284:** 14117–14125.

Fulciniti M, Tassone P, Hideshima T, Vallet S, Nanjappa P, Ettenberg SA, Shen Z, Patel N, Tai YT, Chauhan D, et al. 2009. Anti-DKK1 mAb (BHQ880) as a potential therapeutic agent for multiple myeloma. *Blood* **114:** 371–379.

Galis F, Wagner GP, Jockusch EL. 2003. Why is limb regeneration possible in amphibians but not in reptiles, birds, and mammals? *Evol Dev* **5:** 208–220.

Gardiner DM. 2005. Ontogenetic decline of regenerative ability and the stimulation of human regeneration. *Rejuvenation Res* **8:** 141–153.

Gurley KA, Rink JC, Sanchez Alvarado A. 2008. β-Catenin defines head versus tail identity during planarian regeneration and homeostasis. *Science* **319:** 323–327.

Gurtner GC, Werner S, Barrandon Y, Longaker MT. 2008. Wound repair and regeneration. *Nature* **453:** 314–321.

Gurung A, Uddin F, Hill RP, Ferguson PC, Alman BA. 2009. β-Catenin is a mediator of the response of fibroblasts to irradiation. *Am J Pathol* **174:** 248–255.

Harty M, Neff AW, King MW, Mescher AL. 2003. Regeneration or scarring: An immunologic perspective. *Dev Dyn* **226:** 268–279.

Hashimi ST, Fulcher JA, Chang MH, Gov L, Wang S, Lee B. 2009. MicroRNA profiling identifies miR-34a and miR-21 and their target genes JAG1 and WNT1 in the coordinate regulation of dendritic cell differentiation. *Blood* **114:** 404–414.

He B, You L, Uematsu K, Xu Z, Lee AY, Matsangou M, McCormick F, Jablons DM. 2004. A monoclonal antibody against Wnt-1 induces apoptosis in human cancer cells. *Neoplasia* **6:** 7–14.

Ito M, Yang Z, Andl T, Cui C, Kim N, Millar SE, Cotsarelis G. 2007. Wnt-dependent de novo hair follicle regeneration in adult mouse skin after wounding. *Nature* **447:** 316–320.

Kaiser M, Mieth M, Liebisch P, Oberlander R, Rademacher J, Jakob C, Kleeberg L, Fleissner C, Braendle E, Peters M, et al. 2008. Serum concentrations of DKK-1 correlate with the extent of bone disease in patients with multiple myeloma. *Eur J Haematol* **80:** 490–494.

Kawakami Y, Rodriguez Esteban C, Raya M, Kawakami H, Marti M, Dubova I, Izpisua Belmonte JC. 2006. Wnt/β-catenin signaling regulates vertebrate limb regeneration. *Genes Dev* **20:** 3232–3237.

Kazanskaya O, Glinka A, del Barco Barrantes I, Stannek P, Niehrs C, Wu W. 2004. R-Spondin2 is a secreted activator of Wnt/β-catenin signaling and is required for Xenopus myogenesis. *Dev Cell* **7:** 525–534.

Kersting M, Friedl C, Kraus A, Behn M, Pankow W, Schuermann M. 2000. Differential frequencies of p16(INK4a) promoter hypermethylation, p53 mutation, and K-ras mutation in exfoliative material mark the development of lung cancer in symptomatic chronic smokers. *J Clin Oncol* **18:** 3221–3229.

Kim JB, Leucht P, Lam K, Luppen C, Ten Berge D, Nusse R, Helms JA. 2007. Bone regeneration is regulated by Wnt signaling. *J Bone Miner Res* **22:** 1913–1923.

Kim KA, Wagle M, Tran K, Zhan X, Dixon MA, Liu S, Gros D, Korver W, Yonkovich S, Tomasevic N, et al. 2008. R-Spondin family members regulate the Wnt pathway by a common mechanism. *Molec Biol Cell* **19:** 2588–2596.

Kim J, Kim J, Kim DW, Ha Y, Ihm MH, Kim H, Song K, Lee I. 2010. Wnt5a induces endothelial inflammation via β-catenin-independent signaling. *J Immunol* **185:** 1274–1282.

Koch S, Nava P, Addis C, Kim W, Denning TL, Li L, Parkos CA, Nusrat A. 2011. The Wnt antagonist Dkk1 regulates intestinal epithelial homeostasis and wound repair. *Gastroenterology* **141:** 259–268.

Korpal M, Lee ES, Hu G, Kang Y. 2008. The miR-200 family inhibits epithelial-mesenchymal transition and cancer cell migration by direct targeting of E-cadherin transcriptional repressors ZEB1 and ZEB2. *J Biol Chem* **283:** 14910–14914.

Kragl M, Knapp D, Nacu E, Khattak S, Maden M, Epperlein HH, Tanaka EM. 2009. Cells keep a memory of their tissue origin during axolotl limb regeneration. *Nature* **460:** 60–65.

Kubo F, Takeichi M, Nakagawa S. 2003. Wnt2b controls retinal cell differentiation at the ciliary marginal zone. *Development* **130:** 587–598.

Larson BJ, Longaker MT, Lorenz HP. 2010. Scarless fetal wound healing: A basic science review. *Plast Reconstr Surg* **126:** 1172–1180.

Leucht P, Kim JB, Helms JA. 2008a. β-Catenin-dependent Wnt signaling in mandibular bone regeneration. *J Bone Joint Surg Am* **90** (Suppl 1): 3–8.

Leucht P, Minear S, Ten Berge D, Nusse R, Helms JA. 2008b. Translating insights from development into regenerative medicine: The function of Wnts in bone biology. *Semin Cell Dev Biol* **19:** 434–443.

Levesque M, Villiard E, Roy S. 2010. Skin wound healing in axolotls: A scarless process. *J Exp Zool B Mol Dev Evol* **314:** 684–697.

Liang H, Chen Q, Coles AH, Anderson SJ, Pihan G, Bradley A, Gerstein R, Jurecic R, Jones SN. 2003. Wnt5a inhibits B cell proliferation and functions as a tumor suppressor in hematopoietic tissue. *Cancer Cell* **4:** 349–360.

Liu B, Yu HM, Hsu W. 2007a. Craniosynostosis caused by Axin2 deficiency is mediated through distinct functions of β-catenin in proliferation and differentiation. *Dev Biol* **301:** 298–308.

Liu H, Xu S, Wang Y, Mazerolle C, Thurig S, Coles BL, Ren JC, Taketo MM, van der Kooy D, Wallace VA. 2007b. Ciliary margin transdifferentiation from neural retina is controlled by canonical Wnt signaling. *Dev Biol* **308:** 54–67.

Martin P, Leibovich SJ. 2005. Inflammatory cells during wound repair: The good, the bad and the ugly. *Trends Cell Biol* **15:** 599–607.

Mazumdar J, O'Brien WT, Johnson RS, LaManna JC, Chavez JC, Klein PS, Simon MC. 2010. O2 regulates stem cells through Wnt/β-catenin signalling. *Nat Cell Biol* **12:** 1007–1013.

McKillop IH, Schrum LW. 2005. Alcohol and liver cancer. *Alcohol* **35:** 195–203.

McLean KE, Vickaryous MK. 2011. A novel amniote model of epimorphic regeneration: The leopard gecko, Eublepharis macularius. *BMC Dev Biol* **11:** 50.

Mescher AL, Neff AW. 2005. Regenerative capacity and the developing immune system. *Adv Biochem Eng Biotechnol* **93:** 39–66.

Mescher AL, Neff AW. 2006. Limb regeneration in amphibians: Immunological considerations. *Sci World J.* **6** (Suppl 1): 1–11.

Midwood KS, Williams LV, Schwarzbauer JE. 2004. Tissue repair and the dynamics of the extracellular matrix. *Int J Biochem Cell Biol* **36:** 1031–1037.

Minear S, Leucht P, Jiang J, Liu B, Zeng A, Fuerer C, Nusse R, Helms JA. 2010. Wnt proteins promote bone regeneration. *Sci Transl Med* **2:** 29ra30.

Morrell NT, Leucht P, Zhao L, Kim JB, ten Berge D, Ponnusamy K, Carre AL, Dudek H, Zachlederova M, McElhaney M, et al. 2008. Liposomal packaging generates Wnt protein with in vivo biological activity. *PLoS ONE* **3:** e2930.

Newton PM, Watson JA, Wolowacz RG, Wood EJ. 2004. Macrophages restrain contraction of an in vitro wound healing model. *Inflammation* **28:** 207–214.

Nurden AT, Nurden P, Sanchez M, Andia I, Anitua E. 2008. Platelets and wound healing. *Front Biosci* **13:** 3532–3548.

Pereira CP, Bachli EB, Schoedon G. 2009. The wnt pathway: A macrophage effector molecule that triggers inflammation. *Curr Atheroscler Rep* **11:** 236–242.

Petersen CP, Reddien PW. 2008. Smed-βcatenin-1 is required for anteroposterior blastema polarity in planarian regeneration. *Science* **319:** 327–330.

Petersen CP, Reddien PW. 2009. A wound-induced Wnt expression program controls planarian regeneration polarity. *Proc Natl Acad Sci* **106:** 17061–17066.

Polzer K, Diarra D, Zwerina J, Schett G. 2008. Inflammation and destruction of the joints—The Wnt pathway. *Joint Bone Spine* **75:** 105–107.

Popelut A, Rooker SM, Leucht P, Medio M, Brunski JB, Helms JA. 2010. The acceleration of implant osseointegration by liposomal Wnt3a. *Biomaterials* **31:** 9173–9181.

Poss KD, Shen J, Keating MT. 2000. Induction of lef1 during zebrafish fin regeneration. *Dev Dyn* **219:** 282–286.

Qyang Y, Martin-Puig S, Chiravuri M, Chen S, Xu H, Bu L, Jiang X, Lin L, Granger A, Moretti A, et al. 2007. The renewal and differentiation of $Isl1^{+}$ cardiovascular progenitors are controlled by a Wnt/β-catenin pathway. *Cell Stem Cell* **1:** 165–179.

Rai NK, Tripathi K, Sharma D, Shukla VK. 2005. Apoptosis: A basic physiologic process in wound healing. *Int J Low Extrem Wounds* **4:** 138–144.

Ramachandran R, Zhao XF, Goldman D. 2011. Ascl1a/Dkk/β-catenin signaling pathway is necessary and glycogen synthase kinase-3β inhibition is sufficient for zebrafish retina regeneration. *Proc Natl Acad Sci* **108:** 15858–15863.

Raymond PA, Barthel LK, Bernardos RL, Perkowski JJ. 2006. Molecular characterization of retinal stem cells and their niches in adult zebrafish. *BMC Dev Biol* **6:** 36.

Reh TA, Levine EM. 1998. Multipotential stem cells and progenitors in the vertebrate retina. *J Neurobiol* **36:** 206–220.

Rinkevich Y, Lindau P, Ueno H, Longaker MT, Weissman IL. 2011. Germ-layer and lineage-restricted stem/progenitors regenerate the mouse digit tip. *Nature* **476:** 409–413.

Rubinfeld B, Robbins P, El-Gamil M, Albert I, Porfiri E, Polakis P. 1997. Stabilization of β-catenin by genetic defects in melanoma cell lines. *Science* **275:** 1790–1792.

Sanchez Alvarado A. 2000. Regeneration in the metazoans: Why does it happen? *Bioessays* **22:** 578–590.

Sansom OJ, Meniel VS, Muncan V, Phesse TJ, Wilkins JA, Reed KR, Vass JK, Athineos D, Clevers H, Clarke AR. 2007. Myc deletion rescues Apc deficiency in the small intestine. *Nature* **446:** 676–679.

Sato T, Vries RG, Snippert HJ, van de Wetering M, Barker N, Stange DE, van Es JH, Abo A, Kujala P, Peters PJ, et al. 2009. Single Lgr5 stem cells build crypt-villus structures in vitro without a mesenchymal niche. *Nature* **459:** 262–265.

Satoh A, Cummings GM, Bryant SV, Gardiner DM. 2010. Regulation of proximal-distal intercalation during limb regeneration in the axolotl (Ambystoma mexicanum). *Dev Growth Differ* **52:** 785–798.

Sen M, Lauterbach K, El-Gabalawy H, Firestein GS, Corr M, Carson DA. 2000. Expression and function of wingless and frizzled homologs in rheumatoid arthritis. *Proc Natl Acad Sci* **97:** 2791–2796.

Sezer O. 2005. Myeloma bone disease. *Hematology* **10** (Suppl 1): 19–24.

Silva H, Conboy IM. 2008. Aging and stem cell renewal. In *StemBook*, Cambridge, MA.

Sinnberg T, Menzel M, Kaesler S, Biedermann T, Sauer B, Nahnsen S, Schwarz M, Garbe C, Schittek B. 2010. Suppression of casein kinase 1α in melanoma cells induces a switch in β-catenin signaling to promote metastasis. *Cancer Res* **70:** 6999–7009.

Steele BM, Harper MT, Macaulay IC, Morrell CN, Perez-Tamayo A, Foy M, Habas R, Poole AW, Fitzgerald DJ, Maguire PB. 2009. Canonical Wnt signaling negatively regulates platelet function. *Proc Natl Acad Sci* **106:** 19836–19841.

Stephens WZ, Senecal M, Nguyen M, Piotrowski T. 2010. Loss of adenomatous polyposis coli (apc) results in an expanded ciliary marginal zone in the zebrafish eye. *Dev Dyn* **239:** 2066–2077.

Stewart S, Tsun ZY, Izpisua Belmonte JC. 2009. A histone demethylase is necessary for regeneration in zebrafish. *Proc Natl Acad Sci* **106:** 19889–19894.

Stoick-Cooper CL, Weidinger G, Riehle KJ, Hubbert C, Major MB, Fausto N, Moon RT. 2007. Distinct Wnt signaling pathways have opposing roles in appendage regeneration. *Development* **134:** 479–489.

Straube WL, Brockes JP, Drechsel DN, Tanaka EM. 2004. Plasticity and reprogramming of differentiated cells in amphibian regeneration: Partial purification of a serum factor that triggers cell cycle re-entry in differentiated muscle cells. *Cloning Stem Cells* **6:** 333–344.

Thatcher EJ, Paydar I, Anderson KK, Patton JG. 2008. Regulation of zebrafish fin regeneration by microRNAs. *Proc Natl Acad Sci* **105:** 18384–18389.

Thompson DA. 1884. The regeneration of lost parts in animals. *Mind* **9:** 415–420.

Tian E, Zhan F, Walker R, Rasmussen E, Ma Y, Barlogie B, Shaughnessy JD Jr. 2003. The role of the Wnt-signaling antagonist DKK1 in the development of osteolytic lesions in multiple myeloma. *N Engl J Med* **349:** 2483–2494.

Trembley A. 1744. *Memoires pour servir a l'histoire d'un genre de polypes d'eau douce, a bras en forme de cornes.* Verbeek, Leide.

Tropepe V, Coles BL, Chiasson BJ, Horsford DJ, Elia AJ, McInnes RR, van der Kooy D. 2000. Retinal stem cells in the adult mammalian eye. *Science* **287:** 2032–2036.

Vaid M, Prasad R, Sun Q, Katiyar SK. 2011. Silymarin targets β-catenin signaling in blocking migration/invasion of human melanoma cells. *PLoS ONE* **6:** e23000.

van der Flier LG, Clevers H. 2008. Stem cells, self-renewal, and differentiation in the intestinal epithelium. *Annu Rev Physiol* **71:** 241–260.

Vermeulen L, De Sousa EMF, van der Heijden M, Cameron K, de Jong JH, Borovski T, Tuynman JB, Todaro M, Merz C, Rodermond H, et al. 2010. Wnt activity defines colon cancer stem cells and is regulated by the microenvironment. *Nat Cell Biol* **12:** 468–476.

Villar J, Cabrera NE, Valladares F, Casula M, Flores C, Blanch L, Quilez ME, Santana-Rodriguez N, Kacmarek RM, Slutsky AS. 2011. Activation of the Wnt/β-catenin signaling pathway by mechanical ventilation is associated with ventilator-induced pulmonary fibrosis in healthy lungs. *PLoS ONE* **6:** e23914.

Wehkamp J, Wang G, Kubler I, Nuding S, Gregorieff A, Schnabel A, Kays RJ, Fellermann K, Burk O, Schwab M, et al. 2007. The Paneth cell α-defensin deficiency of ileal Crohn's disease is linked to Wnt/Tcf-4. *J Immunol* **179:** 3109–3118.

West DC, Shaw DM, Lorenz P, Adzick NS, Longaker MT. 1997. Fibrotic healing of adult and late gestation fetal wounds correlates with increased hyaluronidase activity and removal of hyaluronan. *Int J Biochem Cell Biol* **29:** 201–210.

Yaccoby S, Ling W, Zhan F, Walker R, Barlogie B, Shaughnessy JD Jr. 2007. Antibody-based inhibition of DKK1 suppresses tumor-induced bone resorption and multiple myeloma growth in vivo. *Blood* **109:** 2106–2111.

Yazawa S, Umesono Y, Hayashi T, Tarui H, Agata K. 2009. Planarian Hedgehog/Patched establishes anterior-posterior polarity by regulating Wnt signaling. *Proc Natl Acad Sci* **106:** 22329–22334.

Yokoyama H, Maruoka T, Ochi H, Aruga A, Ohgo S, Ogino H, Tamura K. 2011. Different requirement for Wnt/β-catenin signaling in limb regeneration of larval and adult Xenopus. *PLoS ONE* **6:** e21721.

Zhao J, Kim KA, De Vera J, Palencia S, Wagle M, Abo A. 2009a. R-Spondin1 protects mice from chemotherapy or radiation-induced oral mucositis through the canonical Wnt/β-catenin pathway. *Proc Natl Acad Sci* **106:** 2331–2336.

Zhao L, Rooker SM, Morrell N, Leucht P, Simanovskii D, Helms JA. 2009b. Controlling the in vivo activity of Wnt liposomes. *Methods Enzymol* **465:** 331–347.

Targeting Wnt Pathways in Disease

Zachary F. Zimmerman[1,2], **Randall T. Moon**[2,3], **and Andy J. Chien**[2,4]

[1]Department of Medicine, Division of Oncology, University of Washington, Seattle, Washington 98195

[2]Institute for Stem Cell and Regenerative Medicine, University of Washington, Seattle, Washington 98109

[3]Department of Pharmacology and The Howard Hughes Medical Institute, University of Washington, Seattle, Washington 98195

[4]Department of Medicine, Division of Dermatology, University of Washington, Seattle, Washington 98195

Correspondence: andchien@u.washington.edu

Wnt-mediated signal transduction pathways have long been recognized for their roles in regulating embryonic development, and have more recently been linked to cancer, neurologic diseases, inflammatory diseases, and disorders of endocrine function and bone metabolism in adults. Although therapies targeting Wnt signaling are attractive in theory, in practice it has been difficult to obtain specific therapeutics because many components of Wnt signaling pathways are also involved in other cellular processes, thereby reducing the specificity of candidate therapeutics. New technologies, and advances in understanding the mechanisms of Wnt signaling, have improved our understanding of the nuances of Wnt signaling and are leading to promising new strategies to target Wnt signaling pathways.

As covered by other Wnt signaling investigators, the Wnt family of secreted glycoproteins act as ligands to activate multiple signal transduction pathways. The best understood is the Wnt/β-catenin pathway, which activates the function of β-catenin in the nucleus to regulate expression of genes by binding to T-cell factor/lymphoid enhancer factor (TCF/LEF) transcription factors, in addition to its role in regulating cadherin-dependent cell adhesion at the plasma membrane. The Wnt/β-catenin pathway acts in a context-dependent manner to regulate cell proliferation, migration, differentiation, and survival in both embryonic development and in adults.

Perturbations in the levels of Wnt/β-catenin signaling are linked to many disease processes. Elevation of Wnt/β-catenin signaling has been linked to cancer, whereas conversely attenuation of Wnt/β-catenin signaling has been linked to a distinct set of diseases including Alzheimer's disease, familial exudative vitreoretinopathy (FEVR), and disorders of bone formation (Robitaille et al. 2002; reviewed in De Ferrari and Moon 2006; Hoeppner et al. 2009).

Reports have even shown that Wnt/β-catenin signaling can regulate aspects of human immunodeficiency virus (HIV) including gene expression and replication, further highlighting the ubiquitous function of this pathway in

Cite this article as *Cold Spring Harb Perspect Biol* doi: 10.1101/cshperspect.a008086

almost all cell types (Wortman et al. 2002; Carroll-Anzinger et al. 2007; Kumar et al. 2008; Kameoka et al. 2009). As the diverse roles for Wnt/β-catenin signaling in tissue homeostasis and disease continue to be elucidated, interest in therapeutic targeting of this pathway has expanded enormously.

In addition to the Wnt/β-catenin pathway, β-catenin-independent pathways such as the planar cell polarity (PCP) pathway and Wnt/Ca^{2+} pathway have been described. These pathways are less well understood in part owing to the paucity of established reporter assays. In addition, β-catenin-independent Wnt signaling often inhibits the Wnt/β-catenin pathway, making the relative contribution of pathway activation versus inhibition to an observed phenotype difficult to distinguish. Studies in a variety of organisms have established that β-catenin-independent Wnt signaling is involved in regulating cell polarity during gastrulation in embryos, and in the polarized orientation of hair cells in the inner ear, as well as mesenchymal stem cell maintenance and renal development (reviewed in Sugimura and Li 2010). The β-catenin-independent Wnt pathway(s) are also linked to disease processes, notably cancer.

From a therapeutic point of view, the Wnt signaling pathways present several challenges to the development of a targeted drug, so it is not surprising that drug strategies specifically directed at this pathway are in a state of relative infancy. In addition to the existence of 19 Wnt ligands and 10 FZD receptor isoforms, specificity of targeting is further complicated by the convergence of downstream Wnt signaling events on promiscuous enzymes like glycogen synthase kinase 3 (GSK3) and on proteins that are central to fundamental and ubiquitous cellular structures such as the cytoskeleton and cell–cell junctions that are critical to all cells. Predictability of drug effects can be problematic with a pathway as ubiquitous as Wnt, particularly because it is quite likely that the majority of both somatic cells and stem cell niches in the body will show some effect with either Wnt activation or inhibition. Nevertheless, many studies in manipulation of Wnt signaling pathways have shown promise that we will examine in more detail in this review.

WNT/β-CATENIN SIGNALING IN CANCER

The Wnt/β-catenin pathway has been associated with cancer ever since the gene *int-1* was identified as a mammary oncogene in mice (Nusse and Varmus 1982; Rijsewijk 1987). This relationship was solidified with the discovery that the adenomatous polyposis coli (APC) gene associated with familial adenomatous polyposis (FAP) is inactivated in ~85% of colorectal carcinomas leading to constitutive nuclear translocation of β-catenin (Kinzler et al. 1991; Nishisho et al. 1991; Su et al. 1993). As a result, the therapeutic targeting of Wnt/β-catenin signaling has received considerable interest in the context of cancer.

Dysregulated Wnt/β-catenin signaling occurs in many solid tumors and hematologic malignancies even in the absence of documented mutations. Frequently, altered levels of expression of Wnt/β-catenin pathway regulators have been observed without mutations in the coding regions of the respective genes. For example, epigenetic silencing of genes that encode putative extracellular Wnt antagonists such as the secreted frizzled-related proteins (SFRPs) has been described in colon, breast, prostate, lung, and other cancers (Caldwell et al. 2004; Lee et al. 2004a; Suzuki et al. 2004; Fukui et al. 2005; Zou et al. 2005). Increased expression of other Wnt/β-catenin pathway members such as Wnt ligands (Rhee et al. 2002; Wong et al. 2002; Milovanovic et al. 2004) or dishevelled (DVL) have also been described (Okino et al. 2003; Uematsu et al. 2003a,b).

CONTEXT-DEPENDENT CORRELATIONS BETWEEN β-CATENIN AND PATIENT SURVIVAL

The role of Wnt/β-catenin activation via APC mutations in colorectal cancer progression as well as the initial discovery of *int-1* in a mammary tumor screen has led to the dogmatic view that this pathway is oncogenic. Indeed, studies correlating increased Wnt/β-catenin signaling in tumors to worse prognosis for patients with colorectal cancer have helped solidify this viewpoint (Cheah et al. 2002; Miyamoto et al. 2004; Lugli et al. 2007). However, it is important to emphasize that elevated Wnt signaling does not correlate with reduced patient survival in all types of cancer. For example, in melanoma, active Wnt/β-catenin signaling as shown by nuclear β-catenin in tumors is associated with a lower proliferative index and correlates with a more favorable prognosis in melanoma patients (Kageshita et al. 2001; Maelandsmo et al. 2003; Bachmann et al. 2005; Chien et al. 2009b). Consistent with this observation, overexpression of Wnt3a in a murine melanoma model results in a less aggressive disease phenotype as well as an increase in melanocyte differentiation markers, which are often lost during melanoma progression (Chien et al. 2009b). Mechanistically, the effects of elevated β-catenin in melanoma differ from colorectal cancer because in the former the β-catenin promotes expression of the transcription factor *MITF*, which drives differentiation toward a melanocytelike cell fate and reduces cell movements (Dorsky et al. 2000; Takeda et al. 2000; Hornyak et al. 2001).

In some contexts, active Wnt signaling is correlated with less aggressive disease. Recently, discrete molecular subtypes of medulloblastoma were identified that included a subgroup characterized by constitutive Wnt/β-catenin pathway activation. The tumors in this subgroup universally harbor activating mutations in β-catenin (*CTNNB1*), and appear to derive from a distinct anatomic region within the brain. These Wnt/β-catenin-driven tumors are associated with a much better prognosis compared with those without active Wnt/β-catenin signaling (Gibson et al. 2010). Similarly, Wnt activation appears to function as a tumor suppressor in some sarcomas, as inhibition of Wnt in human mesenchymal stem cells leads to high-grade sarcoma formation in nude mice (Matushansky et al. 2007). In mouse models of pancreatic ductal adenocarcinoma (PDAC), the forced expression of a nondegradable constitutively active β-catenin can arrest Kras-driven transformation, resulting in a different tumor type that is much more benign compared with PDAC (Heiser et al. 2008; Morris et al. 2010). Other studies in prostate, ovarian, and even in colorectal cancer have also found that increased Wnt signaling in patient tumors is associated with a better prognosis in at least some stages of disease (Gamallo et al. 1999; Horvath et al. 2005; Elzagheid et al. 2008).

These findings suggest that the effects of Wnt/β-catenin signaling differ widely depending on the stage of disease progression, cell of origin, or context in which the signaling occurs. In our current model, dynamic regulation of Wnt signaling is tightly regulated in cells, resulting in an ongoing modulation of cellular Wnt activation levels by feedback mechanisms, which involve other diverse signaling pathways. The identification of putative Wnt inhibitors including AXIN2, DKK1, and TCF7L2 as direct transcriptional targets of β-catenin signaling supports the model of a tightly regulated pathway. To develop and successfully implement Wnt-directed therapies, several contextual considerations are essential. For one, understanding the mechanism of Wnt dysregulation, whether it be through ligand overexpression or mutational activation, will be central to developing appropriate strategies for targeting the pathway in different diseases. Second, the idea that Wnt activation may be an important part of maintaining homeostasis in normal, noncancerous cells must be considered in the bigger picture. Finally, the effects of targeting this pathway will undoubtedly vary across different cell types, which have broad implications for both predicting the efficacy of treatment as well as for predicting potential side effects in the context of an entire organism.

NONCANONICAL β-CATENIN-INDEPENDENT WNT SIGNALING IN CANCER

Although most studies have focused on the well-described Wnt/β-catenin pathway, Wnts can also activate pathways that act independently of β-catenin. These pathways, often referred to as "noncanonical" Wnt pathways, are currently known collectively as β-catenin-independent Wnt signaling. In both developmental models and in cultured mammalian cells, β-catenin-independent Wnt signaling is a key modulator of cellular adhesion and motility, and several studies have shown the involvement of this pathway in tumor cell migration and metastasis. Perhaps the best-studied activator of β-catenin-independent Wnt signaling is WNT5A, which has been shown to act both as a tumor suppressor as well as a promoter of tumor invasion depending on the model (Weeraratna et al. 2002; Kurayoshi et al. 2006; Pukrop 2006). In melanoma, WNT5A mediates a polarized redistribution of proteins such as myosin IIB to facilitate directional movement along a chemokine gradient (Witze et al. 2008). Other components of β-catenin-independent Wnt signaling such as VANGL1 and VANGL2 have also been implicated in tumor cell migration. These genes are mammalian homologs of the *Drosophila* PCP gene *van gogh/strabismus* and represent putative tetra-membrane-spanning proteins widely expressed in human cancers (reviewed in Katoh 2002). VANGL2 has recently been shown to form a Wnt-induced receptor complex with the receptor ROR2, which acts to sense Wnt dosage gradients in developing limb buds (Shafer et al. 2011). VANGL2 expression also appears to promote tumor cell migration and matrix metalloproteinase-dependent invasion in a human colon cancer line (Cantrell and Jessen 2010). VANGL1 also promotes metastases in murine models of colorectal and squamous cell carcinoma (Lee et al. 2004b, 2009; Kho et al. 2009). Interestingly, other roles for β-catenin-independent Wnt signaling besides migration are emerging. For example, WNT5A was found to potentiate androgen-mediated signaling through the androgen receptor in prostate cancer cells, thereby promoting tumor formation in a murine prostate cancer model (Takahashi et al. 2011).

Currently, the development of drugs targeting β-catenin-independent Wnt signaling has been limited by several key factors. First, assays for measuring activation of β-catenin-independent Wnt signaling are diverse, likely context dependent, and technically involved, thereby limiting the ability to perform large-scale exploratory studies such as high-throughput signaling. Second, our understanding of the mechanisms involved in β-catenin-independent Wnt signaling remains limited, even with regard to basic events such as the identification of receptors involved in ligand binding. Third, β-catenin-independent Wnt signaling can antagonize Wnt/β-catenin signaling, so even specific targeting of ligands such as Wnt5A could conceivably have implications for the broad effects of Wnt/β-catenin signaling in both cells and organisms. Nevertheless, there have been some efforts to target this pathway (Figs. 1 and 2; Tables 1 and 2) that have shown promise in early studies.

WNT PATHWAYS IN NEUROLOGIC DISEASE

Wnt-mediated pathways have a well-established role of the development of the central nervous system in embryonic development (reviewed in Salinas and Zou 2008). Therefore, it is not surprising that aberrations in Wnt signaling are common in neurologic diseases in the adult. One such disease that has received considerable attention is schizophrenia, in which genetic perturbations predicted to impact Wnt regulation have been recently identified. For example, a Scottish family with a high incidence of schizophrenia, depression, and bipolar disorder was found to carry a balanced chromosomal translocation involving disrupted in schizophrenia 1 (*DISC1*), encoding a protein subsequently found to have an important role in brain development and neural progenitor proliferation (Blackwood et al. 2001; Mao et al. 2009). DISC1 directly interacts with and inhibits GSK3β activity, resulting in enhanced β-catenin-mediated signaling (Mao et al. 2009). WNT1 was also observed at higher levels in the hippocampal region of brains of schizophrenic patients compared with controls on postmortem analysis (Miyaoka et al. 1999). Genetic linkage studies have also shown an association with schizophrenia and the Wnt receptor *FZD3* in a subset of schizophrenic Chinese patients (Yang et al. 2003a).

Other neurologic diseases have been linked to aberrant Wnt signaling as well. For example, the presenilin proteins, which have been associated with early-onset Alzheimer's disease, are

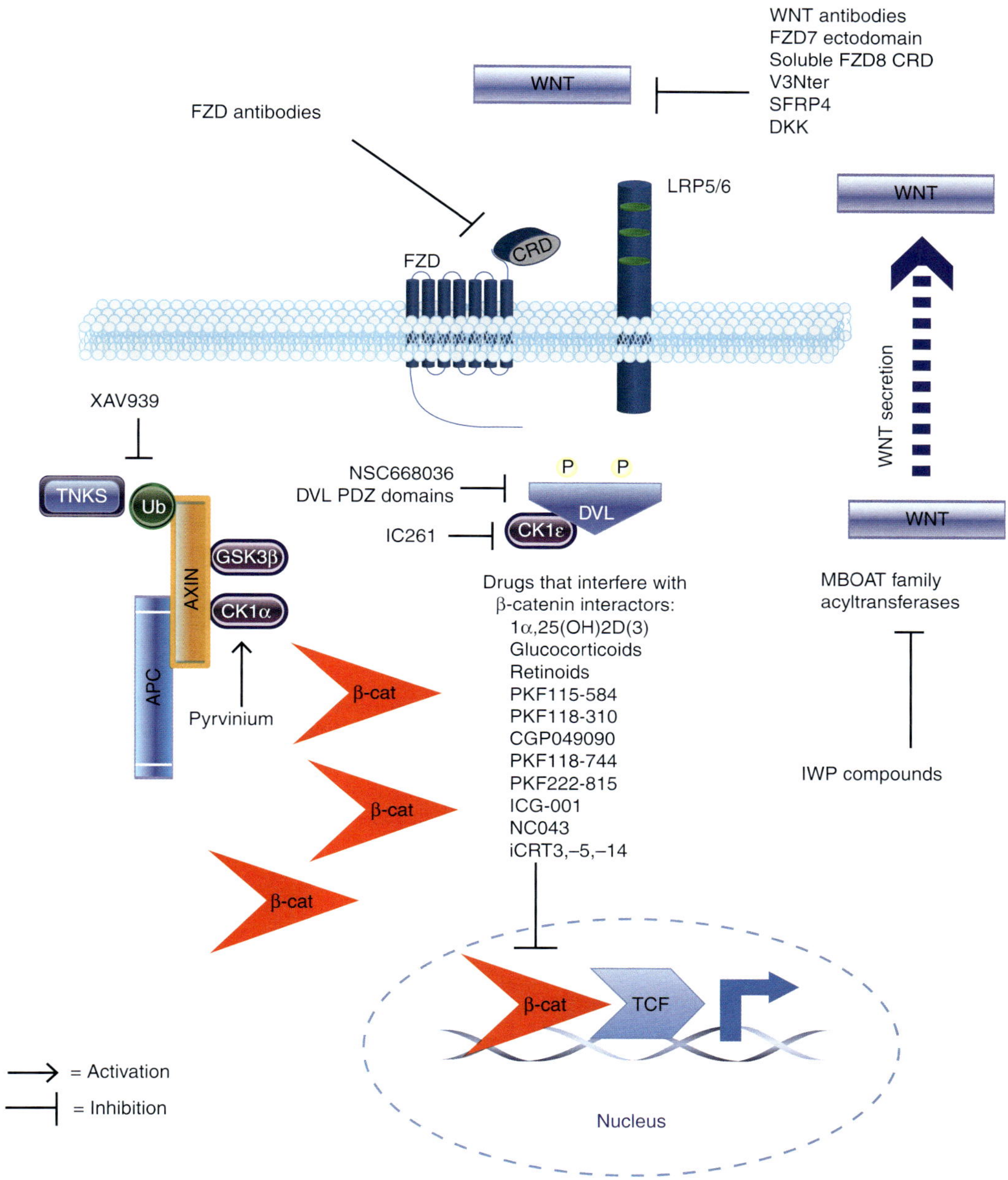

Figure 1. Pharmacologic inhibitors of the Wnt/β-catenin signaling pathway and known interactions with pathway components (see Table 1 for references). This figure depicts the Wnt/β-catenin pathway in its active state, with cytoplasmic accumulation and nuclear translocation of β-catenin, along with the transactivation of target genes in the nucleus mediated by interaction between β-catenin and members of the TCF family of transcription factors. Note that in contexts in which downstream mutations in APC or β-catenin result in constitutive pathway activation, the use of upstream inhibitors that target ligand secretion, ligand binding, and even the APC-AXIN-GSK3β complex may not be sufficient to inhibit signaling. Also, targeting of β-catenin interactors can involve either the disruption of protein interactions or the direct inhibition of transcriptional activity (see also Table 1 for additional details and references).

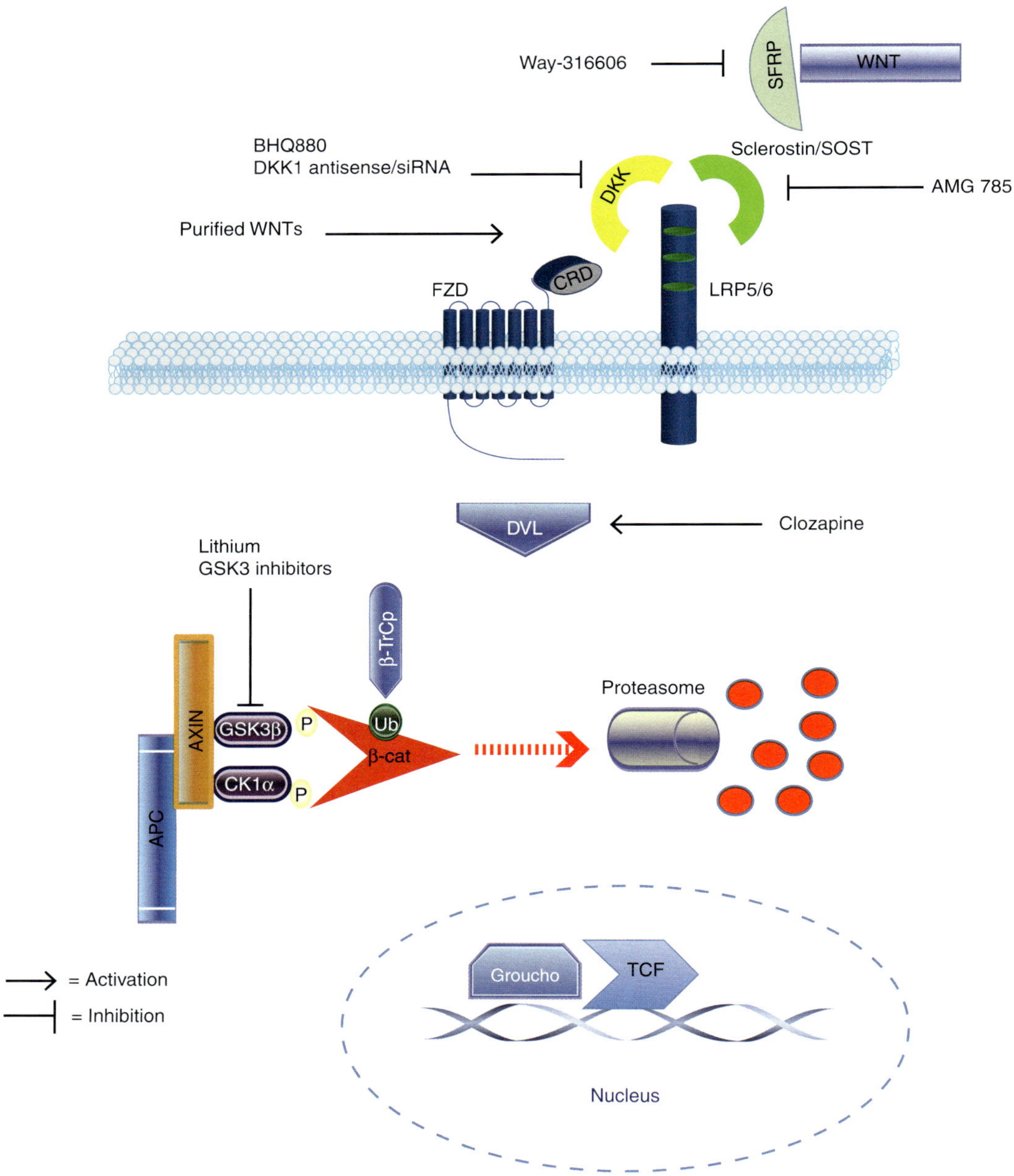

Figure 2. Pharamcologic activators and enhancers of the Wnt/β-catenin signaling pathway and known interactions with pathway components (see Table 2 for references). This figure depicts the pathway in the inactive state, in which constitutive phosphorylation of β-catenin by GSK3β targets the protein for ubiquitylation and subsequent proteasomal degradation. In the absence of nuclear β-catenin, TCF-family transcription factors are bound by members of the Groucho family of transcriptional repressors, thus inhibiting target gene expression.

Cite this article as *Cold Spring Harb Perspect Biol* doi: 10.1101/cshperspect.a008086

Table 1. Drug candidates that inhibit Wnt signaling and diseases/disease models in which they have been studied

Drug candidate	Target	Diseases or disease model(s)	References
Biologics			
Wnt1 antibody	Wnt1	NSCLC, head and neck cancer, sarcoma, colon cancer, breast cancer	He et al. 2004, 2005a,b; Mikami et al. 2005; Wei et al. 2009
Wnt2 antibody	Wnt2	Melanoma, mesothelioma, nonsmall cell lung cancer	You et al. 2004a,b; Mazieres et al. 2005b
Wnt5a antibody	Wnt5a	Rheumatoid arthritis	Sen et al. 2001
Wnt16 antibody	Wnt16	Acute lymphoblastic leukemia	Mazieres et al. 2005a
Fzd5 antibody	Fzd5	Melanoma, rheumatoid arthritis	Sen et al. 2001; Weeraratna et al. 2002
Soluble Fzd7	Fzd7	Hepatocellular carcinoma	Wei et al. 2011
Fzd10 antibody (conjugated with Y90)	Fzd10	Synovial sarcoma	Fukukawa et al. 2008
Fzd7 ectodomain	Fzd7	Colon cancer	Vincan et al. 2005
Soluble Fzd8 CRD	Wnt binding	Teratocarcinoma	DeAlmeida et al. 2007
V3Nter	Wnt binding	Colon cancer	Lavergne et al. 2011
sFRP4	Wnt binding	Renal injury	Surendran et al. 2005
Small molecules			
Glucocorticoids	β-catenin	Osteosarcoma	Takayama et al. 2006
Retinoids	β-catenin	Colon cancer	Xiao et al. 2003
Vitamin D(3)	β-catenin	Colon cancer	Pálmer et al. 2001
IWP compounds	MBOAT family of acyltransferases	Inhibition of Wnt-dependent colon/prostate cancer proliferation	Chen et al. 2009
IWR compounds	Axin stability	Inhibition of Wnt-dependent colon/prostate cancer proliferation	Chen et al. 2009
XAV939	TNKS1, TNKS2	Colon cancer, cardimyogenesis	Huang et al. 2009; Wang et al. 2011a
PKF115-584 PKF118-310 CGP049090 PKF118-744 PKF222-815	TCF4/β-catenin	Colon cancer proliferation	Lepourcelet et al. 2004
ICG-001	CBP	Pulmonary fibrosis, colon cancer	Emami et al. 2004; Henderson et al. 2010
NC043	TCF4/β-catenin	Colon cancer	Wang et al. 2011b
iCRT3,-5,-14	TCF/β-catenin	Breast cancer, colon cancer	Gonsalves et al. 2011
IC261	CKIε	Breast cancer	Kim et al. 2010
Fumagillin, TNP-470	MetAP-2	Angiogenesis	Zhang et al. 2006; Cirone et al. 2008
Phenylmethimazole	Wnt5a expression	Melanoma	Schwartz et al. 2009
Pyrvinium	CK1α	Cardiac repair	Saraswati et al. 2010; Thorne et al. 2010

NSCLC, non-small-cell lung cancer; CRD, cysteine-rich domain; CBP, CREB-binding protein; CKIε, casein kinase-Iε.

Table 2. Drug candidates that enhance Wnt signaling and diseases/disease models in which they have been studied

Drug candidate	Target or class	Diseases or disease model(s)	References
Biologics			
Purified Wnts	LPR5/6/ Fzd	Fracture repair, ES/iPS cell differentiation	Reya et al. 2003; Willert et al. 2003; Nostro et al. 2008; Lengerke et al. 2009; Vijayaragavan et al. 2009; Minear et al. 2010
DKK1 antibody (BHQ880)	DKK1	Osteoporosis, osteoarthritis, fracture healing, multiple myeloma	Wang et al. 2008; Fulciniti et al. 2009; Glantschnig et al. 2010; Komatsu et al. 2010
DKK1 antisense oligonucleotides	DKK1	Brain injury, osteoperosis	Cappuccio et al. 2005; Wang et al. 2007
DKK1 siRNA	DKK1	Alzheimer's disease, brain injury	Caricasole et al. 2004; Cappuccio et al. 2005
Sclerostin/SOST antibodies	Sclerostin	Osteoporosis, fracture healing	Li et al. 2009; Agholme et al. 2010; Ominsky et al. 2010; Padhi et al. 2011
Foxy-5	Wnt5a mimetic	Breast cancer	Säfholm et al. 2006; Ford et al. 2009
Small molecules			
Lithium	GSK3β	Bone formation, ES differentiation, Alzheimer's disease	de Boer et al. 2004; Zamani et al. 2009; Toledo and Inestrosa 2010
GSK3β inhibitors	GSK3β	Pancreatic transplant, bone formation, schizophrenia, multiple myeloma	Kulkarni et al. 2006; Roh et al. 2007; Freyberg et al. 2010; Graham et al. 2010; Gunn et al. 2010
WAY-316606	sFRP1	Bone formation	Moore et al. 2009
Iisoquercitrin and isorhamnetin	Unknown	Adipocyte inhibition	Lee et al. 2011
Riluzole	GRM1	Melanoma, bipolar disorder	Yip et al. 2009; Biechele et al. 2010; Le et al. 2010

ES, embryonic stem; iPS, induced pluripotent stem.

negative regulators of canonical Wnt signaling (Kang et al. 2002). Additionally, variant alleles in the Wnt receptor LPR6 have been associated with Alzheimer's disease in population-based linkage analyses. Functional analysis of one of these common alleles revealed decreased β-catenin signaling when this variant was expressed in HEK293T cells (De Ferrari et al. 2007). Wnt may also have a role in the metabolism of the amyloid precursor protein (APP) and inhibit phosphorylation of the tau protein, two events thought to be critical in the pathophysiology of the disease (Mudher et al. 2001). Wnt signaling may also participate in neuroprotection after brain injury. Enhanced Wnt/β-catenin was shown in newly derived glial progenitors and astrocytes in a model of traumatic brain injury (White et al. 2010). These observations establish the potential rationale for exploring the use of Wnt-directed therapeutics in these disease conditions.

WNT SIGNALING IN AUTOIMMUNE DISEASE AND INFLAMMATION

Wnt signaling has long been implicated in development of the immune system (reviewed in Yu et al. 2010) but the effects of Wnt signaling in adult immunity are only beginning to be appreciated. Studies of adult T cells have shown that the Wnt effector TCF1 appears to be required for the functional generation of memory CD8$^+$ T cells (Gattinoni et al. 2009; Jeannet et al. 2010; Zhao et al. 2010; Zhou et al. 2010). Wnt signaling can also affect the differentiation of naïve CD4$^+$ T lymphocytes. Blockade of Wnt by small interfering RNA (siRNA)-mediated knockdown of β-catenin or by treatment of CD4$^+$ T cells with DKK1 inhibited the expression of Th2 cytokines (Notani et al. 2010). Wnt/β-catenin has also been shown to affect immunosuppressive T regulatory (Treg) cells that have been implicated in the pathogenesis of autoimmune disease, cancer, and infectious diseases (Ding et al. 2008). Enhanced Wnt signaling in gut-associated dendritic cells also led to immune suppression and tolerance in a model of inflammatory bowel disease, the effects of which were partially mediated through the induction of Tregs (Manicassamy et al. 2010).

Aberrant Wnt signaling has also been observed in autoimmune disorders. In rheumatoid arthritis (RA) patients, Wnt and *FZD* gene expression is higher in synovial membranes of affected joints compared with controls without RA (Sen et al. 2002). Increased expression of *WNT7B* has also been shown in RA patients and exogenous expression of *WNT7B* in human synovial cells induced inflammatory cytokines such as interleukin 1β, interleukin-6, and TNFα (Nakamura et al. 2005). β-catenin-independent Wnt signaling has also been implicated in the development of RA. *WNT5A* expression was noted to be higher in cultured synoviocytes from patients with RA compared with controls (Sen et al. 2000). The autoimmune disease systemic lupus erythematosus (SLE) has been associated with aberrant Wnt gene expression as well (Sheng et al. 2011). In a murine model of SLE, increased canonical Wnt pathway activity was shown in kidneys affected by lupus nephritis, suggesting a role for Wnt activity in mediating end organ damage associated with the disease (Tveita and Rekvig 2011).

In summary, manipulation of Wnt signaling may provide a novel means of shaping the immune response in autoimmune disease, tumor immunotherapy, transplantation, and vaccine development. Once again, the in-depth involvement of Wnt signaling pathways in immunological function highlights both the power of these pathways as regulators as well as the potential difficulties involved in the directed targeting of Wnt signaling in disease.

WNT SIGNALING IN TISSUE HOMEOSTASIS AND REGENERATIVE MEDICINE

Given the importance of Wnt pathways in development and stem cell biology, it should not be surprising that manipulating Wnt signaling has proven useful in the burgeoning field of regenerative medicine. One of the central problems that must be addressed in applying stem cell biology to regenerative medicine is how to specifically control stem cell renewal and differentiation to direct stem cells toward specific fates. There are numerous studies implicating Wnts as important components of the stem cell signaling network that can be used to direct specific cell fates (Nusse 2008). For example, Wnt signaling has been implicated in the self-renewal and maintenance of hematopoietic stem cells (Reya et al. 2003; Willert et al. 2003; Wang and Nakayama 2009). Also, canonical Wnt signaling appears to be required for differentiation of embryonic stem (ES) cells toward hematopoietic stem cells and primitive hematopoietic progenitors (Lengerke et al. 2008; Nostro et al. 2008; Wang and Nakayama 2009). β-Catenin-independent Wnt signaling has also been implicated in driving mesoderm specification and exit from the pluripotent state from ES cells (Vijayaragavan et al. 2009).

Wnt is important for the development and maintenance of other tissue types as well. For example, the maintenance of intestinal crypt stem cells requires the Wnt pathway transcription factor TCF7L2/TCF4 (Korinek et al. 1998). Accordingly, Wnt3a is one of a host of factors that can be used to derive human intestinal cells from induced pluripotent stem (iPS) cells (Spence et al. 2011). Wnt/β-catenin signaling has been implicated in cardiomyogenesis from human ES cells in a biphasic manner, promoting cardiomyocyte differentiation when triggered at early stages of differentiation, while inhibiting cardiogenesis at later stages (Paige et al. 2010; Yamauchi et al. 2010).

One of the most active fields in Wnt biology and regenerative medicine is in the field of bone metabolism. The initial discovery that Wnt pathways were related to bone formation in the adult was made when mutations in the Wnt coreceptor LRP5 were linked to the autosomal recessive disorder osteoporosis-pseudoglioma syndrome (OPPG) (Gong et al. 2001). Among other features of the syndrome, individuals with OPPG suffered with low bone mass and a predisposition to skeletal fractures. Carriers of the LRP5 mutation also had lower bone mass compared with age-matched controls. A different mutation in LRP5, G171V has been associated with increased muscle mass and skeletal strength in humans (Boyden et al. 2002). An indirect role for Wnt signaling in osteoblast biology has been suggested by the observation that *Lrp5* deficiency in the duodenum has an endocrine effect on osteoblasts via gut-derived serotonin binding the receptor hydroxytryptamine (Htr1b) expressed on osteoblasts (Yadav et al. 2008), but a more recent study refutes this view and indicates a direct role of LRP5 in osteoblasts (Cui et al. 2011). Several other studies have also shown an important direct role for Wnt/β-catenin signaling in osteoblasts at multiple stages, including promoting osteoblast differentiation from progenitors as well as promoting osteoblast and osteocyte survival in vitro (Day et al. 2005; Rodda and McMahon 2006). In addition, β-catenin was found to regulate osteoprotegerin (OPG/TNFRSF11B) production in mature osteoblasts. Because OPG is an inhibitor of osteoclast differentiation, Wnt/β-catenin signaling in osteoblasts appears to inhibit bone resorption as well (Glass et al. 2005; Holmen et al. 2005).

As new models for regeneration are developed, the role of Wnt/β-catenin signaling as a therapeutic target will be further clarified. Because regeneration is a long-term process, the question of how persistent therapeutic activation or inactivation of Wnt signaling will affect an organism becomes more relevant. However, it should be noted that in developmental models, including embryonic development, the effects of Wnt/β-catenin signaling can be elicited through rather short exposure to Wnt ligand. During embryonic development, Wnt signaling over short time periods may completely alter the specification of cell fate, suggesting that prolonged treatment courses with activators or inhibitors may not be needed (Bakre et al. 2007; Ueno et al. 2007; Trowbridge et al. 2010). In zebrafish models, pulsed exposure to Wnt signaling can result in measurable improvements in fin regeneration, further supporting the idea that long-term, continuous activation or inhibition of the pathway may not be absolutely required to achieve the desired end result (Stoick-Cooper et al. 2007).

DRUGS THAT INHIBIT WNT SIGNALING

Given the aberrant Wnt pathway activation in cancer, it is not surprising that inhibition of Wnt signaling has been an active area of investigation in both academia and the biotechnology sector. One past limitation in identifying clinically useful drugs has been the lack of an obvious enzyme target to inhibit in the pathway. Another has been the complexity of Wnt pathway regulation. At least 19 Wnt isoforms have been identified in humans (Chien et al. 2009a), all of which share significant sequence homology. The mechanisms that determine the specificity of a particular Wnt isoform remain unresolved, even with regard to whether isoforms exclusively activate Wnt/β-catenin signaling or β-catenin-independent Wnt signaling.

Existing Patient-Experienced Drugs with Wnt Inhibitory Activity

Despite the inherent difficulties in developing novel Wnt inhibitors, many existing FDA-approved drugs inhibit Wnt pathways (Table 1). Nonsteroidal anti-inflammatory drugs (NSAIDs) were initially found to inhibit polyp formation both in patients with FAP as well as in the corresponding murine model, APC^{min} (Labayle et al. 1991; Giardiello et al. 1993; Mahmoud et al. 1998; Yang et al. 2003b; Boon et al. 2004). In FAP patients, reduced levels of β-catenin were shown in polyps of patients treated with the NSAID sulindac (Boon et al. 2004). The selective COX2 inhibitor celecoxib has been FDA approved in patients with FAP as it has been shown to reduce polyp formation as well (Steinbach et al. 2000; Phillips et al. 2002). It has also been shown to reduce nuclear β-catenin in colorectal cancer cell lines similarly to other NSAIDs (Steinbach et al. 2000). The R-enatiomer, the NSAID etodolac, whose structure is a nonsuperimposable mirror image to the parent drug but itself does not possess COX inhibitory activity, was found to inhibit Wnt/β-catenin target genes in HEK293 cells and enhance the apoptosis of chronic lymphocytic leukemia (CLL) cells in vitro (Jensen et al. 2008).

Several nuclear receptors have also been found to act, at least partially, through effects on the Wnt/β-catenin pathway. Retinoid-mediated signaling through retinoic acid receptors can compete with several TCF factors for β-catenin binding, thus inhibiting Wnt/β-catenin target genes (Xiao et al. 2003). Vitamin D derivatives can also inhibit Wnt signaling in colorectal cancer. Studies in colorectal carcinoma cell lines have shown that analogs of the active form allows the vitamin D receptor to bind β-catenin and compete with TCF7L2 to reduce Wnt target gene expression in these cells (Pálmer et al. 2001). Additionally, in a human osteosarcoma line that is sensitive to glucocorticoid-induced cell cycle arrest, glucocorticoid-mediated down-regulation of cyclin D1 was found to occur via interactions between the glucocorticoid receptor and the TCF/β-catenin complex (Takayama et al. 2006).

Other classes of drugs have been identified that can inhibit Wnt signaling by screening libraries of FDA-approved drugs. For example, the anti-helminthic drug pyrvinium was identified as an agent that could potentiate the activity of CK1α, thus leading to enhanced β-catenin and coactivator Pygo degradation and diminished signaling (Thorne et al. 2010). Pyrvinium has shown promise as a promoter of tissue repair in a mouse model of myocardial infarction, in which the inhibition of Wnt/β-catenin signaling led to global improvements in cardiac remodeling (Saraswati et al. 2010).

In summary, although the clinical use of drugs for the primary purpose of inhibiting Wnt signaling is still in early development, there are numerous drugs or compounds in current usage that affect Wnt/β-catenin signaling. Whether the effects of these drugs on inhibiting Wnt/β-catenin may contribute to some of the efficacy or side effects seen with the current use of these drugs will likely become more apparent with improvements in our understanding of Wnt signaling across various tissues and diseases.

Approaches to Identifying Novel Inhibitors of Wnt Signaling

Because Wnt pathways are subjected to extensive regulation at various levels, high-throughput screening technologies have greatly facilitated the identification of both biologics and small molecule pathway regulators. One promising approach has been the use of cell-based reporter assays to screen siRNA or chemical libraries. These cell-based reporters provide a rapid, reliable, and quantitative means of interrogating Wnt/β-catenin signaling, and have been critical tools in the identification and subsequent characterization of every Wnt-modulating small molecule identified to date (Biechele and Moon 2008). Genome-wide siRNA and drug library screening approaches are powerful but have many difficulties such as uneven transfection efficiency, the knockdown efficacy of siRNA sequences, or off-target effects leading to false hits. Such screens also offer little in the way of mechanistic data. These complications have been mitigated somewhat by improvements in assay automation and rigorous validation. Another approach to minimize the false hit rate and provide insights into mechanism is to integrate data from different platforms. For example, an integrated analysis from a genome-wide siRNA screen of modulators of Wnt signaling with protein interaction networks derived from mass spectrometry of 23 bait proteins or derived from known protein interactions in the literature identified 38 novel regulators, several of which were involved in chromatin remodeling (Major et al. 2008). Another siRNA-based loss-of-function screen of β-catenin-mediated transcription was integrated with gene copy number analysis in human colon cancer cells and identified cyclin-dependent kinase 8 (CDK8) as a key regulator of Wnt target gene expression (Firestein et al. 2008).

Strategies for Inhibiting Proximal Wnt Signaling

One attractive approach for inhibiting Wnt signaling has been targeting either ligands or extracellular receptors that are readily accessible to antibodies or small molecules (Fig. 1; Table 1). This biologics approach may be particularly relevant for diseases that display aberrant Wnt signaling in the absence of known downstream pathway mutations. In these cases, dysregulation of Wnt signaling is instead attributed to the aberrant expression of Wnt ligands. For example, WNT1 is overexpressed in numerous human cancers and a WNT1-blocking antibody both inhibited proliferation and induced apoptosis in many cancer cell types (He et al. 2004, 2005a,b; Mikami et al. 2005; Wei et al. 2009).

The FZD or LRP coreceptors are also potential targets for therapeutic manipulation. For example, synovial sarcomas show significant overexpression of FZD10 compared with normal tissue. In a mouse synovial sarcoma xenograft model, FZD10 antibody was used to target delivery of the radioisotope Yttrium-90 to tumors resulting in inhibition of tumor growth (Fukukawa et al. 2008). Using the FZD ectodomain to competitively bind Wnt ligand is another strategy that has been used. Overexpression of the FZD7 ectodomain attenuated proliferation in a colon cancer cell line in vitro as well as a murine xenograft assay (Vincan et al., 2005). Additionally, a soluble Wnt receptor consisting of a FZD8 cysteine rich domain (CRD) fused to a human Fc domain showed activity against teratoma lines in vitro and in vivo (DeAlmeida et al. 2007).

Another approach to target proximal events in Wnt signal transduction is to use endogenous extracellular modulators of Wnt such as Dickkopf (DKK), Wnt inhibitory factor-1 (WIF1), or SFRPs. SFRPs contain a cysteine-rich domain (CRD) similar to the extracellular Wnt-binding domain of the frizzled receptors, and have been postulated to act as a decoy receptor by directly binding Wnt ligands. WIF1 also directly binds Wnt ligands, whereas DKK binds to and inhibits transduction via the LRP coreceptor. Dysregulation of these endogenous inhibitors is common in many tumor types, typically by promoter hypermethylation (Suzuki et al. 2004, 2008; Fukui et al. 2005; Zou et al. 2005). Although SFRP is lost early in the progression of colon cancer, restoration of SFRP in colon cancer can attenuate Wnt signaling even in the presence of downstream activating mutations (Suzuki et al. 2004). In another study, a peptide containing a frizzled CRD derived from a cell-surface collagen VIII termed V3Nter was able to act in an SFRP-like fashion and inhibit canonical Wnt signaling in colorectal carcinoma lines as well as inhibit tumor growth in a xenograft model (Lavergne et al. 2011).

In addition to the use of Wnt inhibitors as soluble proteins, forced cellular overexpression has also been tested as a strategy. Forced DKK3 expression in renal cell carcinoma lines resulted in the induction of apoptosis, although this appeared to be mediated by alterations in the noncanonical JUN kinase (JNK) pathway rather than β-catenin inhibition (Ueno et al. 2011). Viral vector-mediated overexpression of DKK3 inhibits prostate and breast cancer growth in mice (Edamura et al. 2007; Kawasaki et al. 2009). Although the use of endogenous Wnt inhibitors seems to be a reasonable strategy, the finding that proteins like SFRPs can potentially enhance Wnt signaling in some contexts, perhaps by acting as chaperones, suggests that the therapeutic use of these particular proteins may be complicated (Uren et al. 2000; Chien et al. 2009a; von Marschall and Fisher 2010). Furthermore, DKK family members can themselves transduce receptor-mediated cellular signals that are not completely understood at this time (Wu et al. 2000; Mukhopadhyay et al. 2006; Yamamoto et al. 2008).

Another approach to inhibit upstream events in Wnt signaling is disruption of the interaction between the receptor FZD and DVL. For example, a structure-based virtual ligand screening identified a small molecule that is capable of binding to the PDZ domain of DVL required for its interaction with FZD receptors. This small molecule is able to inhibit Wnt3A-mediated signaling (Shan et al. 2005). Another study found that peptide binding to the DVL PDZ domain is more permissive compared with PDZ domains in other proteins and used peptide-phage display to identify peptide ligands, which bound to the PDZ domain of human DVL2. These peptide ligands were then shown to disrupt Wnt/β-catenin signaling in cell lines (Zhang et al. 2009).

Drugs that interfere with Wnt secretion have also been used as an approach to inhibit proximal signaling. For example, the drug IWP-2, identified in a screen of a synthetic chemical library, was found to inhibit the activity of Porcupine, a membrane-bound acyltransferase that modifies Wnt ligands with a palmitoyl group that is required for their secretion and signaling activity (Takada et al. 2006; Chen et al. 2009).

Drugs Inhibiting Downstream Wnt Signaling

Given the wide range of biologic effects of Wnt-mediated signaling in the adult, targeting more upstream signaling events may lead to unwanted side effects. Furthermore, in diseases such as colon cancer in which there frequently are mutations in *APC*, upstream inhibition at the level of the receptor may be ineffective for preventing the effects of a downstream activating mutation.

High-throughput screening approaches have been used to identify promising lead compounds that may one day lead to the development of bona fide clinical Wnt inhibitors (Fig. 1; Table 1). For example, HEK293 cells expressing the SuperTOPFLASH Wnt-responsive reporter were screened with a chemical library that led to the identification of XAV939 as a potent inhibitor of Wnt/β-catenin signaling. Mechanistic studies of this drug revealed a novel regulatory pathway by which the stability of Axin, a critical negative regulator of the pathway, was regulated by PARsylation and proteasomal degradation by tankyrases, TNKS1 and TNKS2 (Huang et al. 2009). This drug was able to inhibit colony formation of colorectal cancer cell lines and has also been used to inhibit β-catenin in a manner that promotes cardiomyocyte differentiation from embryonic stem cells (Huang et al. 2009; Wang et al. 2011a). Other substrates of tankyrases must be considered if tankyrase inhibitors are to be used clinically. For example, the dominantly inherited disorder cherubism is characterized by interosseous fibrocystic lesions resulting from hyperactive osteoclasts results from single missense mutations in a region of the adaptor protein 3BP2, which is part of a multiprotein signaling complex containing the SRC family kinases, SYK, and RHO family guanine nucleotide exchange factor VAV (Ueki et al. 2007; Guettler et al. 2011). These mutations have been found to disrupt the binding of tankyrase to 3BP2, inhibiting 3BP2 degradation and resulting in high levels of SRC, SYK, and VAV activity in osteoclasts (Guettler et al. 2011; Levaot et al. 2011). These observations may be relevant if tankyrase inhibitors are to be considered as cancer therapeutics, for instance, owing to not only potential side effects such as osteoporosis, but also concerns about potentiating the SRC pathway, which is already hyperactive in many human tumors.

As the β-catenin TCF/LEF complex is the ultimate downstream effector required for transcription of Wnt target genes, disruption of this complex has been studied as a therapeutic strategy. The ability to rationally design drugs to intervene in this interaction has been facilitated by characterization of the crystal structure of TCF/β-catenin complexes (Graham et al. 2000). One class of small molecules able to disrupt this interaction was identified by a high-throughput screening approach utilizing an assay based on the binding affinity of purified β-catenin to a TCF4 (TCF7L2) fragment (Lepourcelet et al. 2004). These compounds were able to inhibit growth of colorectal carcinoma cells in vitro (Lepourcelet et al. 2004). Another study used a small molecule secondary structure template library, which identified a compound, ICG-001 that was able to down-modulate β-catenin/TCF signaling by binding to the CREB response element-binding protein (CBP) and specifically inhibiting the CBP/β-catenin interaction (Emami et al. 2004). This compound was also shown to inhibit tumor formation in a mouse colon cancer xenotransplant model as well as attenuate pulmonary fibrosis induced by bleomycin, which is thought to be in part mediated by wnt/β-catenin signaling (Emami et al. 2004; Henderson et al. 2010). Another small molecule screen yielded the drug NC043, a diterpine alkaloid that can interrupt the β-catenin/TCF4 interaction and inhibit colorectal cancer growth in vivo (Wang et al. 2011b). Recently, several compounds were identified that inhibited β-catenin/TCF4 interactions utilizing a chemical library screen of *Drosophila* cells to mitigate functional genetic regulatory redundancies in mammalian cells. These compounds were able to inhibit β-catenin-dependent invasion of breast cancer cells and proliferation of colon cancer cell lines (Gonsalves et al. 2011). One conceptual difficulty in utilizing such drugs as therapeutics is that the structural basis for the interaction between β catenin and TCF is almost identical to that with E-cadherin, although there is some evidence for binding specificity mediated by β-catenin modifications (Graham et al. 2000; Huber and Weis 2001; Gottardi and Gumbiner 2004). Consequently, drugs designed to interrupt the β-catenin/TCF interaction may also disrupt the cadherins junction leading to unpredicted and potentially untoward side effects.

Other regulators of Wnt/β-catenin signaling have also emerged as potentially useful targets. For example, casein kinase-Iε (CKIε) was identified based on siRNA screening of breast cancer cell lines as a positive regulator of β-catenin-dependent transcription. Pharmacologic inhibition of CKIε with the drug IC261 inhibited breast cancer cell proliferation (Kim et al. 2010). Many other kinases that positively regulate Wnt have emerged as potential targets including phosphatidylinositol 4-kinase type IIa (PI4KIIa), phosphatidylinositol-4-phosphate 5-kinase type Ib (PIP5KIb), PAR-1, TNIK, and MAP3K1 (Sun et al. 2001; Pan et al. 2008; Mahmoudi et al. 2009; Sue Ng et al. 2010).

As the tools for interrogating Wnt signaling continue to evolve, novel pathways and promising compounds will continue to be identified as modulators of Wnt/β-catenin signaling. Because the effects of most of these compounds will be cell type/context-dependent, the challenge will be to understand the role of Wnt/β-catenin signaling across various tissues in such a way as to develop a practical therapeutic strategy.

STRATEGIES TO ACTIVATE WNT SIGNALING

Although much work has been performed identifying antagonists of Wnt signaling, particularly with regard to cancer therapy, drugs that activate Wnt pathways may be beneficial in many diseases including osteoporosis and certain cancers (Fig. 2; Table 2). Several drugs exist that act, at least partially, by enhancing Wnt pathways. For example, lithium has long been known as a powerful morphogen in embryonic development. Research into the mechanism for this property identified it as an inhibitor of GSK3β, which ultimately results in constitutive Wnt pathway activation (Klein and Melton 1996). Lithium is a well-known psychoactive drug and it has been postulated that its effects in the central nervous system (CNS) may be partially attributable to Wnt enhancement. Interestingly, many anti-psychotic drugs may act, at least to some degree, through interaction with Wnt pathway components. For example, the atypical anti-psychotic drug clozapine has been shown to increase levels of Dvl and phospho-Dvl in the rat brain, thereby enhancing β-catenin-mediated signaling (Roh et al. 2007).

Several studies have linked the use of lithium to increased Wnt/β-catenin signaling. For example, treatment of Lrp5-deficient mice with lithium significantly increased trabecular bone mass. Lithium also increased bone mass in animals with age-related osteoporosis and oophorectomy-induced osteoporosis (Clément-Lacroix et al. 2005). Furthermore, a retrospective clinical study comparing patients receiving lithium to age-matched controls noted increased bone density, lower serum ALP, and osteocalcin (markers of bone turnover) in patients who received lithium (Zamani et al. 2009). The directed clinical use of lithium specifically as a Wnt modulator is complicated by the substantial toxicities in vivo that accompany the very high doses needed to achieve Wnt activation in different contexts. More potent GSK3β inhibitors are available and may also be clinically useful, although these agents are still considered experimental at this time. In vivo effects on bone metabolism comparable to lithium have been seen with an orally available potent dual inhibitor of GSK3α/β (Engler et al. 2004). Pharmacologic inhibition of GSK3β has been shown to potentiate regulatory T-cell activity in vitro and extend allograft survival in a mouse model of pancreatic islet cell transplantation (Graham et al. 2010). Unfortunately, inhibition of an enzyme as ubiquitous as GSK3β has multiple consequences that are independent of the Wnt pathway, leading to significant side effects that have complicated the use of these otherwise potent compounds in human patients.

Other indirect means of enhancing Wnt signaling have been explored. For instance, knockdown of *DKK1* by RNA interference (RNAi) significantly inhibited bone loss in a rat model of estrogen depletion as well as in a model of glucocorticoid-mediated bone loss (Wang et al. 2007, 2008). Investigators have also used anti-DKK1 monoclonal antibodies as a means to increase bone mass and mineralization in several small animal and nonhuman primate models (Glantschnig et al. 2010). DKK1 antibody has also been shown to inhibit degenerative osteoarthritis in the setting of inflammatory arthritis and to facilitate fracture healing (Diarra et al. 2007; Komatsu et al. 2010). Additionally, in a small animal model of ischemic brain injury, expression of DKK-1 was found to be strongly induced during injury. Treatment with antisense oligonucleotides against DKK-1 significantly inhibited neurotoxicity (Cappuccio et al. 2005). Interest in inhibiting the function of DKK has also emerged because it has been appreciated that tumor-derived DKK may inhibit Wnt-induced osteoblastic differentiation, thus facilitating osteolytic bone metastases in diseases such as breast cancer and multiple myeloma (Tian et al. 2003; Bu et al. 2008; Qiang et al. 2008). An anti-DKK1 antibody, BHQ880 was shown to increase osteoblasts and bone density in a (SCID)-hu murine model of human multiple myeloma as well as inhibit the growth of multiple myeloma cells on bone marrow stroma in vitro (Fulciniti et al. 2009).

Another extracellular regulator of Wnt signaling has been targeted as an approach to enhance bone formation. In *Xenopus*, a secreted cysteine-knot protein, Wise, was identified that acted as a context-dependent activator or inhibitor of Wnt signaling by interacting with LPR6 (Itasaki et al. 2003). Deficiency of a related mammalian gene product, sclerostin (SOST), is associated with the inherited disorder sclerosteosis, which results in hyperactive osteoblasts (Brunkow et al. 2001). SOST-deficient mice also showed markedly increased bone formation and bone strength (Li et al. 2008). Sclerostin was confirmed to bind LPR5/6 and inhibit Wnt/β-catenin signaling (Li et al. 2005; Semënov et al. 2005). Several anti-sclerostin antibodies have been developed that have shown significant activity in enhancing bone mass in small animal and nonhuman primate models (Li et al. 2009; Agholme et al. 2010; Ominsky et al. 2010). A single dose, placebo-controlled phase I clinical trial with a humanized sclerostin antibody, AMG 785, has been recently reported. In this study, patients receiving single doses of AMG 785 had increased markers of bone turnover and higher bone mineral density compared with placebo, and treatment was generally well tolerated (Padhi et al. 2011). A phase II study of this agent is currently ongoing.

Another putative secreted inhibitor of Wnt signaling, SFRP1 is also an attractive target to enhance Wnt signaling therapeutically. *Sfrp1*-deficient mice have increased trabecular bone density and mineralization, consistent with the observation that deletion of Sfrp1 inhibited age-associated bone loss in mice (Bodine et al. 2004). In one study, a small molecule inhibitor of SFRP1 was identified via a fluorescent polarization binding assay using purified human SFRP1 protein and a cell-based Wnt reporter assay. This small molecule was capable of stimulating ex vivo murine calvarial bone formation (Moore et al. 2009).

Although there has clearly been more work on identifying inhibitors, small molecule compounds that potentiate Wnt/β-catenin signaling have also been described. One study identified flavonoid components of the *Persicaria hydropiper* plant as enhancers of Wnt/β-catenin signaling that can block the differentiation of adipocytes from 3T3-L1 cells (Lee et al. 2011). In another study, a cell-based screen utilizing the BAR reporter of a chemical library of 1857 human-experienced drugs identified 44 compounds that enhanced Wnt3a-mediated stimulation (Pollock et al. 2003; Namkoong et al. 2007; Biechele et al. 2010). One of these activators, riluzole, is an FDA-approved drug for treating amyotrophic lateral sclerosis (ALS). Riluzole is thought to inhibit the metabotropic glutamate receptor GRM1 and is currently in clinical trials for melanoma based on its activity in melanoma cell lines, xenograft models, and a phase 0 patient trial (Pollock et al. 2003; Namkoong et al. 2007; Yip et al. 2009; Le et al. 2010). As Wnt/β-catenin signaling is associated with an improved prognosis in melanoma, it is intriguing to speculate that the mechanism of action of this drug may be at least partially owing to its effect on enhancing Wnt/β-catenin signaling (Chien et al. 2009b).

STRATEGIES FOR TARGETING β-CATENIN-INDEPENDENT WNT SIGNALING

Despite the mounting evidence for involvement of β-catenin-independent Wnt signaling in the pathogenesis of several disorders, approaches to target noncanonical Wnt signaling have been disappointingly limited. WNT5A antisense, dominant-negative WNT5A, and addition of a WNT5A antibody in vitro were shown to inhibit the production of proinflammatory cytokines such as IL-6, IL-15, and RANKL in cultured synoviocytes from RA patients (Sen et al. 2001). Small molecules represent another viable approach to targeting β-catenin-independent Wnt pathways. For example, the drug fumagillin, originally isolated from the fungus *Aspergillus fumagatus*, was initially identified as a compound that bound the protein methionine aminopeptidase 2 (MetAP-2) (Zhang et al. 2006). Fumagillin was subsequently found to disrupt noncanonical Wnt signaling in zebrafish embryos and to inhibit β-catenin-independent Wnt-mediated endothelial cell migration and angiogenesis (Zhang et al. 2006; Cirone et al. 2008). In another study, phenylmethimazole, an inhibitor of Toll-like receptor-3 (TLR3) was found to inhibit *WNT5A* expression in melanoma and pancreatic cell lines, thus inhibiting their growth and migration in vitro and in a murine xenograft model (Schwartz et al. 2009). A formylated hexapeptide derived from the primary WNT5A peptide sequence mimicked the ability of WNT5A to inhibit breast cancer cell migration, suggesting that peptide-based approaches may also be viable tools for modulating Wnt pathways (Säfholm et al. 2006). As more studies elucidate the specific roles of β-catenin-independent Wnt signaling in disease, more therapeutic targets and strategies are likely to emerge.

CONCLUSION

As research tools to understand the molecular mechanisms of human diseases continue to develop, so does our understanding of the prominent role that aberrant Wnt-mediated signaling plays in an ever-expanding number of human diseases. Apart from the examples described here, a role for Wnt is emerging in a multitude of other disease processes for which pharmacologic manipulation could prove clinically valuable. To fully realize the potential of Wnt manipulation, more must be understood about the complex regulation of both canonical and noncanonical pathways. As research tools to understand the molecular mechanisms of human disease continue to develop, so does our understanding of the nuances of Wnt regulation in disease as well as in the context of normal tissue homeostasis. Our ability to target these pathways for therapeutic benefit is only in its infancy. However, promising new technologies are paving the way to drive more effective therapeutic targeting of Wnt pathways in the future.

ACKNOWLEDGMENTS

Z.Z. is funded by a T32 training grant through the National Institutes of Health (T32 CA00 9515-2). R.T.M. is an Investigator of the Howard Hughes Medical Institute. A.J.C. is funded by a K08 Career Development Award from NIH/ NCI (K08 CA128565). We are indebted to our funding agencies for their continued support of our research endeavors. As with any attempt to review a broad and dynamically changing field, we had to limit our scope and exclude certain studies, so we apologize in advance for any unintended omissions. The views detailed in this publication do not represent the official views of the National Institutes of Health or the Howard Hughes Medical Institute.

REFERENCES

Agholme F, Li X, Isaksson H, Ke HZ, Aspenberg P. 2010. Sclerostin antibody treatment enhances metaphyseal bone healing in rats. *J Bone Miner Res* **25:** 2412–2418.

Bachmann IM, Straume O, Puntervoll HE, Kalvenes MB, Akslen LA. 2005. Importance of P-cadherin, β-catenin, and Wnt5a/frizzled for progression of melanocytic tumors and prognosis in cutaneous melanoma. *Clin Cancer Res* **11:** 8606–8614.

Bakre MM, Hoi A, Mong JC, Koh YY, Wong KY, Stanton LW. 2007. Generation of multipotential mesendodermal progenitors from mouse embryonic stem cells via sustained Wnt pathway activation. *J Biol Chem* **282:** 31703–31712.

Biechele TL, Moon RT. 2008. Assaying β-catenin/TCF transcription with β-catenin/TCF transcription-based reporter constructs. *Methods in molecular biology*, Vol. 468, pp. 99–110. Humana Press, Clifton, NJ.

Biechele TL, Camp ND, Fass DM, Kulikauskas RM, Robin NC, White BD, Taraska CM, Moore EC, Muster J, Karmacharya R, et al. 2010. Chemical-genetic screen identifies riluzole as an enhancer of Wnt/β-catenin signaling in melanoma. *Chem Biol* **17:** 1177–1182.

Blackwood DHR, Fordyce A, Walker MT, St. Clair DM, Porteous DJ, Muir WJ. 2001. Schizophrenia and affective disorders—Cosegregation with a translocation at chromosome 1q42 that directly disrupts brain-expressed genes: Clinical and P300 findings in a family. *Am J Hum Genet* **69:** 428–433.

Bodine PVN, Zhao W, Kharode YP, Bex FJ, Lambert A-J, Goad MB, Gaur T, Stein GS, Lian JB, Komm BS. 2004. The Wnt antagonist secreted frizzled-related protein-1 is a negative regulator of trabecular bone formation in adult mice. *Mol Endocrinol (Baltimore, Md)* **18:** 1222–1237.

Boon EM.J, Keller JJ, Wormhoudt TA.M, Giardiello FM, Offerhaus GJ.A, van der Neut R, Pals ST. 2004. Sulindac targets nuclear β-catenin accumulation and Wnt signalling in adenomas of patients with familial adenomatous polyposis and in human colorectal cancer cell lines. *Brit J Cancer* **90:** 224–229.

Boyden LM, Mao J, Belsky J, Mitzner L, Farhi A, Mitnick MA, Wu D, Insogna K, Lifton RP. 2002. High bone density due to a mutation in LDL-receptor-related protein 5. *N Engl J Med* **346:** 1513–1521.

Brunkow ME, Gardner JC, Van Ness J, Paeper BW, Kovacevich BR, Proll S, Skonier JE, Zhao L, Sabo PJ, Fu Y, et al. 2001. Bone dysplasia sclerosteosis results from loss of the SOST gene product, a novel cystine knot-containing protein. *Am J Hum Genet* **68:** 577–589.

Bu G, Lu W, Liu C-C, Selander K, Yoneda T, Hall C, Keller ET, Li Y. 2008. Breast cancer-derived Dickkopf1 inhibits osteoblast differentiation and osteoprotegerin expression: Implication for breast cancer osteolytic bone metastases. *Int J Cancer* **123:** 1034–1042.

Caldwell GM, Jones C, Gensberg K, Jan S, Hardy RG, Byrd P, Chughtai S, Wallis Y, Matthews GM, Morton DG. 2004. The Wnt antagonist sFRP1 in colorectal tumorigenesis. *Cancer Res* **64:** 883–888.

Cantrell VA, Jessen JR. 2010. The planar cell polarity protein Van Gogh-Like 2 regulates tumor cell migration and matrix metalloproteinase-dependent invasion. *Cancer Lett* **287:** 54–61.

Cappuccio I, Calderone A, Busceti CL, Biagioni F, Pontarelli F, Bruno V, Storto M, Terstappen GT, Gaviraghi G, Fornai F, et al. 2005. Induction of Dickkopf-1, a negative modulator of the Wnt pathway, is required for the development of ischemic neuronal death. *J Neurosci* **25:** 2647–2657.

Caricasole A, Copani A, Caraci F, Aronica E, Rozemuller AJ, Caruso A, Storto M, Gaviraghi G, Terstappen GC, Nicoletti F. 2004. Induction of Dickkopf-1, a negative modulator of the Wnt pathway, is associated with neuronal degeneration in Alzheimer's brain. *J Neurosci* **24:** 6021–6027.

Carroll-Anzinger D, Kumar A, Adarichev V, Kashanchi F, Al-Harthi L. 2007. Human immunodeficiency virus-restricted replication in astrocytes and the ability of γ interferon to modulate this restriction are regulated by a downstream effector of the Wnt signaling pathway. *J Virol* **81:** 5864–5871.

Cheah PY, Choo PH, Yao J, Eu KW, Seow-Choen F. 2002. A survival-stratification model of human colorectal carcinomas with β-catenin and 27kip1. *Cancer* **95:** 2479–2486.

Chen B, Dodge ME, Tang W, Lu J, Ma Z, Fan C-W, Wei S, Hao W, Kilgore J, Williams NS, et al. 2009. Small molecule-mediated disruption of Wnt-dependent signaling in tissue regeneration and cancer. *Nat Chem Biol* **5:** 100–107.

Chien AJ, Conrad WH, Moon RT. 2009a. A Wnt survival guide: From flies to human disease. *J Invest Dermatol* **129:** 1614–1627.

Chien AJ, Moore EC, Lonsdorf AS, Kulikauskas RM, Rothberg BG, Berger AJ, Major MB, Hwang ST, Rimm DL, Moon RT. 2009b. Activated Wnt/β-catenin signaling in melanoma is associated with decreased proliferation in patient tumors and a murine melanoma model. *Proc Natl Acad Sci* **106:** 1193–1198.

Cirone P, Lin S, Griesbach HL, Zhang Y, Slusarski DC, Crews CM. 2008. A role for planar cell polarity signaling in angiogenesis. *Angiogenesis* **11:** 347–360.

Clément-Lacroix P, Ai M, Morvan F, Roman-Roman S, Vayssière B, Belleville C, Estrera K, Warman ML, Baron R, Rawadi G. 2005. Lrp5-independent activation of Wnt signaling by lithium chloride increases bone formation and bone mass in mice. *Proc Natl Acad Sci* **102:** 17406–17411.

Cui Y, Niziolek PJ, MacDonald BT, Zylstra CR, Alenina N, Robinson DR, Zhong Z, Matthes S, Jacobsen CM, Conlon RA, et al. 2011. Lrp5 functions in bone to regulate bone mass. *Nat Med* **17:** 684–691.

Day TF, Guo X, Garrett-Beal L, Yang Y. 2005. Wnt/β-catenin signaling in mesenchymal progenitors controls osteoblast and chondrocyte differentiation during vertebrate skeletogenesis. *Dev Cell* **8:** 739–750.

DeAlmeida VI, Miao L, Ernst JA, Koeppen H, Polakis P, Rubinfeld B. 2007. The soluble wnt receptor Frizzled8CRD-hFc inhibits the growth of teratocarcinomas in vivo. *Cancer Res* **67:** 5371–5379.

de Boer J, Siddappa R, Gaspar C, van Apeldoorn A, Fodde R, van Blitterswijk C. 2004. Wnt signaling inhibits osteogenic differentiation of human mesenchymal stem cells. *Bone* **34:** 818–826.

De Ferrari GV, Moon RT. 2006. The ups and downs of Wnt signaling in prevalent neurological disorders. *Oncogene* **25:** 7545–7553.

De Ferrari GV, Papassotiropoulos A, Biechele T, Wavrant De-Vrieze F, Avila ME, Major MB, Myers A, Sáez K, Henríquez JP, Zhao A, et al. 2007. Common genetic variation within the low-density lipoprotein receptor-related protein 6 and late-onset Alzheimer's disease. *Proc Natl Acad Sci* **104:** 9434–9439.

Diarra D, Stolina M, Polzer K, Zwerina J, Ominsky MS, Dwyer D, Korb A, Smolen J, Hoffmann M, Scheinecker C, et al. 2007. Dickkopf-1 is a master regulator of joint remodeling. *Nat Med* **13:** 156–163.

Ding Y, Shen S, Lino AC, Curotto de Lafaille MA, Lafaille JJ. 2008. β-Catenin stabilization extends regulatory T cell survival and induces anergy in nonregulatory T cells. *Nat Med* **14:** 162–169.

Dorsky RI, Raible DW, Moon RT. 2000. Direct regulation of nacre, a zebrafish MITF homolog required for pigment cell formation, by the Wnt pathway. *Genes Dev* **14:** 158–162.

Edamura K, Nasu Y, Takaishi M, Kobayashi T, Abarzua F, Sakaguchi M, Kashiwakura Y, Ebara S, Saika T, Watanabe M, et al. 2007. Adenovirus-mediated REIC/Dkk-3 gene transfer inhibits tumor growth and metastasis in an orthotopic prostate cancer model. *Cancer Gene Ther* **14:** 765–772.

Elzagheid A, Buhmeida A, Korkeila E, Collan Y, Syrjanen K, Pyrhonen S. 2008. Nuclear β-catenin expression as a prognostic factor in advanced colorectal carcinoma. *World J Gastroentero* **14:** 3866–3871.

Emami KH, Nguyen C, Ma H, Kim DH, Jeong KW, Eguchi M, Moon RT, Teo J-L, Oh SW, Kim HY, et al. 2004. A small molecule inhibitor of β-catenin/CREB-binding protein transcription. *Proc Natl Acad Sci* **101:** 12682–12687.

Engler TA, Henry JR, Malhotra S, Cunningham B, Furness K, Brozinick J, Burkholder TP, Clay MP, Clayton J, Diefenbacher C, et al. 2004. Substituted 3-imidazo[1,2-a]pyridin-3-yl-4-(1,2,3,4-tetrahydro-[1,4]diazepino-[6,7,1-hi]indol-7-yl)pyrrole-2,5-diones as highly selective and potent inhibitors of glycogen synthase kinase-3. *J Med Chem* **47:** 3934–3937.

Firestein R, Bass AJ, Kim SY, Dunn IF, Silver SJ, Guney I, Freed E, Ligon AH, Vena N, Ogino S, et al. 2008. CDK8 is a colorectal cancer oncogene that regulates β-catenin activity. *Nature* **455:** 547–551.

Ford CE, Ekström EJ, Howlin J, Andersson T. 2009. The WNT-5a derived peptide, Foxy-5, possesses dual properties that impair progression of ERα negative breast cancer. *Cell Cycle (Georgetown, Tex)* **8:** 1838–1842.

Freyberg Z, Ferrando SJ, Javitch JA. 2010. Roles of the Akt/GSK-3 and Wnt signaling pathways in schizophrenia and antipsychotic drug action. *Am J Psychiat* **167:** 388–396.

Fukui T, Kondo M, Ito G, Maeda O, Sato N, Yoshioka H, Yokoi K, Ueda Y, Shimokata K, Sekido Y. 2005. Transcriptional silencing of secreted frizzled related protein 1 (SFRP 1) by promoter hypermethylation in non-small-cell lung cancer. *Oncogene* **24:** 6323–6327.

Fukukawa C, Hanaoka H, Nagayama S, Tsunoda T, Toguchida J, Endo K, Nakamura Y, Katagiri T. 2008. Radioimmunotherapy of human synovial sarcoma using a monoclonal antibody against FZD10. *Cancer Sci* **99:** 432–440.

Fulciniti M, Tassone P, Hideshima T, Vallet S, Nanjappa P, Ettenberg SA, Shen Z, Patel N, Tai Y-T, Chauhan D, et al. 2009. Anti-DKK1 mAb (BHQ880) as a potential therapeutic agent for multiple myeloma. *Blood* **114:** 371–379.

Gamallo C, Palacios J, Moreno G, Calvo de Mora J, Suárez A, Armas A. 1999. β-Catenin expression pattern in stage I and II ovarian carcinomas: Relationship with β-catenin gene mutations, clinicopathological features, and clinical outcome. *Am J Pathol* **155:** 527–536.

Gattinoni L, Zhong XS, Palmer DC, Ji Y, Hinrichs CS, Yu Z, Wrzesinski C, Boni A, Cassard L, Garvin LM, et al. 2009. Wnt signaling arrests effector T cell differentiation and generates $CD8^+$ memory stem cells. *Nat Med* **15:** 808–813.

Giardiello FM, Hamilton SR, Krush AJ, Piantadosi S, Hylind LM, Celano P, Booker SV, Robinson CR, Offerhaus GJ. 1993. Treatment of colonic and rectal adenomas with sulindac in familial adenomatous polyposis. *N Engl J Med* **328:** 1313–1316.

Gibson P, Tong Y, Robinson G, Thompson MC, Currle DS, Eden C, Kranenburg TA, Hogg T, Poppleton H, Martin J, et al. 2010. Subtypes of medulloblastoma have distinct developmental origins. *Nature* **468:** 1095–1099.

Glantschnig H, Hampton RA, Lu P, Zhao JZ, Vitelli S, Huang L, Haytko P, Cusick T, Ireland C, Jarantow SW, et al. 2010. Generation and selection of novel fully human monoclonal antibodies that neutralize Dickkopf-1 (DKK1) inhibitory function in vitro and increase bone mass in vivo. *J Biol Chem* **285:** 40135–40147.

Glass DA II, Bialek P, Ahn JD, Starbuck M, Patel MS, Clevers H, Taketo MM, Long F, McMahon AP, Lang RA, et al. 2005. Canonical Wnt signaling in differentiated osteoblasts controls osteoclast differentiation. *Dev Cell* **8:** 751–764.

Gong Y, Slee RB, Fukai N, Rawadi G, Roman-Roman S, Reginato AM, Wang H, Cundy T, Glorieux FH, Lev D, et al. 2001. LDL receptor-related protein 5 (LRP5) affects bone accrual and eye development. *Cell* **107:** 513–523.

Gonsalves FC, Klein K, Carson BB, Katz S, Ekas LA, Evans S, Nagourney R, Cardozo T, Brown AM.C, DasGupta R. 2011. An RNAi-based chemical genetic screen identifies three small-molecule inhibitors of the Wnt/wingless signaling pathway. *Proc Natl Acad Sci* **108:** 5954–5963.

Gottardi CJ, Gumbiner BM. 2004. Distinct molecular forms of β-catenin are targeted to adhesive or transcriptional complexes. *J Cell Biol* **167:** 339–349.

Graham TA, Weaver C, Mao F, Kimelman D, Xu W. 2000. Crystal structure of a β-catenin/Tcf complex. *Cell* **103:** 885–896.

Graham JA, Fray M, de Haseth S, Lee KM, Lian MM, Chase CM, Madsen JC, Markmann J, Benichou G, Colvin RB, et al. 2010. Suppressive regulatory T cell activity is potentiated by glycogen synthase kinase 3β inhibition. *J Biol Chem* **285:** 32852–32859.

Guettler S, Larose J, Petsalaki E, Gish G, Scotter A, Pawson T, Rottapel R, Sicheri F. 2011. Structural basis and sequence rules for substrate recognition by tankyrase explain the basis for cherubism disease. *Cell* **147:** 1340–1354.

Gunn WG, Krause U, Lee N, Gregory CA. 2010. Pharmaceutical inhibition of glycogen-synthetase-kinase-3-β reduces multiple myeloma-induced bone disease in a novel murine plasmacytoma xenograft model. *Blood* **117:** 1641–1651.

He B, You L, Uematsu K, Xu Z, Lee AY, Matsangou M, McCormick F, Jablons DM. 2004. A monoclonal antibody against Wnt-1 induces apoptosis in human cancer cells. *Neoplasia* **6:** 7–14.

He B, Barg RN, You L, Xu Z, Reguart N, Mikami I, Batra S, Rosell R, Jablons DM. 2005a. Wnt signaling in stem cells and non-small-cell lung cancer. *Clin Lung Cancer* **7:** 54–60.

He B, Reguart N, You L, Mazieres J, Xu Z, Lee AY, Mikami I, McCormick F, Jablons DM. 2005b. Blockade of Wnt-1 signaling induces apoptosis in human colorectal cancer cells containing downstream mutations. *Oncogene* **24:** 3054–3058.

Heiser PW, Cano DA, Landsman L, Kim GE, Kench JG, Klimstra DS, Taketo MM, Biankin AV, Hebrok M. 2008. Stabilization of β-catenin induces pancreas tumor formation. *Gastroenterology* **135:** 1288–1300.

Henderson WR Jr, Chi EY, Ye X, Nguyen C, Tien Y-t, Zhou B, Borok Z, Knight DA, Kahn M. 2010. Inhibition of Wnt/β-catenin/CREB binding protein (CBP) signaling reverses pulmonary fibrosis. *Proc Natl Acad Sci* **107:** 14309–14314.

Hoeppner LH, Secreto FJ, Westendorf JJ. 2009. Wnt signaling as a therapeutic target for bone diseases. *Expert Opin Ther Targets* **13:** 485–496.

Holmen SL, Zylstra CR, Mukherjee A, Sigler RE, Faugere MC, Bouxsein ML, Deng L, Clemens TL, Williams BO. 2005. Essential role of β-catenin in postnatal bone acquisition. *J Biol Chem* **280:** 21162–21168.

Hornyak TJ, Hayes DJ, Chiu LY, Ziff EB. 2001. Transcription factors in melanocyte development: Distinct roles for Pax-3 and Mitf. *Mech Dev* **101:** 47–59.

Horvath LG, Henshall SM, Lee C-S, Kench JG, Golovsky D, Brenner PC, O'Neill GF, Kooner R, Stricker PD, Grygiel JJ, et al. 2005. Lower levels of nuclear β-catenin predict for a poorer prognosis in localized prostate cancer. *Int J Cancer* **113:** 415–422.

Huang S-MA, Mishina YM, Liu S, Cheung A, Stegmeier F, Michaud GA, Charlat O, Wiellette E, Zhang Y, Wiessner S, et al. 2009. Tankyrase inhibition stabilizes axin and antagonizes Wnt signalling. *Nature* **461:** 614–620.

Huber AH, Weis WI. 2001. The structure of the β-catenin/E-cadherin complex and the molecular basis of diverse ligand recognition by β-catenin. *Cell* **105:** 391–402.

Itasaki N, Jones CM, Mercurio S, Rowe A, Domingos PM, Smith JC, Krumlauf R. 2003. Wise, a context-dependent activator and inhibitor of Wnt signalling. *Development* **130:** 4295–4305.

Jeannet G, Boudousquié C, Gardiol N, Kang J, Huelsken J, Held W. 2010. Essential role of the Wnt pathway effector Tcf-1 for the establishment of functional CD8 T cell memory. *Proc Natl Acad Sci* **107:** 9777–9782.

Jensen M, Engert A, Weissinger F, Knauf W, Kimby E, Poynton C, Oliff IA, Rummel MJ, Osterborg A. 2008. Phase I study of a novel pro-apoptotic drug R-etodolac in patients with B-cell chronic lymphocytic leukemia. *Invest New Drugs* **26:** 139–149.

Kageshita T, Hamby CV, Ishihara T, Matsumoto K, Saida T, Ono T. 2001. Loss of β-catenin expression associated with disease progression in malignant melanoma. *Brit J Dermatol* **145:** 210–216.

Kameoka M, Kameoka Y, Utachee P, Kurosu T, Sawanpanyalert P, Ikuta K, Auwanit W. 2009. Short communication: RNA interference directed against Axin1 upregulates human immunodeficiency virus type 1 gene expression by activating the Wnt signaling pathway in HeLa-derived J111 cells. *AIDS Res Hum Retroviruses* **25:** 1005–1011.

Kang DE, Soriano S, Xia X, Eberhart CG, De Strooper B, Zheng H, Koo EH. 2002. Presenilin couples the paired phosphorylation of β-catenin independent of axin: Implications for β-catenin activation in tumorigenesis. *Cell* **110:** 751–762.

Katoh M. 2002. Strabismus (STB)/Vang-like (VANGL) gene family (Review). *Int J Mol Med* **10:** 11–15.

Kawasaki K, Watanabe M, Sakaguchi M, Ogasawara Y, Ochiai K, Nasu Y, Doihara H, Kashiwakura Y, Huh NH, Kumon H, et al. 2009. REIC/Dkk-3 overexpression downregulates P-glycoprotein in multidrug-resistant MCF7/ADR cells and induces apoptosis in breast cancer. *Cancer Gene Ther* **16:** 65–72.

Kho DH, Bae JA, Lee JH, Cho HJ, Cho SH, Seo Y-W, Ahn KY, Chung IJ, Kim KK. 2009. KITENIN recruits Dishevelled/PKCδ to form a functional complex and controls the migration and invasiveness of colorectal cancer cells. *Gut* **58:** 509–519.

Kim SY, Dunn IF, Firestein R, Gupta P, Wardwell L, Repich K, Schinzel AC, Wittner B, Silver SJ, Root DE, et al. 2010. CK1epsilon is required for breast cancers dependent on β-catenin activity. *PloS ONE* **5:** e8979.

Kinzler KW, Nilbert MC, Su LK, Vogelstein B, Bryan TM, Levy DB, Smith KJ, Preisinger AC, Hedge P, McKechnie D. 1991. Identification of FAP locus genes from chromosome 5q21. *Science (New York, NY)* **253:** 661–665.

Klein PS, Melton DA. 1996. A molecular mechanism for the effect of lithium on development. *Proc Natl Acad Sci* **93:** 8455–8459.

Komatsu DE, Mary MN, Schroeder RJ, Robling AG, Turner CH, Warden SJ. 2010. Modulation of Wnt signaling influences fracture repair. *J Orthop Res* **28:** 928–936.

Korinek V, Barker N, Moerer P, van Donselaar E, Huls G, Peters PJ, Clevers H. 1998. Depletion of epithelial stem-cell compartments in the small intestine of mice lacking Tcf-4. *Nat Genet* **19:** 379–383.

Kulkarni NH, Onyia JE, Zeng Q, Tian X, Liu M, Halladay DL, Frolik CA, Engler T, Wei T, Kriauciunas A, et al. 2006. Orally bioavailable GSK-3α/β dual inhibitor increases markers of cellular differentiation in vitro and bone mass in vivo. *J Bone Miner Res* **21:** 910–920.

Kumar A, Zloza A, Moon RT, Watts J, Tenorio AR, Al-Harthi L. 2008. Active β-catenin signaling is an inhibitory pathway for human immunodeficiency virus replication in peripheral blood mononuclear cells. *J Virol* **82:** 2813–2820.

Kurayoshi M, Oue N, Yamamoto H, Kishida M, Inoue A, Asahara T, Yasui W, Kikuchi A. 2006. Expression of Wnt-5a is correlated with aggressiveness of gastric cancer by stimulating cell migration and invasion. *Cancer Res* **66:** 10439–10448.

Labayle D, Fischer D, Vielh P, Drouhin F, Pariente A, Bories C, Duhamel O, Trousset M, Attali P. 1991. Sulindac causes regression of rectal polyps in familial adenomatous polyposis. *Gastroenterology* **101:** 635–639.

Lavergne E, Hendaoui I, Coulouarn C, Ribault C, Leseur J, Eliat P-A, Mebarki S, Corlu A, Clément B, Musso O. 2011. Blocking Wnt signaling by SFRP-like molecules inhibits in vivo cell proliferation and tumor growth in cells carrying active β-catenin. *Oncogene* **30:** 423–433.

Le MN, Chan JL-K, Rosenberg SA, Nabatian AS, Merrigan KT, Cohen-Solal KA, Goydos JS. 2010. The glutamate release inhibitor Riluzole decreases migration, invasion, and proliferation of melanoma cells. *J Invest Dermatol* **130:** 2240–2249.

Lee AY, He B, You L, Dadfarmay S, Xu Z, Mazieres J, Mikami I, McCormick F, Jablons DM. 2004a. Expression of the secreted frizzled-related protein gene family is down-regulated in human mesothelioma. *Oncogene* **23:** 6672–6676.

Lee JH, Park SR, Chay K-O, Seo Y-W, Kook H, Ahn KY, Kim YJ, Kim KK. 2004b. KAI1 COOH-terminal interacting tetraspanin (KITENIN), a member of the tetraspanin family, interacts with KAI1, a tumor metastasis suppressor, and enhances metastasis of cancer. *Cancer Res* **64:** 4235–4243.

Lee JK, Bae JA, Sun EG, Kim HD, Yoon TM, Kim K, Lee JH, Lim SC, Kim KK. 2009. KITENIN increases invasion and migration of mouse squamous cancer cells and promotes pulmonary metastasis in a mouse squamous tumor model. *FEBS Lett* **583:** 711–717.

Lee S-H, Kim B, Oh MJ, Yoon J, Kim HY, Lee KJ, Lee JD, Choi K-Y. 2011. Persicaria hydropiper (L.) spach and its flavonoid components, isoquercitrin and isorhamnetin, activate the Wnt/β-catenin pathway and inhibit adipocyte differentiation of 3T3-L1 cells. *Phytother Res* **25:** 1629–1635.

Lengerke C, Schmitt S, Bowman TV, Jang IH, Maouche-Chretien L, McKinney-Freeman S, Davidson AJ, Hammerschmidt M, Rentzsch F, Green JBA, et al. 2008. BMP and Wnt specify hematopoietic fate by activation of the Cdx-Hox pathway. *Cell Stem Cell* **2:** 72–82.

Lengerke C, Grauer M, Niebuhr NI, Riedt T, Kanz L, Park I-H, Daley GQ. 2009. Hematopoietic development from human induced pluripotent stem cells. *Ann NY Acad Sci* **1176:** 219–227.

Lepourcelet M, Chen Y-NP, France DS, Wang H, Crews P, Petersen F, Bruseo C, Wood AW, Shivdasani RA. 2004. Small-molecule antagonists of the oncogenic Tcf/β-catenin protein complex. *Cancer Cell* **5:** 91–102.

Levaot N, Voytyuk O, Dimitriou I, Sircoulomb F, Chandrakumar A, Deckert M, Krzyzanowski PM, Scotter A, Gu S, Janmohamed S, et al. 2011. Loss of tankyrase-mediated destruction of 3BP2 is the underlying pathogenic mechanism of cherubism. *Cell* **147:** 1324–1339.

Li X, Zhang Y, Kang H, Liu W, Liu P, Zhang J, Harris SE, Wu D. 2005. Sclerostin binds to LRP5/6 and antagonizes canonical Wnt signaling. *J Biol Chem* **280:** 19883–19887.

Li X, Ominsky MS, Niu QT, Sun N, Daugherty B, D'Agostin D, Kurahara C, Gao Y, Cao J, Gong J, et al. 2008. Targeted deletion of the sclerostin gene in mice results in increased bone formation and bone strength. *J Bone Miner Res* **23:** 860–869.

Li X, Ominsky MS, Warmington KS, Morony S, Gong J, Cao J, Gao Y, Shalhoub V, Tipton B, Haldankar R, et al. 2009. Sclerostin antibody treatment increases bone formation, bone mass, and bone strength in a rat model of postmenopausal osteoporosis. *J Bone Miner Res* **24:** 578–588.

Lugli A, Zlobec I, Minoo P, Baker K, Tornillo L, Terracciano L, Jass JR. 2007. Prognostic significance of the wnt signalling pathway molecules APC, β-catenin and E-cadherin in colorectal cancer: A tissue microarray-based analysis. *Histopathology* **50:** 453–464.

Maelandsmo GM, Holm R, Nesland JM, Fodstad Ø, Flørenes VA. 2003. Reduced β-catenin expression in the cytoplasm of advanced-stage superficial spreading malignant melanoma. *Clin Cancer Res* **9:** 3383–3388.

Mahmoud NN, Boolbol SK, Dannenberg AJ, Mestre JR, Bilinski RT, Martucci C, Newmark HL, Chadburn A, Bertagnolli MM. 1998. The sulfide metabolite of sulindac prevents tumors and restores enterocyte apoptosis in a murine model of familial adenomatous polyposis. *Carcinogenesis* **19:** 87–91.

Mahmoudi T, Li VSW, Ng SS, Taouatas N, Vries RGJ, Mohammed S, Heck AJ, Clevers H. 2009. The kinase TNIK is an essential activator of Wnt target genes. *EMBO J* **28:** 3329–3340.

Major MB, Roberts BS, Berndt JD, Marine S, Anastas J, Chung N, Ferrer M, Yi X, Stoick-Cooper CL, von Haller PD, et al. 2008. New regulators of Wnt/β-catenin signaling revealed by integrative molecular screening. *Sci Signal* **1:** ra12.

Manicassamy S, Reizis B, Ravindran R, Nakaya H, Salazar-Gonzalez RM, Wang Y-C, Pulendran B. 2010. Activation of β-catenin in dendritic cells regulates immunity versus tolerance in the intestine. *Science (New York, NY)* **329:** 849–853.

Mao Y, Ge X, Frank CL, Madison JM, Koehler AN, Doud MK, Tassa C, Berry EM, Soda T, Singh KK, et al. 2009. Disrupted in schizophrenia 1 regulates neuronal progenitor proliferation via modulation of GSK3β/β-catenin signaling. *Cell* **136:** 1017–1031.

Matushansky I, Hernando E, Socci ND, Mills JE, Matos TA, Edgar MA, Singer S, Maki RG, Cordon-Cardo C. 2007. Derivation of sarcomas from mesenchymal stem cells via inactivation of the Wnt pathway. *J Clin Invest* **117:** 3248–3257.

Mazieres J, You L, He B, Xu Z, Lee AY, Mikami I, McCormick F, Jablons DM. 2005a. Inhibition of Wnt16 in human acute lymphoblastoid leukemia cells containing the t(1;19) translocation induces apoptosis. *Oncogene* **24:** 5396–5400.

Mazieres J, You L, He B, Xu Z, Twogood S, Lee AY, Reguart N, Batra S, Mikami I, Jablons DM. 2005b. Wnt2 as a new therapeutic target in malignant pleural mesothelioma. *Int J Cancer* **117:** 326–332.

Mikami I, You L, He B, Xu Z, Batra S, Lee AY, Mazieres J, Reguart N, Uematsu K, Koizumi K, et al. 2005. Efficacy of Wnt-1 monoclonal antibody in sarcoma cells. *BMC Cancer* **5:** 53.

Milovanovic T, Planutis K, Nguyen A, Marsh JL, Lin F, Hope C, Holcombe RF. 2004. Expression of Wnt genes and frizzled 1 and 2 receptors in normal breast epithelium and infiltrating breast carcinoma. *Int J Oncol* **25:** 1337–1342.

Minear S, Leucht P, Jiang J, Liu B, Zeng A, Fuerer C, Nusse R, Helms JA. 2010. Wnt proteins promote bone regeneration. *Sci Transl Med* **2:** 29ra30.

Miyamoto S, Endoh Y, Hasebe T, Ishii G, Kodama K, Goya M, Ono M, Saitoh N, Chiba T, Ochiai A. 2004. Nuclear β-catenin accumulation as a prognostic factor in Dukes' D human colorectal cancers. *Oncol Rep* **12:** 245–251.

Miyaoka T, Seno H, Ishino H. 1999. Increased expression of Wnt-1 in schizophrenic brains. *Schizophr Res* **38:** 1–6.

Moore WJ, Kern JC, Bhat R, Commons TJ, Fukayama S, Goljer I, Krishnamurthy G, Magolda RL, Nogle L, Pitts K, et al. 2009. Modulation of Wnt signaling through inhibition of secreted frizzled-related protein I (sFRP-1) with N-substituted piperidinyl diphenylsulfonyl sulfonamides. *J Med Chem* **52:** 105–116.

Morris JP IV, Cano DA, Sekine S, Wang SC, Hebrok M. β-Catenin blocks Kras-dependent reprogramming of acini into pancreatic cancer precursor lesions in mice. *J Clin Invest* **120:** 508–520.

Mudher A, Chapman S, Richardson J, Asuni A, Gibb G, Pollard C, Killick R, Iqbal T, Raymond L, Varndell I, et al. 2001. Dishevelled regulates the metabolism of amyloid precursor protein via protein kinase C/mitogen-activated protein kinase and c-Jun terminal kinase. *J Neurosci* **21:** 4987–4995.

Mukhopadhyay M, Gorivodsky M, Shtrom S, Grinberg A, Niehrs C, Morasso MI, Westphal H. 2006. Dkk2 plays an essential role in the corneal fate of the ocular surface epithelium. *Development* **133:** 2149–2154.

Nakamura Y, Nawata M, Wakitani S. 2005. Expression profiles and functional analyses of Wnt-related genes in human joint disorders. *Am J Pathol* **167:** 97–105.

Namkoong J, Shin S-S, Lee HJ, Marín YE, Wall BA, Goydos JS, Chen S. 2007. Metabotropic glutamate receptor 1 and glutamate signaling in human melanoma. *Cancer Res* **67:** 2298–2305.

Nishisho I, Nakamura Y, Miyoshi Y, Miki Y, Ando H, Horii A, Koyama K, Utsunomiya J, Baba S, Hedge P. 1991. Mutations of chromosome 5q21 genes in FAP and colorectal cancer patients. *Science (New York, NY)* **253:** 665–669.

Nostro MC, Cheng X, Keller GM, Gadue P. 2008. Wnt, activin, and BMP signaling regulate distinct stages in the developmental pathway from embryonic stem cells to blood. *Cell Stem Cell* **2:** 60–71.

Notani D, Gottimukkala KP, Jayani RS, Limaye AS, Damle MV, Mehta S, Purbey PK, Joseph J, Galande S. 2010. Global regulator SATB1 recruits β-catenin and regulates T(H)2 differentiation in Wnt-dependent manner. *PLoS Biol* **8:** e1000296.

Nusse R. 2008. Wnt signaling and stem cell control. *Cell Res* **18:** 523–527.

Nusse R, Varmus HE. 1982. Many tumors induced by the mouse mammary tumor virus contain a provirus integrated in the same region of the host genome. *Cell* **31:** 99–109.

Okino K, Nagai H, Hatta M, Nagahata T, Yoneyama K, Ohta Y, Jin E, Kawanami O, Araki T, Emi M. 2003. Up-regulation and overproduction of DVL-1, the human counterpart of the *Drosophila* dishevelled gene, in cervical squamous cell carcinoma. *Oncol Rep* **10:** 1219–1223.

Ominsky MS, Vlasseros F, Jolette J, Smith SY, Stouch B, Doellgast G, Gong J, Gao Y, Cao J, Graham K, et al. 2010. Two doses of sclerostin antibody in cynomolgus monkeys increases bone formation, bone mineral density, and bone strength. *J Bone Miner Res* **25:** 948–959.

Padhi D, Jang G, Stouch B, Fang L, Posvar E. 2011. Single-dose, placebo-controlled, randomized study of AMG 785, a sclerostin monoclonal antibody. *J Bone Miner Res* **26:** 19–26.

Paige SL, Osugi T, Afanasiev OK, Pabon L, Reinecke H, Murry CE. 2010. Endogenous Wnt/β-catenin signaling is required for cardiac differentiation in human embryonic stem cells. *PloS ONE* **5:** e11134.

Pálmer HG, González-Sancho JM, Espada J, Berciano MT, Puig I, Baulida J, Quintanilla M, Cano A, de Herreros AG, Lafarga M, et al. 2001. Vitamin D(3) promotes the differentiation of colon carcinoma cells by the induction of E-cadherin and the inhibition of β-catenin signaling. *J Cell Biol* **154:** 369–387.

Pan W, Choi S-C, Wang H, Qin Y, Volpicelli-Daley L, Swan L, Lucast L, Khoo C, Zhang X, Li L, et al. 2008. Wnt3a-mediated formation of phosphatidylinositol 4,5-bisphosphate regulates LRP6 phosphorylation. *Science (New York, NY)* **321:** 1350–1353.

Phillips RKS, Wallace MH, Lynch PM, Hawk E, Gordon GB, Saunders BP, Wakabayashi N, Shen Y, Zimmerman S, Godio L, et al. 2002. A randomised, double blind, placebo controlled study of celecoxib, a selective cyclooxygenase 2 inhibitor, on duodenal polyposis in familial adenomatous polyposis. *Gut* **50:** 857–860.

Pollock PM, Cohen-Solal K, Sood R, Namkoong J, Martino JJ, Koganti A, Zhu H, Robbins C, Makalowska I, Shin S-S, et al. 2003. Melanoma mouse model implicates metabotropic glutamate signaling in melanocytic neoplasia. *Nat Genet* **34:** 108–112.

Pukrop T. 2006. Wnt 5a signaling is critical for macrophage-induced invasion of breast cancer cell lines. *Proc Natl Acad Sci* **103:** 5454–5459.

Qiang Y-W, Chen Y, Stephens O, Brown N, Chen B, Epstein J, Barlogie B, Shaughnessy JDJ. 2008. Myeloma-derived Dickkopf-1 disrupts Wnt-regulated osteoprotegerin and RANKL production by osteoblasts: A potential mechanism underlying osteolytic bone lesions in multiple myeloma. *Blood* **112:** 196–207.

Reya T, Duncan AW, Ailles L, Domen J, Scherer DC, Willert K, Hintz L, Nusse R, Weissman IL. 2003. A role for Wnt signalling in self-renewal of haematopoietic stem cells. *Nature* **423:** 409–414.

Rhee C-S, Sen M, Lu D, Wu C, Leoni L, Rubin J, Corr M, Carson DA. 2002. Wnt and frizzled receptors as potential targets for immunotherapy in head and neck squamous cell carcinomas. *Oncogene* **21:** 6598–6605.

Rijsewijk F. 1987. The *Drosophila* homology of the mouse mammary oncogene int-1 is identical to the segment polarity gene wingless. *Cell* **50:** 649–657.

Robitaille J, MacDonald ML.E, Kaykas A, Sheldahl LC, Zeisler J, Dubé M-P, Zhang L-H, Singaraja RR, Guernsey DL, Zheng B, et al. 2002. Mutant frizzled-4 disrupts retinal angiogenesis in familial exudative vitreoretinopathy. *Nat Genet* **32:** 326–330.

Rodda SJ, McMahon AP. 2006. Distinct roles for Hedgehog and canonical Wnt signaling in specification, differentiation and maintenance of osteoblast progenitors. *Development (Cambridge, England)* **133:** 3231–3244.

Roh M-S, Seo MS, Kim Y, Kim SH, Jeon WJ, Ahn YM, Kang UG, Juhnn YS, Kim YS. 2007. Haloperidol and clozapine differentially regulate signals upstream of glycogen synthase kinase 3 in the rat frontal cortex. *Exp Molec Med* **39:** 353–360.

Säfholm A, Leandersson K, Dejmek J, Nielsen CK, Villoutreix BO, Andersson T. 2006. A formylated hexapeptide ligand mimics the ability of Wnt-5a to impair migration of human breast epithelial cells. *J Biol Chem* **281:** 2740–2749.

Salinas PC, Zou Y. 2008. Wnt signaling in neural circuit assembly. *Annu Rev Neurosci* **31:** 339–358.

Saraswati S, Alfaro MP, Thorne CA, Atkinson J, Lee E, Young PP. 2010. Pyrvinium, a potent small molecule Wnt inhibitor, promotes wound repair and post-MI cardiac remodeling. *PloS ONE* **5:** e15521.

Schwartz AL, Malgor R, Dickerson E, Weeraratna AT, Slominski A, Wortsman J, Harii N, Kohn AD, Moon RT, Schwartz FL, et al. 2009. Phenylmethimazole decreases Toll-like receptor 3 and noncanonical Wnt5a expression in pancreatic cancer and melanoma together with tumor cell growth and migration. *Clin Cancer Res* **15:** 4114–4122.

Semënov M, Tamai K, He X. 2005. SOST is a ligand for LRP5/LRP6 and a Wnt signaling inhibitor. *J Biol Chem* **280:** 26770–26775.

Sen M, Lauterbach K, El-Gabalawy H, Firestein GS, Corr M, Carson DA. 2000. Expression and function of wingless and frizzled homologs in rheumatoid arthritis. *Proc Natl Acad Sci* **97:** 2791–2796.

Sen M, Chamorro M, Reifert J, Corr M, Carson DA. 2001. Blockade of Wnt-5A/Frizzled 5 signaling inhibits rheumatoid synoviocyte activation. *Arthritis Rheum* **44:** 772–781.

Sen M, Reifert J, Lauterbach K, Wolf V, Rubin JS, Corr M, Carson DA. 2002. Regulation of fibronectin and metalloproteinase expression by Wnt signaling in rheumatoid arthritis synoviocytes. *Arthritis Rheum* **46:** 2867–2877.

Shafer B, Onishi K, Lo C, Colakoglu G, Zou Y. 2011. Vangl2 promotes Wnt/planar cell polarity-like signaling by antagonizing Dvl1-mediated feedback inhibition in growth cone guidance. *Dev Cell* **20:** 177–191.

Shan J, Shi D-L, Wang J, Zheng J. 2005. Identification of a specific inhibitor of the dishevelled PDZ domain. *Biochemistry* **44:** 15495–15503.

Sheng Y-J, Gao J-P, Li J, Han J-W, Xu Q, Hu W-L, Pan T-M, Cheng Y-L, Yu Z-Y, Ni C, et al. 2011. Follow-up study identifies two novel susceptibility loci PRKCB and 8p11.21 for systemic lupus erythematosus. *Rheumatology (Oxford, England)* **50:** 682–688.

Spence JR, Mayhew CN, Rankin SA, Kuhar MF, Vallance JE, Tolle K, Hoskins EE, Kalinichenko VV, Wells SI, Zorn AM, et al. 2011. Directed differentiation of human pluripotent stem cells into intestinal tissue in vitro. *Nature* **470:** 105–109.

Steinbach G, Lynch PM, Phillips RK, Wallace MH, Hawk E, Gordon GB, Wakabayashi N, Saunders B, Shen Y, Fujimura T, et al. 2000. The effect of celecoxib, a cyclooxygenase-2 inhibitor, in familial adenomatous polyposis. *N Engl J Med* **342:** 1946–1952.

Stoick-Cooper CL, Weidinger G, Riehle KJ, Hubbert C, Major MB, Fausto N, Moon RT. 2007. Distinct Wnt signaling pathways have opposing roles in appendage regeneration. *Development (Cambridge, England)* **134:** 479–489.

Su L, Vogelstein B, Kinzler K. 1993. Association of the APC tumor suppressor protein with catenins. *Science* **262:** 1734–1737.

Sue Ng S, Mahmoudi T, Li VSW, Hatzis P, Boersema PJ, Mohammed S, Heck AJ, Clevers H. 2010. MAP3K1 functionally interacts with Axin1 in the canonical Wnt signalling pathway. *Biol Chem* **391:** 171–180.

Sugimura R, Li L. 2010. Noncanonical Wnt signaling in vertebrate development, stem cells, and diseases. *Birth Defects Res C Embryo Today* **90:** 243–256.

Sun TQ, Lu B, Feng JJ, Reinhard C, Jan YN, Fantl WJ, Williams LT. 2001. PAR-1 is a Dishevelled-associated kinase and a positive regulator of Wnt signalling. *Nat Cell Biol* **3:** 628–636.

Surendran K, Schiavi S, Hruska KA. 2005. Wnt-dependent β-catenin signaling is activated after unilateral ureteral obstruction, and recombinant secreted frizzled-related protein 4 alters the progression of renal fibrosis. *J Am Soc Nephrol* **16:** 2373–2384.

Suzuki H, Watkins DN, Jair K-W, Schuebel KE, Markowitz SD, Dong Chen W, Pretlow TP, Yang B, Akiyama Y, van Engeland M, et al. 2004. Epigenetic inactivation of SFRP genes allows constitutive WNT signaling in colorectal cancer. *Nat Genet* **36:** 417–422.

Suzuki H, Toyota M, Carraway H, Caraway H, Gabrielson E, Ohmura T, Fujikane T, Nishikawa N, Sogabe Y, Nojima M, et al. 2008. Frequent epigenetic inactivation of Wnt antagonist genes in breast cancer. *Brit J Cancer* **98:** 1147–1156.

Takada R, Satomi Y, Kurata T, Ueno N, Norioka S, Kondoh H, Takao T, Takada S. 2006. Monounsaturated fatty acid modification of Wnt protein: Its role in Wnt secretion. *Dev Cell* **11:** 791–801.

Takahashi S, Watanabe T, Okada M, Inoue K, Ueda T, Takada I, Watabe T, Yamamoto Y, Fukuda T, Nakamura T, et al. 2011. Noncanonical Wnt signaling mediates androgen-dependent tumor growth in a mouse model of prostate cancer. *Proc Natl Acad Sci* **108:** 4938–4943.

Takayama S, Rogatsky I, Schwarcz LE, Darimont BD. 2006. The glucocorticoid receptor represses cyclin D1 by targeting the Tcf-β-catenin complex. *J Biol Chem* **281:** 17856–17863.

Takeda K, Yasumoto K, Takada R, Takada S, Watanabe K, Udono T, Saito H, Takahashi K, Shibahara S. 2000. Induction of melanocyte-specific microphthalmia-associated transcription factor by Wnt-3a. *J Biol Chem* **275:** 14013–14016.

Thorne CA, Hanson AJ, Schneider J, Tahinci E, Orton D, Cselenyi CS, Jernigan KK, Meyers KC, Hang BI, Waterson AG, et al. 2010. Small-molecule inhibition of Wnt signaling through activation of casein kinase 1α. *Nat Chem Biol* **6:** 829–836.

Tian E, Zhan F, Walker R, Rasmussen E, Ma Y, Barlogie B, Shaughnessy JD Jr. 2003. The role of the Wnt-signaling antagonist DKK1 in the development of osteolytic lesions in multiple myeloma. *N Engl J Med* **349:** 2483–2494.

Toledo EM, Inestrosa NC. 2010. Activation of Wnt signaling by lithium and rosiglitazone reduced spatial memory impairment and neurodegeneration in brains of an APPswe/PSEN1ΔE9 mouse model of Alzheimer's disease. *Mol Psychiatry* **15:** 228, 272–285.

Trowbridge JJ, Guezguez B, Moon RT, Bhatia M. 2010. Wnt3a activates dormant c-Kit$^-$ bone marrow-derived cells with short-term multilineage hematopoietic reconstitution capacity. *Stem Cells (Dayton, Ohio)* **28:** 1379–1389.

Tveita AA, Rekvig OP. 2011. Alterations in Wnt pathway activity in mouse serum and kidneys during lupus development. *Arthritis Rheum* **63:** 513–522.

Ueki Y, Lin CY, Senoo M, Ebihara T, Agata N, Onji M, Saheki Y, Kawai T, Mukherjee PM, Reichenberger E, et al. 2007. Increased myeloid cell responses to M-CSF and RANKL cause bone loss and inflammation in SH3BP2 "cherubism" mice. *Cell* **128:** 71–83.

Uematsu K, He B, You L, Xu Z, McCormick F, Jablons DM. 2003a. Activation of the Wnt pathway in non small cell lung cancer: Evidence of dishevelled overexpression. *Oncogene* **22:** 7218–7221.

Uematsu K, Kanazawa S, You L, He B, Xu Z, Li K, Peterlin BM, McCormick F, Jablons DM. 2003b. Wnt pathway activation in mesothelioma: Evidence of Dishevelled overexpression and transcriptional activity of β-catenin. *Cancer Res* **63:** 4547–4551.

Ueno S, Weidinger G, Osugi T, Kohn AD, Golob JL, Pabon L, Reinecke H, Moon RT, Murry CE. 2007. Biphasic role for Wnt/β-catenin signaling in cardiac specification in zebrafish and embryonic stem cells. *Proc Natl Acad Sci* **104:** 9685–9690.

Ueno K, Hirata H, Majid S, Chen Y, Zaman MS, Tabatabai ZL, Hinoda Y, Dahiya R. 2011. Wnt antagonist DICKKOPF-3 (Dkk-3) induces apoptosis in human renal cell carcinoma. *Mol Carcinogen* **50:** 449–457.

Uren A, Reichsman F, Anest V, Taylor WG, Muraiso K, Bottaro DP, Cumberledge S, Rubin JS. 2000. Secreted frizzled-related protein-1 binds directly to Wingless and is a biphasic modulator of Wnt signaling. *J Biol Chem* **275:** 4374–4382.

Vijayaragavan K, Szabo E, Bossé M, Ramos-Mejia V, Moon RT, Bhatia M. 2009. Noncanonical Wnt signaling orchestrates early developmental events toward hematopoietic cell fate from human embryonic stem cells. *Cell Stem Cell* **4:** 248–262.

Vincan E, Darcy PK, Smyth MJ, Thompson EW, Thomas RJ.S, Phillips WA, Ramsay RG. 2005. Frizzled-7 receptor ectodomain expression in a colon cancer cell line induces morphological change and attenuates tumor growth. *Differentiation* **73:** 142–153.

von Marschall Z, Fisher LW. 2010. Secreted Frizzled-related protein-2 (sFRP2) augments canonical Wnt3a-induced signaling. *Biochem Biophys Res Commun* **400:** 299–304.

Wang Y, Nakayama N. 2009. WNT and BMP signaling are both required for hematopoietic cell development from human ES cells. *Stem Cell Res* **3:** 113–125.

Wang F-S, Ko J-Y, Lin C-L, Wu H-L, Ke H-J, Tai P-J. 2007. Knocking down dickkopf-1 alleviates estrogen deficiency induction of bone loss. A histomorphological study in ovariectomized rats. *Bone* **40:** 485–492.

Wang F-S, Ko J-Y, Yeh D-W, Ke H-C, Wu H-L. 2008. Modulation of Dickkopf-1 attenuates glucocorticoid induction of osteoblast apoptosis, adipocytic differentiation, and bone mass loss. *Endocrinology* **149:** 1793–1801.

Wang H, Hao J, Hong CC. 2011a. Cardiac induction of embryonic stem cells by a small molecule inhibitor of Wnt/β-catenin signaling. *ACS Chem Biol* **6:** 192–197.

Wang W, Liu H, Wang S, Hao X, Li L. 2011b. A diterpenoid derivative 15-oxospiramilactone inhibits Wnt/β-catenin signaling and colon cancer cell tumorigenesis. *Cell Res* **21:** 730–740.

Weeraratna AT, Jiang Y, Hostetter G, Rosenblatt K, Duray P, Bittner M, Trent JM. 2002. Wnt5a signaling directly affects cell motility and invasion of metastatic melanoma. *Cancer Cell* **1:** 279–288.

Wei W, Chua MS, Grepper S, So SK. 2009. Blockade of Wnt-1 signaling leads to anti-tumor effects in hepatocellular carcinoma cells. *Mol Cancer* **8:** 76.

Wei W, Chua M-S, Grepper S, So SK. 2011. Soluble Frizzled-7 receptor inhibits Wnt signaling and sensitizes hepatocellular carcinoma cells towards doxorubicin. *Mol Cancer* **10:** 16.

White BD, Nathe RJ, Maris DO, Nguyen NK, Goodson JM, Moon RT, Horner PJ. 2010. β-Catenin signaling increases in proliferating NG2$^+$ progenitors and astrocytes during post-traumatic gliogenesis in the adult brain. *Stem Cells (Dayton, Ohio)* **28:** 297–307.

Willert K, Brown JD, Danenberg E, Duncan AW, Weissman IL, Reya T, Yates JR III, Nusse R. 2003. Wnt proteins are lipid-modified and can act as stem cell growth factors. *Nature* **423:** 448–452.

Witze ES, Litman ES, Argast GM, Moon RT, Ahn NG. 2008. Wnt5a control of cell polarity and directional movement by polarized redistribution of adhesion receptors. *Science (New York, NY)* **320:** 365–369.

Wong SCC, Lo SFE, Lee KC, Yam JWP, Chan JK.C, Wendy Hsiao WL. 2002. Expression of frizzled-related protein and Wnt-signalling molecules in invasive human breast tumours. *J Pathol* **196:** 145–153.

Wortman B, Darbinian N, Sawaya BE, Khalili K, Amini S. 2002. Evidence for regulation of long terminal repeat transcription by Wnt transcription factor TCF-4 in human astrocytic cells. *J Virol* **76:** 11159–11165.

Wu W, Glinka A, Delius H, Niehrs C. 2000. Mutual antagonism between dickkopf1 and dickkopf2 regulates Wnt/β-catenin signalling. *Curr Biol* **10:** 1611–1614.

Xiao J-H, Ghosn C, Hinchman C, Forbes C, Wang J, Snider N, Cordrey A, Zhao Y, Chandraratna RAS. 2003. Adenomatous polyposis coli (APC)-independent regulation of β-catenin degradation via a retinoid X receptor-mediated pathway. *J Biol Chem* **278:** 29954–29962.

Yadav VK, Ryu J-H, Suda N, Tanaka KF, Gingrich JA, Schütz G, Glorieux FH, Chiang CY, Zajac JD, Insogna KL, et al. 2008. Lrp5 controls bone formation by inhibiting serotonin synthesis in the duodenum. *Cell* **135:** 825–837.

Yamamoto H, Sakane H, Michiue T, Kikuchi A. 2008. Wnt3a and Dkk1 regulate distinct internalization pathways of LRP6 to tune the activation of β-catenin signaling. *Dev Cell* **15:** 37–48.

Yamauchi K, Sumi T, Minami I, Otsuji TG, Kawase E, Nakatsuji N, Suemori H. 2010. Cardiomyocytes develop from anterior primitive streak cells induced by β-catenin activation and the blockage of BMP signaling in hESCs. *Genes Cells* **15:** 1216–1227.

Yang J, Si T, Ling Y, Ruan Y, Han Y, Wang X, Zhang H, Kong Q, Li X, Liu C, et al. 2003a. Association study of the human FZD3 locus with schizophrenia. *Biol Psychiatry* **54:** 1298–1301.

Yang K, Fan K, Kurihara N, Shinozaki H, Rigas B, Augenlicht L, Kopelovich L, Edelmann W, Kucherlapati R, Lipkin M. 2003b. Regional response leading to tumorigenesis after sulindac in small and large intestine of mice with Apc mutations. *Carcinogenesis* **24:** 605–611.

Yip D, Le MN, Chan JL-K, Lee JH, Mehnert JA, Yudd A, Kempf J, Shih WJ, Chen S, Goydos JS. 2009. A phase 0 trial of riluzole in patients with resectable stage III and IV melanoma. *Clin Cancer Res* **15:** 3896–3902.

You L, He B, Xu Z, Uematsu K, Mazieres J, Fujii N, Mikami I, Reguart N, McIntosh JK, Kashani-Sabet M, et al. 2004a. An anti-Wnt-2 monoclonal antibody induces apoptosis in malignant melanoma cells and inhibits tumor growth. *Cancer Res* **64:** 5385–5389.

You L, He B, Xu Z, Uematsu K, Mazieres J, Mikami I, Reguart N, Moody TW, Kitajewski J, McCormick F, et al. 2004b. Inhibition of Wnt-2-mediated signaling induces programmed cell death in non-small-cell lung cancer cells. *Oncogene* **23:** 6170–6174.

Yu Q, Sharma A, Sen JM. 2010. TCF1 and β-catenin regulate T cell development and function. *Immunol Res* **47:** 45–55.

Zamani A, Omrani GR, Nasab MM. 2009. Lithium's effect on bone mineral density. *Bone* **44:** 331–334.

Zhang Y, Yeh JR, Mara A, Ju R, Hines JF, Cirone P, Griesbach HL, Schneider I, Slusarski DC, Holley SA. 2006. A chemical and genetic approach to the mode of action of fumagillin. *Chem Biol* **13:** 1001–1009.

Zhang Y, Appleton BA, Wiesmann C, Lau T, Costa M, Hannoush RN, Sidhu SS. 2009. Inhibition of Wnt signaling by Dishevelled PDZ peptides. *Nat Chem Biol* **5:** 217–219.

Zhao D-M, Yu S, Zhou X, Haring JS, Held W, Badovinac VP, Harty JT, Xue H-H. 2010. Constitutive activation of Wnt signaling favors generation of memory CD8 T cells. *J Immunol (Baltimore, MD, 1950)* **184:** 1191–1199.

Zhou X, Yu S, Zhao D-M, Harty JT, Badovinac VP, Xue H-H. 2010. Differentiation and persistence of memory $CD8^+$ T cells depend on T cell factor 1. *Immunity* **33:** 229–240.

Zou H, Molina JR, Harrington JJ, Osborn NK, Klatt KK, Romero Y, Burgart LJ, Ahlquist DA. 2005. Aberrant methylation of secreted frizzled-related protein genes in esophageal adenocarcinoma and Barrett's esophagus. *Int J Cancer* **116:** 584–591.

Index

G

H

I

J

K

L

M

N

T

V

W

Z